PROTECTED AREA MANAGEMENT
PRINCIPLES AND PRACTICE

Australian Government

SUSTAINABLE
TOURISM
CRC

For past, current, and future protected area managers

Second Edition

PROTECTED AREA MANAGEMENT

PRINCIPLES AND PRACTICE   Graeme L. Worboys, Michael Lockwood, Terry De Lacy

with Christine McNamara, Madeleine Boyd, Mark O'Connor and Michelle Whitemore

OXFORD
UNIVERSITY PRESS

OXFORD
UNIVERSITY PRESS

253 Normanby Road, South Melbourne, Victoria 3205, Australia

Oxford University Press is a department of the University of Oxford.
It furthers the University's objective of excellence in research, scholarship,
and education by publishing worldwide in

Oxford New York

Auckland Cape Town Dar es Salaam Hong Kong Karachi
Kuala Lumpur Madrid Melbourne Mexico City Nairobi
New Delhi Shanghai Taipei Toronto

With offices in

Argentina Austria Brazil Chile Czech Republic France Greece
Guatemala Hungary Italy Japan Poland Portugal Singapore
South Korea Switzerland Thailand Turkey Ukraine Vietnam

OXFORD is a trade mark of Oxford University Press
in the UK and in certain other countries

National Library of Australia
Cataloguing-in-Publication data:

Worboys, Graeme.
     Protected area management: principles and practice.

     2nd ed.
     Bibliography.
     Includes index.
     ISBN 0 19 551728 8.

     1. Nature conservation—Australia. 2 Natural resource management areas—Australia.
     3. Conservation of natural resources—Australia. I. Lockwood, Michael, 1955–.
     II. De Lacy, Terry. III. McNamara, Christine. IV. Title.

     333.780994

Typeset by OUPANZS
Printed by Bookpac Production Services, Singapore

Foreword

Professor Adrian Phillips

Cardiff University, UK and Vice Chair World Heritage,
World Commission on Protected Areas of IUCN

In 1962 there were an estimated 10,000 protected areas worldwide. Now there are more than 100,000. They occupy more than 11.5% of the terrestrial area of our planet, though less than 1% of the surface of our seas. Taken together, protected areas are larger than the whole of South America. But the growing importance of protected areas lies not just in their being a huge and expanding slice of real estate. They are essential for biodiversity conservation, and in implementing the Convention on Biological Diversity. They deliver vital ecosystem services, protecting watersheds and soils, reducing CO_2 levels and shielding human communities from natural disasters. Many protected areas are important to local communities, especially Indigenous peoples, who depend on them for a sustainable supply of resources. They are places for people to find peace in a busy world—places that invigorate the human spirit and challenge the senses. Many protected areas embody important cultural values, and some help support sustainable land use practices in lived-in landscapes.

While the idea that some places are so special that they require protection has deep historic roots and wide appeal, the rapid growth in the number and extent of protected areas in recent years owes much to international initiatives. There are global agreements, like the 1972 World Heritage Convention and the 1992 Convention on Biological Diversity; and regional ones, like the European Habitats and Birds Directives. A common theme has been concern to stem the rapid loss of species and the degradation of ecosystems, habitats, and landscapes.

The progress made in setting up protected areas was celebrated at the Vth World Parks Congress, held in Durban, South Africa in September 2003. The 3000 people present also recognised the many values of protected areas and their role in bringing 'benefits beyond boundaries' to millions of people. But despite this, protected areas are under threat as never before. They are exposed to pollution and climate change, irresponsible tourism, insensitive infrastructure, and ever-increasing demands for land and water. Many protected areas lack political support and are short of financial and other resources. There are still too many gaps in the global protected area system, there is often poor management, and too often local communities are alienated from, and not linked to, protected areas.

So managers of these places face a host of challenges. While those in poorer countries encounter the greatest difficulties, many protected areas in wealthier parts of the world are also under great pressure. Durban also confirmed that protected areas will not survive in a hostile context. So their future depends on a commitment to addressing global challenges, such as climate change. A greater engagement of local people in protected area management—and the sharing of protected area benefits with them—is required. Protected areas need to be planned as part of a country's sustainable development strategy. They need to be physically linked and buffered. And they require the support of all the sectors that affect them.

To promote the cause of protected areas, the world's leaders in this field have established the World Commission on Protected Areas (WCPA). This is one of six commissions of volunteer experts that are part of IUCN (The World Conservation Union). WCPA has an ambitious program—it was, for example, the architect of the Durban Congress. This book was originally generated as a WCPA Australian and New Zealand regional initiative to encourage better protected area management. It has

been widely used by practitioners and is a prescribed text by universities in Australia and internationally.

The second edition has been updated, improved and expanded to include additional park management themes. It incorporates the outcomes from Durban and the WCPA best practice publications. I can think of few comparable initiatives worldwide. This new edition should be even more useful in sharing experience internationally in planning and managing protected areas.

Contents

Foreword   v
List of Figures   xi
List of Tables   xiv
List of Photographs   xvii
List of Backgrounders,
     Case Studies, and Snapshots   xxii
Australian Jurisdictions
     and Protected Area
     Management Agencies   xxviii
Acknowledgments (First Edition)   xxxi
Acknowledgments (Second Edition)   xxxvi
Preface to the Second Edition   xl
Introduction   xli

Part A
Setting the Context   1

Chapter 1
Australia's Natural Heritage   2

1.1   Evolution of Australia's
       natural heritage   3
1.2   Australia today   12
1.3   Climate   13
1.4   Landforms, soils,
       and landscapes   16
1.5   Biological diversity   20
1.6   Terrestrial vegetation   23
1.7   Terrestrial fauna   28
1.8   Marine flora and fauna   34

Chapter 2
Social Context   35

2.1   A brief history of Australian
       protected areas   36
2.2   Contemporary influences   47
2.3   Responses and initiatives   56
2.4   Government in Australia   64
2.5   Legislation   67
2.6   Society, culture, and values   69

Chapter 3
Concept and Purpose
of Protected Areas   76

3.1   Values   78
3.2   International environmental
       protection   85
3.3   Types of protected areas   91
3.4   Extent of Australia's
       protected areas   105

Chapter 4
Process of Management   112
4.1   How organisations manage   114
4.2   Planning   115
4.3   Organising   116
4.4   Leading   124
4.5   Controlling   126
4.6   Managing for performance   126

Part B
Principles and Practice   131

Chapter 5
Establishing Protected Areas   132
5.1   Terrestrial national
       reserve system   133
5.2   Marine national
       reserve system   138
5.3   Scientific reserve
       selection methods   144
5.4   Land-use planning and
       protected area establishment   147
5.5   Management lessons
       and principles   160

Chapter 6
Obtaining and
Managing Information   162
6.1   Scope of information needs   163
6.2   Data and information
       collection methods   167
6.3   Storage, retrieval,
       and presentation   174
6.4   Analysis and application   179
6.5   Information management
       systems   184
6.6   Management lessons
       and principles   186

Chapter 7
Management Planning   188
7.1   Approaches to planning   190
7.2   From approach to process   195
7.3   Protected area
       management plans   199
7.4   Management lessons
       and principles   217

Chapter 8
Finance and Economics   219
8.1   Financing protected areas   220
8.2   Pricing services
       and facilities   223

8.3   Environmental valuation
       and benefit–cost analysis   237
8.4   Management lessons
       and principles   248

Chapter 9
Administration—Making it Work   250
9.1   Administering people   251
9.2   A principled organisation   256
9.3   Administering finances   257
9.4   Administration of assets,
       standards, and systems   262
9.5   Administration policy and
       process considerations   264
9.6   Management lessons
       and principles   267

Chapter 10
Sustainability Management   269
10.1   The need for sustainability   270
10.2   Sustainability and
        protected areas   274
10.3   Management lessons
        and principles   286

Chapter 11
Operations Management   288
11.1   Planning for operations   290
11.2   Operations implementation   300
11.3   Management lessons
        and principles   307

Chapter 12
Natural Heritage Management   309
12.1   A national perspective   310
12.2   Conserving fauna   314
12.3   Conserving flora   328
12.4   Fungi   331
12.5   Managing water   332
12.6   Managing soils and geology   335
12.7   Conserving scenic quality   342
12.8   Managing fire   343
12.9   Management lessons
        and principles   349

Chapter 13
Cultural Heritage Management 351
13.1 An ancient heritage 352
13.2 Types of cultural heritage sites 353
13.3 Conserving cultural heritage 358
13.4 Conserving Aboriginal
cultural heritage sites 363
13.5 Conserving historic sites 367
13.6 Management lessons
and principles 370

Chapter 14
Threats to Protected Areas 371
14.1 Underlying causes 372
14.2 Type and nature of threats 377
14.3 Preventing environmental threats 386
14.4 Minimising environmental
threats 392
14.5 Management lessons
and principles 400

Chapter 15
Incident Management 402
15.1 Organisations with incident
management responsibilities 403
15.2 Managing incident responses 404
15.3 Managing fire incidents 412
15.4 Managing wildlife incidents 417
15.5 Managing incidents arising
from natural phenomena 418
15.6 Management lessons
and principles 422

Chapter 16
Tourism and Recreation 424
16.1 Global tourism and
environmental performance 427
16.2 Tourism and recreation in
Australia's protected areas 429
16.3 Managing tourism
and recreation 442
16.4 Managing for quality 456
16.5 Management lessons
and principles 462

Chapter 17
Working with the Community 465
17.1 A critical working relationship 466
17.2 Making the relationship work 469
17.3 Communicating with
stakeholders 474
17.4 Interpretation: Communicating
with heart and mind 484
17.5 Management lessons
and principles 491

Chapter 18
Indigenous People and
Protected Areas 493
18.1 Who are Indigenous people? 494
18.2 'Caring for country' 495
18.3 Land rights 498
18.4 Indigenous people and
protected areas 499
18.5 Implications for Aboriginal
land management 503
18.6 Management lessons
and principles 507

Chapter 19
Linking the Landscape 509
19.1 Importance of linkages 511
19.2 Conservation at a
regional scale 515
19.3 A menu of policy
instruments 525
19.4 Management lessons
and principles 530

Chapter 20
Marine Protected Areas 532
20.1 Importance of marine
protected areas 533
20.2 Australian marine
environments 534
20.3 Marine management issues 539
20.4 Marine park management 541
20.5 Management lessons
and principles 552

Chapter 21
Evaluating Management
Effectiveness   553
21.1   Purposes of management
        effectiveness evaluation   554
21.2   Developing evaluation systems   558
21.3   Guidelines for evaluating
        management effectiveness   563
21.4   Management lessons
        and principles   567

Chapter 22
Futures and Visions   569
Appendices   575

Appendix 1
        Chronology of Major Events   576

Appendix 2
        Extent of Australian
        protected areas   584

Commonwealth
protected areas   585
Australian Capital Territory
protected areas   587
New South Wales
protected areas   588
Northern Territory
protected areas   590
Queensland protected areas   592
South Australia
protected areas   594
Tasmania protected areas   596
Victoria protected areas   598
Western Australia
protected areas   600
International designations   602

References   603
Index   629

List of Figures

Figure 1 Structure of the book   xlvi

Figure 1.1 Australia—island continent
                                          colour section

Figure 1.2 El Niño effect, drought effects,
and bushfires in Blue Mountains
National Park   15

Figure 1.3 Australia climate   colour section

Figure 1.4 Australia topography   colour section

Figure 1.5 Australia's protected areas with
respect to IBRA Version 5.1 bioregions   23

Figure 1.6 Major vegetation changes
since 1788   colour section

Figure 1.7 Location of threatened
plant species   27

Figure 1.8 Distribution of species
richness for birds   29

Figure 1.9 Location of Australia's
threatened bird species   29

Figure 1.10 Distribution of species
richness for reptiles   30

Figure 1.11 Location of Australia's
threatened reptile species   30

Figure 1.12 Distribution of species
richness for amphibians   31

Figure 1.13 Location of Australia's
threatened amphibian species   31

Figure 1.14 Distribution of species
richness for mammals   32

Figure 1.15 Location of Australia's
threatened mammal species   32

Figure 2.1 Growth in Australian
protected areas 1968–2002   42

Figure 3.1 Classification system for the
values of nature   79

Figure 3.2 Fred Williams's *Waterpond in
a Landscape II*   colour section

Figure 3.3 Extent of the world's
protected areas   colour section

Figure 3.4 Global distribution of
natural and mixed World
Heritage sites   colour section

Figure 3.5 The World Heritage emblem   90

Figure 3.6 Australian protected areas   106

Figure 4.1 Some inputs into the process
of management   114

Figure 4.2 The four functions of
management   115

Figure 5.1 Reservation levels for
IBRA regions   139

Figure 5.2 Reservation priorities for
IBRA regions   139

Figure 5.3 Australia's marine jurisdiction   140

Figure 5.4 Protected area network
in the South Coast Region of
Western Australia   150

Figure 5.5 Key Victorian forest planning
instruments   157

Figure 5.6 Proportion of public land in
forest management zones, plantations,
and conservation reserves   158

Figure 5.7 Process used to develop the forest zoning scheme 159

Figure 6.1 Level of service framework 166

Figure 6.2 State-wide maps used to illustrate variability between parks 182

Figure 6.3 The future role of *State of the Parks* reporting in Parks Victoria's management model 183

Figure 7.1 An adaptive planning process 198

Figure 7.2 Outline of a rational, adaptive, and participatory planning process 199

Figure 7.3 Some Kosciuszko National Park planning documents 201

Figure 7.4 From vision to performance standard 210

Figure 8.1 Undersupply of public goods 221

Figure 8.2 Supply, demand, and market efficiency 229

Figure 8.3 Consumer and producer surplus 230

Figure 8.4 EMC revenue 1993 to 2003 235

Figure 8.5 GBRMPA revenue sources for 2002–03 235

Figure 8.6 Change in consumer surplus following a public infrastructure investment 239

Figure 9.1 Example of layout for a 'typical' budget report for a major region of a state 261

Figure 10.1 The global environmental management context within which protected area management operates 270

Figure 10.2 Protected area management and environmental accountability 271

Figure 10.3 Moreton Island National Park 281

Figure 12.1 Number of Gould's Petrel fledglings produced on Cabbage Tree Island for ten seasons, 1989 to 1998 316

Figure 12.2 Location of Anangu Lands in three Australian jurisdictions 324

Figure 12.3 Carbonate karsts in Australia 337

Figure 13.1 Heritage conservation planning framework 361

Figure 14.1 Distribution of some pest animals in Australia 381

Figure 14.2 Concentration of weeds in Australia 382

Figure 14.3 Changes in the measured rate of muddy estuarine bank erosion in response to management controls 384

Figure 14.4 Great escarpment of eastern Australia 388

Figure 14.5 Australian Alps–Great Escarpment conservation corridors 389

Figure 14.6 Continental-scale corridors of mountain protected areas 390

Figure 15.1 The incident control team 406

Figure 15.2 Average frequency of large bushfires 413

Figure 16.1 Recreation opportunities on public land in Victoria 439

Figure 16.2 RBSim output example: predicted average queuing time at car parks, Twelve Apostles, Port Campbell National Park Victoria, for 2001, 2006, 2011 440

Figure 16.3 Visitor codes of conduct produced for the Australian Alps National Parks 448

Figure 16.4 A planning framework for visitor destinations 449

Figure 19.1 Fitzgerald Biosphere Reserve showing national park core area and surrounding buffer-corridor zone 521

Figure 19.2 Macro-corridor network in the south coast region of Western Australia 522

Figure 21.1 Management cycle for the Tasmanian Wilderness World Heritage Area 556

Figure 21.2 Evaluation in the management cycle 558

Figure 21.3 Process for establishing an evaluation system 561

Figure A2.1 Protected areas under Commonwealth jurisdiction 585

Figure A2.2 Protected areas under ACT jurisdiction 587

Figure A2.3 Protected areas under NSW jurisdiction 588

Figure A2.4 Protected areas under Northern Territory jurisdiction 590

Figure A2.5 Protected areas under Queensland jurisdiction   592

Figure A2.6 Protected areas under South Australian jurisdiction   594

Figure A2.7 Protected areas under Tasmanian jurisdiction   596

Figure A2.8 Protected areas under Victorian jurisdiction   598

Figure A2.9 Protected areas under Western Australian jurisdiction   600

Figure A2.10 World Heritage Areas, Biosphere Reserves, and Ramsar Wetlands   602

List of Tables

Table 1 Management themes and desired outcomes   xliii

Table 1.1 Earth's evolutionary development with particular reference to Australia   4

Table 1.2 Distinctive characteristics of the Australian continent   12

Table 1.3 Some Australian atmospheric (and climate) changes   14

Table 1.4 The diversity of life on Earth   21

Table 1.5 Australian species diversity   22

Table 1.6 The conservation status of Australia's flora   27

Table 1.7 The conservation status of Australia's birds   29

Table 1.8 The conservation status of Australia's reptiles   30

Table 1.9 The conservation status of Australia's amphibians   31

Table 1.10 The conservation status of Australia's mammals   32

Table 2.1 Selected influential events and milestones in the history of Australian protected areas   43

Table 2.2 An economic classification of goods   50

Table 2.3 Key initiatives for sustainable development   51

Table 2.4 Modes of protected area governance in Australia   59

Table 3.1 Australian protected area management departments and agencies   95

Table 3.2 Extent of Australian terrestrial protected areas (by designation)   108

Table 3.3 Extent of Australian marine protected areas (by designation)   110

Table 3.4 Extent of Australian terrestrial protected areas (by IUCN category)   110

Table 3.5 Extent of Australian marine protected areas (by IUCN category)   111

Table 5.1 Conservation status of the broad vegetational structural types in Australia   135

Table 5.2 Characteristics of the RFA and LCC processes   148

Table 6.1 Types of information systems   185

Table 7.1 Kosciuszko National Park planning, 1965–2004   200

Table 7.2 Good practices in protected area planning participation   205

Table 7.3 A classification of ends and means   209

Table 8.1 Selected examples of visitor and commercial use charges for Kangaroo Island National Parks and Wildlife sites   225

Table 8.2 Revenue objectives for facilities, services, and values   231

Table 8.3 Revenue collection methods   233

Table 8.4 Results for a sample of
Australian travel cost studies   239

Table 8.5 Example calculation of net
present value   246

Table 8.6 Benefit–cost analysis of forest
reservation in East Gippsland   247

Table 10.1 Sustainable futures and
strategic considerations   273

Table 10.2 Sustainability policy
and planning   275

Table 10.3 Indicators and measures
utilised in analysing accommodation
providers' performance in comparison
to baseline levels   276

Table 10.4 Average per capita
performance for environmental
management parameters   277

Table 10.5 Estimates of water
consumption relevant to
protected areas   283

Table 11.1 Examples of operational
planning approvals   294

Table 11.2 Selected competency
qualifications required for protected area
operations staff   295

Table 12.1 Fauna inventory: Some field
survey techniques used   317

Table 12.2 An Australian wildlife year:
some fauna management implications   325

Table 12.3 Examples of soil
management actions   336

Table 12.4 Some karst management
actions   339

Table 12.5 Management considerations
in utilising geological materials for roads,
tracks and other purposes   341

Table 12.6 Some measures for
managing scenery   342

Table 13.1 One categorisation of
cultural heritage resources   354

Table 13.2 Some Aboriginal sites   354

Table 13.3 Some significant Aboriginal
sites and their age   355

Table 13.4 Some historic cultural heritage
resources that may be managed by
protected area organisations   355

Table 13.5 Some examples of NSW
'standing structure' (built environment)
historic heritage sites and
their management   357

Table 14.1 Types of threats to
protected areas   377

Table 14.2 Some mammalian pest
animals introduced to Australia and
some control techniques   380

Table 15.1 Skills required for a
fire incident   409

Table 16.1 Visitation statistics, Seal Bay
Conservation Park, 1988–2002   430

Table 16.2 Potential conflicts between
recreation groups in protected areas   434

Table 16.3 Recreation Opportunity
Spectrum as utilised by the NSW NPWS   436

Table 16.4 Indicative tourism and
recreation activities undertaken for
Recreation Opportunity Classes   437

Table 16.5 Environmental threats to
protected areas from tourism   443

Table 16.6 Visitor expectations of
services relative to the nature of
visitor destination settings based
on the ROS   458

Table 17.1 Examples of the many 'public
faces' of protected area organisations
(with related actions and public
interaction skills)   467

Table 17.2 Communication methods   475

Table 17.3 Types of interpretive guides   489

Table 17.4 Positive and negative aspects
of some interpretive media   490

Table 21.1 Summary of the WCPA
framework   559

Table 21.2 Examples of evaluation
assessments undertaken in the Great
Barrier Reef   565

Table A1.1 Chronology of major events   576

Table A2.1 Proportion of land managed as protected area in Australia   584

Table A2.2 Protected areas by type managed by the Commonwealth Government   586

Table A2.3 Commonwealth protected areas by IUCN category   586

Table A2.4 Protected areas by type managed by the ACT Government   587

Table A2.5 ACT protected areas by IUCN category   587

Table A2.6 Proportion of land managed as protected area in the ACT   587

Table A2.7 Protected areas by type managed by the NSW Government plus IPAs   589

Table A2.8 NSW protected areas by IUCN category   589

Table A2.9 Proportion of land managed as protected area in NSW   589

Table A2.10 Protected areas by type managed by the Northern Territory Government plus IPAs   591

Table A2.11 Northern Territory protected areas by IUCN category   591

Table A2.12 Proportion of land managed as protected area in the Northern Territory   591

Table A2.13 Protected areas by type managed by the Queensland Government plus IPAs   593

Table A2.14 Queensland protected areas by IUCN category   593

Table A2.15 Proportion of land managed as protected area in Queensland   593

Table A2.16 Protected areas by type managed by the South Australian Government, plus IPAs   595

Table A2.17 South Australian protected areas by IUCN category   595

Table A2.18 Proportion of land managed as protected area in South Australia   595

Table A2.19 Protected areas by type managed by the Tasmanian Government, plus IPAs   597

Table A2.20 Tasmanian protected areas by IUCN category   597

Table A2.21 Proportion of land managed as protected area in Tasmania   597

Table A2.22 Protected areas by type managed by the Victorian Government   599

Table A2.23 Victorian protected areas by IUCN category   599

Table A2.24 Proportion of land managed as protected area in Victoria   599

Table A2.25 Protected areas by type managed by the Western Australian Government, plus IPAs   601

Table A2.26 Western Australian protected areas by IUCN category   601

Table A2.27 Proportion of land managed as protected area in Western Australia   601

List of Photographs

Wollemi Pine, Wollemi National
Park, NSW   10

Cambrian/Precambrian boundary,
Wilkawillana Gorge, Flinders Ranges
National Park, SA   11

Summer storm clouds over Kosciuszko
National Park, NSW   14

Ribbon Gum (*Eucalyptus viminalis*) and
Alpine Ash (*Eucalyptus delegatensis*),
Kosciuszko National Park   16

Wetland, Yellow Water, Kakadu National
Park, NT   17

Rocky shore and seal colony, Montague
Island Nature Reserve, NSW   18

Termite mounds, Tennant Creek, NT   18

Mt Jagungal, Kosciuszko National
Park, NSW   19

Sunrise over Sydney, NSW   20

Premier Bob Carr, South East Forests
National Park, NSW   25

Uluṟu–Kata Tjuṯa National Park,
NT   colour section

Fan Palms, Wet Tropics World
Heritage Area, Qld   colour section

Jagungal Wilderness, Mt Jagungal,
Kosciuszko National Park,
NSW   colour section

Diving with Whale Sharks (*Rhincodon
typus*) at Ningaloo Reef, WA   colour section

Cave painting, Nourlangi, Kakadu
National Park, NT   colour section

Male Superb Lyrebird (*Menura
novaehollandiae*) in display   colour section

Slender Blue-tongue Lizard
(*Cyclodomorphus venustus*)   30

Corroboree Frog
(*Pseudophryne corroboree*)   31

Australian Sea Lion (*Neophoca cinerea*)
at Seal Bay Conservation Park,
Kangaroo Island, SA   32

Scientist surveying for small mammals,
Kosciuszko National Park   33

Marine flora (kelp)   34

Aboriginal Cave Paintings, Carnarvon
Gorge National Park, Qld   37

The Belair National Park Board of
Commissioners on a park inspection
in the early 1920s, SA   38

Travertine terraces, Mammoth
Hot Springs, Yellowstone National
Park, USA   39

Boatshed and original causeway,
1899, Audley, Royal National
Park, NSW   40

Ranger David Kerr, first uniformed ranger
for NSW, circa 1960   41

Rangers, early 1970s, Smiggin Holes,
Kosciuszko National Park, NSW   46

2003 Vth World Parks Congress, Durban,
South Africa   60

Neville Gare, Superintendent, Kosciuszko
State Park, 1961   75

Wilsons Promontory National Park, Vic   77
Little Tern (*Sterna albifrons*)   83
Central Eastern Rainforest Reserves
    World Heritage Area, Dorrigo National
    Park, NSW   87
David Sheppard and IUCN headquarters,
    Gland, Switzerland   92
Norfolk Island National Park and
    Botanic Garden   96
Southwest National Park, Tas   100
Operational plans were used to design
    the Blue Mountains National Park
    Heritage Centre seen here under
    construction, NSW   116
Field officer setting out information sign
    to be routed   119
Then Chief Executive, NSW NPWS,
    Ms Robyn Kruk and Premier of NSW,
    Mr Bob Carr, launch NSW's 100th
    national park, SE Forests National
    Park, NSW   124
Dune Lake, Great Sandy National Park,
    Fraser Island World Heritage Area, Qld   127
Park staff fire training, Kosciuszko
    National Park, NSW   129
Columnar rhyolite geological feature,
    Sawn Rocks, extension to Mt Kaputar
    National Park, NSW   136
Snorkellers, Great Barrier Reef, Qld   138
South East Forest National Park, NSW   147
Visitor information sign, South East
    Forest National Park   153
Visitor lookout, South East Forest
    National Park   154
Biological survey camp in the Anangu
    Pitjantjatjara lands, SA   170
Charles Sturt University researcher
    Jamie Weber and Little Penguin
    at Montague Island Nature
    Reserve, NSW   172
Catherine Gillies, park management
    library, Kosciuszko National Park, NSW   177
Botanist and endangered rainforest
    species—Daintree Wet Tropics, Qld   178

Ranger rest break, remote area
    monitoring, NSW   187
Ranger on patrol, Kosciuszko National
    Park, NSW   204
Bushwalkers along Ramsay Beach,
    Hinchinbrook Island, Qld   212
Protection works, Mimosa Rocks National
    Park, NSW   218
Information stall at the 1994 'Picnic in
    the Park', Belair National Park, SA   222
Boardwalk, Seal Bay Conservation Park,
    Kangaroo Island, SA   224
Port Arthur Historic Site, Tas   236
Carnarvon Gorge, Carnarvon National
    Park, Qld   241
Warrumbungles National Park, NSW   241
Zoe Bay, Hinchinbrook Island National
    Park, Qld   241
Volunteers for the State Emergency
    Service of NSW assisting fire
    operations at Kosciuszko National
    Park, NSW   254
Management training at Dorrigo National
    Park, NSW   255
Project Manager Iris Paridaens at work   268
Electricity generation, South Gippsland
    wind farm near Toora, Victoria   272
Solar panels, Montague Island Nature
    Reserve, NSW   273
Photo voltaic array, Qld   282
Sydney Olympics recycling   284
Recycling, Point Pelee National
    Park, Canada   284
Road upgrading works, Ben Boyd
    National Park, NSW   289
Soil erosion, Flinders Ranges National
    Park/Gammon Ranges National
    Park, SA   291
Rabbit warren, Flinders Ranges National
    Park/Gammon Ranges National
    Park, SA   292
Planting saltbush, Flinders Ranges
    National Park/Gammon Ranges
    National Park, SA   293

Dr Catherine Pickering inspecting
gravel walking track, Kosciuszko
National Park, NSW   299
Property protection burn, Blue Mountains
National Park, NSW   302
Parks field officer backburning, Blue
Mountains National Park, NSW   302
Banding Gould's Petrel, Cabbage Tree
Island, NSW   315
Hairs collected from hair tubes are
analysed to assess mammal species
attracted by the bait inside   318
Elliot traps are used in small
mammal studies   318
Mountain Pygmy-possum   319
An adult female of *Amphylaeus morosus*
(sub-family Hylaeinae)   319
Tasmanian Cave Spider   320
Radio tracking sea turtles   321
Ginger Wikilyiri aerial spotting and
sharing his knowledge of land and
resource management   324
Surveying mangrove species at
permanent vegetation plots, Tully
River, Qld   331
Fungi   332
Hedley Tarn, Kosciuszko National
Park, NSW   334
Kosciuszko walkway under construction,
Kosciuszko National Park, NSW   335
Jenolan Caves, NSW   337
Speleologist, Yarrangobilly Caves,
Kosciuszko National Park, NSW   338
Thredbo landslip stabilisation works   340
Eucalypt regeneration at Geehi following
the January 2003 Australian Alps
fires, NSW   343
Fuel measurement, Blue Mountains
National Park, NSW   348
Zoologist and the endangered
Mountain Pygmy-possum, Kosciuszko
National Park, NSW   350
Mungo National Park, Willandra Lakes
World Heritage Area, NSW   353

Historic wagon, Barrington Tops National
Park, NSW   356
Old Telegraph Station, Alice Springs, NT   356
Rangers talking with property owner,
Culgoa National Park, NSW   362
Engraving site at Bulgandry, NSW,
scratched by vandals—before
restoration   364
Engraving site at Bulgandry, NSW,
scratched by vandals—after
restoration   364
Artefact site, Namadgi National
Park, ACT   365
Cape Otway Lighthouse, Vic   367
Ship's insignia engraved on rock,
Quarantine Station, Sydney Harbour
National Park, NSW   368
Lighthouse, Montague Island Nature
Reserve, NSW   370
A construction vehicle accident that
caused serious diesel pollution
to a creek, Kosciuszko National
Park, NSW   378
Pig control program in Namadgi National
Park ACT—distributing 'free feed'   379
A cat scavenging on King Penguin
(*Aptenodytes patagonica*) chicks
killed by Southern Giant Petrel
(*Macronectes giganteus*),
Macquarie Island, Tas   379
Weed: Monterey Pine (*Pinus radiata*)   379
Rockface vandalism, Lamington National
Park, Qld   382
The fox, a major pest to Australian fauna   382
Destruction of park signage by
vandals, NSW   383
Habitat clearing   386
Lantana researcher Daniel Stock,
Springfield National Park, Qld   394
The rabbit has had a major impact
on burrow-nesting seabirds on
Macquarie Island, Tas   397
Plant washdown unit for Cinnamon
Fungus control, Kangaroo Island, SA   400

NPWS chainsaw crew (Neville Brogan and colleague) 'mopping up', NSW   401

Search and rescue, Kosciuszko National Park, NSW   404

Incident Controller Dave Darlington, 2003 Australian Alps fires, Jindabyne, NSW   407

Multi-agency incident control team, 2003 Australian Alps fires, Jindabyne, NSW   407

NSW Rural Fire Service tankers and fire-fighters near Smiggin Holes, Australian Alps fires January 2003, NSW   408

Subalpine woodland burning near Dainers Gap, January 2003 Australian Alps fires, Kosciuszko National Park, NSW   410

Helicopter used for firefighting   411

Fire crew awaiting evacuation by helicopter after completing their shift   411

Fire trail used as a fire control line   416

Volunteers often play an important role in marine mammal incidents   417

Rehabilitation of oiled wildlife   419

Clearing up storm damage   420

Heavy snow, Kosciuszko National Park, NSW   421

Crocodile warning signs, Daintree River, Qld   423

Desert garbage—trailer abandoned on the French Line Track, SA   429

Visitors, Seal Bay Conservation Park, SA   430

Bushwalking group, Ben Boyd National Park, NSW   441

Walker impacts, South Coast Track, Southwest National Park, Tas   446

Researcher Carolyn Littlefair collecting data, Lamington National Park, Qld   447

Hardened site at Murramurang National Park, NSW   451

Montague Island Nature Reserve, NSW   451

Construction of boardwalk, Macquarie Island   452

Track profile monitoring, McKeown's Valley Track, Jenolan, NSW   453

Inappropriately placed campsite, Zoe Bay, Hinchinbrook Island, Qld   454

Valley of the Giants, WA   455

Tourists at Minnamurra Rainforest Centre, Budderoo National Park, NSW   456

Geoff Martin, Chief Ranger, Kosciuszko State Park, NSW   457

Ranger-guided walk, Border Ranges National Park, NSW   459

Parks and Wildlife Rangers Chas Delacoeur (left) and Greg Boehme assist an injured walker   459

Visitors on elevated walkway in the high country of Kosciuszko National Park, NSW   460

Boardwalk, Wet Tropics World Heritage Area, Qld   463

Department of Conservation information sign, Mount Cook National Park World Heritage Area, NZ   464

Friends of the Great Victoria Desert Parks, SA, during a desert clean-up in the unnamed conservation park   467

School students on a suspension bridge at Tarra-Bulga National Park, Vic   468

Volunteer Venturer scouts repainting heritage stairs, Montague Island Nature Reserve, NSW   470

Venturer scout Bernadette Dadday describing for television the heritage of Montague Island Native Reserve   478

NSW NPWS communications team Stuart Cohen and Penny Spoelder   478

Information shelter, Bungle Bungle National Park, WA   484

Entrance sign, Keep River National Park, NT   485

Interpretative sign, Keep River National Park, NT   485

Interpreter Pat Hall, leading a group of visitors at Fitzroy Falls, Moreton National Park, NSW   487

Interpretation sign using international
symbols, Mimosa Rocks National
Park, NSW 489
Local advisory committees at the South East
Forests National Park launch, NSW 491
Aboriginal ranger at Nitmiluk National
Park, NT, lights speargrass for an
early dry season burn 494
Parks and Wildlife Ranger Lance Spain
patrols the Katherine River in Nitmiluk
National Park, NT 504
NPW SA and Irrwanyere staff at the end
of work to relocate the camp ground
at Dalhousie Springs, Witjira National
Park, SA 506
Petroglyphs, Mootwingee Historic
Site, NSW 508
Granite landscape: South Bald Rock with
tors, Girraween National Park, Qld 510
Community planting at the opening of
the Burnett Bushcare Support
Centre, Qld 513

School children involved in rainforest
rehabilitation, Norfolk Island
Botanic Gardens 518
Lower Snowy River, Kosciuszko National
Park, NSW 519
Tree nursery, NSW 531
Quicksilver pontoon tourist
destination, Great Barrier Reef
Marine Park, Qld 535
Quicksilver tourist glass bottom boat
observation area, Great Barrier Reef
Marine Park, Qld 538
Governor Island Marine Reserve,
Bicheno, Tas 541
Education sign, Bicheno, Tas 550
Kelp, Bicheno, Tas 551
Penguin researcher Amy Jorgensen (right)
and volunteer monitoring Fairy Penguin
population numbers, Montague Island
Nature Reserve, NSW 563
Nourlangie Rock, Kakadu National
Park, NT 570

List of Backgrounders, Case Studies, and Snapshots

Snapshot 1.1 Australian Fossil Mammal Sites World Heritage Area, Riversleigh (inscribed 1994)   10

Snapshot 1.2 Australian Fossil Mammal Sites World Heritage Area, Naracoorte (inscribed 1994)   10

Snapshot 1.3 The Wollemi Pine (*Wollemia nobilis*), the Greater Blue Mountains Area World Heritage Area (inscribed 2000)   11

Snapshot 1.4 Shark Bay Marine Reserves World Heritage Area (inscribed 1991)   11

Snapshot 1.5 Attitudes to the desert centre: the literary record   18

Backgrounder 1.1 The diversity of life on Earth   21

Backgrounder 1.2 The biosphere   21

Backgrounder 1.3 Interim Biogeographic Regionalisation for Australia   22

Snapshot 1.6 The South East Forest National Park of NSW   25

Snapshot 1.7 Surveying for the elusive Long-footed Potoroo of south-eastern Australia   33

Case Study 2.1 Developing the first national protected area legislation for Australia   44

Snapshot 2.1 Natural Heritage Trust   48

Backgrounder 2.1 Sustainable development initiatives   51

Backgrounder 2.2 Traditional and emerging protected area paradigms   57

Snapshot 2.2 WPC Recommendation 17: Recognising and supporting a diversity of governance types for protected areas   61

Snapshot 2.3 Local government protected area on the Mornington Peninsula   61

Snapshot 2.4 Nitmiluk (Katherine Gorge) National Park   62

Snapshot 2.5 Community management networks   62

Backgrounder 2.3 Protected area governance principles   62

Snapshot 2.6 Using art to change the way we see our heritage   71

Snapshot 2.7 Ethnic visitation to parks in the Sydney region   72

Case Study 2.2 Kosciuszko National Park: A beginning   75

Backgrounder 3.1 Convention for the Protection of the World's Cultural and Natural Heritage 1972 (World Heritage Convention)   86

Backgrounder 3.2 Fifth IUCN World Parks Congress 2003   88

Backgrounder 3.3 IUCN protected area categories   93

Snapshot 3.1 Bookmark Biosphere Reserve   95

Snapshot 3.2 Norfolk Island National Park
and Botanic Garden   96

Snapshot 3.3 Management of the Wet
Tropics World Heritage Area   97

Snapshot 3.4 Tri-National Wetlands
Memorandum of Understanding   98

Backgrounder 3.4 Many guiding hands   99

Case Study 3.1 Protected area
designations in Tasmania   100

Case Study 4.1 New Zealand Department
of Conservation strategic systems   117

Backgrounder 4.1 Evolving protected
area agency organisational structures
within Australia   120

Case Study 4.2 Leadership and pressure
at Brisbane Forest Park   125

Snapshot 4.1 Monitoring and evaluation—
experiences from Fraser Island World
Heritage Area   127

Backgrounder 5.1 Targets for a global
protected area network   137

Case Study 5.1 Global representative
system of marine protected areas   141

Snapshot 5.1 New marine national parks
for Victoria   142

Case Study 5.2 Establishment of MPAs
in Commonwealth Waters: South-East
Marine Region   143

Snapshot 5.2 C-Plan   146

Snapshot 5.3 Identifying gaps and
priorities for reserves in NSW   147

Case Study 5.3 Regional planning for
protected areas in the South Coast
Region, Western Australia   149

Case Study 5.4 The contribution of the
Land Conservation Council to reserve
selection in Victoria   152

Case Study 5.5 Selection of forest
reserves in North East Victoria   156

Case Study 6.1 Sustainable resource
allocation for visitors to Victoria's parks   165

Snapshot 6.1 Monitoring programs   168

Case Study 6.2 Remote sensing in
protected area management   169

Case Study 6.3 Biological survey of
South Australia   170

Snapshot 6.2 Research partnership
between Narooma District of NSW
P&W and Charles Sturt University   172

Case Study 6.4 Using historical long-term
vegetation studies for management
and restoration   175

Case Study 6.5 Establishing and
maintaining a reference collection   176

Backgrounder 6.1 The World Conservation
Monitoring Centre   178

Backgrounder 6.2 Use of digital
information and technology in managing
protected areas   179

Case Study 6.6 *State of the Parks*
reporting: Parks Victoria's experience   181

Case Study 6.7 Management
Information System, Uganda   185

Case Study 7.1 A new plan of
management for Kosciuszko National Park   202

Case Study 7.2 Hinchinbrook Island
National Park Management Plan   212

Case Study 7.3 Zoning in the Great
Australian Bight Marine Park   215

Snapshot 8.1 The National Parks
Foundation of South Australia   222

Snapshot 8.2 Business plan for Fitzroy
Falls Visitor Centre   223

Case Study 8.1 User-pays fees and
regional development on Kangaroo Island   224

Snapshot 8.3 Charging for
ecosystem services   228

Backgrounder 8.1 Supply, demand,
and recreation pricing   229

Case Study 8.2 Great Barrier Reef Marine
Park Environmental Management Charge   234

Backgrounder 8.2 National Competition
Policy   236

Snapshot 8.4 Summary of results and
extract from the CV survey   243

Snapshot 8.5 Regional economic impact
of the Dorrigo and Gibraltar Range
National Parks   244

Backgrounder 8.3 Steps in a benefit–cost analysis   245

Snapshot 8.6 Benefit–cost analysis of reserving East Gippsland forests   247

Snapshot 9.1 Assigning funds to a capital works project   259

Snapshot 9.2 The process of developing a policy   264

Case Study 9.1 Access to genetic resources in Australia's protected areas—intellectual property and patenting issues   265

Snapshot 10.1 Tensions over water in southern Africa   271

Snapshot 10.2 Water plan for the Murray–Darling Basin   272

Backgrounder 10.1 Life cycle assessments   276

Backgrounder 10.2 Green Globe 21 and earthcheck™ benchmarking indicators for tourism operators   277

Case Study 10.1 Sustainable ranger housing   278

Snapshot 10.3 Government department introduces hybrid vehicles   280

Case Study 10.2 Sustainable park management   280

Snapshot 10.4 Landcare Research NZ   286

Case Study 11.1 Operation Bounceback, South Australia   291

Case Study 11.2 Walking track materials research, Kosciuszko National Park   298

Backgrounder 11.1 Managing a prescription burning operation   300

Backgrounder 11.2 Prescription burning tragedy, Ku-ring-Gai Chase National Park   303

Backgrounder 11.3 A day in the life of a ranger   304

Case Study 12.1 The Australian Natural Heritage Charter: A tool for protected area managers   311

Case Study 12.2 Recovery of an endangered species—the Gould's Petrel   314

Snapshot 12.1 The Mountain Pygmy-possum   319

Snapshot 12.2 Keeping invertebrates on the conservation agenda   319

Snapshot 12.3 The Richmond Birdwing Butterfly   319

Snapshot 12.4 Monitoring cave fauna in Tasmania   320

Snapshot 12.5 Conserving the Green and Golden Bell Frog   321

Snapshot 12.6 Looking after miyapunu: Indigenous management of marine turtles   321

Snapshot 12.7 Kuka Kanyini – looking after wildlife preferred by Anangu of the Pitjantjatjara Lands   323

Snapshot 12.8 Protecting endangered ecological communities in the ACT   329

Snapshot 12.9 The Wee Jasper Grevillea   330

Snapshot 12.10 Water in Barmah Forest   333

Snapshot 12.11 Elevated walkway to the summit of Mount Kosciuszko   335

Snapshot 12.12 Fossil site management, Naracoorte Caves   338

Backgrounder 12.1 Guidelines for caves and karst management   338

Backgrounder 12.2 Ecological fire management planning   344

Snapshot 12.13 Fire management—Wilsons Promontory National Park   349

Backgrounder 13.1 Cultural heritage in protected areas   359

Snapshot 13.1 Involving the community to assess cultural heritage   362

Backgrounder 13.2 Nine criteria used for assessing the national estate significance of a place   363

Snapshot 13.2 Cataloguing Aboriginal sites at Namadgi   365

Snapshot 13.3 Rock art at Namadgi   367

Snapshot 13.4 Adaptive reuse of Coolart homestead   368

Snapshot 13.5 Managing mining heritage in Victoria   369

Snapshot 14.1 Urban pressures on the Dandenong Ranges National Park   375

Snapshot 14.2 Chronology of changes to habitats and species in the Bega Valley since first European settlement 378

Snapshot 14.3 Managing Lower Gordon River erosion—geoconservation of landforms affected by tourism 384

Snapshot 14.4 Management of zinc toxicity 385

Backgrounder 14.1 Fragmentation of natural habitat and extinctions 387

Backgrounder 14.2 Environmental impact assessment 391

Snapshot 14.5 Willows—what's the worry? 393

Snapshot 14.6 Management of Lantana in eastern Australian subtropical rainforests 394

Snapshot 14.7 Weed management in central Australia 394

Snapshot 14.8 Western Shield 395

Snapshot 14.9 Eradication planning for invasive alien animal species on islands 396

Snapshot 14.10 Namadgi National Park feral pig control program 396

Snapshot 14.11 Vertebrate pest management in Macquarie Island Nature Reserve 397

Snapshot 14.12 Managing a legend— wild horse management In Kosciuszko National Park 398

Snapshot 14.13 Weeds, tourism, and climate change 399

Snapshot 14.14 Management of Cinnamon Fungus on Kangaroo Island 400

Snapshot 15.1 Safety and remote-area firefighting 410

Snapshot 15.2 The 2003 Australian Alps fires 413

Snapshot 15.3 Fire planning completed by the Southern Regional Fire Association of NSW 414

Snapshot 15.4 A whale rescue—50 False Killer Whales *(Pseudorca crassidens)* stranded at Seal Rocks, July 1992 418

Snapshot 15.5 Wildlife response to the *Iron Baron* oil spill in July 1995 419

Snapshot 15.6 Mystery Creek Cave flooding leads to drownings 420

Snapshot 15.7 Closure of visitor access to Minnamurra Falls 422

Backgrounder 16.1 Definitions of tourism and visitor use 425

Backgrounder 16.2 The origins of the concept of ecotourism 426

Snapshot 16.1 Four-wheel-drives in the desert 429

Snapshot 16.2 Balance on the beach: Sustainable tourism at Seal Bay Conservation Park, Kangaroo Island 430

Case Study 16.1 Binna Burra Mountain Lodge, Lamington National Park, Queensland 432

Backgrounder 16.3 Recreation Opportunity Spectrum 435

Backgrounder 16.4 Visitor-use data 438

Case Study 16.2 Visitor Management Model for Port Campbell National Park and Bay of Islands Coastal Park 439

Snapshot 16.3 The problems of overuse in Tasmanian parks 444

Backgrounder 16.5 ESD and the tourism industry 444

Snapshot 16.4 Sustainable recreational diving in marine protected areas 445

Backgrounder 16.6 Resisting the obvious 445

Snapshot 16.5 The Desert Parks Pass 446

Snapshot 16.6 Using interpretation to manage recreation in Lamington National Park 447

Snapshot 16.7 Australian Alps codes of conduct 448

Backgrounder 16.7 Limits of Acceptable Change, Visitor Impact Management, and other visitor management models 450

Snapshot 16.8 Sustainable visitor-use limits: Montague Island Nature Reserve 451

Snapshot 16.9 Planning visitor use at sub-Antarctic Macquarie Island 452

Snapshot 16.10 Determining carrying capacity and visitor impacts at Jenolan Caves   453

Snapshot 16.11 Balancing seal conservation and sustainable visitor management   454

Snapshot 16.12 Saving the giants   455

Snapshot 16.13 Key destination: Minnamurra Rainforest Centre, Budderoo National Park, southern NSW   456

Snapshot 16.14 Rainforests for the people   457

Case Study 16.3 National visitor asset management   461

Snapshot 17.1 Tarra–Bulga National Park: a little help from some friends   468

Snapshot 17.2 Parks Victoria and tourism following the 2003 Australia Alps fires   468

Snapshot 17.3 Communication plan—Southern Region of NSW NPWS   470

Case Study 17.1 Management of wild dogs and foxes: a nil tenure approach to a landscape issue   471

Snapshot 17.4 Community surveys   474

Backgrounder 17.1 What is communication?   475

Case Study 17.2 Managing public information—Australian Alps bushfires, 2003   476

Snapshot 17.5 The right channels   477

Snapshot 17.6 The Cooma necklace   479

Snapshot 17.7 Doing a good turn for Little Terns   479

Snapshot 17.8 Community involvement in the Great Barrier Reef   480

Snapshot 17.9 Involving local government in conservation—dog and cat control   481

Snapshot 17.10 Managing community participation and consultation programs   481

Case Study 17.3 The use of marketing as a tool in protected area management   482

Snapshot 17.11 Interpretation and marketing—as experienced by a visitor   486

Backgrounder 17.2 Theories of learning and communication   488

Snapshot 18.1 Conservation: A family matter   496

Snapshot 18.2 Traditional Aboriginal land management and biological survey on the Anangu Pitjantjatjara Lands, South Australia   497

Snapshot 18.3 Joint management at Uluru–Kata Tjuta National Park   500

Snapshot 18.4 Bama involvement in the management of the Wet Tropics of Queensland World Heritage Area (WTWHA)   500

Snapshot 18.5 Indigenous peoples managing Indigenous Protected Areas—Nantawarrina, South Australia   501

Snapshot 18.6 A traditional owner's perspective on Indigenous protected areas   502

Snapshot 18.7 Learning about issues for Indigenous people in protected area management   502

Snapshot 18.8 Indigenous peoples and marine protected areas   503

Snapshot 18.9 A walk through time   504

Snapshot 18.10 Empowerment (Djabugay Ranger and Land Management Agency)   505

Snapshot 18.11 Go for it!   505

Case Study 18.1 Looking After the Munda (land)—joint management of Witjira National Park   506

Snapshot 19.1 WildCountry partnerships   516

Case Study 19.1 Stepping outside—a bioregional and landscape approach to nature conservation   520

Snapshot 19.2 Blackwood Biodiversity Program, south-west Western Australia   524

Snapshot 19.3 Woodland Watch—community conservation of eucalypt woodlands, Western Australia   530

Snapshot 20.1 Ningaloo Reef, WA   539

Backgrounder 20.1 Coral bleaching   539

Snapshot 20.2 Mooring policy for WA Marine Conservation Reserves   542

Snapshot 20.3 Environmental impact assessment on the Great Barrier Reef   542

Case Study 20.1 Day-to-day management
of the Great Barrier Reef Marine Park   543
Case Study 20.2 Re-zoning of the Great
Barrier Reef Marine Park   545
Snapshot 20.4 Tasmanian Seamounts
Marine Reserve   547
Snapshot 20.5 Management of whale
shark watching, Ningaloo Marine Park   548
Snapshot 20.6 Management of the
Great Barrier Reef trawl fishery   550
Case Study 21.1 Is the management
plan achieving its objectives?   555

Snapshot 21.1 Evaluating the dingo
education campaign, Fraser Island   558
Backgrounder 21.1 Process for developing
a monitoring and evaluation system   561
Backgrounder 21.2 Internationally applied
systems for evaluating management
effectiveness   562
Snapshot 21.2 Evaluation of effectiveness
of protected area management in India   564
Case Study 21.2 Evaluating management
effectiveness in the Great Barrier Reef
Marine Park   565

Australian Jurisdictions and Protected Area Management Agencies

Jurisdiction	Jurisdiction Abbreviation	Agency	Agency Acronym	Notes	Web Site
Commonwealth		Parks Australia	PA	PA incorporates elements of the former Australian National Parks and Wildlife Service (ANPWS), and now operates within the Commonwealth Department of Environment and Heritage	http://www.deh.gov.au/parks/
		Wet Tropics Management Authority	WTMA		http://www.wettropics.gov.au/
		Great Barrier Reef Marine Park Authority	GBRMPA		http://www.gbrmpa.gov.au/
Australian Capital Territory	ACT	Australian Capital Territory Parks and Conservation Service	ACT PCS	ACT PCS operates within Environment ACT, which is part of the Department of Urban Services	http://www.environment.act.gov.au/
New South Wales	NSW	New South Wales Parks and Wildlife	NSW P&W	NSW P&W is part of the Department of Environment and Conservation	http://www.nationalparks.nsw.gov.au/
Northern Territory	NT	Parks and Wildlife Commission of the Northern Territory	PWCNT	PWCNT is part of the Department of Lands, Planning and Environment	http://www.nt.gov.au/ipe/pwcnt/

Jurisdiction	Jurisdiction Abbreviation	Agency	Agency Acronym	Notes	Web Site
Queensland	Qld	Queensland Parks and Wildlife Service	Qld PWS	Qld PWS is part of the Queensland Environmental Protection Agency	http://www.epa.qld.gov.au/
South Australia	SA	National Parks and Wildlife South Australia	NPWSA	NPWSA is part of the Department for Environment and Heritage	http://www.environment.sa.gov.au/
Tasmania	Tas	Parks and Wildlife Service Parks, Tasmania	PWST	PWST is part of the Department of Tourism, Parks, Heritage and the Arts	http://www.dtpha.tas.gov.au/parks
Victoria	Vic	Parks Victoria	PV	Parks Victoria operates as an independent corporation, but receives its direction from the Department of Sustainability and Environment	http://wwww.parkweb.vic.gov.au/
Western Australia	WA	Department of Conservation and Land Management	CALM		http://www.calm.wa.gov.au/

Related web sites

ABHF	Australian Bush Heritage Fund	http://www.bushheritage.asn.au/
	Greenpeace	http://www.greenpeace.org/
	Landcare Australia	http://www.landcareaustralia.com.au/
ABS	Australian Bureau of Statistics	http://www.abs.gov.au/
ACF	Australian Conservation Foundation	http://www.acfonline.org.au/
ALGA	Australian Local Government Association	http://www.alga.asn.au/
CBD	Convention on Biological Diversity	http://www.biodiv.org/
DC	New Zealand Department of Conservation	http://www.doc.govt.nz/
FAO	Food and Agriculture Organisation of the United Nations	http://www.fao.org/
GA	Greening Australia	http://www.greeningaustralia.org.au/
	Birds Australia	http://www.birdsaustralia.com.au/
GEF	Global Environment Fund – United Nations Development Program	http://www.undp.org/gef/
ICOMOS	International Council on Monuments and Sites	http://www.icomos.org/

IUCN	International Union for the Conservation of Nature (The World Conservation Union)	http://www.iucn.org/
IUCN WCPA	IUCN World Commission on Protected Areas	http://www.iucn.org/themes/wcpa/
MAB	UNESCO Man and the Biosphere (World Biosphere Reserve)	http://www.unesco.org/mab/
OECD	Organisation for Economic Co-Operation and Development	http://www.oecd.org/
Rasmar	Ramsar convention	http://www.ramsar.org/
UN	United Nations	http://www.un.org/
UNEP	United Nations Environment Programme	http://www.unep.org/
UNESCO	United Nations Educational, Scientific and Cultural Organization	http://www.unesco.org/
WCMC	World Conservation Monitoring Centre	http://www.wcmc.org.uk/
WH	World Heritage Areas	http://whc.unesco.org/
WWF	World Wide Fund for Nature	http://www.panda.org/
	World Wide Fund for Nature (Australia)	http://www.wwf.org.au/

Acknowledgments
(First Edition)

The first edition of this book was achieved as a result of a collective effort by many people from around Australia. These contributions were highly valued and helped to ensure the success of the book. Recognition for these contributions is reproduced in this new edition as it was presented in the original text. It should be noted that the name of many organisations and the titles of many individuals will have changed.

The members of the ANZECC Working Group on National Parks and Protected Area Management and our state advisers have been most supportive, efficient and patient. In particular, our thanks go to David Barrington, Peter Bosworth, Doug Brown, Michael Butler, Jeff Francis, Rod Gowans, Colin Ingram, Peter Ogilvie, David Phillips, David Ritchie, John Senior, and Bill Woodruff. These people played the primary role in identifying and obtaining potential Case Study and Snapshot contributions from people in their state.

The protected area maps and protected area status tables in part C were derived from the Collaborative Australian Protected Areas Database (CAPAD). This information is compiled and maintained by Environment Australia. Organisations that contribute to CAPAD include:

- Western Australian Department of Conservation and Land Management
- South Australian Department of Environment and Heritage, National Parks and Wildlife Service
- South Australian Department of Administrative and Information Services, Forestry SA
- Parks and Wildlife Commission of the Northern Territory
- Queensland Department of Environment and Heritage, Environment Protection Agency
- Queensland Department of Natural Resources
- New South Wales National Parks and Wildlife Service
- Australian Capital Territory Department of Urban Services
- Australian Capital Territory Land Information Centre and Environment ACT
- Victorian Department of Natural Resources and Environment, Land Information Group
- Tasmanian Department of Primary Industries, Water and Environment, Resource Management and Conservation.

The IUCN World Commission on Protected Areas (WCPA) network assisted the development of this book. It was enthusiastically supported by the Chair of IUCN WCPA, Adrian Phillips, and Head of the IUCN Protected Areas Program, David Sheppard. It was sponsored by the Australia Region of the IUCN WCPA. It has been prepared consistent with IUCN WCPA strategic planning objectives, and in particular 'the need to strengthen capacity and effectiveness of managers of protected areas through the provision of management guidelines, tools and information'.

Lee Thomas and David Phillips have been our close contacts at Environment Australia. As members of the steering committee they have supported us and provided advice throughout the project. Lee also provided considerable assistance in his role as Vice-chair, IUCN World Commission on Protected Areas Australia Region Committee. David played a major role in facilitating production of the maps and figures. Marc Hockings organised the placement of the supplementary web-based text onto the IUCN web site.

Our appreciation is extended (in no particular order) to:

- Julie Collins for writing Chapter 15
- Peter Bridgewater for his personal support, and the then Australia Nature Conservation Agency for the States Assistance Grant (1996), which permitted this project to proceed
- Peter Taylor, Hillary Sullivan, and Colin Griffith at Environment Australia, who provided critical assistance and encouragement in the initial stages of the project
- Charles Sturt University, University of Queensland, NSW National Parks and Wildlife Service, Environment Australia, and the CRC for Sustainable Tourism for enthusiastically supporting the project
- Ms Robyn Kruk, former Director General NSW NPWS, who strongly supported the concept of the book, and provided the opportunity and support for its development
- Alastair Howard, the then Executive Director Operations NSW NPWS, who strongly supported the concept of the book and assisted the NPWS to cover the absence of Graeme Worboys, then a Senior Manager with the service
- Robert Horrocks, manager of the NSW Premier's Department Executive Development Program, who encouraged and facilitated the development of the book concept
- IUCN/WCPA Australia Region for adopting the book as a special project, and for supporting its development
- Adrian Phillips for writing the Foreword and for his enthusiastic support for the project
- Tony Fleming and Bob Flannigan of Southern Region NSW NPWS for their outstanding support for the project and for editorial contributions
- Ian Close, NSW NPWS for editorial assistance and liaison with Oxford University Press
- Gavin Gatenby, NSW NPWS for developing the illustrations for the book

- Ian Charles, Charles-Walsh Natural Tourism Services, for the preparation of maps and some of the figures
- Ann Hardy, Natural Resources System Program, Environment Australia, for her major contribution to the preparation of the maps and tables in the Appendix
- Robyn Korn for assistance with compiling historical material
- Rosemary Black and Janet Mackay for their valuable contribution 'A day in the life of a ranger', for their involvement in initial discussions about the book, and for editorial contributions
- Bob Beeton for his substantial and generous help in providing guidance and background information
- Marc Hockings, Gregor Manson, Bob Conroy, Tony Fleming, Lee Thomas, David Phillips, Janet Mackay, Sue Feary, and members of the ANZECC Working Group on National Parks and Protected Areas, who provided constructive editorial comments on the manuscript
- Gabrielle Wiltshire, who assisted as Research Assistant in the early stages of the book's development
- Nicole Zammit, who assisted above and beyond the call of duty in the final stages of the project
- Ray Fowke, NSW NPWS for the use of his manuscript on organisational structures (Chapter 4)
- Brian Leahy for his effort in researching and providing historical photographs of Royal National Park
- Brian Gilligan and Sally Barnes, NSW NPWS, for their support for the development of the book
- Mary White for her constructive editorial comments and important improvements to Chapter 1

For their support, advice on specific topics, and valuable contacts, we thank:

Jack Baker, Johannes Bauer, John Benson, Max Beukers, Tony Blanks, Miles Boak, Raoul Boughton, John Briggs, Andrew Claridge, Stuart Cohen, Valda Corrigan, Derrin Davis, James Dawson, Stephen Dovey, Derek Farr, Allan Fox, Elery Hamilton-Smith, Neville Gare, Roger Good, George Grossek, Mark Hallam, Larry Hamilton, Alistair Henchman, Rob Hunt, Omar Ibrahim, Nick Klomp, Frances Knight, Linda Lidbury,

Kreg Lindberg, Dan Lunney, Bob McKercher, Jim Muldoon, Monica Muranyi, Lorraine Oliver, David Pemberton, Rebecca Pirzl, Bob Pressey, Michael Saxon, Bo Slowiak, Warwick Smith, Andy Spate, Amanda Sullivan, and Peter Windle.

We would like to extend our deep appreciation to those people who submitted contributions for consideration as Case Studies or Snapshots. We received an overwhelming number of outstanding contributions. Unfortunately due to space limitations, we could not include them all. Contributors of Case Studies and Snapshots appearing in the book are:

- Roger Armstrong, Department of Conservation and Land Management, WA
- Stephen Arnold, SA National Parks and Wildlife Service
- Lynn Baker, Wallambia Consultants
- David Banner, Heritage Victoria
- Tom Barrett, NSW National Parks and Wildlife Service
- Rosemary Black, Charles Sturt University
- Samantha Bradley, Parks Victoria
- Robert Brandle, Department of Environment, Heritage and Aboriginal Affairs, SA
- John Briggs, NSW National Parks and Wildlife Service
- Linda Broome, NSW National Parks and Wildlife Service
- Craig Campbell, Parks Victoria
- Bill Carter, University of Queensland
- Stuart Cohen, NSW National Parks and Wildlife Service
- Peter Copley, Department of Environment, Heritage and Aboriginal Affairs, SA
- Geof Copson, Parks and Wildlife Service, Tasmania
- Kevin Curran, Parks Victoria
- Bruce Davis, University of Tasmania
- Derrin Davis, Southern Cross University
- Pearce Dougherty, SA National Parks and Wildlife Service
- Michael Driessen, Parks and Wildlife Service, Tasmania
- Anthony English, NSW National Parks and Wildlife Service
- Jeff Foulkes, Department of Environment, Heritage and Aboriginal Affairs, SA
- Robert Furner, Department of Environment, Heritage and Aboriginal Affairs, SA
- Neville Gare, former superintendent, NSW National Parks and Wildlife Service
- Catherine Gillies, NSW National Parks and Wildlife Service
- Roger Good, NSW National Parks and Wildlife Service
- David Greentree, NSW Department of Land and Water Conservation
- John Grinpukel, Parks Victoria
- Angie Gutowski, Parks Victoria
- Nicholas Hall, Australian Heritage Commission
- Barbara Hardy, The National Parks Foundation of South Australia Inc.
- John Hicks, Queensland Parks and Wildlife Service
- Marc Hockings, University of Queensland
- Stuart Johnston, NSW National Parks and Wildlife Service
- Graeme Kelleher, Senior Adviser to IUCN World Commission on Protected Areas
- Rod Kennett, Northern Territory University
- Peter Lavarack, Queensland Parks and Wildlife Service
- Bruce Lawson, Wet Tropics Management Authority
- Vikki Lee, NSW National Parks and Wildlife Service
- David Lloyd, Great Barrier Reef Marine Park Authority
- Trish MacDonald, ACT Parks and Conservation Service
- Janet Mackay, NSW National Parks and Wildlife Service
- David Major, NSW National Parks and Wildlife Service
- Russell Mason, Parks Victoria
- Paul McCluskey, formerly of Parks and Wildlife Commission, NT
- Stuart McMahon, NSW National Parks and Wildlife Service
- Bob Moffatt, NSW National Parks and Wildlife Service
- Bernard Morris, ACT Parks and Conservation Service
- Nanikiya Munungurritj, Dhimurru Land Management Aboriginal Corporation

- Monica Muranyi, ACT Parks and Conservation Service
- Bradley Nesbitt, Wallambia Consultants
- Mark O'Connor, environmental poet and writer
- Tim O'Loughlin, Tasmania Parks and Wildlife Service
- John O'Malley, Parks Victoria
- David Priddel, NSW National Parks and Wildlife Service
- Alison Ramsay, NSW National Parks and Wildlife Service
- Peter Reed, NSW National Parks and Wildlife Service
- Ben Rheinberger, Tasmania Parks and Wildlife Service
- Tony Robinson, Department of Environment, Heritage and Aboriginal Affairs, SA
- Darren Roso, ACT Parks and Conservation Service
- Katharine Sale, NSW National Parks and Wildlife Service
- John Senior, Parks Victoria
- Dermot Smyth, Smyth and Bahrdt Consultants
- Roy Speechley, Parks Victoria
- Allan Spessa, Environment Australia
- Ted Stabb, Forests Service, Department of Natural Resources and Environment
- Steve Szabo, Environment Australia
- Richard Thackway, Environment Australia
- Lee Thomas, Environment Australia
- Keith Twyford, SA National Parks and Wildlife Service
- Sharon Veale, NSW National Parks and Wildlife Service
- Fraser Vickery, SA National Parks and Wildlife Service
- Russell Watkinson, Wet Tropics Management Authority
- John Watson, Department of Conservation and Land Management, WA
- Lynn Webber, NSW National Parks and Wildlife Service
- Jim Whelan, Parks Victoria
- Kylie White, Forests Service, Department of Natural Resources and Environment
- Cliff Winfield, Department of Conservation and Land Management, WA
- Alan Young, NSW National Parks and Wildlife Service
- Frank Young, Añangu Pitjantjatjara Land Management
- Dino Zanon, Parks Victoria

Contributors of Case Studies appearing on the web site[1] associated with the book are:

- Paul Adam, University of New South Wales
- Stephen Arnold, National Parks and Wildlife, South Australia
- Lynn Baker, Wallambia Consultants
- Katharine Betts, Swinburne University
- Samantha Bradley, Parks Victoria
- Peter Bridgewater, Environment Australia
- Cheryl Brown, Yarrawarra Aboriginal Corporation
- Craig Campbell, Parks Victoria
- Stuart Cohen, NSW National Parks and Wildlife Service
- Peter Copley, Department of Environment, Heritage and Aboriginal Affairs, SA
- Dene Cordes, Department of Environment, Heritage and Aboriginal Affairs, SA
- Ian Cresswell, Environment Australia
- Peter Croft, NSW National Parks and Wildlife Service
- Kevin Curran, Parks Victoria
- Bruce Davis, University of Tasmania
- Karen Edyvane, South Australian Research and Development Institute
- Anthony English, NSW National Parks and Wildlife Service
- Neville Gare, formerly of NSW National Parks and Wildlife Service
- Claire Grant, Queensland Parks and Wildlife Service
- Angie Gutowski, Parks Victoria
- Barbara Hardy, Department of Environment, Heritage and Aboriginal Affairs, SA
- Terry Harper, Queensland Parks and Wildlife Service
- Colin Harris, Department of Environment, Heritage and Aboriginal Affairs, SA
- Peter Hensler, Queensland Department of Environment and Heritage
- Graeme Kelleher, Great Barrier Reef Marine Park Authority

- Dave Lambert, NSW National Parks and Wildlife Service
- Bruce Lawson, Wet Tropics Management Authority
- Brian Leahy, NSW National Parks and Wildlife Service
- Bruce Leaver, Tasmanian Resource Planning and Development Commission
- David Lloyd, Great Barrier Reef Marine Park Authority
- Pam Lunnon, NSW National Parks and Wildlife Service
- Cathy Mardell, NSW National Parks and Wildlife Service
- Russell Mason, Parks Victoria
- Christopher McCormack, NSW National Parks and Wildlife Service
- Stuart McMahon, NSW National Parks and Wildlife Service
- Ian Miles, Department of Natural Resources and Environment, Victoria
- Joanne Millar, Queensland Parks and Wildlife Service
- John Mortimer, NSW National Parks and Wildlife Service
- Jim Muldoon, Environment Australia
- Bradley Nesbitt, Wallambia Consultants
- Jan Palmer, North East Catchment Management Authority
- John Senior, Parks Victoria
- Peter Reed, NSW National Parks and Wildlife Service
- Corazon Sinha, University of Western Sydney
- Dermot Smyth, Smyth and Bahrdt Consultants
- Mark Sutton, NSW National Parks and Wildlife Service
- Steve Szabo, Environment Australia
- Richard Thackway, Environment Australia
- Lee Thomas, Environment Australia
- Keith Twyford, Department of Environment, Heritage and Aboriginal Affairs, SA
- Fraser Vickery, Department of Environment, Heritage and Aboriginal Affairs, SA
- Russell Watkinson, Wet Tropics Management Authority
- John Watson, Department of Conservation and Land Management, WA
- Lynn Webber, NSW National Parks and Wildlife Service
- Geoff Wescott, Deakin University
- Brent Williams, Parks and Wildlife Commission of the Northern Territory
- Paul Williams, Queensland Department of Environment and Heritage
- Alexandra Wyatt, NSW National Parks and Wildlife Service

Note

1   This web site has been decommissioned following publication of the second edition.

Acknowledgments
(Second Edition)

This second edition has been prepared thanks to the assistance of many people. Our special thanks are extended to Debra James, Lucy McLoughlin, and other staff of Oxford University Press. We would like to express our appreciation to Dr Karen Edyvane for writing Chapter 20; Dr Marc Hockings, Dr Fiona Leverington, and Robyn James for writing Chapter 21; and Julie Collins for reviewing Chapter 18.

International protected area management experts worked on the 'Principles' sections of the book. In September 2003, 59 experts attended a mountain protected area management workshop convened by IUCN WCPA Mountains Biome Vice-Chair Professor Larry Hamilton as part of the Vth IUCN World Parks Congress. The workshop was held at Didima in the uKhahlamba-Drakensberg World Heritage Site, South Africa. Appreciation is expressed to the following international experts who helped to improve the principles at the workshop:

- Paulina Arroyo, Ecuador Program, The Nature Conservancy, Quito, Ecuador
- Rodney Atkins, Environment Australia, Canberra
- Dr Yuri Badenkov, Russian Academy of Science, Moscow, Russia
- Dr Bill Bainbridge Resource Management, Pietermaritzburg, South Africa
- Silvia Benítez, Director/Ecuador Program, The Nature Conservancy, Quito, Ecuador
- Dr Edwin Bernbaum, The Mountain Institute, Berkeley, California, USA
- Tom Cobb, Association for Protection of the Adirondacks, New York, USA
- Nicholas Conner, Economist, NSW National Parks and Wildlife Service
- Juan Pablo Contreras, Regional Director Protected Areas CONAF, Antofagosta, Chile
- Brent Corcoran, KwaZulu-Natal Nature Conservation Pietermaritzburg, South Africa
- Dr Pitamber Prasad Dhyani, G B Pant Institute of Himalayan Environment and Development, Kosi-Katarmal Almora, India
- Barbara Ehringhaus, Pro Mont-Blanc Mountain Wilderness Suisse, Switzerland
- Elizabeth Fox, Consultant in Ecotourism and Environmental Journalist, Roma, Italy
- Professor Roberto Gambino, Politecnico di Torino, Torino, Italy
- Mervyn Gans, Mountain Club of South Africa, Westville, South Africa
- Roger Good, Parks and Wildlife, Queanbeyan, NSW
- Paul Green, Director, Tongariro/Taupo Conservancy, Turangi, New Zealand
- Dr Chandra P. Gurung, WWF Nepal, Kathmandu, Nepal
- Dr Larry Hamilton, Vice-Chair for Mountains, IUCN WCPA, Vermont, USA
- David Harmon, Executive Director, The George Wright Society, Michigan, USA .
- Enrique Jardel Peláez, Universidad de Guadalajara Autlán, Jalisco, Mexico
- Bruce Jefferies, Conservation Management and Planning Systems, Taupo, New Zealand
- Sonja Krueger, Ezemvelo KwaZulu-Natal Wildlife, Cascades, South Africa

- Dr Monica Kuo, Chinese Culture University, Taipei, Taiwan
- Harvey Locke, Canadian Parks and Wilderness Society, Toronto, Ontario Canada
- Janet Mackay, Director, Planning for People, Berridale, NSW
- Kathy MacKinnon ENVGC, The World Bank, Washington DC, USA
- Gregor Manson, Executive Director, Great Barrier Reef Marine Park Authority, Townsville, Queensland
- Linda McMillan, Vice-President, American Alpine Club, San Rafael, California, USA
- Angeles Mendoza Durán, University of Calgary, Calgary, Alberta, Canada
- Dave Morris, President UIAA Mountain Protection Commission, International Mountaineering and Climbing Federation, Kinross, Scotland
- Bob Moseley, Conservation Science Director, The Nature Conservancy Yunnan Great Rivers Project, Kunming, Yunnan, People's Republic of China
- Tobgay Namgyal, Director, Bhutan Trust Fund for Environmental Conservation, Thimphu, Bhutan
- Krishna Oli, IUCN Country Office, Bakhundole, Lalitpur, Kathmandu, Nepal
- Dr David J. Parsons, Director, Aldo Leopold Wilderness Research Institute, Missoula, Montana, USA
- Dr John Peine, USGS, University of Tennessee, Knoxville, Tennessee USA
- Andrew Plumptre, Director, Albertine Rift Programme, Wildlife Conservation Society, Kampala, Uganda
- Dr Martin Price, Director, Centre for Mountain Studies, University of Highlands and Islands, Perth, United Kingdom
- Miquel Rafa, Fundació Territori Paisatge, Caixa Catalunga, Barcelona, Spain
- Manuel Ramirez U, Director, Southern Mesoamerica Program, Conservation International, Apartado San Jose, Costa Rica
- Professor Bernardino Romano, Università dell'Aquila, Monteluco di Roio, L'Aquila, Italy
- Ian Rushworth, Ecological Advice Coordinator uKhahlamba, Ezemvelo KwaZulu-Natal Wildlife, Cascades, South Africa
- Trevor Sandwith, Coordinator, Cape Action Plan for the Environment, Claremont, South Africa
- Nikhat Sattar, Head, Emerging and Emergency Programmes, IUCN Asia Programme, Karachi, Pakistan
- Mingma Sherpa, Director, Asia and Pacific Programs, World Wildlife Fund—US, Washington DC, USA
- Assoc. Professor D. Scott Slocombe, Wilfrid Laurier University, Waterloo, Canada
- Jordi Sorgatal, Fundació Territori Paisatge, Caixa Catalunga, Barcelona, Spain
- Gary Tabor, Yellowstone-to-Yukon Program Coordinator, Wilberforce Foundation, Bozeman, Montana, USA
- Dr James Thorsell, Parks Country Environmental Consultants, Banff, Canada
- Hernán Torres, Environmental Consultant, Huelén, Santiago, Chile
- Miriam Torres, The Mountain Institute, Huaráz, Peru
- Dr J. Alejandro Velázquez M., Universidad Nacional Autónoma, El Pueblito, Morelia Michoacán, Mexico
- Per Wallsten, Principal Conservation Officer, Swedish Environmental Protection Agency, Stockholm, Sweden
- Sangay Wangchuk, Head, Nature Conservation Section, Forestry Services Division, Thimpu, Bhutan
- John Watson, Manager/South Coast Region, Dept of Conservation/Land Management, Albany, Western Australia
- Sean White, Chief Technical Adviser, Mt Elgon Conservation and Development Project, Muthaiga, Nairobi, Kenya
- Edgard Yerena, Altamira, Caracas, Venezuela
- Dr Pralad Yonzon, Coordinator, Resources Himalaya, Kathmandu, Nepal
- Kevan Zunckel, Ezemvelo KwaZulu-Natal Wildlife, Cascades, South Africa.

New Case Studies, Snapshots, figures and other information have been included in this revised edition. We would like to express our appreciation to the following managers, authors, and organisations that facilitated and provided us with this information:

- Linus Bagley, Binna Burra Mountain Lodge, Lamington National Park, Queensland
- Grant Baker, Department of Conservation, New Zealand
- Tom Barrett, Conservation Information Officer, NSW Department of Environment and Conservation, Queanbeyan, NSW
- Lindsay Best, Department for Environment and Heritage, Adelaide
- Rachael Bevan, Department of Conservation, New Zealand
- Steven Bourne, Department for Environment and Heritage, Naracoorte, South Australia
- John Bradbury, Department of Primary Industries, Water and Environment Tasmania
- Stuart Cohen, Department of Environment and Conservation, NSW P&W, Queanbeyan
- Mark Collins, UNEP-World Conservation Monitoring Centre, Cambridge, United Kingdom
- Mary Cordiner, UNEP-World Conservation Monitoring Centre, Cambridge, United Kingdom
- Harry Creamer, NSW P&W, Port Macquarie
- Pam Cromarty, Department of Conservation, New Zealand
- Dr Jocelyn Davies, School of Earth and Environmental Studies, Adelaide University
- John Day, Great Barrier Reef Marine Park Authority, Queensland
- Lorraine Donne, NSW P&W, Hurstville
- Susan Downing, Griffith University, Nathan Campus, Queensland
- Mike Edgington, Department of Conservation, New Zealand
- Karen Edyvone, University of Tasmania, Hobart
- Alan Feely, Environment Protection Agency, Queensland National Parks and Wildlife Service, Brisbane
- Roger Good, Senior Project Manager, NSW P&W, Queanbeyan
- Yani Grbich, Griffith University, Gold Coast Campus, Queensland
- Linda Greenwood, Parks Victoria, Melbourne
- Andrew Growcock, Griffith University, Gold Coast, Queensland
- Wendy Hill, Griffith University, Gold Coast Campus, Queensland
- Rob Hughes, Environment Protection Agency, Queensland National Parks and Wildlife Service, Brisbane
- Rob Hunt, Department of Environment and Conservation, NSW P&W, Queanbeyan, NSW
- Glenys Jones, Tasmanian Parks and Wildlife Service, Hobart
- Ray Jones, Environment Protection Agency, Sustainable Industries Division, Brisbane, Queensland
- Francis Johnston, Griffith University, Gold Coast Campus, Queensland
- Alex Knight, Anangu Pitjantjatjara Yankunytjatjara Land Management, Umuwa, via Alice Springs, NT
- Charles Lawson, Griffith University, Gold Coast Campus, Queensland, Australia
- Bruce Leaver, Department of Environment and Heritage, Canberra
- Caroline Littlefair, Griffith University, Gold Coast Campus, Queensland
- Lynette Liddle, Uluṟu Kata-Tjuṯa National Park, Yulara, Northern Territory
- Doon McColl, Wet Tropics World Heritage Area Management Authority, Cairns, Queensland
- Ian Miles, Department of Sustainability and Environment, Victoria
- Dr Graeme Moss, Department for Environment and Heritage, Kangaroo Island, South Australia
- Martin O'Connell, Department of Environment and Conservation, NSW P&W, Byron Bay
- Pam O'Brien, NSW P&W, Jindabyne
- Bill O'Connor, Parks Victoria, Melbourne
- John Ombler, Department of Conservation, New Zealand
- David Osborn, Commonwealth Department of Environment and Heritage, Canberra
- Michael Pemberton, Department of Primary Industries, Water and Environment, Hobart
- Dr Catherine Pickering, Griffith University, Gold Coast Campus, Queensland
- Damien Pierce, Department for Environment and Heritage, Adelaide
- Sarah Pizzey, Commonwealth Department of Environment and Heritage
- Julie Richmond, Parks Victoria, Melbourne
- Kate Sanford-Readhead, Commonwealth Department of Environment and Heritage, Canberra

- Pascall Scherrer, Griffith University, Gold Coast Campus, Queensland
- Dianne Smith, Manager Tourism Partnerships, Parks Victoria
- Penny Spoelder, Department of Environment and Conservation, NSW P&W, Queanbeyan
- Paul Stevenson, Parks Australia, Commonwealth Department of Environment and Heritage
- Daniel Stock, Griffith University, Gold Coast Campus, Queensland
- Sharon Sullivan, Sullivan Blazejowski and Associates, Nymboida
- Guy Thomas, Acting Senior Ranger, Queensland National Parks and Wildlife Service, Moreton Bay, Queensland
- Sally Troy, Parks Victoria, Melbourne
- Geoff Vincent, Parks Victoria, Melbourne
- Russell Watkinson, Wet Tropics World Heritage Area Management Authority, Cairns, Queensland
- George Wilson, Australian Wildlife Services
- Kerry Yates, Brisbane, Queensland
- Steve Yorke, The Briars Park, Mornington Peninsula, Victoria
- Frank Young, Añangu Pitjantjatjara Land Management, Umuwa, Northern Territory
- Dino Zanon, Parks Victoria, Melbourne

The photograph of the Royal National Park in Chapter 2 is reproduced courtesy of the NSW Government Printing Office. The material in Chapter 1 and the epigraph introducing Chapter 14 from Mary White's books are used with permission from Mary White. Figures 1.3 and 1.4 were supplied by LANDINFO, the Spatial Division of Sinclair Knight Merz. Table 1.5 and Figures 1.6, 1.7, 1.8, 1.9, 1.10, 1.11, 1.12, 1.13, 1.14, 1.15, 14.1, 14.2, and 15.2 are reproduced with the permission of Environment Australia (now Department of Environment and Heritage). Figures 1.5, 5.1, and 5.2 are reproduced with permission from the Australian Government, Department of Environment and Heritage. Figure 5.3 is reproduced with the permission of the Commonwealth of Australia, and Figure 7.1 is reproduced with the permission of Island Press. The extracts from Discovering Monaro by W.K. Hancock are reproduced with the permission of Cambridge University Press. The epigraph introducing Chapter 7 is reproduced with the permission of Alex Grieg, Peter Fraser, and Dunlop Group Ltd. Figure 12.3 is reproduced with the permission of Mr Andy Spate. Backgrounder 12.1 is reproduced with the permission of IUCN. Figure 13.1 is reproduced with the permission of Melbourne University Press. Table 13.1 is reproduced with the permission of the University of Queensland. Tables 13.2 and 13.4 are reproduced with the permission of Dr D.H.R. Spennemann.

Every effort has been made to trace the original source of copyright material contained in this book. The publisher would be pleased to hear from copyright holders to rectify any errors and omissions.

Preface to the Second Edition

It is four short years and two reprints since the first edition of this book was produced in 2001. This second edition of *Protected Area Management: Principles and Practice* builds on the successful first edition. The book was designed to fill a gap in protected area management educational literature and to help underpin an improvement in management of our magnificent Australian protected area system. The success of the first edition both within Australia and (somewhat unexpectedly) internationally has reflected a very real need for the best protected area management information to be readily available in one compendium. This need is unchanged and the second edition has responded with new chapters, new maps, revised and updated text, and a new schedule of protected area history. A great deal has happened in the protected area management profession since 2001 and improvements to this book capture this.

Global protected area management issues dominated the lead-up to the Vth World Parks Congress in Durban, South Africa in September 2003. World Park Congresses such as Durban are milestone events. They are conducted every 10 years and provide a basis for reviewing past protected area management performance as well as establishing directions for the future. Seven major themes of protected area management and additional cross-cutting themes were dealt with by 3000 delegates from 154 countries. The very latest in protected area management principles and practice were reported.

This second edition captures the most important of this information. Nearly all chapters have been revised or upgraded. The principles at the end of each chapter in Part B have been revised with the help of 59 leading international protected area management professionals at a workshop conducted as part of the Durban World Parks Congress. We have added to these principles some of the major lessons learned from national and international collective experience with protected area management.

Five new chapters have been added: Establishing Protected Areas; Sustainability Management; Operations Management; Marine Protected Areas; and Evaluating Management Effectiveness. Other chapters have been expanded to include new topics such as protected area governance. A new schedule of protected area historical events has been included in an appendix. New national and international protected area management Case Studies and Snapshots have been included. New global protected area maps provided by the World Conservation Monitoring Centre provide an improved context setting source of information for the 100,000 plus protected areas globally.

Introduction

Protected areas are special places. They include the world's highest mountains, its greatest coral reef system, the purest of freshwater lakes, heritage-rich tropical rainforests, ice caps and mountain glaciers, vast wildlife-studded plains, active volcanic islands, and ancient monuments. The idea of protecting special places forever has deep historic roots and is a concept that has almost universal acceptance among the nations of the Earth.

Protected areas are essential for the overall health and well-being of human and other species for the long term. They are areas on land and sea that are established to conserve a representative sample of the world's biological diversity. They protect natural phenomena, cultural and social heritage sites, and areas of significance, ecosystems, landscapes, and natural ecosystem processes. They help to preserve the genetic diversity of the Earth. The definition of protected areas by the International Union for the Conservation of Nature World Commission on Protected Areas has wide acceptance and it defines protected areas as:

> *an area of land and/or sea especially dedicated to the protection and maintenance of biological diversity, and of natural and associated cultural resources, and managed through legal or other effective means (IUCN 1994).*

No other land-use type has been as effective in achieving conservation outcomes in the face of changes impacting the planet. Protected areas help to retain the life-essential ecosystem services,

ecosystem processes, and other life-support systems of the planet (Wilson 1992, Mackinnon et al. 1986). They provide opportunities for a basic human need for interrelationships with nature and they protect areas of special spiritual and cultural significance. They often have an influence well beyond their boundaries and form part of a landscape and ecosystem approach to conservation. As such, they may be a very important part of the sustainable cultural and economic well-being of local communities (Phillips 2003).

Properly designed and managed protected areas offer major sustainable benefits to society. They help ensure that utilisation of species by humans is sustainable. They play a central role in the social and economic development of rural environments and contribute to the economic well-being of urban centres and the quality of life of their inhabitants (Mackinnon et al. 1986). They provide opportunities for outdoor recreation and tourism.

There are formidable expectations for protected areas. They are a principal tool for the Convention on Biological Diversity attaining its global objectives for biodiversity conservation (UNEP-SCBD 2001). They are critical for ecosystem services and life-support systems of the planet and they are a necessary alternative to rapid changes happening to the world. The response by nations of the planet has been remarkable. By the start of the twenty-first century, the greatest ever voluntary land-use transformation accomplished by independent governing countries had been achieved. In 1962, there were 10,000 protected areas worldwide. In 2003, there were more than 100,000 protected areas occupying almost 12% of the surface of the terrestrial area of the planet (Chape et al. 2003). However, significant gaps remain, and much needs to be done to establish additional protected areas, particularly in freshwater and marine environments.

Purpose and structure of this book

The scope of protected area management is broad and diverse. Protected area managers are required to meet a wide range of management objectives, and some of these may be competing. This diversity of management may be illustrated by identifying key management themes for protected areas. Protected area objectives of management help define, guide, and prioritise management actions. An inventory of theme keywords provides a guide to the scope of protected area management. We developed such an inventory by combining the twenty-first century paradigm for protected areas and its management 'objective-themes' described by Phillips (2003) with additional material derived from the 2003 Vth World Parks Congress outputs (IUCN 2003 a,b,c,d). The resulting list of protected area management themes is given in Table 1.

This book has been prepared for all people involved or interested in the management of protected areas. Effective management is essential if the purposes for which protected areas are set aside are to be achieved. This book deals with all aspects of protected areas and their management.

Our focus is practical. This is an Australian text, written in the context of Australia's social and cultural setting, as well as being informed by international experiences. We have included contributions from all over Australia and from a wide range of speciality areas. International examples are also included. We believe the practices we suggest for managing protected areas will be of profound value also in other countries. These practices are not just based on the personal views of the authors. Reserve managers from around Australia and overseas have collectively evolved this know-how. Appropriately, then, throughout the book we use contributions from protected area staff all over Australia and internationally. Longer contributions are presented as Case Studies, and

shorter ones as Snapshots. Together they give the sense of many voices reporting on personal experience of contemporary practices for conservation.

We have written this book to help policy-makers, practitioners, and students (those future practitioners) to improve the ways they manage protected areas, in Australia and elsewhere. It will also interest stakeholders, including park neighbours, rural communities, conservationists, and Indigenous peoples. In as much as Australia is leader in some aspects of protected area management, we would like to share these processes with protected area managers around the globe. The Case Studies demonstrate the stunning variety of Australian protected areas, and help to pass on the wealth of practical knowledge stored up in the experience of its managers. Space allowed us to use only some of these valuable Case Studies, and many have been shortened more than we would like. We have chosen to provide a glimpse of subjects rather than omit them, in the hope that this will inspire readers to discover more about the subject elsewhere.

Nomenclature for plant and animal species mentioned in this book generally follows Harden (1990, 1991, 1992, 1993) for native plants, Parsons and Cuthbertson (1992) for weeds, Strahan (1995) for mammals except cetaceans, Menkhorst (1995) for cetaceans, Simpson and Day (1989) for birds, Cogger (1994) for reptiles and amphibians, and Gommon et al. (1994) for fish. On first mention of a species, both common name and scientific name are given, with common name used thereafter. Common names are capitalised, with the exception of introduced fauna species.

The book is structured in three parts (Figure 1). Part A sets the context for protected area management in Australia. Part B contains 18 chapters that examine in depth the main principles and practices of such management. These chapters cover:

establishing protected areas; using, obtaining, and managing information; planning; finance and economics; administration; sustainability management; operations management; natural heritage; cultural heritage; threats; incidents; tourism and recreation; working with the community; Indigenous people and protected areas; linking the landscape scale; marine management; evaluating management effectiveness and futures. Part C provides a protected area history schedule of key events and a summary of information on the protected areas in each Australian state and territory.

Table 1	**Management themes and desired outcomes** (adapted from Phillips 2003, IUCN 2003 a,b,c,d)
Protected area management themes	Desired outcome(s) for themes
ESTABLISH (Chapter 5)	New protected areas established Viable representative samples of all ecosystems conserved Critically endangered species conserved Threatened species conserved Globally significant areas of species conserved Freshwater systems conserved Marine and coastal systems conserved
GOVERN (Chapters 2, 3)	Coherent national framework for biodiversity conservation achieved Effective conservation governance principles introduced Recognition of diverse knowledge systems Openness, transparency, and accountability in decision-making Inclusive leadership Mobilising support from diverse interests Sharing authority and resources; devolving/decentralising decision-making Conservation policies harmonised with sectoral policies and laws Recognition of a range of governance types achieved
PLAN (Chapter 7)	National strategic plan for protected area system(s) prepared Strategic, tactical, and operational protected area plans developed
RESEARCH (Chapter 6)	Active research in protected areas and protected area systems Changes in biodiversity and key ecological processes researched Climate change effects researched
ADMINISTER, BUILD SKILL, CAPACITY & COMPETENCY (Chapter 9)	Multiskilled competent workforce with the capacity to manage in place Capacity for protected area management increased at all levels Individual and group capacity to manage strengthened Generic competency standards achieved Management standards introduced Self-assessment introduced Learning systems introduced

Table 1	(continued)
Protected area management themes	**Desired outcome(s) for themes**
CONSERVE (Chapters 10, 12, 13, 14, 15, 16, 18, 19, 20)	Biodiversity conservation achieved Natural heritage-natural phenomena conservation achieved Cultural heritage conservation achieved Social heritage conservation achieved Threatening processes eliminated
ADAPT (Chapter 7)	Adaptive management implemented applying lessons learnt Climate change effects mitigated by adaptive management
REHABILITATE (Chapters 11, 14)	Damaged environments are repaired/restored/rehabilitated
MONITOR (Chapters 7, 21)	Protected area management effectiveness monitored
EVALUATE EFFECTIVENESS (Chapter 21)	Protected area management effectiveness evaluation completed as a routine part of management for improved management and more transparent and accountable reporting
FINANCE (Chapter 8)	Finance mechanisms supporting protected areas achieved Diversity of funding sources achieved Stability of funding sources achieved Sustainability of funding sources achieved
BE BUSINESS-LIKE (Chapters 7, 8)	Business guidelines and standards achieved Business guidelines and standards for business guide improved protected area management techniques
SUSTAINABLE USE (Chapter 10)	Sustainable visitor use of protected areas achieved with conservation as the primary objective
RESPECT (Chapter 17, 18)	Rights of Indigenous peoples respected Informed consent of Indigenous peoples obtained with the establishment of new protected areas Formal recognition of the intellectual property of Indigenous peoples
UNDERSTAND, APPRECIATE, INCLUDE (Chapter 17)	Understanding of community social values achieved Appreciation of community aspirations and community needs achieved Inclusion of cultural heritage considerations in management Laws and policies that foster multicultural values adopted Sensitive management of sacred places undertaken
BENEFIT, SHARE (Chapters 8, 17, 21)	Benefits, economic values, costs known Equitable sharing of benefits and costs

Table 1	(continued)
Protected area management themes	**Desired outcome(s) for themes**
PARTNER, SHARE, TRANSFER (Chapters 2, 19)	Effective shared management systems achieved. These may include: Joint management Co-management Shared management; involvement of local communities Managed by local communities
INTEGRATE, NETWORK (Chapter 19)	Integrated conservation objectives across land/sea; regional and sectoral planning levels achieved Protected areas designed with linkages to surrounding ecosystems Transboundary protected area management facilitated
PARTICIPATE, COOPERATE (Chapter 17)	Co-operation with other organisations achieved Partnerships established including for research and training Conservation work by local communities supported Community conserved areas supported Work with communities: pro-poor policies implemented
APPRECIATE (Chapter 17)	Protected area and protected area systems viewed as a community asset and as an international asset Mechanisms for the participation of stakeholders in management in place
BE POLITIC (Chapters 4, 17)	Political considerations managed effectively
EDUCATE (Chapter 17)	Integrate a multi-level communication strategy for protected areas and protected area systems Strengthen knowledge exchange and professional development Support agencies as learning organisations
PROMOTE (Chapters 2, 17)	Importance and benefits of protected areas promoted Benefits of investments in protected areas promoted Stronger support for protected areas from urban constituency achieved
REDUCE EFFECTS OF CONFLICT (Chapters 2, 17, 18, 19)	Peace and conflict impact assessment conducted Training, mediation, and support for local staff for dealing with conflict in protected areas achieved Parties supporting protected areas are neutral Alternatives for local communities to exploitation of protected areas in times of crisis achieved Impacts derived from humanitarian relief efforts minimised Capacity for prompt rehabilitation post conflict organised

Figure 1   Structure of the book

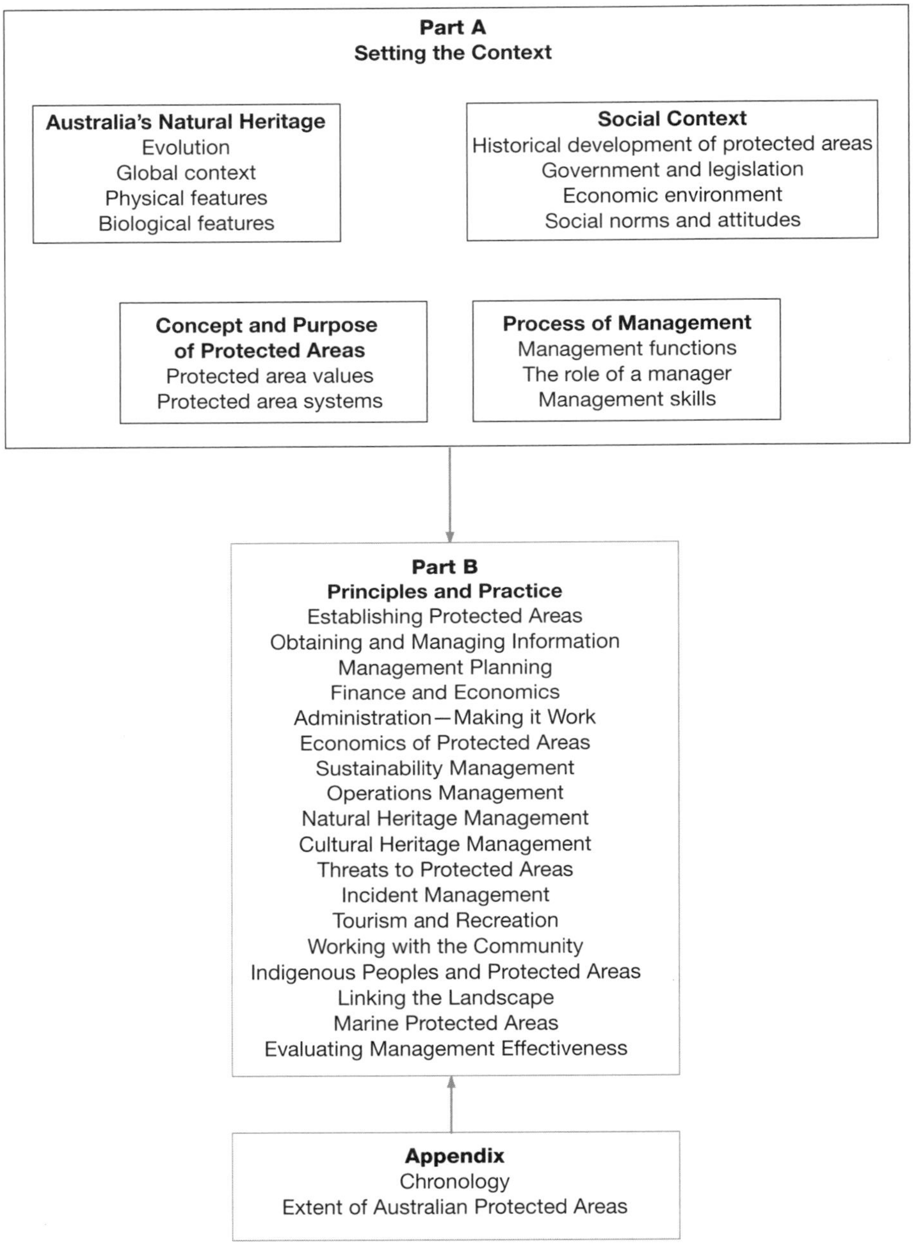

**Part A
Setting the Context**

Australia's Natural Heritage
Evolution
Global context
Physical features
Biological features

Social Context
Historical development of protected areas
Government and legislation
Economic environment
Social norms and attitudes

**Concept and Purpose
of Protected Areas**
Protected area values
Protected area systems

Process of Management
Management functions
The role of a manager
Management skills

**Part B
Principles and Practice**
Establishing Protected Areas
Obtaining and Managing Information
Management Planning
Finance and Economics
Administration—Making it Work
Economics of Protected Areas
Sustainability Management
Operations Management
Natural Heritage Management
Cultural Heritage Management
Threats to Protected Areas
Incident Management
Tourism and Recreation
Working with the Community
Indigenous Peoples and Protected Areas
Linking the Landscape
Marine Protected Areas
Evaluating Management Effectiveness

Appendix
Chronology
Extent of Australian Protected Areas

PART A SETTING THE CONTEXT

01 Australia's Natural Heritage   2
02 Social Context   35
03 Concept and Purpose of Protected Areas   76
04 Process of Management   112

01

Australia's Natural Heritage

> *The Australian continent is flat and dry, has poor soils by world standards, is fire-prone and has long been exposed to erosive forces; yet it is rich with heritage—geologically, biologically and culturally. This ancient land is one of the 12 most biologically diverse countries on Earth. It is a continent on a remarkable planet and is part of a living, changing, renewing, evolving, and interconnected global biosphere. Being geologically stable, Australia has some of the oldest land surfaces on Earth. With an international reputation for its koalas, kangaroos and coral reefs, distinctive eucalypts and wildflowers, vivid red outback soils and clear aqua-green coastal waters, Australia is both diverse and unique. To help manage the finest heritage of this remarkable land is a special privilege, and challenge.*

1.1	Evolution of Australia's natural heritage	3	
1.2	Australia today	12	
1.3	Climate	13	
1.4	Landforms, soils, and landscapes	16	

1.5	Biological diversity	20
1.6	Terrestrial vegetation	23
1.7	Terrestrial fauna	28
1.8	Marine flora and fauna	34

The people of Australia are proud of their country's outstanding natural and cultural heritage and they care about its future. The 7.68 million km² continent of Australia lies in the southern hemisphere between its northern point, Cape York (latitude 10°41' south); its most southern point, South East Cape, Tasmania (43°39' south); its western point, Steep Point in Western Australia (longitude 113°09' east) and its most easterly point Cape Byron, NSW (153°39' east). (See Figure 1.1 in the colour section.)

Australia is the world's smallest continent and one of two continents and some sizeable islands in the southern hemisphere. The nation of Australia is the sixth largest country (by area) in the world. It occupies the entire Australian continent and many outlying islands.

This chapter sets Australia's biophysical environment in a global context. It explains the importance of Australian environments and species, and briefly describes the evolution of Australia's bioregions from the perspective of a conservation manager. We also briefly illustrate this history by describing some outstanding protected areas that conserve samples of Australia's past evolutionary processes, whether as fossils or living relics. We then describe Australia as it is today to provide a context for the more detailed chapters on management practice.

The text for this chapter, particularly the geological evolution of Australia and the importance of the biosphere has been influenced by the outstanding work of Mary White and her books, *The Nature of Hidden Worlds* (1990) and the new edition titled *Reading the Rocks* (1999); *After the Greening: the Browning of Australia* (1994); *Listen … Our Land is Crying* (1997); *The Greening of Gondwana* (1998); *Running Down, Water in a Changing Land* (2000) and, *Earth Alive, From Microbes to a Living Planet* (2003). The reference *Ecology, an Australian Perspective* (Attiwill & Wilson 2003) is also highly recommended reading.

1.1 Evolution of Australia's natural heritage

One of the challenges for protected area managers in Australia is to comprehend the vast time-scale of the heritage they must conserve. The geological time-scale is breathtaking, especially if we consider that the first hominids evolved in the order of four million years ago and modern humans have probably only evolved in the past 100,000 years (Table 1.1) (Gore 1997). Yet humans settled Australia at least 40,000 years ago and maybe even 50,000 years ago or more. Every protected area in Australia shares a part of the planet's evolutionary history, and some of them contain outstanding examples of geological evolution. This heritage warrants conservation, no less than do contemporary life forms and landforms. It provides a context for understanding and managing our current environments. It is a rich source of information for scientific enquiry, and its preservation is essential if we are to educate current and future generations about the richness of our heritage, and its possible futures. We live on a dynamic planet where geologic forces, biosphere life support dynamics, and evolutionary processes continue, despite direct and indirect human interventions. Protected area managers need to have a very clear idea of this context when they set management goals.

Since understanding the past is critical to making such decisions, we will briefly summarise the development of the Australian continent through information provided in Table 1.1 and some Snapshots that help to illustrate the rich history of this continent. Readers are encouraged to consult Mary White's texts for a more detailed treatment of this subject.

Protected area management and Australia's evolutionary heritage

Significant samples of Australia's geological heritage have been permanently protected within the reserve system of Australia. Interpreting the past through this record allows us to prepare for and manage for the future. In addition, there is a sense of pride in understanding how this extraordinary continent has evolved upon the planet (Table 1.1).

Table 1.1	**Earth's evolutionary development with particular reference to Australia** (Vandenbeld 1988, White 1990, Vickers-Rich & Rich 1993, White 1994, Mulvaney & Kamminga 1999, Groombridge & Jenkins 2000, Rich & Vickers-Rich 2000, White 2000a, Woodford 2000, Groombridge & Jenkins 2002, Allen et al. 2002, Burenhult 2003, White 2003)			
Geological era/ period/epoch	Million years ago	Atmosphere, climate, major tectonic events, natural and human phenomena	Flora	Fauna
Holocene Epoch	0 to 0.01	First Fleet reach Australia, 1788 Macassan trade, northern Australia 1700s 1616 Dutch sailor Dirk Hartog landfall, Cape Inscription, WA Approx 4000 years ago Aboriginal 'backed blades' appear in the archaeological record Relatively stable, aridity a major feature, modern fire regimes Last 4000–5000 years climate strongly influenced by the El Niño–Southern Oscillation patterns Henbury meteorite craters (NT) 5000 years ago (remembered in an Aboriginal Dreamtime story) Approx. 5000 years ago, fine shaped spear points emerge in the archaeological record Sea level reaches its modern level about 6000 years ago The modern Great Barrier Reef (Qld) commences forming about 8000 years ago	Modern flora Relic rainforests of Queensland wet tropics area and the cool-temperate forests of Tasmania trace their origins back 60 to 80 million years in Australia (World Heritage Areas)	Fossil evidence of Tasmanian Devil (*Sarcophilus harrisii*) in Victoria 600 years before Europeans Marsupials Monotremes Dingo (*Canus lupus dingo*) arrives in Australia about 5000 years ago
Pleistocene Epoch	0.01 to 1.6	Land bridge connection with Tasmania severs about 13,000 years ago Extreme aridity	Cyclic changes from open/drier vegetation in glacial periods to wetter vegetation in interglacials— recovery and establishment of modern vegetation over the last 16,000 years	Marsupials Megafauna, diprotodonts present Monotremes

Table 1.1	(continued)			
Geological era/ period/epoch	Million years ago	Atmosphere, climate, major tectonic events, natural and human phenomena	Flora	Fauna
Pleistocene Epoch	0.01 to 1.6	Last glacial cycle about 18,000 years ago—sea levels 120 metres below today—Australian mainland connected to PNG and Tasmania, Australian Alps' glacial lake depressions form (World Biosphere Reserve) Tower Hill (Vic) volcanic event 23,000 years ago (protected area) Willandra Lakes full of water and occupied by Aboriginal communities at about 40,000 years ago—ancient burial cremations 25,000 years ago (World Heritage Area) Tasmania Cave occupation sites 35,000 years ago Aborigines arrive in Australia at least 40,000 years ago 91,000 years ago, earliest *Homo sapiens sapiens* fossils found (Israel) 460,000 Early Homo fossils found in China 700,000 Early humans migrated out of Africa across the old world	38,000 years ago, Lynch's Crater, Cape York shows a change from fire sensitive plants to eucalypts 60,000 years ago Lake George NSW fossil record shows changes in vegetation from casuarina woodland to eucalypt-dominated flora with major increases in charcoal from fires	Lord Howe Island Giant Horned Turtle known from 40,000-year-old sediments
Tertiary Period Neogene Epochs: Pliocene, Miocene	1.6 to 23.7	Early ancestor of humans in Africa, 1.6 million years ago Africa, 2.5 million years ago, tool-making by ancestor of humans	Modernisation of Australia's vegetation begins in the Miocene. Spread of eucalypts and acacias—adaptations to drier climates, poorer soils	Primitive platypus, megafauna including diprotodonts, giant kangaroos and wombats, and great running birds

Table 1.1	(continued)			
Geological era/ period/epoch	Million years ago	Atmosphere, climate, major tectonic events, natural and human phenomena	Flora	Fauna
Tertiary Period Neogene Epochs: Pliocene, Miocene	1.6 to 23.7	Cool, dryness starts to appear in upper Tertiary Late Miocene cooling—southern ice cap increases about 6 million years ago Cooler Warm Indo-Australia plate collides with SE Asia about 15 million years ago Artesian basin-fed mound springs exist in central Australia Faulting in the Miocene uplifts Kosciuszko	Nucleus of arid flora develops in drier areas Grasslands in the centre of Australia in the late Miocene	Oligocene-Miocene frogs at Riversleigh Crocodiles, lizards, snakes, pythons, rich bird life, bats, marsupials at Riversleigh (Snapshot 1.1)
Tertiary Period: Palaeogene Epochs: Oligocene, Eocene, Palaeocene	23.7 to 66.4	Periods of relative aridity Warm and wet Australia separates from Antarctica about 45 million years ago and moves northward from high latitudes Deep weathering of soils creates duricrust now prominent in central Australia 60 million years ago, earliest known primate lived in Africa	Evolution of a unique Australian flora starts Mixed Gondwanan forest in central Australia Coal swamps in the south (Eocene)	Evolution of a unique Australian fauna starts Ancestral dasyurids and other marsupial families Birds, first penguins evolve Flightless birds
Cretaceous Period	66.4 to 144	Warm equable climate Australia in high latitude location Uplift of the great divide and great escarpment 60 to 80 million years ago	Mass extinctions in plants, highest loss in angiosperms, lowest in ferns Mixed Gondwanan forest Last period of fern and gymnosperm dominance	Mass extinction of the dinosaurs and many species of animals First bird's feather found in Australia dated to this time

Table 1.1	(continued)			
Geological era/ period/epoch	Million years ago	Atmosphere, climate, major tectonic events, natural and human phenomena	Flora	Fauna
Cretaceous Period	66.4 to 144	Rifting in Gondwana between 96 to 132 million years ago Gondwana fragmenting Lower Cretaceous marine flooding worldwide about 120 million years ago	Wollemi Pine, a living relic of the Cretaceous (Snapshot 1.3) Major diversification of angiosperms worldwide. Mass marine extinctions, and freshwater and terrestrial vertebrates	Fossil ancestor of the Queensland lungfish found at Koonwarra, Victoria Early Cretaceous placental mammal; monotreme; and dinosaur at Dinosaur Cove, Otways, Victoria World's oldest monotreme, a Platypus at Lightning Ridge Turtles present in Australia Crocodiles at Lightning Ridge Pterosaurs Dinosaurs
Jurassic Period	144 to 208	Subtropical Warm and wet No polar ice Australia in high latitudes, warm world, no polar ice, winter darkness Major rifting processes commence 150 million years ago, leading to the breakup of Gondwana Freshwater aquifers of the Great Artesian Basin develop Pangaea comes together about 200 million years ago	Araucarians, Podocarps Age of conifers, cycads, tree ferns	Age of reptiles including the dinosaurs Ancestor of monotremes Cicadas Multituberculates
Triassic Period	208 to 245	Warm to hot Periods of aridity and seasonal rainfall	Conifers, cycads, ginkgoes, ferns, club mosses, and horsetails	Mass extinctions of marine invertebrates; some land vertebrates lost Age of amphibians— Labyrinthodonts

Table 1.1	(continued)			
Geological era/ period/epoch	Million years ago	Atmosphere, climate, major tectonic events, natural and human phenomena	Flora	Fauna
Triassic Period	208 to 245	South Pole situated in western NSW, winter darkness, no polar ice Warm		Reptiles, dinosaurs first appear, but are rare First mammals appear in the northern hemisphere Amphibian fossils in Sydney sandstone national parks
Permian Period	245 to 286	Volcanism Warm to hot Temperate Glacial pavement at Halletts Cove Conservation Park, South Australia Ice cap advances and retreats	Mass extinctions in plants Extensive swampy conditions for the deposition of coal Glossopteris flora	The most severe extinction event, marine and terrestrial Labyrinthodonts (primitive amphibians) common
Carboniferous Period	286 to 360	Glaciation Australia in high southern latitudes Cool Climate changes as Australia moves from tropics to high latitudes Warm	Pro-gymnosperms are replaced in tundra areas Club mosses and horsetails diminish with the cold	First sharks appear First amphibians appear
Devonian Period	360 to 408	Warm and wet Australia in the southern hemisphere, subtropics	First extinction crisis in plants Rich flora in swamps Club mosses are preserved in sediments of the Mimosa Rocks National Park, NSW	Mass extinctions in marine animals Age of fishes—fossil fish deposits of the red beds of Ben Boyd National Park in NSW Massive fossil reef formation, Windjana Gorge National Park, Western Australia

Table 1.1	(continued)			
Geological era/ period/epoch	Million years ago	Atmosphere, climate, major tectonic events, natural and human phenomena	Flora	Fauna
Silurian Period	408 to 436	Sufficient ozone in atmosphere able to screen ultraviolet rays, permitting life outside of the seas Hot, arid Australia in the northern hemisphere, tropics	First land plants evolve from marine algae	Records of one of the first animals to walk on land—Eurypterid tracks of the Tumblagooda Sandstone in the Kalbarri National Park, Western Australia
Ordovician Period	436 to 505	Atmosphere unable to support life on land Warm	Algae in the sea	Marine extinction event First vertebrates appear (fish) Rich trilobite and shellfish fauna, Hatton's Corner Nature Reserve, NSW
Cambrian Period	505 to 570	Atmosphere unable to support life on land Hot and dry	Algae in the sea	Archaeocyathid (sponges) limestone reefs, Flinders Ranges National Park, South Australia
Precambrian Era	570 to 4500	Living organisms steadily release oxygen to the oceans and atmosphere 300 million year ice age that peaks at 620,770 and 940 million years ago Life providing oxygen causes oxidation of ocean minerals and produces 'red beds' 2500 million years ago Ancient atmosphere consists of carbon dioxide, water vapour, methane, ammonia, and hydrogen	Cyanobacteria—living examples of stromatolites, Shark Bay Marine Reserves World Heritage Area (Snapshot 1.4) 3200 million years ago oldest evidence of simple single cells Stromatolites are the most common form of life for 3000 million years 3500 million years ago, stromatolites appear	Ediacaran fauna (jellyfish etc) of the Precambrian, Flinders Ranges National Park 600 million years ago, oldest multi cell animal fossils appear 1500 million years ago, first complex cells appear in the fossil record

Snapshot **1.1**

Australian Fossil Mammal Sites World Heritage Area, Riversleigh
(inscribed 1994)

Riversleigh is part of Lawn Hill National Park, 200 kilometres north-west of Mount Isa in Queensland. It is one of the most significant fossil sites in the world and 10,000 hectares of the park has been recognised as a World Heritage property. It is one of two parts to this World Heritage area, the other being Naracoorte, some 2000 kilometres away. The fossil sites inscribed were consistent with meeting world heritage criteria as 'outstanding examples representing major stages of the Earth's evolutionary history' and as 'outstanding examples representing significant ongoing ecological and biological evolution'. The Riversleigh fossils preserve the remains of a wide cross-section of vertebrate animals from up to 25 million years ago, and have effectively doubled our knowledge of Australia's past fauna. It is one of the world's richest Oligocene-Miocene mammal records, linking the period 15–25 million years ago. The fossils confirm that there was once a tropical rainforest over the Riversleigh site 25 million years ago, but the sequence continues, and shows the profound effects on fauna when Australia's rainforests largely vanished. The fossils include marsupial lions, carnivorous kangaroos, diprotodonts, seven-metre-long pythons, early ancestors of the now extinct Tasmanian Tiger or Thylacine (*Thylacinus cynocephalus*), and primitive platypuses. Evidence from Riversleigh shows that the fauna of the lowland rainforests of 20 to 15 million years ago became the progenitors for almost all of Australia's living animals, from those that still persist in the remnant rainforests to those that live in its grasslands and burrow through its deserts. Riversleigh tells us that Australia's surviving rainforests are more than just beautiful remnants of a once green continent. They contain many of the descendants of the 'seminal' creatures that spawned thousands of new species to rapidly fill a continent that had become 44% arid.

Over the last 3000 million years, only a couple of places on Earth have experienced conditions conducive to preserving representative samples of species and communities. These sites are of immense value. Riversleigh is one of them (Attenborough cited in Archer et al. 1994).

Snapshot **1.2**

Australian Fossil Mammal Sites World Heritage Area, Naracoorte
(inscribed 1994)

Naracoorte Caves are found in south-east of South Australia and they host 300 hectares of lands inscribed as World Heritage within the Naracoorte Caves National Park. The fossil site complements the Riversleigh fossil site. The fossils in the caves illustrate faunal change spanning several ice ages and highlight impacts of climate change and the influence of humans on Australia's mammals from at least 350,000 years before present. Some 99 vertebrate species have been discovered, including exceptionally well-preserved examples of the ice age megafauna, as well as a host of modern species such as the Tasmanian devil, thylacine and others (DEH 2004a).

Wollemi Pine, Wollemi National Park, NSW (P&W Collection, Jamie Plaza)

Snapshot **1.3**

The Wollemi Pine (*Wollemia nobilis*), the Greater Blue Mountains Area World Heritage Area (inscribed 2000)

A new member of the Araucariaceae family, the Wollemi Pine, was discovered in the Wollemi National Park near Sydney in August 1994 (NSW NPWS 1995, Woodford 2000). Described as a 'living fossil' from the age of the dinosaurs, this pine has very close relatives that date back to the Jurassic, Cretaceous, and Tertiary times. The discovery caused great excitement, as Australia's flora is considered to be well known. The pines reach about 35 metres in height, with a main trunk up to one metre in diameter. The discovery of a new species of tree, especially one that grows to such an impressive height, is most unusual. The habitat of the trees—protected steep-sided canyons north-west of Sydney, which act as refuges from fires that frequently burn the adjacent plateaus—contributed to their continued existence. The tree, given the common name of 'pine', is a conifer but is closer to the Norfolk Island Pine (*Araucaria heterophylla*) than to true pines. The discovery is a dramatic demonstration that parts of our biological heritage remain unknown (SEAC 1996). The importance of the Wollemi Pine directly contributed to the inscription of the Greater Blue Mountains Area, which includes the Wollemi National Park, as World Heritage in 2000.

Snapshot **1.4**

Shark Bay Marine Reserves World Heritage Area (inscribed 1991)

The Shark Bay World Heritage Area (WHA) near Carnarvon in Western Australia covers about 2.2 million hectares, of which about 71% is marine. The WHA contains an outstanding example of Earth's evolutionary history in the stromatolites and hypersaline environment of Hamelin Pool. The stromatolites in the Pool are the most diverse and abundant examples found anywhere in the world. These are living representatives of the stromatolites that existed some 3500 million years ago and that are also found in Western Australia's fossil record. Shark Bay is Australia's largest enclosed marine embayment. It contains the largest seagrass meadow in the world; it has a stable population of 10,000 Dugong (*Dogong dugon*); it is the most important Loggerhead Turtle (*Caretta caretta*) nesting site in Western Australia; it is a gathering ground for migratory Humpback Whales (*Megaptera novaeangliae*); and it is rich in other species, including birds (CALM 1996).

Shark Bay was included on the World Heritage List in December 1991 on the basis of its natural values and met all four of what were, at that time, the criteria for World Heritage listing of natural sites (CALM 1996), specifically: outstanding examples representing the major stages of Earth's evolutionary history; outstanding examples representing significant ongoing geological processes, biological evolution and human interaction with the natural environment; unique, rare, or superlative natural phenomena, and features of exceptional natural beauty; and important and significant habitats where threatened species of plants and animals of outstanding worldwide value from the point of view of science and conservation still exist.

Cambrian/Precambrian boundary, Wilkawillana Gorge, Flinders Ranges National Park, SA (Graeme Worboys)

1.2 Australia today

Australia's north–south spread gives rise to a wide range of climates. This, together with its evolutionary history, has produced a rich 'biogeodiversity' with a range of habitats and landscapes. Some qualities of Australia's present day natural heritage are briefly described (Table 1.2).

Table 1.2	Distinctive characteristics of the Australian continent (CSIRO 1986, SEAC 1996, White 1997, White 2000a)
Characteristic	**Notes**
World's lowest continent	Australia has the world's lowest average elevation. Its few mountain ranges are very low by world standards and most of the country is a broad, flat plain.
Geologically stable	Australia has some of the oldest land surfaces on Earth. Much of the surface has been exposed to weathering and erosive forces for long periods of geological time. In comparison, Europe and North America contain some landscapes dating back only to 16,000 to 14,000 years ago, when the great ice sheets melted. Australia is the only continent that has no tectonic mountain building or active volcanoes.
Driest inhabited continent	Australia receives the lowest annual average rainfall for any continent, apart from Antarctica. More than one-third is classified as arid (receiving less than an average of 250 mm per annum) and another third is semi-arid (250 to 500 mm per annum). Australia's climate is highly variable, largely because of changing ocean current temperatures, some of which generate *El Niño* effects.
Fewer sizeable rivers and less runoff than any other continent, except Antarctica	The low rainfall means that for its size, Australia has few large rivers, and accumulates little fresh water on its surface. The Murray–Darling river system carries about 22 billion m^3 of water compared with the Rhine (70 billion m^3) and the Mississippi (593 billion m^3). Australia's rivers have the most variable flow of any in the world. Australia has large internal drainage basins. It has fewer native freshwater fish species as a consequence.
Soils among the most nutrient-poor in the world	Dryness, geological stability, the antiquity of land surfaces, and long exposure to weathering have created poor soils. Less than 10% of Australian soils are reasonably productive and can sustain agriculture or dense vegetation. Australian topsoils are comparatively very thin. Only 6% of soils are of arable quality.
Most fire-prone continent	Large fires occur within Australia at any time of the year. Much of the native vegetation is adapted to fire.
Rich in minerals	Australia has more than 20% of the world's stock of recoverable bauxite, iron ore, uranium, mineral sands, and diamonds.
World's largest island	The coastline of Australia extends for 36,735 kilometres.
World's largest coral reef	The Great Barrier Reef complex extends some 2500 kilometres along the coast of Queensland. The reef is a World Heritage Area.

Table 1.2	(continued)
Characteristic	**Notes**
One of the 12 most biologically diverse countries in the world	The rich diversity is a consequence of Australia's long geological isolation from other continents (following the break-up of Gondwana in the Cretaceous) and its many climate zones.
Vegetation dominated by two genera of trees, *Eucalyptus* and *Acacia*	Acacias are widespread in southern continents, but eucalypts are confined to Australia and New Guinea. Most Australian trees are adapted to dry conditions (sclerophyllous) and low-nutrient soils.
Global significance of the ancient Australian rainforests	Australian rainforests are extremely old and have survived since the early Tertiary in refugia. They contain many species whose origin can be traced back to Gondwanan forests.
Unique mammal fauna	Australia is the world centre for marsupials (mammals whose young develop in a pouch), with over 144 species. Australia has two of the world's three species of monotremes. Monotremes (mammals that lay eggs) are only found in Australia (the Platypus (*Ornithorhynchus anatinus*) and the Short-beaked Echidna (*Tachyglossus aculeatus*)) and New Guinea (an echidna).

1.3   Climate

Photos of our planet from space dramatically illustrate our continents, oceans, and atmosphere and show how finite they are relative to the immensity of space (Figure 1.1). Such images also illustrate the interconnectedness of our atmosphere and reinforce a need to respond to Australia's climatic management responsibilities in a context of our planet's biosphere (White 2003).

The Earth's climate is like a giant solar–powered engine (White 1990). It is a global circulation of hot rising air and cooling and descending air that is complicated by the effects of a tilted and rotating planet, the location of continents, a varied terrain, and day and night differences. There is also a dynamic and an evolutionary aspect to the atmosphere, which is directly linked to Earth's remarkable biosphere. Mary White (2003) describes this interconnectedness in her book *Earth Alive*, which protected area managers are encouraged to read. Life on Earth has helped to create the atmosphere as we know it today. It helps to maintain it.

Important 'stages' in the development of Earth's atmosphere have been described (Table 1.1). Life on land on our planet has not always been possible.

It took until the late Silurian for photosynthesis to build up sufficient atmospheric oxygen to create a thick ozone layer. This was needed to filter out harmful radiation before life on land was possible (White 1990, p. 87).

Throughout geological time, five great extinctions of species have been recognised and have probably been caused by changes to the biosphere. The Late Ordovician extinctions are thought to have been caused by cooling and warming of the Earth, marine transgressions and regressions, and anoxia (Groombridge & Jenkins 2002). The most probable causes of the late Devonian extinctions were marine transgressions and anoxia; the end of Permian extinctions due to volcanism, warming, marine transgression, and anoxia; the end of Triassic extinctions, marine transgression, and the end of the Cretaceous mass extinction due to the impact of a large meteor, volcanism, cooling, and marine regression (Groombridge & Jenkins 2002).

Other perturbations to the biosphere have occurred. In the Palaeocene (55 million years ago) a phenomenon believed to have been linked to a sudden, widespread release of methane from sediments

Table 1.3	Some Australian atmospheric (and climate) changes (Manins et al. 2001)	
Atmosphere condition	Change in condition	Measurement period (1910–2000)
Average surface temperature	Increase 0.760°C	90 years
Average minimum temperature	Increase 0.960°C	90 years
Average maximum temperature	Increase 0.560°C	90 years

caused rapid global warming that took tens of thousands of years to regain equilibrium (White 2003). Ice cores from Greenland reveal a 5 to 10 degrees Celsius warming for a 20-year period just 12,000 years ago.

Humans are now introducing major changes to the biosphere, including the atmosphere. It has been

Summer storm clouds over Kosciuszko National Park, NSW (Graeme Worboys)

estimated that in less than 1000 years, 2 to 4 giga-tonnes of carbon have been added to the atmosphere by humans (White 2003).

Reduction in the thickness of the protective ozone layer over Antarctica occurred slowly from the 1950s to the 1970s, with a rapid decline in the mid 1990s. There was a 60% loss in ozone (Gifford 2003). Chlorofluorocarbon and methyl bromide pollution of the upper atmosphere caused this problem. The area of the 'ozone hole' over the Antarctic was near its maximum size in 2001 (Gifford 2003), which caused ultra-violet stressed phytoplankton over Antarctica (White 2003). Urgent global responses have halted impacts; however full ozone layer recovery to pre 1970 levels is unlikely to occur before 2050 (Manins et al. 2001, Gifford 2003).

Highly publicised changes are happening to the planet's climate as a direct result of human activities (Table 1.3). These changes will directly influence the way in which protected areas are managed for the long term.

The eight warmest years measured globally have occurred in the 1990s and 2000s. Australia's warmest year on record was 1998, the second warmest year was 2002, and the warmest decade ever recorded with measuring instruments was the 1990s. An increase in global average temperatures of 1.5–6 degrees Celsius is forecast for the year 2100 (Howden et al. 2003).

Australia's climate

Beckmann (1996) provides a succinct description of Australia's current average climatic conditions. The following is adapted from his work. Australia's latitude ranges from 10° south to about 43° south. This puts much of it in the zone of high pressure and therefore low rainfall, and this is exacerbated by the absence of high mountains. More than one-third of Australia is classified as arid, which means that it receives an average annual rainfall of less than 250 mm. Another third is semi-arid, with an annual average rainfall between 250 and 500 mm. Despite its aridity, Australia has few areas of true desert. Australia has a wide range of climatic zones: temperate climates in coastal Tasmania and the southern mainland; Mediterranean in the south-west and south-central areas; tropical climates in the north; subtropical along the warm east coast; and a small region of alpine climate in the south-east of the mainland and in central Tasmania (Beckmann 1996).

Characteristics of Australia's climate include:

- extensive areas of arid and semi-arid land
- hot and moist climates to the north, with distinctive wet summer and dry winter seasons
- hot and moist climates with uniform rainfall in the wet tropics around Cairns
- an east coast that receives steady rainfall from south-east wind systems, with a rapid decrease in rainfall inland of the coastal ranges

- a southern coastline, including Tasmania and the south-west of Western Australia, which receives winter rainfall maximums from westerly air systems—these systems can also bring snow to the higher peaks of NSW, Victoria, and Tasmania
- extreme weather conditions that include cyclones during the summer in northern Australia, and winter rain and snow storms and summer thunderstorms in southern Australia
- a wide variation in annual rainfall, with frequent droughts.

Australia is also significantly influenced by the El Niño–Southern Oscillation (a natural, aperiodic ocean-atmosphere fluctuation) (Allan 2003) phenomenon and the cycle of drought and bushfires that it brings (Figure 1.2). ENSO involves fluctuations between two extremes, the El Niño (warm episode) and La Niña (cold episode) events. ENSO events and their severity may be affected by global warming (Allan 2003). Seven broad climatic zones (Figure 1.3, see colour section) have been identified for Australia, based on long-term climatic data.

Figure 1.2   El Niño effect, drought effects, and bushfires in Blue Mountains National Park
(NSW NPWS et al. 1998)

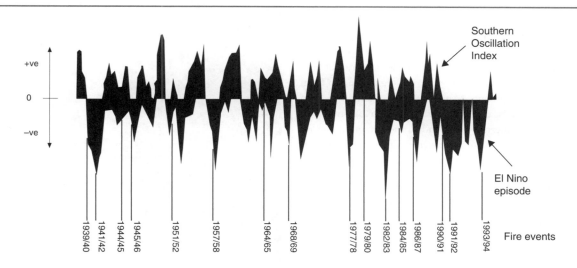

Negative index values coincide with major fire events

1.4 Landforms, soils, and landscapes

Landforms

Australia is the smallest, lowest, and flattest continent (Figure 1.4, see colour section). The average elevation is only 330 metres, with the highest point being Mt Kosciuszko at 2228 metres. Uplands rarely exceed 1200 metres. The continent has three distinct landscapes: the Western Plateau, the Central Lowlands, and the Eastern Highlands (ABS 1995, 1996). The centre of the continent is inwardly draining and the Western Plateau is virtually flat.

The Western Plateau consists of very old rocks (some over 3000 million years old), and much of it has existed as a land mass for over 500 million years. Several sections are identified as individual plateaux (Kimberley, Hamersley, Arnhem Land, Yilgarn). The Nullarbor, a limestone plain about 25 million years old, is an uplifted sea floor (ABS 1995).

The Central Lowlands stretch from the Gulf of Carpentaria through the Great Artesian Basin to the Murray-Darling Plains. Much of central Australia is flat though there are several ranges (for example MacDonnell, Musgrave) and some individual features of which Uluṟu (Ayers Rock) is the best known. In South Australia, fault movements formed a series of ranges (Mt Lofty, Flinders Ranges, Adelaide Hills). The lowlands are either occupied by sea (Spencer Gulf) or plains (lower Murray Plains).

Some areas of the Eastern Highlands are dissected hills. Other parts of the Eastern Highlands rise gently from central Australia towards a series of high plateaus, which at the eastern margins tend to form high escarpments. Many of these are united to form the great escarpment of eastern Australia, stretching from northern Queensland to the Victorian border. In eastern Victoria, however, the old plateau has been eroded into separate high plains.

Soils

Australian soils are generally thin and infertile. This reflects the age of the continent and constant exposure to the processes of erosion and weathering. Over 33% of the continent is made up of unconsolidated sands, saline, and sodic soils. Only about 6% of Australia has reasonably productive soils (White 2000b). Soils of Australia, especially when compared to those of Europe and North America, are fragile and susceptible to wind and water erosion, to loss of structure and nutrients, and to salinisation, sodification, and acidification. Australian soils have been broadly grouped into 14 major types. These can be further grouped into six major classes: shallow stony soils, deep sands, cracking clays, calcareous soils, massive sequioxidic soils, and duplex or 'texture-contrast' soils.

Australian landscapes

Forests and woodlands

Forests cover a very small part of Australia, about 5% of the total land mass. Forests are restricted to parts of the coastal and near-coastal areas with high rainfall and suitable soils. Of the forests that existed when the first white settlers arrived, only one-third remain. Most of this loss has occurred since 1900, leading to major loss of habitat. In a continent restricted in

Ribbon Gum (*Eucalyptus viminalis*) and Alpine Ash (*Eucalyptus delegatensis*), Kosciuszko National Park (Graeme Worboys)

forest resources, rainforest areas are among the most valuable and most restricted. Rainforests have an incomparable richness of flora and fauna and are important reserves of genetic diversity. A 10 km² area of eucalypt forest will typically have some six to eight tree species in the canopy layer. In a rainforest there may be 200 or more.

The wetter environments of the great escarpment of eastern Australia, of the south-west of Western Australia, and of Tasmania, support tall, open eucalypt forests and rainforests. Some of these forests are the tallest hardwood forests in the world. Landscapes are typically hilly to mountainous, with the dominant *Eucalyptus* genus varying from 100-metre giants in favourable sites to stunted woodlands at the tree line in alpine areas. The open forests and woodlands of the drier plains form a gentler landscape, often mixed with native cypress pines.

Wetlands

Wetlands include swamps, marshes, billabongs, lakes, saltmarshes, mudflats, mangroves, coral reefs, fens, peatlands, or bodies of water, whether natural or artificial, permanent or temporary (DEH 2004b). They are defined by the Ramsar Convention (Chapter 3):

> *Wetlands are areas of marsh, fen, peatland or water, whether natural or artificial, permanent or temporary, with water that is static or flowing, fresh or brackish or salt, including areas of marine water the depth of which at low tide does not exceed 6 metres (DEH 2004b).*

Australia's wetlands are diverse and include a range of fauna and flora. There are more than 851 nationally important wetlands recorded for all states and territories (Environment Australia 2001). Sixty-four of these wetlands are registered under the Ramsar Convention (DEH 2004b). There are marine and coastal zone wetlands, inland wetlands, and human-made wetlands. For Australia's river systems, there are 13 drainage divisions recognised for Australia (Neal et al. 2001) and within these, there are 22 drainage basins that contain 10 or more nationally significant wetlands (Environment Australia 2001). Australia's marine, estuarine, and freshwater wetland environments include a rich array of species and habitats.

Wetland, Yellow Water, Kakadu National Park, NT
(Graeme Worboys)

Rivers, streams, and lakes

By comparison with other countries, Australia is deficient in rivers and streams. Not only are there few of them, but they carry much less water in relation to the size of their catchments than comparable river systems elsewhere. Streams, lakes, the short, sharp fast-flowing rivers of the eastern escarpments, the large tidal river systems of northern Australia, and the long meandering inland river systems such as the Murray–Darling, complete a surprisingly diverse list for a generally dry continent. Many rivers are ephemeral and part of large inward-draining systems, such as the Lake Eyre basin. Usually dry, they come alive with wildlife following major rainfall events in distant parts of their catchments. Anabranching of rivers is a characteristic because of the extreme flatness of the land. Floodplains are an integral part of the major river systems. Freshwater rivers and streams include a rich fauna and flora such as biofilms, algae, aquatic plants, reeds, sedges, micro-organisms, waterbirds, fish such as Murray Cod (*Maccullochella peelii*) and Golden Perch (*Macquaria ambigua*), amphibians, freshwater crayfish, and platypus (Young 2001). There are 302 species of freshwater fish belonging to 59 families in Australian rivers and streams (Allen et al. 2002). This is a small number compared to other continents, reflecting the dry conditions of the continent. They are strongly endemic species. The Queensland lungfish is the world's oldest surviving vertebrate species, and has persisted for 150 million years (Allen et al. 2002).

Estuaries, coastlines and islands, coral cays

Australia has an enormous length of coastline, and a wide range of coastal environments. These may be rocky, sandy, muddy, sand dunes, or off-shore coral reefs. Brackish to saline estuaries and mudflats are rich sources of food for many bird and fish species. There are many islands scattered around the coastline of Australia. Some are the remains of drowned coastlines. Others are the creation of island-building corals. Many islands have been formed by sand deposits, and Fraser Island, a World Heritage Area, is the world's largest sand island. Islands may contain rare localised species or remnants of mainland species that have elsewhere disappeared or diverged. This makes them especially significant for the study of genetics, evolution, and species interaction. Other spectacular coastal features include the 'Twelve Apostles' on the south coast of Victoria and the giant carbonate cliffs of parts of the Great Australian Bight.

Arid and semi-arid

The arid lands of Australia occupy nearly three-quarters of the land surface of the continent, an arid area only exceeded by that of North Africa and the

Rocky shore and seal colony, Montague Island Nature Reserve, NSW (Graeme Worboys)

Termite mounds, Tennant Creek, NT (Graeme Worboys)

Middle East. These lands support highly specialised and adapted plants and animals—an invaluable pool for genetic and physiological investigation. Vegetation in arid terrain tends to be sparse, dwarfed, often ephemeral, and in a delicate ecological balance. In large areas the present vegetation has stabilised a pattern of sand dunes that were mobile only some

Snapshot 1.5

Attitudes to the desert centre: the literary record

The way Australians speak and write about 'the centre' reveals a steady movement towards more positive attitudes. The 'hellish country' and 'God-forsaken land' of some early observers, a 'huge abraded rind ... crow-countries, graped with dung' (Kenneth Slessor) where the white cockatoo 'screams with demoniac pain' (James McCauley), has become 'the wide brown land' of which Australians are quietly proud, and even 'the three-quarters of our continent set aside for mystic poetry' (Les Murray). Writers like Eric Rolls and Judith Wright began to see regions with close affection—'part of my blood's country rises that tableland, high delicate outline ... full of old stories that still go walking in my sleep' (Wright). 'Ayers Rock' has changed from a popular curiosity, a rock that turned pink at sunset, to Uluṟu, the mystical marker of Australia's imagined centre point, a place whose Aboriginal significance matters deeply to all groups in Australia.

10,000 to 15,000 years ago, when Australia was much drier than today. The wide flat landscapes of the centre of Australia have a special magic for many people, and are the way most overseas visitors now visualise a typical Australian landscape (Snapshot 1.5.).

Alpine areas

The highest mountains of Australia are low in altitude compared to the great alpine areas of the world. The Australian alpine areas nevertheless include many endemic plants and some mammals, and many features of great scientific interest. There are two main alpine regions: the Australian Alps in the highest part of the Great Dividing Range, and the Central Highlands of Tasmania. The Australian Alps covers a continuous area from Baw Baw National Park east of Melbourne to the northernmost parts of Kosciuszko National Park and Brindabella National Park to the west of Canberra. The Central Highlands of Tasmania are represented by the Central Plateau Region, Mt Field National Park, and Cradle Mountain–Lake St Clair National Park. The alpine and subalpine environments of Australia are small by world standards, since Australia's tallest mountains are only just high enough for snow cover. Commonly there is no snow for parts of the year, with only isolated snow-

Mt Jagungal, Kosciuszko National Park, NSW (Graeme Worboys)

drifts surviving the summer. The tree-line is about 1800 metres in NSW. The twisted, gnarled, and stunted eucalypt tree-lines of Australia are in stark contrast to the conifer-dominated tree-lines of the northern hemisphere.

Heath

Typically heathlands are found in areas of low soil nutrients such as sandstone soils. They are limited in area but important for biodiversity. Many rare and endangered plant and animal species are found within the heath communities of Western Australia and coastal eastern Australia.

Grasslands

Grassland communities dominate the tropical north of Australia. These include the grassy savannahs of the tropical north and the Mitchell Grass plains of the Great Artesian Basin. Grasslands are also found in some areas to the south, such as the frost-hollow basins of the Snowy Mountains in the Australian Capital Territory, and the drier basalt areas of the Monaro in southern NSW. Many of the temperate grassland communities have been impacted by agriculture, and only exist today as small remnants on roadsides, stock routes, and cemeteries.

Marine

The marine landscapes of Australia are extraordinarily diverse, ranging from cool-temperate and sub-Antarctic waters to tropical waters. They host a wide range of fish, mammal, bird, plant, and other species.

Karst landscapes

These specialised landscapes of limestone, dolomite, and other rocks influenced by solution are also limited in Australia. Yet some spectacular karst landscapes do occur, for instance at Jenolan Caves in NSW, the Chillagoe Caves in Queensland, and the spectacular dolines (solution depressions) of the Nullarbor Plain.

Urban

Australia is a highly urbanised country. Major cities such as Sydney, Melbourne, Adelaide, Brisbane, and Perth contain most of the population. The

Sunrise over Sydney, NSW (Graeme Worboys)

better-watered east coast of Australia has the highest density of population. Around the cities, urban landscapes and hobby farms have expanded into native bushlands. This brings a peculiar interaction of natural phenomena (bushfires, plants, and animals) into a human-dominated landscape, and provides special challenges for protected area managers.

Agricultural

Agricultural landscapes dominate many parts of Australia, including large areas of cereal crops, irrigated orchards and vineyards, sheep and cattle stations, and timber plantations. There is a large variety of agricultural output, and Australia is a net exporter of food. Important remnants of the original vegetation are still found scattered through these landscapes. Agricultural lands are increasingly being affected by salinisation and soil degradation.

1.5   Biological diversity

The term biological diversity (or biodiversity) refers to the variety of all living things: the plants, fungi, animals, and micro-organisms, the genetic information they contain, and the ecosystems they form. It was defined by the United Nations Convention on Biological Diversity (UN 1992a) as:

> The variability among living organisms from all sources including, inter alia, terrestrial, marine and other aquatic ecosystems and the ecological complexes of which they are part; this includes diversity within species, between species and of ecosystems.

Australia's biological diversity is part of the rich global biological diversity (Backgrounder 1.1), and which is supported by an interconnected global biosphere (Backgrounder 1.2). The biosphere has evolved over geologic time, and is essential for life on Earth:

> the biosphere is alive and metabolising, everything is forever changing, renewing, evolving, and it is the interconnectedness and symbiotic functioning

> of living matter at all levels that maintains the checks and balances. The natural laws that apply in the natural world, particularly those that balance populations and nutrients, maintain equilibrium (White 2003, p. 174).

It is estimated that Australian ecosystems support about one million species (ABS 1996, ASEC 2001a). Australia is one of the 12 nations, and the only developed country, that contains major repositories of species diversity. Other countries include Indonesia, with its wealth of islands and different habitats, Zaire in equatorial Africa, and Brazil, with its expanses of rich tropical rainforests, rivers and mountains (SEAC 1996). The estimated extent of Australia's species diversity is shown in Table 1.5.

Australia's bioregions

A landmark document identifying biogeographic regions for Australia was researched and developed by Thackway and Cresswell (1995a). This work has assisted in setting priorities for conservation reservation (Backgrounder 1.3).

Backgrounder 1.1   The diversity of life on earth

The diversity of life on earth is remarkable. There is an estimated 14 million species of which 1.75 million have been described (Table 1.4).

Table 1.4	**The diversity of life on Earth** (Groombridge & Jenkins 2003)			
Domains	Eukaryote Kingdoms	Phyla	Species described	Estimated total
Bacteria			10,000	?
Protoctista			80,000	600,000
Eukarya	Animalia	Cranatia (vertebrates) total	52,500	5500
		Mammals	4630	
		Birds	9750	
		Reptiles	8002	
		Amphibians	4950	
		Fishes	25,000	
		Mandibulata (insects and myriapods)	963,000	8,000,000
		Chelicerata (arachnids etc)	75,000	750,000
		Mollusca	70,000	200,000
		Crustacea	40,000	150,000
		Nemotoda	25,000	400,000
	Fungi		72,000	150,000
	Plantae		270,000	320,000
TOTAL			1,750,000	14,000,000

Backgrounder 1.2   The biosphere

The biosphere is the envelope around the Earth's surface that contains all living organisms and the elements they exchange with the non-living environment (Groombridge & Jenkins 2000). It is estimated to be 20 kilometres thick. The top is the top of the atmosphere, and its bottom is continental rock and ocean depths (White 2003). The world's weather is found within it and sunlight provides the energy that maintains the biosphere. It is used by micro-organisms, algae, and plants to produce organic molecules by photosynthesis. The net primary production includes organic material synthesised by photosynthetic organisms and the amount of energy-rich material left to sustain all other life on Earth (Groombridge & Jenkins 2000). The biosphere is a closed ecological system.

It is also an intricately interconnected system. It works on cause and effect and action and reaction and ultimately sets the rules for the survival of species of all sorts, including humans (White 2003). It is the interconnectedness of everything within this system that has influenced the evolution of species and thus of the biosphere itself over billions of years (White 2003).

Table 1.5	**Australian species diversity** (SEAC 1996)	
Major group	Estimated number of species	Percentage described
Micro-organisms		
Protozoans	65,000	40
Fungi	160,000	5
Bacteria	40,000	0.1
Invertebrates		
Arthropods		
Coleoptera (beetles)	30,000	67
Lepidoptera (moths and butterflies)	20,000	53
Hymenoptera (ants, wasps, and bees)	23,000	33
Diptera (flies and mosquitoes)	11,000	75
Other insects	15,000	20
Arachnids (e.g. spiders and mites)	39,000	14
Crustaceans (e.g. crabs and prawns)	18,000	5
Springtails	2500	14
Other arthropods	?	?
Molluscs (e.g. snails, oysters, squid)	19,000	?
Sponges	1400	28
Nematodes	150,000	1
Other invertebrates	?	?
Vertebrates	5588	90+
Plants		
Higher plants	20,000	90+
Algae	22,000	50

?   *For some groups the level of knowledge is so poor that estimates are unavailable.*

Backgrounder 1.3   Interim Biogeographic Regionalisation for Australia

By the 1960s, managers realised that they needed urgently to acquire not just any natural lands that were available but representative samples of the different habitat-types, and especially of rare and threatened ones. Unfortunately there was an inevitable bias in the types of land available. The easiest land to acquire for reserves is land that has remained with the Crown because it is unsuitable for agriculture or forestry. There was also political pressure from lobby groups to conserve the 'tall green end of the spectrum of ecosystems' (Thackway & Cresswell 1995a). This meant that some types of habitat were over-represented in reserves, while others were left unprotected.

Managers began to investigate and actively counter this 'bias' in their own areas, but the job really needed to be done at national level. The will to cooperate was there, but a way to standardise the data and to reconcile the different categories that managers in different regions were using to classify and count types of habitat was needed. The result was the Interim Biogeographic Regionalisation for Australia (IBRA) (Thackway & Cresswell 1995b).

The original IBRA recognised 80 biogeographic regions for Australia's land mass. The biogeographic regions vary in size from 2372 km^2 (Furneaux Island, in Bass Strait) to 423,751 km^2 (Great Victoria Desert). The larger IBRA regions are mostly in flatter arid or semi-arid areas, whereas its regions are smaller near the coast where rainfall and elevation fluctuate more widely. Even so, each IBRA region is a broad category, and contains many smaller habitats.

Thus even when a high proportion of a given IBRA region is protected in reserves, many of its internal environments may not be. It turned out that some types of region suffer more than others from such 'internal bias' in the way that their smaller habitats are represented. Only in western Tasmania and small areas in Victoria and South Australia (less than 1% of Australia's land mass) were regions found whose whole range of internal environments or ecosystems were represented, without 'bias', in reserves. The situation was worst in arid and semi-arid lands (Thackway & Cresswell 1995b).

In 2000, IBRA Version 5 was developed, with 85 bioregions delineated based on the major environmental influences on flora and fauna distribution. Regional and continental scale data on climate, geomorphology, landform, lithology, flora, and fauna were interpreted to develop the bioregional boundaries. The distribution of Australia's protected areas with respect to IBRA Version 5.1 is shown in Figure 1.5. The National Reserve System for Australia is using IBRA as its planning framework (Chapter 5).

Figure 1.5   Australia's protected areas with respect to IBRA Version 5.1 bioregions

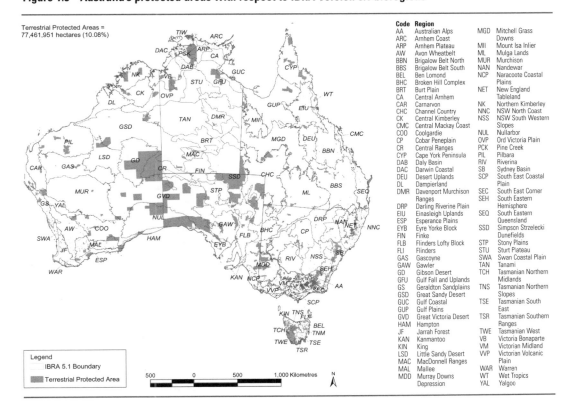

Terrestrial Protected Areas = 77,461,951 hectares (10.08%)

Code	Region		
AA	Australian Alps	MGD	Mitchell Grass Downs
ARC	Arnhem Coast		
ARP	Arnhem Plateau	MII	Mount Isa Inlier
AW	Avon Wheatbelt	ML	Mulga Lands
BBN	Brigalow Belt North	MUR	Murchison
BBS	Brigalow Belt South	NAN	Nandewar
BEL	Ben Lomond	NCP	Naracoote Coastal Plains
BHC	Broken Hill Complex		
BRT	Burt Plain	NET	New England Tableland
CA	Central Arnhem		
CAR	Carnarvon	NK	Northern Kimberley
CHC	Channel Country	NNC	NSW North Coast
CK	Central Kimberley	NSS	NSW South Western Slopes
CMC	Central Mackay Coast		
COO	Coolgardie	NUL	Nullarbor
CP	Cobar Peneplain	OVP	Ord Victoria Plain
CR	Central Ranges	PCK	Pine Creek
CYP	Cape York Peninsula	PIL	Pilbara
DAB	Daly Basin	RIV	Riverina
DAC	Darwin Coastal	SB	Sydney Basin
DEU	Desert Uplands	SCP	South East Coastal Plain
DL	Dampierland		
DMR	Davenport Murchison Ranges	SEC	South East Corner
		SEH	South Eastern Hemisphere
DRP	Darling Riverine Plain		
EIU	Einasleigh Uplands	SEQ	South Eastern Queensland
ESP	Esperance Plains		
EYB	Eyre Yorke Block	SSD	Simpson Strzelecki Dunefields
FIN	Finke		
FLB	Flinders Lofty Block	STP	Stony Plains
FLI	Flinders	STU	Sturt Plateau
GAS	Gascoyne	SWA	Swan Coastal Plain
GAW	Gawler	TAN	Tanami
GD	Gibson Desert	TCH	Tasmanian Northern Midlands
GFU	Gulf Fall and Uplands		
GS	Geraldton Sandplains	TNS	Tasmanian Northern Slopes
GSD	Great Sandy Desert		
GUC	Gulf Coastal	TSE	Tasmanian South East
GUP	Gulf Plains		
GVD	Great Victoria Desert	TSR	Tasmanian Southern Ranges
HAM	Hampton		
JF	Jarrah Forest	TWE	Tasmanian West
KAN	Kanmantoo	VB	Victoria Bonaparte
KIN	King	VM	Victorian Midland
LSD	Little Sandy Desert	VVP	Victorian Volcanic Plain
MAC	MacDonnell Ranges		
MAL	Mallee	WAR	Warren
MDD	Murray Downs Depression	WT	Wet Tropics
		YAL	Yalgoo

Legend
IBRA 5.1 Boundary
Terrestrial Protected Area

500   0   500   1,000 Kilometres   N

1.6   Terrestrial vegetation

The present vegetation of Australia reflects major events over the last two million years such as ice-age climatic fluctuations. In more recent times, the vegetation has been modified by human use of fire as well as large-scale clearing for agriculture.

It is estimated that there are some 20,000 to 25,000 vascular plant species in Australia (Briggs &

Leigh 1995, ASEC 2001a). It is one of the world's six 'floristic realms' or regions supporting a characteristic flora (SEAC 1996). Some of its special characteristics are listed below.

• Australian floras show standard features of the southern hemisphere flora group. For instance, the dominant flowering trees are evergreen, unlike the deciduous trees of the temperate northern hemisphere.

- In contrast to the northern hemisphere floras, Australian floras have relatively few conifers.
- The arid zone has a unique combination of cosmopolitan and ancient Australian plants. It differs from other deserts in having very few succulents. Most of its woody plants are drought resisters with a capacity to endure dry conditions. The deserts have a beautiful ephemeral flora that is rarely seen and includes threatened species.
- The alpine habitat is very limited in extent, and carries a rare combination of ancient southern plants and adapted Australian ones. These occur above a tree-line that is remarkably low.
- The scleromorphic ('hard-leaved') floras of poor soils and seasonal habitats show a bewildering diversity, in which they are matched only by the Cape flora of South Africa. Endemism in these floras is high.
- The rainforests are largely remnants of the ancient Gondwanan flora and harbour some of the world's most important links in plant evolution. There is a rich assemblage of orchids, with 1500 known species.
- Australia is one of the most lichen-rich countries of the world, with about 2275 species.
- Australia has the world's greatest diversity in the family Proteaceae, with 42 genera (860 species) from a world total of 72 (Barlow 1981, DEST 1994).
- The genus Eucalyptus, according to some authorities, consists of about 900 species, all but 13 endemic. Most of the (estimated) 1070 Australian taxa of Acacia occur nowhere else in the world. They are diversified into almost every habitat on the continent (Commonwealth of Australia 1996b).

It is unusual for a landmass as large as Australia to have its vegetation dominated by two genera over seventy-five per cent of its area. The genus Eucalyptus dominates in better-watered regions with a rainfall of at least 300 millimetres, Acacia takes over dominance in the arid centre of the continent where the rainfall is less than 300 millimetres. The two genera share dominance in interzones ... Chenopod shrubland (saltbush, bluebush) and tussock grasses occupy the largest, about equal, areas in the twenty-five per cent of the continent not characterised by Acacia or Eucalyptus. Casuarina dominates in smaller areas in the southern half of the continent. Melaleuca paperbark forests of considerable extent are found in the Carpentaria Basin/Cape York area; and mixed or other floristic groups comprise smaller vegetation patches.

Only the fringes of the continent carry forest vegetation of a type with a closed canopy, or open tall tree type, or with dense medium trees in open layered formation with lower trees and shrubs. The drier woodlands, shrublands and scrub have grass and forb understorey. Rainforests, the last remnants of Gondwanan forest which exist in refuges along the east coast from Tasmania to Cape York, show a variety of structure and content as a result of the sifting and sorting of the original mixed forest of the early Tertiary. Rainforests of northeastern Queensland are the closest to those early warm temperate forests ... Tall open forests have suffered the most from heavy logging ... Open forest and woodland of the better-watered eastern sector of the continent have largely been replaced by pastures and cultivated lands (White 1994, p. 207).

Classification of plant communities

Based on the work of Specht (1970) and Beard & Webb (1974), Australia's vegetation may be classified into the following major structural categories (AUSLIG 1990) (Figure 1.6, see colour section).

Tall Closed Forest (trees >30 m; 75% foliage cover). These rainforests are among the most complex ecosystems in the world. They survive in Australia today as only a few scattered relics in north-eastern Queensland, and they are the Australian equivalents of the tall tropical lowland rainforests of the Amazon and Zaire basins. Their height ranges up to 40 metres or more.

Tall Open Forest (trees > 30 m; 30 to 70% foliage cover). These towering Eucalyptus forests, standing over 30 metres tall and reaching heights of 100 metres in places, are the tallest of Australia's hardwood forests. There are few areas that are untouched by logging during the last 200 years. Tall Open Forest covers about 50,000 km² or 0.65% of Australia. Understoreys range from rainforest and tree-ferns to low trees and small shrubs. These forests are usually found where the rainfall is reliable and greater than 1000 mm per annum (Snapshot 1.6).

Closed Forest (trees 10 to 30 m; >70% foliage cover). These rainforests vary from the cool temperate Southern Beech forests of Tasmania to the tropical vine forests of Cape York Peninsula in Queensland. Small rainforest patches also occur in the top end of the Northern Territory and in the Kimberley region of Western Australia. Originally much more extensive, they presently occupy about 20,000 km² or 0.26% of the Australian land mass. Most of these rainforests occur in areas of high rainfall (> 1200 mm annually) and in altitudes of more than 1200 metres. Four broad climatic groupings of this rainforest are recognised, including tropical rainforests (North Queensland), subtropical and warm temperate

types (Mackay, Queensland, to East Gippsland, Victoria), and the cool temperate rainforests of Victoria and Tasmania.

Open Forest (trees 10 to 30 m; 30 to 70% foliage cover). Open forests form the bulk of Australia's forested country and are the primary resource for the nation's timber industry. Native open forests, with *Eucalyptus* hardwoods predominant, cover about 140,000 km² or 1.8% of Australia. This represents close to half of the original area of these forests. Open forests are generally confined to the coast or nearby ranges.

Woodland (trees 10 to 30 m; 10 to 30% foliage cover). Woodlands form a transitional zone between the higher-rainfall forested margins of the continent and the arid interior. *Eucalyptus* is the most widespread tree component, though there is a wide range of understorey types. Woodlands have almost entirely been removed from the cereal cropping lands in the south-east and the far south-west of Australia. Thinning and removal of shrubby understorey has occurred in areas used for grazing.

Snapshot **1.6**

The South East Forest National Park of NSW

On 29 January 1997, the Premier of NSW, Mr Bob Carr, officially announced—from a podium on the edge of the great escarpment of eastern Australia, surrounded by a magnificent Tall Open Forest—the establishment of the 100th national park for NSW. The 1150 km² South East Forest National Park protected some of the finest of the tall eucalypt forests of NSW and, importantly, achieved conservation links along the great escarpment, creating a continuous 150 kilometre north–south swathe of protected area. The declaration of the park was a climax to more than 20 years of bitter controversy between loggers and conservationists, and achieved the Premier's long-held vision of conserving an adequate sample of these magnificent forests. In 2003, the conservation corridor was extended to 350 kilometres north–south by the State Labor Government Minister for the Environment, Mr Bob Debus. There is an imperative to extend the corridor further to more than 600 kilometres north–south to minimise the loss of biodiversity through fragmentation (Pulsford et al. 2003).

Premier Bob Carr, South East Forest National Park, NSW (Graeme Worboys)

Approximately 538,000 km² (or 7% of Australia) of native woodland remains. Approximately 70,000 km² of other vegetation types has been converted into woodland through activities such as grazing and clearing.

Open Woodland (trees 10 to 30 m; <10% foliage cover). The largest natural occurrences of open woodland are the eucalypt-studded grasslands on the flood-plains of the upper tributaries of the Darling River and on the undulating country extending inland from Townsville. Large-scale tree clearing in the agricultural areas of eastern Australia has resulted in the creation of extensive artificial open woodlands. Approximately 141,000 km² (or 1.8% of Australia) of open woodland remains. Approximately 262,000 km² (or 3.4% of Australia) of open woodland has been created by human activities.

Low Closed Forest (trees <10 m; >70% foliage cover). The low closed forests have a widespread but patchy distribution, especially across northern Australia. The main occurrences are on rich soils or basalt outcrops in Queensland. In monsoonal areas, they are mostly confined to small, fire-protected sites. Approximately 1000 km² (or 0.01% of Australia) of this community remains.

Low Open Forest (trees <10 m; 30 to 70% foliage cover). Approximately 32,000 km² (or 0.4% of Australia) of this community remains in a number of distinct communities throughout the continent. The most extensive are the Lancewood *(Acacia shirleyi)* forests of the escarpment country in the Northern Territory and inland Queensland.

Low Woodland (trees <10 m; 10 to 30% foliage cover). Low woodlands are floristically diverse, and they occur extensively within the sub-humid and semi-arid zones of the continent. Approximately 432,000 km² (or 5.6% of Australia) of this community remains.

Low Open Woodland (trees <10 m; <10% foliage cover). Low open woodlands are found throughout much of inland Australia where the scarcity of water and the poor soils limit the height and density of trees. Eucalypts and acacias commonly dominate the tree layer. Approximately 1,453,000 km² (or 18.9% of Australia) of this community remains.

Open Scrub (shrubs >2 m; 30 to 70% foliage cover). These are dense formations of tall shrubs that are found on low-nutrient or waterlogged soils. Approximately 29,000 km² (or 0.37% of Australia) of this community remain.

Tall Shrubland (shrubs >2 m; 10 to 30% foliage cover). Two of Australia's best known inland vegetation types, mallee and mulga, occur within this category, though they have been extensively modified by human use. Much former mallee land now lies within the wheatbelt, and large areas of mulga shrubland have been reduced to open shrubland by more than a century of grazing. Approximately 724,000 km² (or 9.4% of Australia) of this community remains.

Tall Open Shrubland (shrubs >2 m; <10% foliage cover). This is the single most widespread structural form of vegetation in Australia. It covers a remarkable 82.4% of the continent. The largest areas are found in the desert sand country.

Open Heath (shrubs <2 m; 30 to 70% foliage cover). Heaths are a floristically rich formation of diverse shrubby genera. In Western Australia, for example, the heath flora contains about 2000 species and a number of endemic genera. Heaths are a very important fire-adapted habitat for a variety of native fauna. Heaths have a very patchy distribution in coastal and near-coastal areas throughout Australia. They tolerate annual rainfalls from more than 1600 mm to less than 300 mm. About 8000 km² (or 0.1% of Australia) of this community remains.

Low Shrubland (shrubs <2 m; 10 to 30% foliage cover). This community formerly covered about 5% of Australia. The major part of this area is still dominated by saltbush and bluebush, but shrub density and foliage cover has declined as a result of grazing. (The treeless alpine vegetation is classified as low shrubland.) Approximately 60,000 km² (or 0.78% of Australia) of this community remains.

Low Open Shrubland (shrubs < 2 m; <10% foliage cover). Many of the natural occurrences of this formation are found in the arid interior of Australia. Here they occur in extreme environments such as rocky ranges or the skeletal soils of erosional landscapes. Approximately 92,000 km² (or 1.2% of Australia) of this community remains in a natural condition, while another 308,000 km² (or 4% of Australia) has been created by human activities and natural processes.

Hummock Grassland. While hummock grasslands without an overstorey are quite restricted, other forms of vegetation with a hummock grass understorey cover 25% of the continent. Hummock grasses are members of the genera *Triodia* and *Plectrachne*, found only in Australia.

Tussocky Grassland. Vast plains of tussock grasses are a distinctive Australian vegetation type, and the Mitchell Grass plains of northern NSW, Queensland and Northern Territory cover enormous areas. They are found mainly on cracking clay soils.

European settlement introduced rapid and fundamental changes to the land cover of Australia. The area now used for agriculture (including grazing) and forestry covers more than 5,000,000 km² or 70% of the continent. Grazing lands have been modified by the selective forces imposed by exotic herbivores. European influence on vegetation includes:

- the taking up of most suitable grazing land
- the decrease in Australia's forested lands from 690,000 km² to 390,000 km²
- the decrease in Australia's woodlands from 1,570,000 km² to 1,070,000 km²
- the decrease in Australia's tall shrublands from 1,230,000 km² to 770,000 km²
- an increase in the areas of open woodland, tall open shrubland, low open shrubland, and grassland (AUSLIG 1990).

Rare or threatened Australian plants

Briggs & Leigh (1995) recorded 1009 threatened Australian flora species and 5031 rare and threatened Australian plant species, which was then 15% of the world's flora classified as threatened (Figure 1.7). They also advised that almost 20% of the world's flora classified as presumed extinct came from Australia. The proportion of Australia's flora that has been identified as being rare or threatened is high relative to other countries where comparable detailed studies are available. The world figure

in 1995 was 6%, with South Africa at 9.2%, New Zealand at 9.3%, Europe at 17%, USA at 17%, and Australia at 22.9% (Briggs & Leigh 1995). In 2001, there were 517 endangered and 654 vulnerable vascular plants. This is an increase in numbers of plants of this status (ASEC 2001a). In 2004, the conservation status of Australia's plants showed 560 critically endangered and endangered plants and 670 vulnerable plants (Table 1.6).

Table 1.6	The conservation status of Australia's flora (DEH 2004c)
Conservation status (EPBC Act 1999)	**Number of species**
Extinct	61
Critically endangered	52
Endangered	508
Vulnerable	670
Conservation dependent	–

Figure 1.7   Location of threatened plant species (SEAC 1996)

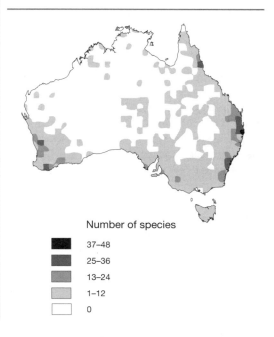

Number of species

- 37–48
- 25–36
- 13–24
- 1–12
- 0

1.7 Terrestrial fauna

Australia has a distinctive fauna. The *Australia State of the Environment Report* (Biodiversity) recognised 369 mammal, 825 bird, 633 reptile, and 176 frog species that had been formally described (ASEC 2001a). Many of Australia's birds, and most reptiles and amphibians, are endemic. Australia's fauna is also highly diverse compared to that of other continents. For example, parrots are more varied in Australia than elsewhere; and the Australian arid zones house two to three times as many lizard species in a given area as desert regions in either Africa or North America.

Australia has a large invertebrate fauna. About one-third of the estimated total species awaits discovery. The insects are extremely diverse, with perhaps 140,000 species. Some families and most species only occur in Australia. Ants are so abundant that there are at least as many genera of ants in part of a Canberra nature reserve as there are *species* in the whole of Britain. Other terrestrial invertebrate groups such as worms have endemic examples such as the Giant Gippsland Earthworm (*Megascolides australis*).

Australia's animals possess a wide range of adaptations to help them cope with arid conditions. Approximately one-third of Australian frogs can burrow and some also form cocoons. Many marsupials produce concentrated urine and some can survive without drinking.

Australia's modern animal assemblage reflects its evolution in this extraordinary country. Heatwole (1987) subdivided Australia's fauna into five categories based on evolution. The oldest contain those taxa that are relicts from former, larger continents and are referred to as the Pangaean (Archaic element) and the Gondwanan Element (Old Southern Element). Those more recent taxa with an Asian origin can be subdivided according to the time of their arrival in Australia. The oldest members of this group began to appear in the Tertiary fossil record and, therefore, have had considerable time to undergo adaptive radiation in Australia. These are designated as the Asian Tertiary Element. Later arrivals, in the Pleistocene and more recently, are called the Modern Element. The fifth category, the Introduced Element, includes those species introduced by humans.

The Pangaean (Archaic Element) includes cockroaches; the relict ant *Nothomyrmecia*; the Tasmanian endemic spider family *Hickmaniidae*; the fly genus *Nemopalpus*; the beetle genera *Cupes* and *Rhysodes*; and the earthworm tribe *Acanthodrilini*.

The Gondwanan Element (Old Southern Element) includes marsupial mammals, ratite birds, chelid turtles, diplodactyline geckos, megascolecine earthworms, certain terrestrial molluscs, certain terrestrial spiders, many insects, and the scorpion genus *Cercophonius*.

The Asian Tertiary Element reflects the Australia tectonic plate colliding with the Asia tectonic plate during the Tertiary, and the faunal exchange that occurred at that time. It includes most families of lizards and snakes, conilurine rodents, many birds, burthid and scorpionid scorpions, theraphosid spiders, and many insects.

The Modern Element taxa have scarcely diverged, if at all, from their New Guinea or Indonesian relatives. Where there is divergence, it is usually only at the specific or subspecific level. Examples of this element include the ranid frog, a turtle, native rats, the Rainbow Bee-eater (*Merops ornatus*), and various spiders and insects.

Birds

Australia's birds show a rich diversity with their greatest concentration being found on the better-watered east coast (Figure 1.8). Some 744 birds were recorded in Australia in the 1977–81 and 1998-2001 Bird Atlas periods (NLWRA 2002). Trends in 497 species were able to be determined from the two Atlas periods from 1.7 million records (NLWRA 2002).

Birds are most numerous along the east coast and in the south-eastern region of Australia, and they diminish in the arid interior (SEAC 1996, NLWRA 2002). The lowest numbers of species were recorded from Tasmania, the arid interior and the western half of the continent. The highest concentration of species only found in Australia are found in southern Australia (particularly inland), while Western Australia has the greatest number of species particularly adapted to Australia's harsh conditions (NLWRA 2002). The highest densities of threatened birds occur on the major offshore islands, Tasmania, the Houtman Abrolhos, the Nullarbor Plain, and the eastern Mallee region. Major threats or reasons for decline are clearance for agriculture, grazing particularly by sheep (*Ovis aries*) and rabbits (*Oryctolagus cuniculus*), altered fire regimes, predation by introduced species including cats (*Felis catus*) or black rats (*Rattus rattus*), hunting by early settlers, and trapping for the bird trade (ANPWS 1992, NLWRA 2002). Formal description has been completed for 825 Australian bird

Table 1.7	The conservation status of Australia's birds (DEH 2004c)

Conservation status (EPBC Act 1999)	Number of species
Extinct	23
Critically Endangered	5
Endangered	34
Vulnerable	63
Conservation dependent	–

Figure 1.8   Distribution of species richness for birds (SEAC 1996 pp. 4–32)

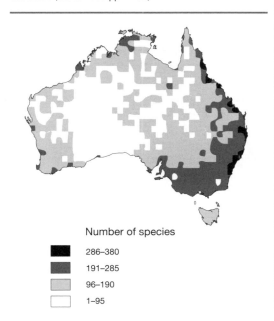

Number of species

- 286–380
- 191–285
- 96–190
- 1–95

Figure 1.9   Location of Australia's threatened bird species (SEAC 1996, pp. 4–36)

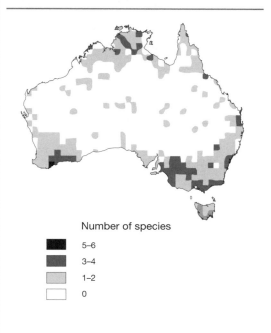

Number of species

- 5–6
- 3–4
- 1–2
- 0

species (ASEC 2001a). The conservation status of birds in 2004 shows 23 birds that are now extinct and 39 that are either critically endangered or endangered (Table 1.7). The distribution and species richness of birds, as well as the location of threatened species in 1996, is shown in Figures 1.8 and 1.9.

Reptiles

A total of some 836 species of Australian reptiles (Wilson & Swan 2003) has been identified from a world total of about 8000 species (Groombridge & Jenkins 2000). Australian reptiles are exceptionally diverse and include 136 genera spanning 17 families (Figure 1.10). Within this fauna, some 270 species (36%) of the total reptilian fauna have been described in the past two decades (ANCA 1993). A remarkable 89% of reptiles are endemic to Australia (SEAC 1996). Species richness of reptiles is high in the arid zone and the tropics. No Australian reptile is known to have become extinct, but there are many species that are endangered or threatened (Table 1.8), most of them in the south-east of Australia (Figure 1.11).

Slender Blue-tongue Lizard (*Cyclodomorphus venustus*)
(Tony Robinson)

Table 1.8	**The conservation status of Australia's reptiles** (DEH 2004c)

Conservation status (EPBC Act 1999)	Number of species
Extinct	–
Critically Endangered	–
Endangered	12
Vulnerable	38
Conservation dependent	–

Figure 1.10   Distribution of species richness for reptiles (SEAC 1996, pp. 4–32)

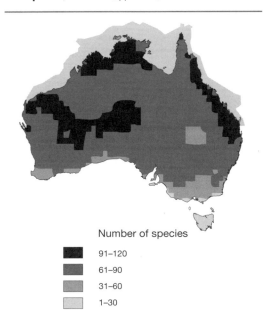

Number of species

- 91–120
- 61–90
- 31–60
- 1–30

Figure 1.11   Location of Australia's threatened reptile species (SEAC 1996, pp. 4–36)

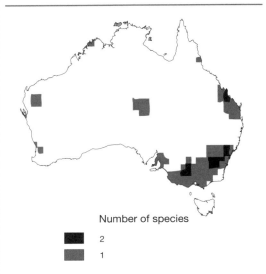

Number of species

- 2
- 1

Amphibians

There are three major groups of living amphibians. The largest of these and the only one found in Australia is the order Salientia to which belong the world's frogs and toads. (The other two orders include newts and salamanders, and Gymnophonia, worm-like burrowing animals (Cogger 1994).) A total of 203 species of Australian amphibians has been identified from a world total of about 4000 species (Figure 1.12). It is estimated that a further 25 to 30 species in Australia await discovery. Major centres of diversity are South America and Africa, while Australia and New Guinea are significant. There are 208 Australian species of frog recognised, with the earliest Australian fossil evidence dating back to the Eocene (Wildlife Australia 1997). In 2001, 176 Australian Frogs had formally been described (ASEC 2001a). Approximately 93% of frogs are endemic to Australia (SEAC 1996). There has been a dramatic and unexplained worldwide decline in frog species in recent years, and several Australian species are also threatened (Figure 1.13). Given that frogs have survived in Australia for over 45 million years, this is of major concern (Wildlife Australia 1997). Species richness of amphibians is highest in parts of the east and across the north, with some pockets in the south-west (Table 1.9).

Corroboree Frog (*Pseudophryne corroboree*) (P&W Collection, Dave Hunter)

Table 1.9	The conservation status of Australia's amphibians (DEH 2004c)

Conservation status (EPBC Act 1999)	Number of species
Extinct	4
Critically Endangered	–
Endangered	15
Vulnerable	12
Conservation dependent	–

Figure 1.12 Distribution of species richness for amphibians (SEAC 1996, pp. 4–32)

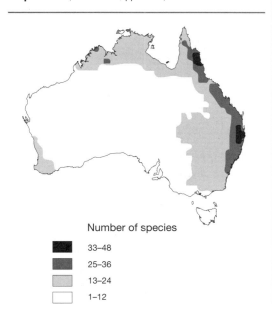

Number of species

- 33–48
- 25–36
- 13–24
- 1–12

Figure 1.13 Location of Australia's threatened amphibian species (SEAC 1996, pp. 4–36)

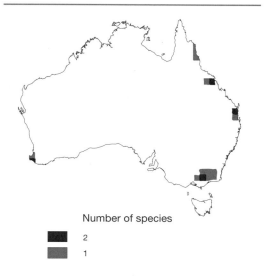

Number of species

- 2
- 1

Mammals

Native Australian mammal species comprise only 5–6% of the world's total, but this is appropriate to a continent that constitutes only 5.7% of the land surface of the Earth. Of the 17 orders of living terrestrial mammals, only four (monotremes, marsupials, rodents, and bats) include species that are native to Australia, whereas seven to ten orders are represented in each of the other continents (Strahan 1995). Australian rodents comprise only 4% of the world total; and Australian bats make up 7% of the world total, but only 4% are endemic (Strahan 1995). In 2001, 369 Australian Mammals had been formally described (ASEC 2001a). There were 305 terrestrial mammals in Australia and its external territories and an additional 26 exotic species (NLWRA 2002). Their distribution (Figure 1.14) shows high mammal numbers along the eastern margin of the continent, in the far north, the south-west, and part of the arid zone. Approximately 82% of Australia's mammals are endemic. The conservation status of Australian native land mammals is given in Table 1.10. Some rare species, such as the endangered Long-footed Potoroo (*Potorous longipes*), have been the subject of considerable scientific research (Snapshot 1.7).

Australian Sea Lion (*Neophoca cinerea*) at Seal Bay Conservation Park, Kangaroo Island, SA (Fraser Vickery)

Table 1.10	The conservation status of Australia's mammals (DEH 2004c)
Conservation status (EPBC Act 1999)	Number of species
Extinct	27
Critically Endangered	1
Endangered	33
Vulnerable	51
Conservation dependent	1

Marsupials and monotremes have not fared well since Captain Cook arrived in Australia a little over 200 years ago. Australia accounts for about one-third of all mammal extinctions worldwide since 1600, and most extinct Australian mammals have been marsupials. Since 1788, 27 species of Australian mammals have become extinct and many more are threatened with extinction (Figure 1.15, Table 1.10) (NLWRA 2002). Fortunately, no species of monotreme is

Figure 1.15   Location of Australia's threatened mammal species (SEAC 1996, pp. 4–36)

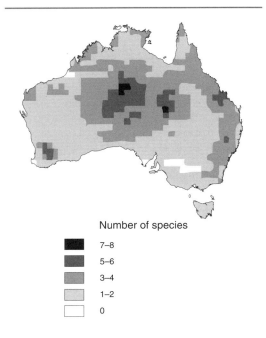

Number of species

- 7–8
- 5–6
- 3–4
- 1–2
- 0

Figure 1.14   Distribution of species richness for mammals (SEAC 1996, pp. 4–32)

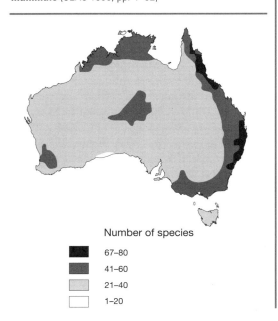

Number of species

- 67–80
- 41–60
- 21–40
- 1–20

threatened. These figures, startling as they are, do not reflect the degree to which increasing numbers of species are under pressure regionally. In the central deserts about one-third of all mammal species have disappeared, while in some heavily cleared agricultural areas, over one-quarter of marsupial species are locally extinct (Wildlife Australia 1996a). There has been a massive contraction in the geographical ranges and species composition of Australia's mammal fauna.

Snapshot **1.7**

Surveying for the elusive Long-footed Potoroo of south-eastern Australia

The endangered Long-footed Potoroo is a remarkably cryptic animal. While Victorian researchers had been successful in the live capture of these animals in Eastern Victoria, little success had been achieved by their researcher colleagues Michael Saxon and Linda Broome using the same techniques in nearby NSW. The presence or absence of the animal was important to land-use decisions associated with the woodchipping and timber industry at the time of the surveys. So a range of survey techniques was used. Live traps were set, hair tubes (narrow feeding tubes with sticky tape on their walls that retain samples of the fur of animals using the tubes) were placed in likely and more obscure habitat, scats were collected (including predator scats), trip cameras were used, and scratchings were recorded. The presence of the animals was confirmed from hair tubes, fox scats, and potoroo scats. Scratchings were only indicative. However, the degree of effort to achieve any results at all was extraordinary. After nine years of survey between 1986 and 1995, some 11,020 hair tubes had been set, and there had been 6618 trap nights completed. This resulted in a total of 17 definite records of potoroo including 11 from hair tubes, and six from predator scats. Two potoroo scats were most probably that of the Long-footed, but could not be confirmed (Broome et al. 1996). Following confirmation of the

Scientist surveying for small mammals, Kosciuszko National Park (Graeme Worboys)

presence of the animals, the majority of the confirmed potoroo habitat was permanently reserved as part of the NSW South East Forest National Park.

This is unparalleled in any other component of biodiversity in Australia or anywhere else in the world (NLWRA 2002).

Some marsupials are very rare. The Northern Hairy-nosed Wombat (*Lasiorhinus krefftii*), for example, has been reduced to about 67 individuals, of which only about 15 are breeding females. The Mala (the central Australian subspecies of the Rufous Hare-wallaby (*Lagorchestes hirsutus*)) became extinct in the wild in 1991 and now exists as about 150 animals in captivity and as one experimental reintroduction of only about ten wild animals (Wildlife Australia 1996a). The key findings of the Australian Terrestrial Biodiversity Assessment were that mammals persist best in high rainfall areas; arid and semi-arid areas have lost a high proportion of their mammals; critical weight range mammals are most susceptible to extinction and bats demonstrate lower levels of decline and extinction (NLWRA 2002).

Terrestrial invertebrates

The conservation status of Australia's terrestrial invertebrates is poorly known. As with other fauna, habitat clearing and disturbance have affected invertebrates. Three species are presumed extinct and 118 are either endangered or vulnerable. Another 291 species are of indeterminate status (SEAC 1996). In 1996, there were 281 species of Australian invertebrates listed on the IUCN Red List of Threatened Animals (ASEC 2001a). Invertebrates are an integral part of all ecosystems, and are critically important as primary food sources for many other animals.

1.8   Marine flora and fauna

Australia's marine environments include a rich array of species and habitats. They are summarised by the Interim Marine and Coastal Regionalisation for Australia (IMCRA).

The coastal and marine environments of Australia extend from the coastline to the boundary of its 200 nautical mile Exclusive Economic Zone and cover about 16 million km^2 of seas. These environments cover a wide range of climates, and geological and biological regions (SEAC 1996). Rocky shores, sandy beaches, algal reefs, and kelp forests dominate the temperate south; and coral reefs, estuaries, bays, seagrass beds, mangrove forests, and coastal saltmarshes dominate the tropical north. There are also the less well-known habitats such as mid-water, outer-shelf and deepwater areas. Australia's marine environments also include external territories in the Indian Ocean, South Pacific Ocean, Southern Ocean, and Antarctica (IMCRA 1998).

The extent and diversity of Australia's coastal and marine environments has resulted in some of the most diverse, and unique marine flora and fauna in the world. Australia has the:

- highest species diversity of tropical and temperate seagrasses
- highest diversity of marine macroalgae
- largest area of coral reefs
- highest mangrove species diversity
- highest global levels of diversity for a range of marine invertebrates (such as bryozoans, ascidians and nudibranchs) (Zann 1995).

Some 4150 species of fish (ASEC 2001a), 43 species of whales and dolphins, and six of the seven world species of marine turtles are recorded in Australian waters (IMCRA 1998). All major groups of marine organisms are represented, and many species are endemic or unique to Australia's waters. In temperate southern Australian waters, which have been geographically and climatically isolated for around 65 million years, most known species (90 to 95%) are endemic or restricted to the area. In the waters of tropical northern Australia, which are connected by currents to the Indian and Pacific Ocean tropics, 85 to 90% of species are shared with the Asia-Pacific region (IMCRA 1998).

In addition, our seas support 1200 species of macroalgae (seaweed), including 30 species of seagrasses; the largest intact coral reef in the world; some of the best preserved mangrove forests in the world; and a moderately rich cetacean fauna by world standards, with eight of the world's 13 cetacean families represented and 54% of the species (SEAC 1996, Wildlife Australia 1996b).

More details on marine ecosystems, habitats, and biodiversity are given in Section 20.1.

Marine flora (kelp) (Michael van Ewijk)

The natural heritage of Australia is one of the most remarkable of any country. Australia is a whole continent. It is also the world's most long-isolated continent apart from Antarctica (some of whose now-extinct species it presumably preserves). This huge heritage is important to Australians, for whom it is tied up with sense of place and with national identity. It is also important to people everywhere. And it is of particular importance to Australia's Aboriginal peoples, many of whom retain deep ties to the land.

A wide range of practical responses is required to ensure conservation of this ancient and diverse heritage. We will be discussing many of these responses in Part B of this book.

Figure 1.1 Australia—island continent

Australia - the global context

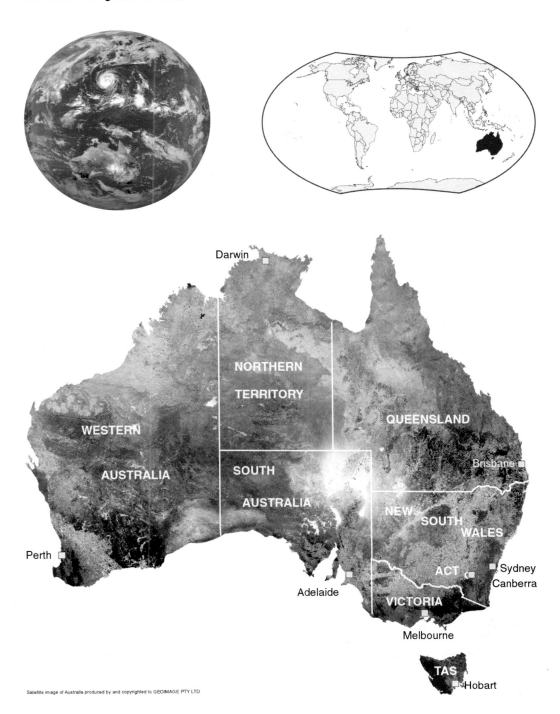

Figure 1.3 Australia climate

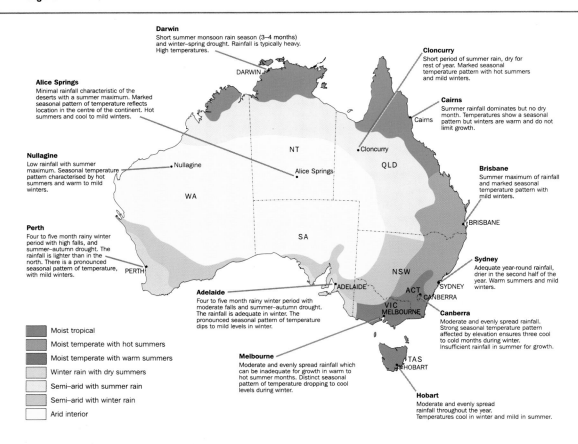

Darwin
Short summer monsoon rain season (3–4 months) and winter–spring drought. Rainfall is typically heavy. High temperatures.

Cloncurry
Short period of summer rain, dry for rest of year. Marked seasonal temperature pattern with hot summers and mild winters.

Alice Springs
Minimal rainfall characteristic of the deserts with a summer maximum. Marked seasonal pattern of temperature reflects location in the centre of the continent. Hot summers and cool to mild winters.

Cairns
Summer rainfall dominates but no dry month. Temperatures show a seasonal pattern but winters are warm and do not limit growth.

Nullagine
Low rainfall with summer maximum. Seasonal temperature pattern characterised by hot summers and warm to mild winters.

Brisbane
Summer maximum of rainfall and marked seasonal temperature pattern with mild winters.

Perth
Four to five month rainy winter period with high falls, and summer–autumn drought. The rainfall is lighter than in the north. There is a pronounced seasonal pattern of temperature, with mild winters.

Sydney
Adequate year-round rainfall, drier in the second half of the year. Warm summers and mild winters.

Adelaide
Four to five month rainy winter period with moderate falls and summer–autumn drought. The rainfall is adequate in winter. The pronounced seasonal pattern of temperature dips to mild levels in winter.

Canberra
Moderate and evenly spread rainfall. Strong seasonal temperature pattern affected by elevation ensures three cool to cold months during winter. Insufficient rainfall in summer for growth.

Melbourne
Moderate and evenly spread rainfall which can be inadequate for growth in warm to hot summer months. Distinct seasonal pattern of temperature dropping to cool levels during winter.

Hobart
Moderate and evenly spread rainfall throughout the year. Temperatures cool in winter and mild in summer.

Legend:
- Moist tropical
- Moist temperate with hot summers
- Moist temperate with warm summers
- Winter rain with dry summers
- Semi–arid with summer rain
- Semi–arid with winter rain
- Arid interior

Uluṟu–Kata Tjuṯa National Park, NT (Graeme Worboys)

Figure 1.4 Australia topography

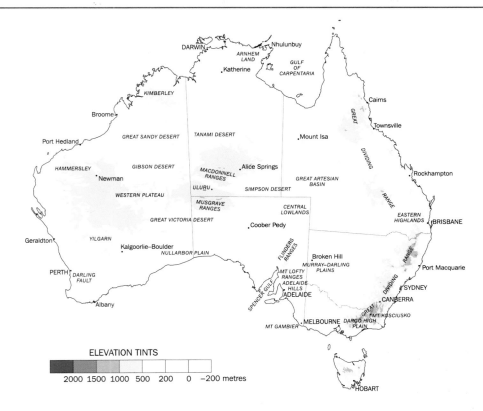

ELEVATION TINTS

2000 1500 1000 500 200 0 −200 metres

Fan Palms, Wet Tropics World Heritage Area, Qld (Graeme Worboys)

Figure 1.6 Major vegetation changes since 1788

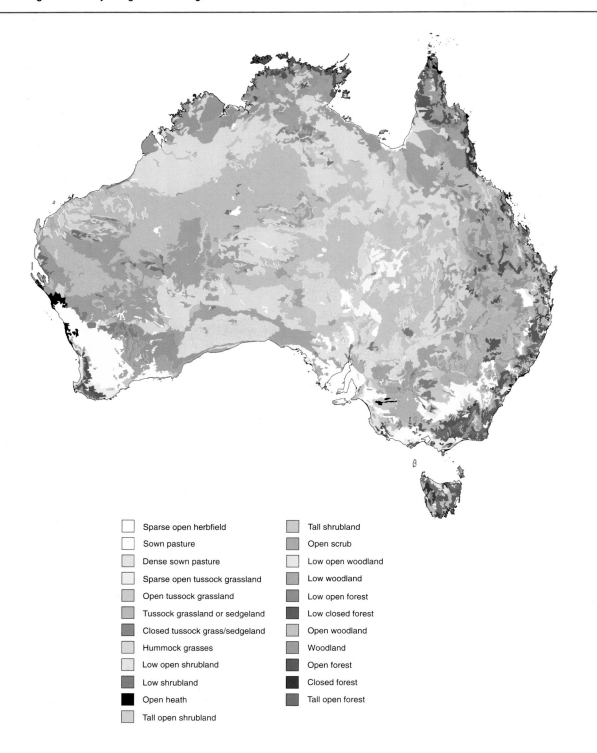

Sparse open herbfield
Sown pasture
Dense sown pasture
Sparse open tussock grassland
Open tussock grassland
Tussock grassland or sedgeland
Closed tussock grass/sedgeland
Hummock grasses
Low open shrubland
Low shrubland
Open heath
Tall open shrubland

Tall shrubland
Open scrub
Low open woodland
Low woodland
Low open forest
Low closed forest
Open woodland
Woodland
Open forest
Closed forest
Tall open forest

Jagungal Wilderness, Mt Jagungal, Kosciuszko National Park, NSW (Graeme Worboys)

Diving with Whale Sharks (*Rhincodon typus*) at Ningaloo Reef, WA (Derrin Davis)

Cave painting, Nourlangi, Kakadu National Park, NT (Graeme Worboys)

Male Superb Lyrebird (*Menura novaehollandiae*) in display (Craig Campbell)

Figure 3.2   Fred Williams's *Waterpond in a Landscape II*, © Fred Williams, licensed by VISCOPY, Austalia, 2004

Figure 3.3   Extent of the world's protected areas

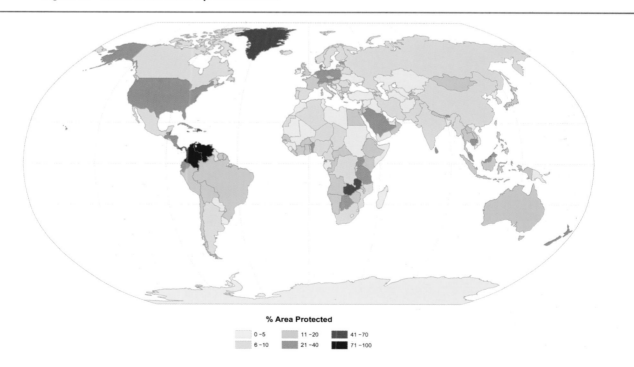

% Area Protected

0 –5   11 –20   41 –70

6 –10   21 –40   71 –100

Figure 3.4   Global distribution of natural and mixed World Heritage sites

Legend
- Natural Sites
- Mixed Natural and Cultural Sites
- Forest (% tree cover > 40%)

Prepared with the kind support from:

UNITED NATIONS FOUNDATION

Map prepared by C. Ravilious, UNEP-WCMC.

List of sites correct as at September 2003

149 properties which the World Heritage Committee has inscribed on the World Heritage List for natural values and 23 properties for both natural and cultural values.

UNEP WCMC

IUCN   The World Conservation Union

Social Context

> Some time back a rare but interesting genus was trapped on a slowly drying plain. Unable to compete, it put forth a cluster of short-lived species, much as an overgrazed bush pushes up spindly shoots. One by one each of these species went extinct, unable to survive fierce predators and energetic competitors. A slight change in the world's weather, or a single volcanic eruption might have made the whole genus extinct; yet just one species from it survived by a fingernail, found good times, proliferated, and got through to the present.
>
> Recently one of its members painted the Sistine Chapel.
>
> *Mark O'Connor*

2.1 A brief history of Australian protected areas   36

2.2 Contemporary influences   47

2.3 Responses and initiatives   56

2.4 Government in Australia   64

2.5 Legislation   67

2.6 Society, culture, and values   69

Managing protected areas is essentially a social process. Throughout our history, environmental problems have resulted largely from human behaviour. Hence solving them is a social issue and long-term social forces need to be considered when we seek to develop conservation policies.

Protected area staff of all kinds, ranging from on-the-ground managers to those who formulate national policy, must grasp the broader context in which their profession is embedded. The meanings, purposes, and management of protected areas are not static, but develop in conjunction with wider social, economic, and cultural influences. Managers must recognise and meet responsibilities concerning regional communities and Indigenous peoples. To relate effectively to tourists and visitors day-to-day, a ranger must understand how an individual's cultural background can cause them to hold certain views or to act in certain ways. When devising policies and strategies a manager must consider, realistically, how various interest groups will respond. To manage a protected area effectively, a manager must take account of politics, the legal system, the internal dynamics of institutions, and the broad social and political structures of our society. Each requires some practical know-how.

Australia's democratic political system is ever changing. At all levels of government, new individuals and new groups constantly replace each other. As they rise to power they bring new policies and priorities. They may change the laws that set the powers of protected area staff. As different environmental issues come to be more or less in favour, funding may become harder or easier to obtain.

The forces of global capital and of consumer markets dominate our economy. Managers must work within these. If possible, they need to harness those market forces to conserve biodiversity. Managers also work within an Australian community that is partly formed (and reformed) by the stories we tell ourselves of our past. It is also formed, of course, by our dominant social institutions, and by the many cultural forces that constantly reshape our lives.

This chapter explores the social context within which protected areas must be managed. First, we give a brief history of protected areas in Australia. We describe the major international trends that are shaping protected area management, and the responses of managers to these trends. We then cover the political and legal system in Australia, with particular reference to environmental controls. Finally, we explore the social and cultural factors that shape Australians' attitudes towards the environment.

2.1   A brief history of Australian protected areas

In Australia, protected areas are not pristine wilderness, if 'wilderness' is taken to mean free from human presence or influence. The natural areas of Australia are cultural landscapes. For thousands of years, Indigenous people have influenced them. Compared with this, European practices have been very recent. Aboriginal people have traditionally related to and managed the country on a broad landscape scale. Protected area managers have recognised only recently how significant this management tradition has been (Chapter 18).

Aboriginal people have long regarded, and continue to identify, certain places as what we now call protected areas. This ancient system was severely disrupted with the coming of Europeans to the continent. However, since the late nineteenth century we have seen the gradual re-emergence of the importance of protected areas, so that now they again have a central place in the way humans relate to and manage the Australian environment.

The original protected areas

The notion of protected areas dates back to the first humans on this continent. Aboriginal people, the first Australians, had, and retain today, a profound reverence for the land—what they now call 'country'.

Country is a place that gives and receives life.
Not just imagined or represented, it is lived in

and lived with … *Each country has its sacred origins, its sacred and dangerous places, its sources of life and its sites of death … In Aboriginal Australia each country is surrounded by other countries. The boundaries are rarely absolute; differences are known, respected and culturally elaborated in many ways … Each nourishing terrain, each promised land, was cared for (Rose 1996, pp. 7, 9).*

The period during which Aboriginal people established themselves on the continent of Australia (at least 40,000 years ago) appears in their legends as the Dreamtime. The spirits from the Dreamtime are the focus for their intimate relationship, mythically and spiritually, with their country. The common ancestry they share with wildlife and with some features of the landscape forms the basis for their tribal laws, taboos, ceremonies, and totemic beliefs (Whitelock 1985). The resulting religious and social order used to restrict the way they used the environment in many ways. For example, they protected certain plants and animals from the Dreaming (Rose 1988). In many areas, sacred sites are protected. Activities such as hunting, fishing, gathering or burning are often not permitted there, or may take place only within prescribed boundaries. Sacred sites may be nesting or breeding places for certain native animals (Rose 1996).

The dreaming lays down the laws concerning the accessing of resources from the environment. The environment relates directly to social organisation, kinship and social obligations, sacred law, offences against property and persons, marriage, and an individual's relationship with the land. Aboriginal land and the meaning behind it passes on information about the environment to each generation, depending on where each person was conceived, when she/he quickened in the womb, and the totemic associations of the father and mother. On this basis, certain resources could be gained in different areas by accessing kinship rights. Conversely, restriction and taboos would apply to other persons in other areas. As a result of these checks and balances, sanctuaries occurred throughout the environment where certain species could be reproduced without threat of destruction (Bayet 1994, p. 28).

Aboriginal Cave Paintings, Carnarvon Gorge National Park, Qld (Graeme Worboys)

Dame Mary Gilmore was a member of one of the first European families to settle in the Wagga Wagga area of NSW. She noted how important protected areas were to local Aboriginal people:

'Beside the fish, where there were deep valleys, running water and much timber, the natives invariably set aside some parts to remain as breeding places or animal sanctuaries. Where there were plains by a river, a part was left undisturbed for birds that nested on the ground. They did the same thing with lagoons, rivers, and billabongs for water birds and fish. There once was a great sanctuary for emus at Eunonyhareenyha, near Wagga Wagga. The name means "The breeding-place of the emus"—the emus' sanctuary' (Rose 1996, p. 50).

Aboriginal groups changed continually in their density and distribution. They were often not so much nomads as seasonal re-settlers. This allowed them to be where food was plentiful in season, and may also have been a conscious way of reducing the impact of their numbers on the land. They manipulated their environment through practices such as burning, sometimes collecting and redistributing plant species, and conducting religious rituals. The rituals were designed to ensure the landscape remained productive (Walsh 1996). They managed their country in terms of sacred sites and other spiritual aspects: '… there's a lot of dreaming trails which cross over, these are really important places. They are so sacred you can't kill animals or even pick plants. And of course you don't burn them. You might burn around them in order to look after them' (Latz cited in Rose 1996, p. 51). The traditional relationships between Aboriginal people and country are further discussed in Chapter 18.

The re-emergence of protected areas

In the second half of the nineteenth century, the first non-Indigenous concerns about conserving Australia's natural resources were raised. This was after a century of European exploration and development. The English nature poets, such as William Wordsworth, and American writers such as Henry David Thoreau and George Marsh, had provoked and directed such thinking. Published in New York in 1864, George Marsh's *Man and Nature* argued that humanity's dominant role over the natural world was having significant, unrecognised, and largely destructive consequences. It was a radical position at the time (Powell 1976). In the 1860s the Australian colonial press popularised Marsh's ideas about the damage caused by clearing forests. *Man and Nature* prompted a turning point in attitudes among naturalists. In its time it was as significant a book as Rachel Carson's *Silent Spring* was a century later (Hutton & Connors 1999).

The attitudes of European Australians towards the environment developed in the context of their strong links with Britain. It reflected their need to address the problems of managing land in an unfamiliar environment. It expressed their growing national identity. Australian writers, poets, and painters played a pivotal role. They evoked a range of images and stories that rapidly advanced from hostility towards the Australian bush, through to mixed attitudes, and finally to appreciation of its special character (Powell 1976). Key artistic figures include the Heidelberg school painters such as Tom Roberts and Frederick McCubbin. Poets include Henry Kendall and Adam Lindsay Gordon.

The Belair National Park Board of Commissioners on a park inspection in the early 1920s, SA (Courtesy of Dene Cordes, SA NPWS)

Naturalists were struck by Australia's unique environments and began to appreciate their vulnerability. They were among the first to call for the protection of areas and species. They conducted their campaigns through various scientific societies, and the public supported them widely. By the 1890s royal societies and/or natural history societies were established in all the Australian colonies. Among them one of the most active conservers of nature was the Royal Society of South Australia. Established in 1853 as the Philosophical Society of South Australia, the organisation changed its name in 1876. In 1883, the Royal Society of South Australia established a section of field naturalists (Whitelock 1985, Hutton & Connors 1999), which played a crucial role in having Belair and Flinders Chase National Parks and other reserves declared.

Scientists also led efforts to establish practical conservation policies. In Victoria, botanist Baron Ferdinand von Mueller regarded forest as a heritage provided by nature, to be honoured reverently, maintained carefully, and used wisely (Hutton & Connors 1999). In Australia the first colony to legislate to protect native fauna was Tasmania. In 1860 it passed laws that protected various game bird species during their breeding season. In 1874, when NSW and Queensland were legislating for eradication of kangaroos and wallabies from properties, a closed season was declared in Tasmania for the hunting of kangaroos during the breeding season (Quarmby 1998).

In the nineteenth century, the moral principle behind the conservation movement's thinking was conservation for future exploitation. They argued for the economic usefulness of protecting birds, the efficiency in terms of timber production for preserving native forests, and the recreational value of public parks (Hutton & Connors 1999). The acclimatisation movement believed that introducing new species would improve nature. They sought laws that supported their importing game species, and which also protected certain 'useful' native fauna (Pettigrew & Lyons 1979). In 1861 the Victorian Acclimatisation Society, for example, fought one of the earliest campaigns for land reservation. It attempted to designate Phillip Island as a zone for acclimatising introduced animals (Hutton & Connors 1999).

In 1866, Jenolan Caves in NSW was declared a water reserve. Its major proponent was John Lucas, MLA, who saw the purpose of the reserve as being protection of the caves. It can be argued that Jenolan Caves, therefore, was Australia's first non-Aboriginal protected area (Hamilton-Smith 1998). A few years later, in 1871, the efforts of explorer and statesman John Forrest led to the establishment of Kings Park, a sizeable area of bushland in the Western Australian capital, Perth (Serventy 1999). The United States Congress dedicated Yellowstone, the world's first 'modern' national park, in 1872.

The idea of national parks as a land use spread to Australia, with the establishment in 1879 of The National Park (later called Royal National Park), near Sydney. However, the purpose and management of the park were closer to the large recreation parks on the outskirts of London (Pettigrew & Lyons 1979) than to our current concept of a national park. In the mid 1800s Sydney was a fast-developing city, with most land owned by private interests. When 7284 hectares of land south of Port Hacking, 22.5 kilometres from the centre of Sydney, was first set aside under the name of national park, a wide range of purposes was proposed. In the words of the

Travertine terraces, Mammoth Hot Springs, Yellowstone National Park, USA (Graeme Worboys)

legislators, these parks were to 'ensure a healthy and consequently vigorous and intelligent community' (Pettigrew & Lyons 1979, p. 15). The park allowed for portions to be used for 'ornamental lawns and gardens … zoological gardens … a racecourse … cricket and other lawful games … bathing places … or any other public amusements' (Pettigrew & Lyons 1979, p. 22). Other uses included acclimatisation of foreign birds and game animals to Australian conditions, and military training.

The park was managed by a board of trustees who were guided by legislation and by public opinion, both of which at that time saw nature primarily as being there for human use. A central issue for the trustees, and one that continues to be central to conservation agencies, was how to fund management. The prevailing views of the times permitted them to exploit the park's natural resources. Over the next 70 years a variety of small-scale enterprises were attempted, though none proved particularly lucrative: gravel was collected for roads and clay for bricks, timber was felled for mine props and bakers' ovens, mature trees were turned to charcoal, and oysters were pulled from the rocks (Pettigrew & Lyons 1979).

Aside from government grants and the occasional gifts from benefactors, tourism was the most sustained source of income. Visitors were catered for with a growing range of infrastructure that included guesthouses, as well as boating and swimming facilities. Several roads were built in the first decades, and as motorcars became cheaper and more common, so their use in the park increased. Soon they were transporting more and more visitors to remote picnic spots. For many years construction of private facilities was also permitted, with minimal planning. Buildings, from shacks to churches, were constructed through the park, especially during the depression. After a prolonged battle the majority of these were removed, but many shacks remain. Temporary residents and visitors to the park created a range of impacts. For example, a popular pastime early in the century was collecting wildflowers. The trustees recognised this was depleting favourite species such as the Waratah (*Telopea speciosissima*) and Gymea Lily (*Doryanthes excelsa*), and in 1908 they imposed a total prohibition on picking flowers (Pettigrew & Lyons 1979).

Reserves were also being established elsewhere in Australia (Appendix 1). In 1866, a 597-hectare

Boatshed and original causeway, 1899, Audley, Royal National Park, NSW (courtesy NSW Government Printing Service)

area at Tower Hill near Warrnambool in western Victoria was set aside as a public park. It was prized for its geological features. In 1892 a special act of parliament gave national park status to the area. In 1878, Queensland pastoralist Robert Collins was inspired by the national park ideals promoted in the United States by Muir and others. With Romeo Lahey, he initiated a lengthy campaign to protect the McPherson Ranges. This eventually led to the Bunya Mountains and Lamington National Parks being declared. In 1898 the Victorian *Lands Act* set aside areas including Wilsons Promontory and Mt Buffalo (Goldstein 1979). Railway posters, magazines, and pictures promoted Mt Buffalo and other scenic areas such as Mt Field and Russell Falls in Tasmania as desirable destinations for recreation (Hutton & Connors 1999). In 1912, the Tasmanian Tourist Association instigated walking tracks and accommodation that enabled people to access the mountains, falls, lakes, and fishing opportunities at Mt Field. In that year as well William Crooke led the establishment of a National Parks Association that campaigned successfully for a national park at Russell Falls and Mt Field. In 1915, an 11,000-hectare park was declared (Goldstein 1979, Quarmby 1998).

From the 1920s, bushwalking clubs became increasingly important in conservation campaigns. Conservationist Myles Dunphy was an especially influential figure. Dunphy was influenced by the philosophy and management approach of · the United States National Parks Service, established in 1916, and by the US Forest Service concept of wilderness areas, which he called *primitive areas* (Strom 1979a). The first successful conservation campaign conducted by the walking clubs of NSW was the Blue Gum Forest Reserve campaign in 1931. In 1957 the National Parks Association of NSW was created under the sponsorship of the NSW Federation of Bushwalking Clubs. The prime objective of the new organisation was to push for the establishment of a National Parks Act in NSW. With the support of Tom Lewis, a minister in the then Liberal Government, the *National Parks and Wildlife Act* was passed in 1967 (Strom 1979b). This paved the way for a professional organisation to be established to manage the park and reserve system (Fox 1979).

Ranger David Kerr, first uniformed ranger for Kosciusko State Park, circa 1960 (Courtesy Neville Gare)

Until the late 1950s, conservationists pursued their objectives mainly through formal deputations and well-researched lobbying. This approach became less effective, however. Shaped by the social turmoil of the 1960s, the conservation movement became more radical. Environmental groups accepted that research and lobbying were important. Increasingly they added other methods: educating the Australian people and taking direct action. An example is the campaign in 1968 to save the Myall Lakes, NSW. They conducted an intensive education program with media releases. They organised tours to the region. They distributed leaflets and car stickers. They lobbied politicians. A campaign by the Fraser Island Defence Organisation was also based on educating the public, lobbying governments and political parties, and taking legal action (Hutton & Connors 1999).

In 1979, the Terania Native Forest Action Group took direct action. They blockaded logging operations in the Terania Creek catchment in northern NSW. The national publicity that they achieved provided a model for other radical interventions by conservationists. Similar blockades were instigated at various sites including Mt Nardi in NSW, the Gordon and Franklin Rivers in Tasmania, and the Errinundra Plateau in Victoria.

In south-west Tasmania there was conflict over whether large natural areas should be protected or whether dams should be built to generate hydro-electric power. This debate helped shape Australians' thinking about conservation issues. In 1955, the protection of Lake Pedder was recommended to the Scenery Preservation Board. Soon after its proclamation, however, walkers became aware of plans for a hydroelectricity scheme in the pristine area. Conservationists mounted a concerted campaign against the lake's flooding. Formed from representatives of various outdoor clubs and conservation groups, the South-West Committee pressed for a large national park. The Lake Pedder Action Committee was also established. Lake Pedder supporters set up their own political party, the United Tasmanian Group. It can be considered the world's first green political party (Hutton & Connors 1999). Against considerable public opposition, assent was given to the hydroelectricity scheme. Lake Pedder was submerged and lost.

In the early 1980s further proposals were put forward for a major hydro-development on the Gordon and Franklin Rivers. Conservationists in Tasmania and across Australia were determined these natural areas would not suffer the same fate as Lake Pedder. The Save the Franklin campaign mobilised large numbers of activists and supporters and brought the issue into the mainstream of Australian politics. It also helped establish a widespread view that nature had rights beyond its use to humans (Hutton & Connors 1999). Lobbying and education campaigns failed to stop the project. Hence in the summer of 1982–83 the Tasmanian Wilderness Society organised a blockade. The publicity that resulted nationally and internationally, along with political and educational activities, led finally to the development being stopped on legal grounds. Bob Brown, later an Australian Greens senator in the federal parliament, was a central figure. Most of south-west Tasmania is now protected as national park.

Men such as Myles Dunphy and more recently Milo Dunphy and Bob Brown have become figureheads in the Australian conservation movement. Many women have also played an important, though less publicised, role in establishing protected areas in Australia. In the 1920s and 30s, conservationists who were active in their respective states included Marie Byles in NSW, Jessie Wakefield in Tasmania, and Irene Longman in Queensland (Hutton & Connors 1999). More recently, poet Judith Wright was a key contributor to the campaign that led to the protection of the Great Barrier Reef. Key figures in the Tasmanian Wilderness Society included Cathy Plowman and Karen Alexander. Penny Figgis played a major role in shaping the policies of the Australian Conservation Foundation.

Figure 2.1   Growth in Australian protected areas 1968–2002 (Cresswell & Thomas 1997, p. 5, updated with CAPAD 1999 and 2002 data)

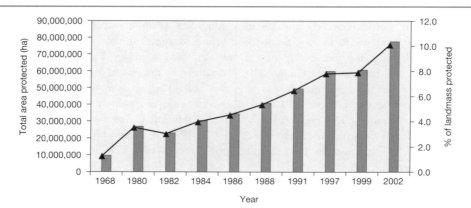

The activities of the conservation movement during the 1960s, 1970s, and 1980s led to significant additions to the network of protected areas. Progressively, parks and wildlife legislation was modernised. Processes for resolving disputes were established (Hutton & Connors 1999), including Victoria's Land Conservation Council (Chapter 3). The growth in protected areas in Australia from 1968 to 2002 is shown in Figure 2.1. Some major events in the history of Australia's protected areas are summarised in Table 2.1 and a more detailed chronology is given in Appendix 1.

Table 2.1	Selected influential events and milestones in the history of Australian protected areas
Year	Event
1864	Marsh's book published on land and water degradation
1866	Jenolan Caves Reserve (NSW) declared
1866	Tower Hill public park (Victoria) declared—upgraded to a national park in 1892
1879	The National Park (Royal) (NSW) declared
1891	Belair National Park (South Australia) declared
1893	The Australasian Association for the Advancement of Science formalised the concept of establishing reserves for the protection of flora and fauna
1898	Mt Buffalo and Wilsons Promontory National Parks (Victoria) declared
1906	Collins and Lahey campaigned for national parks in McPherson Ranges (Queensland) and for park legislation, Mt Tambourine and Bunya Mountains National Parks (Queensland) declared
1909	Wildlife Preservation Society of Australia established, led by David Stead
1916	Mt Field National Park (Tasmania) declared
1931	Blue Gum Forest conservation campaign. Myles Dunphy active in conservation of areas in Blue Mountains and Garawarra Primitive Area campaigns
1944	Kosciuszko State Park declared—management body set up to maintain the then largest single protected area in Australia
1945	United Nations Educational, Scientific and Cultural Organisation (UNESCO) established
1948	International Union for the Conservation of Nature (IUCN or World Conservation Union) established
1952	Victorian National Parks Association formed to lobby for park legislation
1957	National Parks Association of NSW formed
1960s to 1980s	Various controversies and campaigns—large increase in number of protected areas, establishment of protected area management agencies in all jurisdictions
1962	Rachel Carson's influential book *Silent Spring* published
1962	Wildlife Preservation Society of Queensland formed, which led the Great Barrier Reef campaign
1962	The first world congress on national parks and protected areas held in Seattle, USA, organised by the IUCN
1965	Australian Conservation Foundation formed
1967	NSW NPWS established
1968	10 million hectares reserved as protected areas in Australia
1969	Land Conservation Council established in Victoria to plan for balanced land use
1970	UNESCO Man and the Biosphere Program established

Table 2.1	**Selected influential events and milestones in the history of Australian protected areas** (continued)
Year	Event
1972	Concept of sustainable development promulgated at the UN-sponsored conference on the human environment in Stockholm
1972	World Heritage Convention established
1974	Specht Report provided the first national evaluation of the conservation of major plant communities of Australia—highlighted major deficiencies in the protected area system
1974	Lake Pedder flooded
1974	Great Barrier Reef Marine Park (Queensland) declared
1974	Australia became a party to the World Heritage Convention
1975	Australian National Parks and Wildlife Service (now Parks Australia) established (Case Study 2.1)
1975	Australian Heritage Commission established
1976	World Biosphere Reserves established under the UNESCO Man and the Biosphere Program
1977	Uluṟu-Kata Tjuṯa National Park declared
1978	Declaration of Fitzgerald River Biosphere Reserve, Western Australia
1978	Flora and Fauna Guarantee Act 1978 passed by the Victorian parliament—a model for legislation designed to conserve species and communities
1979	Kakadu National Park (Stage I) (Northern Territory) declared
1981	World Wildlife Fund (WWF) established (now known as World Wide Fund for Nature except in the United States and Canada)
1982–83	Franklin River blockade, federal government intervenes to halt the Gordon-below-Franklin dam
1983	Our Common Future and its advocacy for sustainable development generated by a UN commission
1992	UN Conference on Environment and Development (UNCED) held in Rio de Janeiro, resulting in an action plan for sustainable development (Agenda 21), as well as a Convention on Biodiversity Conservation
1996	National Strategy for the Conservation of Australia's Biological Diversity established
1997	60 million hectares of Australia reserved as protected areas
2003	Over 100,000 protected areas cover over 11.4% of the Earth's surface.

Case Study 2.1

Developing the first national protected area legislation for Australia

Neville Gare, former Deputy-Director of the Australian National Parks and Wildlife Service

For many people the world concept of the protected area originated in 1872 with the dedication of the Yellowstone National Park in the USA. In fact the word 'national' was not applied to Yellowstone until several years later, but 'Yellowstone Park' was from the start the creation and the responsibility of the Federal Government of the USA. Other countries picked up the idea and eventually the term 'national park' was widely adopted.

The Australian colonies did not federate until 1901, but the State of NSW had by then already established its own 'National Park', in 1879, and before the end of the century, South Australia, Victoria, and other states had followed suit. Federation gave rise to national attitudes to many things, but nature and landscape conservation did not seem to be an urgent priority. What has been termed the 'Rail Gauge Syndrome' prevailed (different rail gauges in each state), as if the vegetation, fauna, other wildlife, and landforms of the continent existed in six or seven discrete parcels, each strictly contained within state and territory boundaries.

So, 70 years after political federation there was still no truly national protected area, nor were there clear policies or programs designed to give a national thrust to the conservation of landscapes and ecosystems with some sort of leadership from the Commonwealth Government. The principle of 'states' rights' prevailed under the influence of politicians and bureaucrats, jealously guarding their powers dating back to colonial origins and limited in their ability and willingness to see the big picture of a unique island with national systems straddling the artificial political boundaries within its 7.6 million km^2.

In 1972, the Australian House of Representatives Select Committee on Wildlife Conservation completed two years of investigations, hearings, and interviews across the country and reported to the parliament. Among its main proposals was the establishment of a Commonwealth authority to be responsible for national parks and wildlife conservation in Commonwealth territories and the administration of international agreements on migration and endangered species, and to pursue a range of national nature conservation initiatives in cooperation with the states. The Whitlam Government came to power later in 1972 with a bold agenda for change. Among its stated policies was a promise to 'set up a national parks service' to pursue some of the recommendations in the House of Representatives Committee's report.

Towards the end of 1973 I was in Papua New Guinea (PNG) and nearing the conclusion of my contract to establish that country's first national park and a small parks service, when the Federal Minister for Environment and Conservation, Dr Moss Cass, visited Port Moresby. He was accompanied by the head of his department, Dr Don McMichael, a former director of the NSW NPWS, from which I was on loan to PNG through the Federal Government. As a result of discussions with them, I moved to Canberra early in 1974 to take up a post as executive officer with Dr Cass's department, to help draft a bill to establish the first federal nature conservation agency, ultimately to be called the Australian National Parks and Wildlife Service (ANPWS).

For the next year I worked with Ian Turnbull, senior parliamentary counsel in the Attorney-General's Department, in the drafting of the legislation. In the spirit of an innovative and reformist government we had as a guide a joint opinion, dated 23 March 1973, from Attorney-General Lionel Murphy and Solicitor-General Bob Ellicott, on the powers of the national parliament in respect to national parks and wildlife flowing from the federal constitution. By the time we started work Ellicott had resigned from the public service and become a shadow minister for the Opposition. The latter had a majority in the senate, which of course had much to do with the fall of the Whitlam Government in November 1975. It also had quite a bearing on projects such as ours; state governments were feeling threatened by what they saw as a power grab for land and its management within their boundaries and consequently were working hard to defeat legislative proposals in the senate. Former state officers like myself working for the Federal Government were not popular in some states at this time.

The rise to power in the mid 1960s of an ambitious and vigorous state minister, Tom Lewis (NSW), had led to the establishment of that state's National Parks and Wildlife Service and the introduction of annual conferences of state and federal ministers and heads of departments dealing with nature conservation, initially hosted by NSW. Tom Lewis's view of the Canberra initiative of the 1970s was not softened when his NPWS Director resigned to head the Federal Department of Environment and Conservation; nor did he warm to the federal minister's announcement of the formation of a Council of Nature Conservation Ministers (CONCOM), effectively replacing Lewis's annual conference of ministers. Other conservative state ministers agreed, and early CONCOM meetings were somewhat ill-tempered arenas in which to discuss national nature conservation issues.

Rangers, early 1970s, Smiggin Holes, Kosciuszko National Park, NSW (Graeme Worboys)

Despite this climate, Ian Turnbull and I set out with some very positive aims. We drew as widely as possible on the Constitution powers advised by Murphy and Ellicott. We took into consideration other draft federal legislation, including the Great Barrier Reef Marine Park Bill, the Australian Heritage Commission Bill, the Environment Protection (Impact of Proposals) Bill, and the Aboriginal Land Rights (Northern Territory) Bill.

In the latter case we drew on my experience of customary land situations in Papua New Guinea to provide a section in our bill to enable the Director of ANPWS to cooperate with and assist Aboriginal people in the conservation management of Aboriginal land. This was to lead to the great initiative of Kakadu National Park and later Uluru-Kata Tjuta National Park, with Aboriginal boards of management and Aboriginal people involved in day-to-day management activities.

The original cabinet decision had authorised a three-person National Parks and Wildlife Commission, but on a trip to the Kakadu area the Prime Minister told Dr Cass and me that the government was being criticised 'for appointing too many commissions—we cannot have both an Australian Heritage Commission and a National Parks Commission; see what you can do about it!' We knew the Heritage Commission was to be principally an advisory body, and did not want our agency to be a toothless tiger attached to it. So, we proposed a Director of National Parks and Wildlife with wide powers directly responsible to a minister. It was accepted and so we proceeded.

Knowing that in the years ahead governments of all persuasions would be elected and that the 'states' rights' question would always be around, we tried to broaden the Bill's provisions for action by the director as much as possible. He was given wide powers to act both directly and on a cooperative basis with a whole range of people and institutions both nationally and internationally. The purchase by the Whitlam Government of land at Towra Point (NSW) and Lake Cootharaba (Queensland) for nature reserves related to Australia's obligations under international migratory bird treaties upset both states, but in the long run these initiatives spurred more positive nature conservation actions by NSW and Queensland, including the passage of Queensland's own National Parks and Wildlife Act.

Ian Turnbull had been a police prosecutor in Rhodesia (now Zimbabwe). He was keen to ensure that regulations made under the new Act would make it as easy as possible for rangers and other field staff to do their job. Thus, the appropriate section of the Act made comprehensive provisions for the making of regulations. Public participation in the park and reserve planning process was well covered, the director being required to give public notice of his intention before preparing a draft plan. Once accepted by the minister, proposed plans of management had to be laid before both houses of parliament for 20 sitting days of each house.

There was an effort to provide a check and balance on the exercise of power. The director had wide powers but had to exercise them in accordance with any directions given by the minister. In turn the director was required to give particulars of any such ministerial directions in his annual report for presentation to the parliament.

Because of the Opposition's senate majority it was necessary to accept some amendments either in the committee stages or the senate debate. Nevertheless all-party support was given to the Act when it finally was passed in the parliament on 5 March 1975.

Yet the job had only just begun. As the government reeled towards dismissal in November 1975, it was a hard battle for me as acting director of the new service (appointed on April Fool's Day 1975) to get the new outfit afloat, with an initial staffing of six or so people and little support from government bureaucrats concerned about their own survival. An administrative staff review set up by the new Prime Minister Malcolm Fraser, looked set to recommend our abolition, and the untimely death of our supportive minister did not help. But some wiser counsel prevailed and the PM decided we should not be strangled at birth. In the end I retired as Deputy-Director of ANPWS in 1987.

Many good things have resulted, both directly and indirectly, from the ANPWS initiative—Kakadu and later Uluru-Kata Tjuta National Parks with Aboriginal involvement, the Australian Alps Ministerial Memorandum of Understanding for coordinated management of Australia's mainland mountains of the south-east and many other federal–state conservation projects of a cooperative nature, Aboriginal ranger training schemes, marine parks with both state and federal involvement, and many national off-park conservation and biodiversity programs.

Despite some changes implemented by the Federal Government in 1999, hopefully much of the good that has been done will remain and go on to better things in developing a real sense of national identity and identification with the big Gondwana-sourced island that is Australia.

2.2   Contemporary influences

Protected area policy and management is strongly influenced by prevailing social and economic circumstances, as well as cultural and ethical norms. There is a plurality of views about how we should relate to the natural world, why we should protect natural environments, and how we should manage and use them. Broad social concerns and trends that drive these views include changing roles for and organisation of governments; the sustainable development agenda; calls for increased community participation in decisions; and the possibility that we are now living in 'postmodern' times.

Governments and markets

The social, economic, and political structures of western democracies such as Australia include public and private sectors. A public sector, also called 'the state', comprises governments and parliaments, and all their associated departments and agencies, as well as the court system and police forces. The state is responsible for pursuing the common good on behalf of its citizens. A private sector comprises individuals, companies, and the mechanisms for exchanging goods and services. These enable people, within the limits of their economic means, to acquire the necessities of life, and to achieve the standard of living they desire. A key component of the private sector is the market. A market is a formal set of rules and institutions that facilitate the orderly exchange of goods, to the benefits of both the buyers (consumers) and sellers (producers).

A key policy debate throughout the world is, and has been for many years, about a desirable balance between the public sector and private sector. How much power should the public and private sectors have? How should they relate to each other? The debate is crucially important to protected area managers. It influences, among other things:

- who is given the responsibility for managing protected areas
- what resources are allocated for managing protected areas
- who pays for these resources
- who has the power to make decisions
- how decisions are made.

Australia is caught up in a process of international change that is transforming economies and governments. The declining power of the state has been associated with an expansion of market capitalism, such that the role of government is reduced to the efficient operation of markets through micro-economic reform, and provision of a basic level of welfare services (NSW NPWS 1998). Major forces affecting Australia include the internationalisation of capital and markets through the development of an international financial sector; the expansion of free trade agreements; the emergence of dominant transnational corporations; and the development of power blocs based on economic association, such as the European Economic Union and the Asia Pacific Economic Community. These forces have fostered such changes as a reduction in the size of government; corporatisation of public agencies, and the redefinition of the role of the public sector.

The duality between governments and markets has been a defining framework in modern western democracies. Recently, the once clear-cut differentiation between markets and governments has become blurred. The reality now is also that the public and private sectors are not clearly separable, but are increasingly embedded in each other. The boundaries have been blurred by public–private partnerships, corporatisation of public sector agencies, National Competition Policy, and the like. Autonomous public authorities are increasingly acting like hybrids between a public agency and a private firm. Public sector planners use tools developed in the private sector such as strategic planning, as well as market-based analysis such as cost–benefit analysis. Planning and markets can be considered as alternative forms of governance. Many key questions of environmental policy now revolve around deciding on the most effective form of governance, and as discussed in Section 2.3, protected areas are no exception.

Hughes (1998) identified four ways in which the government can intervene in activities occurring within a nation—provision, production, subsidy, and regulation.

Provision. Governments usually provide goods or services through an annual budget cycle. Governments fund the defence forces, social services, education, and protected area agencies. Each

year, for example, the NSW government's budget funds the NSW P&WS.

Production. Governments can produce goods and services for sale. A government might produce and sell maps and brochures that inform the public about the location and features of national parks.

Subsidy. The government may assist individuals or groups in the private economy to provide goods and services that the government desires. Under the *Native Vegetation Act 2003*, for example, the NSW Department of Infrastructure, Planning and Natural Resources provides landholders with financial incentives for the conservation of native vegetation. The Commonwealth Government also uses the Natural Heritage Trust to fund conservation activities across Australia (Snapshot 2.1).

Regulation. Governments have coercive powers—the powers of law. All states and territories have passed legislation that restricts the activities that people may pursue in protected areas.

Snapshot 2.1

Natural Heritage Trust

Prior to the 1995 federal election, both major political parties committed themselves to increase funding for the National Landcare Program (NLP). This included funding for large works on private lands where there were identifiable conservation outcomes. The incoming Liberal government used proceeds from the partial sale of the national communications carrier, Telstra, to increase its funding for Landcare substantially. It did so through a five-year $1.25 billion Natural Heritage Trust (NHT) program. Some major programs in the original NHT (and the dollars allocated for the years 1997 to 2002) included: Bushcare ($328 million); NLP ($264 million); Murray–Darling 2001 ($163 million); Coasts and Clean Seas Program ($106 million); National Rivercare Program ($97 million); and the National Reserves System Program ($80 million).

In 2001 the Commonwealth Government extended the NHT for another five years from 2002–03 to 2006–07, with over $1 billion provided in funding over this period. In response to criticisms of expenditure under NHT1, the Australian Government has adopted a more strategic and targeted approach to allocation of these funds.

Governments carry out these four roles within the frameworks of existing laws. Budget allocations and priorities may vary with government policies at the time. There are a variety of influences on policy development, including:

- political, for example the rise of green independents and their environmental agenda has changed politics and policy in Australia
- social, for example the public are more involved in decision-making for management of public land
- economic, for example introduction of user-pays systems as a result of greater reliance on market-based approaches
- environmental, for example the international recognition of the importance of biodiversity conservation has influenced governments to formulate related policy
- cultural, for example, the Mabo Land Rights decision has required new legislation to deal with land ownership and management (Mercer 2000).

Fenna (1998) described three phases of the public policy-making process—agenda setting, policy formation, and implementation. Agenda setting can include how issues are identified and defined and what priority is given to the issue. Policy formulation involves which solutions are considered, how stakeholders are involved, what policy instruments are chosen, and which department is given responsibility. Implementation involves the resources that are committed to the policy, the time frame over which it is to be implemented, and evaluation of the effectiveness of the policy.

In Australia, the doctrine of economic rationalism has dominated the debate about the relative roles of the public and private sectors. Economic rationalism essentially advocates maximising the market's role in determining how resources are produced and allocated. In theory, perfect markets will tend to be efficient, in that goods will be bought and sold until the point is reached where everyone involved in the market can gain no additional benefit from further exchange—that is, benefits have been maximised. Therefore, the economic rationalists argue, we should strive to create perfect markets for as many goods and services as possible.

An ideal market requires:

- perfect competition between the actors in the market
- availability of full information in relation to the goods being traded and the mechanisms of trade
- allocation of property rights for all goods in the market:
  - that are transferable and secure
  - for which use entitlements are known and enforced
  - that are excludable—benefits and costs accrue only to the holder.

When one or more of these conditions does not apply, market failure occurs. Without effective property rights, some people may choose to 'free-ride' on the payments of others. A free rider is someone who cannot be excluded from enjoying the benefits of a project but who pays nothing (or pays a disproportionately small amount) to cover its costs. While one may devise means of determining preferences for public goods, it is unlikely that any direct way of extracting payment on the basis of these preferences could be devised without encountering free-riding behaviour.

Three sorts of market failure are of particular interest in relation to environmental issues.

1 The excessive production of unwanted 'externalities' can impose costs on other people. For example, clearing native vegetation on one property can cause salinity and thereby decrease the productivity of other properties. Pollution is another example of an externality that imposes large costs on society. In this case, market failure is due to the fact that there is no 'ownership' of the externality. Furthermore, given that there is no market price for the externality, there are no incentives to optimise the level of emissions.

2 The private sector economy will not produce enough public goods. Public goods and services contribute to the general welfare of society, but cannot be 'owned' by individuals—for example, health, education, defence, and parklands. The private sector is not able to efficiently provide these goods and services because they cannot recover all the costs of producing them.

3   Depletion of common property goods occurs due to a lack of any incentive for those extracting the resource to take account of its sustainability—if one user takes less, then there is more available for a competitor. This so called 'tragedy of the commons' occurs when all users take as much as they can, so that the resource is degraded for all. Such unsustainable exploitation has occurred, for example, with respect to commons grazing and fish stocks.

Furthermore, markets do not, in general, guarantee that resources will be used in a sustainable fashion. It can be economically rational to use up an entire stock of resources today, rather than preserving them for future generations.

Table 2.2 shows that, in general, there are four types of good, which are defined according to whether or not they possess the properties of rivalry and exclusivity. A good is rival when one person's use of it automatically precludes others from using it. A good is exclusive when, once the good is provided, only those who pay for it are able to use it. A mixed good is one that has at least two different types of values that are in different categories. For example, a forest is a mixed good that provides private goods such as timber, and public goods such as wildlife habitat.

Markets are very effective at providing an efficient supply of private goods. Public goods such as biodiversity conservation are generally 'produced' by governments through mechanisms such as establishing protected areas. Common property goods such as fishery often require government intervention in the form of regulation and/or allocation of property rights to ensure sustainable management of the resource.

It also needs to be recognised that government policies have also contributed to managing land poorly and to degrading resources. Examples are government's supporting inappropriate irrigation schemes, or taxation incentives that have promoted the clearing of vegetation. Yet the many cases of 'government failure' do not mean that market-based approaches would necessarily have fared any better. Governments must not bow out and expect markets to supply what are essentially 'non-market' services. Rather they must improve their own efficiency and reliability in supplying such services.

Sustainable development

In the 1970s and 1980s the global community realised that supposedly renewable resources of wild stock, for example fish, whale, and timber trees, were being pushed towards extinction. World consumption was increasingly depleting non-renewable resources. The consequences of exploitation, ecologically and socially, were becoming unacceptable. Governments and non-government organisations around the world have responded in a series of sustainable development policy initiatives (Table 2.3, Backgrounder 2.1). Sustainable development is based on three broad goals: environmental integrity, economic efficiency, and equity between present and future generations. The United Nations Conference on Environment and Development (UNCED) was held in Rio de Janeiro, Brazil, in 1992. The conference stimulated debate and action, both nationally and internationally. Two major outcomes were the Rio Declaration and Agenda 21. The Rio Declaration on the Environment and Development is a set of 27 principles designed to guide the economic and environmental behaviour of both nations and individuals. Agenda 21 is an action plan that draws on these principles and addresses the social and economic aspects of the conservation and management of resources.

Table 2.2	An economic classification of goods	
	Rival	Non-rival
Exclusive	Car, house, commercial recreation tour *Private goods*	Road tollway with no delays due to excess traffic *Toll goods*
Non-exclusive	Fishery, crowded park with unrestricted access *Common-property goods*	Uncrowded park with unrestricted access *Public goods*

Table 2.3	Key initiatives for sustainable development
Initiative	**Established**
World Conservation Strategy	IUCN 1980
Brundtland Report	World Commission on Environment and Development, 1987
Agenda 21	UN Conference on Environment and Development, 'Earth Summit', UN 1992
Convention on Biological Diversity	UN Conference on Environment and Development, 'Earth Summit', UN 1992
Kyoto Protocol	Conference of the Parties to the UN Framework Convention on Climate Change, Kyoto 1997
Millennium Development Goals	Millennium Summit, New York, UN 2000
World Summit on Sustainable Development	United Nations, Johannesburg, 2002

Backgrounder 2.1   Sustainable development initiatives

Brundtland Report

The World Commission on Environment and Development was appointed by the United Nations in 1983 to address the apparent conflict between economic and environmental interests, and to propose strategies for sustainable development. Norwegian Prime Minister Gro Harlem Brundtland chaired the Commission and its report *Our Common Future*, published in 1987, was commonly known as *The Brundtland Report*. This landmark report helped trigger a wide range of actions, and brought the concept of sustainable development to the public's attention. This report described sustainable development as '*development which meets the needs of the present without compromising the ability of future generations to meet their own needs*'. There is no standard or legal definition for sustainable development, but a number of organisations have utilised this definition and adapted it to fit their purposes. The major thread is that, in the interests of intergenerational equity, any proposed development should integrate consideration of environmental, economic, and social factors, such that the advancement of one does not degrade the quality of another (Bell & Morse 2003, p. 3).

Agenda 21

Following on from the Brundtland report, the United Nations Conference on Environment and Development (UNCED), or the 'Earth Summit', was held in Rio de Janeiro, Brazil, in June 1992. An outcome of this meeting was Agenda 21, a blueprint for achieving sustainable development (UN 1992b). It provides a framework for how to make development environmentally, socially, and economically sustainable in the twenty-first century. Agenda 21 outlines a plan of action to be taken up by all levels of government and major groups in every area in which humans impact upon the environment. The report was broken into sections covering social and economic dimensions, conservation and management of resources for development, strengthening the role of major groups, and means of implementation. There were 38 issues covered by Agenda 21 and these included topics as varied as: atmosphere, education, forests, health, international law, mountains, science, and waste (UN 2003).

Agenda 21 required each nation state to develop their own National Sustainable Development Strategy, and this led to a succession of national, local, and industry-based versions of Agenda 21 (Stunden 2002). The Commission on Sustainable Development was created in December 1992 to monitor and report on the implementation of sustainable development principles in each country. A special session of the UN General Assembly was called in 1997 to assess progress five years on from the Earth Summit.

Kyoto Protocol

Observations of global climate change commenced in 1957, the International Geophysical Year, when a number of nations set up research stations in Antarctica. In 1985 the hole in the ozone layer was discovered and the shock of this prompted the *Vienna Convention for Protection of the Ozone Layer*. Commitments to reducing emissions of greenhouse gases affecting the ozone layer were produced in the form of the *Montreal Protocol on Substances that Deplete the Ozone Layer*, Montreal 1987 (UNEP–Ozone Secretariat 2003). In 1990 the Intergovernmental Panel on Climate Change's assessment of global warming provided sufficient consensus to underpin the United Nations Framework Convention on Climate Change (UNFCCC), agreed upon in New York, 1992 (Grubb 1999). This Convention provided the political and legal foundations for international action. Five years later, with additional evidence of human-induced climate change, 160 nations met in Kyoto, in 1997, to discuss reducing greenhouse gas emissions. The outcome was the Kyoto Protocol, which required developed nations to limit their greenhouse gas emissions relative to 1990 levels by 2008 to 2012. Many of the countries present signed the Protocol, but a number have since retracted their support (including Australia), and in mid 2004 the Protocol had not yet entered into force.

Australia's obligations under the Protocol were rather generous, with targets to limit greenhouse gas emissions to 8% above 1990 levels (Australian Greenhouse Office 2002). Under the Kyoto Protocol, activities that remove carbon dioxide from the atmosphere, or 'carbon sinks', can be used to meet target commitments, and credits gained from carbon sinks can be traded, often with financial benefits.

Convention on Biological Diversity (CBD 2003)

Biodiversity is the term given to the variety of life on Earth, the plants, animals, microorganisms, and the different gene sequences in each species. At the 1992 Rio Earth Summit one of the agreements adopted was the Convention on Biological Diversity. The Convention establishes three main goals: the conservation of biological diversity, the sustainable use of its components, and the fair and equitable sharing of the benefits from the use of genetic resources.

Protecting biodiversity is in our self-interest. Species have been disappearing at 50 to 100 times the natural rate, and this is predicted to rise dramatically. The convention reminds decision-makers that natural resources are not infinite and sets out a philosophy of sustainable development. The Convention also offers guidance on the precautionary principle. The responsibility for achieving the goals of the Convention rests largely with individual countries, but requires the combined efforts of the world's nations if it is to be successful. Signatories to the Convention periodically meet in a Conference of the Parties to make decisions that advance implementation of the Convention. The most recent meeting was held in Malaysia on 9–20 February 2004. Priority issues discussed at the meeting included the biological diversity of mountain ecosystems and the role of protected areas in the preservation of biological diversity.

Millennium Summit

The Millennium Summit was a special session of the 55th session of the UN General Assembly, held in New York from 6 to 8 September 2000. The purpose of this meeting was to discuss how to strengthen the role of the UN in the twenty-first century. It followed up a report by the UN Secretary General Kofi Annan titled *We the Peoples—the Role of the United Nations in the 21st Century*, commonly known as the Millennium Report. This report outlined the vision for the United Nations in the age of globalisation. One of the outcomes of the Millennium Summit was the Millennium Development Goals. These are eight targets, aimed at reducing poverty and promoting sustainable development, most of which the signatories agreed to meet by 2015.

World Summit on Sustainable Development

Ten years on from Agenda 21, the World Summit on Sustainable Development (WSSD) was held, in Johannesburg 2002, to identify issues that were impeding implementation of Agenda 21. This summit reaffirmed commitment to Agenda 21 and the Millennium Development Goals (UN 2003).

While numerous new commitments and alliances were made, according to Bigg (2003) the Summit fell far short of the criteria set by the UN General Assembly. It did not progress much towards addressing the inadequacies of Agenda 21, and one area in which it really failed was that of sustainable development governance.

The main official outcomes of the Summit were a political statement and the WSSD Plan of Implementation. There were new targets and commitments under the Millennium Development Goals, including a new goal of improving sanitation levels, but there was very little about how any of these goals would be achieved.

In Australia, several major reports, strategies, and conferences have shaped efforts to adopt sustainable development principles and practice. In 1990 the Australian Government produced a proposal to develop a National Strategy for Ecologically Sustainable Development (NSESD) (Commonwealth of Australia 1990). Nine Ecologically Sustainable Development (ESD) working groups were established. Working groups brought together government officials, industry, environment, union, welfare, and consumer groups. They examined sustainability issues in major industry sectors such as agriculture, forests, fisheries, manufacturing, mining, energy, tourism, and transport. The chairs of each working group compiled separate reports on inter-sectoral issues, such as climate change, biodiversity conservation, and urban development. Taken together the reports contained over 500 recommendations for working towards ESD (Environment Australia 1998).

ESD working group reports were used to develop the NSESD. A group comprising the heads of Australia's three tiers of government (Commonwealth, state, and local) established a steering committee to assess the recommendations and what they implied for government policies. On 7 December 1992, a Council of Australian Governments meeting endorsed the NSESD (Commonwealth of Australia 1992a). The establishment of the Natural Heritage Trust (Snapshot 2.1), which implements ESD principles and encourages stakeholder involvement, underscored the NSESD. Implementing ESD in individual jurisdictions would be subject to budget constraints (Environment Australia 1998). Over 150 of Australia's 750 local Councils have now established local sustainability strategies (Environs Australia 2003).

Citizen participation

Citizen involvement is a central component in most if not all statements of sustainability principles. This centrality is a consequence of widespread discontent with the legitimacy and efficacy of representative democratic government; in particular the failure of democratic institutions, on their own, to represent and give expression to citizens' interests and aspirations. Citizen involvement in sustainable development planning and decision-making is manifested in two related but distinct approaches.

Participative approaches involve a shift from representative to participative democracy, in which citizens are actively engaged with the processes of policy development and implementation. Participative methods range from formal enquiries and opportunities to make written submissions, through to informal consultation via face-to-face discussions with participants. Such public engagement can improve the legitimacy of representative democracy by supporting political equality and the rights of citizens to be involved in decisions that affect them, reducing citizen alienation, increasing government accountability, as well as clarifying and representing the diversity of citizen interests and values concerning sustainability policy (Webler & Renn 1995, Selin et al. 2000, Barham 2001). It can improve the efficacy of government policy development and implementation by reducing failure; increasing acceptance; reducing delays; using local knowledge; managing competing interests and mediating conflict; and enhancing public ownership and commitment to solutions (Wondolleck & Yaffee 2000, Pimbert & Pretty 1997, Curtis & Lockwood 1998, Shindler & Brunson 1999).

Guidelines for participation processes include encouragement of all stakeholders to contribute;

opportunities for participation in a manner that best suits the particular understandings, needs and contributions of each participant; and ensuring participants have access to all relevant information (O'Riorden & O'Riorden 1993, Moote et al. 1997). Participation is often marketed by government as a mechanism for giving participants power to influence policy outcomes. However, despite some participatory processes offering opportunities for citizens to express views, and perhaps have an influence at the margin, the core policy agenda and framework largely remain under the control of governments.

More extensive are deliberative democratic processes that provide for collective decision-making through discussion, examination of relevant information, and critical discursive analysis of options. Attempts are made to eliminate the power and advantage afforded by political or economic position, so that participants regard one another as equals, defend and criticise positions in a reasonable manner, and accept the outcomes of such discussions (Dryzek 1997). Deliberative methods include citizen juries, deliberative polls, and consensus conferences. Deliberative democracy is seen by its proponents as more effectively recognising citizens' interests than the more limited participative approaches (Dryzek & Braithwaite 2000). However, in practice deliberative methods remain essentially advisory—the outcomes of deliberative polls and citizen juries are used to aid and inform the policy process, but do not constitute the core of the process itself—that usually remains under the control of governments.

Collaborative approaches constitute a more radical model of participation, which attempts to construct consensus policy decisions in which citizens have a central, not marginal, influence. Collaborative planning, as articulated by leading proponents such as Healey (1997), draws on the theory of communicative rationality, according to which judgments are made about the quality of communication using criteria such as honesty, clarity, sincerity, as well as lack of distortion, manipulation, and deception. With communicative rationality, decisions and actions are valid only if they arise from circumstances where all actors have been able to express themselves without inhibition or constraint, and where outcomes are unconditionally and freely accepted by all parties. In practice, consensus is very difficult to achieve, and collaborative participation largely remains an ideal rather than a reality. Some commentators have also argued that even in apparently well-functioning collaborative processes, it is inevitable that some people will exert undue power and influence (Flyvberg & Richardson 2002). Citizen engagement in relation to protected area decisions generally follows a participative approach (Chapters 7 and 17).

Regional development

The traditional view of protected areas as isolated repositories for natural and cultural heritage ignores the interactions between protected areas and regional and local communities. Protected areas are part of a mosaic of land and natural resource uses that are interdependent with communities and economies. Increasing recognition is being given to the importance of protected areas in furthering regional development. On the other hand, concerns are often raised that reduced access to resources such as timber and grazing adversely affects regional economies and communities. Protected area managers have a responsibility to explain the local and regional benefits that protected areas provide, as well as engaging more fully with local communities to minimise costs and maximise the flow of these benefits.

Over the past 40 years or so, many regional communities in Australia have suffered declining populations, reduced services, employment loss, and economic contraction. While there are many factors underlying this trend, important influences include globalisation of the economy, declining terms of trade for agricultural products, and changing community aspirations and preferences.

Protected areas can play an important role in regional rural development, especially in 'gateway' communities—that is, those communities that serve as primary access points into major protected areas. Important gateway communities in Australia include Cairns (Great Barrier Reef Marine Park and Wet Tropics World Heritage), Alice Springs (Uluṟu–Kata Tjuṯa National Park) and Jindabyne (Kosciuszko National Park).

There is a two-way relationship between regional communities and protected areas. For the values of a protected area to be maintained, it must function as part of its community. Protected areas cannot be divorced from local and regional land uses (Machlis & Field 2000a). Most exist in a matrix of multiple use public lands and private lands devoted to agriculture, private forestry, urban development, and other uses. Regional growth or decline will affect both management and visitor experiences. Some regional communities are changing rapidly, especially in coastal areas, due to the 'sea-change' phenomenon of city dwellers seeking different lifestyle options and the attractiveness of coastal areas for the growing number of retirees.

Local citizen reactions to the social and economic issues confronting a gateway community—such as growth planning, environmental restrictions, and provision of services—can vary, particularly where there has been a recent influx of new residents seeking a rural lifestyle in an aesthetically pleasing environment. Long-term residents may have different views and interests to the newcomers, so that such communities are not homogeneous, but often replete with their own conflicts and competing interests (Jarvis 2000). Protected areas typically require transportation routes, energy grids, water supply, and waste disposal systems. They can create employment, housing needs, and business opportunities, particularly those related to supply of the goods and services needed to support visitor activities. These needs and opportunities in turn trigger development requirements within a region for infrastructure, waste disposal, and natural resources such as water (Machlis & Field 2000a). Management issues ranging from fire protection and prevention to spread of introduced species can arise from such development activity.

This implies that management policies for protected areas should be integrated into the broader context of community sustainability. Strategic planning is required to integrate those concerns within the boundaries of the protected area network (biodiversity conservation, visitor service provision, environmental protection) with wider environmental, economic, and social sustainability. Machlis & Field (2000b) advocate that protected area managers should:

- take responsibility to influence development in rural areas, and aggressively seek to maintain the viability of communities that surround protected areas
- promote a sense of local identity that allows people to determine their own destinies
- create allies among local citizens, especially local leaders, to develop a management capability at a landscape scale
- emphasise the local and regional benefits of protected areas
- adopt a collaborative approach to planning, with citizen participation understood as being crucial to the development of leadership and capacity for sustainable development
- contribute to preserving the overall character and lifestyle adjacent to protected areas while maintaining opportunities for planned growth
- give technical assistance to rural and gateway regions, train staff in rural development and collaboration skills, and assess progress in achieving sustainable rural development.

Postmodern times

It is thought by some academics that over the last three decades or so, a fundamental change has occurred in western society. Since the Enlightenment (1650–1850), western culture has been engaged in 'the project of modernity' (Taylor 1998). Modernity is characterised by a reductionist scientific world view; belief in progress, especially that afforded by technological advances; and a preference for order and classification. The Enlightenment also elevated the importance, rights, and responsibilities of individuals as apart from, or even above, the traditional authority of the church. Democracy, trade based on competitive profit-seeking organisations, together with the technical, economic, and social transformations of the Industrial Revolution are other key aspects of modernity (Friedmann 1987). Positivism is a modernist theory that claims the knowledge generated by rational scientific enquiry is universal and objective truth. In the modern world, technical capacity, positivism, and rational thought enabled humans to maintain the illusion that environments can be controlled.

It is thought by some that we are now moving into postmodern times in which the old 'certainties' and confidence of modernism are being superseded by more complex, diverse, and dynamic understandings and behaviours. The postmodern sensibility is relativistic rather than absolute, pluralistic rather than segregated, richly chaotic rather than ordered. Post-positivist understandings argue that knowledge belongs to particular social and historical contests, and as such is relative and subjective. Truth is replaced by interpretation, and the role of power in establishing interpretations is a central concern of postmodern thinking. Postmodernism rejects any attempt to promote an overall theory or explanation. A post-positivist perspective requires shifting from causal reasoning as a basis for decision-making to discovering and confirming meaning (Allmendinger 2002).

Traditional protected area management grew up during the last phases of modernity, and as such many of its concepts and methods are also modernist in character. If one accepts that there has indeed been a fundamental shift in the nature of Western society, a tension is evident between traditional protected area practice and the postmodern times in which we are now living. For example, postmodernists reject the possibility of a monolithic 'public interest', replacing it with an irreducible plurality of voices and interests (Campbell & Fainstein 2003). Given that one of the traditional roles for protected area management is to serve 'the public interest' such thinking, if accepted, poses significant challenges to the role and place of protected areas in society. The new governance possibilities being considered by protected area proponents (discussed in Section 2.3) can be seen as one way of meeting such challenges.

2.3   Responses and initiatives

Many protected area managers and proponents have been well aware of the trends and influences outlined in the previous section, and have been seeking appropriate initiatives and responses. Protected areas are now conceived as a long-term societal endeavour that goes well beyond the original, limited 'Yellowstone' vision of what a national park should be. Important elements of this endeavour are building a wide constituency that supports protected areas, locating protected areas within the wider agenda of sustainable development, and responding to calls from Indigenous peoples and local communities for more recognition of their rights, needs, and cultures. In sum, these constitute a 'paradigm shift' in thinking about protected areas. These changes are profound, involving as they do a re-orientation of how humans relate to nature. The separation of humans from nature implied by the traditional national parks paradigm, while arguably a necessary stage in the evolution of protected areas, is now maturing into a view in which humans both engage with and show respect for the natural world, together with the values that it affords and embodies.

This shift is summarised in Backgrounder 2.2, where Phillips (2003) characterises the old and new paradigms according to factors such as the objectives of protected areas, their governance, attitudes towards local people, and management. In the rest of this section, we will explore two 'new paradigm' topics in more detail: governance and developing a protected areas constituency.

Governance

Protected area management is more than just a state activity. As part of a wider sphere of environmental planning and management activity, such work is both the privilege and the responsibility of many individuals and organisations (Rydin 2003). To put it another way, protected area management is a mode of governance. Governance modes range from the traditional exercise of government authority, through to a wide variety of partnership, co-management, and informal arrangements involving multiple agencies, interest groups, and individuals (Ostrom 1990, Reeve et al. 2004). Graham et al. (2003, p. 2–3) defined governance as

> *the interactions among structures, processes and traditions that determine how power and responsibilities are exercised, how decisions are taken, and how citizens or other stakeholders have their say. Fundamentally, it is about power,*

Backgrounder 2.2   Traditional and emerging protected area paradigms

Phillips (2003)

Traditional paradigm

Objectives

- 'Set aside' for conservation, in the sense that the land (or water) is seen as taken out of productive use
- Established mainly for scenic protection and spectacular wildlife, with a major emphasis on how things look rather than how natural systems function
- Managed mainly for visitors and tourists, whose interests normally prevail over those of local people
- Placing a high value on wilderness, that is areas believed to be free of human influence
- About protection of existing natural and landscape assets—not about the restoration of lost values

Governance

- Run by central government, or at the very least set up at the instigation only of central government

Local people

- Planned and managed against the impact of people (except for visitors), and especially to exclude local people
- Managed with little regard to the local community, who are rarely consulted on management intentions and might not even be informed of them

Wider context

- Developed separately—that is planned one by one, in an ad hoc manner
- Managed as 'islands'—that is managed without regard to the areas around

Perceptions

- Viewed primarily as a national asset, with national considerations prevailing over local ones
- Viewed exclusively as a national concern, with little or no regard to international obligations

Management technique

- Management of protected areas treated as an essentially technocratic exercise, with little regard to political considerations
- Managed reactively within a short time-scale, with little regard to the need to learn from experience

Finance

- Paid for by the taxpayer

Management skills

- Managed by natural scientists or natural resource experts
- Expert-led

Emerging paradigm

Objectives

- Run also with social and economic objectives, as well as conservation and recreation ones
- Often set up for scientific, economic, and cultural reasons—the rationale for establishing protected areas therefore becoming much more sophisticated

- Managed to help meet the needs of local people, who are increasingly seen as essential beneficiaries of protected area policy, economically and culturally
- Recognises that so-called wilderness areas are often culturally important places
- About restoration and rehabilitation as well as protection, so that lost or eroded values can be recovered

Governance

- Run by many partners, thus different tiers of government, local communities, Indigenous groups, the private sector, NGOs and others are all engaged in protected areas management

Local people

- Run with, for, and in some cases by local people—that is local people are no longer seen as passive recipients of protected areas policy but as active partners, even initiators and leaders in some cases
- Managed to help meet the needs of local people, who are increasingly seen as essential beneficiaries of protected area policy, economically and culturally

Wider context

- Planned as part of national, regional, and international systems, with protected areas developed as part of a family of sites. The CBD makes the development of national protected area systems a requirement
- Developed as 'networks', that is with strictly protected areas, which are buffered and linked by green corridors, and integrated into surrounding land that is managed sustainably by communities

Perceptions

- Viewed as a community asset, balancing the idea of a national heritage
- Management guided by international responsibilities and duties as well as national and local concerns. Result: transboundary protected areas and international protected area systems

Management technique

- Managed adaptively in a long-term perspective, with management being a learning process
- Selection, Planning, and Management viewed as essentially a political exercise, requiring sensitivity, consultations, and astute judgment

Finance

- Paid for through a variety of means to supplement—or replace—government subsidy

Management skills

- Managed by people with a range of skills, especially people-related skills
- Valuing and drawing on the knowledge of local people

---

relationships and accountability: who has influence, who decides, and how decision-makers are held accountable.

Powers exercised by protected area authorities include:

- planning powers relating to designation and management of protected areas
- regulatory and enforcement powers related to the use of land and resources
- spending powers to further conservation and sustainable development
- revenue-generating powers that may be in the form of fees, permits, or taxes
- powers to enter into agreements to share or delegate some of the above powers

- powers to enter into cooperative agreements with other parties with interests or responsibilities germane to protected area concerns (Graham et al. 2003).

In traditional protected area management, governance has been by government, particularly Commonwealth, state, and territory governments. While this is still largely the case, there are now several modes of governance beyond direct management by a government agency. These include various forms of collaborative management, partnership arrangements, delegated authority, and community management. Powers and responsibilities related to protected areas now extend to include Indigenous communities, local community networks, local government, private sector organisations, and individual landholders. A summary of the governance arrangements currently evident in Australian protected area management is given in Table 2.4.

In their *Directions for the National Reserve System: a partnership approach*, the Natural Resource Management Ministerial Council (NRMMC) (2004) recognised that the protected area system cannot meet the goals of a fully comprehensive, adequate, and representative National Reserve System, without the protection of certain ecosystems that are located fully or in part on private lands. This will require development of new partnership and governance arrangements between

Table 2.4	Modes of protected area governance in Australia	
Mode	Type	Example
Government	National	Pulu Keeling National Park (see http://www.deh.gov.au/parks/cocos/index.html)
	State or Territory	Kosciuszko National Park, NSW (see http://www.nationalparks.nsw.gov.au/parks.nsf)
	Local	Briars Park (Snapshot 2.3)
	Delegated (to another government agency)	Heard Island and McDonald Islands Marine Reserve (see http://www.deh.gov.au/parks/index.html)
	Delegated (to statutory authority)	Wellington Park (see http://www.wellingtonpark.tas.gov.au/)
	Delegated (to local government or community group)	Bukkulla Conservation Park in Queensland managed by the Wildlife Land Fund Ltd
Co-management	Collaborative	Kinchega National Park, NSW (see http://www.nationalparks.nsw.gov.au/npws.nsf/Content/Park+management+info)
	Joint	Nitmiluk (Katherine Gorge) National Park (see Snapshot 2.4)
Private	Individual	Winlaton Grassland, Northern Victoria (search http://www.nre.vic.gov.au)
	Not-for-profit organisation	Charles Darwin Reserve (see http://www1.bushheritage.asn.au)
	Commercial organisation	Warrawong Earth Sanctuary (see http://www.warrawong.com/)
Community	Indigenous	Yalata Indigenous Protected Area, South Australia (see http://www.deh.gov.au/indigenous/ipa/declared/yalata.html)
	Local	Grassy Box Woodlands Conservation Management Network (Snapshot 2.5)

governments, communities, and private landowners, including Indigenous landowners. It is important to emphasise that these new modes of governance are complementary to the public network of protected areas. They are a means of extending conservation management across landscapes, thereby providing the scale and extent of conservation management necessary to secure the long-term future of species and ecosystems. The directions recommended by NRMMC (2004) are further considered in Chapter 5.

This expansion of protected area governance matches an international trend, where such developments are even more strongly evident. Governance was a major theme of the Fifth World Parks Congress held in Durban, South Africa, in September 2003. At the congress, Graham et al. (2003) proposed five key principles of good governance for protected areas (Backgrounder 2.3). These principles were endorsed under Recommendation 5.16 of the Congress as a basis for improving protected area governance around the world. The Congress also endorsed the acknowledgment of a range of governance types as a means of expanding the global protected area network and increasing its legitimacy (Snapshot 2.2).

Government management is the traditional mode of protected area governance, and remains the dominant mode in many developed countries, including Australia. Government agencies can be established within the Commonwealth, state/territory or local tiers of government. Governments can also delegate their authority to another government agency, statutory authority, or non-government organisation.

Co-managed protected areas are where authority, responsibility, and accountability are shared among two or more parties, which may include government agencies, Indigenous people, non-governmental organisations, and private interests. There are two types of co-management. With *collaborative management*, authority is held by one party (often a governmental agency) but this party is required to collaborate with other parties. *Joint management* involves true sharing of authority among two or more parties, with none of these parties having ultimate authority in their own right.

2003 Vth World Parks Congress, Durban, South Africa (Graeme Worboys)

Private management of protected areas can be done voluntarily by individuals, not–for profit organisations, or commercial enterprises. Generally the authority of these parties to identify and manage land as a protected area arises from the private property rights they hold over a parcel of land. Protected area designation can be formalised through mechanisms such as a covenant on the title of the property. In some cases, government agencies provide management and financial support to the private owners.

Community managed protected areas (also called community conserved areas) are managed voluntarily by Indigenous or local communities. Management regimes may be established through customary laws and institutions using traditional knowledge, or through partnership agreements amongst consortia of local people. Such areas are generally not formally recognised in national and international conservation systems. For example, in Australia community conserved areas are not registered on CAPAD—the Collaborative Australian Protected Area Database managed by the Commonwealth Department of the Environment and Heritage.

Snapshot 2.2

WPC Recommendation 17: Recognising and supporting a diversity of governance types for protected areas

RECOMMEND governments and civil society:

a. Recognise the legitimacy and importance of a range of governance types for protected areas as a means to strengthen the management and expand the coverage of the world's protected areas, to address gaps in national protected area systems, to promote connectivity at landscape and seascape level, to enhance public support for such areas, and to strengthen the relationship between people and the land, freshwater and the sea; and

b. Promote relationships of mutual respect, communication, and support between and amongst people managing and supporting protected areas under all different governance types

REQUEST the IUCN World Commission on Protected Areas (WCPA) to refine its Protected Area Categorization System to include a governance dimension that recognises the legitimacy and diversity of approaches to protected area establishment and management and makes explicit that a variety of governance types can be used to achieve conservation objectives and other goals.

Snapshot 2.3

Local government protected area on the Mornington Peninsula
Steve Yorke, Team Leader, The Briars Park

The Briars Park is a 220 hectare property owned by the Mornington Peninsula Shire at Mount Martha, comprising a 96 hectare wildlife reserve, 96 hectare farm, eight hectare heritage site jointly owned with the National Trust of Australia (Vic), and several small leased parcels. These include an Outdoor Education camp, vineyard, a rose nursery, a site for the Astronomical Society of Frankston and the caretaker's cottage. One of the heritage listed buildings, a former barn, is leased as a restaurant. There are 15 such buildings, listed on the Register of the National Estate and the Victorian Heritage Register.

A wide range of facilities and activities is provided for a wide cross-section of the community, offering passive recreation, education, and enjoyment. The well-equipped visitor centre provides a hub for education programs and school holiday activities, an aquarium, displays, and audio-visual presentation. Visitors can picnic, walk, watch birds, inspect the 1846 homestead, outbuildings, and historic farm machinery, play tennis or croquet, or attend seminars and workshops.

A historic farming property is the perfect setting for the demonstration of the relationship between human activity, agriculture, and conservation. A whole farm plan, currently being developed, will integrate the management of the farm lease with park objectives and work towards best practice. Already stream lines are fenced and planted, salinity is being monitored in bores and, at the surface, remedial planting has taken place and organic fertilisers are encouraged. The Wildlife Reserve is contained by an electrified vermin-proof fence—eastern grey kangaroos and swamp wallabies are being reintroduced to join the resident koalas. Balcombe Creek has the best water quality of all the streams entering Port Phillip and supports a variety of native fish.

Snapshot 2.4

Nitmiluk (Katherine Gorge) National Park
Adapted from Smyth (2003)

Nitmiluk (Katherine Gorge) National Park came into being as the result of a successful claim by Traditional owners of the former Katherine Gorge National Park under the *Aboriginal Land Rights (Northern Territory) Act 1976*. The main features of joint management arrangements at Nitmiluk include:

- the vesting of the park in a Land Trust on behalf of the Aboriginal Traditional owners
- the lease of the land to the Conservation Land Corporation, on behalf of the Northern Territory Government

- the payment to Traditional owners of an annual rental of $100,000 for the lease of the park; this amount is to be reviewed every three years
- the establishment of a Board of Management comprising 13 members of whom 8 shall be Traditional owners, 4 shall be staff of the Commission and one shall be a local resident appointed by the Mayor of Katherine
- the day to day management of the Park by the Parks and Wildlife Commission of the Northern Territory.

Snapshot 2.5

Community management networks
Adapted from Jaireth (2003) and Figgis (2003)

An innovative governance mechanism is the Conservation Management Network (CMN) that creates a network of vegetation remnants from targeted and threatened ecological community or groups of related communities and draws in landholders and others to manage and grow the network biologically and socially. CMN governance mechanisms are flexible, and may involve government run with a facilitator or technical staff, or may involve a community steering committee. Current CMNs address conservation of two important vegetation types: Grassy Box Woodlands in central western NSW and the Forest Red Gum Grassy Woodlands on the Gippsland Plains in Victoria.

The NSW project was funded by the NHT and implemented jointly by the NSW NPWS and a consultancy firm, Ecological Interactions, Community Solutions. Participants sought out important remnants and provided advice and incentives. Funding up to $60,000 per site was available to assist landholders establish a Conservation Agreement and manage weeds, fencing etc. In 2003 16 properties had Voluntary Conservation Agreements covering 60 sites and more than 2000 hectares.

The Gippsland Plains CMN was formed in 1999 through the purchase by the Trust for Nature of several areas of privately owned Forest Red Gum Grassy Woodlands. The management of these areas was seen as complementing the conservation protection provided by existing public reserves such as the Providence Ponds Flora and Fauna Reserve and other private lands protected through conservation covenants. The effort was originally coordinated by the Trust for Nature, followed by the then Victorian Department of Natural Resources and Environment. By 2003 a specially incorporated membership based CMN organisation had been established that employs a ranger to manage monitoring and restoration works.

Backgrounder 2.3   Protected area governance principles

Modified from Graham et al. (2003)

According to the principles, protected area governance should strive for:

Legitimacy and voice through:

- a supportive democratic and human rights context involving democratic institutions based on free elections, respect for human rights, and promotion of tolerance and social harmony
- an appropriate degree of collaboration and decentralisation in decision-making that involves all affected parties, particularly local and Indigenous people, as well as giving recognition to broader national and international interests

- existence of civil society groups and an independent media to act as a check and balance on the exercise of the powers granted to protected area political leaders and managers
- high levels of trust among the various actors, governmental and non-governmental, national, state and local, involved in the management of protected areas.

Direction through:

- consistency with international conventions, agreements, and best practice guidelines
- legislation (either formal or traditional law) that sets out clear purpose and objectives for protected areas, provides for viable management organisations, and establishes the means by which these organisations will exercise their powers
- national system-wide plans that have quantified objectives for representing values in a protected area network
- preparation of management plans for individual protected areas that reflect citizen participation, have formal approval of the appropriate authorities, set out clear objectives consistent with legislation, set out measurable results to be achieved within specific time frames, are regularly reviewed and updated, and are implemented through annual work plans
- effective leadership that provides an inspiring and consistent vision for the long-term development of the protected area system, mobilises support for this vision, and garners the necessary resources to implement the various plans for the system or individual protected areas.

Performance through:

- cost-effectiveness in achieving objectives
- having the capacity (resources, knowledge) to undertake required functions
- coordination of efforts between the various actors
- public availability of performance information
- responsiveness in dealing with complaints and criticism.
- capacity to undertake regular and comprehensive monitoring and evaluation, and to respond to findings
- ability to provide for learning and adjustment of management actions on the basis of operational experience as part of an adaptive management strategy
- capacity to identify key risks and manage them.

Accountability through:

- clarity in the assignment of responsibilities and the authority
- appropriateness of responsibilities assigned to political leaders as opposed to non-elected officials or semi-independent bodies and the absence of corruption
- effective public institutions of accountability
- transparency in decision-making, resource allocation, and management performance.

Fairness through:

- a supportive judicial context characterised by respect for the rule of law including an independent judiciary, equality before the law, and citizens having the right to seek legal remedies
- impartial and effective enforcement of protected area regulations including transparency, absence of corruption, and rights of appeal
- processes for establishing new protected areas that include public participation and respect for the rights, uses, and traditional knowledge of local and Indigenous peoples related to the area
- management practices that achieve a favourable balance of costs and benefits to local and Indigenous peoples
- mechanisms for sharing or devolving the management decision-making of the protected area with local and Indigenous peoples
- equitable human resource management practices for protected area staff
- processes for recognising and dealing with past injustices resulting from the establishment of protected areas.
    Graham et al. (2003) noted that these principles overlap and may conflict with each other, so applying them requires balance and judgment, as well as recognition of the prevailing social and historical context.

Building a constituency

Constituency-building is a global trend that involves establishing broadly based coalitions and partnerships directed towards sustainable environmental management, including conservation through various forms of protected area. Long-term conservation at a landscape scale requires support and commitment from a wide range of constituencies. Protected area managers must secure widespread community support, both to legitimise their work and to secure the support necessary to expand and strengthen their activities. It is widely acknowledged that achieving satisfactory conservation outcomes will require considerable expenditure of funds—funds that will only be raised if there is community understanding of and support for protected area management objectives. But no matter how much funding is made available, protected area management will not be successful in the long term, unless it is accepted as a core part of a wider social, cultural, economic, and political agenda. Protected areas are already widely supported, and not just among those people who might be identified as 'green'. However, protected areas need to become even more centralised into popular consciousness and acceptance, so that they are recognised as a key element in people's quality of life, linked to their personal identity and aspirations.

Already Uluṟu, the Great Barrier Reef, mountains, deserts, and coastal landscapes are integral to many Australians' self-identity. This connection with Australian natural environments will need to continue to grow and deepen. The full range of benefits provided by protected areas is not well understood or appreciated. Protected area agencies need to clearly identify and articulate the values that protected areas provide to local and regional communities in particular.

2.4 Government in Australia

Federation

The Commonwealth of Australia was formed in 1901 by the federation of the Australian colonies. Australia's federal system involves a division of decision-making powers and activities between the states and a central government (Hughes 1998). Australia has adopted a modified Westminster model, which is similar to the system used in Britain in that:

- parliament is elected by the citizens
- the government is then elected by parliament and is answerable to it
- the executive (prime minister and cabinet) is composed of members of parliament and is answerable to parliament
- the prime minister, while formally appointed by the Crown, is elected by the party with the majority in the lower house.

However, a crucial difference between Australia and Britain is that Australia has a constitution, the *Australian Constitution Act 1901* (Cth), which specifies the powers of the Australian government and establishes the high court to determine what these powers are in matters of dispute. The Australian constitution primarily outlines the separation of powers between the Commonwealth and the states. The constitution specifies the areas of Commonwealth power, whereas the states have power over the residual issues. To carry out their respective constitutional roles the states and Commonwealth each have their own government systems and infrastructures. As well, the Constitution grants the Commonwealth (or federal government) legislative power over the territories, though some of these have evolved self-governments that have some autonomy. A third level of Australian government, one not regulated by the Australian constitution, is that of local government. The states and territories pass local government acts that define their powers.

Commonwealth Government

Since environmental matters were not of public concern at the time, the Constitution says nothing about any powers of the Commonwealth Government in relation to the environment. This means that the state governments have power and can legislate on almost all environmental matters. However, the Commonwealth Government can legislate on environmental issues, by using sections 51 and 52 of the Constitution.

The Commonwealth has exclusive jurisdiction over matters listed in section 52 of the Constitution, including the Australian Capital Territory, Australian external territories, and land owned by the Commonwealth within the states (a total of about 1 million hectares.). The self-governing territories—the Northern Territory, Australian Capital Territory and Norfolk Island—have been granted authority over environmental protection and conservation. The Commonwealth, however, retains ultimate legislative authority (Bates 1995).

Other matters on which the Commonwealth may legislate are listed in section 51 of the Constitution. Section 109 also indicates that where state legislation conflicts with federal legislation, the latter will prevail. The various parts of section 51 are termed 'heads of power', since they provide the Commonwealth Government with the authority to act on the matters specified therein. The key heads of power (subsections of section 51) in terms of environmental matters include:

- Subsection (i) Trade and commerce with other countries
- Subsection (xx) Corporations
- Subsection (xxvi) People of any race, including the Aboriginal people of Australia
- Subsection (xxix) External affairs.

The external affairs power, for example, means that the Commonwealth may legislate on domestic environmental matters where they are of international concern. One way in which bona fide international concern can be demonstrated is via the existence of an international treaty. For example, parts of south-west Tasmania are declared as World Heritage under the *Convention for the Protection of World Cultural and Natural Heritage 1975*. Australia is a signatory to this convention. This enabled the Commonwealth to pass the *World Heritage Properties Conservation Act 1983*, which allowed the Commonwealth Government to stop the Tasmanian Government from proceeding with the Gordon below Franklin dam. Other examples of international conservation agreements include the Ramsar convention for the protection of internationally significant wetlands, and the Japan-Australia and China-Australia migratory birds agreements (JAMBA and CAMBA).

Another section of the constitution that has some environmental relevance is section 96, which allows the Commonwealth to grant financial assistance to the states. The Commonwealth has used this power to provide incentives for environmental protection, as for example through the National Heritage Trust, established from the partial sale of Telstra.

Under international law, the Commonwealth is responsible from the low water mark to 200 nautical miles from this mark. Under an agreement between the Commonwealth and the states, all states and the Northern Territory have property rights on the sea bed within a three nautical mile limit. Similarly, states have control over fisheries within the three-nautical-mile limit.

State governments

State governments have primary responsibility for land use and planning matters on both public and private land. Through their legislative powers, they also control the framework within which local government operates. There is a large body of state legislation, policy, and programs and strategies that focus on environmental planning and management.

Territory governments

The constitution grants the Commonwealth legislative power over the territories. However, the Commonwealth has given self-government to the Northern Territory and the Australian Capital Territory. These territories have their own legislative assemblies and have similar environmental management and planning responsibilities to the states.

Government departments

Commonwealth, state, and territory governments have numerous departments and agencies that assist the development and implementation of policy. These administrative arms of government implement the decisions of parliament. Ministers are responsible for their departments and statutory bodies of the 'public service'. Departmental functions are determined by laws and by executive arrangements (Singleton et al. 1996). However, in 'day-to-day administration … government officers are guided not so much by regulation but by memoranda, guidelines, procedures and directives of all kinds' (Bates 1995, p. 10). The arrangement is not all one-way. Experienced and knowledgable departmental officials also provide advice on policy and legislation to their ministers. The public service is intended to be an impartial body, however, which performs functions and responds to policies set by the government of the time (Hughes 1998). The Australian departments and statutory bodies responsible for managing protected areas are described in Chapter 3.

Local government

Local governments are the third level in the Australian government system but, unlike the other levels, their powers are not regulated by the Australian constitution. Instead, each state and territory has a local government act for this purpose. These acts have been created independently, so the powers of local governments vary between states. Additional powers are generally granted under land-use and planning legislation. Local governments consist of a council made up of part-time members that is chaired by a mayor or shire president. Local government is typically responsible for land-use planning within municipalities, and for providing environmental services such as domestic waste disposal. Some local governments also use instruments such as levies, rate relief and differential rating, grants, and community education to achieve environmental objectives.

In some respects, local governments have substantial potential for capacity building, can be sensitive to local identities and needs, and effectively initiate local action. However, they are often beset by problems such as a lack of jurisdictional autonomy, poor resource availability, and tensions between raising revenue and planning.

Local governments are assuming greater responsibility for managing conservation (Binning 1998). Some have formed alliances with community groups such as Landcare and Greening Australia. Some have provided community incentives for conservation, such as rate rebates. Some local governments have used their land planning powers to create conservation zones, where they have controlled development. Councils are coordinating their efforts because they realise their environmental problems must be solved regionally and cooperatively. Scarce resources also need to be used efficiently. In Tasmania, for example, the Cradle Coast Authority coordinates efforts across nine local governments to achieve regional sustainable development.

Coordinating government action

Efficient and effective cooperation across levels of government and between jurisdictions depends on the extent to which roles and responsibilities are unambiguously defined, duplication is avoided, and effective cooperative processes are established. Australian governments have adopted a policy of cooperative federalism. Under cooperative federalism, the roles of the various governments are negotiated through agreements, national councils, and strategies (Hughes 1998). Some cooperative mechanisms relevant to natural resource management are as follows.

The *Council of Australian Governments* (COAG) is the peak body overseeing closer cooperation between Australian governments. It consists of the prime minister and the leaders of the states and territories and local government and deals with issues concerning roles and responsibilities in areas of shared responsibility, micro-economic reform, regulation, and the environment.

The *InterGovernmental Agreement on the Environment* was signed by the Commonwealth, state and territory governments, and the Australian Local Government Association in 1992. The agreement followed a decade of political controversy and legal argument over specific conservation issues between the Commonwealth and the states. The agreement establishes a cooperative national approach to the environment, and better defines the roles of government (Commonwealth of Australia 1992b). The agreement requires ecologically sustainable development principles and practice to be implemented, including the principle of precaution and the principle of intergenerational equity.

Ministerial councils are made up of Commonwealth, state, and territory ministers with responsibility in a particular area. The Natural Resource Management Ministerial Council comprises all Australian Government, state, and territory ministers with responsibility for natural resources, the environment, water policy, and primary industries. It has a National Reserve System Taskforce, convened under its Land, Water and Biodiversity Committee.

National strategies are developed following input from all Australian governments, industry, and the community. Governments, at least in theory, use national strategies to guide their decision-making. Over recent decades there has been a proliferation of environmental strategies, programs, and initiatives. The National Strategy for the Conservation of Biological Diversity and the National Strategy for Ecologically Sustainable Development set out many of the underlying principles for conservation programs.

2.5 Legislation

The primary way in which parliaments address topical issues, alter the workings of government, and put policy into action, is through the formal enactment of laws approved by established political institutions. Legislation in the form of Acts of Parliament, and regulations under these acts (which often provide details inappropriate to the acts themselves) are the main means by which governments give expression to their policies. In Australia, for bills to become law they must pass through several stages in parliaments. As most parliaments have two levels—called lower and upper houses—they are also known as 'bicameral (two-chamber) legislatures'. The exchanges between the two houses are critical on controversial matters. This is because elected governments dominate lower houses, whereas other parties may hold the balance of power in upper houses.

Many matters that come before parliaments need to be delegated. To do so, parliaments may create statutory (administrative) bodies. Such bodies may also be given powers to create subordinate legislation independent of parliaments. Organisations that manage protected areas are generally statutory bodies. Subordinate legislation includes regulations, such as those made under the *National Parks and Wildlife Act 1974* (NSW) that confer functions, powers, and duties upon park rangers.

Commonwealth environmental legislation

The major legislation at the Commonwealth level is the *Environment Protection and Biodiversity Conser-vation (EPBC) Act 1999*. The EPBC Act came into force on 16 July 2000, and superseded much of the existing Commonwealth environmental legislation. It provides protection for the natural and cultural environment in matters of national environmental significance, and matters involving Commonwealth land or Commonwealth agencies. The goals of the EPBC Act are:

- to protect and conserve nationally significant aspects of the environment
- to provide for a streamlined environmental assessment and approvals process
- to establish an integrated regime for the conservation of biodiversity and the management of protected areas.

The EPBC Act provides protection for:

- listed species and communities in Commonwealth areas
- cetaceans in Commonwealth waters and outside Australian waters
- protected species in the Territories of Christmas Island, Cocos (Keeling) Islands, and Coral Sea Islands
- protected areas (World Heritage properties; Ramsar wetlands; Biosphere reserves; Commonwealth reserves—see Chapter 3).

Under the assessment and approval provisions of the Act, actions that are likely to have a significant impact on a matter of national environmental significance are subject to an assessment and approval process. Matters of national environmental significance are:

- World Heritage properties
- National Heritage places
- Ramsar wetlands of international significance

- nationally listed threatened species and ecological communities
- listed migratory species
- Commonwealth marine areas
- nuclear actions (including uranium mining).

The government states that the *EPBC Act* promotes the conservation of biodiversity by providing strong protection for threatened species and ecological communities; the marine environment; migratory species and internationally significant wetlands; protection from nuclear activity; and regulation of international trade. Through the Act there have been amendments to the list of nationally threatened species, ecological communities, key threatening processes, and recovery plans (DEH 2002). However, the Act has also been criticised for its alleged unduly limited view of Commonwealth powers, marginalisation of public participation, exemption of many forestry and fisheries operations from Commonwealth approval, lack of specific detail on EIA processes, and excessive Ministerial discretion (Mercer 2000).

State and territory environmental legislation

There is a very large body of environmental legislation in each state and territory. The Victorian Department of Sustainability and Environment, for example, lists 53 Acts on its web site (www.dse.vic.gov.au) that have a major influence on its work and functions. The NSW Department of Infrastructure, Planning and Natural Resources is wholly or in part responsible for 30 Acts (DIPNR 2003). There are at least 90 Acts of Parliament in Tasmania that relate to environmental matters (Environment Defenders Office 2001). Such legislation can be grouped into several broad categories.

1. Environmental planning and protection
A. Land-use planning

Land-use controls are the most widely used method of controlling and regulating development activity, and are often achieved through zoning. The routine planning functions on private land are often taken on by local government, with overall control and coordination by state government. In some jurisdictions, such as Victoria, public land-use planning is also enabled through specific legislation.

Examples:

Land Use Planning and Approvals Act 1993 (Tas) is concerned, among other things, with the making and amendment of local government planning schemes.

Victorian Environment Assessment Council Act 2001 (Vic) establishes the Victorian Environment Assess-ment Council to conduct investigations and make recommendations relating to the protection and ecologically sustainable management of public land.

B. Environmental protection

Every state government has accepted the need for some formal legislative procedure to assess the impacts of development proposals. Some states have included these environmental impact provisions in planning legislation (NSW, SA), while others have separate acts (Vic).

Examples:

Environmental Planning and Assessment Act 1979 (NSW) deals with, among other things, procedures for dealing with development applications and for the assessment of environmental impacts.

Environmental Effects Act 1978 (Vic) requires the environmental effects of certain developments to be assessed.

C. Pollution

Controls related to air, water, and noise were developed extensively in the 1970s. Many states set up specialist authorities to address these issues.

Examples:

Environmental Protection Act 1994 (Qld), among other things, identifies levels of 'environmentally relevant activities' based on the risk of environmental harm from released contaminants, and specifies activities that must be licensed or receive an approval.

Environment Protection Act 1970 (Vic) establishes the Environment Protection Authority and makes provision for the Authority's powers, duties, and functions relating to managing waters, control of noise, and control of pollution.

2. Conservation of natural and cultural resources
A. Conservation of areas

Protected areas and other forms of public land reservation are established through legislation.

Examples:

Conservation and Land Management Act 1984 (WA) provides for reservation of national parks, conservation parks, nature reserves, and marine parks, as well as preparation of management plans for these areas.

National Parks and Wildlife Act 1974 (NSW) provides for the designation of several categories of protected area, including national parks, nature reserves, state conservation areas, regional parks, marine parks, Aboriginal areas, and historic sites.

B. Cultural heritage conservation

Heritage sites, buildings, and artefacts require legislation to provide for their conservation and protection.

Examples:

Historical Cultural Heritage Act 1995 (Tas) provides for formal identification of heritage places and ensures consideration is given to their conservation through Tasmania's resource management and planning system.

Heritage Act 1993 (SA) provides for the protection and conservation of places and objects of cultural heritage significance and the registration of such places and objects.

C. Wildlife and flora conservation

Protected areas in themselves are not sufficient to ensure the survival of particular species or ecosystems. Legislation is required to ensure the survival of rare and threatened plants and animals in the wild.

Examples:

National Parks and Wildlife Act 1972 (SA), as well as establishing the state's protected areas, provides protection measures for individual species.

Flora and Fauna Guarantee Act 1988 (Vic) aims to ensure that all Victoria's native species of flora, and its native ecosystems and communities can survive, flourish, and retain their potential for evolutionary development in the wild.

3. Resource allocation

Historically, this type of legislation has been concerned exclusively with the regulation of production of water, minerals, forest products, and agricultural products. Recently many environmental provisions have been included in amendments to the various state Water Acts, Forest Acts, Mining Acts, and so on.

Examples:

Water Act 1989 (Vic) establishes rights and obligations in relation to water resources and provides mechanisms for the allocation of water resources.

Forestry Act 1916 (NSW) provides for the conservation and utilisation of timber on public land.

2.6 Society, culture, and values

Many conservation programs have expressed the need to address human concerns and interests. As well, scientists and philosophers have called for conservationists and other social-change movements to consider patterns of human behaviour more realistically. Singer (1998), for example, remarked that:

> Political philosophers ... have all too often worked out their ideal society ... without knowing much about the human beings who must carry out, and live with, their plans ... those seeking to reshape society must understand the tendencies inherent in human beings.

Similarly, Yaffee (1997) suggested that many authorities have failed to identify how recurring patterns of human behaviour have defeated their environmental policies and decisions. As a result they have failed to find long-term solutions.

As individuals, our experiences are often limited to our own cultural sphere and we tend to generalise about other social groups. We often have shallow conceptions about what motivates them and what pressures colour their lives. This often leads to apparent 'factions' forming, which engender competition rather than cooperation. Examples are the ways environmentalists are often characterised as 'radical, anti-growth preservationists' and timber interests as 'uncaring, profit-hungry forest-rapists' (Yaffee 1997, p. 335)—a damaging situation which some members from both 'sides' have struggled to overcome.

Historically, Australian society has been able to overcome entrenched value conflicts such as those based on racial differences that influenced policies for the first half of our federation. We must struggle to continue this tradition of tolerance and understanding of others. Any successful conservation management activity that does not give due consideration to individuals, their social organisation, and the cultural forces that shape them, will surely fail.

Support for protected areas in Australia is generally high. Nearly 50% of Australians visit a park at least once a year. A survey undertaken in 1999 showed 46% of Queenslanders supported the existing national number of parks and a further 39% believed there should be more parks in the state. However, further education is needed about the role of protected areas in protecting a wide range of landscapes and ecosystems, and the benefits that this protection brings to the local communities and the nation as a whole (NRMMC 2004).

Cultural connections

Collective conscience? One of the most important debates in the social sciences is about the relationship between individuals and the societies of which we are members. Central to the debate is whether we as individuals shape society, its characteristics, and the way it is organised and patterned, or whether (and how) society's structure and systems shape us.

We are all individuals, each with a personal experience of life. The dominant philosophy underlying the Constitution of the USA, a country with similar cultural norms to those in Australia, is personal freedom. Yet individuals also live in societies. We tend to have broadly similar experiences shaped by the dominant social institutions. These social institutions include family, government departments, law

enforcement organisations, political parties, corporations, educational institutions, religions and so on. Our social norms and values are broadly identifiable and distinguishable from those of other societies. We may not realise the extent to which these determine our personal values and actions.

On considering our earlier ways of thinking, we will find that our perceptions were filtered unknowingly through our ideologies. What appeared logical was merely subjective. These questions are central to our achieving ESD. They define how we judge decisions about managing protected areas. How can society be changed towards achieving ESD? To what extent do we need to achieve ESD and conservation outcomes at a landscape level? Can existing social institutions drive changes? Or should we develop new ones?

True blue? What is 'Australian' culture? Is it represented by the bronzed surfer enjoying a beachscape? Or the legend of the heroic pioneer in a bush landscape? The Australian poet Banjo Paterson promoted the legend of the bush pioneer. Writer Henry Lawson questioned the image, although it also received some support in Russel Ward's popular book *The Australian Legend*. At one time *Waltzing Matilda* vied for nomination as Australia's national anthem. Most bush ballads, however, even during their heyday, described life for a certain portion of the population only: in the case of *Waltzing Matilda* it was single, usually white, non-urban men (Bulbeck 1993). Modern Australian poets have demonstrated their wider view of how contemporary Australians relate to the natural world (Snapshot 2.6).

The reality of modern Australia is that it is one of the most urbanised countries in the world. The majority of Australians live in a few large cities. Images presented in glossy magazines and newspaper supplements such as that of the urban Generation X eating vegetarian food and drinking soyaccinos in sidewalk cafes, supporting animal liberation and nature conservation, now loom larger in the national psyche. These points illustrate that while common themes may be created by dominant social institutions, a society may be divided into subgroups that operate according to different social norms. In her social history of Australia,

Snapshot 2.6

Using art to change the way we see our heritage

Mark O'Connor, environmental poet and writer

The arts, including literature, offer direct visual and sensuous experiences, rather than abstractions and generalisations. A painter might present the principle of gravitation, for example, by painting a particular apple falling in an orchard. This starting point might not be so different from Isaac Newton's; but where scientists investigate the particular in the hope of finding general principles, artists seek the perfect particular example. A photographer may try to express the essence of a national park. They will most likely show us a few selected bushes, trees and rocks, in a setting that speaks of larger principles. The arts focus on particulars because they aim to create emotions as well as understanding, and humans find it easier to have emotions about what they can see, hear, smell and touch.

Environmental literature is a crucial art form because it uses words. We use words to debate, with ourselves and with each other, what we should do. When so-called 'scrub' is disappearing under the bulldozer's blade, a writer who can find a more evocative word than 'scrub' for what is disappearing can, literally, halt the clearing. The effect may not be immediate, but it may be permanent. Much of my work is to do with finding tangible sensuous images that give us words with which to celebrate Australia's unique natural and cultural heritage. I don't preach. A poet's business is to find out what a site can and should mean to us. This means making the reader see and feel it vividly. At a waterfall, for instance I might speak of:

The endless soft violence of water
rushing its troubles against the heart,
    or elsewhere of
A shimmer of ants recycling nutrients
through a tussock
through millennia
weather permitting
    or of
Silence as if the snow had washed
the dust of noise out of the air.
A grasshopper's crackling cellophane

in the fox's mouth,
a crow's wings grating on ice-grit air,
ice-slivers breaking loose with a tonk
like the clonkle-tonk of old horse-bells.

    or (of Barramundi Charles at Nourlangie Cave)
painting his people back into the country
under the great faulted cliffs of Arnhem Land
—an old man, despised,
soon to be famous,
renewing the world with the ordinary skills,
while his body rotted here
where religion and increase
had worked for millennia.
    or

Dry hiss of rock-kissing scales
whispers that you trespass here
    or

Sturt on that first boat-trip saw
the giant cod suspended in clear water; mile by mile
each held its place like a solid fisherman's dream,
so many kilos of fighting flesh
just beyond spear's throw, a cloud in clear water,
its tethered shadow rippling underneath.
    or

Everything finished and happened once, back in the Dreamtime.
We live in eternity now.

All of these offer a sensuous image, except the last which works by pure boldness of idea. In each case my aim is to produce the feelings a site could deserve. If I can do that well enough I don't need to say we need to conserve that site. The conservation message is carried obliquely by the poetry.

Bulbeck (1993) pointed out that subgroups—the well-to-do and poor, men and women, migrants, non-Indigenous Australians and Aboriginal people, urban and rural dwellers—can all provide varying narratives of their histories.

A detailed examination of Australia's culture is beyond the scope of this book, but an overview of demographic statistics can suggest the main components, including the social and cultural patterns, of Australian society.

Demographic indicators

The Australian Bureau of Statistics and other research organisations regularly survey the population, looking at such factors as wealth, age, settlement, education, and ethnicity. The surveys act as monitors and impetus for planning and for social policies. These factors are also increasingly related to issues like environmental health and ESD. The broad features of a society can also reveal the predominant social patterns and cultural norms. Some examples are given below.

Population and settlement patterns

In June 2003, the population of Australia was estimated at 19,881,500. The growth rate over the previous year was 1.2%, or 240,500 people (ABS 2003a). The spatial distribution of the population is highly concentrated, with 80% living in eight capital cities (Beckmann 1996). Since the 1970s, internal migration has shown a flow of population towards the warmer coastal regions of New South Wales and Queensland. Where people live may have a direct effect on their perceptions of environmental priorities. In general, air pollution and waste management issues are considered more important by city dwellers, while rural dwellers are more sensitive to land degradation (ABS 1996, EPA 1997).

Ethnicity

Present-day Australian culture and its vitality is a product of the exchanges and discourse among many traditions and ethnic groupings. Patterns of use and activity for different ethnic communities may have implications for park management in relation to interpretation and facility needs, as well as use patterns. For example, large groups of people attend national parks in the Sydney region each year for cultural festivals (Snapshot 2.7).

Snapshot 2.7

Ethnic visitation to parks in the Sydney region
Alison Ramsay, NSW NPWS

Surveys conducted in 1997 by the NSW NPWS looked at the ethnicity of visitors who arrived by car at national parks near Sydney. Staff at the parks asked the question 'Do you speak a language other than English at home?' Responses at Bobbin Head and West Head in Ku-Ring-Gai Chase National Park and in Lane Cove National Park showed that 20%, 13%, and 26% answered yes, naming a great variety of languages. Chinese was the most common. These figures may be an underestimate, since those with poor English were less likely to fill in the questionnaire. Overseas tourists were not distinguished from local residents, but in one of the three surveys it was noted that among those who answered yes, they outnumbered locals. Surveys conducted in non-urban parks are likely to produce very different results.

Education

Education is one of the most important social processes by which culture is transmitted. Level of formal education has been shown to affect the extent of people's environmental concern. The environment tends to be a higher priority for professionals and people with university qualifications, while those without more often prioritise unemployment (EPA 1997). The need for more information about environmental protection measures is frequently cited as a barrier to changing behaviour (Mainieri et al. 1997). Indeed, education is believed by many Australians to be one of the most important initiatives for government to take in addressing environmental problems (EPA 1997). Others debate this point, believing that the level of environmental awareness in the population is now at a point where provision of infrastructure and other support for program implementation is more important. Environmental education and nature appreciation activities are an integral part of management duties in many protected areas and have broader conservation significance through the raising of awareness throughout the community.

Occupation

As a nation rich in natural resources, Australia's economy has been historically built upon primary industries such as mining, agriculture, and logging. Unemployment has consistently been a greater social priority than the environment in Australia (Lothian 1997), particularly in rural towns (EPA 1997) that largely depend upon declining primary industries. The threat of job losses has been an emotive component of many forest and mining conservation campaigns and the challenge to promote alternative industries increases the complexity of government policies. However, employment in this sector has decreased throughout the twentieth century and service industries now account for a far greater proportion of the workforce. For example, in 2002 tourism directly employed 550,000 people and indirectly a further 397,000 (Commonwealth of Australia 2003).

Religion

The majority of the Australian population identifies with some religion. Various forms of Christianity are popular. Other prominent religions include Buddhism, Hinduism, Islam, and Judaism. Australians overall would be considered moderately religious, however, in comparison with many other nations. This is because religious practices play a relatively small role in the day-to-day functioning of Australian society.

Strength of environmental concern

As suggested earlier, general attitudes to environment have been evolving ever since industrial Europeans first encountered the wild and harsh continent of Australia. The land at first overwhelmed them. A prominent attitude was to 'clear and conquer', which still lingers today. Since the 1970s, however, our concern about how the environment is degrading has been high. It had peaks in the early 1970s, in the late 1980s, and again in the early 1990s (Lothian 1997). Environmental issues continue to be among the five most important social issues—crime, health, education, unemployment, and environment—however the percentage of Australians concerned about the environment fell from 75% in 1992 to 62% in 2001 (ABS

2003b). Nonetheless, in NSW at least, people have more sophisticated environmental knowledge; are able to discuss environmental issues in more detail; value the environment as part of their lives; are willing to pay to protect it; and support compulsory environmental regulation (DEC 2004).

The reasons for this lessening of overall concern for environmental problems (or at least of the sense that they are urgent) is not clear, but suggestions include: the environment is now mainstream and already being addressed in many sectors of society (Pakulski et al. 1998); the marketing 'greenwash' and high level lip-service given to environmental issues has resulted in a false sense of security; or other more traditional social issues may be considered more directly relevant to quality of life. Australians today are increasingly aware of the links between the environment and the economy (EPA 1997). In NSW a 2003 survey showed that the environment was ranked sixth behind health, education, public transport, law and order, and roads and traffic, but ahead of unemployment, housing, economic growth, and terrorism/security (DEC 2004).

Not all environmental issues are considered of equal concern. In the past, water pollution and air pollution (including greenhouse and ozone effects) have consistently rated highest. Loss of forests and/or ecosystems have usually been of more concern than extinction of species. Urban sprawl and overpopulation have generally ranked fairly low (ABS 1996, ABS 1997, EPA 1997, Lothian 1997).

A more recent ranking indicates that water is the most important environmental issue, followed by air, flora and fauna protection, waste, urban management, land degradation, and pollution (DEC 2004).

A critical component of investigating the perception of environmental issues in Australia is to consider the relationship between ideas and action. This is discussed below with respect to changing social norms.

Changing social norms

Agents of change

People's attitudes and behaviours are affected by the dominant institutions of their society and the social norms associated with them. To change social norms requires proactive 'agents of change'. These agents may be existing longstanding social institutions, such as political parties and universities; or they may be newer institutions, such as national environmental groups (such as the Australian Conservation Foundation and Greenpeace) and local community action groups. These agents operate at all social levels. Individuals can also be motivators.

A few volunteers may initiate community gardens in inner-city areas, or begin replanting native bushlands. They then provide a nucleus for others. School groups, for instance, may enjoy the chance to 'connect' with nature, an opportunity not generally available to them in the city. The success of programs may depend on how well community leaders convince others of their worth (Wilson 1997, Binning 1998). At the local level, for example, Landcare leaders may have a major role, while government ministers and international figures such as David Suzuki can also be influential. Increasingly eco-friendly activities have become socially acceptable. Involving influential people and institutions has been effective. In urban areas, recycling, for example, was successfully introduced by leaders who initiated the practice on their block (Bryce et al. 1997). Agencies managing protected areas may operate in a similar way.

Individual action

Today we see environmental issues in a broader context. We see the links between the environment and the economy. We see what individuals can do and their broader effects. NSW residents were asked how much they felt they were doing in their everyday lives to help protect the environment. Only 2% felt they were doing nothing, with 60% doing a few things or at least some things, 30% doing quite a lot, and 8% doing a great deal (DEC 2004). Twenty per cent of Australian adults gave time or money towards environmental protection in 2001, compared with 28% in 1992. In contrast, membership of environmental protection groups has been relatively stable since 1998 at 4% of adults. Landcare and catchment management groups were the most popular, with 220,000 members (ABS 2003a).

The gap between ideology and implementing that ideology, which hinders governments, also affects individuals. Those who state their pro-environment attitudes are more likely to behave in accordance with them, although the relationship is not necessary a strong one (Mainieri et al. 1997). Several factors may come into play. More education may be needed to show people how to live according to ESD principles, and how easy the task is. Kerbside recycling is more successful when there are scattered collection points (Bryce et al. 1997).

Political activism takes more than changing your shopping habits. Economic factors arise, such as your ability to pay for more costly eco-items. Different factors may operate in households, depending on who has what powers. Many still believe they cannot do anything to solve environmental problems. Numerous sociological studies have studied the gap between attitudes and behaviours, and we recommend a study of them.

One example is the theory of 'reasoned action'. In the model, behaviour is determined by intention, which is itself a function of attitude and subjective norms (Ajzen & Peterson 1988). Given these relationships, behaviour is predictable from attitude only if there is a good correlation between intention and behaviour. Where the behaviour requires possession of resources, such as money, the degree of volitional control the person has over the situation can intrude between intention and behaviour. High correspondence between attitude, intention, and behaviour relies on consistency of four situational conditions: the action involved, the target of the action, the context in which it occurs, and the time of its occurrence (Ajzen & Fishbein 1977, Ajzen & Peterson 1988).

Moving forward

The roles of protected area managers are changing internationally. Once they were seen as protectors of the 'jewel in the crown' of natural resources. Now they are seen as active supporters and facilitators for ESD not just in their own reserves but also at a landscape level. Conserving protected areas in the long term is not likely to succeed if society or communities do not support them. Protected area managers are not social revolutionaries, but they can further the conservation agenda. To do so they will need to resolve conflicts (Chapter 17). They will also need to set targets for positive outcomes (Chapter 7), and they must do so while remaining sensitive to the social context—that is, to individuals, communities, governments, and industries. In Case Study 2.2, Neville Gare, Kosciuszko National Park's first superintend-

ent, shows the sort of goals that can be achieved by one motivated individual who knows how to engage the community and address local issues.

Case Study 2.2

Kosciuszko National Park: A beginning

Neville Gare, former superintendent, NSW NPWS

In 1944 a visionary Premier of the state of NSW, William McKell, responded to increasing concern for the way in which grazing of sheep and cattle in the high country was degrading the snow-fed watershed of the Southern Alps of NSW. His solution to the land use conflict was the *Kosciuszko State Park Act 1944*. It established a high level trust of eight members, chaired by the NSW Minister for Lands, to have 'care, control and management' of a park of some 526,000 hectares.

Five years after the underfinanced trust started work, the huge Snowy Mountains hydroelectric scheme hit the park like an avalanche. By 1958 the Trust was demoralised and almost overwhelmed by the Snowy scheme and its multiple impacts on the park. The trust decided to recruit a professional land manager to fill the new post of park superintendent. As a forester, I wrote to the Forestry Commission's representative on the Trust, Baldur Byles, to enquire about the job. He sent a six-page reply, which really threw down the challenge to a young red-blooded forester getting a bit frustrated with the NSW public service

Neville Gare, Superintendent, Kosciuszko State Park, 1961 (Courtesy Neville Gare)

system. When my application was successful I resigned from the Forestry Commission and the public service to commence the task. We had a staff of four park workers and a budget of $40,000.

One of my first tasks was to enforce a new government ban on grazing in the higher parts of the park, after 130 years of grazing. This was obviously not going to make me popular with at least some of the locals, and it was not easy gaining acceptance, at either a personal or institutional level. The local shire council consisted largely of graziers and some of them combined with the local member of parliament to make life difficult for me in various ways.

But slowly we made progress, by persisting and doing things. Part of the success was due to establishing personal places in the district community. I played rugby league for Jindabyne, my wife played tennis, and we were active in the Lions Club and Country Women's Association. These social contacts are of vital importance for anyone trying to establish protected area management in a potentially hostile environment.

On the work front I tried to live up to a belief originating in previous hands-on experience in the bush—country people judge you on what you do, not on what you say. On arrival, I had said to local people that one day they and/or their kids would be earning a living largely because of the great national park in the mountains. Now I had to make it happen. I persuaded the trust to build a small information centre at the Kosciuszko Road park entrance to put us on the map for all those people going past on guided tours of the Snowy scheme. We held the first campfire program alongside the Snowy River at Waste Point and gave ranger talks at the Man from Snowy River Hotel in Perisher Valley. I was also busy writing media releases about the park for the local paper and starting work on a master plan to develop a scheme to zone the park and strike a balance between its natural resource base and a range of appropriate uses and activities. We gave wide publicity to these things and employed local people to help make them happen.

I hope that the early efforts at Kosciuszko to build community recognition of the unique values of the protected area on its doorstop will pay off.

03

Concept and Purpose of Protected Areas

3.1 Values   78
3.2 International environmental protection 85
3.3 Types of protected areas   91
3.4 Extent of Australia's protected areas 105

National park—what image do these words conjure in your mind? Many Australians would probably envisage green forest and clear flowing water, mountain peaks, rippling deserts, sweeping plains, and perhaps a crashing ocean and a soaring White-bellied Sea-Eagle (*Haliaeetus leucogaster*). These images of wild sea and landscapes are deeply ingrained in our cultures—cultures that have also generated an integral part of the heritage that is protected in parks. These are places whose continued existence we desire, yet today they are hard to find outside areas that are protected from the impacts of development.

The concept of protected areas has many underlying facets. What is being protected? Why do these areas need to be kept, and why are they unable to exist without specific acts of protection? Some of the answers are biological, others are sociological, and some are personal.

Humans are a part of nature. Our activities have always impacted on the environment in various ways: hunting and gathering, promoting the growth of favoured foods and game, tending fields, domesticating animals, and building shelter. The primary issue in the modern world today is not impact per se. It is that human demands exceed the environment's sustained productive and assimilative capability. For example, at present, over 40% of the planet's photosynthetic activity has been appropriated by the human species and the assimilative capacity of our ecosystems is stretched to the limit (Young 1992). This reflects more than the sheer magnitude of the biological demands of 6 billion people. It also expresses the often wasteful methods by which we meet our needs, wants, and desires. While the magnitude of demand is certainly an issue, equally important are the ways in which we satisfy this demand. There are numerous opportunities to more effectively and efficiently meet our desires, by changing the ways we undertake our recreation, forestry, and agriculture. Moving towards sustainable resource management will also feed back into the way we see ourselves and the relationships between nature, society, and culture—our demands themselves change as our understandings and management capacities mature. For example, increasingly people are willing to pay a premium to purchase products that are certified as having been produced using sustainable methods.

Wilsons Promontory National Park, Vic (Graeme Worboys)

This chapter opens by considering the ways in which nature is valued, and the services nature provides to humans. But protected areas are much more than repositories of natural value—they also provide protection for cultural, social, and even economic values. They are an essential component of ensuring sustainability and diversity of our environment, economy, culture, and society. These values are also considered. The chapter's second section looks at various means of protecting nature at the international level. We then specifically address Australian protected areas: their designations, management, and extent.

3.1   Values

Before considering the 'how' of conservation, it is useful to consider the 'why'. The following discussion is an introduction to a range of ways of looking at nature. Some of these are based on human needs and desires (anthropocentric). Others are independent of humans (biocentric).

Through our art, history, archaeological artefacts, and traditions we discover ourselves. We connect with our forebears and ancestors. We gain pleasure from the existence of these valuable objects and artefacts. Nature is also imbued with immense cultural significance. We often extend protection to noteworthy cultural features and culturally significant locations. Maintaining local cultural traditions and lifestyles is part of the modern protected area ethos. The International Union for the Conservation of Nature (IUCN), for example, recognises the legitimacy of protecting landscapes and seascapes where people and nature have interacted to produce areas of distinct character, areas having significant aesthetic, ecological, or cultural values.

Visions of future human civilisations vary widely. They range from a return to less consumer-oriented ways of life to technological utopias. Within the scope of human existence there will always be a patchwork of landscapes and seascapes that reflect various degrees of human use and involvement. We find some land is valued highly for its agricultural productivity, other land is reserved for housing, industry, and the production of natural resources, while some places are preserved in as near to their natural state as possible. In these places minimal interference is most highly valued. Protected areas can be understood as standing at one end of a spectrum of conservation strategies.

But protected areas also exhibit a range of levels of human-nature interaction. In some, sustainable resource extraction is appropriate. Others are highly 'protected' and few people may visit them. There is a range of potential contributions that protected areas make to our lives. These include:

- allowing other life forms and natural landscapes to exist with minimal human interference
- providing opportunities for traditional cultures to continue and develop on their own paths
- providing contact with nature, and nature education
- being places to find peace, solitude, and spirituality
- providing opportunities for developing ecologically sensitive technologies, industries, and lifestyles
- providing income to local communities through alternative means such as eco-tourism
- offering recreational opportunities
- preserving genetic and chemical wild stocks
- continuing ecological processes, for example, providing clean air and water
- stimulating scientific studies.

Such lists provide an impressive summary of why we need protected areas. A more formal categorisation of protected area values is also helpful to give a comprehensive understanding of the scope and complexity of protected area values, and to avoid confusion associated with comparing value types across classificatory boundaries. However, trying to express the full range of protected area values in a simple typology is a hazardous undertaking. Many values are multi-faceted, and could be located within several categories. Some of the most treasured and important values are difficult to conceptualise and express. So with these cautions in mind, we offer the following overview of protected

Figure 3.1   Classification system for the values of nature

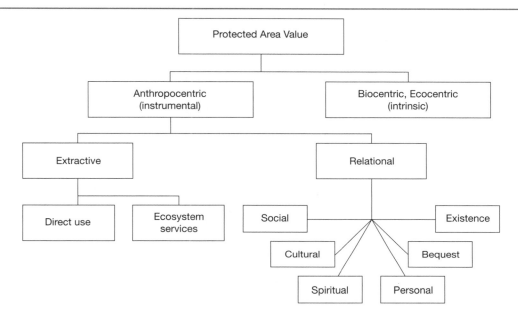

area values, drawing on sources such as Lockwood (1999) and Harmon and Putney (2003). A summary of our classification is given in Figure 3.1.

Intrinsic worth

For something to be intrinsically valuable, it must be an end in itself—valuable only for its own sake regardless of anything else. That humans benefit from nature is a powerful reason for its conservation. However, it can be argued that nature is valuable quite apart from any benefits that humans might gain from it. Natural areas, or parts thereof, can be conceived of as having intrinsic value—that is, value in their own right, regardless of humans. Many people have a strong sense that it is wrong for humans to harm other species. Some religions have placed nature in their codes of morality. Buddhists profess compassion and loving-kindness for all beings. Jainists believe every living organism is inhabited by an immaterial soul, and adhere to a doctrine of non-harm to all living beings. Traditional philosophers of ethics often treated human beings as the only species with any right to existence or to well-being; but such attitudes are

changing. Today it has become much more difficult to ignore complaints of injustice from human minorities, and, in much the same way, many people now regard it as unacceptable arrogance for humans to claim the right to wipe out other species for their own convenience. The intrinsic value of non-human species is a 'widely shared intuition' (Callicott 1986, p. 140). Many people believe in such a value. Intrinsic value in nature means that we cannot simply treat nature as a resource to satisfy human wants and needs

> *things are not merely to be* valued for me and my kind *(as resources), not even as goods of* my kind ... *but as* goods of their kind *(Rolston 1983, p. 191).*

Such beliefs may concern individual animals, and perhaps plants as well—these views are termed *biocentric*. As ecological science has developed, so too has our appreciation of the significance of interactions between plants, animals, and the physical environment—air, water, and earth. This has led some people to believe that ecosystems also have intrinsic value, apart from, or additional to, the intrinsic value of individual plants and animals that live within them. This is an *ecocentric* position.

Various philosophies, such as deep ecology and animal liberation, have placed the view of intrinsic worth on different ethical bases. An ethic is a set of principles of obligation and value. There is an explicit connection between the notion that something has intrinsic value, and the consequence that this imposes obligations in relation to it. The presence of intrinsic value implies moral considerability.

Animal liberationists consider that the sentience of higher animals means that humans have a moral obligation to avoid causing them harm. Deep ecology demands a change in the basic ideas underlying civilisation so that nature will be respected as valuable in itself and also as part of human activity. Deep ecology is a term first coined by Norwegian philosopher Arne Naess in 1972. He stated that the environmental movement could go in the direction of shallow ecology, which is the use of quick-fix solutions to pollution and resource depletion where the problems are abated and their causes are covered. An alternative deep ecology approach would look for fundamental facets of our culture that lead to environmental degradation, including the implications of an intrinsic value in nature for human behaviour and interaction with the natural world. Similarly, Aldo Leopold's notion of the land ethic extends moral consideration beyond humans into nature:

> It is inconceivable to me that an ethical relation to land can exist without love, respect, and admiration for land, and a high regard for its value. By value, I of course mean something far broader than mere economic value; I mean value in the philosophical sense … A thing is right when it tends to preserve the integrity, stability, and beauty of the biotic community. It is wrong when it tends otherwise (Leopold 1949, pp. 223, 224–5).

Such views and how they might be applied have been debated extensively by environmental philosophers. Yet in practice acceptance of an intrinsic value in nature and the obligations that such a belief imposes on humans is widely acknowledged in international fora. The Biodiversity Convention affirmed 'the intrinsic value of biodiversity'. The United Nations World Charter for Nature, signed by over 100 nations and adopted in 1982, stated:

> Every form of life is unique, warranting respect regardless of its worth to man [sic], and, to accord other organisms such recognition, man [sic] must be guided by a moral code of action.

Instrumental values

Instrumental values refer to some 'higher' purpose. There can be a chain of instrumental values that leads finally to an intrinsic value. Intrinsic values therefore provide meaning to instrumental values. For example, food is of instrumental value to human beings because it sustains the *intrinsic* value of human life. Such values are *instrumental* to human needs and wants. Arguments that support the instrumental use of nature for human purposes are known as 'anthropocentric' or 'human-centred'. Instrumental values can be divided into two types: *extractive* and *relational*. Extractive values, either directly or indirectly, involve on-site use or extraction of goods and services from protected areas. Relational values do not necessarily depend on use or consumption of protected area resources, nor do they necessarily require a person to be on-site to gain these values —they are based on emotional, psychological, community, or spiritual relationships between people and one or more protected areas.

Direct use values

Many valuable goods and services are extracted through direct engagement with protected areas. Nature-based recreation is an important part of many people's lives—swimming at the beach or in lakes, mountain biking or walking in the forest, bird watching and picnicking by a stream, photography, skiing, camping, rock climbing, and general sight seeing are just a few of the activities that parks support. Members of naturalist societies and hiking clubs, such as Myles Dunphy, were among the first people to push an agenda for conservation of Australian flora, fauna, and landscapes. Visitors to protected areas also generate considerable economic benefits to regional areas (Chapter 8).

An important part of a visit to a protected area is often an aesthetic appreciation of a red desert plain, a mountain landscape, a waterfall, an 80-metre tall

300-year-old Mountain Ash (*Eucalyptus regnans*), or a tiny Sky Lily (*Herpolirion novae-zelandiae*) in an alpine herbfield. Books, photographs, and films can be enjoyed by many people even if they do not actually go to a park. Books and photographs of Tasmania's south-west wilderness for example, had such an impact that many people who never visited the area became strong supporters for its conservation.

Many artists and writers have found in protected areas qualities that inspired imagination and creative expression. The painter Fred Williams developed his own abstract aesthetic to express his understanding of the Australian landscape (Figure 3.2, see colour section). The poems of Judith Wright often draw on Australian landscapes and wildlife for their subjects and inspiration:

> Carnarvon Creek
> and cliffs of Carnarvon,
> your tribes are silent;
> I will sing for you –
> each phrase
> the size of a stone;
> a red stone,
> a white stone,
> a grey
> and a purple;
> a parrot's cry
> from a blossoming tree,
> a scale of water
> and wavering light … (Wright 1971).

Philosophers have also found inspiration in wild natural places as sources, as well as resources:

> We pass the gate and pay the admission fee; we are inside the park's official boundaries. But politics and society soon fade, and the natural history commands the scene. And the first commandment is: Survive. Adapt. Eat or be eaten. Life or death. Our first observation is: Life goes on—protected in the parks but on its own, wild and free. … Forests and soil, sunshine and rain, rivers and sky, the everlasting hills, the rolling prairies, the cycling seasons—these are superficially just

> pleasant scenes in which to recreate. At depth, they are surrounding creation that supports life. If one insists on the word, they are resources, but now it seems inadequate to call them recreation resources. They are the sources that define life (Rolston 2003, pp. 104–5).

Protected areas are important resources for education and research. They provide a living laboratory in which we can learn more about nature, culture, and ourselves. In some types of protected area, various forms of resource extraction are permitted. Relatively low-impact uses include harvesting of seed or collecting of traditional foods by Indigenous communities. Even higher impact activities such as extraction of forest products and fishing may be compatible with some protected areas (IUCN Category VI—see Section 3.3) as long as they are conducted on an ecologically sustainable basis.

Biodiversity has become increasingly relevant to our agriculture and medicine. Stocks of native plants and animals can provide valuable genetic material and chemical compounds for developing, for example, disease resistant strains of important agricultural crop species or new pharmaceuticals.

All of the above direct use values also have temporal dimension. We can choose to consume them in the current time period, or we can delay consumption to some future time. In some cases, simply having the option to enjoy a particular use in the future can be considered a value, even if that use never actually occurs. For example, someone living in Sydney may value the option of having an opportunity to visit Kakadu National Park in the Northern Territory sometime during their life, even if they never actually do so. It is likely that some protected areas contain resource potential that we are not even aware of. Scientific discoveries in the future may indicate, for example, the medicinal value of a chemical found only in a rare plant species whose population has been conserved in a protected area. Such potential future use values provide yet another reason for establishing and maintaining a comprehensive network of protected areas.

Ecosystem services—nature's lifelines

Every living organism relies upon a network of abiotic and biotic activities known as ecological processes. These include biogeochemical and hydro-logical cycles over long periods. When forests are felled, wetlands filled in, watercourses altered, and species are over-harvested, ecological processes are damaged or destroyed. As a result, many of nature's 'free' services to humanity are lost—the filtering of air and water, the assimilating of waste, the cycling of nutrients, and the creation of building materials. Humanity's survival depends on the survival of nature. *Ecosystem services* flow from natural assets (soil, biota, water systems, and atmosphere) to support human activities and lifestyles that are generated outside natural areas, but indirectly dependent on them.

Ecological functions are used indirectly by everyone among other things to meet their basic needs for food and water. Ecosystem services indirectly support the production of many goods (food and fibre, process and manufacturing etc). For example, wetlands can help remove pollutants from water, thereby ensuring that downstream flow can be extracted for domestic uses. The agricultural industry depends heavily on many ecological processes, including soil formation and nutrient cycling. Common instances of agricultural failure after ecosystem alteration include soil salinisation and soil loss after excessive tree removal, as well as population explosions of pest species once the habitat for their predators has been removed.

Non-use values

Existence values are based upon the enjoyment people get from knowing that an area is being preserved, regardless of whether they will see it or directly use its resources. *Bequest values* are derived from the belief that natural resources should be retained for future generations to appreciate and enjoy. Both existence and bequest values are a source of satisfaction to the holder of such values. I feel enriched and fulfilled simply knowing that places such as Ningaloo Marine Park in Western Australia are protected even though I have no particular desire or intention to visit the area. Similarly, I care that the opportunity for such appreciation and satisfaction is also available to my children.

Spirit, culture, and identity

When we look at any natural object or scene, we assess more than its immediate physical form. Our connections with nature are also spiritual. There are qualities of protected areas that give rise to reverence; to awareness of the sacred. The force of spirituality, while it cannot be measured easily, has been recognised in many fora. In Australia, wilderness areas were declared, in part, to '… provide opportunities for solitude and … spiritual activities' (Cresswell & Thomas 1997, p. 35). Our cultural and spiritual attachments play a central role in conservation. Passions derived from beliefs and feelings, rather than logical reasoning, have led many to protest when development has marched over pristine landscapes.

We interpret and imbue it with cultural significance based on our personal beliefs and associations. In modern industrialised Australia, the human and the natural environments are starkly contrasted. Environmental management has tended to maintain a sharp distinction between 'natural' and 'cultural' heritage. Ongoing connections between people and land have not been well recognised. This is despite the fact that whole environments may embody evidence of past land management practices. Some forests in south-eastern Australia, for example, while they possess natural values, can also be considered as cultural landscapes arising from early timber harvesting and clearing activities. 'Cultural landscapes' is seen by some as a more appropriate term for the vast majority of Australian places (Ghimire & Pimbert 1997, ACF 1998). Cultural evidence in land now reserved as protected area can include mining and logging equipment, roads and tracks, railways, fences, sheds, yards, tree stumps, particular vegetation ages and structures, and so on.

> *Cultural heritage is the term used to refer to qualities and attributes possessed by places that have aesthetic, historic, scientific or social value for past, present and future generations. These values may be seen in a place's physical features, but can also be associated with intangible qualities such as people's associations with, or feelings for, a place (Lennon et al. 1999, p. 8).*

For many Indigenous Australians and Indigenous peoples all over the world, cultural and natural heritage are inseparable. Four of Australia's World Heritage properties, for example, are listed for both natural and cultural values: Tasmanian

Little Tern (*Sterna albifrons*) (P&W Collection, Andrew Brown)

Wilderness, Willandra Lakes, Kakadu National Park, and Uluṟu-Kata Tjuṯa National Park. In 1994, Uluṟu-Kata Tjuṯa became the second national park in the world to be listed as a cultural landscape. Uluṟu has immense cultural significance for the Aboriginal people in its region, as do similar Indigenous sites across the continent.

Non-Indigenous Australians also have strong cultural connections to nature. During the first waves of colonial settlement, the association was often negative. Today, however, many profess their strong personal attachment to the bush and the outback. Historical events since 1788 have made many sites, landscapes, and seascapes particularly significant.

Protected areas provide opportunities to keep alive or conserve cultural values that give a sense of identity, connection or meaning, related to traditional cultures and lifestyles. For many people, personal identity is in part formed by connection with places, including protected areas. To people living in or around protected areas such as Kosciuszko National Park in NSW or Seal Bay Conservation Park on Kangaroo Island in South Australia, such places often engender feelings of pride and belonging. Personal, family, and community histories are often ingrained with stories and memories in which such places are a central element. For many of the people who spent time on the Gordon River in Tasmania protesting against the proposed Gordon-below-Franklin dam, the Gordon–Franklin Wild Rivers National Park is an ongoing source of pride in their achievements.

Protected area managers are increasingly recognising their responsibilities in relation to cultural heritage conservation. Decisions often need to be made where natural, recreation, and cultural objectives are in conflict. Nature conservation values can be adversely affected, or perceived to be affected, by historic features related to past European activities, current activities that have links to traditional European practices, or activities of Indigenous peoples. Cultural heritage values can be degraded or destroyed by attempts to protect or enhance natural ecosystems. Exotic plants with cultural significance can be removed, timber or metal artefacts displaced or destroyed, past building sites rehabilitated. Cultural perspectives come into play, for example, in relation to introduced plant species in national parks. While an introduced plant is nothing but a plant growing 'out of place' the term immediately conjures up the mental image of 'pest species' coupled with a need for 'containment' and ideally 'eradication.' What is often forgotten is that these introduced plant species may well be tracer plants of prior human occupation. The Australian Alps include isolated apple, plum, and walnut trees, as well as jonquils and daffodils. It is usually parks policy to remove such species unless they are historically significant.

Conflicts can arise because of different stakeholder aspirations and behaviours. Protected area management agencies routinely engage with a range of stakeholders, including environmentalists, recreationists, traditional owners, traditional users, and those who have links with previous uses and sites. These stakeholders often make conflicting demands. Recreationists may desire access to sites that Indigenous people consider inappropriate. Environmentalists may object to traditional owners hunting or taking medicinal plants. Recreation activities can also damage sensitive cultural sites such as sand middens. Protected area management agencies have tended to be more responsive to those stakeholders who supported the establishment of protected areas in the first place. It must be noted that this emphasis is also consistent with the charter that most agencies have been given by their respective governments. Not surprisingly, some agency staff can have a primary interest in the conservation of natural heritage, so that cultural heritage concerns may be marginalised within the organisation (Lockwood & Spennemann 1999).

Protected areas can also have negative impacts on local and Indigenous cultures and lifestyles. In many parts of the world, there is a disturbing and unacceptable history of Indigenous people forcibly removed from their lands. For some communities with strong traditional associations and practices, sense of place and cultural connection can give rise to considerable hostility towards protected areas and the injustices arising from their establishment. The first national park, Yellowstone was established in 1872 in the USA. The inhabitants of the area, mainly Crow and Shoshone Indians, either left for reservations after intense pressure, or were driven out (Ghimire & Pimbert 1997). Internationally, protected area managers are now acknowledging and taking responsibility for such costs by discontinuing further alienation of Indigenous peoples from their lands, as well as beginning a process of reconciliation and reparation for those peoples who have been adversely affected. The recent World Parks Congress was a significant step forward in this regard (Section 3.2).

Graziers and timber workers, whose relationships with the land were established through several generations living and working in a particular place, often challenge the value of establishing protected areas in their localities, in part because of the reduced access to resources that form part of their livelihood, but also because of their cultural relationship with preferred land-use practices and activities. Removal or reduction of high country grazing in protected areas in NSW, Victoria, and Tasmania, for example, has engendered sustained hostility from some local communities towards protected areas. On the other hand, many park users and conservationists have welcomed such decisions.

Such difference of views and conflicts need to be acknowledged, and more appropriately addressed by managers and decision-makers. Depending on their category, protected areas can exclude some or all extractive uses. This exclusion can disadvantage local and regional communities, as well as having wider economic impacts. In some areas, such impacts may not be fully offset by gains associated with tourism activities or ecosystem services. While in many cases removal or reduction in traditional European uses are justified, protected area managers need to attempt reconciliation with such communities, as well as more fully acknowledging their responsibilities in assisting regional development and re-establishment of vibrant, healthy communities.

The personal and the social

Protected areas contribute to personal development by providing opportunities for transformative experiences through connection with wild nature. Involvement in activities such as bushwalking can lead to development of leadership skills, building of self-reliance and confidence, and acceptance of responsibility. Protected areas can contribute to bringing people together to work cooperatively. Recreation and management involvement can reduce alienation; as well as facilitate bonding, reciprocity, understanding, and sharing between people. Physical activity in natural settings gives rise to therapeutic values by creating a potential for healing and enhancing psychological well-being (sense of wellness, stress management, prevention and reduction of depression, anxiety, and anger) and physical well-being (cardiovascular, weight, strength, increased life expectancy) (Shultis 2003).

Values and change

In this section we have attempted to indicate the wide and deep extent of protected area values, as well as acknowledge some of their costs. Protected area managers have a responsibility to current generations to ensure, as far as is within their power, that protected areas continue to provide these benefits to humans, as well as securing the protection of that natural world for its own sake. It is also important to recognise that values are not static. Our understanding and appreciation of natural and cultural values is significantly different from that of 50 years ago; of 100 years ago. No doubt future generations will have different values to our own. The present generation has a responsibility to pass on the natural world and cultural heritage in such a condition that future generations have the opportunity to enjoy the many benefits that protected areas offer, however those values are conceived of in the future. Protected area managers are on solid ground when they uphold public interests in natural and cultural heritage conservation against short-sighted or narrow sectional interests.

3.2 International environmental protection

International protected area organisations and programs play a central role in developing best-practice standards and strategies for conserving nature. International efforts have also increased national governments' awareness of the need for protected areas. The extent of the world's protected areas is summarised in Figure 3.3 (in the colour section). Sharing successes and difficulties experienced by various national programs through established communication networks has been beneficial. International congresses, publications, and reports are major fora for discussion and dissemination of information. The significant outcome of such discourse on protected areas has been several major international institutions, conventions and, agreements that have assisted in developing worldwide conservation efforts.

Institutions

The last century has been an era of globalisation. In 1788 the sea passage from England to Australia took a gruelling 11 months. Now email messages travel almost instantaneously around the world. By the end of two world wars, global interactions were shaping national politics. In 1945 the United Nations (UN) was established to deliver peace and security along with global social and economic development. The first partnerships between nations to protect species, and the first international conservation organisations, appeared around this time. The UN has initiated international policies, strategies, conventions, and programs for conserving and managing environments.

Since the UN holds a key position in international politics, it has taken an important role in promoting environmental policies. Several UN agencies are relevant to protected areas.

1   The United Nations Educational, Scientific and Cultural Organisation (UNESCO) was formed in 1945. Its main objective is to bring 'peace and security in the world by promoting collaboration among nations through education, science, culture and communication.' UNESCO negotiated the World Heritage Convention in 1972 and administers the Man and the Biosphere program. The International Council on Monuments and Sites (ICOMOS) is an international organisation linked to UNESCO. It brings together people concerned with the conservation and study of places of cultural significance. ICOMOS has established national committees in some 60 countries, including Australia.

2   The United Nations Environment Program (UNEP), formed in 1972, has its headquarters in Nairobi. It aims to develop international strategies and to fund programs that conserve the world's environment.

3   The United Nations Development Program (UNDP) was established in 1965. Its objectives include 'focusing its own resources on a series of objectives central to sustainable human development: poverty eradication, environmental regeneration, job creation, and advancement of women.'

4   The Food and Agriculture Organisation (FAO) was established in 1945 to promote improvements in agricultural techniques, forestry, fisheries, and human nutrition in developing countries.

5   The Global Environment Facility (GEF), established in 1991, serves as a mechanism for international cooperation. It provides grants and concessional funds to meet agreed global environmental benefits. The focal areas that the GEF supports are: biological diversity, climate change, international waters, and ozone layer depletion. Supported by UNDP, UNEP, and the World Bank, the GEF was the original funding mechanism for the Convention on Biological Diversity.

6   The World Bank administers funds that developed nations provide to support developing nations, economically and socially, and to improve their standard of living.

As well as setting up issue-focused programs and working groups, the UN has sponsored a number of key conferences and policies. The first was the United Nations Conference on the Human Environment held at Stockholm in 1972.

It had a number of significant outcomes. It formally recognised the relationship between humans and the environment. The conference established UNEP, and, negotiated by UNESCO, the World Heritage Convention (Backgrounder 3.1). The World Heritage Convention was one of the first international conservation conventions. Its reference to the 'common heritage of mankind' was established in international law. The concept of sustainable development was in its infancy at the Stockholm conference. Sustainable development was further articulated by UNEP, which instigated the World Commission on Environment and Development (WCED). Headed by the Prime Minister of Norway, Gro Brundtland, the WCED operated from 1983 to 1987, when it produced the sustainable development manifesto *Our Common Future* (Chapter 2).

Backgrounder 3.1 Convention for the Protection of the World's Cultural and Natural Heritage 1972 (World Heritage Convention)

Modified from Thorsell (1997)

Australia (the Commonwealth) is a state party to the UNESCO 1972 World Heritage Convention. When Australia signed the convention in August 1974, like all other signatories, it pledged to conserve the sites and monuments within its borders. These are sites recognised to be of outstanding universal value. The Convention took effect in 1975 in the form of the World Heritage List. In 1995, the World Heritage Convention, with 142 state parties, was the most universally supported of all conservation conventions (McDowell 1995). In 2003, the World Heritage List included 754 properties (582 cultural, 149 natural, and 23 mixed properties) located in 129 countries (Figure 3.4 in colour section).

The Convention established many international conservation standards. Properties listed on the World Heritage List, for example, are considered to be of such exceptional interest and universal value that all humanity assumes responsibility for protecting them. The Convention assumes the existence, therefore, of a world heritage, which belongs to all humans (UNESCO 1989). For a site to merit being inscribed on the World Heritage List, it must have what the Convention calls 'outstanding universal value'. For a natural site to have 'outstanding universal value' it must satisfy one or more of four criteria.

- A site is a unique landform or presents an outstanding example of major stages of the Earth's evolution. An example is the extensive calcite deposits in Huanglong, China. Another example is the Giant's Causeway, Northern Ireland. The site is renowned for its contribution to the science of geology.
- A site presents an outstanding example of significant ecological or evolutionary processes. The processes are also ongoing. An example is Ecuador's Galapagos Islands, where Darwin found living evidence of evolution. Another example is Fraser Island, Australia, which demonstrates exceptionally well the way sand is deposited on coastlines.
- A site contains superlative natural phenomena, formations, or features of outstanding natural beauty. An example is the Grand Canyon National Park in the USA. Its spectacular landscape contains geological exposures that present over two billion years of the Earth's history. A second example is the Iguazu Falls on the border of Argentina and Brazil.
- A site contains exceptional biological diversity or habitats where threatened species of outstanding universal value still survive. An example is Banc d'Arguin National Park in Mauritania, with its huge numbers of Palaearctic birds. Another example is Manu National Park in Peru. Manu is the most biodiverse of all Amazonian basin parks, with 10% of the world's species of birds.

To be listed on the World Heritage List, a natural area must do more than meet one or more of the above criteria. It must also fulfil what the World Heritage Committee's operational guidelines call the 'conditions of integrity'. These specify the longer term requirements that a site must meet. A site must be ecologically viable and protectable. The World Heritage List is not intended to include sites of national importance. The site must be universally or globally significant. The criteria used to help determine such significance include: distinctiveness, integrity, naturalness, dependency, and diversity. Cultural sites have a different set of criteria.

Central Eastern Rainforest Reserves World Heritage Area, Dorrigo National Park, NSW (Graeme Worboys)

Sites on the World Heritage List may be included on a second list—a 'World Heritage in Danger' list—if the values they were selected for are threatened. The committee may be alerted to the presence of dangers through many channels. It may be individuals, non-governmental organisations, or other groups who inform them. Article 11(4) of the Convention outlines the nature of threatening processes warranting concern as: 'serious and specific dangers, such as the threat of disappearance caused by accelerated deterioration, large-scale public or private projects or rapid urban or tourist development projects; destruction caused by changes in the use or ownership of the land; major alterations due to unknown causes; abandonment for any reason whatsoever; the outbreak or the threat of an armed conflict; calamities and cataclysms; serious fires, earthquakes, landslides; volcanic eruptions; changes in water level, floods and tidal waves' (UNESCO 1999). As of 2003, 35 properties were listed as 'World Heritage in Danger'. The 'in danger' list includes such places as the Everglades National Park in the United States, the Rio Platano Biosphere Reserve in Honduras, and the cultural landscape and archaeological remains of the Bamiyan Valley in Afghanistan.

The International Union for the Conservation of Nature (IUCN) (the World Conservation Union) was established in 1947. As a non-government organisation held in high esteem, it enjoys strong support from most national governments. The World Wildlife Fund (WWF) was established in 1961 to raise private sector funds for IUCN and to promote conservation. It now operates independently. In 1986 the organisation changed its name to the World Wide Fund for Nature, but kept the acronym WWF. The role of these and other international organisations has been to push for conservation to be included on international political agendas, improve coordination between nations, and assist in putting strategies into action. Primary mechanisms include:

- initiating international conferences and meetings that bring together experts from around the world
- placing the issue of conservation in political and legal arenas by formulating and administering conventions, agreements, and treaties
- establishing commissions and working groups for collating information and research

- monitoring the state of the global environment and disseminating the data
- assisting national programs directly and indirectly
- attracting funding for all these functions.

In terms of its influence and action, the leading international protected area organisation is the IUCN's World Commission on Protected Areas (WCPA), previously known as the Commission for National Parks and Protected Areas. Many individuals from protected area management agencies and non-governmental organisations involved with protected areas are members of WCPA. They commit their time and energies voluntarily to the global conservation effort. The four goals of the WCPA are:

- to strengthen the capacity and effectiveness of protected area managers through provision of management guidelines, tools, and information
- to integrate protected areas with sustainable development and biodiversity conservation by provision of strategic advice to policy-makers
- to increase investment in protected areas by persuading public and corporate sources of their value

- to strengthen the capacity of the WCPA to implement its program through collaboration with IUCN and partners.

As of September 2003, the WCPA membership comprised 1300 individuals from over 140 countries. WCPA membership is by invitation, and includes managers of protected areas; experts in relation to the fields of WCPA's interests; academic specialists in areas relating to protected areas, resource economics, biogeography wildlife management, marine conservation, and other related fields; officials from relevant non-government organisations involved with protected areas; and members from key partner organisations. Australia is a part of the Australian and New Zealand Region of WCPA, and has its own vice-chair. Many IUCN publications are essential tools for managers and we recommend use of them to supplement the material in this book. This book was inspired by the IUCN regional action plan and indeed the development of this book was an Australian IUCN regional project.

The IUCN also hosts world congresses on national parks and protected areas once every decade. To date there have been five conferences (Appendix 1). Over the 50-year span of the congresses, the feedback and discourse have helped to evolve the strategies and philosophies of protected area management (Whitehouse 1992). The first was held in Seattle, USA, in 1962. The second took place at Yellowstone National Park, USA, in 1972. The third was at Bali, Indonesia, in 1982, and the most recent conference was at Caracas, Venezuela, in 1992, shortly before the Rio Earth Summit. The fifth congress was held in South Africa in 2003. Some of the major outcomes from the South Africa congress are summarised in Backgrounder 3.2.

The World Conservation Monitoring Centre (WCMC) is a non-profit organisation established by IUCN, WWF, and UNEP. It is now part of UNEP. It provides information services on the conservation and sustainable use of the world's living resources. It helps others to develop information systems of their own. The protected areas unit of WCMC publishes regular international inventories of protected areas.

Backgrounder 3.2   Fifth IUCN World Parks Congress 2003

Adapted from IISD (2003)

The Fifth IUCN World Congress on Protected Areas, or World Parks Congress, was held in Durban, South Africa, from 8-17 September 2003. Some 3000 participants attended, representing governments and public agencies, international organisations, the private sector, academic and research institutions, non-governmental organisations, and community and Indigenous organisations. The theme of the 2003 WPC was 'Benefits beyond Boundaries'.

The congress included seven workshop streams. *Linkages in the landscape and seascape* examined ecological and socio-cultural linkages at different scales, and investigated the application of the ecosystem approach to protected area management and governance. *Developing the capacity to manage* focused on the skills, attributes, and support systems needed for protected area institutions, decision-makers, and practitioners. *Building broader support* addressed building cultural support for protected areas; working with neighbours and local communities; and building support from new constituencies. *Evaluating management effectiveness* examined the status of tools for evaluating management effectiveness, including principles, methods, and current issues. *Governance* reviewed different protected area governance models, discussed key governance issues, evaluated good governance, and provided guidance for decision-makers. *Building a secure financial future* addressed a range of financial arrangements and options for generating revenue, with emphasis on the development of a business approach to protected area management and applications of sustainable protected area financing. *Building comprehensive protected area systems* assessed the status of global protected area coverage with a focus on poorly represented biomes; identified gaps in protected area systems and ways to address them; and addressed global change factors and best practice for protected area design.

There were also three 'cross-cutting' workshop themes. *Marine protected areas* (MPAs) emphasised improving MPA management effectiveness; building resilient MPA networks; integrating MPAs in marine and coastal governance;

and expanding MPAs in the high seas and exclusive economic zones. *World heritage* identified ways to capitalise on these areas of outstanding value to build awareness and support, and assessed their characteristics, needs, and potential. *Communities and equity* focused on Indigenous and local communities' rights and responsibilities in protected area management.

The main congress outputs were the Durban Accord and Action Plan, consisting of a high-level vision statement for protected areas, and an outline of implementation mechanisms; 32 recommendations approved by the workshop streams; and the Message to the Convention on Biological Diversity.

The *Durban Accord* is the principal message from the Congress to the world. The development of the Accord began in Albany, Western Australia, in 1997, and continued at other international and regional conservation events. The Accord proposes a new paradigm for protected areas (Backgrounder 2.2) that integrates conservation goals with sustainable development in an equitable way. The Accord highlights a number of concerns, including: inadequate protected area coverage, particularly for marine and freshwater ecosystems; a lack of recognition of the conservation efforts of local communities and mobile and Indigenous peoples; a decline in wild areas outside protected areas; and threats from human-induced climate change. The Accord notes an annual funding gap of US$25 billion, and urges a commitment to: promote the multiple roles of protected areas; achieve adequate representation; promote stakeholder participation in decision-making; and involve local communities, Indigenous and mobile peoples in protected area establishment and management.

The *Durban Action Plan* provides a checklist of the activities needed to increase protected areas' benefits to society and to improve their coverage and management. Key outcomes include: the implementation of protected areas' fundamental role in sustainable development; a global system of protected areas linked to landscapes and seascapes; improved quality, effectiveness, and reporting of protected area management; recognition of the rights of local communities, and Indigenous and mobile peoples; the empowerment of younger generations; increased support for protected areas from other constituencies; improved forms of governance, recognising both traditional and innovative approaches; and increased resources for protected areas. Under each outcome, key targets and specific actions are identified.

Recommendations cover topics such as strengthening institutional and societal capacities; building comprehensive and effective protected area systems; climate change; financial security; transboundary conservation initiatives; tourism, cultural and spiritual values, cities and protected areas; peace, conflict and protected areas; governance; management effectiveness evaluation; IUCN protected area management categories; building a global system of marine and coastal protected areas networks; Indigenous peoples and protected areas; Africa's protected areas; and communication, education, and public awareness. Brief summaries of two example recommendations follow.

Integrated landscape management to support protected areas. With Recommendation 5.09, WPC participants recommend: adopting protected area design principles that emphasise linkages to surrounding ecosystems; restoring ecological processes in degraded areas within protected areas and in their surrounding landscapes; reflecting the presence and needs of human populations in overall protected area design and management; recognising participatory processes; applying principles of adaptive management; and adopting a policy framework to encourage local communities' active involvement in biodiversity stewardship. They also call on international organisations to: build relationships between biodiversity conservation, protected area management, and sustainable development; regenerate cultural landscapes and revitalise rural communities; and promote integrated earthscape management in relevant international agreements.

Poverty and protected areas. With Recommendation 5.29, WPC participants note that protected areas should contribute to poverty reduction, and call for: integrating protected areas into broader sustainable development planning agendas; conserving biodiversity both for its value as a local livelihood resource and as a public good; equitable benefit sharing; fully compensating affected communities; and incorporating a gender perspective in protected area governance. Participants also recommend developing inclusive government for protected area management, based on: building partnerships with poor communities and empowering them to participate in decision-making; developing pro-poor mechanisms to reward environmental stewardship; respecting customary ownership and access rights; and improving accountability and transparency in decision-making.

Numerous other international non-governmental organisations play a significant role in the international conservation movement. They organise projects, disseminate information, and put plans into action at a 'grassroots' level. They also make submissions to international commissions, conferences, and assemblies. Greenpeace, for example, has been vocal on a wide range of issues, from the dumping of toxic waste to the hunting of whales. The Nature Conservancy, based in the USA, together with US Aid, pursues conservation projects around the world.

Conventions

International conventions form a framework for interactions between nations. They document the mutually agreed obligations to work towards common goals. Conventions are formulated through submissions from various interested parties. The final stages of their ratification are often controversial when underlying political issues such as equity are raised. Conventions are generally finalised at internationally convened meetings that are attended by representatives from interested nations and organisations. They are then put forward to be ratified by nations. They come into force as international law when a specified number of nations have ratified them.

Recognising that documents in themselves have limited value, conventions also establish administering bodies. These have a similar role to national bureaucracies. The Convention on Biological Diversity, for example, established a Conference of the Parties to make formal decisions; a secretariat for administrative functions; a Subsidiary Body on Scientific, Technical and Technological Advice to advise the Conference of the Parties; and a financial institution to distribute the funds. At the national and sub-national levels, governments are free to choose whether they enact legislation that pursues the principles and strategies of such conventions.

Australia is a party to several international conventions that encourage reservation of protected areas or directly affect their management, including:

- International Convention for the Regulation of Whaling
- Convention on International Trade in Endangered Species of Wild Flora and Fauna (CITES)
- Agreement between the Government of Japan and the Government of Australia for the Protection of Migratory Birds and their Environment (JAMBA)
- Agreement between the Government of the People's Republic of China and the Government of Australia for the Protection of Migratory Birds and their Environment (CAMBA)
- Convention on the Conservation of Migratory Species of Wild Animals (Bonn Convention)
- World Heritage Convention
- United Nations Convention on the Law of The Sea (UNCLOS).

The Central Eastern Rainforest Reserves in NSW, for example, were successfully listed as World Heritage in 1986, with extensions into Queensland approved in 1994:

Only sites which had secure conservation status would be in the nomination … sites were chosen so as to encompass the range of variation in NSW rainforests and include outstanding examples of rainforest types … Collectively these sites proposed met all four criteria for inclusion … The very distribution of rainforest

Figure 3.5   The World Heritage emblem
This emblem symbolises the interdependence of cultural and natural properties. The central square is a form created by humans, while the circle represents nature—the two are intimately linked. The emblem is round like the world, and is also a symbol of protection (UNESCO 1989).

patches is a reflection of geological history, the once extensive cover having been reduced to isolated refugia during progressive drying of much of the continent in the late Tertiary. The survival of the rainforest was linked to two outstanding geomorphological/geological phenomena—the great escarpment [Chapter 1] and the series of Tertiary volcanic events along the eastern margin of the continent. The nomination was probably the first occasion in which the biogeographic significance of the great escarpment was realised (Adams 1999).

3.3 Types of protected areas

The internationally accepted definition of a protected area is:

An area of land and/or sea especially dedicated to the protection and maintenance of biological diversity, and of natural and associated cultural resources, and managed through legal or other effective means (IUCN 1994, p. 7).

Two main points are highlighted in this definition. The first relates to what is being protected—biological diversity and natural and cultural resources. The second is that management is integral to the concept of protected areas.

Management goals are in essence what distinguish protected areas from other types of land/sea use; in particular, the goal of 'protecting biological diversity and natural and associated cultural resources'. Furthermore, protected areas are often islands in a sea of landscape dominated by environmentally disruptive human activities, the effects of which must be actively mitigated. Activities such as control of weeds and feral animals are part of the day-to-day process of protected area management. Likewise, the variety of conservation strategies described in the previous sections are part of the scope of a protected area manager's job. There is an increasing tendency for their activities to include off-reserve programs for achieving on-reserve goals.

To manage protected areas effectively requires organisations that operate under a defined set of policies and/or legal powers, and which employ appropriately skilled people. Various protected area management organisations exist for this purpose. International coordinating bodies also exist to promote conventions and other means of establishing protected areas. They develop and disseminate effective management standards, strategies, and skills.

An international system of protected areas

Management principles for modern protected area systems around the world have been significantly influenced by the 'Yellowstone National Park Model' initiated in 1872 and more recently the international conventions and agreements described above. National systems have also developed uniquely according to their individual socio-political histories. There are differences in the type and amount of human activities permitted in reserves. There are also differences in the mechanisms of protection.

Existing national conservation strategies vary widely in their effectiveness, and reserve nomenclature is not necessarily indicative of the protected status of an area. Examples are national parks in developing countries. On paper the parks may be afforded top level protection. In reality they may be threatened by a range of degrading processes, often because there is a lack of funding for proper management. This problem is being addressed through the establishment of an international system of protected areas. Cohesiveness is provided by the support and monitoring activities of international organisations. To improve monitoring capabilities and clarify the definition of a 'protected area', IUCN has produced guidelines for protected area classification. Six categories of protected areas (Backgrounder 3.3) are defined in the classification. A central principle of the guidelines for the selection and management of each category is that categories should be defined by the objectives of management, rather than the title of an area or the effectiveness of management in meeting those objectives (IUCN 1994). For each category the IUCN guidelines indicate the following elements.

A *definition* that outlines the broad biophysical and cultural characteristics and the overall management objectives for each category as distinct from the other five.

Management objectives that give more detail on specific management issues, such as Indigenous and local use, resource use, public access, and recreation.

Guidance for selection that specifies the parameters that should be considered when designating a protected area to a category, such as size and naturalness.

Organisational responsibility that outlines which types of organisations could have the necessary resources, skills, and power to manage each category. Government agencies are the organisations best placed to manage protected areas by 'legal or other effective means'. Other organisations include: councils of Indigenous people, private foundations, universities or institutions with established research or conservation functions, non-profit trusts, and non-governmental organisations.

All Australian jurisdictions use IUCN categories to classify their terrestrial and marine protected area systems. Although each of the states, territories, and the Commonwealth continues to name protected areas according to their own individual protected area legislation, their IUCN designations facilitate uniform national reporting and inter-jurisdictional comparisons. Bruce Leaver, as the Executive Commissioner of the Tasmanian Resource Planning and Development Commission utilised the IUCN categories to reduce 138 land-use categories in Tasmania down to 13. He also noted that:

> *In the Commission's view, one of the most important features of the IUCN system is its non-hierarchical nature—it does not represent a gradation of conservation importance from categories I to VI ... The difference between the categories lies in the application of their management objectives—these provide for a gradation of human intervention. The recommended Tasmanian system embodies the principle of non-hierarchical categories (Leaver 1999).*

Of course, before an area is considered for classification it should satisfy the IUCN definition of a protected area. The definition contains two key points. First, the area must be dedicated to protecting

David Sheppard and IUCN headquarters, Gland, Switzerland (Graeme Worboys)

and maintaining biological diversity, and its natural and cultural resources. Second, the area must be managed effectively through legal or other means (ANCA 1996). This is why catchments protected for the supply of potable water will not necessarily be classified as protected areas. The same argument applies to state forests managed principally for commercial timber production.

Backgrounder 3.3   IUCN protected area categories

Category Ia: Strict nature reserve

Purpose: protected area managed mainly for science

Definition: area of land and/or sea possessing some outstanding or representative ecosystems, geological or physiological features and/or species, available primarily for scientific research and/or environmental monitoring.

Category Ib: Wilderness area

Purpose: protected area managed mainly for wilderness protection

Definition: large area of unmodified or slightly modified land and/or sea, retaining its natural character and influence, without permanent or significant habitation, which is protected and managed so as to preserve its natural condition.

Category II: National Park

Purpose: protected area managed mainly for ecosystem protection and recreation

Definition: natural area of land and/or sea, designated to (a) protect the ecological integrity of one or more ecosystems for this and future generations, (b) exclude exploitation or occupation inimical to the purposes of designation of the area, and (c) provide a foundation for spiritual, scientific, educational, recreational and visitor opportunities, all of which must be environmentally and culturally compatible.

Category III: Natural monument

Purpose: protected area managed mainly for conservation of specific natural features

Definition: area containing one or more specific natural/cultural feature that is of outstanding or unique value because of its inherent rarity, representative or aesthetic qualities, or cultural significance.

Category IV: Habitat/species management area

Purpose: protected area managed mainly for conservation through management intervention

Definition: area of land and/or sea subject to active intervention for management purposes so as to ensure the maintenance of habitats and/or to meet the requirements of specific species.

Category V: Protected landscape/seascape

Purpose: protected area managed mainly for landscape/seascape conservation and recreation

Definition: an area of land, with coast and sea as appropriate, where the interaction of people and nature over time has produced an area of distinct character with significant aesthetic, ecological, and/or cultural values, and often with high biological diversity. Safeguarding the integrity of this traditional interaction is vital to the protection, maintenance, and evolution of such an area.

Category VI: Managed resource protected areas

Purpose: protected areas managed mainly for the sustainable use of natural ecosystems

Definition: area containing predominantly unmodified natural systems, managed to ensure long-term protection and maintenance of biological diversity, while providing at the same time a sustainable flow of natural products and services to meet community needs.

International designations

Australia manages World Heritage Areas, biosphere reserves, and Ramsar wetlands in fulfilment of its international obligations. Our international responsibilities entail effective protection. The level of protection and the style of management vary between each of Australia's states, territories, and the Commonwealth. Areas with a supplementary international designation are not management categories in their own right. IUCN suggests such areas should be recorded and classified under the appropriate IUCN management category according to its management objectives with the extra international designation added as an additional description (IUCN 1994).

World Heritage is an international category under the World Heritage Convention adopted by the general assembly of UNESCO. Australia is signatory to the World Heritage Convention (Backgrounder 3.1, Figure 3.5). The following 15 places in Australia have been registered on the World Heritage List under the convention:

- Fossil mammal sites at Riversleigh (Queensland) and Naracoorte (SA)
- Central eastern rainforest reserves (Queensland and NSW)
- Blue Mountains (NSW)
- Fraser Island (Queensland)
- Great Barrier Reef (Queensland)
- Kakadu National Park (NT)
- Lord Howe Island Group (NSW)
- Shark Bay (WA)
- Tasmanian Wilderness (Tasmania)
- Uluṟu–Kata Tjuṯa National Park (NT)
- Wet Tropics (Queensland)
- Willandra Lakes (NSW)
- Purnululu National Park (WA)
- Heard and Macdonald Islands (Commonwealth)
- Macquarie Island (Tasmania).

World Heritage Areas are designated over a range of land tenures. This does not change the ownership or application of state and local laws. It does limit the scope of activities permitted in those areas. These must not threaten the universal, natural, and cultural values of the area (Cresswell & Thomas 1997). In the cases of Uluṟu–Kata Tjuṯa and Kakadu, where joint management plans have been negotiated between the traditional owners and the Commonwealth, activities within the scope of those plans are not restricted by world heritage designation.

Ramsar wetlands are established under the Ramsar Convention, which was initiated at the small Iranian town of Ramsar in 1971. Australia was the first country to become a party to this Convention on 8 May 1974. In 2004 Australia had 64 Ramsar wetlands covering approximately 7.3 million hectares. The Convention provides a framework for international cooperation in the conservation and use of wetlands. The four main commitments for contracting parties are:

- nominating suitable sites as Wetlands of International Importance and thereby ensuring that they are managed so as to maintain their ecological character
- formulating and implementing national land-use planning to include wetland conservation considerations, and as possible, to promote the wise use of all wetlands within their territory
- developing national systems of wetlands reserves, facilitating the exchange of data and publications, and to promote training in wetlands research and management
- cooperating with other nations in promoting the wise use of wetlands, where wetlands and their resources, such as migratory birds, are shared.

World Biosphere Reserves are terrestrial or coastal areas that are internationally recognised within UNESCO's Man and the Biosphere (MAB) Program for promoting and demonstrating a balanced relationship between people and nature. The biosphere reserve network was launched in 1976, and by 2003 had grown to include 440 reserves in 97 countries. Individual countries propose sites within their territories that meet a given set of criteria for this designation. For Australia, the MAB Committee of the Australian National Commission for UNESCO is responsible for evaluating and sending nominations to the International Coordinating Council of MAB in Paris. Biosphere reserves serve to combine three functions:

- a conservation function, to preserve genetic resources, species, ecosystems, and landscapes
- a development function, to foster sustainable economic and human development
- a logistic support function, to support demonstration projects, environmental education and training, and research and monitoring related to local, national, and global issues of conservation and sustainable development.

Physically, each biosphere reserve should contain three elements: one or more core areas devoted to long-term conservation of nature; a clearly identified buffer zone in which activities compatible with the conservation objectives may occur; and an outer transition area devoted to the promotion and practice of sustainable development. They may contain a variety of agricultural activities, settlements, or other activities. The strength of biosphere reserves is the emphasis of their management objectives on the integration of human and natural systems.

Australia's 12 biosphere reserves are the responsibility of the Department of the Environment and Heritage. The Director of National Parks is also the lessee of Calperum and Taylorville Stations, which are part of the Bookmark Biosphere Reserve in South Australia (Snapshot 3.1)

Snapshot 3.1

Bookmark Biosphere Reserve
DEH (2004f)

Bookmark Biosphere Reserve is located in the Riverland area of South Australia. 'Bookmark' is derived from the Aboriginal word Pukumako meaning flint stone axe or sandstone grit hole. The reserve includes the Calperum and Taylorville Stations, which are leased to the Director of National Parks. Calperum Station has significant wetlands, and was purchased by the Australian Government in 1993 in partnership with the Chicago Zoological Society. Taylorville Station was purchased by the Australian Government in partnership with the Australian Landscape Trust in 2000, because of its large areas of intact mallee habitat and the presence of the Black-eared miner (*Manorina melanotis*), a nationally endangered bird species. The Mallee scrubland of Australia is one of the most endangered vegetation types in the world, with approximately 80% cleared for agriculture in the past 150 years.

Table 3.1	Australian protected area management departments and agencies	
Jurisdictions	Departments	Agencies
Commonwealth	Department of the Environment and Heritage	Parks Australia, Wet Tropics Management Authority (Snapshot 3.3), Great Barrier Reef Marine Park Authority
Australian Capital Territory	Environment ACT	Australian Capital Territory Parks and Conservation Service
New South Wales	Department of Environment and Conservation	New South Wales Parks and Wildlife
Northern Territory	Department of Lands, Planning and Environment	Parks and Wildlife Commission of the Northern Territory
Queensland	Queensland Environmental Protection Agency	Queensland Parks and Wildlife Service
South Australia	Department for Water, Land and Biodiversity Conservation	National Parks and Wildlife South Australia
Tasmania	Department of Tourism, Parks, Heritage and the Arts	Tasmania Parks and Wildlife Service
Victoria	Department of Sustainability and Environment	Parks Victoria
Western Australia	Department of Conservation and Land Management	

Protected areas within the nine Australian jurisdictions

International protected area organisations provide valuable guidance for managing protected areas in Australia. However, Australia's system of protected areas also reflects its particular historical, social, legal, and government context (Chapter 2). Australia comprises the land and inland waters of six states, two territories, the external territories (Snapshot 3.2), Commonwealth land, and the marine waters of the exclusive economic zone. Nine different systems (six states, two territories, and the Australian Government) are used to designate and manage protected areas. Although they have common traits, the Commonwealth and each state and territory have evolved different legislation, policy, institutions, nomenclature, and strategies to manage their protected area system. Each government has the potential to make decisions relating to land and water protection in its own jurisdiction. The major Commonwealth, state, and territory protected area management departments and agencies are listed in Table 3.1 (see p. 95). Typically agencies in each jurisdiction have undergone several changes of name and/or departmental location. A brief history of such changes in Tasmania is given as part of Case Study 3.1. The following section is intended to introduce the designation, management, and extent of protected areas in Australia.

Snapshot 3.2

Norfolk Island National Park and Botanic Garden
Lee Thomas, formerly of Environment Australia

Norfolk Island (3455 hectares) in the South Pacific is an external Australian Territory. It is approximately 1700 km from Sydney. Within the territory are the small uninhabited Nepean Island, Phillip Island, and numerous rocky islets dotted about the coastline. Norfolk Island National Park (650 hectares) was first declared in 1986. It now comprises the Mt Pitt section (460 hectares) and neighbouring Phillip Island (190 hectares added in 1996). The Norfolk Island Botanic Garden (5.4 hectares) was declared also in 1986, and has been extended. These areas, proclaimed under both Commonwealth and Norfolk Island Government legislation, are managed by Environment Australia.

Prior to European settlement, Norfolk Island was largely covered by dense subtropical rainforest. Today the national park, which covers just 12% of the island's total area, includes nearly all of the remaining natural vegetation. Weed infestation is significant. Only about 1% of the island's pristine natural vegetation remains. Of the 178 species of plants native to Norfolk, at least 42 are believed in danger of extinction. The outlook for endemic species is also of concern. Of the 14 species and subspecies of native birds, only seven species remain. Of these, the Red-fronted Parrot (*Cyanoramphus novaezelandiae cookii*) and Norfolk Island Boobook Owl (*Ninox novaeseelandiae royana*) are listed as endangered. Norfolk Island, Phillip Island, and the surrounding islets provide significant nesting and roosting habitats for 11 species of seabirds. On Phillip Island, the harvesting of Whale Bird (or Sooty Tern

Norfolk Island National Park and Botanic Garden (ANBG Collection, C. Totterdell)

(*Sterna fuscata*)) eggs is a contentious issue for wildlife managers. The main management work focuses on restoring and managing habitats, and controlling woody weeds and feral animals. Introduced animals include rosellas, cats, and rodents.

Snapshot **3.3**

Management of the Wet Tropics World Heritage Area
Russell Watkinson, Wet Tropics Management Authority

The Wet Tropics Management Authority (WTMA) was formed in late 1990. Its role is to ensure that Australia's obligations under the World Heritage Convention are met. Its primary focus since its inception has been to prepare the Wet Tropics management plan, and to build positive relations with the region's communities. The management plan is subordinate legislation under the *Wet Tropics World Heritage Protection and Management Act 1993*. It provides the authority with a regulatory 'stick' to assist with conserving the values of a World Heritage Area. However, the authority prefers to manage through forming partnerships with stakeholder groups.

Preparation of the plan commenced in 1992, by commissioning a survey on community attitudes. In 1995, a draft plan was released for public comment. The Wet Tropics Ministerial

Council approved a final plan and the Governor of Queensland gazetted it on 8 August 1997. Following an appeal by Aboriginal Land Councils operating in the WHA, the management plan was subject to a judicial review. As a result, the Supreme Court of Queensland (4 November 1997) determined that the plan was 'of no force or effect'. It ruled that the Wet Tropics Ministerial Council had the power to either endorse or reject the plan. It did not have the power to amend the plan. Following further consultations with Indigenous people, a revised management plan was developed that was endorsed by ministerial council and gazetted by the Governor of Queensland on 22 May 1998. The plan has now been fully implemented and regulates land use across the Wet Tropics WHA.

Land tenure

Control over land is determined according to tenure. Australian land tenures include freehold, which is privately owned land managed subject to planning regulations and any covenants; crown land that is controlled by state or Common-wealth Government; and crown land over which leases have been granted (leasehold) for commercial uses such as grazing and mining. Leasehold land is managed by private interests subject to lease and other regulatory conditions. Subsequent to the Mabo and Wik High Court decisions regarding Aboriginal land rights, Commonwealth and state native title laws created sub-categories of Aboriginal freehold and leasehold land. Conservation on Aboriginal land will be discussed in Chapter 18.

Three broad categories of protected area are defined by the NRMMC (2004). *Public Protected Areas* (also known as Dedicated Reserves) are usually reserved under specific legislation and managed by the nature conservation agency in each jurisdiction. *Private Protected Areas* include privately or publicly owned and leased lands managed by private individuals, incorporated groups, companies, and local governments, which are protected through legislation or a legally binding instrument on the land title. For example, the Birds Australia

property, Gluepot (SA), and the Trust for Nature (Victoria) property, Naringaningalook Grassland, are protected by nature conservation covenants under state legislation. *Indigenous Protected Areas* are owned and managed by Aboriginal communities such as the Yalata Aboriginal lands in South Australia, in accordance with management plans prepared by the Indigenous landholders.

Commonwealth protected areas (DEH 2004e)

The Australian Government manages protected areas in Australian territories and within Australian waters beyond the three nautical mile state limit of state authority. Such protected areas are declared under the *Environment Protection and Biodiversity Conservation (EPBC) Act 1999*. As of June 2004, there were 21 reserves declared under the EPBC Act, comprising six national parks, 13 marine protected areas and two botanic gardens. Three of the national parks, Kakadu, and Uluru–Kata Tjuta and Booderee, are managed jointly with their Aboriginal traditional owners. The other three national parks are the Cocos (Keeling) and Christmas Islands in the Indian Ocean, and Norfolk Island in the South Pacific.

These protected areas are managed by Director of National Parks, a position established under the

EPBC Act. The Director is supported by the staff of Parks Australia, a division of the Department of the Environment and Heritage. Management Boards and Committees for specific reserves have been established to advise the Director on management policies and priorities. The responsibilities of the Director include:

- the administration, management, and control of Commonwealth reserves and conservation zones
- assisting with the provision of research and training in the knowledge and skills relevant to the establishment and management of national parks and nature reserves
- cooperating with another country in matters relating to the establishment and management of protected areas (Snapshot 3.4)
- making recommendations to the minister in relation to the establishment and management of Commonwealth reserves.

Management of marine reserves has been delegated by the Director to the Marine and Water Division of the Department of the Environment and Heritage, with the exception of the Heard Island and McDonald Islands Marine Reserve that is managed by the Australian Antarctic Division, again under delegation from the Director. The

Great Barrier Reef Marine Park is managed by the Great Barrier Reef Marine Park Authority under separate legislation. The *Commonwealth Historic Shipwrecks Act 1976* has the primary purpose of protecting shipwrecks, but marine fauna also benefits from the regulation of human activity and the reserves are categorised as IUCN 1a.

In Australia, protected areas with international designations—World Heritage Areas, biosphere reserves, and Ramsar wetlands—are the responsibility of the protected area agencies in whose jurisdictions they fall. World Heritage Areas, for example, are managed according to a number of different models. Some are the sole responsibility of the state or Commonwealth. Others are jointly managed (Lane & McDonald 1997). In the case of protected areas that span state and territory borders, joint management agreements may be negotiated. An example is the Australian Alps, where agencies have integrated their efforts to achieve a cooperative approach to management through the Australian Alps Liaison Committee. Parks with both terrestrial and marine components often also require cooperative management by more than one agency. Both Commonwealth and state governments have responsibilities with respect to the Great Barrier Reef Marine Park. Aboriginal groups manage protected areas in partnership with government agencies (Chapter 18).

In addition to managing those protected areas for which it is directly responsible, Parks Australia cooperates with state and territory agencies to address management issues of common concern. An important part of this cooperation is through a benchmarking and best practice program undertaken by the National Parks and Protected Area Management Committee of the Natural Resource Management Ministerial Council (NRMMC). Reports have been published on topics such as community involvement, user pays, and management planning. The NRMMC, established in 2001 by an agreement between Australian federal, state, and territory governments, is made up of all Australian and New Zealand government ministers responsible for natural resource management. Papua New Guinea and the Australian Local Government Association participate in meetings as observers.

Snapshot 3.4

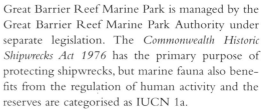

Tri-National Wetlands Memorandum of Understanding
DEH (2004b)

The Director of National Parks signed a Tri-National Wetlands Memorandum of Understanding in June 2002 between Australia, Indonesia, and Papua New Guinea. This Memorandum sets out how the three countries are working together to share and build their expertise and understanding in conserving three internationally significant wetland conservation areas: Kakadu National Park; Wasur National Park in Indonesia; and the Tonda Wildlife Management Area in Papua New Guinea. These areas share very similar ecosystems and management issues, and each is home to Indigenous communities that are dependent on their natural resource base. This project is facilitated by the World Wide Fund for Nature (WWF), and involves staff exchanges, workshops, and capacity building projects.

National estate is documented by the Australian Heritage Council (AHC), which was established under the *Australian Heritage Council Act 2003*, replacing the Australian Heritage Commission. One of the primary roles of the AHC is compilation and management of a register of the national estate that contains:

- a representative list of those places that demonstrate the main stages and processes of Australian geological and biological history
- rare or outstanding natural phenomena, formations, features, including landscapes and seascapes
- habitats of endangered species of plants or animals
- wilderness, forests, and selected habitats and phenomena that, being readily accessible to populated areas, are as valuable as the rarer yet less accessible places in the same category
- significant rock art galleries, ceremonial grounds and sacred sites, quarries and shell mounds, rock and earth arrangements, and important historical and archaeological sites of the Aboriginal peoples
- representatives of the main stages of Australia's architectural and building history
- monuments and historical landmarks, buildings, urban conservation areas, and precincts possessing architectural, social, cultural, aesthetic, historic, or biographical importance or other special values

- buildings, bridges, roads, fences, urban and rural settings, and other structures or ruins that especially illuminate past ways of living, working, or travelling.

The AHC also assesses nominations for the National Heritage List and the Commonwealth Heritage List. This latter list identifies places that come under the environmental protection and assessment provisions of the EPBC Act.

State and territory protected areas

State and territory protected areas are created under legislation, with each jurisdiction having conservation-focused legislation for the creation of national parks and other types of reserves. Protected areas are assigned to different categories according to their specific management objectives, conservation values, land tenures, and political contingencies. Once reserved under a particular act of parliament, management of the protected area can then be influenced by a range of other legislation. The extent of such legislation is exemplified with respect to Western Australia in Backgrounder 3.4.

There are 56 different types of terrestrial protected areas and 17 types of marine protected areas across the nine jurisdictions in Australia (Section 3.4). The diversity of nomenclature is indicative of the various purposes for which these areas have

---

Backgrounder 3.4 Many guiding hands

In Western Australia, marine and terrestrial protected areas are generally established under the *Conservation and Land Management Act 1984* or the *Fish Resources Management Act 1989*. The Department of Conservation and Land Management and the Department of Fisheries are the primary managers of these lands. These agencies are guided not only by the points covered in the establishing legislation but also by other legislation which refers to particular activities or broader laws which also apply in protected areas; for example:

- *Land Act 1933*
- *Wildlife Conservation Act 1950*
- *Sandalwood Act 1929*
- *Aboriginal Heritage Act 1972*
- *Control of Vehicles (Off-Road Areas) Act 1978*
- *Financial Administration and Audit Act 1985*
- *Freedom of Information Act 1992*
- *Reserves Act 1994*

- *Land Acquisition and Public Works Act 1902*
- *Beekeepers Act 1963*
- *Environmental Protection Act 1986*
- *Fish Resources Management Act 1994*
- *Heritage of Western Australia Act 1990*
- *Soil and Land Conservation Act 1945*
- *Public Sector Management Act 1994.*

Southwest National Park, Tas (Graeme Worboys)

been designated. The IUCN category system has been applied to these designations to enable a unified classification (DEH 2002, Appendix 2). Some areas have high conservation value, while others are more noted for their scenic attributes or provide leisure opportunities. Legislative guidelines for each type of protected area generally stipulate:

- management objectives of each protected area category
- requirements for declaration and management
- conditions for altering or revoking the protected status

- activities that may be permitted in the area and under which conditions.

An example of the types of protected area established in Tasmania under the *Nature Conservation Act 2002* is given in Case Study 3.1. As noted, similar arrangements are in place for the other states and territories. There are also resource-based agencies that incorporate in their portfolios the managing of reserves. Forestry Tasmania, for example (Case Study 3.1), manage reserves that are recognised as being part of the National Reserve System (Section 3.4).

Case Study 3.1

Protected area designations in Tasmania

The Tasmanian National Parks and Wildlife Service commenced operations on 1 November 1971 with a staff of 59. The *National Parks and Wildlife Act 1970* repealed the *Scenery Preservation Act 1915* and the *Animals and Birds Protection Act 1928*. Under the Act, provisions were made for the establishment and management of national parks and other reserves and the conservation of flora and fauna. On 1 May 1987 the National Parks and Wildlife Service was amalgamated with the Department of Lands to form the Department of Lands, Parks and Wildlife (TPWS 2004).

The Department of Land, Parks and Wildlife was altered in 1989 to create the Department of Environment and Planning and the Department of Parks, Wildlife and Heritage. This latter Department managed not just land reserved under the National Parks and Wildlife Act but also under the Crown Lands Act 1976, as well as providing for the conservation of wildlife and of Aboriginal and historic heritage. The Royal Tasmanian Botanical Gardens and the Port Arthur Historic Site Management Authority were also part of the Department (PWST 2004).

In 1993, the Department of Parks, Wildlife and Heritage amalgamated with the Department of Environment and Land Management. Recently, responsibility for state-wide flora and fauna management and research has been shifted to the Resource Management and Conservation Division of the Department of Primary Industries, Water and Environment (DPIWE). In August 2002 responsibility for the Parks and Wildlife Service and the Cultural Heritage Branch was transferred from DPIWE to the newly formed Department of Tourism, Parks, Heritage and the Arts (PWST 2004).

Protected areas in Tasmania are established under the *Nature Conservation Act 2002* and managed under the *National Parks and Reserves Management Act 2002*. Crown land can be reserved under the *Nature Conservation Act 2002* as either:

- national park
- state reserve
- nature reserve
- game reserve
- conservation area
- nature recreation area
- regional reserve
- historic site.

To qualify for designation as a particular category, land must possess the values indicated in Column 2, Schedule 1 of the Act (see below). Parliamentary approval is required for declaring land to be reserved land in the class of national park, state reserve, nature reserve or game reserve, or historic site.

Private land can be reserved, if the owner of the land has consented, as a private sanctuary or a private nature reserve. Land vested in public authority, excluding state forest or a public reserve, can be reserved if the public authority in which the land is vested consents.

Land can also be acquired by the government for a conservation purpose, and declared to be reserved land in the class of conservation area. Alternatively, an area can be leased for the term, and on the covenants and conditions, approved by the minister.

Forest reserves are established under the *Forestry Act 1920*. The government may dedicate or revoke any area of land within state forest as a forest reserve for one or more of the following purposes:

- public recreational use
- the preservation or protection of features of the land of aesthetic, scientific, or other value
- the preservation or protection of a species of flora or fauna.

The Forestry Corporation (called Forestry Tasmania) must manage land that is dedicated as a forest reserve according to the following management objectives:

- to conserve natural biological diversity
- to conserve geological diversity
- to preserve the quality of water and protect catchments
- to conserve sites or areas of cultural significance
- to encourage education based on the reserve's purpose and significance

- to encourage research, particularly that which furthers the purpose of reservation
- to protect the reserve against, and rehabilitate the reserve following, adverse impacts of fire, introduced species, diseases and soil erosion on the reserve's natural and cultural values and on assets within and adjacent to the reserve
- to encourage appropriate tourism, recreational use, and enjoyment
- to encourage cooperative management programs with Aboriginal people in areas of significance to them in a manner consistent with the reserve's purpose and other reserve management objectives
- to provide for the controlled use of natural resources
- to provide for exploration activities and utilisation of mineral resources
- to provide for the taking on of an ecologically sustainable basis of designated game species for commercial and private purposes.

Nature Conservation Act 2002 Schedule 1—determination of class of reserved land

Column 1	Column 2
Class	Values of land
National park	A large natural area of land containing a representative or outstanding sample of major natural regions, features or scenery
State reserve	An area of land containing any of the following: a) significant natural landscapes b) natural features c) sites, objects or places of significance to Aboriginal people
Nature reserve	An area of land that contains natural values that a) contribute to the natural biological diversity or geological diversity of the area of land, or both and b) are unique, important or have representative value
Game reserve	An area of land containing natural values that are unique, important or have representative value particularly with respect to game species
Conservation area	An area of land predominantly in a natural state
Nature recreation area	An area of land a) predominantly in a natural state or b) containing sensitive natural sites of significance for recreation.
Regional reserve	An area of land a) with high mineral potential or prospectivity and b) predominantly in a natural state
Historic site	An area of land of significance for historic cultural heritage
Private sanctuary	An area of land that has significant natural or cultural values, or both
Private nature reserve	An area of land that contains natural values that a) contribute to the natural biological diversity or geological diversity of the area of land, or both and b) are unique, important or have representative value.

Non-governmental protected areas

As indicated in Chapter 2, a range of governance mechanisms can be used to establish and manage protected areas. The National Reserve System Program (Chapter 5) recognises that unique ecosystems occur on privately owned and managed land, both leasehold and freehold. Many of these ecosystems have not been adequately protected. The program calls for cost-effective measures so that 'the optimal reserve configuration can be established with the minimal economic and social cost to the community'. Acquisition of private land is a direct method of expanding the public protected area system. However a large number of ecosystems warrant protection, making acquisition by government financially unviable. An alternative is for private individuals and NGOs to acquire and manage these protected areas. This may be done on freehold land by using covenants and easements. They are legal mechanisms that allow interested parties, other than the owner, to enforce proprietary rights over land. Easements include right-of-way to carry out activities on other people's land. Covenants typically restrict the owner's land management rights (Bates 1995). Covenants and easements relating to the protection of biodiversity and natural heritage may be attached to land titles. This ensures legally enforceable protection in the long term, as required by the IUCN protected area definition.

Although still only a relatively minor part of the protected areas system, private conservation reserves are becoming increasingly important in Australia. Until recently, the effort to conserve biodiversity has largely concentrated on reserving areas of public land. While these protected areas on public land are clearly the keystone of any conservation effort, native vegetation on private property is also of importance, particularly for those vegetation types that are not well represented on public land. Over two-thirds of Australia is managed by private landholders. Conservation reserves, comprising just over 10% of Australia, are insufficient to provide an adequate representation of this country's biological diversity.

A growing number of landholders appreciate the value of maintaining native vegetation on their properties. However, these pro-conservation attitudes may not be translated into conservation-oriented decisions and activities. There are a number of factors that may prevent or discourage a landholder from looking after their native vegetation. Barriers to conservation and management include economic pressures to earn a living, the need for landholders to break with their traditional style of land management, and the social pressure in some rural communities not to be seen to be a 'greenie'. Furthermore, the causes of biodiversity loss on private land can be traced back to a number of social, economic, and institutional factors, including the failure of markets to value all biodiversity considerations, incomplete specification of property rights, poor institutional arrangements, failure to distribute information, inadequate resources allocated for biodiversity conservation, and a general lack of awareness of the value of biodiversity.

Given this complexity, it is unrealistic to expect a single policy instrument to be effective. A range of complementary instruments must be employed. The challenge for governments and management agencies is to develop integrated packages that may incorporate:

- legislation or regulations that can be used to create an institutional framework for management, set aside areas of land, and enforce standards and prohibitions
- formal agreements between agencies and landholders
- economic measures such as subsidies and tradeable permits to encourage landholders to provide conservation benefits to the community, assist efficient allocation of resources, and allow for the equitable distribution of costs and benefits of conservation activities.

There are also private sector organisations set up specifically to establish protected areas. The Australian Bush Heritage Fund, for example, is a national, independent, non-profit organisation committed to preserving Australia's biodiversity through the creation of reserves on private land. Since 1990, the Fund has been raising money from the community to create a network of reserves across Australia. This has been achieved by buying land of high conservation value. Bush Heritage Reserves are protecting areas of conservation significance from the tropical rainforests of the Daintree River area of North Queensland to the woodlands of south-western Western Australia. The Nature Conservancy in the USA and the National Trust in the United Kingdom are the most prominent international organisations of this type.

The founder of Earth Sanctuaries, Dr John Walmsley, has taken a more market-based approach. Earth Sanctuaries is a publicly listed conservation company with approximately 6800 shareholders, which purchases land, fences the areas to exclude feral animals, and undertakes any necessary revegetation. Indigenous native species are then reintroduced. Earth Sanctuaries also undertakes ecotourism and education activities.

Multiple use in protected areas

Any area of land or sea may have several potential uses. Conservation often has to compete with one or more short-term, economically profitable, land uses. Conservation and commercial activities have often been considered incompatible. On the one hand, the developers may lament profits, while on the other hand many conservationists staunchly oppose extracting resources in protected areas. Integrating these interests in some cases, however, has had positive outcomes. So long as the primary objectives remain protecting biodiversity and natural heritage, the multiple use of protected areas may provide an impetus for industries to develop best-practice environmental management. This is the conceptual basis for IUCN category VI 'managed resource protected areas' and the transition zone of biosphere reserves.

In Australia there are significant tracts of crown land that have been reserved for future resource extraction. Establishing multiple-use reserves allows both for resource extraction and a level of environmental management. For example, in the mid 1980s an amendment to the South Australian *National Parks and Wildlife Act 1972* created the new designation of regional reserve. In these areas the central management goal remains conserving the wildlife and natural and historic features. Mining and grazing is allowed when the minister recommends it. Exploration for mining is permitted in all regional reserves declared under the Act since its amendment. Wilderness protection areas and zones can be declared. Through applying this strategy, vacant crown land with future mining potential (land that would have otherwise remained unprotected) has

been taken into the protected area system. In 2002, protected areas covered about 25.4% of the state—significantly higher than the national average. In Queensland, state land that has high conservation value but is subject to limited resource use can be declared a resources reserve. Controlled resource use is also allowed in nature refuges and coordinated conservation areas, which can be declared over private land.

Multiple use and marine protected areas

The ocean and coast play large roles in our lives. We use them in many ways. The beach is the most popular 'environmental' site for recreation (ABS 1996). We enjoy surfing. Seafood is an integral part of the Australian diet. The spectacular coral reefs of northern Queensland and Western Australia are major tourist destinations.

The ocean nurtures us with its wealth of biodiversity. The medium of water connects distant ecosystems; circulating sea currents facilitate the dispersal and migration of larvae, sea life, and pollutants alike. Thus, even more so than in terrestrial environments, environmental management must be considered on a broad scale. It is often difficult to conceptualise the delineation of protected areas of sea—it cannot be fenced in. Additionally, some commercial fish range over wide areas and those protected in one area may be caught when they migrate elsewhere. However, many species of marine life are restricted to small territories or are sessile, and sites critical to the life cycle of wide-ranging species, from whales to Southern Bluefin Tuna *(Thunnus maccoyii),* can be identified in Australian waters. For example, estuaries are 'nurseries' for many species; a variety of reef fish defend small territories; and deep sea underwater mountains are home to a range of unique fauna.

Our seas are common property—that is, individuals do not own them. Our scientific research and knowledge of ocean systems has lagged behind our knowledge of terrestrial ecosystems. Overuse of fish resources is rapidly becoming a major problem, as is marine pollution from boating and runoff from coastal activities. Multiple-use marine protected area designation is increasingly being used as one means of conserving marine resources. Marine protected areas receive a variety of designations in the states/territories and Commonwealth (Table 3.3). All are managed for multiple uses. If regulations apply, they usually place limits on resource extraction. Zoning is also being used as a management tool. An example is the zoning of the Great Barrier Reef Marine Park (Case Study 20.2).

3.4 Extent of Australia's protected areas

National parks and other protected areas are an integral part of the landscape for many Australians. Kosciuszko, Kakadu, Uluṟu–Kata Tjuṯa, and the Great Barrier Reef Marine Park are national icons. Bushland reserves provide recreational retreats in many suburban areas. In 1968, the average proportion of land devoted to parks and reserves in Australia was only 1.2% (Cresswell & Thomas 1997). Pressured by conservationists, scientific bodies, and growing public interest, the governments of Australia increased their commitment. By 2002, the total terrestrial area managed for nature conservation was over 760,000,000 hectares or a little over 10% of the Australian land mass, with over 6700 protected areas (Table 3.2), and around 70% by area in IUCN Categories I to IV (Table 3.3). By 2002, nearly 65 million hectares of marine environment was reserved (Table 3.4), with around 45% in IUCN Categories I to IV (Table 3.5).

Most Australian protected areas are considered to be part of the National Reserve System (NRS)—the exceptions are discussed below. NRS data is compiled into the Collaborative Australian Protected Area Database (CAPAD). CAPAD is a record of the jurisdiction, location, area, and IUCN category for all protected areas recognised under the NRS.

There are 56 different types of terrestrial protected areas and 17 types of marine protected areas across the nine jurisdictions in Australia (Figure 3.6, Tables 3.2 to 3.5). Protected areas have also been established on private property. To be included in the NRS, an area must meet the IUCN definition of a protected area given above, as well as:

- being dedicated for the primary purpose of protection and maintenance of biological diversity, with other uses in keeping with this primary purpose
- being able to be classified into one or more of the six IUCN categories
- being secured and managed for the long term by legal or other effective means (NRMMC 2004).

Legal means is defined as land is brought under control of an Act of Parliament, specialising in land conservation practices, and requires a parliamentary process to extinguish the protected area or excise portions from it. Other effective means requires that the protection provide for long-term management, ideally in perpetuity, but at least 99 years, and be secured through a legal means such as a covenant on the title of the land. Nonetheless, there is some concern about the long-term security of private protected areas, and in some cases the lack of public scrutiny of the contract and its implementation (NRMMC 2004). As a result, private protected areas are not yet included in CAPAD, or counted as part of the NRS. While recognising the importance of covenanting for nature conservation, Victoria, Western Australia, and NSW consider their covenants are insufficient to qualify for inclusion in the NRS, whereas the Northern Territory, South Australia, Tasmania, and Queensland consider their covenants to be adequate.

Figure 3.6   Australian protected areas

Terrestrial Protected Areas — 77,461,951 hectares (10.08%)
External Territory and Oceanic Islands Protected Areas — 25,333 hectares
Marine Protected Areas — 64,615,554 hectares

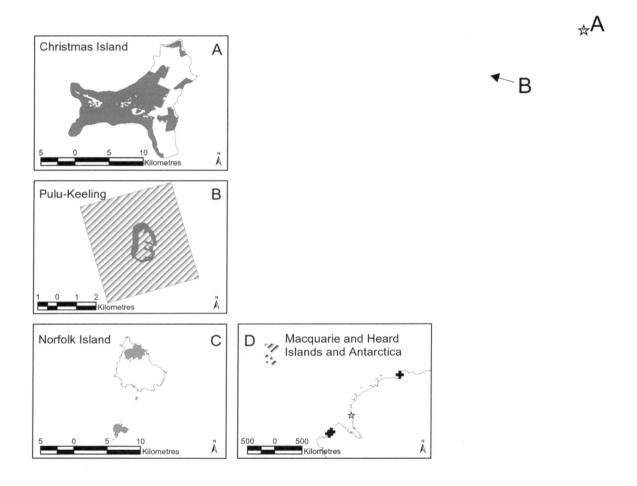

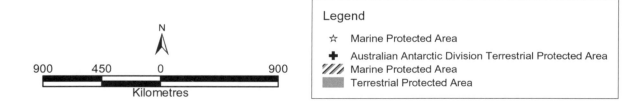

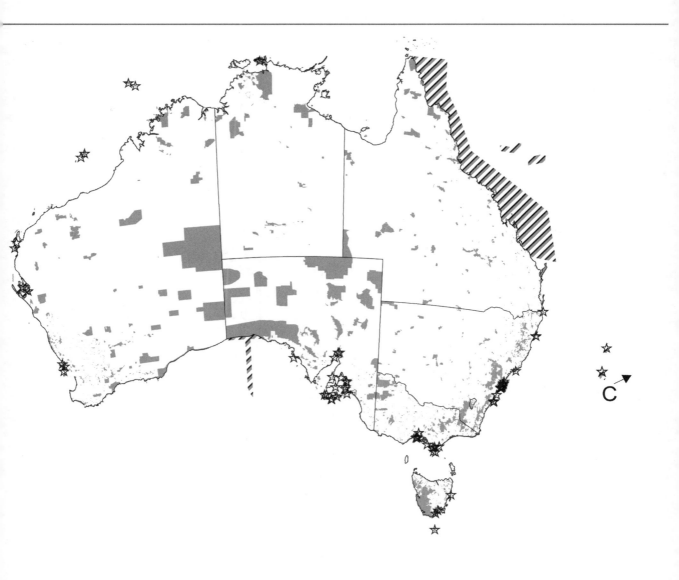

Table 3.2	**Extent of Australian terrestrial protected areas** (by designation) (CAPAD 2002)			
Designation	Number	Area (ha)	Jurisdiction	% Australia
Botanic Garden	1	90	Commonwealth	0.00
Coastal Reserve	1	12,300	NT	0.00
Conservation Area	160	531,282	TAS	0.07
Conservation Covenant	67	9343	TAS	0.00
Conservation Park	417	6,587,151	QLD, SA, WA	0.86
Conservation Reserve	66	305,808	NT, SA	0.04
Feature Protection Area	25	1253	QLD	0.00
Flora Reserve (no spatial data)	103	35,394	NSW	0.00
Forest Reserve	190	174,724	TAS	0.02
Game Reserve	21	37,489	SA, TAS	0.00
Heritage Agreement Area	1193	564,682	SA	0.07
Historic Site	15	16,057	TAS	0.00
Historical Reserve	15	7916	NT	0.00
Hunting Reserve	1	1605	NT	0.00
Karst Conservation Reserve	4	4409	NSW	0.00
Management Agreement Area	5	26,228	NT	0.00
Miscellaneous Conservation Reserve	74	299,923	WA	0.04
National Park	533	28,159,653	All	3.66
National Park (Aboriginal)	4	575,814	NT	0.07
National Park (Commonwealth)	3	2,119,278	Commonwealth	0.28
National Park (Scientific)	7	52,181	QLD	0.01
Native Forest Reserve	58	14,187	SA	0.00
Natural Features Reserve	1	2	VIC	0.00
Natural Features Reserve—Bushland Reserve	1072	45,607	VIC	0.01
Natural Features Reserve—Cave Reserve	6	376	VIC	0.00
Natural Features Reserve—Geological Reserve	10	474	VIC	0.00
Natural Features Reserve—Gippsland Lakes Reserve	29	6318	VIC	0.00
Natural Features Reserve—Natural Features and Scenic Reserves	24	25,385	VIC	0.00
Natural Features Reserve—River Murray Reserve	1	20,313	VIC	0.00

Table 3.2	(continued)			
Designation	Number	Area (ha)	Jurisdiction	% Australia
Natural Features Reserve—Scenic Reserve	52	11,061	VIC	0.00
Natural Features Reserve—Streamside Reserve	156	7740	VIC	0.00
Natural Features Reserve—Wildlife Reserve (hunting)	167	51,601	VIC	0.01
Nature Conservation Reserve	53	15,684	VIC	0.00
Nature Conservation Reserve—Flora and Fauna Reserve	90	144,314	VIC	0.02
Nature Conservation Reserve—Flora Reserve	135	32,023	VIC	0.00
Nature Conservation Reserve—Wildlife Reserve (no hunting)	63	11,602	VIC	0.00
Nature Park	13	24,155	NT	0.00
Nature Park (Aboriginal)	1	3038	NT	0.00
Nature Recreation Area	18	59,223	TAS	0.01
Nature Reserve	1567	11,668,389	ACT, NSW, TAS, WA	1.52
Other Conservation Area	32	190,630	NT, TAS	0.02
Other Park	9	52,479	VIC	0.01
Phillip Island Nature Park	1	2118	VIC	0.00
Private Nature Reserve	1	120	TAS	0.00
Protected Area	3	12,710	NT	0.00
Recreation Park	13	3152	SA	0.00
Reference Area	35	20,459	VIC	0.00
Regional Reserve	28	10,830,360	SA, TAS	1.41
Reserve	9	21,662	NSW	0.00
Resource Reserve	36	348,678	QLD	0.05
Scientific Area	51	21,139	QLD	0.00
State Park	31	183,696	VIC	0.02
State Reserve	60	44,302	TAS	0.01
Wilderness Park	3	202,050	VIC	0.03
Wilderness Protection Area	5	70,069	SA	0.01
Indigenous Protected Area	17	13,794,255	NSW, NT, QLD, SA, TAS, VIC, WA	1.79
Total land area of Australia		**768,826,956**		
Total Terrestrial Protected Areas (2002)	**6755**	**77,461,951**		**10.08**

Table 3.3	**Extent of Australian marine protected areas** (by designation) (CAPAD 2002)		
Type	Number	Area (ha)	Jurisdiction
Antarctic Special Protection Area	1	2088	Commonwealth
Aquatic Reserve	29	17,866	NSW, SA
Conservation Park	2	202	SA
Dugong Protection Area	16	637,298	QLD
Fish Habitat Area	78	729,150	QLD
Fisheries Reserve	5	3357	VIC
Historic Shipwreck	14	827	Commonwealth
Historic Shipwreck Protection Zone	1	78	SA
Marine & Coastal Park	3	62,143	VIC
Marine National Nature Reserve	2	241,714	Commonwealth
Marine National Park	1	126,292	SA
Marine Nature Reserve	4	114,716	TAS, WA
Marine Park	25	59,731,953	Commonwealth, NSW, NT, QLD, SA, VIC, WA
Marine Reserve	5	6,532,748	Commonwealth, VIC
National Nature Reserve	3	1,786,500	Commonwealth
National Park	2	3307	Commonwealth
Nature Reserve	1	81,538	TAS
Total (2002)	**192**	**64,615,554***	

* *Total is not a sum of all protected area type areas due to overlapping of Queensland marine protected areas. To avoid double-counting total area figure is calculated based on a union of Great Barrier Reef Marine Park, Dugong Protection Areas, Fish Habitat Areas, and Queensland State Marine Parks.*

Table 3.4	**Extent of Australian terrestrial protected areas** (by IUCN category) (CAPAD 2002)	
IUCN Category	Number	Area (ha)
IA	2006	18,103,255
IA (Heritage Agreement Areas)	1193	564,682
IB	32	3,963,356
II	642	28,766,907
III	696	390,948
IV	1527	2,225,208
I-IV subtotal	**6096**	**54,014,356**
V	172	788,779
VI	452	22,635,792
V-VI subtotal	**624**	**23,424,571**
Not specified	35	23,024
Total (2002)	**6755**	**77,461,951**

Table 3.5	**Extent of Australian marine protected areas** (by IUCN category) (CAPAD 2002)	
IUCN Category	Number	Area (ha)
IA	12	15,192,688
IB	2	202
II	20	2,140,279
III	0	0
IV	106	12,045,534
I-IV subtotal	**140**	**29,378,703**
V	0	0
VI	38	35,236,024
V-VI subtotal	**38**	**35,236,024**
Total (2002)	**178**	**64,614,727**

04

Process of Management

Hire the best. Pay them fairly. Communicate frequently. Provide challenges and rewards. Believe in them. Get out of their way and they'll knock your socks off.

Mary Ann Allison

4.1 How organisations manage	114	4.4 Leading	124
4.2 Planning	115	4.5 Controlling	126
4.3 Organising	116	4.6 Managing for performance	126

An understanding of management processes is fundamental to successful protected area management. Management is about people. It is a process through which goals are achieved. It involves coordinating all resources (both human and technical) to accomplish specific results. Management is an entirely logical process. We manage instinctively in our everyday lives by planning our day, setting deadlines, budgeting, allocating time and priority to various tasks, and delegating tasks to peers, children, or partners. Organisations are essentially the same but more complex, in that more people are involved and the goals and the work involved are on a larger scale.

Protected area organisation staff manage their area of responsibility every working day. The basic functions and processes of management apply to all staff. If this was not the case, there would be real problems for our rangers managing our parks, who are required to deal with operational matters every 24-hour period. To be reliant on the next communiqué from a senior officer clearly would not work!

The contemporary purpose of protected areas is well defined (Chapter 3). Properly designed and managed, protected areas are areas on land and sea that are established to conserve a representative sample of the world's biological diversity: natural phenomena; cultural and social heritage sites and areas of cultural significance; ecosystems; landscapes and natural ecosystem processes (Mackinnon et al. 1986, IUCN 1994, Phillips 2003). They help preserve genetic diversity. They are a principal tool (Article 8) for the Convention on Biological Diversity attaining its global objectives related to biodiversity conservation and the means for measuring the global achievements of the Convention (UNEP-SCBD 2001).

Protected areas are a critical part of retaining essential ecosystem services, ecosystem processes, and life support systems of the planet (Mackinnon et al. 1986, Wilson 1992). They provide opportunities for a basic human need for interrelationships with nature (biophylia) (Wilson 1992) and protect areas of special spiritual and cultural significance. Protected areas are part of a landscape and ecosystem approach to conservation and they are an important part of the sustainable ecological, cultural, and economic well-being of local communities (Phillips 2003). They

help ensure that utilisation of species by humans is sustainable. They play a central role in the social and economic development of rural environments and contribute to the economic well-being of urban centres and the quality of life of their inhabitants (Mackinnon et al. 1986). They provide opportunities for outdoor recreation and tourism. Properly designed and managed protected areas provide these wide-ranging benefits and services.

Protected area establishment (Chapter 5) is just the start of the process for achieving the objectives for which they were reserved. Active management is required. There is a multiplicity of threats and other actions that need to be dealt with to maintain the purpose and integrity of protected areas (Mackinnon et al. 1986, Brandon et al. 1998b, Dudley et al. 1999a, Van Schaik et al. 2002, Chape et al. 2003, Du Toit et al. 2003). The phenomenon of 'paper parks' where protected areas are designated but never managed is recognised as a serious issue (Dudley et al. 1999a). Simply designating protected areas does not ensure their survival, nor guarantee that social and economic benefits are derived from them (Chape et al. 2003).

This chapter introduces the basic principles of protected area management. It describes a process of how management is undertaken for organisational tasks. It then discusses the specific skills and knowledge needed by those who manage tasks at different levels within an organisation. Too often, the roles of management positions within a protected area agency are underestimated or poorly understood by others within the same organisation. We also describe in some detail the four basic management *functions*: planning, organising, leading, and controlling (Bartol et al. 1998, Robbins et al. 2003). (For a more comprehensive description of these terms and their application we refer our readers to the excellent texts prepared by Robbins et al. (2003), Bartol et al. (1998) and Ivancevich et al. 1997)). We also introduce 'managing for performance', which means seeking maximum efficiency through the optimum blend of these four *functions*. To achieve protected area conservation outcomes, protected area executive leaders with rich protected area knowledge and a depth of relevant experience are essential. Competent protected area staff whose skills involve a blend of theoretical knowledge about management and practical experience of it are also critical. We will discuss these points. The chapter describes the functions of management and why managing protected areas has different requirements to traditional bureaucratic organisational management.

4.1 How organisations manage

In Chapters 2 and 3 we described the social context in which an organisation operates. Inputs from government are part of this social context and help to determine the way organisations operate. Some inputs to the process of management are shown in Figure 4.1.

All of these inputs are important in the process of management. Typically, organisations will manage by undertaking at least four basic functions (Figure 4.2). First, they *plan*. Then they *organise* the available resources that will be needed to achieve the planning goals determined. Following this, *leaders* will motivate and assist staff to achieve the work and the goals. Finally, organisations will be interested in how well they have performed, so they will *monitor* their progress.

Figure 4.1   Some inputs into the process of management

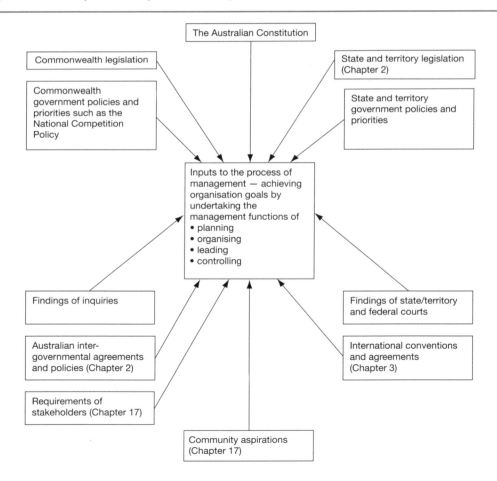

Figure 4.2 The four functions of management (adapted from Bartol et al. 1998, Robbins et al. 2003)

4.2 Planning

Planning is one of the first management functions undertaken by any organisation. First the organisation must determine its goals. Its planners should also look at the range of issues they may encounter and where strategic improvements can be made. Such planning is critical to achieving conservation goals. While this section deals briefly with planning in general, Chapter 7 looks at the specific types of planning commonly associated with protected areas.

Three levels of planning

The function of planning is commonly undertaken at three levels of detail within an organisation. An organisation cannot achieve its primary goal unless each level of management carries out the appropriate level of planning.

Theorists of management often prescribe a top-down system whereby senior executives turn the organisation's goals into a series of high-level 'strategic' plans. These plans, as they pass down the hierarchy, are translated first into a series of 'tactical' and then 'operational' plans, which finally become the instructions to the front-line staff (Bartol et al. 1998). Such a system can only work if each level in the agency clearly understands its role and is provided with the freedom to manage.

Long-term conservation management is dependent on understanding as well as administering bureaucratic processes. People who have managerial skills but who have also themselves done the jobs they now ask of others are the most valuable. With these reservations, let us now look at the 'strategic–tactical–operational' model.

Strategic plans

To achieve its principal goal, an organisation identifies what major strategic goals must first be attained. Strategic planning finds ways to achieve these strategic goals. Typically, strategic plans are prepared by senior executives. Such plans have ramifications for the whole of an organisation and have a long-term time frame. In protected area management, examples of strategic planning include corporate planning; organisational policies (Chapter 9); organisational planning and budget systems and business planning (Chapters 7 and 8); management effectiveness evaluation systems including monitoring (Chapter 21); organisational baseline sustainable performance measures (Chapter 10); and operational procedural systems and statements (Chapter 11).

Tactical plans

Tactical plans set tactical goals that help in the implementation of a strategic plan. They prescribe how elements of a strategic plan are to be achieved.

Operational plans were used to design the Blue Mountains National Park Heritage Centre seen here under construction, NSW (Graeme Worboys)

Tactical plans typically establish a set of steps to achieve each tactical goal. Such plans are usually developed by middle-level managers and staff and have an intermediate time frame. They represent a critical level of planning. Good tactical planning ensures efficient and effective allocation of an organisation's internal resources.

Tactical planning is becoming more important with the evolution of concepts such as regional ecosystem management and landscape management (Chapter 19). A tactical plan, for instance, might deal with tourism and the sharing of the visitor load across a region, and might be developed cooperatively by a number of organisations.

Operational plans

Operational plans are directed towards short-term goals derived from the strategic and tactical plans. They are usually implemented by an organisation's front-line managers (or managers who work at the 'delivery end' of an organisation). Operational plans may be developed as a consequence of tactical plans. As such, they contribute to achieving tactical goals. In the example of the tactical plan for distributing the visitor load across a region, one of its operational plans could be to set up a staff-training course that focuses on customer service for visitors.

Planning a planning project

The project of developing a plan (at strategic, tactical, or operational level) is no different from any other project. Linking the planning, budgeting, and operational systems within an organisation into one unified system is quite critical. The excellent system implemented by the Department of Conservation in New Zealand is described here (Case Study 4.1).

4.3   Organising

As a management function, organising is concerned with how managers allocate and arrange human and other resources to enable plans to be implemented (Bartol et al. 1998). It involves managers determining the range of tasks to be performed and allocating the available resources to obtain the best results most efficiently. Organising is a process that never stops. In a fast-changing world, managers and staff are constantly refining how their organisations work towards required goals.

Case Study 4.1

New Zealand Department of Conservation strategic systems

Grant Baker, Department of Conservation, Wellington, New Zealand

The Department of Conservation's annual business planning system is one of the core business systems of the department. It is integrated with other systems and has three phases: directions and expectations setting, business planning, and work planning. The purpose of the process is to allocate resources to departmental priorities and achieve conservation outcomes.

Directions and expectations

Annual business planning starts from setting directions and expectations (D&Es). This is a 'top down' process (Chapter 7). The D&Es set the directions and expectations of what are aimed to be achieved for the coming year. It guides the department's business planning and work planning. The Chief Executive sets up the D&Es for the whole department at a high strategic and outcome level. The D&Es are then cascaded down to General Managers and other managers who then interpret the Chief Executive's D&Es and set up their D&Es for their divisions, units, and their staff.

The D&Es setting phase opens the discussion among managers at different levels on what the stakeholders such as the minister and New Zealand public expect the department to achieve, the department's performance, strategic risks, options, and alternatives to meet the stakeholder expectations, and the capability requirement. The D&Es setting is supported by the department's environmental scan and department strategies and policies.

Business planning

During the business planning stage the managers develop their business plans based on the D&Es of their managers to prioritise work, allocate resources such as staff time and financial resource, and set up performance target estimates. Business planning is a bottom-up process. Unit and divisional business plans are consolidated and reprioritised to best use the resources and achieve conservation outcomes.

The business planning process is integrated into the government budgeting process, such as the budget bids process. It is supported by the department's performance measurement system, which covers the conservation outcomes, intervention logic, and outputs. The result of the business planning process contributes to the department's Statement of Intent, the definitive planning document for the next three to five years, with a focus on the year one.

Work planning

Work planning starts after the department's budget and Statement of Intent have been approved by the government and signed off by the minister. The work planning sets up details of work programs and projects, which include tasks, time, dollars, and performance targets. They identify work that will be carried out by staff day by day.

During the year, business plans and work plans may be modified and performance targets may be reforecasted based on the circumstance. The business plans and work plans provide the base for performance monitoring and reporting. It also provides the criteria for staff performance appraisal and remuneration.

As to how organisations should be organised, there are many approaches, particularly in the public sector. The organisational structure (the way in which an organisation arranges its staff) can be defined technically as 'the formal pattern of interactions and coordination designed to link the tasks of individuals and groups to achieve organisational goals' (Bartol et al. 1998, p. 875). Usually this is portrayed as an organisational chart that shows the organisation's major positions or departments and the reporting arrangements (chain of command) from higher to lower levels.

Typically, such structures are organised into divisions, with units and individuals reporting in a systematic way. Organisational activities should be coordinated both horizontally and vertically, or across and up and down the organisation.

Demands on protected area agencies are somewhat different from those facing most organisations. There is a need to ensure other public and private sector organisations are aware of these differences, and that 'standard' organisational models are not inappropriately applied to undertake protected area management. Some of the special characteristics are as follows.

- Protected area lands and waters are dynamic, living systems, and the dynamics of natural events are superimposed on the routine bureaucratic timetable of events.
- Protected areas are often rough, rugged, and remote, giving rise to special management needs related to organisational time and resource allocation, as well as staff competencies and capacities.
- Protected areas are a 24 hours a day, 7 days a week operation and with operational matters that arise on protected area lands or waters often needing a rapid response.
- Terrestrial protected areas are usually surrounded by neighbours, and again a round-the-clock response capability is usually required.
- Protected areas are used by a wide range of recreational and other users, with peak use periods often clashing with peak incident periods.
- Unplanned incidents such as fires, search and rescues, whale strandings, and other events (Chapter 15) are normal occurrences, and they may cut across bureaucratic process timetable events.
- The practical and experiential knowledge accumulated by protected area staff is crucial for wise protected area decision-making—such 'corporate memory' is rarely accessible from bureaucratic files or data bases (Stevenson 2004).
- Protected areas need planning and management investments that are continuous and long term—much longer than election and budget cycles, for example.

Organising work

Protected area management organisations are typically under funded and have too few staff for the work they need to do. Every staff position is vital. Every resource that will help an organisation achieve its goals is critical. How these resources are best organised is fundamental to an organisation's success. This is why there is regular change within most organisations. There is a constant need to position the available staff, funding, and resources in the best way to achieve goals. Adaptive management (Chapter 7) practices mean that organisations will need to be designed for flexibility.

Most agencies have mechanisms to ensure that all staff are aware of how they are contributing to their organisation's primary goals. However, these mechanisms sometimes fail. Staff may not always appreciate the roles that their colleagues play in other parts of an organisation. Grumbling that 'head office staff have it easy' or that 'rangers have a wonderful life working in paradise' reflects such ignorance. Other comments such as 'the only real work for conservation is here at the coal face' reflect at best a misguided view of how organisations work and, at worst, a dangerously divisive attitude that could jeopardise the work of an organisation. It may also give the wrong message to external stakeholders. Internal cohesiveness and teamwork is critical.

So, what do different levels of an organisation do? Typically there are three levels of management for protected area organisations: top level, middle level, and front-line staff (Bartol et al. 1998, Robbins et al. 2003).

Top level

Top-level managers are ultimately responsible for the entire organisation. The executive provides leadership for an organisation including long-term strategic planning, and the monitoring of the organisation's performance. Typically there is a chief executive and a small team of senior staff. Top-level managers usually have exceptional conceptual and human management skills but fewer technical skills (Bartol et al. 1998, Robbins et al. 2003). The chief executive is important both as a figurehead and as the lead spokesperson. Top-level management sets the strategic directions. This level also devises, with input from staff, organisational restructures necessary to meet new strategic directions.

Middle level

Middle-level managers and staff develop plans to advance corporate goals. For large organisations, there may be a number of layers of middle managers, depending on needs. The trend of many organisations is to 'flatten' structures so as to reduce the number of middle managers. This may cut costs and may streamline communication. If the cuts are too great, it may impact on the organisation's performance. The limit to such steps is governed by the effective span of control (that is, by the number of staff one manager can effectively supervise) and the volume, sensitivity, and complexity of work. The common result of having fewer hierarchical layers is that the remaining middle management levels gain greater autonomy and responsibility. Typically, middle-level managers will have a balanced approach to dealing with the four functions of management, giving each function roughly the same attention. Technical skills, conceptual skills, and people-management skills are important for people in these positions (Bartol et al. 1998). Middle-level managers are expected to be entrepreneurial and exhibit leadership in their roles. Middle-level managers in a conservation agency may be responsible for functional tasks such as human resource management, legal services, financial management, research, community relations, and policy. Organisations may need to delegate tasks on a geographic basis. Middle-level managers lead such units. Considerable effort is needed by managers and staff to work horizontally or across an organisation in order to achieve its goals, as well as managing up and managing down. They undertake a range of work tasks. For a public sector protected area management organisation, for example, work completed by middle-level managers includes an array of tasks.

Front line

Front-line managers and staff are at the operational level in the hierarchy. They ensure that the day-to-day operations of a protected area run smoothly (Chapter 11). In undertaking the four functions of management, a front-line manager is typically involved in less planning and organising and more leading. Controlling is an important function. The position usually requires strong human and technical skills, with conceptual skills being less important (Bartol et al. 1998, Robbins et al. 2003). Front-line managers are increasingly being involved with whole-of-organisation tasks, as well as with increasingly sophisticated management control systems.

Organising work vertically and horizontally

Typically, protected area organisations (especially in the public sector) are spread across wide areas of Australia. It is very easy for isolated units to work at

Field officer setting out information sign to be routed
(Graeme Worboys)

variance with the primary goals of an organisation. Even when units are in the same building, strong-minded managers or poor systems of coordination may lead to problems such as units or individuals 'doing their own thing' or concentrating on lower priority tasks. Effective coordination of work effort is required up, down, and across an organisation. Some systems and techniques are available to achieve this and include:

- preparing policies and procedures
- providing effective delegations
- a span of control that lets managers deal effectively with their responsibilities
- clear operational policies in relation to centralised accountabilities and decentralised responsibilities.

Contemporary public sector organisation structures

Australia's protected area organisations are structured in a variety of ways. Some have a hierarchical structure and report directly to the minister. Other organisations report through a Head of Department and may involve a Board. Contemporary Australian protected area organisational structures are described on protected area agency web sites. The most important consideration is not how an organisation is structured, but whether or not it is delivering conservation outcomes as a consequence of an organisational structure that is in place. A discussion paper prepared by the NSW NPWS (Backgrounder 4.1) provides an insight into how protected area agencies could be or have been organised.

| Backgrounder 4.1 | Evolving protected area agency organisational structures within Australia |

Adapted from material provided by Ray Fowke, NSW NPWS (1998)

Conservation agencies have structured themselves in various ways to achieve their primary goals. Structure matters, since an organisation needs to be efficient and effective to achieve conservation goals. Organisations are continually evolving. Some such structures for conservation agencies are discussed below. They include market-based, legislative, administrative, and ecologically based issues. Note that some of the options reviewed are still theoretical.

Such structures are currently the subject of considerable debate, from the global to the national and local levels. This debate has been spurred on by the continuing degradation of the environment and by the limited resources of governments. The debate also reflects broader arguments about the evolving role of government. It is especially influenced by ongoing pressures to increase efficiency and accountability.

A trend in the late 1990s was to expose conservation agencies to what can generally be considered 'market-based' mechanisms (including the contracting out of services, corporatisation of agencies, and so on). This approach was largely founded on a belief that private sector efficiency and competition are answers to the many perceived problems of government bureaucracy, waste, and overspending.

Governments of all political persuasions broadly supported such 'reform' of the public sector (particularly through the National Competition Policy (NCP) (Chapter 8)), which meant that conservation agencies were not immune to these trends.

Government conservation agencies are not the only organisations to provide nature conservation management services. They are also increasingly involved in both on- and off-park conservation initiatives. Despite this growth in responsibilities, the criticism remains that they have become 'stranded' in an inflexible institutional structure largely created in the 1970s that is unable to cope with the emerging complex array of social, political, ecological, and economic demands. One needs to make sure that any new system is more efficient, not just in some few areas, but across the whole desired range of ecological, economic, and social outcomes for the community. We need to give priority to maintaining heritage conservation, including biodiversity.

Market-based approaches. Market-based approaches to the delivery of government services are currently the subject of much interest. This stems in large part from the reform process set in train by the NCP, which establishes an agreed and consistent basis for ensuring competition in all areas of the economy. At the most basic level, the NCP will open up profitable areas of public service and provision to competition and privatisation. Importantly, both the NCP and the subsequent Competition Principles Agreement note that the application of competition principles should be selective and should consider issues such as ecologically sustainable development, social welfare and equity, and the general interests of consumers. Market-based approaches to natural resource management include a number of models.

Contracting-out and competitive tendering. Typically, contracting-out means relying on suppliers or contractors for goods or services that cannot be provided in-house. This has been done for many years, especially for 'non-core' business. Competitive tendering, on the other hand, is an extension of competition into areas previously undertaken solely by protected area agency staff. In effect, this means that agency divisions or 'business units' are expected to compete for contracts with other potential contractors. Contracting-out and competitive tendering are both presented as having advantages over the 'public sector monopoly' approach to natural resource management. In particular, they are seen as mechanisms to improve the efficiency and quality of service delivery. Given the demands of the community for a better quality environment with more recreation facilities and improved accessibility, it is argued that these mechanisms can deliver better outcomes at less cost—a primary consideration given the increasing limits on government resources.

However, the advantages are not always clear-cut. For example, there is some evidence that competitive tendering is more often than not done primarily as a cost-cutting exercise, rather than to improve the quality of services. Both contracting-out and competitive tendering can also result in the eventual privatisation of services. In addition, international research indicates that they reduce employment conditions, mostly through casualisation of the workforce, while providing no guarantee of greater consumer choice or quality of service. Contracts are usually long term and create a single private rather than a single public provider. More significantly, overseas experience indicates that on occasions contractors are unable to go on providing services at the negotiated price. Hence the initial cost savings are later eroded as competition for contracts declines.

Purchaser-provider. The fundamental feature of the purchaser-provider organisational model is to separate regulatory and policy functions from service delivery and operational functions. The model acknowledges that while it is the responsibility of government to fund 'public good' activities, such as the management of protected areas, these should be purchased from the most efficient provider, whether that is a private or public organisation. The model also seeks to increase the range of choices available to 'customers' and 'consumers' of services and to ensure that agencies do not compete unfairly with the private sector in the delivery of services. In addition, by proposing the splitting of functions, the model attempts to counter the 'gamekeeper-poacher' dilemma by ensuring that regulatory and operational tasks are separated.

Much of the impetus for the purchaser-provider model flows from the NCP. This model has become increasingly prevalent in Australia within the field of natural resource management (as well as other areas such as electricity and water supply). The purchaser-provider model offers a range of options, from establishment of separate units within agencies with separate operating accounts through to full legal separation of units and exposure to corporations law. Protected area management organisations in Victoria and the ACT have moved to separate policy and regulatory functions from operational functions as they relate to the provision of conservation and recreation management goals and actions. It is not yet clear how well this will work in Australia, and especially whether conservation will be compromised.

The outcomes of using the purchaser–provider model in natural resource management within Australia have not been adequately assessed. In particular, there is little evidence available at this stage to enable assessment of the conservation outcomes of the model. In addition, there are concerns that there will be considerable difficulty in developing and applying meaningful performance indicators for environmental management to determine whether the provider has delivered the required services (ecological outcomes are obviously more intangible and difficult to measure than financial achievements), although judging environmental outcomes is also problematic for other natural resource management models.

Corporatisation and privatisation. In recent years Commonwealth, state, and territory governments have sought either to corporatise or privatise various public agencies (for example, Qantas, Telstra, Sydney Water, and so on). Corporatisation (or commercialisation) is primarily designed to improve the public sector's financial management and to deliver services more efficiently. In Australia, it has generally taken the form of an increased emphasis on managerialism (for example corporate planning, restructuring), charging government agencies for services provided by other agencies, increased competition with the private sector, and contracting out some functions such as information technology maintenance. Corporatised agencies are often required to continue to provide services to meet explicitly identified 'non-business' community service obligations that, in theory, are funded from consolidated revenue. Corporatisation is considered to have several advantages.

- It results in more efficient and economic service delivery.
- There is greater focus on the customer.
- 'User pays' concepts can expose the true cost of services.
- Identification of the full cost of inputs allows a more rational use of resources.

The potential disadvantages of contracting-out and competitive tendering have been discussed above. In addition, there may be potential problems in balancing the autonomy and commercial responsibilities of corporatised organisations with their community obligations and with the broader policy objectives of government.

Privatisation, on the other hand, involves the total or partial transfer of public assets, goods, or services to the private sector, and is usually preceded by corporatisation. It provides government with a one-off revenue gain. It has also been used by a number of governments as an instrument of structural change (for example, reform of the energy sector). Reasons for privatisation include: improved budgetary position, better management and efficiency of enterprises, and notions of the 'appropriate' role of government. A specific advantage of privatisation is the scope it provides to reduce debt or fund recurrent expenditure. In general, privatisation only occurs where the service being privatised is profitable. While this may initially seem to exclude protected area management there is no reason to assume that this is the case. There are now several private protected area commercial enterprises that are not only achieving a profit, but are apparently achieving solid results in threatened species recovery.

However, experience raises doubts about the effectiveness of privatisation. Short-term revenue gains may be offset by the long-term loss of income from profitable agencies. As well, privatisation may simply create new private sector monopolies that provide fewer, cheaper, or inferior services; and, extensive government regulation may be necessary to prevent private monopolies exploiting their control of electricity, water supply, or even national parks.

Private sector organisations. It is generally recognised that there are limitations in the traditional approach to natural resource management, which divides responsibility among specific-purpose agencies and which has long relied solely on government funding. Government funding is not coping well with increasingly expensive management requirements. This issue is further complicated by the growing pressures from increasing tourism.

Hence the private sector is increasingly promoting its willingness and desire to participate. On this there appear to be two schools of thought. Some commentators argue that as governments will never be able to achieve all the community's demands for preserving and rehabilitating the natural environment, the private sector therefore has a very real role to play.

The alternative viewpoint generally opposes the laissez-faire implications of the former, but concedes there is some limited role for the private sector. The main disagreement concerns exactly where this role begins and ends. Some observers argue that private sector involvement, with particular reference to protected areas, should be restricted to appropriate accommodation and tourism. Others contend that the public and private sectors have more in common than they realise, and that scope exists for more cooperative relationships, ranging from a partnership approach, through to direct involvement of the private sector in day-to-day management of natural resources, including protected areas.

Legislation and administration. Laws and administrative rules have long been used to control protected areas and natural resources. This includes controls on water and land use, development, pollution, and so on. Yet in many cases rules have overlapped inconsistently and even contradicted one another. In addition, the plethora of regulations that now exist is unwieldy and has been criticised for emphasising process over outcomes. Many governments have looked at reforming the regulations. In New Zealand the preferred model has been to streamline and condense the body of environmental regulations. Elsewhere, a more gradual process has sometimes been favoured. Several regulatory and administrative models are discussed below.

Integrated legislation. In New Zealand the *Resource Management Act 1991* repealed 12 primary statutes, amended 53 other acts, and revoked or amended 21 regulations and orders. The Act integrated the management of land, air, and water resources into one piece of legislation governed by a common purpose ('sustainable management') and provided a consistent hierarchical structure for policy-making, plan development, consent-giving, and enforcement. Both Queensland and the ACT appear to favour this approach.

In general, the fundamental goal of regulatory reform has been to improve economic performance, particularly to reduce the cost of 'red tape' to business, rather than to achieve better environmental outcomes. It is too early to assess the environmental effects, though some claim the major benefit will be a more holistic approach to environmental management.

Integration of agencies

A further organisational model has been to combine the full range of environmental agencies (for example fisheries, forestry, parks, and so on) into one 'mega-agency'. This has been undertaken in WA and in Victoria. As a result of the amalgamation process, Victoria now has a single agency responsible for management of natural resources, including land identification and management, resource development, use and protection, and conservation and environmental management. Parks Victoria, which was set up in 1998, was the 'provider' to the then Department of Natural Resources and Environment (now Department of Sustainability and Environment, DSE) for management of the reserve system. DSE owns the product, sets priorities and policies, provides funding, and monitors outcomes. Parks Victoria is contracted to maintain parks and build assets. Parks Victoria does not own any of the assets it manages, nor those it builds.

The main potential advantages are better coordination of efforts, more effective use of scarce resources and potential to overcome existing inter-agency conflicts and 'turf-wars'. The disadvantage is that it may simply internalise conflict and prevent examination of different perspectives in resource management. In addition, unless functions are integrated there is no guarantee that one large agency will be more resource-efficient and less bureaucratic than a number of smaller, specialist agencies.

Whole-of-government responsibilities. An emerging organisational model for natural resource management is one that tries to develop a system whereby all relevant government agencies assume environmental responsibilities, but without the system being 'owned' by any single agency. In many ways, this approach is essentially a 'reformed status quo' model that seeks to realign the way agencies currently operate.

The whole-of-government model for sharing environmental responsibilities has perhaps been undertaken most comprehensively in Canada. The Canadian Government, for example, has created a Commissioner of the Environment and Sustainable Development, within the Office of the Auditor-General of Canada, which is responsible for ensuring agencies 'green' their policies, operations, and programs. Each agency must have a sustainable development strategy in place within two years and these strategies are to be presented to the Canadian parliament by the relevant minister. This approach is aimed at placing sustainable development among the mainstream operations of all agencies with minimal disruption. It provides 'teeth' through independent monitoring by the commissioner.

The whole-of-government model has a number of potential advantages: a provision for shared responsibility across government (thereby facilitating bureaucratic cultural change), a focus on outcomes rather than processes, less inter-agency conflict, and ease of introduction. In essence, the model appears to provide a framework for reform that is backed by legislation, but without needing to fundamentally alter the scope of existing regulatory responsibilities. However, unless the system is supported by adequate mechanisms to enforce it, it may encourage only minor amendments to current practices. In addition, it is not clear how the system would prevent the duplication of roles across agencies or clearly delineate accountabilities for particular outcomes.

4.4   Leading

Leading involves influencing others' work behaviour towards achieving organisational goals (Bartol et al. 1998). In the process of leading, effective managers become catalysts in encouraging innovation. Leaders kindle the dynamic spirit needed for success. How well an organisation performs depends on the motivation and commitment of staff. Most protected area organisations are made up of people who are committed to the ideals of conservation and who are prepared to work long hours in support of this commitment. Staff who are supported by positive leaders can harness extraordinary energy to achieve conservation goals. A lack of leadership can influence the development of less productive behaviour by staff. We now discuss some aspects of the function of leadership.

Chief executives and senior staff

The roles of chief executives and other senior staff of protected area agencies are pivotal if conservation outcomes are to be achieved. Protected area agencies within Australia and elsewhere have a depth of more than 30 years' professional development and experience, and organisations need to recognise and use such 'corporate capital'. Appointment of senior staff with extensive protected area management competencies is crucial for the success of protected area management.

A common organisational practice of encouraging transfer of chief executives and senior staff between agencies, on the grounds that they have generic executive skills, is a less suitable approach for protected areas. A core of professional protected area management skills and experience is essential for executive parks staff. Protected area management is too big, too complex, too dynamic, and directly impacts too many people for governments to regard senior staff recruiting as just another executive government appointment.

Content knowledge is critical. Operational experience is fundamental. At times of crisis and incidents, leadership decision-making needs to be instinctive, immediate, and underpinned by confidence derived from experience and knowledge. A protected area organisation simply cannot afford to

Then Chief Executive, NSW NPWS, Ms Robyn Kruk and Premier of NSW, Mr Bob Carr, launch NSW's 100th national park, SE Forests National Park, NSW (Graeme Worboys)

wait for a new chief executive to 'learn the game'. The dynamic of biodiversity conservation needs constant and immediate attention. There is no time to learn on the job and advisers can only assist to a level of their own competency and communication ability. Restructures will always be needed, but the challenge is to maintain conservation outputs and outcomes while continuously streamlining management. Chief executives with core professional skills and the loyalty of staff have a greater chance of achieving continuous conservation improvement.

Executive leadership for protected areas ultimately involves decision-making about natural ecosystem processes, biodiversity, and cultural conservation. Wise executive decision-making ensures that organisational systems and budgetary management, staff competencies and capacities, management performance evaluation, and other essential systems are in place, supporting priority conservation outcomes. To this end, we assert the following four understandings related to effective protected area leadership.

1   All protected area managers need to have a basic understanding of the four functions of management and how the management process works.

2   Organisational structures are important, but not as important as professional, experienced, and competent leadership and organisational loyalty achieving conservation outcomes.

3   Chief executives and senior staff with professional protected area management experience and executive competencies are best equipped to lead protected area organisations and secure conservation outcomes for the long term.

4   Protected area corporate knowledge and professional competencies gained by individuals over many years of field-based and policy-based protected area management are important assets for protected area organisations.

Working with and motivating staff

The performance of staff has been shown to be a function of ability, motivation, and environmental conditions. There are many theories of motivation that need to be understood. Whereas natural instincts and life skills are important, knowledge of motivation theory will ensure that 'natural managers' are even better at their work, and will equip them for more senior management roles (Case Study 4.2). Managers are expected to achieve a positive and motivating workplace. If they understand theories such as Maslow's hierarchy-of-needs (Robbins et al. 2003, Ivancevich et al. 1997) and a range of other motivational theories, managers can positively assist staff in many ways.

Case Study 4.2

Leadership and pressure at Brisbane Forest Park

Bill Carter, University of Queensland

Brisbane Forest Park (BFP) was established under its own legislation (*Brisbane Forest Park Act 1978*), with a brief that emphasised nature-based recreation. I was appointed as the first manager, with a mandate from the chair of the Board (Sir Wallace Rae) and the Authority (Hon. Bill Glasson) to manage the park with a 'people focus' through interpretation, positive facilitation, and in partnership with the private sector as well as government departments. It was also made clear that the government had high expectations for the park and that there would be support, but only a short 'honeymoon' period.

I put much thought into establishing a vision and a direction for the park. To prioritise tasks, I devised a management model that recognised we were starting with no resource knowledge, visitors, or services. The model established how we would need to apply our resources to reach an ideal management status. I realised that to meet the expectations of the Authority and the government we would need to develop a major new facility every second year and an upgrade every other year.

Public relations were crucial. Improvising, we abbreviated the annual report and turned it into leaflets and pocket diaries. Volunteer rangers became an important bridge to the community. We decided to go after a clientele for whom we could effectively provide and manage. We would not be able to handle the 1.5 million visitors per year who came to the park unless they were predisposed to the product we were offering. We decided that all media contact was to include the words 'nature- and family-based recreation'. The client focus became established and the park entered the world of tourism, marketing, and entrepreneurship.

We had specialist consultants do a thorough 'resource inventory' of the park. These studies became vital in several instances where we were able to argue from a position of information strength against proposed actions that might damage both the recreation experience and the resource. The results were also published, which gave us credibility in academic circles. The park became noticed as a place for research. Another sector of the community was thus brought in to assist the park's program; and by default BFP began to be recognised as more than a bit of bush suitable for picnics.

My first development effort was Jolly's Lookout. I enlisted the design skills of a landscape architect. He returned with a design that hardened and modified the site to the extreme, which was not the national park tradition, but it did fit the brief, and the reality. Despite more than 10 years of hosting 500,000 visitors per year, the one-hectare site retains its laid turf. The design worked. I learnt to design for outcomes, capacity, and experiences, and adopted the rule: *Do it once, do it well, and it is cheap. Cost lies in maintenance, not capital.*

I was lucky to have attracted highly competent, innovative, motivated, and creative staff. To manage them, my principle was: *Let go, have confidence, set the guidelines and direction and then don't interfere.* Once I had signed off, responsibility for getting the job done was theirs alone.

During this period a number of park management philosophies were born that separated BFP from other park managing agencies:

- one person, one job, with total responsibility
- no one dabbles in someone else's area, including the manager
- a sign-off means just that
- it is not a stuff-up if it can be fixed
- no recriminations.

We learnt that specialisation saves money. Our trained rangers did their own jobs best, but skilled contractors were able to quote for maintenance at a third of what it was costing us. We were lucky with funding. Our performance spoke loudly, but there is little doubt that the goodwill of senior government members and the behind-the-scenes support and lobbying of an influential board and chair ensured that funds could be obtained and effectively spent. If we often bent the rules, we made a point of discussing it first with treasury officers in the case of fiscal matters, and board members for administrative and operational issues.

Brisbane Forest Park remains an exception to much established wisdom. Success is due not to following rules but to intelligent adaptation to circumstances.

4.5 Controlling

Controlling is concerned with monitoring the performance of an organisation against management benchmarks (Bartol et al. 1998, Robbins et al. 2003). Managers need to set performance measures and the criteria for how they will be evaluated (Chapter 21). Controls help managers and staff cope with uncertainty, detect irregularities, identify opportunities, handle complex situations, and decentralise authority (Bartol et al. 1998, Robbins et al. 2003). The basic process involves establishing standards, measuring performance and comparing performance to those standards. It also involves responding with corrective actions (Snapshot 4.1).

4.6 Managing for performance

Management performance comprises two important dimensions: effectiveness and efficiency. Effectiveness is the ability to choose appropriate goals and achieve them, whereas efficiency is the ability to make best use of the available resources in the process of achieving goals (Bartol et al. 1998). In undertaking the process of management, managers and staff will employ, in their estimate, the optimum balance of planning , organising, leading, and controlling to achieve management goals. This

Snapshot **4.1**

Monitoring and evaluation—experiences from Fraser Island World Heritage Area
Marc Hockings, University of Queensland, and Keith Twyford, formerly of Queensland Parks and Wildlife Service

The notion of performance indicators and program evaluation took the public sector (or at least its rhetoric) by storm in the 1980s. A section on monitoring and evaluation, and a list of performance indicators, became mandatory in most planning documents. But how many indicators were ever measured, and how many monitoring programs were put in place? In our experience there were very few.

In 1994, we teamed up to see if we could devise an effective way of monitoring the management of one park: Fraser Island World Heritage Area. We had the great advantage that its new management plan had clearly laid down the desired outcomes of management. We could thus measure progress towards stated goals, instead of trying to measure efficiency in the abstract. We had a second advantage. One of us (Marc Hockings) was an academic interested in the issue of monitoring, and could thus bring a certain objectivity to the project,

while the other (Keith Twyford) was a manager working inside the system. This combination meant we could devise monitoring tests that were useful and meaningful to managers. We could get their cooperation while offering (and receiving) ongoing feedback. Our aim was a monitoring program that would be precise, rigorous, practical, and useful to managers. It had to measure progress towards all of the stated desired outcomes, yet remain affordable. Having the help of an 'insider' made this possible and, we believe, more than made up for any risk of losing objectivity. We decided to set up monitoring programs for the effects of beach camping, of people and vehicles disturbing wading birds, of fire altering flora and fauna, and of roads. We also monitored threatened fauna and water quality. (Sometimes the same study plot could monitor two things at once.) The result was a partnership between researcher and manager that helps managers manage better.

Dune Lake, Great Sandy National Park, Fraser Island World Heritage Area, Qld (Graeme Worboys)

---

balance will vary in relation to organisational level and purpose of the management position.

Attributes of a competent protected area manager

It is recognised that the mix of technical, human, and conceptual skills are fundamental for managers. Technical skills are more important for front-line

staff, and conceptual skills are more important for top-level managers (Bartol et al. 1998).

Technical skills. Protected area staff must have a cross-section of technical skills and practical skills to undertake conservation management of heritage areas. Conserving species, preserving heritage buildings, protecting significant Aboriginal and

European heritage sites and making critical decisions during incidents such as oil spills and bushfires, all need sound judgment as well as a fundamental understanding of conservation issues. Whether staff are at top, middle, or front-line levels, they must also understand the ramifications of decisions being made. There needs to be a continuous process of learning, and systems available to achieve this. Staff need to be open-minded, perceptive, and sensitive to such information.

Human skills. Protected area managers must be able to work with people and draw on the fundamental skills of organising, motivating and inspiring their staff. Managers are advocates for their agency's policies when interacting with local communities, neighbours, local government, and local members of parliament. They are partners in local cooperative projects. They assist in a range of projects that involve volunteers. They work with the central agencies of government or with commercial or business partners in facilitating conservation. Managers in more senior, supervisory positions are expected to be the trusted confidants of staff, to be ethical in their behaviour, and to work to invest in the professional careers of people.

Conceptual skills. Managers may be selected in competitive interview processes for their demonstrated ability to understand the contextual relationship of their managerial role, to visualise beneficial futures for the organisation, and to proactively pursue opportunities with internal and external partners towards conservation goals. Conceptual skills provide the basis for goal-focused active leadership, for efficiency and effectiveness, and are particularly important for staff in senior positions. Managers who have the combined skills of conceptual insight, hands-on, and social skills are highly sought after.

Competent managers do the right thing, at the right time, and in the right way. Managers can be trained, but they must be willing to learn from experience. Good managers have a range of competencies (Bartol et al. 1998).

1   They are willing to learn from experience.
2   They clearly understand the organisation's goals.
3   They actively pursue excellence and best practice.

4   They are sensitive to trends and conditions inside and outside of their organisation.
5   They have analytical, problem-solving, and decision-making skills.
6   They possess emotional resilience, and can work effectively under pressure.
7   They work ethically.
8   They understand the impact on others of their use of power.
9   They can see the bigger picture.
10 They are able to be innovative.

These principles apply equally to all protected area management staff. Even a temporary staff member brought in to do a small job of piecework ought to grasp the aims of the organisation, and proceed to plan, organise, control, and even lead (as appropriate for their area and task). This is not to deny that some people in the agency—those commonly referred to as executive managers or senior managers—will have wider responsibilities. It might seem that of the four main functions of management (plan, organise, lead, control), the function 'lead' is not appropriate to an officer who has no one 'under' them. Yet this is to make the mistake of assuming that instruction and knowledge must come down from above. It is often the person on the ground who leads those 'above' them to a more concrete understanding of what can and should be done. The most effective, dynamic, and innovative conservation agency will be one in which each officer is a team player, yet is also his or her own main manager in helping to achieve the organisation's mission.

Decision-making

Decisions made by managers and staff are critical to achieving conservation outcomes. Such decisions need to be right. They need to ensure that the best decision-making processes have led to the right management responses. Long-term conservation outcomes are dependent on competent and wise management decisions at every level of management. Protected area managers and staff are constantly being asked to make judgments on how to best allocate their resources. What will achieve the greatest contribution to conservation for the long term? It is a balancing act where the decisions made need to be the best possible. Managers need to be alert to problems in decision-making such as limits to rational decision-making, perceptions, personal biases, motivational factors, issues associated with group decision-making, and the stifling of creativity.

Scanning the external 'environment' is a critical part of decision-making, and managers are repeatedly required to consider internal and external environments. In particular, managers need to be aware of at least six critical decision-making considerations: environmental/ecological, economic, social, political, legal, and managerial.

Environmental/ecological considerations

Life on the planet is dependent on the retention of natural systems and processes. The richness of Australian society is reflected by its history and diverse cultural heritage. The primary purpose for the reservation of protected areas is to conserve environmental heritage (including natural and cultural heritage) and ecological processes and values. Management decisions need to be made in

the context of this primary purpose. They need to be made consistent with the principles of precaution. Most protected areas, for example, will not exist in an unmodified state without active management intervention.

Economic considerations

Protected areas play an important part in the community, providing important economic benefits. From an economic perspective, efficient and effective protected area management is critical to the long-term economic well-being of communities. Protected area management decisions may include economic considerations such as those appraised in benefit–cost analysis (Chapter 8).

Social considerations

The community cares about the environment and expects that protected areas will be well managed. The process of management for conservation includes the whole community. Protected areas are only part of a total conservation effort. Conservation of Australia's heritage will fail if it

Park staff fire training, Kosciuszko National Park, NSW (Graeme Worboys)

relies on the government-established protected areas alone. Working with the community is critical. Community needs (social considerations) are important inputs to the management decision-making process. The community needs to be involved in the management of 'their' protected areas. They need to be aware of the critical importance and benefits of protected areas.

Political considerations

Executive government can intervene in the process of management of organisations at any time and can change the ground rules for how protected area management goals are to be achieved. They can also change the very goals themselves. Politicians, in addition to ideology, are strongly influenced by community attitudes and aspirations in providing leadership at the state/territory or national level. Managers need to be sensitive to the political process. In making decisions, they need to be politically astute in relation to their organisational responsibilities.

Legal considerations

The courts can provide direction that is contradictory to the policies and priorities of executive government. Legislation enacted by parliament may not be the legislation of executive government. Managers are required to make decisions within this environment. They are also required to make decisions in the context of all relevant legislation.

Managerial considerations

People within organisations, their commitment, their enthusiasm, their corporate knowledge and wisdom, their competence, and their teamwork determine the difference between the efficient and effective conservation of protected areas and the loss of Australia's heritage. How organisations are structured, how they operate, their culture, their goals, and how leaders perform are all dependent on effective and efficient management. Management decisions influence how organisations work. They impact on people. They need to be made carefully and with great judgment if conservation goals are to be realised.

PART B PRINCIPLES AND PRACTICE

05	Establishing Protected Areas	132
06	Obtaining and Managing Information	162
07	Management Planning	188
08	Finance and Economics	219
09	Administration—Making it Work	250
10	Sustainability Management	269
11	Operations Management	288
12	Natural Heritage Management	309
13	Cultural Heritage Management	351
14	Threats to Protected Areas	371
15	Incident Management	402
16	Tourism and Recreation	424
17	Working with the Community	465
18	Indigenous People and Protected Areas	493
19	Linking the Landscape	509
20	Marine Protected Areas	532
21	Evaluating Management Effectiveness	553
22	Futures and Visions	569

05

Establishing Protected Areas

We, the 3,000 participants of the Vth World Parks Congress, celebrate, voice concern and call for urgent action on protected areas. We bear witness to those places most inspirational and spiritual, most critical to the survival of species and ecosystems, most crucial in safeguarding food, air and water, most essential in stabilizing climate, most unique in cultural and natural heritage and therefore most deserving of humankind's special care.

The 2003 Durban Accord

5.1 Terrestrial national reserve system   133
5.2 Marine national reserve system   138
5.3 Scientific reserve selection methods 144

5.4 Land-use planning and protected
     area establishment   147
5.5 Management lessons and principles 160

Australia is a signatory to the *Convention on Biodiversity* that asks countries to develop guidelines for the selection and management of a protected area system. State, territory, and Commonwealth governments have demonstrated their commitment to increasing the extent of Australia's protected areas. They have done so by endorsing several national polices and strategies. Objective 10.1 of the National Strategy for Ecologically Sustainable Development states that the objective of a nature conservation system is to:

> establish across the nation a comprehensive system of protected areas which includes representative samples of all major ecosystems, both terrestrial and aquatic; manage the overall impacts of human use on protected areas; and restore habitats and ameliorate existing impacts such that nature conservation values are maintained and enhanced (Commonwealth of Australia 1992a, p. 54).

Item 13 of the Intergovernmental Agreement on the Environment schedule on nature conservation stated:

> the parties agree that a representative system of protected areas encompassing terrestrial, freshwater, estuarine and marine environments is a significant component in maintaining ecological processes and systems. It also provides a valuable basis for environmental education and environmental monitoring. Such a system will be enhanced by the development and application where appropriate of nationally consistent principles for management of reserves (Commonwealth of Australia 1992b, p. 40).

The *National Strategy for the Conservation of Australia's Biological Diversity* aims to achieve the ecologically sustainable use of Australia's natural resources. Protected areas are to be integrated with other measures to achieve this goal:

> Central to the conservation of Australia's biological diversity is the establishment of a comprehensive, representative and adequate system of ecologically viable protected areas integrated with the sympathetic management of all other areas, including agricultural and other resource production systems (Commonwealth of Australia 1996b, Principle 8).

Protected areas can be established by governments under legislation, by Indigenous communities, and by private landowners (Chapter 3). This chapter considers the principles and processes that guide protected area identification and selection in Australia. The current system is of relatively recent origin. Prior to the 1990s, reservation decisions were largely made based on conservationists' pressure and interest together with case-by-case identification of scenic, recreation, and nature conservation values. One notable exception was Victoria, which through the work of the Land Conservation Council and its successors, has administered a formal state-wide process of reserve identification on public land. Researchers such as Dr Bob Pressey from the NSW P&W have also been longstanding advocates for more formal scientifically based processes of protected area selection.

We firstly introduce system-wide principles for identifying and establishing national protected area networks in terrestrial and marine environments. Following consideration of formal scientific reserve selection methods, we then outline two processes for identifying land for reservation as protected areas— the historic work of the Land Conservation Council in Victoria and the more recent reservations under Regional Forest Agreements. We conclude the chapter with a summary of the principles that should guide reserve selection in the future.

5.1   Terrestrial national reserve system

National planning for a system of protected areas includes consideration of:

- defining the priority of protected areas as a worthwhile national concern

- defining the relationships between various categories of protected area
- defining the relationships between protected areas and other land use and tenure categories
- identifying gaps in protected area coverage

- habitat requirements of rare or other species and their minimum viable population sizes
- connectivity between units (corridors) to permit wildlife migration
- perimeter/area relationships
- natural system linkages and boundaries
- traditional use, occupancy, and sustainability
- cost of achieving protected area status (Davey 1998).

The selection criteria have not always been so sophisticated. When national parks were first formed, the conservation of nature per se was not a primary aim. In 1918 the USA was a leader in the field of protected areas. In that year the US secretary of the interior, Franklin Lane, instructed the National Parks Service Director, Stephen Mather, to select national parks on the basis of their having 'scenery of supreme and distinctive quality so extraordinary or unique as to be of national interest and importance'. Scenery was the main selection criterion. The first national park in Australia, Royal National Park, was designated in 1879 originally for recreation and public enjoyment (Section 2.1). In 1894 the Australasian Association for the Advancement of Science called for flora and fauna protection reserves to be designated (Cresswell & Thomas 1997). At that time, government policy and the economic climate were intensively pro-development.

In Australia, as in other countries, protected area systems reflect spatial patterns. These patterns, which emerge on retrospection, derive from various factors. Historically, selection criteria have not been based on sound ecological data. Profitable uses of land have out-competed conservation. A time lag arises largely because conservation steps are usually taken after concerns are revealed over the effects of development (Pouliquen-Young 1997).

One outcome of these factors is a certain bias towards some types of landscapes and ecosystems rather than others. Many of the large parks, and indeed much remnant bushland, are mountainous (Lamington National Park, Blue Mountains, Dandenong Ranges National Park, Border Ranges National Park, and the Australian Alps, to name a few prominent examples) or in areas of low productivity, which are difficult to access and less suitable for other uses. The classic example in Australia is the Western Australian wheatbelt, in which most of the remaining native vegetation available for protection has only been incidentally preserved for road corridors, town sites, water supply, and railway reserves (Pouliquen-Young 1997). Similarly, in the pastoral regions of NSW, travelling stock routes are some of the few repositories of critical vegetation remnants including grasslands, which are among Australia's most endangered ecosystem types.

This lack of comprehensive protection for Australian flora and fauna was first brought to light with the release of the study *Conservation of major plant communities in Australia and Papua New Guinea* by Specht et al. (1974). Since that time various studies have been undertaken to both delineate the variety of landscape types in Australia and to determine their conservation status. Status is evaluated by indicators such as: amount cleared, thinned, and uncleared; rate of disturbance and loss; primary threatening processes (Chapter 14); representation in reserves; and tenure status of land on which it occurs. Table 5.1 contains recent disturbance and reservation data for the broadly classified vegetation structural types (according to the classification by Specht et al. (1974)), which illustrates the range in status and protection.

Biodiversity can be classified at the genetic, species, and ecosystem levels. The protection of genetic variation for a species or for a population is important for the maintenance of their long-term viability. A species approach to conservation planning provides the ability to focus on restricted or threatened species. An ecosystem-based approach to conservation aims to protect a range of habitats, their species, and ecological processes within a particular geographic area. However, there is no generally accepted ecosystem classification and available data on ecosystem distribution is often poor. Consequently, a 'surrogate' is often used to represent changes in biodiversity. The most commonly used surrogates are vegetation communities (Smart et al. 2000).

In Chapter 1 the Interim Biogeographic Regionalisation of Australia (IBRA) program was described. The primary purpose of IBRA is to provide a sound ecological basis for the expansion of the protected area system. For this purpose the

Table 5.1 **Conservation status of the broad vegetational structural types in Australia**
(adapted from Graetz et al. 1995)

Vegetation structural type (percentage of continent covered)	Dominant overstorey genus	T	C	U	I	R
Closed forests: tall, medium, and low (0.6)	Other than Eucalyptus or Acacia	17	9	74	0	23
Forests: tall, medium, and low (6.4)	Eucalyptus	17	20	62	1	10
Forests: tall, medium, and low (1.6)	Acacia	37	43	20	0	1
Open forests: medium (12.7)	Eucalyptus	23	42	27	8	4
Open forests: low (2.9)	Eucalyptus	10	29	19	42	6
Open forests: low (2.7)	Acacia	21	65	14	0	8
Open forests: low (1.8)	Other than Eucalyptus or Acacia	2	8	90	0	5
Sparse forests: medium (2.3)	Eucalyptus	43	49	8	0	1
Sparse forests: low (12)	Eucalyptus	11	22	20	47	6
Tall shrublands: (0.3)	Eucalyptus	2	97	1	0	2
Tall shrublands: (0.6)	Acacia	18	44	38	0	8
Tall open shrublands: (3.7)	Eucalyptus	19	63	18	0	23
Tall open shrublands: (0.7)	Other than Eucalyptus or Acacia	18	43	39	0	17
Tall sparse shrublands: (1.9)	Eucalyptus	12	74	14	0	7
Low open shrublands: (4.9)	Other than Eucalyptus or Acacia	34[1]	66[2]	10		
Grasslands: (4.8)	Variable	5[1]	95[2]			1

Present disturbance measured as percentage thinned (T), cleared (C), uncleared (U), and indeterminate (I). Future potential for disturbance indicated by percentage reserved for conservation (R).

1   *slight grazing disturbance*

2   *substantial grazing disturbance*

reservation status of each bioregion and the ecosystem bias, or degree to which the most extensive ecosystem types (sub-regions) are comprehensively represented in protected areas, has been determined (Thackway & Cresswell 1995a).

Chapter 1 illustrated the range of unique and diverse types of Australia's ecosystems. The development of IBRA and the increasing focus on ecosystem diversity in environmental management has stemmed from the recognition that all ecosystem types, from wetlands to grasslands, are repositories of biological diversity and play a critical role in the wider picture of planetary health. For example, wetlands are no longer perceived to be stinking swamps to be filled in and developed; they are highly productive ecosystems often critical to the life cycle of migratory birds and other fauna, and act as vast water filters.

The conservation status of biodiversity and natural heritage is determined not only by the present level of protection in reserves, but also by the potential for increasing protection. For example, Thackway and Cresswell (1995a) identified three major issues that act to constrain the development of a representative reserve system.

• Economic considerations: the high cost of establishing the national reserve system.
• Social and cultural considerations: nature conservation as a land use and a development decision is not widely understood or accepted and may be viewed as either a restricted opportunity or a lost opportunity for gaining revenue.

• Data quality considerations: appropriate indicators of ecosystem characteristics at various scales are required for bioregional planning.

The aim of the National Reserve System (NRS) Program is to develop a comprehensive, adequate, and representative reserve system for all Australia's bioregions, including both the land and sea environments. It was established in 1992 as a collaborative program between the Commonwealth, states, and territories. State and territory conservation agencies have the primary responsibility for selecting and managing new protected areas. The NRS is now administered by Parks Australia and is one of the five programs funded under the Natural Heritage Trust. The NRS is Australia's contribution to a wider effort aimed at establishing a comprehensive global network of protected areas (Backgrounder 5.1).

The NRS seeks to address the gaps in the existing reserve network at a national scale for terrestrial ecosystems other than forest ecosystems that are considered under the Regional Forest Agreement (RFA) process. Marine ecosystems are considered in a separate process, Section 5.2). Gaps in the representation of ecosystems are identified on a regional basis using IBRA. Consideration is also given to rare or threatened species and ecosystems, particular those listed under the *Commonwealth Environment Protection and Biodiversity Conservation Act 1999* or state, territory, and local government legislation or policy. Species with specialised habitat requirements, wide-ranging or migratory species, and species vulnerable to threatening processes are also taken into account (NRMMC 2004). A strategic approach has been developed to identifying the requirements of a NRS, based on a number of criteria and principles, including the following.

Comprehensiveness. The NRS aims to include the full range of regional ecosystems within each IBRA region.

Adequacy. The NRS aims to provide reservation of each ecosystem to the level necessary to provide ecological viability and integrity.

Representativeness. Areas selected for inclusion in the NRS should reflect the variability of the ecosystems they represent.

Columnar rhyolite geological feature, Sawn Rocks, extension to Mt Kaputar National Park, NSW (Graeme Worboys)

Backgrounder 5.1 Targets for a global protected area network

IUCN (2003b)

The action plan developed in 2003 as part of the Vth World Parks Congress urged governments, non-government organisations, and local communities to maximise representation and persistence of biodiversity in comprehensive protected area networks in all ecoregions by 2012, focusing especially on threatened and under-protected ecosystems and those species that qualify as globally threatened with extinction under the IUCN criteria. This will require that:

- all globally threatened species are effectively conserved in situ with the following immediate targets:
    i.   all Critically Endangered and Endangered species globally confined to single sites are effectively conserved in situ by 2006
    ii. all other globally Critically Endangered and Endangered species are effectively conserved in situ by 2008
    iii. all other globally threatened species are effectively conserved in situ by 2010
    iv. sites that support internationally important populations of congregatory and/or restricted-range species are adequately conserved by 2010
- viable representations of every terrestrial, freshwater, and marine ecosystem are effectively conserved within protected areas, with the following immediate targets:
    i.   a common global framework for classifying and assessing the status of ecosystems established by 2006
    ii. quantitative targets for each ecosystem type identified by 2008
    iii. viable representations of every threatened or under-protected ecosystem conserved by 2010
    iv. changes in biodiversity and key ecological processes affecting biodiversity in and around protected areas are identified and managed
- systematic conservation planning tools that use information on species, habitats, and ecological processes to identify gaps in the existing system are applied to assist in the selection of new protected areas at the national level
- regional landscape and seascape planning considers locally generated maps, and incorporates zoning and management planning processes to assist in designing and enhancing comprehensive protected area networks that conserve wide-ranging and migratory species and sustain ecosystem services
- protected area systems are established by 2006 that adequately cover all large intact ecosystems that hold globally significant assemblages of species and/or provide ecosystem services and processes
- coverage of protected areas in freshwater ecosystems is increased as proposed by the *Convention on Biological Diversity* Recommendation VIII/2 to establish and maintain a comprehensive, adequate, and representative system of protected inland water ecosystems … using integrated catchment/watershed/river basin management by 2012
- a representative network of marine protected areas is created by 2012.

Threat. Priority should be given to ecosystems where there is a high risk of loss (irreplaceability).

Precautionary Principle. The absence of scientific certainty is not a reason to postpone measures to establish protected areas that contribute to a comprehensive, adequate, and representative protected area system.

Landscape context. Biodiversity conservation outcomes should be as far as possible optimised through the application of scientifically established protected area design principles.

Highly protected areas. The NRS aims to have some highly protected areas (IUCN Categories I and II) located in each IBRA region.

Public land. Priority should be given, in the first instance, to meeting the above criteria from public land, with private land targeted to fill any gaps that remain.

Least cost approach. Selection should consider long-term and short-term environmental, economic, social, and equity implications so that an optimal protected area network is established with minimal economic and social cost.

Consultation. Consultation should be undertaken with community and interest groups to address social, economic, and cultural issues.

Indigenous involvement. The biodiversity conservation interests of Australia's Indigenous peoples should be recognised and incorporated in decision-making (NRMMC 2004).

Gaps in the current protected areas network have been identified based on:

- percentage area of reservation within each IBRA (<1%, 1-5%, 5-10%, >10%) (Figure 5.1)
- level of bias between regions in terms of comprehensiveness (nil, low, moderate, high, no reserves)
- level of threat to biodiversity within each IBRA according to degree of modification, existing land-uses, species extinctions, and abundance of introduced species.

Based on these data, each IBRA has been given a priority classification in regard to its priority for the inclusion of land within the NRS (Figure 5.2).

Snorkellers, Great Barrier Reef, Qld (Graeme Worboys)

5.2   Marine national reserve system

A marine protected area (MPA) is an area of sea (which may include land, the seabed, and subsoil under the sea) established by law for the protection and maintenance of biological diversity and of natural and cultural resources (Environment Australia et al. 2003). In Case Study 5.1 progress towards expanding the MPA system at an international level is described. In Australia, while it is over a century since the first terrestrial protected areas were declared (Section 2.1), establishment of such reserves in marine environments is a relatively recent initiative. MPAs began to receive more attention during the 1990s in light of the Commonwealth Ocean Rescue 2000 program, launched in 1991, and the National Oceans Policy established in 1998.

The United Nations Convention on the Law of the Sea establishes Australia's rights and responsibilities over about 9 million km² of seas, which is 16% larger than the land area (IMCRA 1998) (Figure

5.3). Under this convention, in force since 1994, Australia is required to protect and conserve the marine environment. In 1992 a Taskforce on Marine Protected Areas was established to provide a mechanism for the states, the Northern Territory, and the Commonwealth to collaborate on development of a national representative system of MPAs. A principal aim is the establishment of a National Representative System of Marine Protected Areas (NRSMPA). A NRSMPA is also an obligation under the *Convention on Biological Diversity*, the *National Strategy for Ecologically Sustainable Development*, and the *National Strategy for the Conservation of Australia's Biological Diversity*.

MPAs can be declared under Commonwealth, state (Snapshot 5.1), or Northern Territory legislation. State and Northern Territory governments have primary responsibility for marine environments up to three nautical miles out from the territorial sea

Figure 5.1   Reservation levels for IBRA regions (NRMMC 2004, p. 14)*

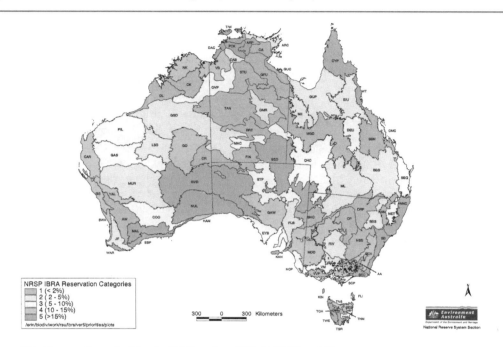

Figure 5.2   Reservation priorities for IBRA regions (NRMMC 2004, p. 14)*

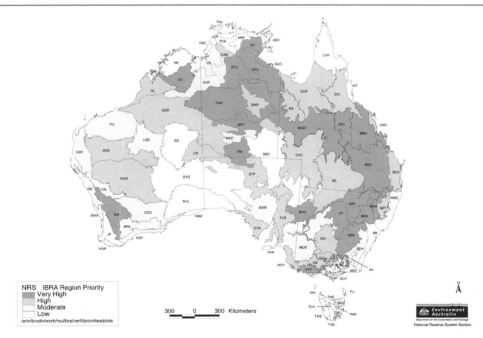

*   *Note that both levels and priorities change over time. Note also that most bioregions, including those with 'low priority', contain particular regional ecosystems that are poorly conserved*

baseline, except for the Great Barrier Reef Marine Park, which extends seaward from the low water mark. Along most of the coast, the territorial sea baseline is the low water mark, but in some areas is up to 60 nautical miles offshore. In general, the Commonwealth Government manages oceans from the state or Northern Territory limit to the edge of Australia's marine jurisdiction, which is 200 nautical miles out to sea. Since the declaration of the Great Barrier Reef Marine Park in 1975, an additional 13 Commonwealth MPAs have been established ranging from strict nature reserves (IUCN category Ia) to managed resource protected areas (IUCN category VI). The Great Australian Bight Marine Park, Ningaloo Marine Park, and the Solitary Islands Marine Reserve, which encompass both state and Commonwealth waters are managed under special cross-jurisdictional arrangements (Environment Australia 2003a, Environment Australia et al. 2003). For example, while the primary management agency for the Great Australian Bight Marine Park is Parks

Australia, management plans include provision for multiple use, and are being developed in conjunction with a range of stakeholders.

Similar to terrestrial environments, the NRSMPA system is attempting to be:

- comprehensive, in that it will include MPAs that sample the full range of Australia's ecosystems
- adequate, in that it will include MPAs of appropriate size and configuration to ensure the conservation of marine biodiversity and integrity of ecological processes
- representative, in that it will include MPAs that reflect the marine life and habitats of the area they are chosen to represent (Environment Australia 2003a).

It is hoped that the system will:

- contribute to the long-term ecological viability of marine and estuarine systems
- maintain ecological processes and systems

Figure 5.3   Australia's marine jurisdiction (Environment Australia 2003a)

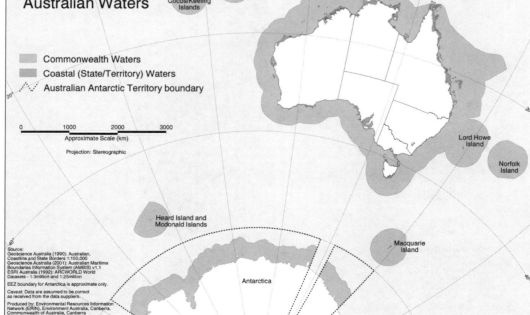

- protect marine and estuarine biodiversity
- provide a management framework for the broad range of human activities in marine protected areas
- provide for recreational, aesthetic, and cultural values (Environment Australia 2003a).

As of 2002, over 64.6 million hectares of Australia's waters (comprising States, Territories, External Territories, and Commonwealth waters) have been reserved in 192 MPAs. While there are no specific targets for setting aside a certain percentage of Australia's marine jurisdiction, the system aims to include some highly protected areas in each bioregion (IUCN Category I and II), as well as establishing MPAs in categories IV and V. There is a need to accelerate development of the NRSMPA both for conservation and to give security of access to industries reliant on ocean resources (Commonwealth of Australia 1998). Significant gaps are evident in the distribution of MPAs, especially in deepwater and cooler temperate systems. The Interim Marine and Coastal Regionalisation for Australia (Chapter 1) is the main tool that is being used to help guide the identification of areas for inclusion in the NRSMPA and ensure that it is representative of the full range of Australia's marine environments (IMCRA 1998). The priority is to establish MPAs in large-scale bioregions that are not already well represented within the NRSMPA. The first reserve selection planning project is being undertaken in the South-east Marine Region (Case Study 5.2), and will build on existing Commonwealth MPAs in the region (Tasmanian Seamounts Marine Reserve and Macquarie Island Marine Park) as well as complementing MPAs in adjoining state waters (Environment Australia et al. 2003).

Case Study 5.1

Global representative system of marine protected areas

Graeme Kelleher, Senior Adviser to IUCN's World Commission on Protected Areas

IUCN and the World Wilderness Congress have defined an MPA as:

Any area of intertidal or subtidal terrain, together with its overlying water and associated flora, fauna, historical and cultural features, which has been reserved by law or other effective means to protect part or all of the enclosed environment (Kelleher & Kenchington 1992).

In 1990 the 17th General Assembly of IUCN adopted a primary goal of reserving and conserving marine environments. Resolution 17.38 was as follows:

To provide for the protection, restoration, wise use, understanding and enjoyment of the marine heritage of the world in perpetuity through the creation of a global, representative system of marine protected areas and through the management in accordance with the principles of the World Conservation Strategy of human activities that use or affect the marine environment (Kelleher & Kenchington 1992).

To move towards the attainment of this goal, the following actions have been taken:

- the writing, publication and wide distribution of simple, inexpensive guidelines that describe the approaches that have been successful in establishing and managing MPAs in various social and ecological situations (Kelleher & Kenchington 1992)
- the division of the world's coastal seas into 18 major biogeographical regions, the recruitment of working group leaders for these regions and one for the high seas, and the establishment of regional working groups, consisting of scientists and managers, government, and non-governmental people

- the establishment of working groups in the countries of each region
- the establishment and empowerment of networks of professionals and activists concerned with MPAs, to work with the working groups
- the provision of assistance and encouragement to working groups in their work.

Considerable progress has been made, but there is still a very long way to go. Thousands of MPAs have been established. Many more are proposed. Some are well managed. Some exist in theory only. In 1994 the Convention on Biological Diversity and the UN Convention on the Law of the Sea gave MPAs an international base for the first time. The field is an exciting one, as some fundamental principles have been accepted generally, such as the need for community ownership of MPAs. There are grounds for optimism.

In 1995 the IUCN, the World Bank and the Great Barrier Reef Marine Park Authority published *A Global Representative System of Marine Protected Areas*. This four-volume report listed the existing MPAs in each country. It identified the highest priorities for establishing new MPAs or for converting paper parks into effective MPAs. It proposed a series of actions necessary to achieve the primary conservation goal, a goal not confined only to MPAs. Progress since then has been highly variable. Some countries, such as Canada, have produced beautifully crafted MPA systems plans, but they have had limited success in implementing these. Others, such as in South-East Asia and the Baltic, are moving with determination and commitment to involve local communities. They are establishing MPA systems that bring ecological and economic benefits. Canada's difficulties arose very likely because it did not integrate socio-economic factors in its process for assessing and making decisions about its MPAs from the start. Australia can learn from Canada's example.

Snapshot **5.1**

New marine national parks for Victoria
DSE (2004)

In 2002 the Victorian Government declared 13 marine national parks and 11 marine sanctuaries totalling about 54,000 hectares or 5.3% of Victoria's marine jurisdiction. Before the establishment of these marine national parks and reserves, only 0.05% of Victoria's marine environment was in highly protected conservation reserves (compared with 16% of the land area). The Parks were established as a result of a review process carried out by the Environment Conservation Council (ECC) (Section 5.4) with the aim of establishing a comprehensive, adequate, and representative marine national parks system in Victoria. There was extensive community and industry consultation over with over 4500 submissions and letters and numerous meetings with a wide range of interest groups.

The largest of the marine protected areas established in Victoria is the Wilsons Promontory Marine National Park. The park encompasses 15,500 hectares and extends along 70 kilometres of coast. It is contiguous with the southern portion of the Wilsons Promontory National Park. Being at the boundary of two major ocean currents, Wilsons Promontory Marine National Park contains species at the limits of their distribution. For example, species of colder waters such as the Leafy Seadragon (*Phycodurus eques*), Red Velvetfish (*Gnathanacanthus goetzeei*), and the intertidal Pheasant Snail (*Phasianella australis*) are at their eastern limit in the park, whereas warmer species such as the Blue-lined Goatfish (*Upeneichthys lineatus*) and the Brittle Star (*Ophionthrix spongicola*) are at the western limit of their distribution. Habitats in the marine park include subtidal rocky reefs covered in kelps and other seaweeds; seagrass beds in some of the sheltered bays; intertidal granite rocky shores; soft sediment areas among offshore reefs; 'sponge gardens' in deeper areas under ledges or in caves, and offshore islands that support colonies of Australian Fur Seal.

Case Study 5.2

Establishment of MPAs in Commonwealth Waters: South-East Marine Region

Karen Edyvane, University of Tasmania

Commonwealth MPAs are established and managed as Commonwealth reserves under the *Environment Protection and Biodiversity Conservation Act 1999*. MPAs may be established for two purposes: as representative MPAs (established on a regional basis under the regional marine planning process); and as MPAs that protect special places or threatened species (which are considered on an individual case-by-case basis, at any time). Under Australia's Oceans Policy (1998) regional marine plans provide the framework for integrated ocean governance, and also the protection and integrated, sustainable ecosystem-based management of Australia's Exclusive Economic Zone and its natural resources. A key outcome of the regional marine plans is the development of representative MPA proposals under the Commonwealth's NRSMPA program. Significantly, the integration of the regional marine planning and MPA processes enables other conservation measures to be considered when designing the MPA system, and ensures that MPAs are not identified in isolation from the management of sustainable resource use (Environment Australia et al. 2003).

Potential representative areas and areas of known outstanding conservation significance go through a similar process prior to a decision being made whether to proceed with the Commonwealth reserve declaration process: scientific assessment, stakeholder consultation, socio-economic assessment, and consideration of conservation objectives. The results of these processes are reported to the Minister for the Environment and Heritage, who decides whether to proceed to the statutory declaration process.

The South-East Regional Marine Plan (2004) is the first of the Commonwealth's regional marine plans and takes in more than 2 million km^2 of Australia's ocean territory around Victoria, Tasmania, eastern South Australia, and southern New South Wales, as well as the sub-Antarctic Macquarie Island. The region encompasses Australia's largest and richest offshore gas and petroleum fields (Gippsland, Otway, and Bass Basins) and major commercial fisheries (lobster, abalone, shark, scalefish), with over 270,000 people working in marine-related industries in the region, valued at $19 billion annually. These same cold temperate waters contain Australia's most important cetacean, seal, and seabird foraging areas and migratory routes. They also contain Australia's most rare and unique marine habitats and species (>90% endemism). With some of the highest levels of marine uses in Australia, they are also, undoubtedly, some of the most threatened.

The South-East Regional Marine Plan (2004) represents the first attempt to identify and establish representative MPAs across a large-scale deep offshore marine region. The identification and selection of MPAs in the region has been facilitated by:

- a list of operational criteria for identifying and selecting a CAR system of MPAs within the South-east Marine Region
- the identification of 11 Broad Areas of Interest (BAOI) that provide a focus in identifying candidate options for MPAs—these areas contain the greatest diversity of bioregions and sea floor features (for example the shelf, shelf edge, slope, abyssal plain, seamounts, and rotated continental blocks) for sampling within candidate MPAs
- MPA guidelines and ecological specifications to assist stakeholders to design options for candidate MPAs within each BAOI
- data on biological values such as seabird breeding colonies, seal foraging areas, whale migrating routes, shark residence areas, and spawning areas of threatened fish (Environment Australia et al. 2003).

The approach is also the first attempt to use key stakeholders to develop candidate MPA options.

Stakeholder engagement is essential to develop MPA options that minimise social and economic impacts of MPA establishment, and to secure ongoing commitment and support for subsequent MPA management. This requires, at the earliest stage of MPA design, the involvement and consideration of the needs of stakeholders, including knowledge of cultural, social and economic impacts (SERMP 2004, p. 79).

Under the South-East Regional Marine Plan, MPAs have been identified for two of the 11 BAOIs: Murray (south-east South Australia) and Zeehan (north-west Tasmania), with the remaining MPAs to be identified later. Following this, a scientific assessment of the entire MPA system within the region will be conducted. Candidate MPA options were developed with input from leaders of peak industry associations from the oil and gas and commercial fishing sectors, as well as conservation and Indigenous groups, scientific organisations, and government agencies. Another approach is to have scientists first identifying candidate protected areas according to scientific criteria, before a selection process is undertaken that incorporates wider stakeholder participation, including consideration of social and economic values. This may clearly reveal the requirements for a comprehensive MPA system. On the other hand, early involvement of stakeholders in MPA planning processes may facilitate ongoing support for management.

The current MPA proposals will continue to be progressed through ongoing stakeholder consultations and scientific peer review. This could result in some fine-tuning of candidate MPAs to address either biodiversity conservation objectives or socio-economic interests (Leanne Wilks, Department of Environment and Heritage, pers. comm.). It is to be hoped that the final outcome will be a system that protects major oceanographic features (upwellings) and areas of high productivity and biodiversity, including key foraging areas of seals and seabirds, shark residence areas, and known spawning areas of threatened fish.

5.3 Scientific reserve selection methods

Most protected areas in Australia have been established through the political process. That is, government agencies and/or interest groups have supported the reservation of an area, and this support has ultimately been manifested in declaration of the area under appropriate legislation. All Australian governments have enacted legislation to enable the reservation of land as protected areas. The political process is discussed in more detail in Chapter 2.

The political approach to selecting protected areas is often ad hoc or opportunistic, is heavily influenced by threat and availability, and primarily determined by economic and cultural factors (Margules 1989, Pressey et al. 1994a). While many important natural areas have been protected in this manner, regional conservation of biodiversity and consideration of other significant conservation values are not guaranteed. As we have seen in the previous section, the distribution of protected areas across the various biogeographic regions of Australia is very uneven, with good representation in some areas, and very poor representation in others (Figure 5.1). Formal selection procedures, while not a substitute for the political process, can allow for more informed land-use decisions based on key biological and social criteria. The NRS provides a set of criteria for reserve selection, but in itself does not immediately identify which specific areas are priorities for inclusion in the protected area network. A number of formal reserve selection procedures have been developed that can potentially perform this task.

A procedure for the selection of protected areas should be explicit, systematic, and straightforward, and should consider the extent to which the options for reservation are lost if a particular site is not preserved, while also recognising the values of efficiency and flexibility (Pressey et al. 1994b). Systematic approaches to protected area selection are characterised as being:

1 data-driven, using *features* such as species, vegetation types, reserve size, or connectivity; and *selection units* that are divisions of the landscape that are to be evaluated for their contribution to satisfying some objectives (point 2)
2 objective-led, based on a set of criteria that have quantitative targets for each feature
3 efficient, in that they attempt to achieve the goals at a minimum cost in terms of other potential land uses
4 transparent, in that reasons behind selection of each reserve are explicit
5 flexible, because features and targets can be varied to explore how changing these parameters influences the configuration and extent of the selected reserve network (Pressey 1998).

As noted in point 1, such methods first divide a study area into selection units. Units are elements of the landscape that are assigned values and form the pool of land areas from which a protected area network is constructed. A unit can be any spatially defined area such as a catchment, environmental domain (the classification of an area according to climatic, terrain, and substrate attributes) or grid square. Units should be small enough to build a reserve with precision, but not so large that a vast number of units is required.

Criteria are then selected, which are used to assess whether each unit should be included in the reserve network. A useful criterion will reflect a significant aspect of reserve selection. Criteria can be divided into four categories: biophysical, social, planning, and reserve design.

Biophysical criteria include factors such as rarity of species, representativeness of ecosystems, diversity of habitat, and naturalness. Social criteria include threat of human interference, community appeal, aesthetics, education value, and recreation and tourism. Planning criteria include adherence to catchment principles, bioregional boundaries, natural boundaries, fire control, and availability of the land. Reserve design criteria are concerned with the spatial placement and characteristics of protected area networks and individual units, including their size, boundaries, shape, connectivity, and geographic relationship to other units. The use of these criteria reflects the importance of considering the relationship of individual units to a network as a whole.

Reserve design criteria are also often employed to combat the problems associated with ecosystem fragmentation or isolation. The way in which a reserve is designed can influence the protection of conservation values and the effectiveness of management. Reserve design criteria recommended as part of the RFA process, for example, included the following.

• Boundaries should be set in a landscape context with strong ecological integrity, such as catchments.
• Large reserved areas are preferable to small reserved areas.
• Boundary-area ratios should be minimised and linear reserves should be avoided where possible except for riverine systems.
• Reserve design should aim to minimise the impact of threatening processes, particularly from adjoining areas.
• Reserves should be linked through a variety of mechanisms, wherever practicable, across the landscape (Commonwealth of Australia 1998b).

Targets and/or thresholds are then specified for each criterion. Targets concern the number of units required to satisfy a given selection criterion. A threshold is a level above which a unit is accepted as contributing to a criterion—that is, a minimum acceptable value that must be met so that the unit is considered to possess a particular feature. For example, the criterion rarity could have the target: 'where possible, each rare species must be found in at least two units'.

Once the criteria to be used have been selected and each unit measured according to each criterion, the resulting data must be analysed to determine which units perform the best, and so should be included in the reserve network. Several systematic procedures incorporating biological selection criteria have been developed (for example Kirkpatrick 1983, Margules et al. 1988, Bedward et al. 1992, Pressey et al. 1994a).

A simple way of testing the units against the criteria is to combine several criteria into a single index, and select the sites that score most highly according to this index. However, often there is no obvious way of weighting and combining criteria. This approach is also inefficient, because there is no way to minimise the number of sites required to satisfy each criterion. A better, but more complicated, approach is to use an iterative algorithm. An algorithm is a rule or series of rules that are applied to each unit to determine whether it should be part of the reserve system. Algorithms can efficiently select a reserve system. However, algorithms that incorporate several scientific, social, and management criteria can become very complex.

As has already been noted, advantages of scientific selection methods include their explicit recognition of why each unit was recommended for the reserve network; consistency of assessment—the same method can be used to recommend reserve networks in different regions; ability to ensure that different types of environments such as deserts and woodlands, as well as the more popular rainforests and mountain areas, are adequately represented in the reserve system; and ability to minimise the area required to adequately satisfy a given set of selection criteria. A weakness of such methods is the heavy information requirement and associated expense in gathering this information.

C-Plan (Pressey & Logan 1995) is an important example of a reserve selection program that has been used for conservation assessment in NSW. The algorithm used in C-Plan is designed to achieve a set of conservation goals for as many features as possible in the minimum area (NSW NPWS 1999b). Included in C-Plan is the concept of irreplaceability. Irreplaceability measures how essential the site is, based on how much of a feature is contained in other sites and how many times a given site is essential, in combination with other sites, to meet a target (Ferrier et al. 2000, Snapshot 5.2). C-Plan was used, for example, in a conservation assessment of the Cobar Peneplain Biogeographic Region in central western NSW. The Cobar Peneplain is a semi-arid area of approximately 73,500 km^2 (9% of NSW), bounded by the Darling and Bogan Rivers. The project involved the use of C-Plan to assess the relative conservation values of land across the region, as well as investigation into ways of incorporating traditional Aboriginal ecological knowledge (Chapter 18) into conservation assessments (Smart et al. 2000). Another example of a quantitative approach to identifying reserve priorities is given in Snapshot 5.3.

Snapshot 5.2

C-Plan
Tom Barrett, NSW P&W

Conservation-Plan (C-Plan) is a software system developed by NSW P&W to support conservation planning decisions. C-Plan was primarily developed for use in the NSW Comprehensive Regional Assessment component of the National Forests Policy. In this process, interest groups (conservationists, timber industry, Indigenous peoples) negotiate with the aim of designing a reserve network that protects the natural and cultural features on NSW public land while maintaining a viable timber industry.

Used with a GIS, C-Plan:

- maps the options for achieving conservation goals in a region
- allows users to decide which sites (areas of land or water) should be placed under some form of conservation management

- accepts and displays these decisions and then:
- lays out the new pattern of options that result.

C-Plan looks at regional biodiversity using the 'irreplaceability' measure. Highly irreplaceable sites are most important for achieving the goal. Sites with a low irreplaceability measure are relatively unimportant for achieving the goal, although some of them must be allocated to conservation management if the regional goal is to be achieved.

C-Plan also facilitates the use of resource, cultural, and social data and allows the user to design decision rules to select areas based on these data. C-Plan allows decision-makers to identify areas for conservation that will have the least impact on industries that depend on forest resources.

Snapshot **5.3**

Identifying gaps and priorities for reserves in NSW
Hubbard et al. (2001)

The so-called 'State Project' provided a quantitative overview of the effectiveness of the NSW reserve system as it was at the end of 1998. Landscapes were used as a surrogate for biodiversity—1486 landscape types were derived by combining data on geology, climate, and environmental provinces. Gaps in the NSW reserve system were identified based on an indicative conservation target of 15% reservation of each landscape type. The project found that 756 (51%) of landscapes had zero reservation, 380 (25.5%) had between zero and 15% reservation, and only 350 (23.5%) of landscapes were adequately reserved. It was also found that 94 landscapes had insufficient vegetation for the 15% target to be met.

Priority areas for reservation were then identified based on irreplaceability and vulnerability. Irreplaceabilty was measured as the percentage of the remaining vegetation needing reservation to reach the 15% target, and vulnerability was the percentage of cleared private and leasehold land within each landscape. The results show a concentration of high priority landscapes in the NSW South Western Slopes IBRA Region, with 85% of the high priority landscapes located on private land, 14% on unreserved public land, and 1% on leasehold land.

South East Forest National Park, NSW (Graeme Worboys)

5.4 Land-use planning and protected area establishment

As noted above, most protected areas in Australia have been established through the political process, rather than using a formal planning framework. There are, however, some notable exceptions. We will consider three of these—regional planning in south-west Western Australia; the Victorian Land Conservation Council (LCC), and subsequent public land-use planning processes; and the Regional Forest Agreement (RFA) process. Regional planning in south-west Western Australia is described in Case Study 5.3. Some key characteristics of the RFA and LCC processes are summarised in Table 5.2 and each process is described more fully below.

LCC and its successors

The LCC process operated in Victoria from 1970 through to 1997, when it was replaced by the Environment Conservation Council (ECC). During the 27 years of its operation, the LCC made recommendations to government concerning the establishment of hundreds of protected areas, including national parks, state parks, wilderness areas, crown land reserves, and several other categories of protected area. No other Australian state or territory has had such an influential or wide-ranging planning process specifically dealing with public land use.

Table 5.2	Characteristics of the RFA and LCC processes	
	RFA	**LCC**
Involvement of stakeholders	Potentially high throughout the process, both formal and informal, though a high level of disputation persisted in some regions	Formal and largely based on written submissions at key stages, although some stakeholders are represented on the council itself
Use of resource data	Extensive gathering and use of biophysical and social data	Extensive gathering and use of biophysical and social data
Use of formal tools and procedures to assess options	Establishment of measurable targets for key features; some (failed) attempts to use scientific reserve selection methods	Minimal
Principal mode of decision-making	Negotiation between Commonwealth and state government representatives	Council of public servants and stakeholders, the latter in the minority
Influence of major lobby groups on decisions	Moderate	Moderate
Main contributions to land-use planning	Integration of land-use objectives across Commonwealth, state, and regional scales	One of the first formal processes to explicitly involve stakeholders
	Use of agreed targets for key biodiversity indicators	First state-wide process for public land-use planning
	Assisting establishment of a national comprehensive, adequate, and representative reserve system	Considered all categories of public land
		Attempted to develop a state-wide comprehensive and representative reserve system
Major criticisms	Targets may not be high enough to secure biodiversity conservation	Process tended to be dominated by major pressure groups
	A so-called 'flexibility' criterion meant that targets did not have to be met in all cases, allowing instances where timber values outweighed important biodiversity concerns	No systematic attempt to reserve adequate representations of all ecosystem types
	Targets could be met in part through a patchwork of small exclusion zones (stream-sides, steep slopes, etc) within logged areas	

Case Study 5.3

Regional planning for protected areas in the South Coast Region, Western Australia

John Watson, South Coast Regional Manager, WA Department of Conservation and Land Management, Albany

Note: Several of the concepts and procedures detailed here have evolved from discussions with existing and former colleagues, especially Ian Herford and Richard May.

Regional planning is a valuable tool for setting a guiding framework for more detailed planning levels. In particular it can provide direction on priorities for subsequent area management plans, and can identify a 'spectrum' of protected area types within a single IUCN category (Chapter 3). Regional planning also provides a sound basis for strategic operational plans, for recognising management capability limitations, and for setting, financing and evaluating individual works programs.

In the South Coast Region of Western Australia there is a network of a dozen or so major national parks (IUCN category II), over 100 nature reserves (mainly IUCN category I), and a small number of timber reserves and state forest areas (category V), totalling over 2.4 million hectares and extending along some 1500 kilometres of coastline across the region (Figure 5.4). There is a legislative requirement under the *Conservation and Land Management Act 1984* (WA) for management plans to be prepared for all the category I and II areas. When we began this task in the mid 1980s we decided to use a broad regional approach through the preparation of a Regional Management Plan. A draft regional plan was released for public comment in 1989 and the final statutory document, duly modified in response to comments made, came out two years later. The regional approach also enabled five key aspects of protected area planning to be addressed:

- overall review and enhancement of the protected area system i.e. bioregional planning
- indicative priorities for more detailed area management plans
- application of the Recreation Opportunity Spectrum for national parks (see Chapter 16)
- operational priorities and funding
- a strategy for addressing management capability.

Review of the physical and biological attributes of a region is a key basis for assessment of the existing and proposed protected area network. Hence climate, geology, landform and soils, vegetation, and fauna are considered. Our Regional Management Plan recommended an improved regional network of protected areas based on biological and physical attributes, juxtaposition with respect to other parts of the network, and the potential for major corridor linkages, particularly along uncleared river systems through the eastern agricultural zone (Watson 1991). There were around 150 changes of protected area tenure or purpose proposed with some quite large areas being proposed to change from Category I to Category II or vice versa.

In recent years, the concept of geodiversity has been increasingly articulated (Kiernan 1996). Whereas geology, landforms, and soils did form a part of the basis of our regional overview, and indeed assessment of individual areas, at the first 10-year review of the South Coast Regional Management Plan in 2002, greater attention was given to this concept, ensuring that a geologically representative system of protected areas is also achieved. Although the vegetation is overall an excellent indicator of geology, landform, and soils, and hence may have ensured a reasonable degree of geodiversity in our protected area system by default, a conscious review is nevertheless required to check for completeness.

Interestingly our regional planning approach has many parallels with what is more recently referred to as 'system planning' for protected areas (Davey 1998). This is possibly a reflection of the relatively large size of our region and its protected area estate.

Our national park network in the South Coast Region comprises about a dozen major areas spread across some 700 kilometres from east to west. Furthermore, some four parks are located within a one- to two-hour drive from Esperance and eight within a similar distance from Albany. As Esperance and Albany are the two major regional centres of population, and both are key tourist towns, we proposed through the mechanism of the regional plan a conceptual 'recreation opportunity spectrum' of parks at the 'macro' level. Thus, parks with major wilderness potential or, conversely, parks with existing or potential major site developments were identified.

This approach has provided a powerful tool when individual area management plans are subsequently prepared (Chapter 7). Typically, during the management planning process there is some community pressure for a 'bit of everything' in each separate area. However, by viewing each park in its regional context we have been able to set it roughly in a position on the conceptual spectrum. For example, where some members of the community have sought a wilderness zone in each park, planners have been able to argue that wilderness doesn't really fit in all areas and is far better catered for in another national park within the local network (Herford et al. 1995).

For all our management plans, we develop implementation programs. The Regional Management Plan is no exception in this regard. The implementation program lists all recommendations or actions from the management plan and identifies those which are 'completed', those which are 'ongoing', those which will be initiated in the next three years ('new') and those which will be 'deferred' beyond three years. For the 'ongoing' and 'new' prescription we indicate who is responsible for the action, how it may be resourced (such as local staff, volunteers, external funding, sponsorship, and so on) and for 'new' prescriptions whether proposed for years 1, 2, or 3.

In recent years there has been increased discussion on the issue of 'paper parks'—those protected areas that may be formally established but not adequately managed or functioning in practice (Dudley et al. 1999a). Lack of management capacity and inadequate resources are among reasons identified (Potter 1998). In our Regional Management Plan we recognised that due to the size of the region and the small number of staff and their location, our ability to effectively manage protected areas and undertake other responsibilities varied markedly across the region. For example, we have no staff presence

Figure 5.4   Protected area network in the South Coast Region of Western Australia (1999)

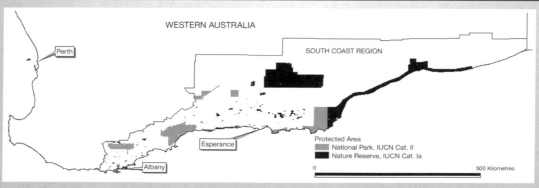

in the eastern half of our region that includes the Southern Nullarbor Plain. We therefore defined three broad zones of management.

- Zone A. Management capability good. Ability to plan operations programs such as fire protection and to respond to emergencies such as wild fires. Generally high public use.
- Zone B. No immediate staff presence, but generally accessible within two hours' drive. Limited ability to plan operations programs and to respond to emergencies. Generally medium public use.
- Zone C. Occasional staff presence only. More than two hours' drive from nearest CALM office or out-station. Can only plan operations programs in exceptional circumstances and at high financial and staff time cost. No ability to respond to emergencies. Generally, low public use except in the vicinity of major highways.

Whereas such an approach does not address the problem associated with the 'zone C' category of 'paper parks' per se, it does however provide a first level of formal recognition of the practical limitations of what in our case is essentially a remote area management issue.

We have found that for our extensive and diverse protected area system it has been crucial to have in place a regional overview and a systematic method of then addressing more detailed management plans for individual protected areas or groups of areas within the region. The use of a hierarchy of planning is particularly valuable as it enables the setting of overall regional priorities and forms an overview basis for a bioregional and geo-regional approach to a protected area system. A regional plan can be particularly valuable in helping to set subsequent planning priorities, in helping to set the 'type' of national park within a user opportunity spectrum, and also to address implementation issues such as priorities, funding, and the management capability zoning of remote areas.

The basic work of the LCC involved systematic assessment of all Victorian public land. The state was divided into study areas, and land-use recommendations developed for each area. This process was repeated for most of the study areas after about a 10-year period. The LCC also undertook special land-use investigations as directed by the government. Major special investigations were conducted on topics such as wilderness, pine plantations, and rivers and streams. The history of the LCC is further elaborated in Case Study 5.4. It should be noted that, despite its achievements, conservationists consider that the LCC process was too strongly influenced by opponents of protected areas, especially those representing forestry, agriculture, and mining interests. As shown in Figures 5.1 and 5.2, the process also did not deliver a comprehensive, adequate, and representative protected area network across all bioregions, although in part this is due to the need for reserves on private land—consideration of which was outside the LCC's jurisdiction.

The *Land Conservation Act 1970* was repealed in 1997 and replaced by the *Environment Conservation Council Act 1997*. The new act established the ECC with a role similar to the LCC, in that it conducts investigations into the balanced use or development of public land within the state. The ECC had three members appointed by the Governor in Council on the recommendation of the minister. The act specified that the ECC must have regard to:

- the ability of any existing or proposed development or use of the land to be ecologically sustainable and economically viable
- the economic and social value of any existing or proposed development or use of the land or resources
- the need to conserve and protect any areas of ecological, historical, cultural, or recreational value or areas of landscape significance
- the need for the creation and preservation of a comprehensive, adequate, and representative system of parks and reserves within the state

Case Study 5.4

The contribution of the Land Conservation Council to reserve selection in Victoria

Ian Miles, Department of Sustainability and Environment

Prior to the twentieth century, the predominant view about public land in Victoria, and indeed other parts of Australia, was that until the land was cleared for farming or pastoral purposes, it was wasteland and that any commercial values should be utilised as soon as possible. This view resulted in the alienation of some two thirds of Victoria by the mid 1950s and significant parts of the remaining public land had been badly affected by human activity.

There was no attempt to systematically retain examples of different types of land in its natural state. Crown reserves were set aside for purposes complementary to agricultural or other human uses, that is, for roads, water supply, timber, and for recreation and public purposes. Utilisation of public land continued apace into the twentieth century.

In the late 1960s considerable controversy arose over government proposals to subdivide public land in the Little Desert in western Victoria for new farms. While the proposal included the establishment of a new national park in part of the Little Desert, the remainder comprising 81,000 hectares was to become agricultural land.

In the public debate that followed, it became clear that conservation, community, and government bodies questioned the merit of such a proposal. Agricultural economists publicly queried the profitability and the effectiveness of its contribution to decentralisation strategies, and voiced strong opposition to the expenditure of scarce capital for farm development in an area of low agricultural potential. Existing procedures for determining the use of public land were perceived to be controlled by an individual minister and public reaction indicated that this was no longer acceptable.

In response to these concerns the *Land Conservation Act 1970* was passed and the LCC was established. The legislation recognised the need for an independent body with relevant expertise to examine competing uses of public land and recommend to the government the most appropriate uses in order to accommodate those people with legitimate interests and the needs of the general community.

The strengths of the LCC model lie in its independence from government, in terms of its chairperson, budget, and staff; the inclusion of experts with diverse interests from outside government as members of the Council; the legislative requirement that the public have the opportunity to participate before any final decision is made; publication of all LCC reports and recommendations to be tabled in parliament; and the legislative requirement that approved recommendations must be given effect by government departments. The *Land Conservation Act* was the first piece of legislation in Australia to contain a provision requiring public participation in land-use decision-making and one of the first such bodies to include members from outside government.

The legislation emphasised that the role of the LCC is to provide scientific and expert advice and that recommendations only be made after all relevant available information had been assessed. These important elements led to the broad community acceptance of the LCC and its processes, even though some sections of the community disagreed with certain decisions and outcomes. Most of the 6000 or so recommendations made by the LCC since its inception were approved by successive Victorian governments and those relating to the establishment of the protected area system have received bipartisan support in the Victorian parliament.

At the time the LCC commenced its work in 1970 the total area of Victoria set aside in the protected area network was approximately 3%. Today, it is more than 15%, largely as the result of LCC recommendations to the Victorian government. Most natural environments occurring on public land in Victoria are well represented in the protected area system, stretching from the semi arid landscapes and deserts in north-western Victoria to the alpine country straddling the Great Dividing Range in the east of the State. The varied forest and woodland environments and the coastal landscapes are a feature of the Victorian protected area system.

The LCC developed a systematic framework for the protected area network, based on land systems mapped across the whole state that identify areas with similar landform, native vegetation, climate, and geology. The types of reserves making up the protected area network include national and state parks, reference areas (set aside as undisturbed examples of natural systems), a network of wilderness areas, and numerous flora and fauna reserves.

The establishment of the protected area network has not been without some cost. Previous uses of public land, including timber production, domestic stock grazing, broombush harvesting, beekeeping, and some recreational activities have been restricted with the result that employment opportunities and economic potential have been foregone. However, the LCC, wherever possible, minimised the impact of its recommendations by adopting a range of strategies to achieve socially acceptable outcomes. For example, providing time to adjust to land-use changes before they are implemented has been an important aspect of the LCC's work. This was critical to the acceptance by the broader community of the necessary changes.

The relative comprehensiveness of the protected area system on public land in Victoria is a legacy of the LCC and has enabled Victoria to participate effectively in the Regional Forest Agreement process, knowing that its reserve system already substantially meets nationally agreed conservation reserve criteria.

Visitor information sign, South East Forest National Park (Graeme Worboys)

Visitor lookout, South East Forest National Park (Graeme Worboys)

- any international obligations entered into by the Commonwealth and any national agreements with the Commonwealth or other states and territories
- the need to protect and conserve biodiversity.

The ECC has now been replaced by the Victorian Environmental Assessment Council, appointed under the *Environmental Assessment Council Act 2001*. The Council provides advice to the government on environmental protection and ecologically sustainable management, including the creation of a comprehensive, adequate, and representative reserve system.

National Forest Policy and Regional Forest Agreements

Australia's National Forest Policy of 1992 established agreement between the Commonwealth, state, and territory governments regarding the broad conservation and industry goals for the management of Australia's forests. RFAs aim to provide long-term stability of timber supplies, within the limitations imposed by a comprehensive, adequate,

and representative reserve system. They address such matters as specifying land-use boundaries, forest management guidelines, and consultative arrangements between governments. The duration of each agreement is subject to negotiation between governments and may vary, but typically they operate for 10 to 20 years, and are subject to review.

Assessment reports were produced for each region, covering environmental, heritage, economic, and social values. These Comprehensive Regional Assessments included consideration of biodiversity, old growth forest, wilderness, national estate values, world heritage values, social values and impacts, resource availability and demand, and regional economic development. Options for forest use based on resource demands and allocations were developed in consultation with industry and the community. Implementation of RFAs is largely the responsibility of state governments (Commonwealth of Australia 1998b).

RFAs can establish a regional forest management plan dealing with detailed operational and implementation matters, or endorse a plan prepared under a state planning process. Agreements and their associated management plans contain a mix of specific and general provisions. In keeping with the National Forest Policy Statement, plans under an RFA incorporate management for sustainable yield, the application and reporting of codes of practice, and the protection of rare and endangered species, national estate values, and biodiversity. Plans may also specify acceptable levels and types of disturbance for particular forest communities, as well as specific performance indicators and monitoring and enforcement arrangements.

The first agreement, for the East Gippsland region of Victoria, was signed in February 1997. Subsequently, RFAs were finalised for:

- Tasmania in November 1997
- Central Highlands (Victoria) in March 1998
- South West forest region of Western Australia in May 1999 (now abandoned by the WA government)
- Eden (NSW) in August 1999
- North-east Victoria in August 1999
- Gippsland (Victoria) in March 2000
- West Victoria in March 2000
- North East New South Wales in March 2000.

The then Australian and New Zealand Environment and Conservation Council (ANZECC), together with the Ministerial Council on Forestry, Fisheries and Aquaculture (MCFFA), established the Joint ANZECC/MCFFA National Forest Policy Statement Implementation Sub-committee (JANIS) to assist with implementation of the National Forest Policy. A technical working group comprising representatives from state forestry and conservation agencies and the CSIRO was established by JANIS to develop criteria for a national forest reserve system. The resulting JANIS report was released in 1996 for public comment. The criteria were finalised and endorsed by the ministers from ANZECC and MCFFA.

These forest reserve criteria (also known as the JANIS criteria) include consideration of biodiversity, old-growth forest, wilderness, and other natural and cultural values of forests. The criteria recognise that both public and private land have a role in meeting conservation objectives. Forests outside reserves will be available for wood production, subject to codes of practice. The national forest reserve system is to be:

- comprehensive, in that it will include the full range of forest communities recognised by an agreed scientific classification at appropriate hierarchical levels
- adequate, in that the ecological viability and integrity of populations, species, and communities will be maintained
- representative, in that the areas of the forest that are selected for inclusion in reserves will reflect the biotic diversity of the communities (Commonwealth of Australia 1998b).

More specifically, the national reserve system will aim to incorporate:

- 15% of the distribution of each forest ecosystem that existed prior to Europeans arriving in Australia
- 60% or more of existing old-growth forest
- 90%, or more, of high quality wilderness (Commonwealth of Australia 1998b).

Old growth forest is defined as forest that is ecologically mature and has been subjected to negligible unnatural disturbance such as logging, roading, and clearing. Where old growth forest is rare or depleted within a forest ecosystem, the aim was to protect all viable examples. Wilderness is land that is in a state that has not been substantially modified by, and is remote from, the influences of European settlement; is of sufficient size to make its maintenance in such a state feasible; and is capable of providing opportunities for solitude and self-reliant recreation (Commonwealth of Australia 1998b).

There are additional targets for vulnerable and endangered ecosystems. At least 60% of vulnerable ecosystems should be reserved. A vulnerable forest ecosystem is one that is approaching a reduction in areal extent of 70% within a bioregional context and which remains subject to threatening processes; or not depleted but subject to continuing and significant threatening processes that may reduce its extent. An example of a vulnerable forest ecosystem that is not depleted but that is sensitive to threatening processes is King Billy Pine (*Athrotaxis selaginoides*) rainforest in Tasmania. This rainforest is highly susceptible to further destruction by fire and it has limited ability to invade burned areas. Although 60% of the pre-1750 distribution of this ecosystem remains, its vulnerability justifies higher levels of reservation.

In addition, all remaining occurrences of rare and endangered forest ecosystems should be reserved or protected by other means as far as is practicable. A rare ecosystem is one where its geographic distribution involves a total range of generally less than 10,000 hectares, a total area of generally less than 1000 hectares, or patch sizes of generally less than 100 hectares, where such patches do not aggregate to significant areas. An endangered ecosystem is one where its distribution has contracted to less than 10% of its former range or the total area has contracted to less than 10% of its former area, or where 90% of its area is in small patches that are subject to threatening processes and unlikely to persist (Commonwealth of Australia 1998b).

The aim of the RFA process was to meet these targets subject to a 'flexibility criterion' that allowed the JANIS criteria to be set aside if necessary to secure 'acceptable environmental, social and economic outcomes'. The approach taken by the then Victorian Department of Conservation and Natural Resources to meet the RFA targets in the north-east of the state, as well as other requirements under the Victorian forest planning process, are detailed in Case Study 5.5.

The target-based approach has been criticised by some eminent scientists as being too simplistic and without sound scientific justification (Figgis 1999). The flexibility criterion also means that targets did not have to be met. For example, in the Tasmanian RFA, the targets were not met for several types of old growth forest, including high timber value Messmate Stringybark (*Eucalyptus obliqua*) and Mountain Ash (*Eucalyptus regnans*) forests. Also of concern is the allowance that small patches of forest excluded from logging under harvesting prescriptions can count towards meeting the criteria. In Case Study 5.5, for example, areas excluded from harvesting under the Code of Forest Practice, such as steep slopes and streamside reserves, contribute to the targets even though they are generally very small, fragmented areas that do not meet reserve design principles concerning spatial placement of protected areas, including their size, boundaries, shape, and connectivity.

Case Study 5.5

Selection of forest reserves in North East Victoria

Kylie White and Ted Stabb, Forests Service, Department of Natural Resources and Environment (now Department of Sustainability and Environment)

A key component of the management planning process for Victorian forests is the preparation of a forest management plan. The state has been divided into Forest Management Areas (FMAs) and Regional Forest Agreement Areas. The North East Forest Management Plan area covers over 2.3 million hectares in the north-east of the state and includes the Central FMA (in part), Benalla-Mansfield FMA, Wangaratta FMA, and Wodonga FMA (in part). The Regional Forest Agreement for the North East uses the same boundaries. Public land comprises 1,253,800 hectares, or 54% of the area, and is mostly native forest. The forest management plan applies to state forest, which covers approximately 30% (700,400 hectares) of the planning area and 56% of the public land. The purpose of the plan is:

to ensure that State forest is managed in an environmentally sensitive, sustainable and economically viable manner. Forest management plans also seek to ensure that planning is a continuing process, responsive to changing community expectations and expanding knowledge of the forest ecosystem (DNRE 1999a, p. 1).

To achieve this, the plan established strategies for integrating the use of state forest for wood production and other purposes with the conservation of natural, aesthetic, and cultural values across the whole planning area. The plan applies for 10 years unless a substantial change of circumstances (such as a major wildfire) warrants a review before then. Flexible strategies will, however, enable progressive refinement in response to new information.

Forest management plans are prepared in the context of state-wide policies and legislation. The most important of these are the *Forest Act 1958,* which establishes a maximum sustainable volume of timber that may be harvested from each FMA; the *Conservation, Forests and Lands Act 1987,* which establishes an overall forest planning framework; and the Code of Forest Practices for Timber, which details a range of harvesting and roading conditions that must be met. For example, the code requires the exclusion of timber harvesting from a minimum 20-metre buffer on permanent streams, and a general exclusion of timber harvesting from slopes with a gradient of more than 30°.

Forest management plans may be prepared in conjunction with RFAs, as well as a number of other state planning instruments (Figure 5.5). Details of exactly which areas are to be harvested each year are set out in wood utilisation plans. A coupe plan is prepared for each area that is logged, to show the exact locations of roads, log landings, streamside reserves, and other areas from which logging is excluded. There are numerous opportunities throughout the process for public comment.

In the North East, the planning team undertook a detailed review of current management issues and gathered information from a wide variety of sources about the resources, uses, and values. In addition, the North East Victoria Comprehensive Regional Assessment reported on the natural, cultural, social, resource, and economic values of the forests in the North East. The extensive biological, social, and cultural data presented in the North East Victoria Comprehensive Regional Assessment has been used in the

Figure 5.5   Key Victorian forest planning instruments

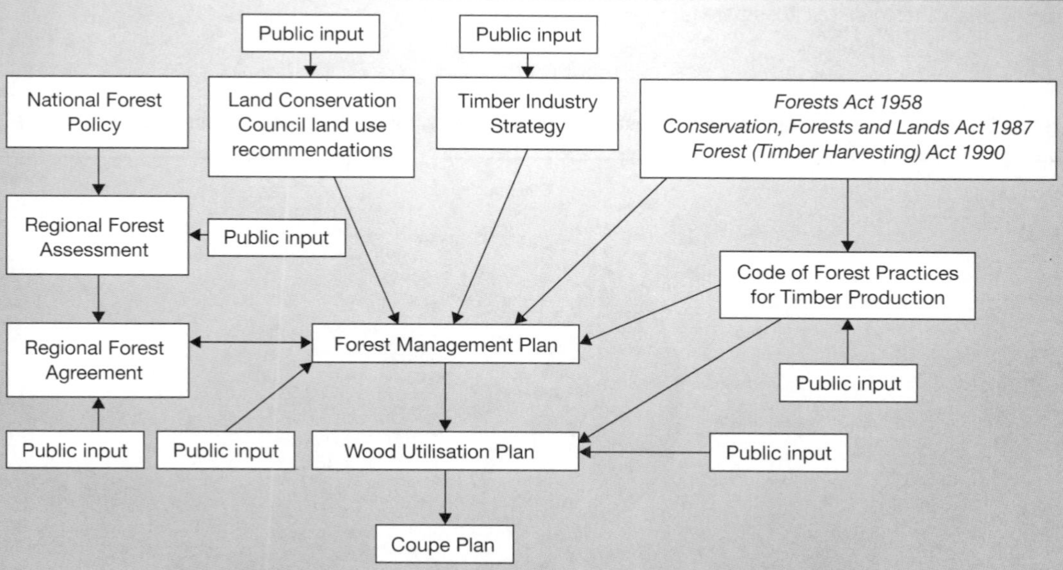

preparation of the plan.

To achieve the aims of the plan, the state forest was divided into three management zones to enable management strategies to be developed for specified values (Figure 5.6).

Special Protection Zones (SPZs) are managed for conservation and timber harvesting is excluded. SPZs form a network designed to link and complement conservation reserves. Larger components of the zone are based on representative examples of vegetation communities and localities of key threatened and sensitive fauna. Several smaller areas identified as sites of biological significance and some research sites are also included in SPZs. Each component of this zone is managed to minimise disturbances or processes that threaten their respective values, and timber harvesting will be excluded.

Special Management Zones (SMZs) are managed to conserve specific features, while catering for timber production and other activities under certain conditions. The areas included in this zone cover a range of natural or cultural values, the protection or enhancement of which require modification to timber harvesting or other land-use practices rather than their exclusion.

General Management Zones (GMZs) are managed for a range of uses and values, with timber production a major use. Forest in this zone will be managed for the sustainable production of timber and other forest products in accordance with the *Code of Forest Practice* and more detailed local management prescriptions.

The zoning system consolidated and integrated information and management requirements from many sources. One of the aims in developing the zoning system was to establish a comprehensive, adequate, and representative forest reserve system in accordance with nationally agreed criteria, while limiting resource withdrawals as far as possible to minimise economic disruption in the region. The process used to develop the zoning scheme is summarised in Figure 5.7. The principal elements of the process are described below.

Remote and Natural Areas. The Land Conservation Council, in its Wilderness Study Final Recommendations (Land Conservation Council 1991), identified Remote and Natural Areas and made recommendations about their management. Implementation of the recommendations requires the exclusion of most forms of disturbance and, accordingly, these areas have been included in SPZs.

Figure 5.6   Proportion of public land in forest management zones, plantations, and conservation reserves

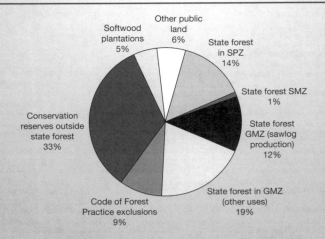

Figure 5.7   Process used to develop the forest zoning scheme

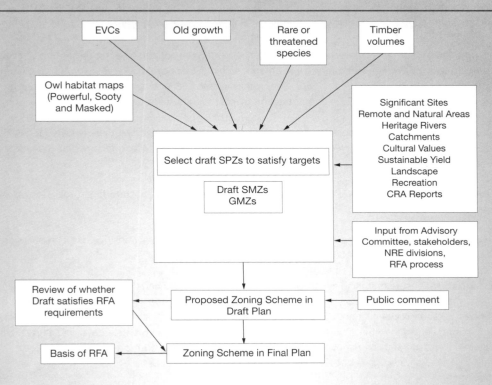

Heritage River Areas. The *Heritage Rivers Act 1992* defines Heritage River Areas and Natural Catchment Areas. Where these areas occur in state forest, they have been included in SPZ.

Flora and Fauna Guarantee Act action statements. Action statements specify the management require-ments for species, communities, or potentially threatening processes that are listed in schedules of the *Flora and Fauna Guarantee Act 1988*. Action statements are developed by multi-disciplinary teams that consider ecological and economic issues relevant to the subject of the statement. Some action state-ments establish management strategies that translate into forest zoning decisions. In these cases, the forest management plan is the primary instrument for implementing the action statements in state forest.

Forest owl conservation. Forest owls breed and hunt over large areas of mature forest. Implementing man-agement strategies developed for Powerful Owl (*Ninox strenua*), Sooty Owl (*Tyto tenebricosa*), and Masked Owl (*Tyto novaehollandiae*) required the establishment of extensive areas of state forest SPZ to complement the habitat provided in existing parks and reserves.

Ecological Vegetation Class protection. Fifty-eight Ecological Vegetation Classes (EVCs) have been identified as currently occurring in the North East. Each EVC represents one or more plant communities that occur in sim-ilar types of environment. Within the state forest, 24 EVCs have been identified. SPZs have been established to contribute to conservation targets for EVCs in accordance with the national reserve criteria. As far as practicable, all areas of rare EVCs in state forest have been included in SPZs. The plan ensures that at least 15% of the pre-1750 extent of common EVCs is protected in either conservation reserves or state forest SPZs.

Old growth protection. SPZs have been established to contribute to conservation targets for old growth forest in accordance with the national reserve criteria. As far as practicable, all areas of rare EVC old growth in state forest have been included in SPZs. To protect at least 60% of the area of common old growth EVCs, conservation reserves, SPZs, and protection by prescription were used.

Landscape. Several areas considered to have high sensitivity to landscape disturbance have been included in SPZs and SMZs. These areas are managed to minimise the medium- to long-term visual impact of management activities.

Recreation sites. SPZ buffers have been established around several major state forest recreation sites to maintain the aesthetic qualities of the sites.

Designated catchments. Several small catchments that may be particularly sensitive to the impacts of timber harvesting and road construction have been included in SMZs.

In addition, to ensure that the diversity of landforms and vegetation, and the range of land uses and management practices are adequately considered over the breadth of the North East, the area has been subdivided into 19 geographic units. These units have been used principally as a basis for analysis of the degree of representation of the various vegetation communities, for establishing targets for featured species conservation, and to establish targets for the provision of recreation opportunities and facilities.

5.5   Management lessons and principles

Lessons

1   The establishment of protected areas can be a long and drawn-out process. It can take many years, involve numerous inquiries, and witness multiple conflicting land-use debates and even protests and campaigns. The fight for the conservation of the northern NSW rainforests, the conservation of Fraser Island, Queensland and south-west Tasmania are examples of hard fought but successful conservation campaigns. All of these areas are now inscribed on the World Heritage List.

2   Formal processes of land-use allocation such as the RFA still require negotiation and hard-fought campaigning to achieve workable protected area outcomes.

3   IUCN Category V and Category VI protected areas hold promise for landscape scale conservation initiatives, but Categories I to IV must be relied upon to ensure adequate biodiversity conservation outcomes.

Principles

Given the processes and criteria described in this chapter, together with considerations of governance and community participation (Chapter 2), and the IUCN classification system (Chapter 3), the guiding principles for selection of protected areas are as follows:

1   A comprehensive, adequate, and representative network of protected areas should be established for terrestrial ecosystems and species

(including freshwater systems) within each IBRA region.

2 Marine protected areas should be established based on adequate representation of ecosystems and species within each IMCRA region.

3 Reserve networks should be designed to take into account the location of protected areas relative to each other and to other land uses, including issues of ecologically optimal boundaries and connectivity.

4 Decisions to establish protected areas should involve participation of those with an interest and those affected by any potential reservation.

5 The interests and concerns of Australia's Indigenous peoples and local communities should be recognised and incorporated in reserve selection decisions.

6 Each IBRA and IMCRA region should contain representative and adequate areas of IUCN category I, II, III, and IV reserves.

7 Protected areas on public land should be complemented by reserves on private lands.

8 The National Reserve System should recognise and incorporate IUCN Categories I to V, so as to create a landscape-scale matrix of protected areas with varying uses and emphases.

9 The National Reserve System should recognise the range of governance mechanisms outlined in Section 2.3, provided an acceptable level of reservation security is afforded.

06

Obtaining and Managing Information

Knowledge is power. Information is liberating.

Kofi Annan, UN Secretary General, Global Knowledge 97 Conference, Toronto, Canada

6.1 Scope of information needs   163

6.2 Data and information
     collection methods   167

6.3 Storage, retrieval, and presentation   174

6.4 Analysis and application   179

6.5 Information management systems   184

6.6 Management lessons
     and principles   186

Good management leads to wise decisions. Wise decisions depend on knowing all of the relevant facts. Both the Burra Charter (Chapter 13) and the Australian Natural Heritage Charter (Chapter 12) warn against making premature decisions in protected area management. Too often, data has been available but has not been used and mistakes made in the past have often been repeated. We live in an information world. It has never been so easy to access so much data and information, and communication technology will only continue to make this easier. Geographic Information Systems (GIS) are a good example of where computer systems have significantly enhanced the capability of protected area managers. In this chapter we look at obtaining and managing data as a basis for competent protected area management.

Access to and use of the most relevant, recent, and cutting-edge knowledge can be beneficial to protected area organisations. Knowledge is synthesised from information, which in turn is derived from analysis of data. Data is collected and stored, for example, on visitor numbers, behaviour, and attitudes. This data is analysed to provide information about comparative visitor use of resources and responses to management actions. This information can be used to help managers to prioritise investment decisions in relation to the provision of infrastructure and services for visitors. Vital to this process is developing a system to manage information, which in turn provides a framework for collecting and analysing data of importance to protected area management. This is not a simple process and often considerable resources and expert knowledge needs to be invested into information management systems. This chapter focuses on issues relating to the scope of information, data collection, storage, analysis, and presentation required at a practical level for protected area managers.

Where full, correct, and relevant data already exists, managers must know how to find and organise it. Where it does not exist, managers must arrange or commission the research that will produce it. Managers need skills in sorting and organising (or managing) data to identify the facts relevant to a given decision. They must be able to spot the gaps where more research is needed and they need skills in interpreting data, especially where it does not lead to 'black and white' conclusions. Managers should be familiar with the different types of data, the different ways it may be accessed or organised and the different places where it may be collected and stored (from the heads of old-timers or skilled tradespeople to books and databases).

6.1 Scope of information needs

There is a range of information requirements for managing protected areas, from detailed scientific knowledge of flora and fauna to visitation figures and financial records. We outline some of the information requirements below.

Physical inventory

Depending on the type and location of the protected area, information will be required on geology, soils, surface water, ground water, ocean currents, geomorphology, and climate. For example, information on flow regimes in the Murray River is vital for management of the Barmah State Park in Victoria. An understanding of geological features is central to management of the Chillagoe-Mungana Caves National Park in Queensland.

Information on geomorphologic processes is crucial for management of coastal protected areas.

Biological inventory

In-depth information about the ecosystems and species within a protected area is vital in order to conserve them. For some species there may be extensive amounts of data, and storing and analysing this data may be the greatest challenge. For other species there may be very little known and primary research will need to be conducted, either by the management organisation or by a consultant. Data may include species composition, diversity, distribution, habitat, and vulnerability, or it may be time series data tracking the effects of climate changes in the region.

Environmental condition

Continual monitoring of park resources is vital for effective conservation management. Indeed many state governments require protected area agencies to provide reports on the state of the parks. This requires collecting, storing, and presenting information on various environmental indicators; for example, wildlife populations, water quality, environmental needs, and so on. Regular benchmarking of these indicators is required. Effective, yet simple survey techniques will need to be put in place for this purpose.

New developments within or adjacent to protected areas will have an environmental impact and research will need to be conducted to assess this impact. Ongoing monitoring of the impacts will be required.

Cultural inventory

Protected areas are designated on cultural values as well so it is important to establish and maintain data on cultural artefacts, sites, beliefs, and rituals. Maps indicating sites of significance should be maintained to assist in the planning of activities and infrastructure and to ensure that these sites are not inappropriately intruded upon. Detailed descriptions of these sites need to be annotated and maintained. Information needs to be provided to front-line managers to assist them with management and to provide information for visitor groups. Cultural information and artefacts have special significance to the local population as well as being of interest to visitors.

Social and land-use history

Knowledge of the human or social history of the area is extremely valuable to protected area managers. In combination with the environmental history this information can provide a complete picture of what has occurred in the area, for example the variety of land use, which may help to explain the composition of the landscape. Like the cultural resources the social history can be utilised to provide interpretation for visitors.

Visitor use

It is important to monitor the level of visitor use at protected areas. Visitation figures may be utilised to calculate the regional economic benefit of parks and they are necessary for estimating environmental impacts, carrying capacity, making decisions about infrastructure investments, and monitoring visitor satisfaction (Case Study 6.1). Protected areas are valuable tourist attractions and visitor monitoring programs need to be implemented.

Non-recreational uses

Depending on their IUCN category (Chapter 3), protected areas can be subject to a range of uses other than recreation. Examples include scientific research, honey production, and seed collecting. Information is needed to ensure effective and appropriate regulation and management of such uses, including minimising their effects on other park values.

Socio-economic costs and benefits

Managers have a responsibility to recognise and, where possible, influence the impacts, both positive and negative, of protected areas on local and regional communities and economies. As shown in Chapter 8, protected areas can make a considerable contribution to local communities. They can also impose costs. Information on such costs and benefits can be used by managers to guide strategic decisions about appropriate relationships with, and contributions to, local communities.

Infrastructure and facilities

Protected areas often contain a diverse array of structures and equipment. Some of these relate to visitor use—walking tracks, visitor centres, signage, car parks; or for marine areas jetties, pontoons, marker buoys, and so on. Others may be incidental to the main mission of the protected area, but still need to be considered by management—transmission lines, water storages, and so on. Infrastructure is often required to support management activities and may include staff accommodation, power supply, helipads, and telecommunications facilities. In terrestrial areas there is usually an access network of roads and tracks. Information is needed on the location, condition, and management of all this infrastructure and facilities.

Management process

Effective protected area management requires efficient administrative systems. Information needs to be stored in regard to human resources, payroll, financial accounts, assets, policies, and procedures etc. Relational database management systems that link, for example financial management and control with visitor management and park infrastructure, enable the efficient connection of different data sets.

Strategic use of information

Different levels within protected area management organisations require different types of information and data. The top (strategic) level of an organisation (Chapter 4) requires information that helps to:

Case Study 6.1

Sustainable resource allocation for visitors to Victoria's parks

Bill O'Connor, Parks Victoria

Appropriate visitor access and service delivery, in parks and reserves, to foster public recreation, education, and understanding of natural systems, is an obligation for park management organisations. With finite resources it is impossible to keep all of Victoria's existing visitor services and assets in optimum condition. This case study focuses on Parks Victoria's *Levels of Service* (LOS) framework, a tool that objectively and effectively supports resource prioritisation decisions, and aids management decision-making towards achieving a financially sustainable service offer.

The LOS framework is a key management tool used by Parks Victoria to establish the 'optimum' quantity and mix of visitor services across the state, given forecast user demand and the availability of resources. The LOS framework utilises all available visitor, asset, and resource data to:

- define consistent, transparent, and objective service standards (including definitions of the resources and infrastructure needed to meet such service standards) across diverse park settings
- ensure that resourcing decisions have a customer focus and balance the management challenges of growing visitor demand/ expectations/ priorities, against the capacity of the organisation to meet such demand (based on asset condition and current inconsistencies in resourcing levels)
- identify state-wide optimum scenarios of the best range of affordable and sustainable recreational activities, and mix of services in appropriate settings, in the context of required service standards and given financial parameters.

In applying the LOS framework, five distinct steps are followed for each visitor node in each park. Figure 6.1 sets out the five steps in more detail.

Sites that are diverse in geography, condition, and type of visitor services offered can be equitably compared in a transparent, consistent manner that is based on current and future customer demand. Coupled with the asset management and visitor research tools, the LOS framework provides a rational and objective basis for applying limited resources to the most valuable sites and visitor services across the state. This maximises Parks Victoria's state-wide performance in relation to visitor services and provides the best return to the community.

It also supports and nurtures cultural change within the organisation. Resistance to individual decisions regarding direction and resourcing of specific parks is minimised because:

- consistent techniques, standards, and scoring methods facilitate objectivity
- a wide range of people are involved in developing and applying the tools that support such decision-making.

Figure 6.1   Level of service framework

STEP 1

Establish model levels of service for each of the five discrete customer profiles
Model service requirements of the five discrete park-users/customer profiles are catagorised from very basic to very high against the 30 most important factors that influence visitor satisfaction (e.g. access; information, interpretation and education; facilities and management services).

STEP 2

Quantify existing levels of service
Assessments are based on the existing condition and functionality of site facilities. Graded scores are allocated to each site, based on the presence or absence of facilities and the type of services provided, compared to those demanded by visitors to the site.

STEP 3

Assess extent of gap between existing and proposed levels of service
Proposed levels of service are based on projected numbers and types of visitor, and the model level of service for each visitor profile and the difference between the existing and proposed level of service based on a standardised score.

STEP 4

Determine relative site importance
In addition to the above influences, site importance assessments are also based on a range of factors including natural/cultural values and the extent of biodiversity in the park, to determine the relative merit of additional investment (both capital and recurrent). A standardised score is also used to measure site importance.

STEP 5

Establish appropriate, affordable and sustainable statewide optimum service level scenarios, given available resourcing
The Service Level Gap is plotted against the Site Importance rating for each park. Each park's relative position in the statewide scattergraph determines what type of resourcing approach will be applied to the park, and the priority of this approach. Cost estimation models are developed and 'ground-truthed' to provide a financial quantification.

Parks Victoria has embraced a program of management reform to achieve higher degrees of accountability, to be more effective and productive, and to address increasing public expectations for better quality public services, all within a finite and limited resource base.

The LOS framework and other performance evaluation tools combine to not only ensure that Parks Victoria's services are in line with government outcomes and longer term organisational objectives, but that these tools enhance the organisation's capacity for planning, resourcing, delivering, and evaluating in an iterative cycle that nurtures continuous improvement.

- determine the adequacy and representativeness of the reserve system
- monitor national and global heritage resources, including world heritage, Ramsar and biosphere reserves (Chapter 3)
- assess whether operations are meeting ESD objectives (Chapter 2)
- set conservation priorities at a state or territory level and thus help allocate finances and staff within an organisation
- manage the flow of visitors at a state-wide level
- prepare audit reports, such as a 'State of the Parks' annual report.

The middle (tactical) level managers need information that helps to:

- set priorities for heritage conservation at a bioregional or landscape level (Chapters 5 and 19) and make sure their policies fit state-wide and national conservation priorities
- identify conservation actions that should have priority at a landscape level, such as replanting projects
- allocate resources precisely within a region, especially for combating threatening processes (Chapter 14) and for dealing with visitors

- educate the community about heritage resources.

The front-line (operational) managers of an organisation need information to:

- understand the context in which they are managing for conservation and the state-wide and national importance of their areas
- cooperate with the local community for conservation
- deal with park visitors and other operational functions on a day-to-day basis
- allocate resources between competing priorities.

6.2 Data and information collection methods

A core component in information management is obtaining the data. It is important to structure data collection to ensure that all of the necessary questions are answered but time is not wasted collecting superfluous data. There are numerous methods of collecting data either from primary research or secondary sources. Primary methods include conducting geological, vegetation (Snapshot 6.1), wildlife, visitor, or neighbour surveys, undertaking experiments, case studies, and interviews. Sources of secondary data include literature surveys, web searches, manuals, databases, media articles, photographs, and reference material such as plant and animal collections, books, journals, and conference proceedings. Data is also collected when conducting evaluation and monitoring, utilising both primary and secondary methods.

Reference data is typically held at a local level, but could also be retrieved from other organisations, and may include:

- published material in libraries and reference collections (includes books, reports, and journal articles)
- published data that gives a social, economic, or political context
- research reports on natural/cultural heritage
- annual reports
- slide and image collections
- video and audio collections
- museum collections, including cultural items and specimens of plants, animals, and minerals

- computer-stored data, CD-ROM and internet sources
- maps, GIS, and planning data
- aerial photographs and satellite images (Case Study 6.2).

The following issues need to be considered in data collection.

1 Data collection needs to be strategic. Attention must be given to what information is needed to plan and manage protected areas and what data must be collected to supply this information. One needs to ask questions such as: 'What data is most critical to the organisation's goals; what is the best and most efficient way to collect it; and will it be presented in ways that tell us how well we are achieving our goals?'

2 Top-level managers need to ask: 'What data needs to be collected from each section (internal monitoring data) to measure the organisation's performance? Can we afford the long-term cost of the proposed system?'

3 How much collecting of data should the organisation conduct in-house and how much should it contract out? Is it worth setting up in-house units and systems?

4 How is the data to be collected and stored? What systems need to be put in place to manage it and to ensure it is compatible for use at a whole-of-government level, nationally and internationally?

Snapshot **6.1**

Monitoring programs
Andrew Growcock, Griffith University

Established monitoring programs can provide an objective record of resource conditions within a changing management organisation. Such programs assist in detecting and evaluating trends between past and present assessments and allow deteriorating conditions to be identified before severe or permanent damage occurs. They also help in assigning priorities for maintenance works, evaluating the success of management approaches, and planning responses for sites demonstrating similar conditions as the monitored site (Marion & Lueng 1997).

In surveys of vegetation and soil condition there are a number of approaches that can be used. Each varies in precision, reliability, cost and the amount and type of information they provide (Marion & Lueng 1997, Hammitt & Cole 1998). Examples include:

- photo point systems—using photos to visually identify condition changes
- condition class systems—using descriptive visual criteria of general site conditions with sites being assigned a numeric value of 1 to 5
- multiple parameter rating systems—allowing individual parameters to be measured, assessed, and scaled with all parameters totalled to provide an overall impact rating.

---

5 How is the data to be analysed to supply the information required?
6 How is the information to be disseminated and to whom?

Operational data is important for day-to-day activities and for dealing with incidents and with visitor safety and may include:

- conditions at specific sites, such as ambient temperatures, wind speeds, relative humidities, rainfall, days since rain, fuel build-up, carbon dioxide levels in tourist caves
- recorded incidents, such as accidents involving visitors
- maintenance requirements
- staffing and finances.

Monitoring data. Monitoring involves comparing current data with benchmark data. This can include:

- maintaining fixed photographic points
- maintaining scientific transects for such things as vegetation histories
- using fixed inventory sites to check fuel loads
- fauna censuses (against benchmarks), especially as part of recovery programs for endangered species
- checking changes in the environment at visitor sites
- counting visitors and their activities
- assessing the numbers of pest animals
- measuring the extent of weeds

- measuring the effect of culling programs on native fauna
- collecting oral histories of areas and events.

Data for decisions on land use. Large amounts of heritage, social and economic data are collected for decision-making processes. For instance, the inventory of data collected for the Regional Forest Assessment process (Chapter 5) included:

- identifying vegetation communities from air photos and 'ground-truthing' this data
- fauna and flora surveys
- plotting topographic data with GIS
- collecting climatic data
- plotting geological and soil data
- collecting social and economic data such as visitor perceptions and behaviour and community attitudes
- multivariate data analyses based on computerised decision-support models.

In South Australia a long-term biological survey has been conducted by the South Australian National Parks and Wildlife Service with considerable success, as described in Case Study 6.3.

Research

Research is a critical means of collecting and analysing data in order to provide information to advise management. Research identifies and assesses

Case Study 6.2

Remote sensing in protected area management

Roger Good, NSW P&W

Many remote sensing techniques are now available for the acquisition of natural resource data for use in the planning and management of protected areas. These techniques enable detailed mapping of resources and the analysis and assessment of data, providing the opportunity to model and develop management alternatives and outcomes.

Prior to the 1970s, aerial photography was most commonly used to map soils, vegetation, habitat, erosion, and land tenure and, while still used in protected area management and planning projects, it has largely been overtaken by the use of remotely sensed satellite imagery and data analysis.

The projects in which satellite imagery has been used in protected area planning and management are numerous but most commonly it is used in detailed vegetation mapping, the determination of 'commence to fill' levels and the extent of riverine wetland flooding, the planning and development of protected area infrastructure such as walking trails and management access, the location of fuel breaks and fire trails, the planning of recreation and tourist management, ecosystem rehabilitation, revegetation and erosion restoration, wildfire mapping and impact assessment, and the monitoring of predicted climate change impacts, particularly as a result of greenhouse gas emissions from wildfires.

For wildfires the most widely used sensors have been those of the National Oceanic and Atmospheric Administration's Advanced Very High Resolution Radiometer, the Landsat Multispectral Scanner and Landsat Thematic Mapper (TM).

During the very extensive wildfires that burnt across south-eastern Australia in January 2003, Landsat imagery was used continuously in the day-to-day plotting of fire fronts and as an aid in planning the deployment of fire suppression personnel and resources. Following the suppression of the fires the satellite imagery and data was used to determine the total burnt area, the day-to-day fire behaviour and the resulting impacts on the vegetation and habitats through which the fires burnt.

While the use of Landsat data and imagery cannot provide for mapping of fire intensities it can, and has, provided an opportunity to map the severity of impact of the Australian Alps 2003 fires, which correlates closely with the range of fire intensities. The process involves image processing to derive a Normalised Burn Ratio (NBR) difference or a fire severity index. The NBR difference index is influenced by changes resulting from the passage of a fire, these being a decrease in green biomass and vegetative cover, a decrease in the cover of ground fuels (increased consumption), an increase in ash and mineral soil exposure, and a decrease in soil moisture.

The NBR technique was applied to the entire burnt area of some 2.7 million hectares in New South Wales, Victoria, and the ACT. From the subsequent mapping of the severity of the impacts of the fires, a number of post-fire management issues, research programs, and rehabilitation projects were identified. The severity mapping provided a basis for ranking the most significant sites in terms of vegetation and habitat damage, the extent of erosion, the degree of stream entrenchment, and the urgency (priority) for the implementation of post-fire rehabilitation and restoration programs. Landsat TM data and imagery was used to map the burnt and unburnt bogs and fens remotely located in the alpine and subalpine areas of Kosciuszko National Park. The TM imagery contributed to the planning and development of rehabilitation techniques for these very important communities.

The use of remotely sensed satellite data and imagery following the extensive fires in south-eastern Australia in 2003 provides but one example of how such techniques are adding to the capacity of protected area planning and management staff to better understand natural resources and systems and to plan and implement innovative and appropriate management programs.

Case Study 6.3

Biological survey of South Australia

Tony Robinson, Jeff Foulkes and Robert Brandle, Department for Environment and Heritage, SA

The South Australian National Parks and Wildlife Service (SA NPWS) was established (by amalgamation) in 1972. By this stage in the development of the state, over 80% of land in the agricultural districts had been cleared and much of the semi-arid and arid pastoral land was severely degraded. We knew that many species of both plants and animals had been lost forever and many more were surviving in only a tiny fragment of the ranges they had inhabited prior to European settlement.

The SA NPWS realised that knowledge of the remaining plants and animals of the state had to be increased to encourage their conservation. SA NPWS chose to concentrate on two groups: vascular plants (ferns and their allies, conifers, grasses and their allies, and flowering plants) and vertebrates (mammals, birds, reptiles, and frogs). These two groups were well known and most of their species can be identified with some confidence. We hoped that if we could conserve a representative sample of them, then most of the other groups, such as insects and fungi, would have a good chance of surviving.

We realised that before beginning we needed a baseline set of data and that this should be collected in a systematic and repeatable manner at sites that could be found again. It should establish the plant and animal species that were 'originally' located at these sites. With this, we could keep track of long-term trends.

The result was the Biological Survey of South Australia. This took shape largely during surveys of the state's offshore islands from 1976 to 1982. The next step was a document that outlined a program of systematic biological surveys that covered the whole of mainland South Australia.

Biological survey camp in the Anangu Pitjantjatjara lands, SA (Tony Robinson)

The first such survey covered all of the vast Nullarbor Plain. This was done in cooperation with the WA Department of Conservation and Land Management. The fieldwork was carried out in 1984 and a report published in 1987. During this survey we learnt a lot about field logistics, establishing computer databases and analysing survey data. We have continued to refine our approach since then. On the survey's recommendation, the South Australian Government proclaimed the present Nullarbor Regional Reserve and, with the purchase of Koonalda Station, extended the Nullarbor National Park westwards to the SA/WA border. On the Western Australian side of the Nullarbor, there was a network of representative Nature Reserves proclaimed in accordance with the survey's recommendations. Subsequent biological surveys continue to lead to further reserves being declared.

The biological survey databases were established following the Nullarbor survey. They contain in excess of 400,000 quadrat-based records of plants and animals. These data were sourced from over 100 discrete surveys that were conducted throughout the state. As well, the 'opportune system', which is non-site based, contains approximately 25,000 records of plants and animals.

By December 1999, the Biological Survey of South Australia had covered just over half of the state at this level of detail. This involved sampling over 15,500 vegetation quadrats, of which 4200 had also been sampled for vertebrates. All of these vertebrate survey quadrats are permanently marked in the field. Hence, they represent a vital baseline against which biologists in the future can begin to measure long-term ecological changes in this state.

Clearly, this level of commitment to a survey is not possible without considerable planning and coordination. A Biological Survey Coordinating Committee formed in 1984 and met quarterly. It had representatives from the Departments of Environment and Heritage (National Parks and Wildlife, Pastoral Management, Wildlife Management, Coastal Management, State Herbarium), Transport, Urban Affairs (Geographic Analysis and Research), and Primary Industries and Energy (Marine Conservation, Soil Conservation). This committee ensured the most efficient use of the increasingly scarce resources within government.

These surveys do not happen without financial support. The early days of the survey were dependent on obtaining large Commonwealth grants from bodies such as the Australian Heritage Commission and other agencies of the Commonwealth Department of Environment. This is still necessary to some extent, but the level of state government support dramatically increased since 1994. By 1999 the annual total funding had risen to $500,000 per annum and there were three full-time staff.

Yet the major reason for the success of the Biological Survey of South Australia to date is our use of hundreds of enthusiastic volunteers in our field sampling program: mostly young biological science graduates or talented field naturalists. This dedicated bunch have endured extremes of weather ranging from a snow storm in the Flinders Ranges to tornado-like winds in the far north of the state.

People often ask 'Have you found any new or endangered species?' and invariably the answer is 'Yes we have'. Two new species are a yellow-flowered *Lechenaultia* from the Yellabinna sand dunes (since named *L. aphylla*) and the Slender Blue-tongue Lizard from the Stony Deserts (since named *Cyclodomorphus venustus*). There have of course been many new records for the state. Threatened species, such as the Sandhill Dunnart (*Sminthopsis psammophila*), have been found on the surveys; and as a result, research programs have been established to study their ecology in more detail. Some plants and animals thought to have been rare have been found to be quite common. Others have been shown to be genuinely restricted and often in need of special management to ensure they survive.

The Biological Survey of South Australia hopes to complete its coverage of the state by about the year 2015. Already, we have improved our ability to conserve our natural heritage into the future.

the presence, significance, functioning, and interdependencies of natural, cultural, social, and economic resources and ecosystems. It reveals our rich cultural heritage and shows the antiquity and diversity of Aboriginal culture. It helps manage fires, make plans to recover endangered species, manage visitors, understand communities, and improve operational systems. Managers need to constantly interact with researchers and facilitate their vital work.

Our focus here is on the manager's viewpoint. How can managers encourage researchers to work in their area and to address salient issues? How can managers influence the design of research projects, to make sure they do not damage the area? How can managers help reduce the risks researchers sometimes run in the field? Why should at least some research be done in-house?

Facilitating research

Managers rarely have the resources for collecting large amounts of data, but they have ways of facilitating such work. It can be beneficial to all parties involved if protected area agencies form partnerships with other research institutions. They may pay consultants, or team up with universities or other government organisations such as the CSIRO or cooperative research centres. Volunteer and community groups can sometimes be very helpful and the private sector may sometimes sponsor research programs. A good example of

Snapshot 6.2

Research partnership between Narooma District of NSW P&W and Charles Sturt University

The partnership began with a simple research project on Montague Island Nature Reserve near Narooma. The district's early requirement was to closely monitor populations of seabirds on the island. This was part of managing a new ecotourism venture that was attracting visitors to the district. Lecturers and students from Charles Sturt University matched the district's need with an enthusiastic commitment to understand the ecology of the Short-tailed Shearwater (*Puffinus tenuirostris*) and the Little Penguin *(Eudyptula minor novaehollandiae)* populations of Montague Island. It was a fruitful partnership. P&W provided logistic support (boat access and accommodation) and some assistance with costs. In return it received annual updates on the dynamics of the seabird populations. With the Narooma District Advisory Committee, the University co-hosted a scientific seminar that highlighted the benefits of the collaborative research work on Montague Island. This received considerable, and positive, media coverage. P&W provided further logistic support to the researchers when a vacant room on the island was converted into a small laboratory.

The partnership evolved beyond Montague Island. A formal agreement (Memorandum of Understanding) was drawn up between the University and P&W. This has produced a major series of baseline studies (social and ecological) for a newly declared national park within the same district, the Eurobodalla National Park. The studies contributed directly to the preparation of the plan of management for the new park.

Charles Sturt University researcher Jamie Weber and Little Penguin at Montague Island Nature Reserve, NSW (Graeme Worboys)

cooperation between managers and a research organisation comes from Narooma in NSW (Snapshot 6.2). As the Snapshot suggests, managers can actively attract research to their area by liaising with research institutions and potential researchers (or their supervisors). An area can gain a reputation as being a good place for postgraduate students and others to do research. Managers may be able to provide some seed money, which would be gratefully received by students and may sometimes be able to fund a PhD or a consultancy project.

Logistical support can attract and keep researchers in an area. Simple matters such as reliable access, communications, safety backup, and accommodation are all important for researchers. In this way, managers can greatly increase a researcher's chance of completing a project. Researchers can then focus more on their own work, to the benefit of both.

Working with researchers—research design. Unless managers invest some of their funds and energy into data-gathering they limit their chances of conserving species and ecosystems for the long term. Of course, when managers sponsor research, it needs to be focused on relevant issues. Managers can often influence the design of research projects within their area and their advice can often save a great deal of stress and unnecessary cost. Researchers from other institutions should consult protected area managers when designing their research program. Sometimes their work, while of interest in itself, may contribute little to solving management problems. Managers should appreciate the need for pure research; yet when consulted they should endeavour to steer researchers towards projects that produce knowledge that helps conservation.

There are a number of questions that managers may raise when consulted by a researcher about a new project.

- Is this research appropriate for a protected area and what environmental impacts will it cause?
- How relevant is this research to good management of the area and what priority should it have?
- Does it duplicate previous work?
- What will be the costs of supporting the project, both directly and in kind?

- What risks and what safety issues does it involve?
- Is the researcher insured, especially if operating in remote or dangerous locations?
- Is the researcher suitably equipped and trained to deal with the terrain?
- Could the research have economic and/or political ramifications, and who must approve publication or public statements?
- Have the required ethical and legal clearances been obtained?
- Will data be collected in a format that standard data systems can use?
- Will all materials and equipment be cleaned away, even if the project is abandoned? What happens if the researcher fails to clean them up?
- Is the researcher familiar with the current management plan for the area and with any relevant codes of conduct?
- What supplies, equipment, or loans of equipment are the managers asked to provide? Can they keep up these commitments for the length of the project?
- Have intellectual property issues been addressed—that is, who owns the knowledge produced by the research?

These important questions would normally be raised on the application form for a research permit, though this in itself may not be enough. A face-to-face meeting is best. Poorly designed research projects have created problems for managers in the past, as shown in the following examples.

- Researchers can demand special access that contradicts a plan of management, such as four-wheel-drive access to wilderness areas where all other visitors are required to walk.
- Researchers may request use of a pristine area as an experimental site, precisely because it is pristine. One example from Kosciuszko National Park illustrates: a bungled experiment where domestic bees were introduced to the alpine area, which is normally free of introduced bees.
- Researchers may abandon a project without cleaning up the site, or they may fail to remove apparatus such as fences, star-pickets, marker-tapes, hides, and shelters until well after the research project has finished.

- Researchers may complete the project, but not keep their promise to compile, analyse, and pass on the data such that it contributes to improved conservation management.

Research—risk management

Managers need to work very closely with those researchers who plan to work in remote locations. There should be a contingency plan for such things as severe weather, floods, impassable roads, sickness, and injury. The plan should not rely on evacuating researchers by helicopter in an emergency. This is costly and should only be used as a last resort. Managers may need to link the researchers to the management's own two-way radio system and set up a series of scheduled radio contacts. Researchers should be skilled in using the relevant communications system, as well as four-wheel-drives, chainsaws, winches, boats, and other crucial equipment.

All researchers must work under a signed research agreement, which clearly sets out all issues relating to safety and risk, including adequate insurance cover.

Research capacity

Managers need their own in-house specialists whose research helps define future methods and priorities for conserving natural and cultural heritage. Such experts give an organisation a base level of expertise for dealing with complex issues and are a critical part of a balanced organisation. They often set the conservation agenda by contributing to high-quality published papers, conferences, and seminars. Whether it is for cultural heritage, social, economic, recreation, threatened species, wetlands, managed fauna, or other research work, in-house scientists and professionals provide a leadership role in defining heritage conservation actions for organisations. Case Study 6.4 describes the valuable role that long-term vegetation studies can have for conservation management.

Monitoring

Monitoring our heritage is a crucial part of any long-term management of a protected area. The National Strategy for Ecologically Sustainable Development (Commonwealth of Australia 1992a) called for regular reporting on the state of the environment. It was recommended that a system be introduced at the national level, so that it could provide feedback and measure success in achieving ecologically sustainable development (ESD). If the reports show that environmental indicators are falling, then ESD is not being achieved.

The task was to develop a nationally agreed set of environmental indicators. Saunders et al. (1998) prepared a consultancy report titled 'Environmental Indicators for National State of the Environment Reporting'. The study was one of a number of expert reports commissioned by Environment Australia and it illustrates the complexity of this work. It recommended that reporting be based on some 53 environmental indicators for reporting at a national scale. Such research must be done at various levels: state or territory, bioregion, and protected area.

The Regional Forest Agreement process (Section 5.4) requires local managers to monitor the conservation condition of forests, though the data are used at a national level. Species recovery plans must be monitored so they can be improved. Such monitoring leads directly to sound management of natural and cultural heritage, as described in Chapters 12 and 13. Monitoring can also focus on the organisation's operational methods and its services to the community.

Systematic monitoring is necessary to assess and improve upon benchmarking performance. This forms the basis for adaptive management, whereby procedures that are not producing the desired outcomes can be amended and more innovative techniques applied.

6.3   Storage, retrieval, and presentation

Accurate and comprehensive data are crucial. So too is being able to store and retrieve them quickly and simply. This is true at a local, state, national, and international level. Local systems are just as important as the more sophisticated information systems that cover national and international areas.

Case Study 6.4

Using historical long-term vegetation studies for management and restoration

Pascal Scherrer and Catherine Pickering, Griffith University

Protected area managers can access a wealth of information from historical long-term vegetation studies that, if maintained, could provide answers to current problems. The studies provide information about recovery from past disturbance, response to natural events, and introduction of feral animals and plants as well as indicate what natural cyclic patterns may be in operation.

The value of long-term data about ecosystem processes is likely to outweigh the design of the original experiment, incomplete data sets, and varying sampling frequencies. Constraints of funding and logistics often limit the establishment of new and broader studies. Thus it is important to maintain and utilise, where possible, existing long-term studies.

An example of an existing monitoring program is the Kosciuszko alpine long-term vegetation transects. Alpine ecologists Alec Costin and Dane Wimbush established one of the most important monitoring programs in the Australian Alps in 1959, in what is now Kosciuszko National Park. It is widely recognised that the data accrued from their studies have provided ecological support and direction to the management of the alpine and subalpine zones for the past 40 years. The results of the long-term transect study were published in 1979 (Wimbush & Costin 1979), and monitoring continued, with data from 1959 to 2002 recently analysed by the authors (Scherrer 2003).

Wimbush and Costin's study consists of a series of six 152.5-metre transect lines in two alpine valleys, one near Mt Kosciuszko and the other near Mt Gungartan. The transects were established to monitor vegetation change and to assess the rate of recovery, and the process of re-establishment, of native vegetation on denuded or eroded alpine humus soils following the removal of livestock grazing. The two areas were selected as representative samples of alpine vegetation with different histories of land use (Wimbush & Costin 1979).

Data from the permanent transects at Gungartan and Kosciuszko provide a detailed account of changes in the cover of individual plant taxa (species, genera, and life forms) at these sites. The long-term data set provides information on the severity of climatic effects (such as drought) on the vegetation and gives an indication of the time frames of recovery from disturbance, vegetation growth, and natural patterns. The study provides information for the selection of native species, which could potentially be used in future restoration programs. It also documents the valuable role of individual species in the colonisation of bare ground and may provide an important baseline to monitor potential shifts in species composition, such as the predicted upward shift of species and the increase in weed establishment and distribution with human-induced climate change. Historical data sets, such as Wimbush and Costin's study, are invaluable sources of information to assist managers of natural areas in the planning and management of land-use activities and the development of successful restoration programs.

Research collections, even in the simplest form, are valuable aids for managers. The larger systems are used more for setting priorities and for making closely analysed comparisons. By contrast, local data is used directly by local managers as a basis for the actions they take. The development of a local information system and the establishment of databases on which they depend, is of vital importance. Simple yet powerful proprietary packages such as Microsoft Access can be used. But a word

Case Study 6.5

Establishing and maintaining a reference collection

Catherine Gillies, NSW P&W

In late 1996 we were due to move to new offices in Jindabyne, complete with a large, fireproof room to house the reference collection. This would be the perfect opportunity to give the Snowy Mountains Region a new library. A vision began to form: a clean library with adequate space, a Dewey shelving system, a computer catalogue with keywords, and not a rat, silverfish, or flood in sight. In short, I wanted to create a specialised library where the staff of Kosciuszko National Park could find good information on park management and on the Snowy Mountains Region. It would be a library that was easy to manage and where data was easy to find. Above all, it would be a fitting home for all the unique materials held in the region.

The process. Having established that a major overhaul of the collection was needed, we went looking for advice from the staff and for funding. A submission to the regional executive secured a $10,000 budget to employ a librarian and buy software. An informal survey sent to staff told us exactly what people wanted their library to be. I was lucky that so many staff saw the enormous potential of the collection and made useful suggestions that determined the main aims of the library project.

As soon as the collection was moved, we contracted a professional librarian for two months. Together we set up the framework of the library. We made a priority of the catalogue, making sure it held the data we wanted and that it was easy to use. The system was to operate for many years to come, so it had to be done properly. All the theory aside, the business of actually creating a library can be daunting. Using Dewey numbers to find books is easy, but assigning those numbers to uncatalogued books is another ball game: If I shelve it here, will people be able to find it? Should a book on park management go under a parks number or a management number? Do walking tracks go under recreation or facilities management? Keywords can be tricky too—how many is enough? Are these the best we can use?

In the end we stuck to three basic principles: make the keywords and the tools of Dewey work the way you want them to work, keep it consistent, and always keep in mind what you are trying to achieve.

One of the best aspects of the project was the creativity involved: we built a computer for the catalogue from bits of old broken computers, we begged copies of old editions of Dewey books and of subject-heading microfiche from the National Library, and we talked to other people who ran small specialist libraries. The most satisfying initiative was arranging for the staff who worked on the park's visitor-entry stations to help with the cataloguing of pamphlets during their quiet spells. They appreciated having productive work to do, they learnt new skills, and their contribution was valuable work that I simply did not have time to do.

Once the system was in place, the key to keeping it going was constant attention. I spent countless days entering the details of books, maps, journals, pamphlets, and photographs into the catalogue, with little to inspire or reward me other than the increasing number at the bottom of the screen, which reminded me of how far we had come. Feedback from staff was required to make sure we were on the right path. To get feedback I sent out a newsletter, offered to run a tour, sent a progress report to the regional executive, and sent around an email asking for comments and suggestions.

In terms of maintenance, if a system is well set up, there shouldn't be too much to do. I checked loans in and out once a week and did a shelf check (stocktake) every 12 months. New items were catalogued

Catherine Gillies, park management library, Kosciuszko National Park, NSW (Liz Wren)

and shelved once a fortnight. Whenever I was in the library, I checked that it was tidy, that the computer was working, and that the desk was clean and conducive to being used.

The success story. A few years later, the computer catalogue had over 3000 entries, with author, title, date, and keywords stored for each. The Dewey-based shelving system made things easy to group and easy to find. The new surroundings provided room for reading and working. The shelving was clean and there was plenty of it. A recent shelf check showed that all is in good order.

Staff use the library regularly. There are more than 300 loans recorded, and more than 1000 visits by staff, students, and researchers annually. Feedback shows that people realise there is useful information in the library and that they can easily find what they want. Cooperative work with other park offices means that they hold copies of our catalogue and we of theirs, which helps exchange of information between offices.

And the moral is … I think the secret of our success was to stay focused on the objective and to keep staff involved. If it is their library and they have input in creating it, they will respect it. If it works for them and is easy to use, they will keep using it and value it. The final thing is not to give up. Even two years after creating the library system I still had 15,000 slides, 60 photo albums, 50 journal collections, 100 boxes of journal articles, and 1000 maps to go. But I kept looking at that number on the screen that said '3000 records' and thought of the old reference collection; and then I *knew* I had achieved something worthwhile.

of warning—be sure to identify the specific uses for the database before it is established.

Protected area agencies generally keep a multitude of reports produced by their own agency staff or contractors; the problem is usually finding what you are searching for. Catherine Gillies relates her experiences with developing and managing a reference collection in Case Study 6.5. It is important to have in place a good reference collection linked to a regularly maintained database or catalogue system to make sure that all material is easy to locate.

Electronic systems for storing and retrieving data range from the simple to the sophisticated. Information can be stored using a range of computer programs and hardware, and there are as many ways of retrieving information, especially with the use of the Internet and electronic library and journal catalogues. The trick is to know how to search these databases, compile the data, and turn it into meaningful information. Researchers have developed a range of systems for accessing this information, including national systems such as the Environmental Research Information Network (ERIN). At the international level, the UNEP World Conservation Monitoring Centre (WCMC) works closely with the IUCN (the World Conservation Union). It has set up a global database to which Australia is a contributor (Backgrounder 6.1).

GIS is a powerful storage, analysis, and presentation tool and the use of GIS in protected areas is discussed in detail in Backgrounder 6.2.

Botanist and endangered rainforest species—Daintree Wet Tropics, Qld (Graeme Worboys)

Backgrounder 6.1 The World Conservation Monitoring Centre

Modified from WCMC (1999)

The World Conservation Monitoring Centre (WCMC) is internationally renowned as a provider of a wide range of information services on conservation and sustainable use of the world's living resources. Based in the United Kingdom, the WCMC locates, compiles, and manages data on species, habitats, and sites. It does this through global networks of experts and through published and unpublished literature. WCMC disseminates its information both electronically (Internet, CD-ROM) and via hard copy (reports/publications). The WCMC accesses and manages data covering:

- globally threatened plant and animal species
- habitats whose conservation is of special concern, in particular tropical forests, coral reefs, and wetlands
- sensitive sites that are of critical international importance for nature conservation
- the global network of protected areas
- the trade in wildlife and its impacts on populations in the wild.

The WCMC provides:

- national biodiversity profiles tailored to particular needs or users
- information ranging from maps and descriptions of protected areas to lists of key contacts
- summaries of what is known about particular species or habitats and about programs to conserve them
- lists of where and how further information can be found, from published or unpublished sources.

| **Backgrounder 6.2** | Use of digital information and technology in managing protected areas |

Tom Barrett, NSW Department of Environment and Conservation

Geographic Information Systems (GIS) are computer programs that facilitate the display, manipulation, and analysis of spatial data such as digital maps (or 'layers'). GIS layers must be geo-referenced; that is the location of each feature must be described using a coordinate system. For example, the 'Geographic' projection uses a latitude and longitude coordinate system to describe locations on the Earth's surface.

One of the most powerful capabilities of a GIS is the ability to overlay two or more GIS layers to explore the relationship between them or generate a new layer by combining them. For example, by overlaying a vegetation community layer with a protected areas layer it is possible to assess the amount of each community that is protected.

A plan of management has to be developed for all national parks and nature reserves in NSW under the NSW *National Parks and Wildlife Act 1974*. An integral part of these plans is the description of the park's geology, fauna and flora, and Aboriginal and European cultural heritage. The distribution of these natural and cultural features can be defined and quantified using GIS analysis of digital information. A GIS allows these features to be compiled and presented on a map, which facilitates the planning of management actions.

Fire management plans have to be developed for all national parks and nature reserves in NSW. Two of the primary objectives for fire management in protected areas (as stated in the *National Parks and Wildlife Act 1974*) are to avoid the extinction of native species and to protect all Aboriginal sites and culturally significant features known to exist within the reserves. These objectives can only be met if there is sufficient knowledge on how native species respond to fire in combination with distribution information describing where these species and associated communities occur in the reserve. GIS mapping allows managers to access this information in combination with fire frequency mapping, compiled over many years of record keeping.

Mapping the distribution of weeds and pests with the use of Global Positioning Systems (GPS) builds layers of information that can be used by a GIS. These layers can be linked to databases that record management actions on the ground and over time. These systems allow pest management officers to monitor the effectiveness of the control programs.

In NSW the Department of Environment and Conservation maintains a centralised database, called the NSW Wildlife Atlas, which records fauna and flora sightings. In recent years the use of GPS has proven to be an invaluable tool for the collection of accurate location information when a species is sighted. These records of sightings are especially important for determining the status of threatened species across their range. Species records allow us to characterise the types of habitat that the species are likely to occur in through their association with mapped environmental variables. It is only by protecting these habitats across the landscape that we will prevent the extinction of these vulnerable species.

6.4   Analysis and application

Documents on individual heritage resources may not mean much by themselves. One needs to be able to collate, compare, and analyse them so as to see overall patterns and draw conclusions.

Strategic level management is likely to require data analyses that show:

- the actual cost of conserving individual heritage sites, species, habitats, and ecosystems
- the likely economic benefits to the community of conserving ecological systems and processes
- the success of heritage conservation programs in reaching their targets

- how such programs have performed, by national and international benchmarks and how well they have fulfilled national and international agreements
- the long-term trends of conservation management.

Tactical management needs data analyses that show:

- conservation priorities at a bioregional level, including data that is required for whole-of-government conservation programs and the effectiveness and cost-effectiveness of regional programs

- the status of the environment
- whether ESD targets are being met
- how well particular programs are working
- staffing details.

Operational management needs data analyses that show:

- the conservation priorities for the local area, which sets them in a national and state-wide context
- the effectiveness of local conservation programs
- the case for various land-use options (including conservation versus development) for all sites in the area
- how other organisations achieve ESD targets
- ecological data needed for recovery of endangered species.

Good data analysis can assist managers to do first what needs doing first—right down to operational level. It means that they can put forward the very best arguments when they seek funds from treasury. Staff in customer service positions need to be kept up to date on developments within the park and of any new discoveries in regard to local flora and fauna so that they can keep the public informed.

Disseminating information

The community needs to be given as much information as possible about natural heritage values and about the benefits of conservation. That way they are more likely to support the efforts of managers, directly and indirectly. In particular, managers should provide the very latest natural heritage information to those who make decisions at state and local government level. This should include an analysis of what needs to be done to conserve local bioregions. These priorities should be updated annually.

Such information makes politicians and other decision-makers more likely to focus on the effects their decisions will have on conservation. If managers hope to conserve natural areas in perpetuity, they must become skilled in providing clear and impressive data to those whose decisions determine the fate of these areas.

There are various methods of disseminating information to staff, the general community, and government officials. Information can be distributed to the community via newsletters, fliers sent in the mail, posting information on a noticeboard or web site, or through fact sheets, either web-based or hard copy, which also may be available to visitors to protected areas. During the 2003 Kosciuszko bushfires, NSW P&W utilised a variety of methods, including email, to disseminate information to the community. The Internet provides a fast and effective way of disseminating up-to-date information to the wider community. For example, the NSW P&W site (www.nationalparks.nsw.gov.au) has a link to information in regard to fires, floods, and park closures.

Fact sheets may take on a number of roles, instructing visitors about developments in the park, providing guidelines on behaviour within the area, or informing visitors about the local wildlife or features of the park. On-site fact sheets may complement interpretation facilities, including visitor centres, signage, and guides. Information can be visual as well as textual and hence can include physical displays or images. Protected area managers might make use of the media—print, radio, television—in order to communicate this information to the general public, particularly if there are health warnings regarding protected areas, for example, flood or fire warnings, or if specific events are planned.

Government departments often have specific reporting requirements such as annual reports, and state of the environment or state of the park reports (Case Study 6.6) in order to assess park performance and to meet their own reporting requirements. These reports are an effective way of informing the general public and partner organisations about the environmental, cultural, and financial achievements of the protected area, and provide a way of checking accountability. An accumulation of internal organisation reports on topics such as the status of financial and staffing records, the effectiveness of management programs, visitation levels, the conservation status of species, and performance against other benchmarking criteria, may go towards compiling an annual report.

Case Study 6.6

State of the Parks reporting: Parks Victoria's experience

Julie Richmond, Linda Greenwood, and Sally Troy, Parks Victoria

Parks Victoria, the custodian of Victoria's 3.8-million-hectare parks and reserves network ('the parks network') produced the inaugural *State of the Parks* report in 2000. It was the first time that consistent and comparable information was compiled for a parks network within Australia and was only the second report of its kind in the world. This first edition presented a snapshot of natural attributes protected in parks and reserves, outlined key conservation issues, and established priorities for natural values management.

This case study describes Parks Victoria's approach to the production of the inaugural report, the lessons learnt from that experience, and how those lessons are being used to guide the development of future editions of *State of the Parks* reports.

The Auditor-General of Victoria (1995) identified a critical lack of consolidated state-wide information relating to environmental attributes, and concluded that it was not possible to determine whether the management of protected areas was satisfying the legislated obligations of the *National Parks Act 1975* (Vic). He recognised that it was necessary to document the state of the parks network in order to develop a clear basis for allocating resources to environmental management. Parks Victoria's production of *State of the Parks 2000* was one of several major initiatives in response to the Auditor-General's report and is part of an ongoing process of reporting to the community on progress in conserving Victoria's natural heritage within parks.

State of the Parks 2000 was based on an environmental management framework using principles of risk-management. The framework was used to develop management programs, which initially require an understanding of the way that environmental attributes are distributed within parks across the State and how that distribution changed with time. The next step was understanding the extent of environmental threats and the risks they pose to environmental attributes. *State of the Parks 2000* presented the analysis of attribute and threat information and used that analysis to establish strategic priorities for park management. The approach was broadly consistent with the pressure-state-response model used nationally for State of the Environment reporting in 1996.

State of the Parks 2000 presented information on the environmental attributes of more than 90% of the then 3.6-million-hectare protected area network managed by Parks Victoria. The report was published as a two volume set. Volume 1: *The Parks System* described the landscape context of the parks network, the crucial role the network plays in protecting biodiversity attributes, and the impact of certain threats on their long-term viability. This information was grouped into categories to enable comparisons between parks across Victoria (Figure 6.2). Volume 2: *Park Profiles* presented this information for a set of key measures for individual parks.

Elements contributing to the landscape context included physical attributes such as park size, shape and connection with other areas with Indigenous vegetation, and historical use of individual parks. These attributes were considered major influences on the susceptibility of parks to impacts from threatening processes and surrounding land uses. Much of the information in the report was presented as maps to enable comparison of the variation across the network.

The potential impacts of a range of threatening processes are also presented, focusing on a network-wide examination of pest plants and some pest animals, fragmentation, salinity, and water flows.

Detailed information only existed for some parks and reserves; hence the report contained the best available data on the whole parks network at the time of publication. A combination of state-wide datasets

Figure 6.2   State-wide maps used to illustrate variability between parks

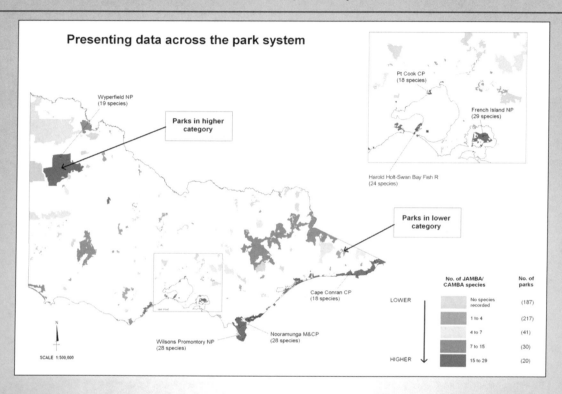

and results of a staff questionnaire formed the basis of the report, and this was supplemented by expert modelling of the latest data on specific issues. An example of the latter was an assessment of the adequacy of environmental water flows throughout the parks network.

The questionnaire enabled staff to verify environmental attribute information and provide detailed on-ground knowledge of threats to those environmental attributes. The results were highly varied, as were staff levels of confidence in the information they provided. Only information available for the majority of the parks network was used in the report.

The report was intended for use by informed individuals and groups with an interest in Victoria's parks network. Environmental attributes and threatening processes were illustrated using numerous coloured photographs and state-wide maps in an A3 size book (Volume 1) and alphabetically listed by park in an A4 size, 347-page book (Volume 2). A separate overview brochure summarising the content and findings of the report and highlighting some of the major issues was disseminated widely to community groups and schools.

The full report was distributed to a wide range of environmental management agencies and stakeholders. Community groups and interested individuals were notified of the report's release and a number of information sessions were held to communicate the report's content and purpose. Copies were available for a fee or could be downloaded free-of-charge via the Parks Victoria web site (www.parkweb.vic.gov.au). Copies were also widely circulated among Parks Victoria staff.

Members of the public were invited to respond to the release of *State of the Parks 2000* by submitting written comments via the Parks Victoria web site. An external consultant conducted a more formal review of the report and a cross-section of stakeholder groups and individuals were surveyed to assess their acceptance and understanding of the report.

The review revealed that the report was welcomed as a major achievement in an area where there were few models or accepted standards; however there were a number of suggested improvements. Suggestions included the following.

- Clarify the purpose and audience of the report.
- Provide information on cultural heritage (Indigenous and non-Indigenous) and visitor services and impacts, not just environmental attributes.
- Provide more information on the condition of the parks network, beyond an inventory of the environmental attributes contained within it.
- Clearly identify benchmarks (and targets) for parks from which future trends can be identified.
- Review and clearly outline methods used throughout the report to classify and rank parks.
- Present detailed data where it exists for parks.
- Provide a complete list of parks and reserves.

Figure 6.3   The future role of *State of the Parks* reporting in Parks Victoria's management model

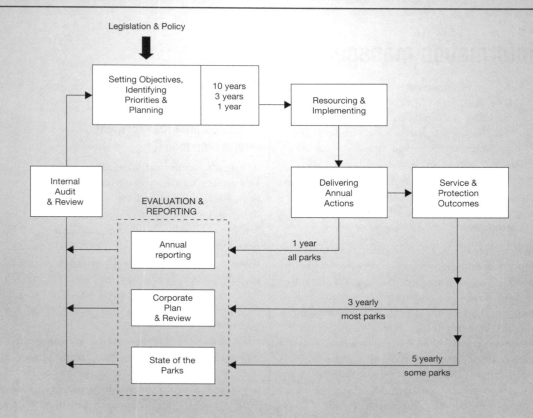

- Establish a consultation/reference group to guide the development and structure of future editions.
- Review the A3 size and colour format.

These suggestions reflect the wide range of expectations, both within and external to Parks Victoria, of what the report should do. Encouraging input and clearly communicating the purpose of the report are critical to managing these expectations. The next edition, *State of the Parks 2005,* is currently being developed, informed by community and stakeholder views with preliminary advice leading to a more clearly defined role in Parks Victoria's management model.

The report will form a critical component of a 'whole-of-organisation' approach to park management in Victoria (Figure 6.3) and will be broadened to include cultural heritage and visitor service elements. The report will no longer set strategic management priorities but will retrospectively assess the effectiveness of Parks Victoria's management of the network against longer-term objectives and consequently inform management. The format of the report will also change with the publication of park profiles on the Parks Victoria web site. This will facilitate regular electronic updates to the information presented in the profiles for all parks.

A project structure has been established that includes an internal reporting and communications framework and an external consulting process. An executive level control group provides direction, and an experienced team of staff from across the organisation is developing the report. Representatives of other management agencies, key stakeholder groups, and expert scientists are providing independent advice on the content of and methods to be used in the report.

6.5 Information management systems

As much as 80 per cent of the typical executive's day is dedicated to information—receiving, communicating and using it in a wide variety of tasks. Because information is the basis for virtually all activities performed in an organization, systems must be developed to produce and manage it. The objective of such systems is to ensure that reliable and accurate information is available when it is needed and that it is presented in a useable form. (Senn 1990, p. 8)

When developing information management systems it is important to consider who will be the users of the information; how to make the information easily accessible; what information is required; how to manage it; and how to control the use of terminology (Orna & Pettitt 1998). Once this is confirmed it needs to be documented and a policy for information management developed. Senn (1990) describes six different types of information system, each of which are aimed at processing data to either capture details of transactions, enable people to make decisions, or to communicate information between people and locations (Table 6.1).

Protected area management information model

The kinds of information required and hence the best systems to store and retrieve data change with time. Today, institutions such as the CSIRO, universities, and museums have reorganised their data systems to enable much more information to be drawn from their collections. They can then make it widely available through the Internet. An information-hungry public is now aware of and has developed an appetite for such data. This in turn puts pressure on managers to acquire, store, and release more data than in the past.

A comprehensive protected area management information model, if developed, would provide a framework for collecting and analysing data of importance to protected area management agencies. Such a model would not reflect the range of data that is available, but the total range of important

Table 6.1	**Types of information systems** (adapted from Senn 1990, p.13)
Type of system	System purposes
Transaction processing systems	Process data about transactions, for classification, calculation, sorting, summarising, and storage
Management information systems	Provide information for decision support where information requirements are regular and can be identified in advance
Decision support systems	Assist managers with unique (nonrecurring) strategic decisions that are relatively unstructured
Executive support systems	Assist top-level executives in acquiring and using information needed to run the organisation—brief them on day-to-day activities and provide information to identify emerging problems
Work group support systems	Assist and support managers, staff, and employees in carrying out day-to-day activities—combine computer processing, data communications, electronic message transmission, and image processing
Expert support systems	Use computer programs to store facts and files to mimic the decisions of a human expert

data. In this way, data gaps could be identified. The model would be equally relevant at all organisational levels within an agency and for information about individual sites, as well as compound information. The model would serve as a basis for database development, information collection strategies, and data analysis, allowing data collected from different sites in different ways to be compared. This would considerably assist macro-level management, but also facilitate effective information management at the micro level.

The protected area management information model would integrate details about the entity (person, community, organisation, or heritage resource) upon which information is collected with their related domain (demography, economy, socio-culture, environment, polity, and description). Within these, information could be managed into identifying classifications (such as atmospheric sciences), which can then be broken down into specific subject areas (such as weather).

An example of an information management system that could be adapted by protected areas is MIST in Uganda (Case Study 6.7). If a number of protected area organisations utilised a system such as MIST they could integrate their systems of data collection, storage, and processing, allowing for information exchange across organisations.

Case Study 6.7

Management Information System, Uganda

Adapted from Schmitt & Sallee (2002)

As part of its advisory services to the Uganda Wildlife Authority (UWA), the German Technical Cooperation has developed a Management Information System, termed MIST, to provide managers and planners at all levels with timely and up-to-date information for planning, decision-making, and evaluation.

When designing the system all staff were consulted about what they wanted out of the system, and this determined data collection, analysis, and output. The organisational structure of UWA and the behavioural principles and technical factors were also considered. The system is very user-friendly and all users have easy access to a central database, a meta-database (provides details about the information, such as location, date, and methods used for collection) and all documents of relevance to UWA, through the local area network, or through using digital data transfer or zip-disks.

MIST has been placed in an Information Management Unit to ensure effective data management and flow. The system integrates information on ecological, social, and economic dimensions of wildlife conservation as well as tourism data, and literature and address databases. Through MIST access is provided to digital archives of documents stored at UWA headquarters.

MIST incorporates data collected by front-line staff, air surveillance, communities, and researchers. Only data that can be transformed into useful information is collected, stored, and processed in MIST. Standardised data collection is an integral part of MIST. Easy to use data sheets have been generated for use by ranger law-enforcement patrols and by communities. This opportunistically collected data is used to calculate indices for monitoring of trends in wildlife populations and illegal activities, and for monitoring of the sustainability of resource harvests. MIST processes visitor data and systematic monitoring and research data. All data collected is geo-referenced using Global Positioning System.

Outputs from MIST include monthly/quarterly/annual reports, and routine or specific requests for information. MIST can present information in the form of reports, maps/diagrams, figures/tables, and charts/graphs. It can work in conjunction with accounting software and financial management systems, and handles data entry, management, and evaluation of the annual operations plans. MIST also maintains its own staff and salary database tables.

MIST improves management and measures management effectiveness by providing baseline data for planning and information for decision-making, as well as monitoring and evaluation of annual operations and management plans, and creating a culture of information exchange. MIST is capable of measuring how individuals adapt to new developments by monitoring changes in implementation of activities and in the uptake of MIST.

MIST is a sustainable system due to the following factors: it does not require a highly trained GIS expert; it is easy to use, maintain, and customise; and it provides different levels of access for users. UWA is currently working with the Wildlife Conservation Society to develop a monitoring and research plan for all of their protected areas, and they are utilising MIST to compile and analyse the data. MIST has also been adapted for use in two national parks in Cambodia.

6.6 Management lessons and principles

Lessons

1   The accessibility of information on the Internet has transformed opportunities for protected area staff and stakeholders in remote and other locations in Australia to access the very latest information on subjects of interest. It is a powerful tool that has already helped many practitioners.

2   Investment in the continuous education of staff through attendance at conferences and workshops of relevance is fundamental to the successful management of protected areas. Also of benefit is staff documenting and publishing their experiences and knowledge and contributing to the debate on best practice protected area management.

3   Keeping up with the literature on protected area management subjects can be a formidable task when the daily grind of operational management takes up so much time. Special 'one-off' or regular in-house workshops on protected area management subjects can assist very busy managers by making it easy to be brought up to date on areas of immediate operational relevance.

4   The design of information systems to service rapid turnaround times and potentially life-threatening circumstances (such as wildfire management) as well as routine information services is crucial.

Principles

1   Effective stewardship requires the best available information on all aspects of protected areas and their surrounding environments, including natural heritage, cultural heritage, economics, and social aspects such as visitor values, attitudes, and behaviour. It is critical to understand the limitations of the data.

2   Access to and the ability to use the most relevant, recent, and cutting-edge information is essential to achieve management objectives.

3   A systematic approach to collecting, organising, storing, accessing, and analysing data is fundamental to delivering useful information. Recent advances, such as GIS and electronic databases are essential tools.

4   Research is a core function of protected area management and should be facilitated by protected area organisations. Research priorities should be clearly documented. Research partnerships should be developed with universities, science organisations, and other research providers.

Ranger rest break, remote area monitoring, NSW
(Graeme Worboys)

5   Monitoring (including the appropriate selection of indicators) provides critical information for evaluating progress, understanding the consequences of management actions, and establishing the basis for adaptive management.

6   Processes should be in place to ensure that information is easily accessible (at no cost) to all interested parties. It needs to be recognised that those accessing the data have different levels of skill and access and hence the information needs to be provided in different formats.

7   Agencies should ensure that staff have the capability to access, understand, interpret, and apply information, made available from research, monitoring, and other sources.

07

Management Planning

Sir, I have a cunning plan.

Baldric

7.1 Approaches to planning   190
7.2 From approach to process   195

7.3 Protected area management plans   199
7.4 Management lessons and principles   217

Planning is something most of us do in one form or another every day. It is also a specialised skill practised by corporate managers, town planners, and natural resource managers. In essence, planning is concerned with the future, and, in particular, future courses of action. Planning is a process for determining 'what should be' (usually defined by a series of objectives), and for selecting actions which can help achieve these objectives. Other definitions of planning include:

Planning is intervention with an intention to alter the existing course of events (Campbell & Fainstein 2003).

Planning is the generic activity of purposeful anticipation of, and provision for, the future (Selman 2000).

Planning can occur at various geographic scales. *Land use planning* (Chapter 5) is the process of deciding in a broad sense which areas of land will be used for what purpose—for example which areas will be national park and which areas will be state forest. This may be undertaken at a national, state, or regional scale. *Area management planning* is concerned with how to manage these areas once their land-use designation has been determined. A park management plan for a national park is an example of an area management plan. Both land-use and area management planning typically deal with a wide range of management issues.

Site planning deals with design details associated with, for example, the development of a visitor facility. A park management plan might recommend the establishment of a camping area of a certain standard in a particular location to provide for a specified number of people. A separate and subsidiary site plan will specify the location and design of access, barriers, campsites, toilets, and so on within the camping area. *Functional planning* focuses on a particular issue, for example fire management or conserving a significant species. *Organisational planning* is concerned with the purpose, structure, and procedures of a management agency. Within an organisation responsible for managing natural areas there may be several levels and types of management planning documents and activities. If the organisation is working well, all these activities and documents should be coordinated and integrated. For example, the objectives of a plan for an individual park should relate to, and be consistent with, a plan at a higher level such as a regional, tactical, or corporate plan. A corporate plan identifies an organisation's collective goals, objectives, policies, and activities, and provides a context and guidelines for area management and functional plans.

There are many other types of planning and related activities associated with establishing and managing protected areas. Examples include *impact assessment, economic planning, financial planning, business planning* (Chapter 8), *operational planning* (Chapter 11), *species recovery planning* (Chapter 12), and *incident planning* (Chapter 15). This chapter will concentrate on area management planning.

There are several reasons why one needs to plan for the management of protected areas. In general, planning can help conserve a resource while providing for its appropriate use. More specific reasons for embarking on a planning project include:

- meeting global responsibilities under such agreements as the Convention on Biological Diversity
- meeting statutory obligations (for example, the *National Parks and Wildlife Act 1974* (NSW) requires management plans to be prepared for areas reserved under the act)
- directing management towards achieving the goals established in legislation or elsewhere
- refining broad goals into specific, achievable objectives
- facilitating the making of sound decisions
- facilitating the resolution of conflicts over resource management
- aiding communication between different levels within a hierarchical organisation—for example, between top-level staff and front-line staff such as rangers who are often responsible for on-ground implementation of actions
- providing continuity of management despite staff changes
- making explicit decisions and the means by which they were arrived at—important components of management that might otherwise remain hidden
- giving the community, interested groups, and individuals an opportunity to take part in decisions
- providing for public accountability.

In our view, this impressive list justifies governments and management agencies placing a high priority on achieving high quality planning. In this chapter we describe the different ways in which a planner might approach his or her task, the process that might develop from the chosen approach, and an account of a typical area planning project. We conclude the chapter by identifying what we consider to be the principles of high quality planning.

7.1 Approaches to planning

Before looking at the specifics of land use and management planning in Australia, it is important to consider how one might, in theory, approach a planning problem. Most planning practitioners disregard theory.

> *There has always been a gap between what academics think planners should do, and what planners actually do … (Sorensen & Auster 1999).*

Why then, should we study planning theory? Although in their day-to-day work many planners rely on professional experience, this cumulative professional knowledge can be understood as assimilated theory. Even without conscious adoption, theory is latent in planning practice, and is implicitly used to guide and establish frameworks for this practice. Theory allows us to see the assumptions and value judgments that underpin planning practice. Good theorising can motivate, define, contextualise, drive forward, and inform practice (Campbell & Fainstein 2003). Theory also provides a means for practitioners to understand planning processes in a way that is outside experience, intuition, or common sense.

Historically, planning as a profession has been dominated by planners working in urban contexts. Much of the theory and practice of planning has been, and continues to be, heavily influenced by this heritage. Since the 1970s, planning has also emerged as an important activity for non-urban land uses such as protected areas, landscape-scale regions such as catchments, and natural resources such as forests, fisheries, and water. In contrast with urban planning, these more 'natural' and 'rural' areas of planning practice initially were heavily influenced by the biophysical problems they sought to address, principally protection of biodiversity and prevention of resource depletion. Planners working to address such issues typically were trained in the natural and physical sciences, and so they tended to adopt planning practices and processes that were consistent with their systematic, scientific understandings.

Early environmental planning focused on natural systems and protection of natural values, working in a 'top-down' fashion. Top-down planning is initiated and conducted by planners or a small group of 'experts', such that there is little or no opportunity for staff at other levels of the organisation, or for other stakeholders, to exert an influence on its outcomes. In contrast, 'bottom-up' planning includes extensive stakeholder and non-planning staff involvement in planning processes. It is now recognised that planners need to consider:

- integration of social, cultural, economic and natural concerns
- development of social and cultural values as well as maintaining natural values
- sharing or devolution of decision-making power
- interdependence of conservation and development
- managing ecosystems in a human context (Maltby 1997, Mercer 2000, Selman 2000).

> *The earlier separation of ecology and economics and of 'nature' and human society (at least in western cultures) favoured still by some conservationists is not helpful. Instead it is necessary to break down such dualities and ensure that we manage natural resources consistently for sustainable benefits to both wildlife and people (Maltby 1997, pp. 3–4).*

Protected area management planning has gone through several phases. Plans in the 1970s and early 1980s tended to be dominated by extensive inventories of natural and cultural resources. They were developed with little community participation and the data collection effort tended to be at the expense of strategic considerations and substantive management decisions. In the mid 1980s until the early 1990s, plans were more focused on specific management objectives and actions, often framed by a zoning scheme. Community participation also became an important component of planning processes. While these plans provided more management guidance than the earlier plans, they often quickly became out of date, and were generally written with little regard to available management resources. They tended to be 'wish lists' rather than realistic management prescriptions. Such rigidity and implementation difficulties meant that they often suffered from the 'sitting on the shelf' syndrome and consequently did little to guide day-to-day management.

As a reaction against these failings, and under the influence of wider trends such as the increasing popularity of strategic planning derived from business management, plans from the mid 1990s were typically much leaner documents. They articulated a strategic direction, but often did not detail specific outcomes or management decisions. Such plans were politically expedient, in that in the absence of any performance measures, agencies could not be held to account. Their lack of specificity meant that they were also of little use in guiding management. Of course specific decisions were still needed—these tended to be made in within-agency operational planning processes that took place out of the public gaze.

We are now entering an era where plans are attempting to address these various limitations. State-of-the-art planning now seeks to produce relatively short strategic documents that none-the-less contain a realistic set of objectives to enable performance evaluation, as well as actions that, in the immediate future, are considered the best options to meet the objectives. Ideally, the plans are also flexible enough to allow modification of actions on the basis of experience and new information, as well as some adjustment of objectives and performance measures.

Land use and area management planning inevitably involve many stakeholders, often with widely diverging values and opinions. This raises a number of questions that planners need to address when designing their planning project.

- How should people be organised to facilitate the planning process?
- Who should have the power to make decisions?
- What planning methods or procedures should be used?
- Who should decide what the planning objectives should be?
- What criteria should be used to select the best courses of action?

The answers to these will largely depend on the approach adopted by the people initiating the planning activity. However, it is important to recognise that most planners do not consciously select a particular approach or mixture of approaches. Furthermore, approaches are often prescribed by management agencies. Important influences on the approaches that are adopted include agency traditions, the prevailing mode of public policy development, institutional structures, and the intellectual traditions most influencing those people directing the planning process. There are four major approaches to a planning project:

- rational comprehensive
- incremental
- adaptive
- participatory.

Note that these approaches are rarely, if ever, used in their pure form—in general, planning projects can be described in terms of mixtures of these approaches. The particular mix implicit in a planning project is reflected by the relative contributions made by the underlying pure approaches.

Rational comprehensive planning

Rational comprehensive planning is a top-down approach that attempts an objective and exhaustive inventory of current conditions, analyses these conditions, develops possible solutions to issues based on these descriptions and analysis, and selects a preferred solution according to a set of measurable criteria (Briassoulis 1989). For example, as part of a land-use planning project, the following process might be adopted under a rational comprehensive approach:

- map current land uses in the study area
- determine the population distribution of all rare or threatened species in the study area

- establish targets for each species, based on minimum viable population estimates, and then
- select protected areas such that at least the minimum viable population of each rare or threatened species is protected, while minimising the impact of reservation on other land uses.

Of course, many other factors would also need to be considered in the selection of protected areas, but this simplified example illustrates some of the key characteristics of rational comprehensive planning. It is a scientific and technically demanding approach, in which experts assume a key role. It relies on high quality data, and often makes use of mathematical models.

Such formality and rigour are both a strength and a weakness. On the strength side, rational comprehensive planning should produce decisions that can be clearly explained and justified. Debate about the decisions tends to focus on technical issues such as the reliability of the data used or the validity of the models used to process the data. The often inefficient processes of public decision-making can be avoided, and political bias minimised. However, the weaknesses of the rational comprehensive approach are that it is inflexible and is disposed to ignoring social and political factors. The rational comprehensive approach to planning tends to give rise to a static planning process, in which a particular set of objectives is established and a number of decisions made that will apply for some specified period of time. Uncertainty and risk are very difficult to accommodate. In the example above, the values built into process through the setting of targets and the minimisation of land-use conflicts would tend to be made by the experts conducting the analysis. The role of other stakeholders in the decision-making process tends to be minimal. The recommendations from a rational comprehensive process may therefore not reflect community values or aspirations, and ignore political and institutional limitations.

Incremental planning

Incrementalism, also known as crisis management or ad hoc planning, uses small incremental changes to deal with problems in an essentially uncoordinated manner. Decisions are made without reference to specific objectives. The dominant factor in selecting a particular course of action tends to be the ease of implementation in terms of political, administrative, and economic feasibility. No explicit attempt is made to consider the combined impact of individual decisions. Incremental decisions generally only have a marginal effect on an existing situation. However, long-term adoption of an incremental approach can result in significant change—the aggregate effect of a large number of incremental decisions can be considerable, and perhaps undesirable. Incrementalism makes it difficult for stakeholders to determine the reasons why a particular decision has been taken. Powerful elites and lobby groups can have a major influence on the process.

Despite, or perhaps because of, these characteristics, incrementalism is widespread in environmental planning (Briassoulis 1989). Incrementalism also has some influential supporters, most notably Charles Lindblom. Lindblom (1979) pointed out that we are kidding ourselves if we think we can ever fully come to grips with complex policy issues—we can never perform a truly complete or holistic analysis. He then argued that it is better to undertake a partial analysis by simplifying the problem to a few familiar policy alternatives. This is a strategic incrementalism—a conscious and skilful recognition of our limitations in dealing with complex problems.

Adaptive planning

Adaptive planning treats management as an iterative process of review and revision, not as a series of fixed prescriptions to be implemented (as in the rational comprehensive approach). Management interventions are seen as a series of successive and continuous adaptations to variable conditions. The approach emphasises flexibility, requires willingness to learn through experience, and may require sacrificing present or short-term gains for longer term objectives (Briassoulis 1989). The emphasis is on learning how the system works through management interventions that are both issue-oriented and experimental (Dovers & Mobbs 1997).

Some adaptive planning places considerable emphasis on the use of predictive models to guide actions. Where outcomes are not as expected, the models are modified to improve their predictive power. In this respect, adaptive planning has some similarities with the rational comprehensive approach. The essential difference is that adaptive planning is flexible and responsive to changing circumstances, whereas rational comprehensive planning tends to be rigid and prescriptive. Adaptive planning recognises that there is often considerable uncertainty about the outcomes of any particular action. This uncertainty is built into plans, so that information about the actual results of actions is used to inform and, where necessary, modify management practices.

Adaptive management in its earlier forms concentrated on ecological modelling that aimed to put possible management interventions in terms of scientific hypotheses, capable of being tested. It was an attempt to combine the methods of scientific research with the practicalities and realities of management. More recently, the approach has been expanded to integrate social and institutional aspects with the ecological and managerial dimensions (Dovers 1998). The main features of this expanded notion of adaptive planning are:

- recognition of the contribution that the natural and social sciences can make to dealing with management problems
- recognition of uncertainty, complexity, and long time-scales
- acknowledgment that management interventions are essentially experimental, and while directed towards improving environmental and human conditions, also allow for testing and improving understanding and capabilities along the way
- design and maintenance of sophisticated mechanisms (institutions and processes) to allow feedback and communication between theory, policy and practice, and across different situations (Dovers 1998).

It is a process of learning by experience. This means that mistakes may, and probably will, be made. It is not possible to predict accurately the precise outcome of any course of action—there are always too many variables involved, and the interactions between the variables are too complex. A planner can develop models to help forecast and predict the future, but these predictions will only be as good as the information on which they are based and the model that processes this information. This means that planning will inevitably involve uncertainty. In this sense a plan is not a blueprint. While one might reasonably expect a house constructed on the basis of a blueprint plan to be accurately represented by that plan, the same cannot be said for plans addressing the future of protected areas.

There is always the risk that a predicted result will not occur and that the consequences of a recommended action will not be those desired. A planner needs to recognise this fact, and include an awareness of uncertainty and risk into the planning process. This is particularly important for irreversible decisions. However, the difficulty of making decisions, especially when they are irreversible in nature, should not be used as an excuse to avoid making a decision altogether. 'Non-decisions' also have consequences.

As with rational comprehensive planning, some attempts at implementing the adaptive approach have suffered from excessive reliance on mathematical models, neglect of qualitative data, and inadequate attention to the institutional and policy frameworks in which the planning takes place (McLain & Lee 1996). Furthermore, integrating effective stakeholder participation with the adaptive approach is not straightforward. In the past, area management planning often maintained a separation between the planning phase and implementation of the plan. Evaluation and monitoring, though specified in many plans and policies, have often not been implemented. Stakeholder participation, particularly in the management of public lands, has generally been confined to the initial planning phase. In contrast, effective stakeholder participation in the adaptive approach demands an ongoing and long-term involvement (Dovers & Mobbs 1997).

Such extensive and open-ended commitment places considerable demands on all stakeholders—demands that are often impossible to meet. It is probable that the only stakeholders to maintain engagement with an adaptive process would be those with the most to gain (or lose). Stakeholders such as urban residents in regional centres and city-based environmental groups often find it difficult to make a meaningful contribution to such processes. Any approach that disadvantages certain stakeholders will pose problems of legitimacy and credibility for the outcomes.

Conventional plans tend to be inflexible, and have limited ability to adapt to changing circumstances and a developing knowledge base. Though review processes are usually built into such processes, the opportunity for making major new management decisions is limited to specified review periods, usually held at three- to five-year intervals. This means that many of these management plans are perceived to become quickly out of date and tend to be ignored. We suspect that sometimes, however, the 'out-of-date' accusation hides more substantive reasons why an expensive management plan ends up sitting on a shelf gathering dust. Sometimes governments simply do not provide enough resources to enable the plan to be implemented. A perfectly good plan can also die of neglect because the planning team who constructed it has moved on to the next planning project, leaving no one with a commitment to the plan to ensure that it remains the basis for the works plans that govern the day-to-day disposition of staff and resources. This is why many agencies directly involve local staff in the development of a plan.

Here an adaptive process may have an advantage. If the planning team has an ongoing responsibility for management planning, then the contact with the implementation staff will also of necessity be ongoing. There is no 'draft' or 'final' plan, but simply a continuously evolving document, which is being continually developed by a planner or planning team, concurrent with day-to-day management activities. However, this type of planning can very easily slip from adaptive to incremental.

Participatory planning

The participatory and adaptive approaches are emerging as particularly influential components of modern natural resource planning and management (Dovers 1997). Demands for greater community participation reflect concerns about the legitimacy and efficacy of modern systems of representative government. Perceptions of community participation vary, largely in terms of the extent to which the community exercises decision-making power, with notions of participation ranging from the provision of information through to local control of decision-making (Arnstein 1969).

There are ethical and pragmatic reasons for involving the public in decision-making. Public participation is believed to legitimise planning outcomes, reduce citizen alienation, avoid conflict, give meaning to legislation, build support for agency programs, tap into local knowledge, provide feedback on program outcomes, contribute to community education, and enhance democratic processes by increasing government accountability (Creighton 1981, Daneke 1983, Lyden et al. 1990).

Creighton (1981, p. 3) defined community participation as:

> *a process, or processes, by which interested and affected individuals, organisations, agencies and government entities are consulted and included in the decision-making of a government agency or corporate entity.*

The participatory approach has its philosophical and political roots in liberal–democratic notions of equality of persons and rights of the individual. If the principle of equality of persons is accepted, then the objectives and outcomes of protected area planning should reflect a synthesis of the interests of stakeholders and not relate to the interests of a single individual or subgroup of individuals. Stakeholder participation is a mechanism for improving the efficacy of representative democracy. Indeed, increased stakeholder involvement in decisions reflects a shift from a purely representative model of democracy to a participative democracy in which there is an expectation among citizens that they will not just be represented by elected officials, but actively and continually engage with the processes of policy development and implementation (Daneke 1983).

The legitimacy of government decisions rests on the consent of the governed. The challenge for government is that it must be able to demonstrate that it has gone to all reasonable lengths to command legitimacy, and through that, credibility. Legitimacy requires that public participation processes, among other things, permit all stakeholders to contribute; ensure these contributions are respected; enable participation to occur in a manner that best suits the particular talents and contributions of the various stakeholders; allow stakeholders to express judgments honestly and sincerely; provide adequate notice of all relevant information and events; make available all relevant documentation to all stakeholders; and ensure that documents are understandable to all stakeholders (O'Riorden & O'Riorden 1993).

If stakeholders are adequately represented in decision-making, and if decision-making processes are adopted that allow stakeholders to cooperate in an honest and open exchange of views, stakeholders can develop empathy for the positions of others and it is possible for agreed positions to be reached that are accepted as fair to all parties (Kaplan & Kaplan 1989, Landre & Knuth 1993). Ostrom (1990) believed this 'social capital' would allow stakeholders to develop cooperative mechanisms to resolve common pool resource dilemmas as alternatives to reliance on market forces or a central authority.

So, involvement of a wide range of groups and individuals throughout the planning process has the following advantages:

- providing the planner with access to a range of information and advice that might otherwise be difficult to obtain
- enabling early identification of major issues and an ongoing check of any further issues that arise
- generating more creative solutions to problems

- reducing implementation failure
- increasing plan acceptance
- managing competing interests and mediating conflict
- enhancing public ownership and commitment to solutions
- supporting the rights of citizens to be involved in decisions that affect them
- increasing government accountability
- articulating and representing the diversity of interests and values involved in a decision (Pimbert & Pretty 1997, Curtis & Lockwood 1998, Tuler & Webler 1999, Wondolleck & Yaffee 2000, Barham 2001).

Disadvantages include:

- the time-consuming nature of a genuine public participation program
- the potentially high financial cost
- the difficulty of obtaining constructive debate when interest groups have entrenched and opposing views
- the impossibility of accommodating conflicting interests under circumstances when only one view can prevail.

7.2 From approach to process

Planning is often connected with the word 'process'. This means that planning is not simply an event or an outcome. Planning is best seen as an interrelated sequence of stages. These stages are linked in a dynamic fashion—the interactions between them may occur in one or more directions and change over time. In addition, while there may be a clearly defined starting point to the process, it is often difficult to define an end point. Indeed, many planning practitioners emphasise the ongoing nature of planning, with the need to regularly review the success and relevance of both a particular plan and even the planning process itself.

The approach or mixture of approaches adopted by a planner will determine the particular stages undertaken in the planning process, as well as the relative importance given to each stage. Most of the approaches would only very rarely be found in their pure form in an actual planning project. More

commonly, a project is made up of a combination of approaches.

Of course, planners are usually required to follow a particular series of steps (that is, a process). There is generally little opportunity for a planner to change the nature of this process. In most agencies, the planning process has been developed by senior planning staff, based on such factors as the types of issues that must be addressed, the prevailing attitude to the importance of planning, political acceptability, and the academic disciplines with which they are most familiar. There is generally little awareness of the possible approaches that form the platform for planning processes. The strengths, weakness, and applicability of the various approaches are not generally considered in the development of a planning process. Many protected area planners remain unaware of the approaches that are implicit in the process that has been adopted. This is unfortunate, because it limits the ability of planners to constructively address problems with existing planning processes. In our view, a good understanding of planning approaches is a prerequisite for high quality planning.

Most protected area planning in Australia has been undertaken using some mix of participative, rational comprehensive, and incremental approaches. This mix has generally not been arrived at through any consciously deliberative process. It simply reflects the imperatives of dealing with the issues at hand and the nature of the organisations responsible for planning and policy development. The move towards adaptive management is in part recognition of the failure of the incremental and comprehensive rational approaches. Incrementalism has been particularly evident in the development of environmental policy, which has had a history of 'lurching, myopic "ad hocery" ' (Dovers & Mobbs 1997, p. 47). Inadequacies of current natural resource management plans as identified by Fallding (2000) include:

- plans that are either too long and scientific or too short and general
- plans that have unrealistic expectations
- objectives that do not have effective implementation mechanisms
- the absence of an adaptive framework.

The range of issues and environments in Australia, together with uncertainty surrounding the effectiveness of the various approaches, means that there is no 'perfect' mix of approaches, nor a 'perfect' process arising from such a mix. We can, however, identify several elements that need to be incorporated into the selection of planning approaches and the development of an effective process. In general, we recommend that planning processes be based on the adaptive and participatory approaches. The requirements necessary to support an intelligent mix of the adaptive and participatory approaches are considerable:

- sophisticated and accessible systems of research, monitoring and communication
- integration across disciplines and professions
- commitment to persistence and accountability
- having 'spare capacity' in natural and human systems so that managers can honestly adapt and make adjustments

- democratised, open, and accessible processes, with participation structured so as to be clear and to persist over time
- political, stakeholder, and community will to engage in difficult, long-term processes
- persistent yet flexible institutional arrangements to allow fulfilment of all the other requirements (Dovers & Mobbs 1997, Dovers 1998).

In addition, successful planning processes have or develop:

- clear articulation of the process
- links to larger-scale strategies, providing a strategic focus
- a clear understanding of the issues
- explicit measurable objectives
- decisions that are justified and transparent
- explicit linkages between objectives and actions
- actions that allow for consistent interpretation and application
- explicit links between actions, available resources, and budgets
- explicit lines of responsibility regarding implementation and evaluation
- availability of suitably trained staff to guide the process and implement the plan.

We will look in more detail at how a management planning project might implement the adaptive and participatory approaches in the context of preparing a plan for a protected area. Before doing so, we first consider three general types of planning processes: strategic planning, systems planning, and adaptive planning.

Strategic planning processes

David (2001) described strategic management as the art and science of formulating, implementing, and evaluating cross-functional decisions that enable an organisation to achieve its objectives. This process requires that organisations continually monitor internal and external events and trends so that they can adapt effectively to change.

Strategic planning was developed in a business management context, and largely absorbed and superseded the previously popular approach called Management by Objectives (MBO). MBO was a somewhat rigid, prescriptive approach, which tended to lock organisations into a fixed set of prescribed outcomes in a manner that stifled creativity, reinforced top-down management and did not facilitate effective responses to changing conditions. Strategic planning emerged as these limitations became evident, and as a response to increased complexity and uncertainty in the business environment.

Strategic planning has been used to describe a number of different planning styles. However, there are some common elements that constitute the core of strategic planning processes. Three essential stages are formulation, implementation, and evaluation (David 2001).

Strategy formulation involves several substages.

1   Develop a vision and mission. A vision statement answers the question 'What do we want to become?' A mission statement is an enduring statement of purpose that addresses the question 'What is our purpose?'

2   Identify an organisation's external opportunities and threats, and determine internal strengths and weaknesses. Strategies are needed to take advantage of external opportunities and avoid or reduce the impact of external threats. These factors are largely beyond an organisation's control. The process of researching and assimilating information on opportunities and threats is sometimes called environmental scanning or industry analysis. Internal strengths and weaknesses are an organisation's controllable activities. SWOT (Strengths, Weaknesses, Opportunities, Threats) analysis is a traditional technique used by planners as part of a strategic planning exercise. Depending on the scope of the planning and the budget, such an analysis may be undertaken by consultants or as an in-house planning exercise. Characteristics of a SWOT analysis include an assessment of an organisation's environment and organisational factors that influence the ability of an organisation to achieve its goals. For the purposes of a SWOT analysis:

   •   a strength is an internal aspect that can improve an organisation's competitive situation

   •   a weakness is an internal aspect where the organisation is potentially vulnerable to a competitor's strategic moves

   •   an opportunity is an environmental condition that can significantly improve an organisation's situation relative to that of competitors

   •   a threat is an environmental condition that can significantly undermine an organisation's competitive situation (Bartol et al. 1998).

3   Establish objectives. Objectives are specific results that an organisation seeks to achieve. They are essential for success because they state direction, aid evaluation, reveal priorities, and provide a basis for effective planning, organising, motivating, and controlling activities. Objectives should be challenging, measurable, consistent, reasonable, clear, and prioritised.

4   Generate alternative strategies, and choose particular strategies to pursue. Strategies are the means by which long-term objectives will be achieved.

Strategy implementation means mobilising employees and managers to put formulated strategies into action. Strategy evaluation involves reviewing the external and internal factors that are the bases for current strategies, measuring performance, and taking corrective actions. Benefits claimed for strategic planning are that it:

•   identifies, prioritises, and exploits opportunities
•   develops a shared vision
•   provides a framework for coordination and control of actions
•   minimises effects of adverse conditions
•   allows decisions to better support objectives
•   provides a framework for internal communication
•   integrates individual efforts into a total effort
•   encourages forward thinking
•   gives work a degree of discipline and formality
•   empowers staff—involving, learning, educating, supporting, owning, and enabling
•   reduces resistance to change
•   enhances problem-prevention capabilities (Kaufman & Jacobs 1996, Koteen 1997, David 2001, Hunger & Wheelan 2001).

Systems planning processes

Systems planning considers the object of planning (the environment) to be a system of interconnected parts. Such systems have a coherence and identity imparted by the relationships and interconnectedness among their constituent parts. Systems planning requires an understanding of the system, allows a planner to forecast the effects of interventions on system components, and facilitates incorporation of strategic thinking and flexibility. While early applications tended to overlook the value-based political nature of planning, attempts have been and continue to be made to incorporate social and economic factors into a systematic series of rational steps. Thus systems-based planning processes are characterised by:

- logical connections between means and ends
- a central role for professional planning expertise and in some cases technical expertise
- heavy reliance on scientific information
- integration of all values and issues into one framework.

Most systems planning attempts to be rational/comprehensive in approach, while including participatory and adaptive elements. However, it is rare that serious attention is given to technical aspects of the process such as forecasting potential effects of interventions. Natural, social, and political systems often prove to be so complex that it is very difficult if not impossible to successfully implement a comprehensive systems process. Elements of systems thinking, however, are evident in most protected area planning processes.

Adaptive planning

As noted above, adaptive planning treats management as an iterative process of review and revision. The approach requires willingness to learn through experience, with an emphasis on learning how the system works through management interventions that are both issue-oriented and experimental. Adaptive management processes systematically test options and assumptions in order to learn and thereby improve outcomes. It is best applied to complex systems that are constantly and unpredictably changing, where full information is not available, and for which immediate action is required (Salafsky et al. 2001). One model for an adaptive planning process is given in Figure 7.1. Examples of adaptive planning processes include:

- fisheries planning in the Great Barrier Reef Marine Park
- tropical savanna management in northern Australia
- environmental flows in the Murray Darling Basin (Allen 2002, Ladson & Argent 2002).

Figure 7.1   An adaptive planning process (Salafsky et al. 2001)

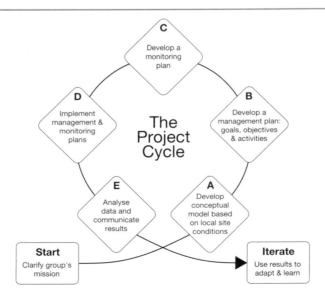

7.3 Protected area management plans

Once land-use planning decisions have been made regarding the tenure and broad purpose of a particular area, decisions still need to be made about how the area will be managed. This is the function of an area management plan. Typically, writing a plan involves the compilation and consolidation of material arising from numerous planning process stages into a coherent document. There are several potentially important questions to consider in this regard.

- Who are the audiences for the plan (and therefore what should be the writing and presentation style)?
- How much, if any, background 'resource inventory' information should be included? Should it be located at the beginning of the plan, the end, or integrated into particular sections?

- Should the process used to develop the plan be explained in the plan itself?
- Should methods used to make the decisions articulated in the plan be described?
- Should community involvement that helped establish the plan content be described, and if so how?
- Will the plan be written at a broad strategic level, or will it include measurable objectives, performance indicators, and standards?
- Will specific actions be given for each goal/ objective, and will time frames and responsibilities be attached to these actions? Some further commentary on this point is provided below.
- What, if any, spatial representations of planning decisions will be required (zoning maps and the like)?

Figure 7.2   Outline of a rational, adaptive, and participatory planning process

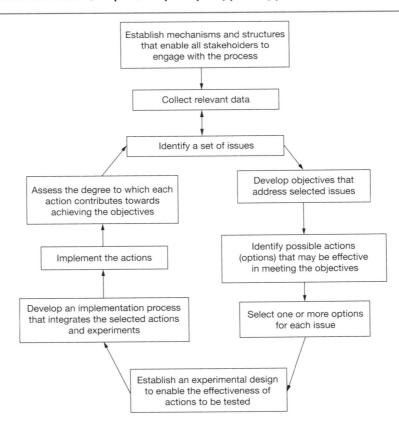

Most plans go though many drafts both in their formative stages and as a result of approvals, public participation, and editorial processes. A planner must be prepared for a long and often tedious series of writing and rewriting sessions. Preparation of a protected area management plan generally involves consideration of the following topics:

- conservation of native flora and fauna
- protection and enhancement of landscape quality
- water quality management
- visitor use management
- fire protection and management
- management of threats, including control of introduced plant and animal species
- management of other uses compatible with the park objectives
- management of other authorised uses, including concessions and leases.

Table 7.1	Kosciuszko National Park planning, 1965–2004	
Year	**Action taken**	**Notes**
1965	Kosciuszko State Park plan	First plan for the park prepared for the State Park Trust— not formally adopted
1974	Kosciuszko National Park management plan	First plan for the new national park, prepared in Head Office, Sydney
1979	Decision to upgrade and amend the plan	Planning team assembled and budget provided
1980	Community information brochure released	Marketing leaflet advising the public that the plan was to be amended
	Key issues identified, and the need to obtain additional community input recognised	Issue statements circulated on fire management, resorts, summit area, huts
	Extensive consultation taken, both within the NPWS and among stakeholders	Leaflet outlining how people can get involved in the process, media interviews, articles, and meetings— submissions received on key issues
1981	Draft plan published	Leaflet encouraging comment and advising where the draft plan can be obtained
1982	Final plan published	
1984	Plan amended to deal with construction of the ski tube	Amendments linked to an Environment Impact Statement
1987	Supplementary plan published	Objectives and actions related to the Cooleman Plain karst area
1988	Plan amended, ski resorts	Proposed expansion of ski resorts to the Ramshead Range and Twin Valleys not approved
1994	Plan amended, ski resorts	
1999	Plan amended, ski resorts	Expansion of Perisher Valley bed limits following a Commission of Inquiry
2001	Commencement of a major plan revision	See Case Study 7.1
2004	Draft plan of management placed on public exhibition	See Case Study 7.1

An outline of a planning process that incorporates rational, adaptive and participatory elements is given in Figure 7.2 (see p. 199). The extended nature of the planning process is illustrated in Table 7.1, Figure 7.3, and the latest process of plan revision (Case Study 7.1).

We do not believe it is useful to prescribe a 'cookbook' approach to protected area planning. However, there are some steps that will be common to most planning processes. When reading the following descriptions of these steps, keep in mind that the particular mix of approaches guiding the planning process will influence exactly how

these steps are carried out, their relative importance, and their relationship to each other—that is how they fit in to the overall planning process.

Establish participatory mechanisms and structures

As we have already discussed, the type and degree of public participation adopted in a planning project will depend on the approach taken by the planner and management agency. Stakeholders may come together in a formal setting such as an advisory committee meeting, or a planner may receive an informal

Figure 7.3   Some Kosciuszko National Park planning documents

Case Study 7.1

A new plan of management for Kosciuszko National Park

Penny Spoelder, NSW Department of Environment and Conservation

Kosciuszko National Park, which encompasses 690 425 hectares, is the largest national park in NSW and one of the largest and most complex conservation reserves in Australia. It is one link in a chain of protected areas that stretches along the spine of the Australian Alps for a distance of some 340 kilometres.

The park attracts over 1 million visitors per year who enjoy alpine and cross-country skiing, sightseeing, rock climbing, ice climbing, horse riding, bushwalking, fishing, camping, caving, and canoeing. The park contains glacial landforms and possesses an exceptional diversity of alpine plant communities and species that provide habitats for a number of rare and unusual animal species. Elsewhere, the park contains significant karst systems, deep river valleys and frost hollows, and vegetation communities ranging from snow gum woodlands and subalpine grasslands, to extensive eucalypt forests, pockets of cool temperate rainforest and stands of native cypress pines. The snow-fed rivers of the mountains provide some of Australia's most important water catchments. The park contains major commercial interests in the form of alpine resorts and the Snowy Mountains Hydro-Electric Scheme, which contribute significantly to state and regional economies. The park is also rich with evidence of, or associations with, Aboriginal culture and the phases of historic land uses, scientific endeavour, and recreation and conservation efforts of many generations. The park's cultural heritage value resides as much in intangible values as it does in physical form. Places within the park have been the scenes of innumerable human experiences. Some of these have survived as legends or anecdotes; others are remembered within place names, songs, literature, art, customs, symbolism, or spiritual observance.

In February 2001, the NSW Government announced that the *Kosciuszko National Park Plan of Management* would be reviewed. The review commenced in January 2002 involving wide public consultation. A team of people were employed with expertise in protected area management from all over Australia to prepare the new plan. At the outset they agreed that the new plan would provide a management framework for NPWS and the community for the next 10 to 20 years, protecting the park's values for future generations. As a first step a number of plans of management prepared for reserves throughout Australia and overseas were reviewed, and park-planning specialists were consulted for advice on what they believed to be the strengths and weaknesses of each plan. Other key documents such as the ANZECC Best Practice in Protected Area Management Planning (Tasmanian Parks and Wildlife Service 2000) and NPWS Plan of Management Manual (NPWS 2001) were also used as reference documents for the development of the project plan.

A process was designed that centred around encouraging the involvement of the general public as well as park users, neighbours, scientists, interest groups, individuals, and local communities in the review of the plan of management. The process was developed following detailed consultation with these groups. This resulted in some modifications to the original design, but ensured that there was general agreement that the planning process would be acceptable to all and involve the relevant interests. The process represented a shift from consulting the public to asking them to share the responsibility for making decisions about their park. This approach was considered the cornerstone to help build public understanding, ensure sound decision-making, and increase the probability that the plan would be supported by them. The following consultation mechanisms were established and to date have proven to be a successful way of involving relevant groups in plan making:

- establishment of an Independent Scientific Committee (ISC) made up of recognised scientists and management experts in various disciplines
- establishment of an independently facilitated and chaired Community Forum consisting of representatives from various interest groups to discuss and recommend strategies that address the key threats identified by the ISC
- establishment of an Aboriginal Working Group representing Aboriginal communities and families who have connections to country that is now known as Kosciuszko National Park.

An inter-agency government working group as well as a staff working group were also established to assist with the plan review. A representative from each of these four groups attended the meetings of the other groups as a way of ensuring strong communication links between them.

The ISC found that, while many values were in good and stable condition, various pressures could lead to degradation of significant values if not adequately managed. Such pressures include the expansion of development, imposition of inappropriate fire regimes, increase in summer visitation, possible climate change, and introduced plants and animals. The ISC prepared an interim report on their findings for community input. Interested organisations and individuals were invited to provide comments, thoughts, and suggestions to the ISC. The submissions were then reviewed and incorporated as appropriate into the final ISC report.

A series of workshops were held in local communities in and around the park as well as Sydney and Canberra. Media releases, brochures, radio announcements, and advertisements in local and national press were also used to inform the public that the review process had commenced. A free call number was established for enquiries and the agency web site was also used as a key source of information about the process with links to other relevant sites. The strong cultural connection that some of the communities held with the park was very apparent. The communities also identified key issues that needed addressing in the new plan. They included, among others, the protection of natural and cultural values, control of weeds and pest animals, sustainable use and access, ski resorts, and fire management.

Following the workshops the Community Forum was established, and met fourteen times over two years to work through the issues raised by the community at the workshops. Meetings were generally held over two days and involved a combination of general discussion, field visits to specific parts of the park, presentations from NPWS staff, stakeholders, and specialists, and small-group work where specific questions and issues were worked through in more detail. The Community Forum developed a series of principles to assist them in the development of strategies associated with the management of the park. The principles related to the protection of Kosciuszko's natural values; recognition and celebration of cultural heritage values; respect for Aboriginal culture; interpretation, education and awareness; maintaining the economic importance of the park; the need for partnerships and participation by all involved in using and managing the park; and the need for ongoing strategic research and monitoring.

The Forum also tackled some difficult issues such as the management of:

- Kosciuszko's huts, particularly following the loss of 17 huts from bushfires in January 2003
- increasing visitation at the summit area
- alpine resorts with growing development pressures and possible implications of climate change
- recreation activities such as horse riding and mountain bike riding
- control of weeds and pest animals.

The Community Forum prepared a summary of these principles and their thoughts on the key issues. Interested organisations and individuals were invited to provide comments, thoughts, and suggestions to the Forum. The submissions were then reviewed and considered by the Community Forum at their subsequent meetings.

The Aboriginal Working Group advised NPWS that the new plan should among other things:

- recognise, acknowledge, and celebrate Aboriginal people's connections to their country now known as Kosciuszko National Park
- strengthen participation by Aboriginal people in looking after the country by working in partnership with NPWS
- recognise the importance of recording history and knowledge, and sharing this with people
- include opportunities for young people in managing the park through employment, education, and training
- rename places in the park with Aboriginal names.

They agreed that the section of the new plan relating to the management of Aboriginal cultural heritage should be primarily written by them.

The advice received from the Community Forum, Aboriginal Working Group, the ISC, other government agencies, and NPWS staff has been invaluable in the development of the new plan of management for Kosciuszko National Park. Greater recognition has been given to the cultural values of the park, community involvement in park management, greater emphasis on environmental stewardship by all agencies and organisations operating in the park, simple zoning schemes, and management strategies identified that ensure sustainable use. The new plan acknowledges the importance of the park's cultural and social values, and the need to protect these values from key threats such as inappropriate fire regimes, climate change, introduced plants and animals, inappropriate development, and unmanaged increases in visitation. The strategies in the plan are commensurate with Kosciuszko's status of one of the greatest national parks in the world.

Ranger on patrol, Kosciuszko National Park, NSW (Geoff Martin)

deputation or phone call from an interested and sometimes irate individual. In many planning projects, the planner is in the position of being a facilitator or leader of a group of people. For example, the planner might be the convenor of a departmental steering committee set up to direct the planning project as well as an advisory committee made up of representatives of key interest groups.

Table 7.2	**Good practices in protected area planning participation** (Parks and Wildlife Commission of the Northern Territory 2002 pp. 16–18)		
Level of Participation Agency/Community	Best/Good Practices	Participation Techniques	Performance Indicators
Inform/Comply Agency informs community e.g. No dogs allowed in park. Community is required to comply with agency requirement.	• Be proactive. • Apply a bottom-up approach. • Research the ways people get their information. • Ensure people are aware of reasons for decisions. • Establish feedback loop to enable the community opportunity to have their say.	• Public meetings • Presentations • Internet and mass media • Communication plans • Press releases • Standard operating procedures • Signs • Internet • Education campaigns • Printed brochures and newsletters	• Level of participation in agency education programs • Number of requests for information • Number of informed people (survey results) • Number of infringement notices issued • Number of complaints • Number of Ministerials • Number of Internet hits
Consult/Cooperate Agency seeks input into decision making process e.g. In developing a plan of management for a park, the community is encouraged to provide input into the planning process. Community agrees to support decisions and becomes involved in programs and activities e.g. 'Friends of the Park' group.	• Consultation takes time and resources—successful outcomes may be undermined where these are insufficient. • Be clear about the basis for involvement. • Value people's contributions. • Promote the inclusion of a diverse range of people and interest groups. • Use language that is inclusive of the community.	• Workshops • Stakeholder meetings • Surveys • Plans of management • Letters to stakeholders • Advertisements in the media • Public displays • Internet	• Quality of submissions • Number of issues raised • Number of stakeholders reached • Diversity of stakeholder input • Level of customer/client satisfaction (measured through surveys/ customer feedback) • Number of staff trained in consultation techniques (e.g. facilitation, conflict resolution)

Table 7.2	(continued)		
Level of Participation Agency/Community	Best/Good Practices	Participation Techniques	Performance Indicators
Consult/Cooperate (continued) Agrees to undertake planting program on park in accordance with agency requirements	• Ensure community is fully aware of issues and what they are asked to do • Acknowledge stakeholder/community input/cooperation • Be very clear if there is no opportunity for people to have a say in the program		• Number of volunteer days and quality of conservation outcomes • Number of volunteer hours • Number of people attending community education programs
Collaborate/Participate Agency invites community to share in decision-making process e.g. Nomination of new marine and terrestrial parks by the community. Community has a formal role in decision-making process.	• Maintain integrity/honesty. • Be open to new ideas. • Respect cultural diversity. • Identify areas of common interest. • Don't make commitments that can't be kept. • Provide opportunities for real involvement. • Be clear about the powers and functions of advisory groups	• Advisory councils • Task forces • Stakeholder feedback • Conservation partnerships with the community, landholders and industry	• Level and type of participation • Level of integration of regional planning decisions with agency management decisions • Number of resolutions • Number of people nominating for advisory consultative groups • Support for decisions
Partner/Participate The agency and community (stakeholders) share responsibility for decision-making e.g. Aboriginal-owned land leased to Government for management as national park	• Maintain dialogue—ensure that all issues are open to discussion. • Provide legislative framework for participation. • Ensure ongoing management of participation. • Set clear outcomes/outputs	• Joint management • Statutory Boards of Management	• Number of partnerships agreements • Quality of relationships • Number of jointly managed protected areas

Table 7.2	(continued)		
Level of Participation Agency/Community	Best/Good Practices	Participation Techniques	Performance Indicators
Handover/ Self-directed Action The agency hands over control and decision-making to the community. The agency may facilitate management by the community through the provision of resources and expertise. Community/ stakeholder has autonomy in decision-making and may seek agency management input, e.g. Landowner wishes to contribute important privately owned land to national reserve system.	• Establish mutual benefits, trust and support • Establish transparent process • Support projects that have good conservation outcomes	• Where government agencies sit on community boards • Provide advice and other resources that result in conservation outcomes • Indigenous Protected Areas • Voluntary Conservation Agreements • Review mechanisms	• Number of private conservation reserves and quality of conservation outcomes • Number of conservation agreements and quality of conservation outcomes • Number and quality of covenants • Area of private land added to the NRS

Groups and individuals who might be included in a participation program can be drawn from within the department itself, from other government departments and agencies, and from the general public. Some of the participation methods used in protected area planning, and linkages with level of participation, best practice guidelines, and performance measures are summarised in Table 7.2.

Collect relevant data

High quality information is an important basis for many aspects of protected area management (Chapter 6), and area management planning is no exception. Incorrect, insufficient, or inadequate resource data can severely hamper the effectiveness and quality of a management plan. However, complete knowledge of a resource is, of course, unobtainable. Collecting and compiling information takes time and costs money, and both of these factors usually place stringent limits on the data collection effort. It is therefore particularly important to concentrate on collecting relevant data. Planning is not about collecting information for its own sake. Comprehensive statistics on visitor activ-

ities may be essential for an area with a significant recreation component, while another area may require only general impressionistic information on visitor activities.

After pursuing all potential sources of information, a planner may find that a key area has not been covered adequately. For example, the distribution and requirements of endangered plant species recorded in the planning area may not be known. In this case the planner has two options. The required information can be gathered in the course of the planning project so that it is available to assist management decisions contained in the plan, or the plan can simply specify an action in relation to the collection of this information.

The first option is by far superior, because it will enable a management decision to be made regarding (in this case) the endangered species, which can then be integrated into the planning process. This means that there can be public input and discussion of the issue, the options for managing the species, and decisions regarding the preferred management actions. However, time and money may preclude selection of this option and the planner may simply have to include a recommendation in the plan regarding future research on the species.

In some cases it may be appropriate to publish the resource data collected as a separate document—a resource inventory. This can be particularly valuable in two respects. First, if there has been very little published information available on the planning area, it gives interested parties access to relevant information early in the planning process. This can considerably improve the quality and utility of input received from people and groups outside the planning team. Second, it can help the planner avoid cluttering up the management plan with a large volume of background information. An excess of such information can distract the reader from the plan itself and make the document too long and unwieldy. The sort of information which may be useful for a planning project includes:

Legal, political, and social

- location, boundaries, and land tenure
- consideration of any international commitments or treaties that relate to the planning area—for example World Heritage listing or protection of migratory birds
- a review of relevant legislation, including a summary of particular limitations or requirements imposed by the legislation
- a description and review of any relevant government policies or strategies.
- a description and review of any relevant policies or strategies of the agency responsible for managing the planning area
- the current legal status of the planning area
- any current planning or zoning schemes that relate to the area, or any previous planning work that has been done
- lists of all parties that may have an interest in the planning area, including neighbouring land holders, local communities, user groups, scientists, and government departments and agencies

Economic

- contribution of the area to the local and regional economy
- income from fees, licences, and so on

Physical

- climate
- geology
- geomorphology
- hydrology
- soils
- fire history
- landscape—an assessment and classification of the visual quality in the planning area

Biological

- general description of the vegetation, including:
  - the locations, requirements, and status of any significant species and communities
  - the location and significance of introduced species
- a list of the vertebrate faunal species and their habitats, including:
  - notes on the particular locations and requirements of any significant species
  - the population and distribution of introduced species
- information on invertebrates, though this is often sketchy or non-existent
- special habitats and their management

Historical and cultural

- Aboriginal history and culture of the area, including a review of Aboriginal sites and their significance.
- European history of the area, including a review of all European historical sites and their significance

Access

- the location and standard of access into and around the planning area
- the location of all entry points (including boat access and walking tracks)
- the location and standard of all access within the planning area, including roads, management tracks, walking tracks, off-road and off-track access, and water access

Visitation

- description of the regional recreation opportunities, including those within the planning area

- the number of visitors, their activities, and the location of these activities within the planning area
- the nature and location of any facilities such as picnic tables, information shelters, and so on
- commercial tours that use the planning area
- forecasts of future visitor numbers and use patterns

Other uses

- timber harvesting
- cattle grazing
- mining
- infrastructure such as power lines, telecommunications installations, dams, pumping stations, and so on
- occupancies or concessions such as cattlemen's huts, sawmills, and kiosks

Management resources

- current staff available, their areas of responsibility and expertise
- cost of providing and maintaining services and infrastructure
- buildings and equipment currently used for the management of the area.

Identify and analyse the issues

The process of compiling a resource inventory should provide a good basis for identifying and analysing problems and issues associated with the planning area. The public participation component of the planning process is also used to identify issues. Issues may involve conflicts between:

- various uses and conservation of natural values—for example, between cattle grazing and conservation of a significant species, or building of a major visitor access route and preservation of scenic quality
- one resource component and another—for example, between an introduced species and a native one
- various uses and the resource on which they depend—for example, the quality of a bush camping experience can be diminished by problems of vegetation depletion, rubbish, disposal of toilet wastes, and so on that are the result of camping

- one use and another—for example, bushwalkers and trail bike riders, or water-skiers and swimmers.

Classifying issues according to this framework can assist the planner to understand the nature and context of the issue. This can be the first step in identifying the underlying cause of the issue or problem. In addition, tackling a problem involving two resource components is likely to require a very different approach in comparison to one involving two conflicting uses.

Another product of analysing problems and issues could be a list of all possible topics for which management objectives and related actions may be required. Typically, the subject matter of objectives and actions covers flora and fauna conservation, various aspects of visitor management, management for other authorised uses, and regulation of use to minimise impacts.

Establish goals and objectives

Planners have used a confusing array of terminology when dealing with means and ends. Words such as 'goal', 'aim', 'objective', 'strategy', 'policy', and 'action' tend to be used loosely and in some cases interchangeably. This makes it rather confusing for a professional not involved in the plan preparation, the staff who have to implement the plan, and the general public who wish to interpret and understand a planning document. Clearly, a common usage and understanding of these terms is desirable. To this end we have adopted the terminology indicated in Table 7.3.

Statements of ends are either goals or objectives, depending on their level of specificity. Statements of means may be general (guidelines) or specific (actions).

A *goal* is a general statement of ends. It is not necessarily achievable in the planning period, but indicates the broad ends to which management

Table 7.3	A classification of ends and means	
	Ends	Means
General	Goal	Guideline
Specific	Objective	Action

aspires. Examples of goals that might appear in a park management plan are:

- to conserve native plants and animals
- to provide a range of recreation opportunities
- to control pest plants and animals.

Goals, because of their very general nature, are by themselves insufficient for directing management. However, it is still important to specify these broad statements of direction in a management plan. They can indicate which goals established in legislation or by the government and the managing agency are particularly relevant for the planning area. A statement of goals is important for establishing the links between broader national, state, or regional planning considerations and a particular management plan. They can also provide a level of detail not included in a broad strategic vision (Section 7.2).

An *objective* is a statement of realistic, measurable, and specific ends to be achieved within a specified period of time. Objectives are required for effective evaluation of a plan, since if it is unclear what a plan intends to achieve, it is not possible to determine its success or failure. Without objectives, a manager cannot know when a particular action achieved the desired result (and

so move on to achieving other objectives). Nor can the manager discover if a particular action is in fact not achieving the desired result, and whether another action should be tried instead. Ideally, an objective should be:

- specific
- clearly stated
- measurable
- realistic
- where appropriate, time limited.

In many cases it may not be possible to formulate an objective that satisfies all of these points. In particular, there is often insufficient information to craft a measurable objective. For example, an ideal objective with respect to an endangered species might read:

> *to increase the population of Leadbeater's Possum (Gymnobelideus leadbeateri) in the areas shown on Map X by 20% over the next ten years.*

However, if there is no inventory of the current location and numbers of the species within the planning area, it is not possible to formulate such an objective. Nonetheless, many objectives in Australian protected area management are what we would call goals, and could be considerably

Figure 7.4   From vision to performance standard

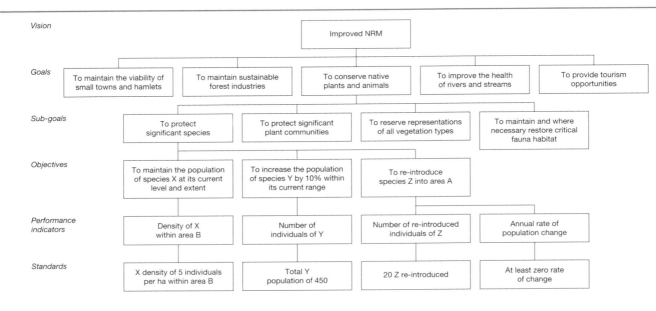

improved particularly through greater specificity and measurability.

Two other terms that are also becoming popular in planning parlance are performance indicator and performance standard. *Performance indicators* are scales that are used to assess the degree to which a desired outcome has been achieved. *Standards* specify the required level of a performance indicator in order for an objective to be met. Generally there is a hierarchy from vision, goal, and objective through to performance indicator and standard, as shown in the example in Figure 7.4.

We use the term *guideline* to refer to a general statement of means. Guidelines, like plans, can be formulated at different levels. A government may have a guideline on conserving plants and animals, a forest service and a parks service may have more specific guidelines on the same topic, and even an individual park may have a flora and fauna guideline. The number and range of guidelines may introduce some confusion as to how the various guidelines relate to each other, and how guidelines relate to actions. The major point about guidelines is that they are generated without specific reference to particular circumstances or locations. *Actions*, on the other hand, do have such specificity and detail.

Case Study 7.2 raises several aspects of developing a park management plan.

Develop options (actions) for achieving objectives

Once goals and specific objectives have been established for each management issue, a planner must explore the possible options for achieving these objectives. Some options will be evident to the planning team from their own professional experience and knowledge of the planning area. Others may be generated through stakeholder and agency staff participation in the planning process. A useful way of getting all these ideas down on paper is to hold a 'brainstorming' session.

In a brainstorm session the planning team considers each objective in turn and lists all ideas that come to mind, no matter how apparently impractical or crazy. This helps prevent premature narrowing of the range of options being considered. Narrowing options too early in the planning

process can stifle lateral thinking and make it more difficult for novel and creative solutions to emerge.

Sometimes the range of options and their possible implications are such that the planning team may decide to prepare and publish a separate issues paper (Table 7.1). These papers can be particularly valuable in a major planning project that involves a number of major and complex issues and is of considerable political significance. An issues paper can facilitate extensive formal and informal public involvement in establishing and evaluating management options. It can also help avoid major conflicts by bringing potentially contentious options into the open early in the process, and allowing plenty of opportunity for their ramifications to be debated.

Select actions

Once the range of possible options for achieving each objective has been established, some basis is required for selecting the best option or combination of options. There is a wide range of methods that could be used to test the options:

* professional judgment
* dialogue involving planners and stakeholders through informal discussion or formal proceedings such as inquiries or conferences
* systematic application of procedures such as benefit–cost analysis (Chapter 8), multicriteria analysis, impact assessment, or voting.

In much protected area planning in Australia, evaluation of options is done implicitly. That is, the planner, planning team, or stakeholder committee discuss and assess the options using their professional judgment. In these assessments the criteria against which the options are being judged are often unstated, and the reasoning behind the testing process is not articulated. The disadvantage of this process is that people who are not directly involved will not have an appreciation of the reasons or justification for the superiority of some actions over others. The decision-making process remains hidden.

One way of making explicit the testing and evaluation steps is to develop a formal assessment procedure. This enables the planner to communicate to interested parties the process by which preferred options were selected, and permits justification of

Case Study 7.2

Hinchinbrook Island National Park Management Plan

Peter Lavarack and John Hicks, Queensland Parks and Wildlife Service

Hinchinbrook Island National Park, which lies offshore from Cardwell and Ingham in North Queensland, is about 35 kilometres long and about 40,000 hectares in area. It is a very rugged, mountainous island with several peaks at or near to 1000 metres. It is separated from the mainland by the narrow Hinchinbrook Channel. The island and channel make up some of the most attractive scenery on the Queensland coast and have attracted much interest from the tourism industry over many years. Some sites such as Zoe Bay, Cape Richards, and Macushla are very popular both with commercial operators and private tourists in their own boats. There is a small tourist resort at Cape Richards that is on a peninsula, isolated from the bulk of the park. The island is part of the Great Barrier Reef World Heritage Area. Access to the island is limited by the exposed nature of the east coast and by the mangroves that line almost all of the west and northern coasts.

With the anticipated development of the Port Hinchinbrook Resort there is renewed interest in Hinchinbrook Island as a major destination for tourists. Several commercial operators have long provided day tours and dropped off campers from Cardwell and Lucinda. Up to the present, use of the island has been managed to retain the 'wilderness' setting of the island, but there is potential for increased pressure due to substantially increased visitor numbers.

A management plan for Hinchinbrook Island National Park had been under development for some time. It became clear that Hinchinbrook Island should not be managed in isolation, but must be considered in the

Bushwalkers along Ramsay Beach, Hinchinbrook Island, Qld (Peter Lavarack)

broader regional context of several other islands in the Rockingham Bay area. These include Dunk Island and Bedarra Island, both of which have resorts. These resorts represent one end of the spectrum of recreational opportunities available in the area. The other end is represented by most of Hinchinbrook Island, with its generally inaccessible coastline and rugged interior. Except for the small resort at Cape Richards, Hinchinbrook Island has not experienced the rapid development that has affected most coastal parts of the wet tropics.

Today Hinchinbrook Island remains a substantial wilderness area, even though it is close to the major population centres of Cairns and Townsville. Being only a few hundred metres off-shore, it suits self-reliant activities such as bushwalking, camping, mountain climbing, and nature appreciation. With the exception of the low-key resort, the only developments on the island are low-key trails, camping, and picnicking facilities at five sites. It has been a major goal of the management plan to preserve this setting of an accessible wilderness area close to population centres. Concurrent with this, is the goal of keeping Hinchinbrook Island at current levels of use, thus preserving a sample of the southern coastal part of the Wet Tropics. These goals are an important part of fulfilling the obligations of managing a World Heritage Area.

To achieve this at a time when tourism operators were very aware of the important scenic attractions of Hinchinbrook Island, we decided to limit the number of visitors at key locations on the island. As more than 90% of visitors reach the island with commercial operators, placing finite limits on the operators was essential, but likely to be met with resistance. A study showed that existing use levels were close to, but not exceeding, sustainable levels, but that permits already granted allowed for a large increase in use. Levels of acceptable use were set for each key area based on the judgment of experienced field staff. The criteria included physical capacity, social capacity, capacity of facilities, and biological carrying capacity. Adjustments were made as required. For example, the initial limit on the Thorsborne trail was 80 people. However, this was reduced to 40 when the effects of the initial level of use were assessed. Consistent with this controlled level of use, a policy was adopted that the Thorsborne Trail, which runs along the east coast for some 32 kilometres, would be managed as a rough bush trail. If monitoring showed that erosion or other problems were developing, the trail would be re-routed to a better site, rather than undertaking extensive construction. Developing or hardening of the trail would be acceptable only in a few key areas such as where it was impossible to avoid steep slopes. In this way, the wilderness character of the trail would remain and impacts would be manageable. This principle was applied to most other visitor destinations, although some site hardening was accepted at a few day-visitor sites.

The decision to limit numbers and to embed limits in the management plan has resulted in a more prescriptive management plan than is now generally undertaken for Queensland protected areas. This level of detail has proved extremely useful (i) in assessing tourism permits; and (ii) in offering the public an open, detailed appraisal of the management proposals in the draft plan.

Management principles from the draft plan were adopted as policy and used to manage the island for several years, with few problems arising. Applications for permits to use the island were assessed on the basis of this policy. A survey of visitor satisfaction showed that most users were satisfied with the experience that the island offered them.

these selections. The decision-making process thus becomes more transparent.

Most formal evaluation systems require that the decision-makers identify criteria that can be used to judge the worth of different options. Multicriteria analysis (MCA), for example, is a widely used approach to assisting decision-making when a range of options need to be assessed according to several criteria—a common circumstance in protected area planning. MCA is a general term used to describe a number of procedures that organise information relevant to the decision-making process. The basic element common to all MCA is an effects table that indicates the performance of each management option in relation to a set of selected criteria. MCA can be used to choose one or more superior alternatives, generate a complete or partial ranking of alternatives, or analyse the acceptability of each alternative (Lahdelma et al. 2000). At its most basic, no attempt is made to formally aggregate across the different criteria to determine the best option. In this case, MCA serves simply as a means of organising and presenting the value of implications involved. The decision-maker(s) can use the MCA effects table as a means of assisting choice and clarifying the nature of the options, but some professional judgment must be explicitly applied to select a preferred alternative. This method has been used, for example, in comparison of road access options (Cape Woolamai Steering Committee 1989).

If the analyst wants to compare options more formally, they can be scaled against a qualitative index. This scaling typically proceeds by determining the performance of each alternative against each criterion using some common measure, multiplying this performance score by a weighting that reflects the relative importance of the criterion, and aggregating across criteria to produce an overall score for the option. This method has been used, for example, in comparison of riparian revegetation options in North Queensland (Qureshi & Harrison 2001) and regional priority setting in Queensland (Hajkowicz 2002).

Integrate actions into a cohesive plan, and implement

Most planning processes have as a major outcome the production of a written management plan. A plan typically incorporates elements such as:

- a description of values and resources
- identification of issues
- goals
- a zoning scheme
- objectives and performance measures
- actions, for which priorities may be established.

We have discussed these elements earlier, with the exception of zoning. Zoning is a technique that involves spatially organising a planning area to facilitate the achievement of management goals and/or objectives. Zoning can direct management towards achieving specific objectives in certain sub-areas of the overall planning area. It can also provide the basis for partitioning the planning area in order to separate incompatible uses and to exclude inappropriate uses from certain areas. An example of a zoning scheme is given in Case Study 7.3.

Ideally, a zoning scheme should be developed from a wide range of spatial resource information, including:

- land capability factors such as slope, soil type, and hydrology
- a general description of vegetation communities
- sites of botanical and zoological significance
- sites of cultural and historical significance
- landscape values
- visitor activities and opportunities
- current land uses
- timber resources (for a forest management plan)
- government decisions regarding land use.

Spatially representing some of these resource factors can be done with the aid of other planning tools. For example, the Recreation Opportunity Spectrum (Chapter 16) can be used to define recreational opportunities, and the Visual Management System (a technique devised by landscape planners) can be used to derive a zoning map depicting areas of different visual quality. By combining all this information, a planner should be able to develop a zoning scheme that takes into account both these resource factors and the goals and objectives for the planning area.

To be useful, zoning must reflect real and significant differences in management emphasis. In some plans there is very little difference between the various zones, and it seems the planners have either been unable to make some hard decisions, for example to exclude certain uses from 'preservation' zones, or they have simply used zoning because it is standard practice, without considering whether it really contributed to the particular planning problem at hand.

Descriptions of values and resources, issues, zoning, goals, objectives, and actions need to be integrated into a written document, which may be a published draft plan (under a conventional approach) or a 'loose-leaf' document that is

Case Study 7.3

Zoning in the Great Australian Bight Marine Park

Karen Edyvane, University of Tasmania

The Great Australian Bight region is increasingly becoming recognised as an area of global conservation significance, particularly for species of rare and endangered marine mammals (Edyvane 2000). Specific areas such as the head of the Great Australian Bight contain the most significant critical breeding and calving sites of the endangered Southern Right Whale (*Eubalaena australis*) in Australia, and the world. More recently, the discovery of globally significant mainland breeding populations of the rare Australian Sea Lion and the Great White Shark *(Carcharadon carcharias)* within the proposed park has added to the national and international conservation significance of the region.

The Great Australian Bight Marine Park comprises two adjoining marine parks, established by the South Australian and Commonwealth Governments. The State waters of the park (total area, 1,683 km^2), extend from the coast to 3 nautical miles offshore and include the Great Australian Bight Marine Park Whale Sanctuary comprising 43,587 hectares (managed under the *Fisheries Act 1982* (SA)), and the Great Australian Bight Marine National Park comprising 124,732 hectares (managed under the *National Parks and Wildlife Act 1972* (SA)). The Commonwealth waters of the park (total area, 19,395 km^2) extend south to the limit of the Australian Exclusive Economic Zone, approximately 200 nautical miles offshore and include the Great Australian Bight Marine Park (Commonwealth Waters) managed under the *Environment Protection and Biodiversity Conservation Act 1999*.

In principle the management of the Great Australian Bight Marine Park is envisaged as one of 'multiple-use', which will provide a unique opportunity for the integrated management of a range of existing and future sustainable uses and activities within the region (that is, mining, commercial fisheries, recreational fisheries, tourism, recreation, research), while protecting the critical breeding and calving areas of the whales and sea lions and also representative marine habitats and ecosystems (Edyvane & Andrews 1995). The integrated management of uses or activities within the Great Australian Bight Marine Park was developed through the preparation of two management plans (one for State waters and one for Commonwealth waters) and achieved through a process of zoning.

Zoning separates a marine park into discrete management units or zones and provides levels of protection that reflect the characteristics of natural resources, biodiversity, and traditional use. Most importantly, by separating potentially conflicting uses and activities into different areas or zones, zoning minimises conflicts that may arise between the different user groups. The levels of protection are achieved through prescribing or regulating activities or uses within zones. The zoning framework was influenced by the relative needs of the community as a whole and was achieved through a formal process of community and industry consultation and participation. While some activities are defined and regulated within each zone, it is important to note that the zoning provisions for uses or activities generally determine issues of access. Four management zones have been defined for the Great Australian Bight Marine Park: two for state waters, Sanctuary and Conservation Zones; and two for Commonwealth waters, Marine Mammal Protection and Benthic Protection Zones.

Sanctuary Zone

The Sanctuary Zone provides the highest level of habitat protection. This zone encompasses the critical breeding and calving areas of the endangered Southern Right Whale and also the breeding colonies and haulout sites of the rare Australian Sea Lions. Potential threats to these species include fisheries net entanglements, vessel strikes, vessel crowding, and acoustic disturbance from boat engines, seismic blasting, and low flying aircraft. Consequently, activities that potentially threaten or disturb these species in the area, such as public access, fishing, mining, and mineral and petroleum exploration are prohibited in this zone. To protect a representative example of the marine habitats in this region, habitat disturbance and the removal of natural resources are also prohibited. Adjacent marine and terrestrial areas should be managed in keeping with the general principles of this zone. Specifically, visible impacts and infrastructure along the terrestrial border should be kept to a minimum so as to maintain the high coastal wilderness values of the region.

Conservation Zone

The Conservation Zone provides a buffer for the Sanctuary Zone and, importantly, provides for unimpeded passage and protection of migrating whales moving along the coast during the breeding season. Priority is given to managing the area to protect natural and cultural values, while allowing for the sustainable use of resources. The zone encompasses the area immediately adjacent to the critical breeding and calving areas of the endangered Southern Right Whale and also the breeding colonies of the rare Australian Sea Lions. Seasonal restrictions (May to November) have been placed on public access, fishing, and mining and exploration activities, in order to keep disturbances to migrating whales to a minimum. Adjacent marine and terrestrial areas should be managed in keeping with the general principles of this zone. Infrastructure and visual impacts should be kept low to maintain the strong coastal wilderness appeal of this area. Mining and exploration activities are prohibited.

Marine Mammal Protection Zone

This Zone also protects the Southern Right Whale and Australian Sea Lion and their habitats. The zone extends from the adjacent South Australian component of the Marine Park (3 nautical miles) out to approximately 12 nautical miles offshore, and is roughly parallel to the coastline of the Great Australian Bight. The management objective for the area is to provide a buffer for the whale and sea lion sanctuaries in the State park through coordination of all activities in the area that may affect these mammals. Hence, demersal (sea floor) trawling and mining (including exploration) is prohibited in this zone, with restrictions on public vessel access from May to November.

Benthic Protection Zone

This zone protects a cross-shelf transect as a representative strip of the sea-floor environment of the Great Australian Bight. The zone encompasses a 20 nautical mile wide strip, extending out to 200 nautical miles offshore and is roughly perpendicular to the coastline of the Great Australian Bight. Sea floor trawling is prohibited in this zone, while mining and exploration may be considered on a case-by-case basis.

continually modified according to a formal process (under an adaptive approach). In an adaptive approach, there is also the additional challenge of incorporating the proposed actions into an experimental design that enables their effectiveness to be tested against the relevant objectives. Other content and structural issues that often confront the writers of management plans include whether to present resource information, and, if so, how much should be provided, and where it should be located.

Budgets and works programs provide a link between the actions in a management plan and their implementation. They indicate the allocation of time, staff, and money required to accomplish each task. Works programs detail who will be carrying out what tasks on what day.

Review effectiveness of actions in achieving the objectives

Monitoring the consequences of actions recommended in a management plan enables a planner to determine whether the actions are in fact achieving the objectives set out in the plan (Chapter 21). Once the actions have been completed and the corresponding action achieved, management can proceed to deal with a new objective. If the action is not making adequate progress towards achieving the related objective, then a new action or series of actions may need to be developed.

Management agencies often have a stated intention to review plans every five years. Such a review enables a plan to be updated to take into account changing circumstances or the availability of new information. It also provides an opportunity for a major review of the objectives and the success of the actions in achieving them. However, if an adaptive approach is adopted, the process of monitoring and revision can be a continuous one. There is not necessarily any 'draft' or 'final' plan at all. There is simply an ongoing working plan that is adapted and modified as objectives are achieved, problems are identified with existing objectives and/or actions, or some external change (such as a major natural event) forces reconsideration of the entire plan.

7.4 Management lessons and principles

Lessons

1 Planning pays off. Whether it is a small operational task or a major management planning process, clarity of purpose and clear direction always wins over ad hoc arrangements.

2 Planning can be very demanding. The preparation of a management plan for a complex and politically sensitive reserve needs to be treated as a substantial project. Such plans are often developed in an environment of multiple opinions, lobbyists, numerous consultations and meetings, a formidable array of information, policy decision-making, and plan documentation. Planning staff need to be carefully rostered and rested, respecting the 'long-haul' nature of such planning tasks.

Principles

There is no consensus among planners as to the best planning approaches and processes. There is also no single best way to undertake a planning project. This means that, in our view, it is not very useful to offer rigid prescriptions on how to do planning. Nonetheless, we can suggest some principles of high quality land-use and management planning.

1 Planners should consciously adopt a suitable mix of planning approaches that are:

- participatory at a level that matches the interests and concerns of stakeholders
- participatory in the identification of issues
- cognisant of the multi-value, multicultural context of protected area management
- rational and participatory in the collection and identification of information to inform management
- rational in the application of formal procedures to assess any changes in land-use or major investment issues
- rational and participatory in the assessment of action options and selection of preferred actions
- adaptive in the implementation, assessment, refinement, and modifications of objectives and actions
- incremental in addressing urgent or minor management requirements that, given information, organisational, or resource constraints, cannot be dealt with in any other way.

2 Effective linkages should be established across planning levels such that:

- strategic planning occurs at the organisational and regional levels, including specification of goals and guidelines
- specific planning occurs at the local level, including development of measurable and

Protection works, Mimosa Rocks National Park, NSW (Graeme Worboys)

realistic objectives that are framed in the context of strategic goals and have clear performance indicators
- explicit linkages are present between objectives and actions and outcomes
- actions are consistent with strategic guidelines, and at a level of detail that allows for consistent interpretation and application.

3 Effective implementation of actions arises from:
- availability of suitably trained staff to guide the planning process and implement the plan
- links between actions, available resources, the budget process, and performance evaluation
- definitions of roles and lines of responsibility in the managing agency regarding implementation of particular actions

- works programs that are linked with the plan, contain dates for completion of actions, and are fed back into the performance evaluation.

4 Formal evaluation of success is an essential part of a successful planning process and involves:
- lines of responsibility in the managing agency regarding evaluating performance against objectives
- mechanisms for formal recognition (and removal from the plan) of objectives that have been met and completed
- mechanisms for addressing objectives and/or actions that have not been met, including, where appropriate, their modification
- clear guidelines for reviewing plans, objectives, and actions, including participants, responsibilities, and periodicity of revisions.

Finance and Economics

There are two problems in my life. The political ones are insoluble and the economic ones are incomprehensible.

Alec Douglas-Home, Prime Minister of Great Britain, New York Times, *9 January 1964*

8.1 Financing protected areas   220

8.2 Pricing services
     and facilities   223

8.3 Environmental valuation and
     benefit–cost analysis   237

8.4 Management lessons and principles 248

High-quality planning and management of protected areas requires input from a wide variety of
disciplines. In the past, the expertise of protected
area managers was concentrated in the biological
sciences. More recently, management has benefited
by drawing from such social science disciplines as
public sector administration, leisure and recreation,
and sociology. While public sector agencies have
always had staff with financial expertise, it is only in
the last decade or so that protected area management agencies have begun to also apply economic
understandings. Finance is concerned with securing
and managing funds, whereas economics addresses
wider issues of resource allocation, benefit assessment, and pricing policy.

In this chapter, we first consider the problem of
funding protected area management and the technique of business planning. We then outline the
contributions that environmental economics can
make to protected area management, including:

- identifying appropriate pricing polices for visitor services and other uses of protected areas
- measuring the economic value of protected
  areas
- assessing the economic benefits arising from
  recreation use, investment of public funds, or
  from creating new protected areas
- documenting the contribution protected areas
  make to regional economies.

8.1 Financing protected areas

Most protected areas are managed by government
agencies and have thus tended to rely heavily on
government funds derived from taxation revenue. In
general, this situation should continue. Governments
must fund protected areas because of the public good
benefits that they provide, and to maintain the
intrinsic values of natural areas. Funding to government departments is provided through annual
appropriations from the state, territory, or federal
treasury, depending on the department involved.
These appropriations are usually divided into
recurrent and capital expenditure components.
Departments may also be able to attract project-
specific funding through various state, territory, and
Commonwealth government grants programs.

The private sector, while making a contribution, cannot and should not be expected to meet
many of the costs associated with protected area
management. Non-use values of natural areas
(Chapter 3) are pure public goods. They reflect the
value people place on the existence of such an area,
regardless of the importance of other values related
to consumption, either of products (such as timber)
or experiences (such as recreation). Such values
would be undersupplied by private nature reserves,
since management would be orientated only
towards providing consumptive activities from
which revenue could be generated, and the area

required to do this would, in general, not be
enough to satisfy the demand for non-use values.

The private sector cannot efficiently provide
public goods and services because benefits arising
from them do not directly accrue to specific individuals. Individuals may use a public good, but may
not be willing to contribute to its cost of production
or maintenance; or there may be simply no mechanism in place to capture individuals' willingness to
pay, or to fund the cost. For example, even though
people are willing to pay for biodiversity protection,
it is rarely possible to sell such protection.

Consider the situation shown in Figure 8.1.
The demand for private goods (such as commercial
recreation tours) from protected areas is shown by
the lower demand curve. A market in protected
areas would produce an efficient level of supply of
these goods from Q_1 protected areas at price P_1.
However, the total demand for protected areas (private plus public good components) is given by the
upper demand curve. The combined demand for
private and public values for protected areas is
much greater than the private demand alone. The
efficient level of supply based on this total demand
is Q_2. Thus a market will undersupply protected
areas by an amount equal to $Q_2 - Q_1$.

This provides a rationale for government funding. The private goods associated with protected

Figure 8.1   Undersupply of public goods

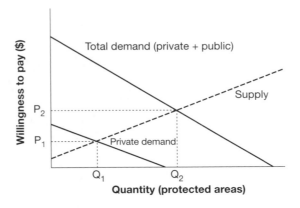

areas can be bought and sold through markets, but the public good values must be funded by a government subsidy equal to the amount $P_2 - P_1$. Markets undersupply public goods, such as biodiversity conservation, to which individual property rights cannot meaningfully be allocated. From an economic perspective, undersupply of public goods constitutes a failure to maximise social economic welfare (Backgrounder 8.1).

However, political and fiscal realities mean it is unlikely that the funding needed to satisfactorily meet all protected area acquisition and management requirements will ever be made available by governments. Financial resources often constrain effective management of protected areas and fall well short of needs. Increasing taxes is always politically difficult, even with community support for additional conservation expenditure, and there are always many other calls on government from health education, social welfare, and so on. In fact, the proportion of public funding going into protected areas is in decline in many countries (IUCN 2000a).

There are opportunities to expand on this public funding base and generate further revenue to meet agency needs. Both public and private revenue need to be optimised, with public revenue linked to public goods and private revenues to private goods. While governments will continue to have a primary role in ensuring the supply of pure public goods, the private sector is becoming increasingly important for providing visitor services and facilities, and for contributing to resource management and the restoration

of sites. The provision of incentives to support conservation activities on private property is addressed in Chapter 19. Further opportunities exist for protected area management agencies to develop constructive partnerships with the private sector. Carter (1996, p. 28) characterised the ideal relationship between the public and private sectors as follows:

> *For resource managing agencies, the private sector can be used to significantly decrease the level of financial resources needed to provide equivalent levels of service. The private sector should thus be viewed as cost-effective human resources available to the public sector for achieving goals of management. For the private sector, resource managers (the public sector) should be viewed as staff responsible for the care and maintenance of the core business resource. Collectively the relationship should be one of joint and cooperative business partners in conserving the natural resource, and providing a public service at minimal cost to each other and the public.*

Protected area managers can sometimes raise considerable sums of money through corporate sponsorship, or from private donors and bequests. Snapshot 8.1 provides an example of what can be done by a private foundation to acquire funding for protected area management. In Section 8.2 we consider how revenue can be raised through adopting 'user pays' policies. Here we first show how business planning is an important step in securing the funding necessary for effective protected area management.

Snapshot 8.1

The National Parks Foundation of South Australia
Barbara Hardy, Hon. Secretary, National Parks Foundation of South Australia Inc.

The National Parks Foundation of South Australia Inc. is an incorporated association, a voluntary non-profit organisation formed in South Australia to support national parks, aid wildlife, and protect the natural environment. The foundation is managed by a 16-member council, comprising experienced South Australian conservationists, business and professional people, and scientists. It meets regularly, and all members donate their services.

The foundation became an incorporated association early in 1982, and later in the year it was granted tax deductibility for donations of $2 and upwards under section 78(i)(a) of the *Income Tax Assessment Act*. In 1998 the foundation had at least 450 regular supporters across South Australia, who assist with donations (from $5 to several thousand dollars). It holds an appeal at the end of the financial year, and a Christmas appeal. The South Australian government matches all donations with a dollar-for-dollar subsidy (up to a maximum of $35,000). All donations are tax deductible.

The foundation also receives income from a special Mastercard credit card known as the green card, which can be applied for at any BankSA Branch. All purchases by green-card holders bring in a small payment to the foundation from BankSA at no cost to the holder.

The foundation has raised more than $1,000,000 since 1983, when it held its first appeal. These funds are invested in

Information stall at the 1994 'Picnic in the Park', Belair National Park, SA (Barbara Hardy)

South Australia's natural heritage. The funds assist in the purchase of land to add to parks (protected areas), or to acquire new parks. The foundation also supports projects to improve park management, sponsors research projects concerned with endangered species and habitats, and supports education about parks and conservation.

Business planning

Business plans are used to guide business development activities. They are being more widely adopted by conservation agencies. Business plans must be developed in the context of a wider management plan that has clearly defined goals and objectives (Chapter 7). This ensures that generating revenue is a means towards the end of more effective management, and does not become an end in itself (IUCN 2000a). A typical business plan contains the following sections:

- an executive summary that outlines the mission and objectives of the business
- a summary that provides an overview of the business, including an assessment of resources, location and facilities, ownership structures, and so on

- identification of the goods and services that will be provided and the business's comparative advantage in providing these
- a market analysis that describes market segments, needs, competitors, trends, and growth, and then uses this to develop a market strategy that outlines which customers will be targeted and why
- a marketing strategy that identifies strategic alliances and explains how the business will position and promote itself and price and distribute its goods and services, together with a sales strategy with forecasts
- an indication of the organisational structure, staff, and decision-making structures needed to implement the business plan
- a financial plan that specifies:

- key financial indicators that will be used to track performance and the amount of financing required to accomplish the goals
- viable funding sources to meet these needs
- projected cash flow and balance sheet (IUCN 2000a).

Importantly, these documents are brief, to the point, and constantly reviewed. One would expect the board to assess the performance of the venture on an annual basis (Snapshot 8.2). As well, there would be at least monthly assessments of business performance throughout the year. Usually this would be undertaken by an executive committee.

As indicated in the list of typical business plan contents, a key component is a financial plan. The financial plan determines the amount and timing of funding required to achieve management objectives, and identifies income sources to meet these needs. Financial planning differs from budgeting (Chapter 9) in that it is more focused on forecasting required funding, as well as the best potential sources to meet short, medium, and long-term needs:

> *Different sources of funding have different characteristics. Some are more reliable than others, some sources are easier to raise than others, and some can be used freely according to management priorities while others come with strings attached. Some funding mechanisms take a long time and a lot of effort to establish; they therefore do not provide a short-term return, but over the longer term they offer the possibility of steady, reliable financing to meet recurrent costs. Some sources of funding have short-term time horizons (such as a bank overdraft) and others have longer-term horizons (such as a mortgage). A good financial plan identifies these characteristics, and builds a revenue stream which matches both the short and long-term requirements of the protected area, or protected area system (IUCN 2000a).*

Snapshot 8.2

Business plan for Fitzroy Falls Visitor Centre

Fitzroy Falls Visitor Centre receives some 400,000 visitors a year. It is near Moss Vale in southern NSW at the site of a very attractive waterfall in sandstone terrain. The visitor centre was constructed at the site during the mid 1990s and a business plan was generated prior to the development. The business plan is updated annually and the centre's performance is measured against projected targets. The plan distinguishes the strictly commercial costs (and profits) of the visitor centre from a different category of costs that are due to providing certain services to the public. These are considered 'social justice contributions'. Because NSW P&W undertakes education and caters for non-paying visitors at the site, some of the operation cost has been deducted from the commercial operating costs. In the past three years the site has operated at a slight loss, then at a slight profit, and has taken corrective actions to improve profit. These actions include devolving the café business to private enterprise and refining the non-commercial aspects. The business plan tool has been effective in improving management of the Fitzroy Falls Visitor Centre.

8.2 Pricing services and facilities

Resource managers are under increasing pressure to adopt user-pays approaches and, where possible, to recover the costs of providing recreation and other services. Managers should be able to justify their pricing of recreation goods and services so that decisions are neither arbitrary nor inequitable (Loomis & Walsh 1997). Some form of user-pays system has been adopted by all Australian protected area management agencies, with most charging for entry to at least some protected areas. Case Study 8.1 describes the user-pays system (and its contribution to regional development through tourism) on Kangaroo Island, South Australia.

Data collected by the Queensland Parks & Wildlife Service and Department of Environment (QPWS 2000) show that the extent of user-pays in national parks around Australia varies greatly. An estimated $54 million in park entry fees was paid to protected area management agencies in 2001–02 (Commonwealth of Australia 2003). This revenue

Case Study 8.1

User-pays fees and regional development on Kangaroo Island

Keith Twyford and Fraser Vickery, National Parks and Wildlife Service, South Australia

Kangaroo Island, at 4500 km² the third largest of Australia's islands, is situated 15 kilometres off the mainland coast of South Australia. The island is promoted as a world-class nature-based tourist destination and is the key ecotourism destination in South Australia.

Over 116,000 hectares or 25% of Kangaroo Island is included within 24 protected areas managed by National Parks and Wildlife South Australia (NPWSA). The agency supports a large infrastructure and staff to facilitate the management of these protected areas, park visitors, and a range of high profile wildlife conservation programs. NPWSA plays a significant role in the Kangaroo Island community, actively participating in a number of government and community boards and committees relating to tourism, development, Landcare, soil conservation, fire, native vegetation, and wildlife management.

The Kangaroo Island economy has traditionally been based upon primary industries that generated $39 million in 1992–93, mainly from wool. Estimates of the economic impact of tourism indicate that in 1996 the industry supported 800 jobs and generated approximately $41 million on gross state product alone. In addition to this, expenditure by day trippers injects significant additional revenue into the regional and state economy. A recent estimate of tourism revenue generation is $80 million.

Kangaroo Island was the first region in South Australia to introduce a comprehensive range of park fees and charges. The Kangaroo Island region of NPWSA has a well established user-pays program that was initially developed in 1987 and in 2000 contributed over $1.6 million towards management of parks and wildlife within the region. Table 8.1 provides some examples of visitor and commercial use charges currently in place.

Boardwalk, Seal Bay Conservation Park, Kangaroo Island, SA (Fraser Vickery)

As well as the 'pay-as-you-go' option outlined above, visitors and residents also have the alternative of purchasing an annual Island Parks Pass entitling the holder to park entry and guided tours at all sites apart from the Discovering Penguins tours. The cost of the pass was $24.00 and $17.00 for an adult and child respectively or $65.00 for a family.

Funds generated from user-pays and commercial-use charges are retained within the Kangaroo Island budget through the NPWSA General Reserves Trust (GRT) and used for park management operations with an emphasis on the provision of visitor services and facilities. Income generation through the Kangaroo Island GRT has increased markedly since fees were introduced over 10 years ago. Income generation in 1996–97 was $1.66 million and $1.71 million in 1997–98. On this basis, Kangaroo Island is the largest NPWSA business enterprise in the state.

The slump in the world wool market has meant that the traditional role of the agricultural industry in sustaining the regional economy and acting as the key employer has been significantly reduced. In comparison, the tourism industry has played an increasingly important role in providing jobs for the local community. In 1999, NPWSA is among the three largest employers on Kangaroo Island, having more than 80 people on its payroll equating to approximately 43 full-time equivalent staff. Generally staff are

Table 8.1	Selected examples of visitor and commercial use charges for Kangaroo Island National Parks and Wildlife sites (as at January 2000)		
Fee type	**Fee ($)**		
Guided tour and entry fees (per day)	Adult	Child/conc.	Family
Seal Bay: Beach tour and boardwalk access	8.50	6.00	20.00
Seal Bay: Boardwalk access	6.00	4.50	16.50
Kelly Hill Caves tour	6.00	4.50	16.50
Cape Borda Lighthouse tour	6.00	4.50	16.50
Cape Willoughby Lighthouse tour	6.00	4.50	16.50
Discovering Penguins tour	6.00	4.50	16.50
	Car	Motorcycle	
Flinders Chase National Park entry	8.00	3.50	
Camping fees	Car (per day)	Motorcycle (per day)	No vehicle (per person)
Flinders Chase National Park			
• Rocky River	12.00	6.00	3.00
• Snake Lagoon, West Bay, Harveys Return	5.00	3.00	3.00
Cape Gantheaume Conservation Park	5.00	3.00	3.00
Commercial filming fees	First day	Each subsequent day	
Video or motion picture filming	350	100	
Still photography	100	100	

involved in directly servicing the tourism industry through guided tours, facility and road construction and maintenance, and on-site visitor management. This level of employment makes an important contribution to the health of the local economy. It has meant that NPWSA has been able to employ large numbers of local people (particularly farmers and their families) on a casual basis to provide the services that are a fundamental aspect of the agency's management of tourism. This alternative employment has been increasingly important to the local economy, particularly in the context of the downturn of the mainstream agricultural sectors.

The protected areas of Kangaroo Island provide the fundamental basis for development of the regional tourism economy while NPWSA itself acts as a major employer and direct source of income for the regional economy. The challenge for Kangaroo Island park managers is to ensure that nature-based tourism continues to be developed and managed in a sustainable manner so that those very values that attract visitors (cleanliness, safety, solitude, wildlife, and wilderness) are maintained and protected into the future.

falls far short of the real cost of servicing tourism and maintaining the recreation values of protected areas. Most Australian protected area agencies have insufficient funds to adequately carry out both natural resource management and visitor infrastructure management simultaneously (TTF 2004). Some agencies charge a fixed fee for all parks, some charge for only certain parks, and some have fees for particular uses or value-added services including:

- admission to particular attractions (for example, National Parks and Wildlife South Australia charge for accessing the boardwalk at Seal Bay Conservation Park)
- use of a specific site or opportunity (for example most agencies charge camping fees at developed sites)
- instruction and education (for example, ACT Parks and Conservation Service charge visitors for ranger-guided activities)
- a licence or permit to undertake an activity (for example, the Queensland Parks and Wildlife Service charge for permits to undertake commercial filming or photography)
- a licence or permit to offer a commercial service to visitors (for example, most agencies charge licence fees to commercial tour operators using parks)
- direct purchase of goods such as maps, books and so on.

Supply costs

Recreation supply costs that may need to be considered in relation to a pricing policy include:

- capital costs to acquire land, develop access roads and facilities
- environmental resource protection costs
- agency operation maintenance and replacement costs
- administrative overhead costs
- congestion costs for users
- opportunity costs of foregone resource development (Loomis & Walsh 1997).

Economists classify costs into two broad categories—fixed and variable. Fixed costs are those that do not vary with changes in the use of a site in the short run. Fixed costs include such things as staff salaries and investment in infrastructure maintenance that is not dependent on visitor numbers. Variable costs change with recreation use, and include various operational and maintenance expenses. Another important distinction is between average cost and marginal cost. Average cost is the cost per visit, for a given number of visits. Average cost includes both fixed and variable costs. Marginal cost is increase in total cost resulting from an increase in production by one visit. Marginal costs are necessarily variable costs. Marginal costs may rise, fall, or remain constant depending on economies or diseconomies of scale.

Measuring expenditure on facilities that are for the sole benefit of visitors is relatively straightforward. However, for many costs it may be difficult to separate out recreation expenditure from other management costs. How much, if any, of the costs associated with running the management agency should be allocated to recreation? How does one separate out the staff costs associated with biodiversity conservation with those associated with recreation? The latter question may be further complicated if some of the biodiversity conservation activities are associated with migrating the effects of visitor use. The exact proportion of total management cost spent on recreation-related services and facilities is often unclear (Beal & Harrison 1997). In the end, some judgment must be made based on a reasonable assessment of factors such as approximate staff time spent on conservation versus recreation-related activities.

There are also various accounting approaches to assessing the supply costs of recreation. Cross-sectional studies summarise average costs for one point in time. Historical line-item accounting procedures show types of expenses such as labour, material, and maintenance. Functional accounting shows expenses by type of program, such as fire control, safety, biodiversity conservation, and so on. Neither line-item nor functional accounting reveal costs for each recreation activity, nor how these costs vary with changes in facility design or level of service (Loomis & Walsh 1997).

Two procedures that can be used to assist recreation economic decisions are cost-effectiveness analysis and break-even charts. Cost-effectiveness is a way of measuring the most effective way of achieving objectives. It simply involves comparing the costs of achieving specified outcomes by alternative methods. For example, options for upgrading visitor access can be assessed in terms of the cost per visitor. A break-even chart is a graphical representation of the relationship between total cost and total revenue for all levels of output (Loomis & Walsh 1997). This method is only useful when there is an identifiable revenue stream that is directly related to a particular facility. The break-even level of facility provision is given by the total fixed cost divided by the price per unit output minus the average variable cost. This relationship can be used to determine a break-even price to cover the cost of increased investment in recreation facilities or the number of new users required to cover the cost of the facility.

Justification for user-pays

Demand for the recreation opportunities afforded by protected areas is likely to continue to rise (Chapter 16). This growth is promoted by, among other things, enhanced information availability about the attractions of protected areas, improved access and transport connections, together with a growing consumer preference for 'quality-of-life experiences', including outdoor recreation. Increased visitor numbers will impose additional costs on protected area management agencies. Services and facilities (car parks, walking tracks, toilets, visitor centres, and so on) will require upgrading and expansion. Environmental damage, and therefore the need to expend resources on rehabilitation, will increase. Costs may also be imposed on visitors in high use areas, as congestion diminishes the quality of recreational experiences (Cotgrove 2004).

These increased costs make the problem of who should pay for them particularly pressing. Governments must remain the principals source of funding, via tax revenue, for the public good and intrinsic values provided by protected areas (Chapter 3). Conservation of natural and cultural resources is rightly regarded as a community service obligation for government agencies and a user-pays system is not applicable to secure the continued supply of these values (QPWS 2000). However the costs of providing appropriate infrastructure, facilities and services, repairing environmental damage, and limiting congestion are generated by 'private' consumption of protected area values. The beneficiary and polluter-pays principles suggest that these costs should not be borne by the taxpayer, but by users who either gain benefits from the infrastructure, facilities, and services (beneficiaries pay) or impose environmental or congestion costs on others (polluter pays).

Non-users effectively subsidise users when fees are not charged. Subsidies may be justified to enable low-income earners to visit natural areas. However, at sites primarily visited by high-income earners, the poor may be worse off as they subsidise the free entry of rich visitors through their taxes. A related issue arises when sites have a significant number of foreign visitors who are more wealthy than the local taxpayers. This is particularly an issue when visitors from developed countries visit developing countries (Lindberg 1998).

Recreation activities are not the only uses that impose environmental costs. Some protected areas are subject to honey production, fishing, cattle grazing, and other extractive uses. Again, the 'polluter pays' principle has application here. Similarly, visitors are not the only beneficiaries from protected areas. Protected areas provide a range of ecosystem services (Chapter 3) that benefit people some distance away. For example, the quality of Melbourne's water supply is in part due to the catchment protection afforded by national parks and other reserves some distance from the city. In this case, applying the beneficiary-pays principle is not easy, but there are examples elsewhere where a mechanism has been developed (Snapshot 8.3). Local communities also benefit (Case Study 8.1),

Snapshot 8.3

Charging for ecosystem services
IUCN (1998)

In 1998 Inversiones La Manguera Sociedad Anonima (INMAN), a Costa Rican hydroelectric company, signed a contract with the Monteverde Conservation League (MCL) to pay for ecological services provided by the protected area managed by MCL. The Bosque Eterno de los Niños (Children's Eternal Rain Forest) is a 22,000-hectare private reserve managed by MCL. Approximately 3000 hectares of the protected forest is part of a watershed that is used by INMAN for generating electric power. Recognising the benefits they receive from protection of this watershed, INMAN entered into an agreement with MCL to pay for the protection of the ecological services provided by Bosque Eterno de los Niños. The contract recognises services such as stabilisation of land, soil protection, humidity and nutrient retention, water protection, protection of species biodiversity, and more. INMAN pays MCL $10 per hectare (a negotiated price) x (a factor that accounts for the amount of energy generated and sold by the hydroelectric plant) x 3000 (for the hectares in the watershed). The money from this tax is used directly to pay for reserve protection programs.

but they often also have to forgo potential benefits from alternative uses such as timber production or mining, and local support can be crucial for achieving successful management outcomes. Such equity and strategic considerations make it generally inappropriate to impose additional costs on locals.

Setting a price

The level of charges in a user-pays system should be determined by a clear set of objectives. An agency's choice of revenue objectives can vary according to the type of value and the beneficiary. Objectives for developing a user pays policy may include:

- equitable allocation of costs
- cost recovery (Backgrounder 8.1)
- economic efficiency through identification of a 'market rate' (Backgrounder 8.1)
- extraction of a 'scarcity rent' (Backgrounder 8.1)
- generation of revenue in excess of costs so that other activities such as biodiversity conservation can be financed
- improving facilities and management
- generation of foreign exchange and/or tax revenues from tourist purchases
- demand management—that is, using fees to limit or redistribute the number of visitors, in order to reduce environmental damage, congestion, or user conflicts (Lindberg 1998, QPWS 2000).

Some potential relationships between the type of facility, service, or value being provided by a protected area, an associated revenue objective, and other factors that may justify modification of this objective, are summarised in Table 8.2.

For walking tracks, camping areas and the like, revenue objectives are usually limited to partial or full cost-recovery (QPWS 2000). As indicated in Backgrounder 8.1, this generally will not constitute the same level of charges as would arise from an objective of economic efficiency. Setting rates below cost recovery can be justified on the basis of the 'external' benefits created by recreation, such as improved health, that reduce costs for other publicly funded welfare programs. On the other hand, fees could be set above cost recovery level if there is a desire to use price as a mechanism for reducing or redistributing demand (see below). The equity

Backgrounder 8.1   Supply, demand, and recreation pricing

Market economic values are determined by the exchange of goods and services in organised markets through the price mechanism. A market will function efficiently only when certain conditions are met. As noted in Chapter 2, an ideal market requires perfect competition between the actors in the market; availability of full information in relation to the goods being traded and the mechanisms of trade; and allocation of property rights such that all goods in the market can be exclusively owned by individuals, and 'non-paying customers' can be excluded.

Under these circumstances, a market will 'automatically' result in a situation in which supply equals demand. Economists use graphs to show how supply and demand change with the price of a good. Figure 8.2 shows a demand curve for recreation that is sloping downwards. This means that as price increases, fewer visits are made. Prices in this context can include all costs associated with making a visit to the protected area in question. These costs can include fuel, vehicle running costs, any charges levied by the management agency, and even an estimate of the dollar value of travel time. The demand curve can be interpreted as representing visitors' willingness to pay (WTP) to 'consume' various 'amounts' of recreational experience. To make the discussion simpler, for the rest of this backgrounder we will consider only that component of demand derived from willingness to pay costs levied at the site itself. We will come back to willingness to pay travel costs when we consider the travel cost method for measuring recreational demand (Section 8.3).

The supply curve shows that as the number of visits increase, the costs faced by the management agency also increase. In general, costs can be calculated to include both the direct costs of providing facilities and services, and externalities associated with rectifying any environmental effects of the recreation activities. Marginal costs are the increase in cost for each additional visit. The average cost per visit for any given level of visitation is given by the total cost (the area under the supply curve) divided by the number of visits.

At price P_1, the demand for visits is Q_1. However, assuming the management agency can capture visitors' willingness to pay, they can still cover costs even if more people visit the area. At Q_2 visits, however, willingness to pay is less than the costs of providing this number of visits, so that at this level of visitation the manager would suffer an economic loss. In a perfect market what tends to happen is that an equilibrium is approached between supply and demand so that Q_3 visits are made and are produced at price P_3.

This level of visitation is efficient, because it maximises the benefits to both the 'producer' (the park agency) and the consumers (visitors). To have one less visit would mean that the agency would miss some potential revenue. To have one more would mean that they would make a loss on this additional visit. From the consumer's point of view, the efficient level of consumption maximises what is known as consumer surplus. In Figure 8.3, the price paid for each visit is P_1, but some consumers are actually willing to pay more for most of their visits, as indicated by the demand curve being above the price

Figure 8.2   Supply, demand, and market efficiency

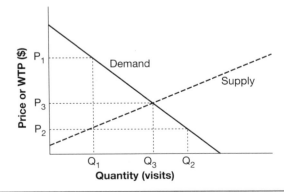

Figure 8.3   Consumer and producer surplus

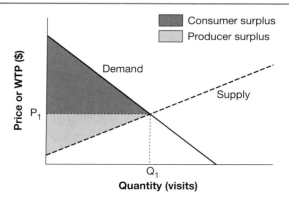

line between zero and Q_1. This difference between the actual price paid and the maximum willingness to pay, is called consumer surplus, because it is a measure of the extra value consumers receive from a purchase over and above the price they pay. There is also a producer surplus generated by Q_1 visits, since extra revenue is obtained over and above 'production' costs (Figure 8.3).

Now, back to objectives for a user-pays policy. First, an agency could set fees per visit at the average cost. This approach would ensure full recovery of costs. At present, most agencies set fees below the level required for full cost recovery.

Second, an agency could set fees at the 'market price'—that is at P. This would ensure economic efficiency, in that benefits (the sum of producer and consumer surplus) would be maximised[1]. At these prices, the capacity of the park to supply recreational opportunities will just equal demand. An illustration of how such a price could be calculated is shown in a study by Beal and Harrison (1997) from Girraween National Park in Queensland. The travel-cost method (Section 8.3) was used to estimate demand functions for the park for the two main user groups: campers and day visitors:

Demand curve for day visits: [Price] = 47.23 − 0.00045[VISITS]
Demand curve for camping: [Price] = 9.08 − 0.0002[VISITS]

VISITS in both cases were expressed in day-visitor equivalents. An aggregate demand curve was calculated by adding the demand for both day visits and camping:

Aggregate demand curve: [Price] = 20.84 − 0.00014[VISITS]

A marginal cost curve was calculated from agency records. Costs included labour, operating, infrastructure, road maintenance, and estimates of additional expenditure required to repair damage caused by recreation activities:

Supply curve: [Marginal Total Cost] = 0.92 + 0.000003[VISITS]

For economic efficiency, supply equals demand, so:

20.84 − 0.00014[VISITS] = 0.92 + 0.000003[VISITS]

From this relationship, the efficient number of visits was calculated to be 139,230. At this level of demand, the efficient price is:

[Price] = 20.84 − 0.00014[139,230] = $1.34

At this price, an efficient demand for VISITS (day visitors) = 38,000, and VISITS (campers) = 102,000. The optimal price per day visit equivalent converts to about $8 for the average camper. Currently, there are no charges for day visitors, and campers are charged at a rate of $7.50 per site, where up to six people can occupy one site. This price structure means that the management agency was not covering the direct costs of providing recreation services, nor charging economically efficient prices.

Third, an agency could attempt to capture a larger proportion of visitors' WTP. They may be in a position to do this because they control a unique resource, for which they are the sole provider. In perfect competitive markets a surplus accrues to the consumer. However, where a supplier has exclusive control over a particular resource, such as access to a protected area, they can potentially charge a higher fee so that the provider appropriates some of what would otherwise be consumer surplus. Note that economists do not regard this capture of a 'scarcity rent' as an efficient outcome, because overall net benefits are not maximised.

1   However, in general, the shape and disposition of the supply and demand curves may be such that an efficient price may not cover the average total costs faced by an agency. Effective competition among private suppliers tends to drive user fees to minimum cost levels, but park agencies do not operate in such a competitive environment and may have a different cost structure. Charging a price equal to the average total cost will ensure that an agency does not make a loss, but this may not produce an 'efficient' amount of recreation (that is, net benefits may not be maximised). For details, see Loomis and Walsh (1997).

implications of charges should also be considered, with concessions available to low-income earners.

Charging to enter a protected area can be problematic. While this may be the most efficient and cost-effective method of collecting fees, visitors should actually be charged for facilities and services, rather than for access. A Senate Committee of Inquiry into park user fees, for example, found widespread support for the principle that access to such areas is a right of all Australians. This emphasises the need to regard fees as payment for a product rather than for the right to be on the land. Entry charges should not be set at such

Table 8.2	Revenue objectives for facilities, services, and values	
Facility, service, or value	Potential revenue objectives	Mitigating considerations
Infrastructure use: e.g. access roads, visitor centre, car parks, walking tracks, campsites	Cost recovery	Equity, competitive neutrality (Backgrounder 8.2), demand management, transaction costs, external benefits
Entry to special attractions: e.g. historic buildings, caves	Cost recovery	Equity, competitive neutrality, demand management transaction costs, external benefits
Commercial products: e.g. on-site accommodation, books, maps, food	Market rate	External benefits
Extractable resources: e.g. honey, fodder, other plant products, fish	Market rate, scarcity rent	Local beneficiaries
Ecosystem services: e.g. clean water, carbon sequestration	Market rate, scarcity rent	Local beneficiaries feasibility, transaction costs
Rental items: e.g. recreation equipment	Market rate	External benefits
Non-extractive occupancies: e.g. power, water, and communications infrastructure	Market rate, scarcity rent	Public service, fairness
Non-commercial non-recreation uses: e.g. scientific research	Cost recovery	Value for informing management
Other commercial non-extractive uses: e.g. filming	Market rate	Education or promotion value

a level as to inhibit the public use. The committee argued that free entry must be retained as a real option, and charges for value-adding services must not become a proxy for entry fees (SECARC 1998).

Activities and infrastructure in protected areas such as resorts, shops, and commercial tours are usually licensed and a fee is often charged. The magnitude of such fees should ensure that, at minimum, the full costs associated with the activity are recovered, and any commercial opportunities are taken advantage of. Commercial products, rental equipment, and other commercial services should generally be priced at a market rate. However, if such goods and services have associated external benefits such as educational or safety value, consideration can be given to charging less than a market price.

Many protected areas include residences (lighthouse cottages, shearers' quarters, homesteads, and so on) that could be utilised to generate a financial return. However, in some cases, it may be more appropriate for an agency to encourage provision of services outside a protected area. Such a strategy can minimise environmental costs, pass on the risk to the private sector, and allow the agency to concentrate on the core business of managing biodiversity, cultural heritage, and recreation. Most protected area agencies provide some basic visitor facilities at destinations, but rely on regional towns (such as Jabiru near Kakadu National Park) to provide services like accommodation, food, and fuel. Some agencies do provide accommodation within parks, such as at Kings Canyon National Park in the Northern Territory, Tidal River at Wilsons Promontory National Park in Victoria, and at Tasmania's Cradle Mountain–Lake St Clair National Park. Some agencies manage these facilities themselves (such as Parks Victoria managing Tidal River) but most lease such facilities for a fee. Hence accommodation within parks usually brings some financial return to agencies. However as noted above they also incur costs, and there is a need to evaluate carefully the real costs of providing and managing such developments.

Many protected areas include power transmission lines or other utilities such as gas pipelines. In some cases protected area agencies may fund the access roads required to service these facilities. These costs should be recovered. It may also be appropriate to license such utilities and to charge a licence fee.

As we saw in Snapshot 8.3, it is possible for a private park provider to generate revenue through pricing of ecosystem service values. While such an idea may also seem attractive for public park agencies, it will generally be very difficult to negotiate an arrangement such as that established between the Costa Rican hydroelectric company and the Monteverde Conservation League. For most protected areas, provision of ecosystem services such as water supply, for example, are assured because of the mandated management regime applied by the park agency. Another authority or company has no need therefore to enter into a contract with the park manager, because they know that they will continue to obtain the ecosystem services for free. However, it is possible to envisage circumstances when revenue might be generated from ecosystem service values. Environmental degradation that occurred prior to establishment of a protected area may still be compromising potential ecosystem service values. In this case, it may be possible for the management agency to recover some or all of the rehabilitation costs from an organisation that will gain ecosystem service benefits from such rehabilitation. It may even be possible to negotiate a market rate or resource rental for the enhanced availability of the valued service. Again, regional development concerns may militate against such a 'hard-nosed' business strategy.

For commercial activities such as cattle grazing that have little or no connection with an agency's mandate, an appropriate objective should be at least total cost recovery, if not market rate or even extraction of the maximum economic rent. The decision as to which of these three approaches to adopt will be influenced by political acceptability and the need to build good relationships with neighbours and an obligation to assist regional economic development.

The cost of collecting user fees is an important factor in establishing a pricing policy. Costs associated with the implementation and administration of a user-pays system are called transaction costs. There is no point charging user fees if the transaction costs are such that they substantially offset the revenue collected. For a park with many

Table 8.3	**Revenue collection methods** (modified from QPWS 2000)	
Fee collection method	Advantages	Disadvantages/constraints
Payment through the post	Administrative convenience, information can be sent	Delay for clients
Credit card over phone or Internet	Speed, administrative convenience	Credit card security, staffing telephones, or maintaining an automated system
Over the counter at park offices	Face-to-face staff contact, client briefing, high compliance	Costs of offices and staff, security of cash transactions
Roving rangers	Staff contact, compliance	Auditing problems, security, time-consuming, staff costs
Park entry stations	Staff contact	Costs of construction and staffing, security
Self-registration stations	Cheap to operate (e.g. in South Australia 5–10% of revenue raised)	Compliance and enforcement, less staff contact, vandalism
Fixed location automatic machines	Computerised records, low labour costs, security	As for self-registration stations, plus costs of installation, power supply
Third-party outlets (shops etc)	Externalised labour costs and security risks, involvement of local community	Revenue shared with the provider, may be no knowledge of parks at point of sale

entrances, the transaction costs associated with establishing numerous fee collection stations would be high. For a park with low annual use, the revenue generated would be low. In both cases, transaction costs are likely to be a high proportion of total costs. Full recovery of these costs is difficult to justify, relative to the value of the damage being caused and/or the services being provided. Of course, transaction costs are also dependent on the collection method employed, and with changing technology, opportunities may arise to significantly reduce transaction costs.

Typical methods of collection include ticketing in advance, tollbooths, roving staff, and honesty boxes. Collection of fees can be done on-site. This has the advantage of making a direct connection between the payment and the service provided and can also facilitate informing visitors about particular activities or regulations. However, such methods are costly in terms of staff salaries, and may not be practical for sites with multiple entry points.

Alternatives include an honour system with drop boxes for payment, or a pass system such as that used in Tasmania where visitors can purchase a pass to visit any of the state's national parks within a specified period of time. Spot checks may be necessary to give such a system credibility. Advantages and disadvantages of various revenue collection mechanisms are summarised in Table 8.3.

If demand management is the objective, peak load pricing can be used to control visitor numbers or redistribute them over different time periods. Peak load pricing refers to the practice of charging different prices over time for the same service. The cost of having excess capacity during off-peak periods can be covered by increasing the amount charged to peak users. Charging higher fees for prime camping sites can help to spread use more evenly. Higher peak-period prices can also be used to perform a rationing function. Variable charges can be used to cover two cost components associated with peak-use periods:

- increased operating costs of providing services to a large number of visitors
- capital costs of providing adequate facilities to meet peak demand (such as sufficient car park capacity).

Another common practice is price discrimination—that is, charging different prices for the same goods or services where the price differences are not proportional to differences in costs. There are a number of reasons why price discrimination may be used. For equity reasons, certain individuals may be charged low prices, or given goods or services free of charge. Such equity-based price discrimination may apply to the very old or very young, local

residents, or low-income earners. Access may be free to traditional Indigenous owners of land. Different prices may also be charged to the same person for consuming large amounts of the same good—as in daily versus weekly rates for ski lift tickets (Loomis & Walsh 1987). Most Australian agencies offer concessions for low-income earners, concession card holders, children, and families (QPWS 2000).

An important factor to consider is that users are more likely to accept the legitimacy of fees if revenue is retained in the local area, and if an explicit connection is made with improvements to services and facilities. Care must be taken, however, that facility improvements, and the potential increase in

Case Study 8.2

Great Barrier Reef Marine Park Environmental Management Charge

QPWS (2000), Skeat & Skeat (2003)

Commercial tourism operators in the Great Barrier Reef Marine Park are required to pay an Environmental Management Charge (EMC). Most commercial operations in the Marine Park are subject to the charge, including tourist operations, aquaculture activities, vessel chartering, vending operations, discharge of sewage, and resorts. Users who require a permit are required to pay permit application assessment fees (including the costs of environmental impact assessments). The growth in revenue over the period 1993 to 2003 is shown in Figure 8.4, and the relative importance of the EMC compared with other income sources is shown in Figure 8.5.

Commercial operations exempt from the charge are private navigational aids, commercial fishing operations, and direct transfer operations from one part of Queensland to another. Commercial fishing does not attract a charge because one is already levied by Queensland fish management organisations. Transfer trips between islands, or islands and the mainland, are exempt on the basis that such passengers are transiting the Marine Park, not taking part in tourist excursions.

For most tourist operations the fee is $4.50 per day for each tourist carried. There are some discounts available. All charges are indexed annually to the Consumer Price Index. The total income from the charge in the 2002–03 financial year was $6.7 million, approximately 20% of the budget of the Great Barrier Reef Marine Park Authority. It was originally proposed to introduce the charge on a formula basis. This was not supported by tourist operators as many believed that the actual numbers of visitors undertaking a tourist program are a more accurate measure of an operator's use of the park. This system has been adopted and involves the addition of new logbooks in which data on park use are recorded. Aggregate data relating to trends in park use provide valuable information on trends and possible problems emerging with increased human activity.

A range of activities that may adversely affect the environment within the park require a permit. Applicants are required to pay a Permit Application Assessment Fee (PAAF) prior to the assessment of any

Figure 8.4   EMC revenue 1993 to 2003 (Skeat & Skeat 2003)

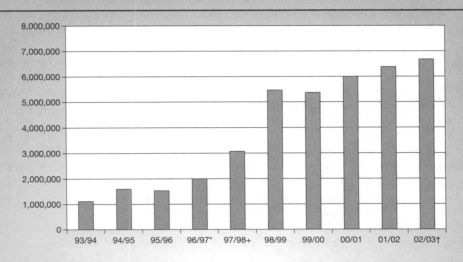

* Standard charge increase – $A1 to $A2
+ Standard charge increase – $A2 to $A4
† Standard charge increase – $A4 to $A4.50

application for a permit. The fees range from $510 for a small tourist program or a vessel mooring to $3660 for a tourist program carrying more than 150 passengers. Where significant impact assessment is required, higher fees can be levied, up to $79,120 when an Environmental Impact Statement is required. A tourism facility such as a new tourist pontoon is usually charged a PAAF of $29,300. Generally these fees are set at cost recovery or below.

Figure 8.5   GBRMPA revenue sources for 2002–03 (Skeat & Skeat 2003)

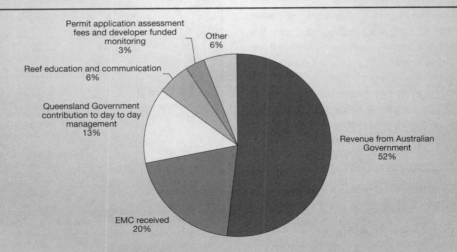

visitation that will arise, are consistent with strategic objectives for the park (QPWS 2000).

Communication with affected businesses can also be crucial to successful implementation of a user-pays system. Initial industry opposition to increased fees for people undertaking commercial tours to the Great Barrier Reef (Case Study 8.2) may have arisen in part from the fact that the introduction of the new fee did not allow operators to incorporate price changes into pre-sold package tours. A common industry recommendation is for 18 months' notice to be given of price changes. It must be noted, however, that many other industries do not enjoy similar forewarning of cost changes. Implicit in industry opposition to fees is that taxpayers should subsidise their businesses. In practice, who actually receives the subsidy depends on how much the business can pass cost savings on to clients (Lindberg 1998). No

or low fees at public sites may also disadvantage private providers who must compete with cheaper public alternatives—a situation that would be at odds with the requirements of the National Competition Policy (Backgrounder 8.2).

Clearly, establishing a recreation pricing policy is a complex task. It is not surprising therefore, that many agencies are having difficulty grappling with these complexities. Best practice guidelines recommended by QPWS (2000) include the following.

* There must be clear definition of revenue-raising objectives for different types of charges.
* Fees should be set to reflect the level of service, the revenue objectives, estimated public willingness to pay, and comparative charges in the market-place.
* Charges should be presented as a fee for services provided, not a fee for entry or access rights.

Backgrounder 8.2   National Competition Policy

QPWS (2000)

National Competition Policy and related legislative amendments to the *Trade Practices Act 1974* (Cth), which became applicable to state government agencies from July 1996, have had an impact on user-pays systems for protected areas. These changes arose from the Hilmer report in 1992, which recommended that there should be national application of a set of competitive conduct rules, greater scrutiny of government processes, and actions that restrict competition and 'competitive neutrality' between government and private firms when they compete in the market. With respect to protected areas' competitive neutrality, for example, it means that protected area agencies should not offer commercial visitor services at such a price as to disadvantage potential private sector providers of such services. Government agencies can charge for services and compete with private operators where the services relate to their 'core business'. Community service obligations are also recognised, such that public benefits are not subject to the same competitive forces, and therefore not required to be managed on a full cost-recovery basis.

Port Arthur Historic Site, Tasmania (Graeme Worboys)

- Agencies should have the ability to adjust fees, at least to the level of the Consumer Price Index, and the public should be informed to expect these small rises at regular intervals.
- Discounts should be available for children and concession card holders.
- Accounting procedures should enable estimation of cost-effectiveness or profitability of user-pays enterprises.
- Good relationships with and controls over all types of commercial operators should be established.
- All conditions of permits, leases, and other agreements should be adequate and enforceable.
- New or increased fees for commercial operators should be advised 12 to 18 months in advance.

- Full revenue retention should be achieved by the management agency, with retention of at least a proportion of funds within the local area.
- Disbursement of funds should be in keeping with management objectives.
- The user-pays system should also be used to collect data to assist management, such as visitor numbers and preferences etc.
- Core management objectives should not be overridden by commercial interests.
- Where necessary, a decline in revenue in the interests of environmental protection should be accepted.
- High compliance levels should be achieved through workable infrastructure, regular compliance checks, and good public support.

8.3 Environmental valuation and benefit–cost analysis

Managers are often in the position of having to justify continued investment in protected areas, argue for additional funds to undertake new projects, or demonstrate the benefits arising from parks as a counter to competing or hostile uses and interests. Assessing the economic benefits arising from recreation use, investment of public funds or protected area establishment has two aspects: first, the relevant economic values need to be measured, and then these values must be incorporated into an assessment methodology. In this section we will consider several valuation methods, and then show how the results can be used in a benefit–cost analysis.

Value measurement

As discussed in Chapter 2, not all goods can be exchanged in markets. Public goods and services contribute to the general welfare of society, but cannot be 'owned' by individuals. Public goods are often discussed in terms of the non-market values they provide. Unlike market values, non-market values cannot be readily quantified, and hence many environmental assets and ecological functions are unpriced and perceived to be 'free'. However, this does not mean that they do not have an exchange (economic) value, or that this value cannot be translated into monetary terms and compared with other things that are valued and priced. The exchange value of a good is measured by the amount an individual is willing to pay for it, or willing to take in compensation for giving it up. If resources are not individually owned or are unpriced, they tend not to be recognised like other assets and there is no economic incentive to protect them. Consequently, they tend to be overused or abused, thereby resulting in environmental damage both at regional and global scales (Young 1992). Environmental economists see a partial solution to environmental problems in ensuring that the environment is properly valued to reflect the relative scarcity of natural resource benefits.

A range of people derive economic benefits from protected areas:

- visitors who only derive use value
- visitors who derive both use and non-use value.
- non-visitors who derive non-use value
- businesses that offer services within the protected area, such as commercial tours, skiing facilities, food, and accommodation

- businesses and local communities that gain indirect benefits from the presence of the protected area, through visitors purchasing fuel, food, accommodation, and other services outside the protected area
- owners of properties in the vicinity of the park that are worth more due to the presence of the park
- owners or users of resources outside the park that are maintained in quantity or quality due to the presence of the park—for example, fisheries outside a park may depend in part on the natural environment being maintained inside the park.

There are also economic costs associated with protected areas that accrue to those who:

- are precluded from gaining some direct use value from the area (such as timber harvesters and graziers)—this is an opportunity cost associated with reservation of the area
- would have gained indirect benefits from the economic activity associated with a precluded use such as logging or grazing
- are owners of properties in the vicinity of the park that are worth less due to the presence of the park
- receive damage from feral animals, weeds, or bushfires that may emerge from some parks.

Economists use two classes of techniques to measure these various economic values—revealed preference and stated preference methods. Conventional revealed preference approaches have relied on measurements based on behavioural expressions of value. People reveal the value they place on a good or service through transactions they make in a market. For some goods, such as recreation undertaken in protected areas, direct markets may not exist, but visitors still reveal their value though their willingness to spend time and money in order to gain access to a site. Such revealed preferences for recreation can be measured using indirect market methods based on travel cost.

Recently economists have also developed methods based on what people say about, for example, their willingness to pay for nature conservation, rather than what they reveal through their behaviour. Such stated preference methods are particularly

important with respect to natural areas, because many of the potential benefits provided by such areas are not revealed in markets, and cannot be recovered through indirect market techniques. At present, the most significant stated preference technique is contingent valuation (CV). Other stated preference techniques that have been explored include contingent rating, contingent ranking, paired comparisons, and choice modelling (Mitchell & Carson 1989, Morrison et al. 1996).

Travel cost method

The major factor in the production of visitor services in protected areas is that individuals must transport themselves to the recreation site to consume the service, rather than have the commodity transported to them. Time and transportation services are scarce resources that are central to the individual's decision whether (and how much) to use the resource. The recreation consumer relates the time and money costs associated with various outdoor activities to their own resources before making the decision about what activities to undertake and where to pursue them. These features of the visitor experience can be analysed to determine the value of a recreation site.

The travel cost method (TCM) is an indirect market technique that is used for estimating the economic value of recreation. The method uses travel costs to measure how much people are willing to pay to come to an area. The method presumes that, as with other economic transactions, people will make repeated trips to a site until the marginal value of the last trip is worth just what they have paid to get there. Assumptions also need to be made regarding the estimated cost of travel, the unit of observation (individual or group), and how costs are allocated when people visit more than one site.

The method has been used to evaluate recreation in a wide range of settings, including national parks, ski resorts, urban parks, and reservoirs. Results from some Australian TCM studies are shown in Table 8.4.

The TCM may also be used to estimate changes to consumer surplus resulting from changes to visitor opportunities or the addition of new recreation sites. Suppose, for example, that a government department is examining the merits of investing in

Figure 8.6   Change in consumer surplus following a public infrastructure investment

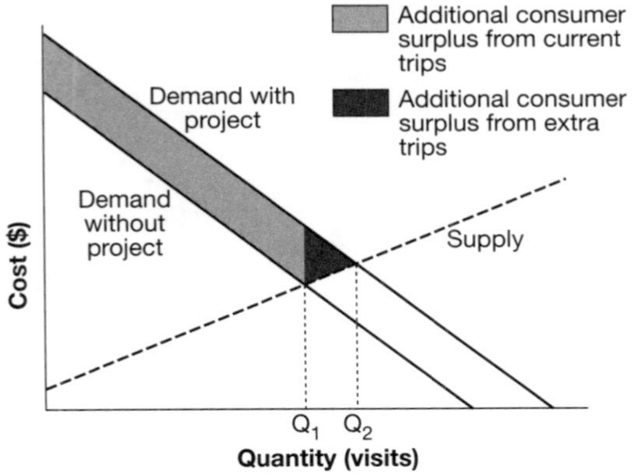

some new tourism infrastructure (such as a visitor centre) in a protected area. A TCM can be used to generate a demand curve for the site under current conditions. A survey can also be conducted to determine how many extra trips might be taken if the visitor centre was built. The resulting demand curves might look something like those shown in Figure 8.6. Each visit is now worth more because of the increased opportunities afforded by the new visitor centre. This yields an increase in consumer surplus for each current trip. There is also an additional surplus from extra trips that are made due to the presence of the new facility. These two sources of additional surplus can be compared with the costs of building the facility to see if the investment is worthwhile, at least from an economic point of view.

There are two basic forms of the TCM method—zonal TCM (ZTCM) and individual TCM (ITCM). Other more sophisticated variants include the hedonic and multiple-site TCM approaches. In ZTCM visitors are grouped into zones of origin and a demand function is estimated from the statistical relationship between the aggregate trips from a zone and the cost of travel. In the ITCM, values are derived from the individual's total trips and the distance the individual travels.

In a basic ZTCM, the catchment area of the recreation resource is divided into zones and users are grouped into residential locations based on zones. The numbers of visits per user are converted to the number of visits per head of population. If entry to the recreation area is free, then the average

Table 8.4	**Results for a sample of Australian travel cost studies** (adapted from Read Sturgess & Associates 1999)				
Author (date)	Location	Vehicle cost	Unit	Result
Beal (1995)	Carnarvon Gorge NP, Queensland	$0.15 per km, 1993 data	$ per person, per day	$34, as a mean of campers and day visitors
Bennett (1995)	Dorrigo NP, Gibraltar Range NP, NSW	$0.15 per km, 1995 data	$ per person, per day	$34 for Dorrigo, and $19 for Gibraltar Range
Ferguson & Greig (1973)	Mt Macedon Forest Park, Victoria	$0.052 per km, 1969 data	$ per group per day, and $ per person per day	$15 per group and $4 per person

Table 8.4 (continued)

Author (date)	Location	Vehicle cost	Unit	Result
Gillespie (1997)	Minnamurra Rainforest centre, NSW	$0.15 per km, 1995 data	$ per person per day	$28 to $48, depending on assumptions made
Greig (1977)	Grampians State Forest, Victoria	1970 data	$ per group, per day	$2
Hundloe et al. (1990)	Fraser Island, Queensland		$ per day per person	$13 for permit holders, $11 for tour visitors
Knapman & Stanley (1991)	Kakadu National Park, NT	$0.17 per km, 1990 data	$ per person per whole visit	$142 per person for the whole visit, or $37 per person per day
Lockwood & Tracy (1995)	Centennial Park, Sydney	$0.33 per km, 1992 data	$ per person per day visit	$7 to $11, depending on assumptions made
Pitt (1992)	Coastal land, NSW	$0.52, 1991 data	(a) $ per person per total visit, (b) $ per person per day	$47 to $151, depending on assumptions made
Resource Assessment Commission (1992)	National Estate Forests of SE Australia		$ per person per day visit	$9
Read Sturgess and Associates (1994)	Grampians National Park, Victoria	$0.17 per km, 1994 data	$ per person, per day	$23 (travel costs only) to $113 (travel costs + time costs + on-site costs)
Sinden (1978)	Colo Shire, NSW	$0.05 per km, 1974 data	$ per person per day	$5 land picnic, $3 driving for pleasure, $2 river picnic, 20c water ski
Sinden & Worrell (1979)	Oxley Wild Rivers NP, NSW	$0.12 per km, 1990 data	$ per person per day for day users, and $ per person per whole visit for campers	$6 per day visit, and $10 per camping visit
Sinden (1990)	Ovens and King Valleys, Victoria	1979 data	$ per group visit, per day	$9
Stoekl (1994)	Hinchinbrook Island NP, Queensland	$0.16 per km, 1992 data	$ per person per day, over all kinds of visit	$57
Thomas (1982)	Murray Catchment, WA	$0.08 per km, 1973 data	$ person per day	$1.10 to $5.60, depending on time costs and other assumptions
Ulph & Reynolds (1978)	Warrumbungle NP, NSW		$ per person day	$100

Carnarvon Gorge, Carnarvon National Park, Qld
(Graeme Worboys)

Warrumbungles National Park, NSW
(Graeme Worboys)

Zoe Bay, Hinchinbrook Island National Park, Qld (Peter Lavarack)

cost of travelling from each zone is an indicator of the price to be paid for visiting the recreational site.

The ZTCM estimates the total value of a site using a two-stage process. In the first stage a statistical estimate of the relationship between visits (per head of population) and cost is derived. In the second stage, a demand curve is estimated by calculating total visitor use at each of several (hypothetical) added costs. Cost is increased until the estimated number of visits from all distance zones is predicted to fall to zero. For each added cost, the expected visitation rate from each travel origin zone is calculated and the values summed across all zones to find the predicted total number of visitors. The area under the resulting demand curve (of added cost versus predicted number of visits) measures the visitor use value (consumer surplus—Backgrounder 8.1) attributed to the resource.

Contingent valuation

CV is a survey technique that, in its simplest form, asks people how much they are willing to pay (WTP) for some change in the provision of an amenity, usually a non-market good. The WTP valuations are determined in the context of a hypothetical market that is constructed in the survey. Participants are asked for their WTP contingent upon the existence of the hypothetical market as described in the survey instrument (Wilks 1990). This hypothetical market typically comprises:

- a description of the amenity
- the change in its provision
- the means (payment vehicle) by which the participant can purchase a particular allocation of the amenity.

CV surveys usually also contain a range of other questions, which have three main purposes:

- to obtain additional information about participants' attitudes and opinions—this information can be useful in its own right, but its main function is to assist in interpretation of the WTP results
- to 'frame' the hypothetical market in such a way that participants will be made aware of other factors that may affect their WTP—for example, personal budget constraints, or other non-market goods that might also potentially make claims for a share of the participant's WTP
- to ascertain participants' major socio-economic characteristics in order to determine the representativeness of the sample, as well as factors influencing WTP.

Over 1000 CV surveys have been conducted in the USA on a wide range of goods including recreation, wetlands, lake preservation, deer hunting, wildlife preservation, aesthetic and health benefits of air and water quality, and wilderness. Australian applications, although considerably less numerous, include:

- eucalypt dieback (Sinden et al. 1982)
- non-use values of a nature reserve (Bennett 1984)
- support for soil conservation (Sinden 1987)
- control of Crown of Thorns Starfish (Carter et al. 1989)
- benefits of forestry research (Young & Carter 1990)
- forest preservation on Fraser Island (Hundloe et al. 1990)
- mining in Kakadu Conservation Zone (Imber et al. 1991)
- preservation of national estate forests in East Gippsland (Lockwood et al. 1993)
- protection of wetlands (Bennett et al. 1997).

Various methods have been used to elicit valuations from participants.

1 *Bidding games.* These resemble an auction in which the interviewer raises or lowers the bid until the respondent decides to make a purchase.
2 *Open-ended questions.* The maximum willingness to pay is simply stated by the respondent without prompting.
3 *Dichotomous choices.* Respondents are required to answer 'yes' or 'no' to the question whether they would be willing to pay a specified dollar amount to preserve the environmental amenity in question. Such questions are easier for respondents to answer than open-ended questions. Dollar amounts are varied among respondents, and statistical techniques are used to estimate average WTP.

Snapshot 8.4 gives an extract from the survey instrument used by Lockwood et al. (1993). This extract illustrates many of the features of establishing a contingent market and using dichotomous choice and open-ended formats.

The CV survey can be conducted as a face-to-face interview, a mail survey, or a telephone interview. The choice of survey technique has implications for the length and complexity of the survey, and is influenced by the good being valued, sample size, and budget.

For a CV survey to produce valid and unbiased data on non-market economic values, participants must:

1   have the opportunity to develop their preferences in the course of answering the survey, since the good may be unfamiliar to them
2   have the ability to express their preferences for non-use value as a willingness to pay.
3   clearly apprehend the good they are to value in a manner that accords with the intentions of the survey
4   consider their budget constraint
5   consider the availability of substitutes, and their potential expenditure on substitutes, for the good being valued

6   accept the justification given in the CV scenario for why payment is required
7   accept the legitimacy of the payment vehicle used in the survey—that is, have no 'in principle' objections to making a payment using this vehicle
8   believe that if they do not pay for the good, it will not be provided
9   trust that any payment they might make would actually be used in the manner specified in the survey (Lockwood 1998).

There has been considerable controversy in the literature concerning the ability of practitioners to design CV surveys that meet all these conditions. The arguments both for and against CV are complex, and at times very technical; we will not detail them here—those interested could try Mitchell and Carson (1989) as a starting point. Suffice to say that CV is very well known, but unlike in the USA, is not yet widely used in Australia. Efforts to further test and refine the method are continuing.

Input–output analysis

Economic valuation techniques such as TCM and CV are directed towards estimating the impacts a particular protected area or policy proposal has on economic welfare. Another approach is to assess the

---

Snapshot 8.4

Summary of results and extract from the CV survey
Lockwood et al. (1993)

A CV survey was conducted to determine Victorians' willingness to pay to reserve unprotected East Gippsland national estate forests in national parks. Such reservation would preclude utilisation of these forests for timber production. The survey was designed to be sent by mail, and included questions concerning attitudes to the forests, WTP for their reservation as national parks, and respondents' demographic characteristics. Several questions such as the importance of forests in relation to other issues were also included, both to provide a realistic context for the valuation exercise and to enable respondents to indicate motivations for their WTP answers.

The survey instrument briefly described and illustrated the general appearance of two major vegetation types, which for the sake of simplicity were termed wet and dry eucalypt forest. Examples of significant animals associated with the forests

and the end use of logging products were also given. The existing logging activities were described by a sequence of pictures depicting the general appearance of forest after logging. Maps and charts were provided to enable the respondent to gauge the size of the areas in question, as well as the proportions currently reserved and available for logging.

Attitudinal questions included in the survey revealed that Victorians consider the non-market values associated with these forests to be more important than market values. The median WTP per respondent's household for reserving the forests was $52, which corresponds to an annual aggregate non-market value to the people of Victoria of $41 million. Data from a subsample of Gippsland residents indicated that people living in or adjacent to East Gippsland placed relatively more emphasis on market values and had a significantly lower WTP for reserving the forests.

contribution a park makes to a regional economy. Economic activity associated with park management and tourism expenditure is a significant component of employment and economic activity in some areas. The technique used to perform such an assessment is called Input–Output (I–O) analysis. I–O analysis measures how an allocation of resources would affect regional income, expenditures, and employment.

The NSW P&W has used I–O analysis to evaluate the direct and flow-on effects on regional economies from agency and tourist expenditure. Direct effects are simply the amount of money spent on managing and using protected areas, including construction supplies, food, transport, commercial tours, and so on. Indirect effects are additional expenditure arising from activity induced by direct expenditure. This includes additional household spending on goods and services enabled by the extra money flowing through the local and regional economy. I–O Studies have been done on areas such as Montague Island Nature Reserve, Fitzroy Falls Visitor Centre, and Warrumbungle National Park. For example the annual contribution of Warrumbungle National Park to the Coonabarabran regional economy was estimated to be $2.08m per year of 'value-added' activity, with $1.37 million of this spent on wages and salaries—the equivalent of 66 full-time jobs (Conner & Gilligan 2003). Snapshot 8.5 gives another example of an I–O analysis that assessed the contribution two national parks made to the regional economy of northern NSW.

Snapshot **8.5**

Regional economic impact of the Dorrigo and Gibraltar Range National Parks
Modified from Powell & Chalmers (1995)

The regional economic impacts of the Dorrigo and Gibraltar Range National Parks in northern NSW were assessed using input–output analysis. These two parks are quite different in the way they are managed and the services they provide, while their location relative to nearby commercial centres is also different. This study therefore provides an indication of how regional economic impacts of national parks might vary, depending on the type of park and its location within a region.

Estimates of visitor expenditure were based on data collected via a visitor survey undertaken as part of a companion travel-cost study into the economic values of the recreation use of Gibraltar Range and Dorrigo National Parks. This indicated that the annual visitor expenditure in the local region associated with trips to the parks was $3.2 million in the Dorrigo region and $1.0 million in the Gibraltar Range region. Estimates of expenditure in the local region by the NPWS in managing the parks were $342,000 for the operation of Dorrigo National Park and $137,000 for the operation of Gibraltar Range National Park.

In total, it was estimated that Dorrigo National Park (with an estimated 160,000 visits per annum) contributed almost 7% of gross regional output, 7.5% of household income, 7.4% of value added (or gross regional product), and 8.4% of regional employment. This represented almost $4 million in output, $2.3 million in value added, including $1.5 million in household payments to 71 people.

The contributions were not as great for the Gibraltar Range National Park (with an estimated 40,000 visits per year). In total, it was estimated that the Gibraltar Range National Park contributed less than 1% of all estimated measures. In value terms this was $1.5 million in gross output, and $0.9 million in value added, including $0.6 million in household income payments to 30 people.

This study demonstrates that national parks, such as Dorrigo and Gibraltar Range, do make a contribution to economic activity in the regions within which they are located and that different parks may have substantially different absolute and relative impacts on regional economies. The study also demonstrated that for both the Dorrigo and Gibraltar Range National Parks the impact of visitor expenditure on regional economic activity exceeded the impact of management expenditure. The total economic impact of every 1000 visitors to the regions ranged from $22,000 to $30,000 for business turnover, from $13,000 to $17,000 in value added, including $8000 to $12,000 in income for households in the area. Total employment impacts ranged from four to six jobs generated for every 10,000 visitors.

Benefit–cost analysis

Benefit–cost analysis (BCA) is the standard economic technique used to assess the economic benefits and costs of public decisions, including those affecting protected areas. BCAs have been conducted, for example, on issues such as the proposed construction of the Gordon-below-Franklin dam in Tasmania (Saddler et al. 1980) and a pulp mill in south-eastern Australia (Streeting & Hamilton 1991).

BCA is an application of modern welfare economics. If a project is to be judged worthwhile according to BCA, its benefits must exceed its costs, where benefits and costs are defined to include any welfare gain or loss that occurs as a result of the project. Cost is often thought of as an opportunity cost (the benefits forgone by proceeding with a project) and the benefits are measured by the consumer surplus arising from

---

Backgrounder 8.3   Steps in a benefit–cost analysis

The basic steps in BCA are:

1   Identify the alternative proposals to be examined, and the 'lifetime' of each proposal—for example, the productive lifetime of a visitor centre might be 20 years.
2   Identify the values associated with each proposal that can be measured in economic terms.
3   Identify each of these values as either a benefit or a cost.
4   Quantify these benefits and costs in dollar terms, using a suitable economic valuation methodology.
5   Assess the project according to a decision rule. Two commonly used decision rules are the benefit/cost ratio (a project has a positive net social benefit if Benefits/Costs > 1); and net present value (NPV) (an activity is economically beneficial if NPV > 0).

The NPV for a proposal is calculated by adding up the net benefits over the lifetimes of the proposal. If the NPV is positive, then benefits outweigh costs and the project makes economic sense. If the NPV is negative, then from an economic point of view the project should not proceed. The higher the NPV, the more economically valuable the project. NPV is given by the following formula:

$$NPV = \sum_{t=1}^{t=n} \frac{B_t - C_t}{(1+r)^t}$$

where:

- t is a particular year of the project, which ranges from 1 at the start of the project to the lifetime of the project at n years
- $\sum_{t=1}^{t=n}$ means 'the sum of, over the years between and including year 1 and year n'
- B_t are the benefits in year t
- C_t are the costs in year t
- r is the discount rate—for a discount rate of 10%, for example, $r = 0.1$

The formula therefore means that NPV is given by the discounted value of benefits minus costs, added up for each of the years during the life of the project. To calculate NPV we must work out the net value for each time period, apply the discount rate, and then add up the results for all time periods. Table 8.5 shows how the NPV for a hypothetical visitor centre in a national park can be calculated. To simplify the example, a time frame of five years has been assumed. Major benefits of the project include:

Table 8.5	Example calculation of net present value					
		Year 1	Year 2	Year 3	Year 4	Year 5
Benefit						
Additional visitors to park		0	50,000	70,000	100,000	120,000
Income from sales and tours		0	50,000	55,000	60,000	65,000
Costs						
Building establishment, maintenance operating		150,000	10,000	10,000	10,000	10,000
Staff		0	50,000	50,000	50,000	50,000
Environmental impacts		0	20,000	28,000	40,000	48,000
Net benefit (B − C)		−150,000	20,000	37,000	60,000	77,000
Discounted benefit $\dfrac{B_t - C_t}{(1 + r)^t}$		−140,187	17,469	30,203	45,774	54,900
NPV						8159

- the consumer surplus associated with additional numbers of visits to the park (measured by TCM)
- income from the sale of books, posters, and guided activities.

   Major costs include:

- building, maintaining, and operating the centre
- staff salaries
- environmental damage caused by increased visitation (measured using CV).

   The net benefit row shows the benefits minus the costs. The discounted benefit row divides these net benefits by the discount rate raised to the power of the year. A 7% discount rate has been used. The last column adds up all the discounted benefits to give the NPV. The NPV is positive, so the project yields a net economic benefit.

Accounting for time

The idea of using a discount rate needs some brief explanation. A complicating factor in BCA is the temporal nature both of costs and benefits. Comparing the benefit stream with the cost stream over time requires that all present and future values be put into a common frame of reference called present value (PV). The determination of PV is done through the use of a discount rate.

   Using a discount rate, the value of a dollar in five years' time is considered to be less than the value of a dollar today. To see this, consider your own attitude to having $100 in your hand today, and $100 in your hand at the end of the year—which would you prefer? If you prefer the money today, then the value of the money in a year's time must be less than it is today. The discount rate r is a measure of how much the value of money decreases over time. The longer the time, the less the value. To be precise, the net value for each year is divided by a factor of $(1 + r)^t$ where t is the time period. Note that we are not talking about inflation here—the value decreases even in the absence of inflation.

the project (Sugden & Williams 1978). A rational social decision is one in which projects are approved if the aggregated benefits to society exceed the aggregated costs.

BCA is concerned with public benefits and costs. Public costs may include impacts on recreation values (measured using the travel cost method), or impacts on the non-use values of a natural area (measured using contingent valuation). On the other hand, an economist conducting a financial analysis of a private investment for a company will use many of the same principles as BCA, but will only consider benefits and costs that directly impact on the company. Furthermore, company managers will be interested in what the analysis says about the return on their investment,

Snapshot 8.6

Benefit–cost analysis of reserving East Gippsland forests
Modified from Lockwood et al. (1993)

Lockwood et al. (1993) conducted a BCA to assess the economic impact of placing the then unreserved national estate forests in East Gippsland, Victoria, into national parks. Streeting & Hamilton (1991) described the benefits and costs of reserving south-east Australian forests as follows:

Benefits
1 saving of capital and operating expenses associated with logging
2 preservation of the natural environment
3 maintenance of visitor opportunities
4 reduced damage to roads caused by logging trucks
5 maintenance of a forest structure that is suitable for non-timber uses such as honey production

Costs
1 value of forgone sawlogs
2 economic costs of increased unemployment
3 social costs of unemployment

4 loss of visitor opportunities that may result from improved access created by roading works associated with logging.

Streeting and Hamilton (1991) estimated the magnitude of Benefit 1 and Costs 1, 2, and 3 for south-eastern Australia. Lockwood et al. (1993) extracted the East Gippsland component of these estimates, and used these data together with their CV value for Benefit 2 (Snapshot 8.4) and the non-market component of Benefit 3 to compute the NPV of reserving the East Gippsland forests. Benefits 4, 5, any market component of Benefit 3, and Cost 4 were not included in the analysis.

A discount rate of 7% was used in the NPV calculations because, at the time, rates of between 4% and 10% were typically used by government agencies in Australia. As shown in Table 8.6, the analysis indicated that reservation of forests in parks would produce a net economic benefit of $543 million. Note that the validity of this estimate is strongly dependent on the validity of the non-market value determined in the CV survey. As noted in the discussion of CV, some economists remain sceptical about CV.

Table 8.6	Benefit–cost analysis of forest reservation in East Gippsland	
		Discounted net benefit[1] @ 7% ($ million)
Saving of logging capital and operating expenses[2]		16
Non-market preservation and recreation value		564
Value of forgone sawlogs[2]		−21
Socio-economic costs of unemployment[3]		−16
Net Present Value		543

1 *Discounted over 49 years, all values rounded to the nearest million dollars ($1 million)*
2 *Values derived from Streeting and Hamilton (1991)*
3 *Based on a compensation package of $18.6 million over 3 years*

rather than using a measure such as net present value (NPV—see below), which is commonly employed in the assessment of public policy. A private company will also consider such matters as the impact of an investment on their share price, their strategic position in the market place, and taxation implications.

BCA is, in effect, a democratic process in which the 'votes' are translated into monetary values. However, the numbers of votes (dollars) are not distributed evenly among individuals within society. Since BCA is an economic technique, it will produce results that reflect the distribution of economic power within society. BCA must assume that this distribution is optimal in order to claim the ability to identify socially optimal decisions.

Furthermore, just as a minority opinion may become swamped by the majority (unless there is a willingness on the part of the majority to recognise and act on that opinion), an individual or group of individuals may suffer considerable costs as a result of a project, and yet on balance, the project can still be regarded as a good thing if benefits to another section of the community are greater.

In other words, BCA cannot provide a consensus measure of a project's value to society. It is possible to envisage a series of projects being implemented, each of which has a net social benefit and each of which engenders costs to a particular section of the community so that that section becomes increasingly disadvantaged. Where differential distribution of benefits and costs is at issue, moral judgment must be applied.

Despite these limitations, Abelson (1979) concluded that BCA performs better than any other evaluation method against the criteria of comprehensiveness, philosophical coherence, ethical acceptability, ability to take into account individual valuations, and practicality. Jacobs (1995) was more critical of the method in terms of its institutional assumptions, ability to identify those projects that satisfy the requirements of ecologically sustainable development, and the ability of economic techniques such as contingent valuation to measure public good values. Some of the basic elements of a BCA are described in Backgrounder 8.3 (see pp 245–6). An example of a BCA, for the reservation of East Gippsland forests in national parks, is given in Snapshot 8.6 (see p. 247).

8.4 Management lessons and principles

Lessons

1   Some aspects of protected area management need to be run like a business.
2   There are opportunities for managers to achieve improved revenue returns by using economic principles to inform pricing policies.
3   By delegating opportunities to secure (appropriate) revenue return to local managers and allowing managers to retain the greater proportion of those funds for works (on sites where the revenue was collected) local managers are empowered and have the incentive to collect revenue. They also have the ability to demonstrate directly to visitors the benefits of the revenue collection.

Principles

As well as the best-practice guidelines listed at the end of Section 8.2, we have established the following key principles for protected area finance and economics.

1   Public funding is essential if conservation agencies are to go on supplying most of their core services.
2   The private sector also has a role in financing protected area management.
3   Protected area managers need to adopt business and financial management techniques to secure the funds necessary to achieve management objectives.
4   Protected areas provide a range of economic benefits to individuals, businesses, and regional communities.
5   Economic principles can be used to help develop pricing policies for visitor services and other protected area uses.
6   Economics can help an agency to justify investments and make decisions that maximise net benefits.

7 Information about the economic benefits of protected areas can be used in planning decisions about park management and investment strategies.

8 Protected area managers need to be aware of the different economic impacts of protected area management and expenditure on different groups.

9 Economic concepts and techniques can be used to measure the economic values and benefits provided by protected areas.

10 Assessment of the costs and benefits of protected area management and investment strategies should take full account of the relevant market and non-market economic values.

09

Administration— Making it Work

9.1 Administering people   251

9.2 A principled organisation   256

9.3 Administering finances   257

9.4 Administration of assets, standards,
     and systems   262

9.5 Administration policy and
     process considerations   264

9.6 Management lessons
     and principles   267

'Administration' for many people is a dirty word. It is often perceived as paperwork and processes that get in the way of 'real work'. Many think that saving animals or putting out fires are the sorts of action that really matter. Some protected area staff probably hold such views; but those with a view of the 'big picture' in conservation must disagree. Most managers know how administration, if done well, helps establish and maintain a successful conservation organisation. They also know that when administration fails, the effectiveness of an organisation is compromised, and that conservation is the ultimate loser.

Administration lies at the heart of a protected area organisation's capacity to operate. People are needed to implement an organisation's primary mission. Staff (and contractors) need to be hired and paid. They need a base from which to operate. Hence offices and workshops must be either purchased, constructed, or leased. People need to be mobile and to have access to equipment and materials. This requires the hire or purchase of vehicles, plant, and other equipment. Staff also need a supportive operating framework, which ranges from employment contracts to workplace safety rules and skills training.

All of this requires well-designed administration systems. Budgets need to be secured and managed. Bills need to be paid. Staff need to be treated fairly. Workplaces need to be safe. Systems need to be in place to evaluate and monitor the staff's performance so that professional standards remain high. Numerous routine administrative tasks and systems are needed to support the conservation of a protected area. Organisations need to operate fairly and equitably, and to be accountable.

In this chapter we give an applied perspective on those basic administration systems that a ranger or conservation manager must understand. Those who require a more extensive theoretical analysis can readily find it in such books as those by Wanna et al. (1992), Corbett (1996), Ivancevich et al. (1997), Fenna (1998), and Robbins et al. (2003).

We now look at the main areas of administration as they apply to protected area organisations.

9.1   Administering people

People make the difference when it comes to achieving conservation outcomes. They do the work, and they make organisations work. They need to be managed well. Investing in people occurs on a daily basis and is one of the great challenges for any manager. Some managers have a natural skill for dealing with people; others need to work on their skills. Many organisations prepare a workforce management plan. This deals with elements such as the structure and flexibility of the workforce; recruitment; performance management; training; skills development; and succession planning.

A workforce management plan is important for the long-term running of a conservation organisation. Having a positive internal culture and a happy staff are critical to its success. A good workforce plan can enable staff to be well organised and comfortable, to work to clear guidelines and purposes, and to use efficient systems. We now set out a series of detailed insights into the administration of people in the workplace, from the day they are recruited to the day they leave an agency. These insights are relevant for any workforce plan.

Recruitment

Managers want the best staff possible for new or vacant positions, so recruitment systems need to be first-class. Job advertisements must not unfairly exclude or advantage certain groups, yet should be clear about the qualities needed for a position. Potential applicants have a right to know the skills and background sought. Staff given the job of responding to calls from applicants need to be thoroughly briefed on the nature of the position.

Just as important is the conduct of the recruitment panel. In the public sector, such panels typically include future supervisors as well as staff from other sections. This helps to ensure fairness. The chair of the recruitment panel will normally have been trained in recruitment procedures, and should ensure that the panel's work and findings are correctly recorded. Independent panel members need to be comprehensively briefed on the nature of the position. For instance, the locations for many jobs in protected area management are in remote parts of Australia, so the temperament of applicants needs to be checked carefully.

Temporary staff

Temporary staff provide assistance to full-time employees of an organisation during, for example, the fire season or peak visitor periods. They may be selected by an interview panel or, if only required for short periods, be directly appointed by a manager. Temporary staff may be employed as a consequence of grants or of revenue earned within protected areas. Managers should take care to spell out and record the terms of employment.

Contract staff

Contract staff are not salaried staff, and are paid for actual work completed, for example carrying out plumbing repairs, or conducting an expert report on a cultural heritage site. Typically they are contracted for a short period to do a specific task. Administration of contractors includes preparing a specific local order or contract. Contracts are usually standard legal documents that are supplemented with specific task descriptions and performance measures. Managers need to ensure that the terms of reference are absolutely clear, including specification of the standards that contractors are required to adopt while working in protected areas. Progress payments are made on the basis of reaching specified 'milestones', with a final payment made upon successful completion of the project. Issues such as who owns intellectual property (for reports and research work) and mechanisms for terminating the contract are specified in the contract.

Induction

Most organisations have a staff induction program which is described in an 'induction manual'. The importance of the induction process is often underestimated by managers. When staff are new they are highly receptive to new ideas and ways of doing things. They tend to closely observe their new environment and appraise the strengths and weaknesses of the organisation. Managers need to ensure that at this time new staff get the attention, skills,

and training they need. Often mentors are assigned to assist in the induction process.

The induction program may comprise a formal course or consist of a more casual process facilitated by an experienced staff member. An induction manual should be provided. It should describe the organisation's structure and mission, as well as:

- strategic priorities
- code of conduct
- personnel policies and systems—such as work and family policy, leave policy, staff accommodation policy, uniform policy, and smoking in the workplace policy
- industrial awards
- competency required (for a given position)
- training and skill development opportunities
- customer service policies and protocols
- how and where to access important corporate information
- how to use internal information systems (such as an intranet).

As one part of an induction program, many managers may set up a familiarisation scheme for their new rangers. They may be required, for example, to inspect, within a fixed time, an extensive list of destinations as part of their orientation. They may also be trained, prior to the fire season or the busy tourist season, in the handling of plant and equipment or in dealing with incidents.

Defining the nature of work and of workplace skills

All staff, and each position in the organisation, should have a precise job description. This should be derived (ultimately) from the mission statement and from current strategic plans. It would describe the key 'accountabilities', the key areas of work to be performed, their work context, their degree of difficulty, and the skills required. Such 'position descriptions' help not only with writing job advertisements but also with establishing the correct salary.

Industrial awards and enterprise agreements

The salaries for staff (other than contract staff) are usually set by industrial awards or workplace enterprise agreements. These specify normal and overtime payments, annual leave, and other entitlements, and work hours (for example, shiftwork, five days a week work). These agreements are legal statements and their negotiation follows clearly defined legal processes. They define how people are expected to operate in the workplace, and their entitlements and wages. Managers may be directly involved in their negotiation. In applying such agreements, managers need to be attentive to the detail. During an incident, for example, managers should try to roster their staff for shifts that fit within work agreements. Otherwise, financial penalties such as double or triple pay might apply.

Competency standards

Organisations may evaluate their staff's competency, and require a certain minimum operating standard for certain positions. Competency standards specify the theoretical and practical knowledge and skill required in the workplace. Salary bonuses are often linked to enhanced competencies, and may involve evaluating a number of relevant skills. Competency standards may be recognised in industrial awards. They are typically linked to performance agreements. Competency standards provide a way to compare qualifications and skills for staff of the same name (such as 'ranger') between the states and territories. This matters when staff move or are exchanged between agencies. Performance management systems (Section 9.5) link competency agreements, performance agreements, and performance evaluation.

Delegations of authority

Most Australian protected area organisations are highly decentralised. Their branches must take on a whole range of duties without direct supervision from head office. To do so, their managers must be given a formal 'delegation of authority'. These delegations may be financial, legal, or as otherwise specified. Organisations typically match such delegations with training as part of their management development programs.

Accountability

When managers speak of the 'accountability' of a given position, they refer not only to the responsibilities of the role, but also to its influence on policy and decisions, its level of independence, and its general stature or importance. Accountability carries an expectation of achievement (including the ability to offer quality advice). It also implies the right to call on a corresponding level of administrative support. Managers with high accountability normally possess a significant delegation of authority.

Staff rosters

Managing a protected area may require having a suitable cross-section of staff (and skills) on duty or at least on call at all times. For this, carefully planned staff-rosters will be needed. Rosters should be prepared some two to four weeks in advance—if possible longer, so that staff can plan their own lives. Managers need to consider many issues. The roster should comply with workplace agreements. There are equity issues. Opportunities for overtime pay need to be fairly distributed. Operational needs will require additional staff on duty during peak holiday periods or during periods of high potential for incidents (such as bushfires). There are also periods when the wildlife may need extra attention or protection. Staff may have special needs for leave such as sick leave, maternity leave, and study leave. Managers need to be modern magicians to balance rosters perfectly, but their best efforts are appreciated.

It is especially important to maintain staff rosters during incidents. In the chaotic 'order' that is brought about by incident control systems (Chapter 15) it is very easy to lose track of how long someone has been on duty. Attention to detail and perseverance with the administration systems is essential for effective management. Managers simply cannot afford the threat of dangerously overtired staff remaining on duty. Furthermore, excessive overtime claims caused by poor administration systems can blow out the budget.

Paying salaries

Salaries are one of the major expenses of an organisation. Most managers have a fixed budget for salaries, which needs to be planned wisely to account for an annual cycle of staffing needs. Managers in the north of Australia tend to need more seasonal staff during the winter dry (tourist) season. Those in the south tend to require more during the summer fire season. Decisions about

Volunteers for the State Emergency Service of NSW assisting fire operations at Kosciuszko National Park, NSW (Graeme Worboys)

salary levels need to be made carefully. Usually managers are guided by the nature of the position and by formal evidence of the competencies, qualifications and experience of a new staff member. Salaries must always be paid promptly.

Volunteers

Volunteers perform a host of invaluable services. These include the removal of weeds; the maintenance of tracks; guiding services; the protection and conservation of mountain huts through the Australian alps; firefighting and bushfire management; working bees for Clean-up Australia Day; and seasonal or rostered staffing of many visitor centres throughout the country. Mixing volunteers with paid staff requires some sensitivity. Agreements with unions usually require that volunteer work does not displace paid staff. Volunteers are an integral part of protected area management. They need management support, which may include training (usually) and uniforms.

Capacity building

For long-term success, an organisation must invest in capacity building and development of its staff. Staff across an organisation need to be up to date with advances in computer software, legislation,

project management techniques, accounting systems, and other organisational aspects. Training helps create an internal culture focused on constant improvement. It can also be used to give staff background information on the history of the organisation. Training is usually administered through the organisation's human resources section.

Capacity building for a protected area organisation needs to be strategic and long term. It should be a systems approach linked to organisational needs and the demography of the workforce. Every protected area organisation needs to be continuously developing its staff in order to achieve conservation outcomes and meet its corporate responsibilities. It always needs to have replacement staff with sufficient skills and experience to achieve smooth successional capacities as staff retire or leave an organisation. This approach is relevant to front-line, middle-level, and top-level staff (Chapter 4). An organisation needs to be actively grooming its senior and experienced personnel for potential executive positions including the Chief Executive Officer position. Front-line staff need to have competencies to operate at the front line of protected areas for all operations. They also need to have opportunities for working in more senior positions.

University graduate qualifications provide ranger staff with basic protected area management skills. They are an essential grounding, but such knowledge and skill need to be augmented by on-the-job training, professional experience, and continuous professional development. Best practice standards for management are constantly being developed, and professional skills are constantly improving. There is a need for protected area agencies to facilitate regular training programs to ensure the currency of their staff in core disciplines of protected area management. These could be developed as postgraduate qualifications (in partnership with universities). In addition, there need to be opportunities where the in-house wisdom and experience of senior managers is formally transferred to ranger and other staff. This has the benefit of reinforcing the value of experienced staff, and helps to build corporate loyalty and commitment to a protected area organisation.

Training needs to be directed towards those competencies that an organisation's corporate mission requires. The first step is to determine the current competencies of the staff, to identify any gaps, and to design a training scheme to bridge these. Then resources (money, staff, and facilities) must be programmed to implement the training. Training needs should be constantly monitored. The human resources section of an organisation may also arrange formal accreditation for its training with a college or university. This ensures that all parties are clear on the levels of proficiency achieved. Other types of training may be delegated or contracted to universities and technical colleges.

Local area managers may also run their own training programs, for example to train new staff in basic operational skills such as using a chainsaw, operating a four-wheel drive, or conducting customer service. Training needs of staff should be recognised in 'performance development' agreements or other similar arrangements with their supervisors. Most organisations foster such an environment of continuous learning, and they reward or explicitly recognise their staff's vocational training. Staff may also benefit from time-release schemes that allow them to be seconded to other organisations or undertake specialist study or project work.

Safety and health in the workplace

Managers need to be eternally vigilant about safety. They must make sure workplaces meet at least the prescribed standards for occupational health and safety. Chemicals or fuel must be stored at a safe distance from workers. Welding and many other engineering tasks need special and safe workplace environments. Office workers need ergonomically suitable furniture and an office layout that is safe. The office should permit rapid evacuation in the event of a fire. Fire evacuation plans and training drills need to have been completed. There should be a trained first aider (and a first aid kit) in each workplace. Other simple arrangements may include a sick room (or bunk), and perhaps somewhere for children who need a safe environment between the end of school and knock-off time for their parent. The safety of visitors to protected areas is also a prime concern, and is discussed in Chapter 16.

Staff housing

Staff housing is commonly provided on parks that are very remote or where staff need to be present to provide special security for a protected area. More commonly today, staff are encouraged to live in nearby towns. On-site housing may be rented at the market rate, or it may be subsidised, especially where the quid pro quo is that staff are nearby and thus potentially on call for serious incidents. Managers will need to ensure that such housing is

Management training at Dorrigo National Park, NSW
(Graeme Worboys)

adequate and in good repair. Much of the rental income may be required for maintaining and occasionally upgrading or replacing the housing. Sometimes staff that live in protected areas but near major metropolitan areas need special security. This could be arranged with local police, or managed through contract security services.

Employee assistance programs

For staff that have personal problems, either in the workplace or at home, some organisations provide personal counselling services. Often this is extended to the immediate family as well. Apart from being a responsible action, the cost is justified by the fact that staff with unresolved troubles are unlikely to work well. Employee assistance programs that offer counselling services can help with marriage and family problems, alcohol and other drug problems, emotional stress, legal problems, interpersonal conflicts, financial problems, and work-related difficulties. These services are confidential. They are normally provided by independent consultants whom staff are invited to contact free of charge. This is better than having managers becoming involved in complex personal issues, as might happen if they felt solely responsible for the welfare of troubled staff members.

Disputes in the workplace

Regrettably, disputes do occur in the workplace. It is often best for managers to discuss the issues with staff directly. Doing this over a cup of coffee, or over a billy of tea in the bush, is a tried-and-true method for resolving grievances. Often though, disputes become entrenched. Managers may then need to address them through formal processes of 'grievance resolution'. (There are similar mechanisms for dealing with industrial issues.) These may include counselling sessions or other resolution actions. Harsher arrangements involve transfer of officers, disciplinary action, or even dismissal.

Maximising conservation outcomes through best use of a workforce

Protected area management is about managing for the long term. It is about instigating conservation work programs that may be in place for generations of managers. It is also about innovative and cutting-edge decisions made by well-informed and dynamic managers. It is about organisations that are adaptive and that have a capacity for continuous improvement. Those organisations that respect, retain, and utilise their depth of talent and experience will have most success in conserving our heritage.

Such mature (in outlook) organisations will continuously train and improve all staff. In particular they will retain their older staff so they can pass on their experience and acquired knowledge to a new generation of managers. They will have a cross-section of staff from new recruits to old-timers, with a balance of genders and a diversity of ethnic backgrounds. They will have a solid corporate culture, and a strong sense of history and of the organisation's prior achievements.

Goodwill

The type of person employed by protected area agencies throughout Australia is typically highly motivated and committed to conservation principles and practice. Many staff members work, on average, 15 hours per week in excess of their 35- to 40-hour-week paid time. This additional workload is equivalent to having 30% more staff members. If this were extrapolated to an organisation with 1000 staff members, the total workload would be equivalent to 1300 people working a 35-hour week. Should management antagonise or undermine such goodwill, then the productivity of organisations is highly likely to substantially reduce, despite a continued professional and responsible performance by staff. Senior management need to be aware of the moral commitment of their staff to the conservation effort, and never underestimate goodwill or take it for granted.

9.2 A principled organisation

Most organisations responsible for managing protected areas have a clear statement of what they stand for. This is usually set out in a corporate plan, and is a group of values and principles the organisation considers important. These principles and values set the scene for how an organisation will deal with its staff members, its customers, its external stakeholders, and how it will go about the business of conservation. For instance, the 2000–03 NSW P&W corporate plan (NSW NPWS 2004d) provides statements of value for conservation, respect for Aboriginal culture and heritage, social cohesion, active community involvement, fairness and equity, professionalism, and ethical conduct. The corporate plan then goes on to describe how these values will be addressed.

Ethics

Managers are at the forefront of their workplace. They are continually being judged on the way they tackle their daily tasks. They are expected to show honesty, integrity, fairness, and balanced judgment. A manager's behaviour and standards will set an example for how the workplace operates. They should be guided—especially in times of stress—by the organisation's code of conduct and by its internal policies on ethical management.

Code of conduct

A code of conduct establishes how employees are expected to behave. It provides guidance to help people to work together comfortably and with mutual respect, despite individual differences of personality or style. Usually the code aims to assist employees when they are faced with ethical issues that may arise during the performance of their duties. It would typically deal with subjects such as professional behaviour; conflicts of interest; gifts, gratuities, and hospitality; personal use of organisational resources; giving and accepting direction; dress; dealings with the public; public comment; confidentiality; other employment; and notification of dishonest or unethical behaviour.

Principles of equity and their practice

There is a basic expectation in Australia that people will be dealt with fairly and equally. Equity means fair and impartial access to community services and to job opportunities. This is reinforced by anti-discrimination laws that are echoed in internal policies. It is unlawful, in all areas of public life, to discriminate against an otherwise suitable person on grounds of their sex, pregnancy, race, colour, nationality, descent, ethnic or ethno-religious background, marital status, disability, sexual preference, or age. Managers can make a huge difference by treating their staff considerately. For instance, they may offer flexible work hours to those with young families, or help accommodate the needs of people with disabilities.

Some organisations have prepared an 'equity management plan' to deal with equity in the workplace. Recruitment is one such issue, and is well dealt with in a report by the NSW Independent Commission against Corruption titled 'Best practice, best person: integrity in public sector recruitment and selection' (ICAC 1999). This report stresses the need for impartiality, accountability, competition, openness, and integrity.

Guarantee of service

Organisations may also prepare a 'Service Charter' or a 'Guarantee of Service', which sets out the standards that citizens and customers can expect from them. It may invite feedback from the public, as well as giving background on the organisation's history, structure, mission, and values.

9.3 Administering finances

Organisations are funded to undertake works. They typically have an annual budget allocation plus some secondary sources of revenue. Managers must account for all the monies they deal with. Budgets must be spent as planned, and regular reports on the progress of works must be provided. Financial management systems help managers achieve this. Audit systems are employed to ensure compliance and to contribute to annual reports.

To manage finances wisely is a critical part of any protected area manager's job. This will mean that important conservation works are achieved efficiently and as needed, and that there is funding also for staff, support infrastructure, equipment, and materials. The mismanagement or fraudulent use of finances, apart from bringing an organisation into disrepute, can lead to staff being reprimanded, dismissed, or even charged with a criminal offence. Since the management of finances is a serious business, it is linked to a series of accountabilities and controls, some of which are described below.

Developing a budget

Apart from an organisation's main budget, there are individual budgets for many specific tasks. These include: special projects, incidents, and proposed projects for which the grant application requires a budget. All budgets must include the following basic elements.

1   There is a clear plan for a project at hand, with a series of objectives and a program of tasks over time (sometimes referred to as a Gant chart).
2   There is a clearly identified source of funds. This may be from a specific source of revenue, or from the organisation's normal budget allocation from treasury. A treasury allocation is typically made up of recurrent funds (funds that are provided for the annual operating costs) and capital funds. Capital funds are provided for the purchase of land, vehicles, equipment, buildings, or other fixed assets, and also for undertaking new works. (Note that many grant applications require the organisation to provide a percentage of the funds for the project or at least in-kind funding.)
3   There is an estimate of staffing costs. Managers will need to ask how many personnel are required, and of what types and skills? Will the work be done in-house or by contract? When and for how long are specific tradespeople required? Will there need to be double shifts or a continuous 24-hour operation to finish on time? Some tasks may need to be undertaken outside office hours to minimise disruption to normal business, and this will usually cost more. Innovative managers may use a special project to train their staff by assigning them the special project task and by putting on temporary staff to do the staff-members' normal work. Additional administration costs (on-costs) should usually be budgeted for. Costs per visitor and costs per hectare could be calculated as a mechanism for benchmarking management performance.

4   There is a costing of project support services. This may include specialist advice and technical designs, for instance the architect's and engineer's plans.
5   There is a costing of materials. For large projects, quantity surveyors may need to supply these details on contract.
6   There is a cost estimate for logistic support arrangements. This covers all temporary equipment, from offices at a construction site to hire of vehicles to transport additional staff, and the hire of supplementary plant such as generators and compressors.
7   There is a costing for basic items such as electricity, water, waste disposal, office supplies, postage, telephones, and other consumables.

Major budgets are normally programmed across a financial year. More advanced budgeting systems allow for project management of funds across two to three years. Once budgets are approved, most organisations expend funds only in accord with firm and clear operating instructions, which are defined by law or by internal procedures. The status of the budget is reviewed against forecast budget performance, usually once a month. Most organisations recognise a budget cycle, and this is linked to their annual planning cycle.

Linking budgets to organisational goals

In an ideal organisation, a corporate plan or equivalent document will clearly define the vision and mission of an organisation. The plan defines a series of actions based on this vision, along with broad time frames for their completion. These actions are assigned to different sections of the organisation to achieve. The budget's funding explicitly recognises the same sections or categories, and thus can be closely linked to the organisation's goals.

The budget process expresses the policies and priorities of government. It also follows an annual cycle. A typical budget calendar for a state-based park service might be as follows.

• October–November. An operational plan is drawn up. It reflects tasks for which the government demands priority, as well as routine tasks for the following financial year (and potentially succeeding years).

- December. The organisational budget is finalised. This follows an internal peer review and a setting of priorities.
- January–February. An organisational budget is submitted to treasury.
- March. Detailed negotiations take place between senior departmental officials and treasury officials.
- April–May. Treasury provides departments with indicative budgets for the new financial year.
- May–June. Detailed planning for the budget implementation is completed in anticipation of the new financial year.
- 30 June. The old financial year finishes, with business transactions against the old budget terminated.
- 1 July. The new financial year begins. Business is conducted on the basis of the supply provisions of parliament until parliament formally approves the new budget. Expenditure is guided by the indicative budget. A review of the previous year's performance is completed.
- August. The premiers' conference, federal budget, and other Commonwealth–state negotiations determine the nature of the state's budget allocations. An annual report for the previous financial year is published.
- August–September. State or territory parliaments formally approve the new budget.
- September. The organisation's budget is finalised and distributed internally, consistent with corporate priorities.

Subsidiary budgets (with funding) are allocated to sections of organisations that have been assigned corporate tasks. This does not always occur without some internal drama. There are always competing priorities for the insufficient funds available. Project champions will jostle for advantage. Competition for the available resources is normal, even though people are working in a collegiate manner to achieve the same vision. You could say that it is part of a corporate culture and part of living in our society. Managers will find that they quickly adapt to this culture, or they will fail in achieving their specific mandate.

The allocation of budgets within an organisation usually follows a careful process that is often centralised. Its stages may include:

- receipt of the budget from state or territory treasury
- removal of all fixed costs (including salaries, rents and rates, insurances, fees, and perhaps contingency funds)
- disbursement of the balance of the budget to priority projects, including some discretionary funds.

Note the roles of the different layers of an organisation in the budget process. Top-level management ensures that budgets are assigned to the key corporate goals that are being targeted. They authorise and adopt the proposed organisational budget. Middle-level management ensures that regional priorities are carried out, and that funds are assigned accordingly. At the front-line, budgets are directly assigned to individual tasks in hand for their implementation.

It is clear that at each stage, social and political priorities influence the way funds are divided. There are also regional issues. A district or local area's funding could depend on the number, value, and condition of its assets; the total area of its protected areas; the number of visitors received; and the number of processes that threaten its heritage values. When allocating funds to a given site, managers need to take into account the safety of staff

Snapshot 9.1

Assigning funds to a capital works project

John O'Malley, Parks Victoria

It is not enough for scarce funds to go to solving the most urgent problems; funds must go to projects, each of which is the best or most cost-effective solution to its problem. It is not enough to get a bridge over the river. It must be the right bridge and in the right place, made or overseen by the right staff, and at the right price. We have distributed to all staff a pocket *Field Guide* for managing capital works and a guide to design standards; also a manual on the process of *Project Initiation*. This insists on clear lines of responsibility for a project, and a proper evaluation process. We have also devised a spreadsheet-based *Investment Prioritisation* model to help us weigh up the claims of competing projects. While not yet perfected, the model provides a transparent, logical, and flexible rationale for tricky decisions.

and the general public; the health of staff and the general public; environment protection needs; the value and condition of assets; and the funds saved by timely rather than by tardy maintenance.

Some of the considerations used in Victoria as part of budgeting and project management of capital works are described in Snapshot 9.1.

Tracking a budget. Managers typically have a 'global' (overall) budget allocation for each major task. To track it they commonly use computerised financial management systems. These systems are supported by manuals that set out the accounting systems used (such as accrual accounting). The manuals guide staff on how to maintain accounts in a way that is consistent with statutory accounting guidelines and internal policies. Usually these global budgets are 'coded' or given a unique number within the overall budget system of an organisation. Managers also have estimates of their budget expenditure across a year. Typically, they use computer support systems to compare forecast expenditure with actual budget performance. This tracking of a budget is critical. It is very easy to over-expend; and it is very easy for estimates of expenditure to be incorrect for one reason or another. When they are, the budget must be revised or fine-tuned. This may mean that managers find themselves in the position of finishing the project with more (or fewer) funds than planned.

Budgets are typically reviewed monthly. This review checks such aspects as revenue received, expenditure completed, and financial commitments. In any financial statement there will be columns that list these items. For instance, equipment or supplies that have been ordered but not yet received would be placed in a commitments column. Actual expenditure would also be shown, as would the balance of the budget that is unspent (Figure 9.1).

Budget management can be severely affected by unplanned events such as incidents (Chapter 15). Experienced managers anticipate this 'incident downtime', and program their year to make sure that tasks, wherever possible, are advanced ahead of schedule.

Audits. The law requires an organisation and its managers to undertake an annual financial audit. Other administrative and management audits may also be taken from time to time. Typically, an internal audit occurs at a local cost-centre office that manages the budget of a given area or section. Audit checks look for compliance between actual and planned or budgeted expenditure. They also examine documents such as invoices and authorisations to check that proper processes have been followed. Internal audits may extend beyond finances to include checks on salary payments, the use of vehicles or equipment, and time sheets. They are a mechanism to help guarantee that funds have been properly spent against the items prescribed in the budget.

Annual report

Annual reports cover all aspects of management, especially the responsible management of finances. They must contain a breakdown of how the various parts of the budget have been spent. To help develop an annual report, local managers must submit a balanced financial report for their areas, as well as a statement of their achievements.

Preventing fraud and corruption

Whether we like it or not, fraud and corruption do occur in our society, and protected area organisations are not immune from this. Examples of corrupt conduct include:

- outright theft
- unauthorised use of an organisation's vehicles or other equipment
- misuse of official or inside information for personal gain
- illicit use for private gain of official powers, or of computers or records
- sale or disposal of assets or services at less-than-fair value
- arranging payment for goods or services that were either not received or not worth so much
- falsifying records or computer programs so as to commit or conceal fraud
- appointing a friend or relative to a position for which there are better qualified applicants
- providing or receiving a bribe to facilitate a contract.

Administrative areas most at risk from fraud include personnel management, information systems, tenders and contracts, licensing and regulation, financial systems and procedures, and arrangements for the use of equipment. Managers

Figure 9.1 Example of layout for a 'typical' budget report for a major region of a state

FICTITIOUS PROTECTED AREA: FINANCIAL REPORT FOR A REGION (SUMMARY)

1998/99 ASSET ACQUISITION PROGRAM MONTHLY ACTIVITY REPORT AS AT 30TH JUNE, 1999

PROJECT CODE	PROJECT NAME	ALLOCATION	ACTUAL EXPENDITURE Y.T.D.	ACTUAL COMMITMENTS Y.T.D.	TOTAL UTILISED	BALANCE OF FUNDS	ESTIMATED COMPLETION DATE MM/YY	PROJECTED EXPENDITURE TO 30TH JUNE	PROJECTED BALANCE AT 30TH JUNE
	REGIONAL SUMMARY								
		$	$	$	$	$		$	$
[1]	[2]	[3]	[7]	[8]	[9] [7]+[8]	[10] [6]–[9]	[11]	[12]	[13] [6]–[12]
	MAJOR WORKS-IN-PROGRESS								
6145/46	GALAH NAT.PARK—NEW HQ	616 000	492 582	4180	496 762	119 238	JUNE 99	498 000	118 000
	TOTAL	**616 000**	**492 582**	**4180**	**496 762**	**119 238**		**498 000**	**118 000**
	R.P.A. PLANT & EQUIPMENT	0	0	0	0	0		0	0
7731	PEST SPECIES MGT RPA	81 000	81 039	0	81 039	–39	MAY 99	81 039	–39
7831	HEAVY FIRE MGT EQPMNT RPA	130 000	119 892	7700	127 592	2 408	MAY 99	128 185	1815
	TOTAL	**211 000**	**200 931**	**7700**	**208 631**	**2 369**		**209 224**	**1776**
	NEW PARK INITIATIVES	0	0	0	0	0		0	0
8090/92	QUOLL CREEK—NEW PARKS	176 000	176 000	0	176 000	0	MAY 99	176 000	0
8100	DINGO STAGE 2—NEW PARKS	871 000	757 667	56 815	814 482	56 518	JUNE 99	821 713	49 287
8110	DOGSHEAD BAY MARINE PARK	269 000	85 557	0	85 557	183 443	MAY 99	88 000	181 000
8270/72	WETLANDS—NEW PARKS	479 000	391 781	26 506	418 287	60 713	JUNE 99	421 000	58 000
	TOTAL	**1 795 000**	**1411 005**	**83 321**	**1 494 326**	**300 674**		**1 506 713**	**288 287**
	CABINET MINUTE PARKS							0	
6020	BANJINE WILDERNESS	2 000	0	2000	2 000	0	MAY 99	2 000	0
6021	NARDOO WILDERNESS	25 000	23 351	0	23 351	1649	MAY 99	23 351	1649
	TOTAL	**27 000**	**23 351**	**2000**	**25 351**	**1649**		**25 351**	**1649**
ANNUAL PROVISIONS—MOU PROJECTS		0	0	0	0	0		0	0
6143	BUNYA-95/96 ACQUISN	150 000	147 395	0	147 395	2605	JUNE 99	148 000	2000
6371	BUDGIE N.P. LOTS 1–7	120 000	117 626	0	117 626	2374	MARCH 99	118 000	2000
6372	BILBY N.P.	70 000	70 848	0	70 848	–848	MAY 99	71 000	–1000
	TOTAL	**340 000**	**335 869**	**0**	**335 869**	**4131**		**337 000**	**3000**
PLANT & EQUIPMENT									
6431	MAJOR PLANT PURCH	230 000	336 407	332	336 739	–106 739	JUNE 99	336 739	–106 739
6431	MAJOR PLANT—SALES		–109 658	0	–109 658	109 658	JUNE 99	–109 658	109 658
	TOTAL	**230 000**	**226 749**	**332**	**227 081**	**2919**		**227 081**	**2919**
GRAND TOTALS		**3 219 000**	**2 690 487**	**97 533**	**2 788 020**	**430 980**		**2 80 3369**	**415 631**

normally prepare a plan to guard against fraud and corruption. Such guidelines must be firm. Audits and tip-offs can help bring fraud to light, but managers should themselves be alert to irregularities. When serious fraud is detected or suspected, criminal investigations and charges, or at least internal disciplinary actions, may follow. Upon suspicion of corrupt behaviour, managers may need to move swiftly to impound all documents and other evidence that investigators would require. Managers have a responsibility to maintain an organisational culture of propriety.

Protected disclosure

It can be difficult for staff to inform on colleagues they suspect of fraud. Some states and territories have passed laws that protect staff who come forward. Such laws protect whistle-blowers from reprisals and intimidation, including harassment and violence. Their disclosures must be kept confidential, and are usually exempt from provisions such as freedom-of-information. These laws help organisations avoid a culture of corruption.

9.4   Administration of assets, standards, and systems

Assets are items of value that an organisation owns or controls. Assets include constructed items such as roads, sewer lines, bridges, buildings, trails, and various cultural heritage structures, as well as tools, vehicles, or even intellectual property. Most organisations have a range of assets to manage, and typically these are inventoried. Asset management systems allow managers to predict when assets will need to be refurbished or replaced ('maintenance cycles'). They can allow for these expenses in their annual budget. They can also keep track of the total value of assets, which is important in 'accrual accounting'.

Asset management should be part of an integrated management system. This system should include data management and IT support systems, integrated organisational management programs for development projects, annual maintenance programs, performance review and assessment, and financial management.

Administration within organisations has many support systems and protocols. They are constantly being improved (and reinvented!). Some of these systems are briefly discussed below.

Best practice and benchmarking

Administrators constantly seek ways to achieve more for less, or to obtain the same effect more efficiently. They try to set standards of efficiency for common administration tasks. In critical areas such as 'information turnaround times' for reports to the minister's office, these standards must be rigorously observed. Many organisations also undertake 'benchmarking' exercises to gauge their performance. Benchmarking enables an organisation to compare its performance with that of similar organisations, and seek improvements. 'Best practice' models developed by governments or commercial firms can serve the same purpose. There are also some established design standards and codes, such as:

- Australian design standards and codes for engineering, construction, and equipment
- corporate designs and standards for such things as letterheads, signs, park furniture, and uniforms
- landscape design codes and rehabilitation manuals
- safety and health codes, and manuals for such things as dangerous goods and chemical safety
- contract management manuals, and leasing and licensing manuals.

Such standards help ensure the safety and health of visitors and staff as well as protecting the environment. For protected area managers, very useful benchmarking and best-practice work has already been done by a highly committed group of managers: the Australia and New Zealand Environment and Conservation Council (ANZECC) Working Group on National Parks and Protected Area Management. This working group has produced a series of reports on best practice in park management and they have been made available on the Department of the Environment and Heritage web site:

- use of fire for ecological purposes (was due in 2003)
- current approaches to performance measurement in protected area management (2002)
- public participation in protected area management (2002)
- cultural heritage management (2001)
- user-pays revenue (2000)
- protected area management planning (2000)
- audit of uptake of best-practice reports (2000)

- park interpretation and education (1999)
- commercial management processes in the delivery of park services (1999)
- visitor risk management and public liability (1999)
- asset management (1997)
- stakeholder management: neighbour relations (1997)
- performance reporting in natural resource management (1996)
- staff training processes (1996)
- data standards (1996).

Environmental best practice

Organisations should practise what they preach. Administrative procedures should ensure that protected area agencies adopt environmental best practice and ecologically sustainable technologies in their use of power, water, paper, and other resources.

The office filing system

Despite the swift advance in electronic systems, the traditional paper-based filing system still prevails today. Computer-aided tracking systems and computer-assisted key-word filing systems are, however, commonplace for larger organisations. It is important to maintain a logical and up-to-date system for filing paperwork and records and a 'paper trail' that tracks management decisions and other procedural tasks. Incidents, coroner inquiries, and Freedom of Information requests do occur and staff can be held personally accountable.

Leasing and licensing systems

Licences and leases are legal contracts between organisations and external individuals or companies. They provide certain rights of occupancy or use of protected areas, usually for a fee. Systems are in place for providing leases and licences that fit in with laws, with planning documents, and with environmental planning requirements. The process needs to be managed carefully and with expert legal advice.

Information tracking systems—managing 'ministerials'

A very important system for senior bureaucrats is the 'ministerial' management system. This deals with urgent requests for information from the minister's office, from parliament, from central government, and from other stakeholders. It ensures that all urgent requests are dealt with as quickly and effectively as possible.

Using clear English for better conservation outcomes

All professions have their own special jargon or technical language. Protected area managers are no exception. Knowing these terms allows skilled people to understand each other more quickly and clearly, and avoids costly misunderstandings. But we need to remember that such jargon tends to exclude the layperson. Jargon also, when it runs to long words or long sentences, makes prose harder and slower to read. Good managers are clear communicators, not only in speech but also in writing, including formal documents. An effective prose style is essential. The rules for good English prose have long been known, and are widely agreed upon. They include the following.

- Avoid using a long word where a short one will do: 'start', 'stop', 'fix', and 'if' are usually better than 'commence', 'discontinue', 'rectify', and 'in the event of'.
- Generally delete unnecessary words.
- Avoid using the passive where you can use the active ('we decided' rather than 'a decision was reached by our group').
- Avoid clichés and stock phrases.

Many managers need to make a conscious effort to break the habit of abstraction. A problem for some with science degrees is that they have been trained to write in a way that breaks most of these rules. The result can be a style that is full of passive verbs and heavily padded with abstract polysyllables. H.W. Fowler called this style 'abstractitis'. It is a disease that produces such sentences as 'Early expectation of a vacancy is indicated by the management' instead of 'The managers say they expect to have a vacancy soon'; or 'Management has a responsibility of intervening to minimise the influences of extinction processes within habitats' instead of 'Managers should act to reduce the risk of extinctions'. To such a person, a 'plane' is a 'fixed wing aircraft'.

Watch the jargon: nests of nouns. All professions have their jargon, but users should note when it is not understood by others. The language of administration is often ridiculed as abstract and pretentious, especially by those who do not understand it. An administrator who says 'undertake' for 'do', and 'implement' for 'carry out' may be making valid distinctions, but risks seeming pretentious to laypersons.

The above is only a brief introduction to the issues as they may affect managers. There are many notable books on the arts of speaking and writing—we recommend Ashe (1972) or a recent edition of Fowler (1926).

Systems for evaluating staff performance

'Performance management and development systems' are important ways of measuring and improving the contributions that staff members make. These systems can also empower staff to plan their own careers, by giving them confidence that they and their supervisors are using the same standards to judge their performance. Such systems link training and development with the requirements of the job and the mission of the organisation. Typically such systems work in three stages.

1   A work plan is developed between a supervisor and an employee. It is clearly linked to the organisation's corporate plan and to any other relevant strategic plans.
2   From this work plan and the personal needs of the employee, they develop a personal plan (a 'staff development plan') for the employee to gain the necessary skills.
3   Progress is reviewed at regular, agreed intervals. At least once a year there is a major review and a report is prepared.

The performance of an administration needs to be measured and monitored at all levels (Chapter 21).

9.5   Administration policy and process considerations

Developing administration policy

Organisations commonly have a policy group that helps to achieve consistency of operations across a state or territory through the development of policies. It may update and propose policies for such matters as fire management, visitor access, wilderness management, recreation management, and commercial use. Such policies may be hotly debated in the community. If so, the advice and support of the organisation's media unit will be important. Allan Young, a senior policy officer of the NSW P&W, describes a process followed to develop a policy (Snapshot 9.2). Parks Australia (2002) has also developed a Policy Development Guide to assist staff.

Sustainability

Administration can greatly assist an aim of a sustainable organisation (Chapter 10). Purchasing policies that favour environmentally friendly products and suppliers, accounting practices that record environmental performance measures (such as energy and water consumption), local purchasing

Snapshot 9.2

The process of developing a policy
Allan Young, NSW P&W

There are at least a dozen steps in developing a policy for the NSW P&W.
1   An issue emerges that requires a policy.
2   We decide if it is local or if it concerns the whole organisation.
3   We ask how soon is a new policy needed. If soon, we may need an interim policy.
4   We identify key contacts and experts within the service, and set up a working group.
5   We research the issue.
6   We check if other organisations already have a policy for it.
7   We discuss and try to balance the views of different sections within the agency.
8   We consult external stakeholders.
9   We draft a policy and seek comments widely.
10   The working group submits its final proposal.
11   Senior management considers, revises, and endorses the document.
12   We print and disseminate the new policy.

policies, and, office practices that minimise consumption of resources all make a difference.

Legal issues: The parliamentary and legislative processes

There are many legal issues in protected area management that need to be managed and administered. Any aspect of protected area management may become a legal matter given the right circumstances. Administrative record systems and access to information are critical. Legal staff generally provide guidelines on the nature of records to be kept and stored. Law enforcement is a good example where records, supported by an excellent records system, are critical. Good record keeping systems will also ensure licences, leases, and other legal agreements are available and that legal commitments are honoured.

The parliamentary process and ministerial information request systems need to be administered efficiently and effectively. Once again record systems administration is critical. Legislative processes include the development or amendment of legislation. Administration of agency responsibilities related to such processes are a major task, with new issues being brought forward regularly. Records of legislation and amendments must be maintained and made available to staff. There is a constant need for improving legislation.

A topical issue in the new millennium is intellectual property issues associated with the genetic resources of protected areas. Legislation may need to be upgraded and administrative processes developed to deal with such rights. This topic is discussed by Charles Lawson and co-authors (Case Study 9.1)

Case Study 9.1

Access to genetic resources in Australia's protected areas— intellectual property and patenting issues

Charles Lawson, Catherine Pickering, and Susan Downing, Griffith University

The *National Strategy for the Conservation of Australia's Biological Diversity* recognised that Australian genetic resources were valuable and that the social and economic benefits should accrue to Australia (Objective 2.8). Implementing this objective is no easy task given Australia's federal system requiring regulation at the Commonwealth, State, and Territory tiers of government. To date, a comprehensive access scheme covering all protected areas in Australia remains unfulfilled. Further, some key issues about the preferred approach to regulating access to genetic resources in protected areas have the potential to undermine the likely economic and other benefits from access to these valuable resources.

In dealing with 'genetic resources' in all areas, including protected areas, Australian law-makers have adopted a broad meaning for this term so that in effect it includes all manner of living organisms and their parts and components. This includes whole organisms, parts of organisms, organs, fluids, information macromolecules (such as DNA and polypeptides) and biochemicals, and any other living materials or derivatives of those materials sourced within the Australia land area and its recognised ocean boundaries.

Any access scheme to protected areas in Australia must be consistent with Australia's commitments to international agreements, such as the United Nations' *Convention on Biological Diversity* (CBD), the United Nations' *Convention on the Law of the Sea* (UNCLOS) and the World Trade Organisation's *Agreement on Trade Related Aspects of Intellectual Property Rights* (TRIPs).

In Australia, proposed Commonwealth regulations under the *Environment Protection and Biodiversity Conservation Act 1999* (Cth) (*EPBC Act*) are consistent with the preferred approach under the CBD for contracts for access between the resource holder and the bioprospector. Some protected areas are within the scope of this legislation where they are 'Commonwealth areas' for the purposes of the *EPBC* Act. The *EPBC Act's* s 301 empowers the making of regulations for the equitable sharing of the benefits arising from the use of biological resources, the facilitation of access to such resources, the right to deny access to such resources, and the granting of access to such resources and the terms and conditions of such access. The concept of access adopted by the proposed regulation deals with the collection of samples from individual organisms and then the determination of their genetic, biochemical, and other attributes together with their potential uses. This will include the taking of native species for conservation, commercial uses, or industrial applications, such as collecting living material, analysing and sampling stored material, and exporting material for purposes such as conservation, research, and potential commercial product development. The *EPBC Act* scheme proposes access permits and a model contract that may then be negotiated between the resource holder and the bioprospector.

These proposed *EPBC Act* regulations hope that the contracting parties will negotiate an adequate level of benefit sharing and access to the technology to both exploit and conserve the genetic resources. The great weakness in this approach is the burdens of responsibility on those negotiating the individual contracts to properly value the resources accessed and ensure the transfer of adequate benefits and technology. This will continue to be a difficult area for protected area management where there is some expectation that access to genetic resources will deliver real and tangible financial and conservation benefits.

While the scope of these *EPBC Act* regulations will capture most protected areas under the control of the Commonwealth, there are significant protected areas outside the scope of these regulations and subject to control by State governments and private landowners. Some State governments are developing access schemes consistent with the Commonwealth's contract model and the requirements of the CBD, such as the Queensland Biodiscovery Bill 2003. Private landowners are being encouraged to adopt the same or similar arrangements to the Commonwealth and States. Again, however, negotiating contracts that equitably share benefits and access technology are likely to be a significant hurdle for managers expecting financial and conservation benefits from access to the resources under their management.

Patenting genetic resources

In Australia, as a generalisation, the *Patents Act 1990* (Cth) grants various exclusive rights for any inventions that satisfy the threshold criteria of being new, not obvious, useful, and described in a way that can be followed by others. This includes inventions involving whole organisms, and parts or components of organisms such as organs, organelles, biological molecules, and the workings of these parts and components. The consequence of using language to define the scope of a patent is the potential for broad claims. As a result of this, broadly claimed patents can be enforced against later inventors across a broad range of products, processes, and uses.

The exact limits of patenting are difficult to predict in detail, as the language of the claims determines the scope of the patent, and these depend in large part on the circumstances in which the patent claims are made and then processed by the Patent Office. The imperative for those drafting the patent claims is to seek to claim as broadly as possible the 'invention' and deliver to the inventor the widest possible scope of 'exclusive rights'. Further, the ingenuity of those drafting patent claims to use language to make broad claims should not be dismissed lightly as the breadth and inclusiveness of language is considerable.

Our concern and criticism of the proposed schemes are addressed at the undermining effects of the internationally agreed minimum standards intellectual property requirements imposed by TRIPs, implemented in Australia under the *Patents Act 1990* (Cth). This is significant as patents are considered to be one of the main mechanisms of benefit-sharing and valuing genetic resources. Such patents provide the prospect of a royalty stream from the commercialisation of accessed genetic resources. The exclusive rights allow the patent holder to commercialise the invention without competition and so capture economic and other benefits as the reward for investing in developing new and useful inventions. Our principal concerns are as follows.

1. The royalty and other benefits flowing from future patents are negotiated at the time access is agreed as a term of the contract between the resource holder and the bioprospector. At this stage the likely economic benefits from access to the genetic resources are very difficult to assess and are likely to be significantly undervalued.
2. Patents may be claimed for inventions independently of the proposed access schemes and without access to the genetic resources. For example, a claim to genes or gene sequences may extend to a range of related genetic materials overlooking the unique diversity that makes Australia's genetic resources useful and valuable. The effect of these claims is to prevent future claims to those materials even though they may not have been known or identified at the time the patent was claimed.
3. Most patents in Australia are granted to non-residents (more than 90%) so the economic benefits from access to Australia's genetic resources are likely to be captured by non-residents with very little opportunity to realise the secondary benefits (such as economic activity, employment, and so on) in Australia.

In our view this is a difficult problem that may only be resolved through genetic resource holders carefully negotiating access contracts that deal in detail, and with some sophistication, with ways to share benefits and access appropriate technology. We also consider there is a role for governments in ensuring adequate technology transfer as the CBD (that includes the access schemes proposed by the United Nations' *Convention on the Law of the Sea* (UNCLOS) and the United Nations' Food and Agriculture Organisation's *International Treaty for Plant Genetic Resources for Food and Agriculture*) was addressed to national benefits. These national benefits include outcomes that may not necessarily be in the immediate interests of individuals. For example, access to a new technology might benefit all Australians (through increased economic activity in Australia and employment), whereas an individual resource holder might consider a minor royalty adequate.

and provides an example of how new imperatives constantly need to be addressed by protected area managers.

Risk management

The systematic treatment of risks for all operations is an essential part of management. Typically a risk management strategic plan is developed for a protected area organisation consistent with Australian and New Zealand Standards (Chapter 11). Risk management updates are usually conducted regularly (Parks Australia in 2003, for example, conducted their review quarterly). Risk management effectively managed is a tool for reducing insurance premiums for agencies as well as the frequency of avoidable incidents.

9.6 Management lessons and principles

Lessons

Administration for protected areas is fundamentally important for the success of protected area agencies. Two key lessons from administrative practice are:

1 Protected area programs that are efficiently and effectively administered more easily attract additional financial and support resources.
2 Poor administration has led to the closure of projects, and even worse, the dismissal of staff and the demise of organisations.

Project Manager Iris Paridaens at work (Graeme Worboys)

Principles

1 Administration is at the heart of a protected area organisation's capacity to operate effectively. Organisations need to strive for continuous improvement in their administration structures and systems.

2 Administration of protected area organisations has a set of values that define ethical behaviour, code of conduct, and professional integrity. Capacity and skills development of staff are an essential component of the human resource management systems.

3 People make the difference when it comes to achieving conservation outcomes. Effective administration of recruitment, induction, and other staff-related processes is essential. People need to be managed fairly, equitably, and have a clear sense of purpose. They need to understand their role, delegations, authority, and opportunities to improve.

4 Acting ethically is part of good administration.

5 Managing finances wisely is a crucial part of any protected area manager's job. Financial management must be consistent with statutory requirements and audit processes.

6 To achieve organisational goals, procedures must have sound budget, financial monitoring, and performance systems and an overall annual business plan (Chapter 8).

7 Standards should be set for administration systems and regular monitoring should be carried out. To conserve continuous improvement in performance the systems should be benchmarked.

8 Protected area organisations should have administrative systems that are: accountable, transparent, auditable, well documented, and founded on written and public policies.

9 Protected area organisations should adopt environmental best practice and ecologically sustainable technologies in their use of power, water, paper, and other resources.

10 Administration system, policies, and procedures are in direct support of a protected area organisation's conservation objectives.

11 Administration systems reflect the program-related outcomes to ensure effective implementation and coordination of program activities.

Sustainability Management

They call her a young country, but they lie:

She is the last of lands…..

A.D. Hope, from Australia

10.1 The need for sustainability   270
10.2 Sustainability and protected areas   274

10.3 Management lessons
      and principles   286

The major objectives of protected area managers are to ensure biodiversity and cultural heritage conservation. At the same time, sustainable management principles need to be adhered to, as the very process of conservation management consumes energy and natural resources and produces wastes, thus impacting upon the global environment. Sustainable protected area management considers these impacts and focuses on reducing greenhouse gas emissions, water and energy consumption, minimising waste production, and ensuring maximum benefits to local communities. Protected area managers operate within the wider context of environmental management and as such there are a number of international environmental policies that govern their operations (Figure 10.1).

This chapter advocates that protected area organisations should be leaders in the field of sustainable management practice (Figure 10.2). Issues of sustainable development and best practice environmental management are discussed along with examples of techniques to minimise resource consumption and waste production. Methods for benchmarking, monitoring, and reporting on compliance with sustainability indicators are described.

10.1 The need for sustainability

The need to implement sustainable development, especially biodiversity conservation, has been outlined in previous chapters. Critically we are all consuming resources and producing more waste than our ecosystems can assimilate. Loss of biodiversity, resource depletion and pollution, and sustainable community development are immense global problems.

Water

Natural resources, including many of those considered renewable, are being rapidly depleted and degraded.

Figure 10.1   The global environmental management context within which protected area management operates

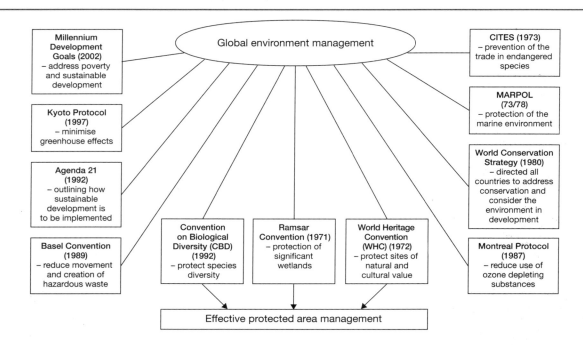

Figure 10.2   Protected area management and environmental accountability

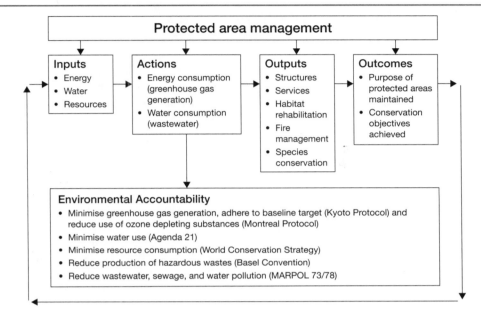

Potable water availability is developing as one of the major environmental concerns of the planet. In 2002, about 30% of the population of the Earth lived in countries suffering from moderate to high potable water supply stress (UNEP 2002). By 2020, it is estimated that 66% of the world's people will be living in water-stressed countries (CSD 1997 cited in UNEP 2002) (Snapshot 10.1). Water is especially important in Australia, the world's driest continent after Antarctica. Intensive development of river systems in the Murray–Darling Basin for irrigation (Snapshot 10.2) and in coastal river systems near major urban centres for potable and industrial water have resulted in water extractions exceeding sustainable yields (Commonwealth of Australia 2001b).

Climate change

The effects of global climate change are becoming increasingly apparent (Table 10.1). To stabilise levels of carbon dioxide (CO_2) (the major greenhouse gas) in the atmosphere, there needs to be a dramatic reduction in deforestation and the consumption of fossil fuels. Evidence of climate change can be found in Australia's terrestrial and marine parks. There has been a reduction in snow cover in Kosciuszko National Park by 30% over 45 years, with the least average snow cover recorded in the last five years (Green 2003). Higher than average sea temperatures have contributed to two recent bleaching episodes on the Great Barrier Reef, estimated to have affected 50% of the reef (Hoegh-Guldberg 2003).

Snapshot 10.1

Tensions over water in southern Africa

In southern Africa, De Villiers (2001, pp. 5–6) describes how there has been tension between Namibia and Botswana (a 'water war') over access to Botswana's Okavango Delta, since 1996. Namibia, one of the driest countries in the world, proposed building a pipeline to the Okavango system. However, the people of Botswana also face water shortages and the Okavango still has not recovered from a severe drought in the 1990s. As water resources continue to dwindle the potential for heightened tensions will become more common among some nations.

Snapshot **10.2**

Water plan for the Murray–Darling Basin

The Murray-Darling Basin is the largest water catchment area in Australia. Irrigation (particularly for pasture) is the greatest user of water in Australia and the Murray-Darling Basin provides the majority of this water. The health of the Murray-Darling has long been suffering from over-exploitation, resulting in severe salinity and erosion and a reduction in native fish and plant stocks.

On 29 August 2003, when much of Australia was in drought, the Council of Australian Governments released a national water plan. A significant component of the plan was the commitment of $500 million to buy back water allocations and divert the water to environmental flows. Whilst these funds would provide for the

re-allocation of 350 gigalitres, this is a long way off the 1500 gigalitres that the Murray–Darling Basin Commission recommended to improve the health of the river system.

The Wentworth Group of Concerned Scientists (2003), an esteemed group of senior researchers, released a Blueprint for a National Water Plan several weeks prior to the government's announcement. In this document they requested an overhaul of how farmers are granted water rights and called for more water to be allocated towards the health of the environment. They proposed the creation of a water bank, and an environmental trust, which could co-ordinate the trading of water rights.

Electricity generation, South Gippsland wind farm near Toora, Victoria (Graeme Worboys)

Additional effects of global warming will include changes in species structure, in which alpine regions will be particularly susceptible as predator species encroach upon higher altitudes. Incursion of feral animals and weeds will increase in many areas as habitats change. Alteration in the foliage structure in temperate regions will likely increase fire frequency and intensity, which will affect Australia's biodiversity. The January 2003 bushfires in the Australian Alps occurred at a time when most of Australia was in severe drought and there were unprecedented by high daytime temperatures. Rainforests, typically

sites of high biodiversity, will be severely affected by increases in average temperatures, as species in these regions are sensitive to even minor temperature changes (Williams 2003).

Dependence on liquid fossil fuels is rapidly coming to a crunch point, as this geofinite resource becomes a supplier's market (Foran & Poldy 2002). We are now in the later stages of the finite age of low-cost liquid fuels, and establishing alternative energy infrastructure now, for the longer term, is a strategic and essential investment (Mason 2003) (Table 10.1).

Table 10.1	Sustainable futures and strategic considerations	
Forecast	Strategic Implications	Source
Shortage of liquid fossil fuels (2015–30): implications of demand exceeding supply for liquid fossil fuels high cost of liquid fossil fuels in the future	Vehicle fleet considerations changing from liquid fossil fuels to other energy sources Heavy plant and equipment considerations given the higher cost of diesel fuel Aircraft use considerations given the higher cost of aviation fuel Rehabilitation and construction cost considerations Increased accountability for greenhouse gas generation	Foran & Poldy (2002), Mason (2003)
Loss of snow (2030–70): forecast reduction in snow season change in the flora and fauna living in alpine and subalpine regions	Monitor feral and weed species intrusion into alpine and subalpine areas Design and construction policies to develop buildings that are adaptable to the changing environment and which have minimal impact and energy consumption Licence agreements, especially in relation to ski resorts, need to reflect climate forecasts	Green (2003)
Impacts to the Great Barrier Reef (2030–50): shift from coral to algal dominance implications for Queensland tourism and fishing industries	Policies to reduce greenhouse gas emissions Consideration of activities in the office and in the field to reduce energy consumption Adoption of renewable energy Reduce wastewater runoff	Hoegh-Guldberg (2003)

Solid waste

Waste production is increasing at alarming rates, aligned with increases in GDP and population. The capacity of many ecosystems to assimilate these wastes is already stretched. Many developing countries have limited methods of dealing with solid waste, but while developed countries have systems in place, waste generation increases with economic growth and so even these countries are having trouble coping with the quantities of waste being generated (UNEP 2002).

Mason (2003) predicted that the cumulative effects of global circumstances, such as limitations on fuel, increased population and poverty, global warming, potable water shortages, excessive waste production, and potential political instability will reach a peak in the decade from 2030.

Solar panels, Montague Island Nature Reserve, NSW (Graeme Worboys)

10.2 Sustainability and protected areas

Sustainable environmental management needs to be part of the daily operations of protected area management. Managers have a responsibility to address environmental issues, provide leadership, and be accountable to the community. Reduction in the use of fossil-based energy decreases the amount of greenhouse gases generated, consuming less water will assist in maintaining the health of our catchment and river systems, and creating less waste helps preserve our ecosystems. Figure 10.2 illustrates how the process of protected area management requires specific environmental management strategies. Each of the activities undertaken in protected areas utilises resources and emits pollutants. Protected area managers are accountable for the resources that they utilise and they have a responsibility to limit the environmental impacts of their activities.

Strategies to reduce greenhouse gases and ensure sustainability outcomes need to be developed and implemented for park management operations. An important component of this is environmental performance assessment and monitoring. Energy, water, and other resource use, waste production, and greenhouse gas emissions need to be assessed for operations. These can then be benchmarked and continual improvement systems implemented. Such 'sustainability assessment' needs to be an integral part of park management planning and operations in the future.

Sustainable development criteria need to be part of the planning, design and construction of new facilities. Issues considered include design for natural lighting, ventilation, and heating; the use of renewable energy sources, the use of recycled materials, water minimisation, recycling and retention systems; and life cycle assessments of building products to reduce the ecological footprint of a development and its continued operation.

Environmental performance reporting on a regular basis will ensure that management continues to operate at the highest sustainability standards and that protected areas assist in educating the community on sustainability principles and practices. Such performance achievements should be made publicly available.

Ensuring that sustainability management practices are incorporated into policy, planning, operations, monitoring, and reporting on performance is the responsibility of all levels of protected area organisations. De Lacy et al. (2002) established a model for addressing each of the aspects required for sustainable management. The model includes:

1   sustainability policy and planning
2   performance assessment and monitoring
3   continual performance improvement
4   sustainability reporting.

Sustainability policy and planning

A protected area organisation needs, at a whole-of-organisation systems level, to establish a policy framework based on international sustainability principles and national, state, and local environmental legislation and policies (Chapter 2).

The application of sustainability principles begins with senior management implementing core sustainability policy statements for the planning and strategic operation of protected area systems. These policies need to be integrated into all levels of plans, including corporate plans, regional plans, management plans, business plans, and individual operations plans (Chapter 7). This planning then governs all activities carried out within protected areas (Table 10.2).

Protected area management agencies need to be able to adapt to environmental changes and should be proactive in identifying problems before they occur. Every operation carried out by protected area managers needs to be assessed for its long-term sustainability. For example, liquid fossil fuel shortages have been forecast in the next 15 to 20 years. Strategies need to be developed now that will allow protected area organisations to deal with such events. Global warming is predicted to have serious consequences for some protected areas requiring strategic responses (Table 10.1).

Where, when, and how these activities are undertaken needs to be developed in strategic plans. For example, when planning road maintenance works, the timing and number of staff to be

Table 10.2	Sustainability policy and planning
Organisation strategic consideration	Protected area agency strategic action
Sustainability policy	Organisational commitment statement for sustainable management (senior manager sign-off)
	Organisational policy that external operators must meet environmental standards
Sustainable targets	Systems level sustainability targets for energy emissions, water consumption, solid waste generation, and other environmental management targets
Inputs management	Systems level purchasing policies target confirmed 'green' products and services
	Systems level designs minimise the need for inputs
Process management	Operation management processes are designed consistent with sustainability principles—systems level
Outputs management	Tasks completed and services provided meet pre-determined sustainability performance targets
	Environmental management consideration for operations that contribute to greenhouse gas generation (such as hazard reduction burning)
Outcomes management	Organisational level performance for key environmental performance areas achieved
	Organisational targets achieved in areas such as greenhouse gas generation, water consumption, solid waste, liquid waste, and habitat rehabilitation

involved, the type of transport and fuel used, and the techniques utilised all need to be considered as the resources consumed and carbon dioxide emitted will vary depending on the approach taken (Chapter 11).

Performance assessment and monitoring

The only way to ensure sustainable outputs and outcomes is to assess and systematically monitor performance. Operations need to be assessed for their environmental (and socio-economic) performance in order to determine what improvements can be made towards sustainability (Chapter 21).

Sustainability performance assessment involves an assessment of the products and services to be used for an operation, including a life cycle assessment (Backgrounder 10.1). Life cycle assessment provides a tool with which to determine the most sustainable option for alternative materials and processes, for example, a decision between wood and steel as building components.

Performance monitoring can be achieved through establishing sustainability indicators and

performance benchmarking (De Lacy et al. 2002). Monitoring should be conducted on a systematic basis and across strategic aspects of an operation, in order to determine the success of previous actions and where and how new improvements can be made. Monitoring will be guided by an organisation's information systems (Chapter 6) and its evaluation strategy (Chapter 21).

The Sustainable Tourism CRC has developed environmental management system software that facilitates performance reporting. The system uses a series of earthcheck™ indicators (Table 10.3) with relevant baseline data from which to compare the performance of individual operations.

> *Indicators are quantitative or qualitative forms of feedback that deliver concise, scientifically credible information in a manner that can be readily understood and used by various audiences to assess the achievement of goals and objectives (Schuh & Thompson 2002, p. 190).*

Indicators provide a framework to monitor, benchmark, and enhance environmental, social, and

Backgrounder 10.1 Life cycle assessments

Life cycle assessment is:

a method for assessing the biophysical and health impacts and resource consumption of a product over its entire life cycle (from raw materials to final disposal) and identifying opportunities for reducing those impacts (Higgins & Thompson 2002, p. 293).

Life cycle assessment is a specific methodology, defined within the ISO 14000 standards, for discerning the environmental impact of products and processes. As indicated by Trusty (2003), assessing the merits of standard building materials, such as wood, plastic, concrete, or steel, can be complex because consideration must be given to each of the composite materials and processes used in manufacturing. A comparative example might be comparing whether to purchase china cups or disposable cups. On the one hand, china cups utilise considerable energy and resources in manufacturing; they need to be transported from the factory to the warehouse and out to individual stores and because they are quite heavy this would involve considerable fuel consumption; once purchased they require washing after each use, consuming water and the use of potentially nutrient-rich detergents. Using an estimate per cup of coffee the resources consumed could be calculated. Alternatively, a disposable paper cup requires energy and resources for manufacturing, but they are much lighter than china cups and hence would not use as much fuel during transportation. Following each cup of coffee these cups are disposed of, contributing to waste production and pollution of the atmosphere. This simple example demonstrates that a life cycle assessment can be quite complex and even give results that differ from normal expectations.

Table 10.3	**Indicators and measures utilised in analysing accommodation providers' performance in comparison to baseline levels** (De Lacy et al. 2002)
Environmental and social performance areas	Benchmarking (earthcheck™) indicators and measures
Policy and planning	Sustainability policy: *policy in place*
Energy management	Energy consumption: *energy consumed/guest night or area under roof*
Freshwater resources	Potable water consumption: *water consumed/guest night or area under roof*
Wastewater management	Cleaning chemicals used: *biodegradables used/total chemicals used*
Waste minimisation	Solid waste production: *volume of waste landfilled/guest night or area under roof*
Ecosystem conservation	Resource conservation: *ecolabel products purchased/products purchased*
Social and cultural impact	Social commitment: *employees living within 20 km/total employees*
Optional indicators	Value of products purchased locally: *funds donated to species conservation*

economic performance. Baseline data is founded on accepted minimum performance standards compiled and averaged from a range of sources, and these are utilised to set the benchmark. Baseline performance figures will vary depending on the location, its climatic conditions, and its socio-economic status. If the average per capita performance for a number of environmental management considerations is used as a baseline performance measure, then this can be a starting point for comparative quantitative environmental performance (Table 10.4). Establishing baseline performance levels becomes a policy decision by organisations. It may vary between aspects of operations. Operators are expected to meet or exceed baseline standards and consistently improve upon these over time. Green Globe 21 (Backgrounder 10.2) is a certification scheme for the travel and tourism industry for improving sustainability performance. Green Globe 21 utilises earthcheck™ benchmarking baseline performance levels and indicators.

Performance improvement

Environmental management performance improvement targets the steady reduction of resource consumption, greenhouse gas production and waste generation in response to baseline environmental management performance criteria. There are methods to assist performance improvement. One common approach is establishing an Environmental Management System (EMS) for operations. For further information on EMS refer to Marguglio (1991), Hillary (1997), Zutshi & Sohal (2001), Thompson (2002), or ISO 14001 (Standards Australia) among others (Chapter 11). We now briefly give an

Table 10.4	Average per capita performance for environmental management parameters (Green Globe Asia Pacific 2003)				
Country	Energy use (MJ/person/day)	CO_2 emissions (kg/person/day)	Water withdrawals (L/person/day)	Solid waste (kg/person/day)	Solid waste (m³/person/day)
Australia	629	47	2299	1.9	0.00291
Canada	910	38	4447	1.3	0.00207
China	104	8	1203	1.5	0.00231
United Kingdom	443	24	438	1.3	0.00202
United States	926	54	4595	2	0.00303

Backgrounder 10.2   Green Globe 21 and earthcheck™ benchmarking indicators for tourism operators

Green Globe 21 assists tourism enterprises and communities with monitoring, improving, and reporting on their environmental management performance (De Lacy et al. 2002). Launched by the World Travel and Tourism Council in 1994 and based on the principles of Agenda 21 (WTTC 1997), Green Globe 21 is the global, environmental, benchmarking, and certification program for the travel and tourism industry. The Sustainable Tourism CRC developed the benchmarking system, earthcheck™, utilised by Green Globe 21. All operators applying for benchmarked status by supplying environmental management performance data are assessed against baseline performance standards by earthcheck™ Pty Ltd.

There are four types of Green Globe 21 registration and there are currently certification standards developed for companies, destinations, design and construct infrastructure, and ecotourism, with one being developed for tourism precincts. In order to be Benchmarked operators must address the requirements of the relevant Green Globe 21 Standard, as well as annually measure and submit information in regard to a number of benchmarking key performance indicators. To achieve certified status operators have to complete the benchmarking process and following approval at that stage they are submitted to an independent audit, to assess and ensure compliance. This is the only known travel and tourism certification system that utilises quantified environmental management performance criteria.

overview with example case studies for making performance improvements in infrastructure, energy consumption, water consumption, waste production, and social and cultural commitment.

Infrastructure design

Detailed infrastructure design can positively contribute to the achievement of sustainability objectives. As well as delivering improved environmental performance it can lower construction costs, enhance aesthetics, and lower long-term maintenance costs. There are a number of guidance documents and references (McHarg 1992, Van Der Ryn & Cowan 1995, Benyus 1998, Hyde & Law 2001). These authors also refer to the importance of learning from nature in design. By mimicking nature, buildings can be designed that fit in better with the surrounding environment, are more durable and suitable for the conditions, and operate more efficiently. Design considerations include:

- development evaluated by an environmental impact assessment process and ensuring that appropriate approvals are obtained
- design with nature principles or biomimicry (Benyus 1998)—examples of where these design principles have been utilised include WA CALM, SA Department for Environment and Heritage, and US National Parks Service
- natural lighting and low-cost, long-term maintenance requirements
- low-energy consumption operating design (Karolides 2002)
- sustainable energy sources built into the development
- use of recycled construction materials
- use of natural materials from sustainable sources, with limited environmental impacts and life cycle assessment considerations (Backgrounder 10.1)
- use of low-impact natural paints and furnishings from local sources

Case Study 10.1

Sustainable ranger housing

Robert Hughes and Ray Jones, Queensland EPA

The Queensland Environmental Protection Agency, incorporating the Queensland Parks and Wildlife Service (EPA/QPWS), provides housing for rangers living in remote and isolated national parks. Providing accommodation enables a Ranger presence to be maintained for the delivery of essential park management services. The type of housing provided varies, from shared 'barracks style' accommodation to purpose-built family residences. A challenge for EPA/QPWS is providing accommodation that meets the needs of a more socially diverse and mobile workforce, while ensuring that EPA/QPWS meets its obligations for sustainable management.

The main consideration was to keep the design as simple as possible and capable of being adapted for a range of circumstances, including terrain and climate. A number of passive (non-mechanical) and active strategies were used to ensure the buildings were comfortable, while reducing reliance on air conditioning.

The buildings have been oriented so that the living spaces have a north orientation, for the houses south of the Tropic of Capricorn and a south orientation for the houses north of the Tropic of Capricorn. This is to reduce the potential for the sun to radiate into the building in summer, yet allow winter sun to penetrate. The buildings have been designed with insulated external walls to minimise the radiation of heat into the houses during summer and retain heat during winter.

Louvered windows and moveable walls have been incorporated into the buildings for maximum ventilation and a central hallway, which acts as a breezeway, has been constructed between the building modules enhancing the cooling properties of the buildings. To counterbalance the potential for heat penetration through the windows, the houses have been designed with large roof overhangs providing further shade to the windows. Bushes and trees planted around the houses will further screen the houses from solar penetration and reduce the outside temperature in summer.

A further innovation has been the construction of a superstructure or 'fly roof' over the main building. This design system provides for a composite roof structure that utilises a well-insulated roof material as the primary surface with a fly roof above to deflect direct sunlight. The space between the two roofs is well ventilated to minimise the radiation of heat onto the main roof.

The houses were designed as modular units that were easy to assemble. This meant that they were easy to transport and could be relocated/reused, without the need for specialist skills or significant construction work. Prefabricated buildings provide substantial cost benefits, reduce building waste during construction, and minimise site disturbance during installation.

A further consideration during the design phase was the choice of materials for the housing. The brief for the housing specified:

- non-toxic materials for improved indoor air quality
- materials selected for minimum 20-year durability
- timber used to be from sustainable sources
- inclusion of energy-efficient appliances and lighting
- fire and termite resistant materials
- rainwater tanks.

The provision of appropriate remote area ranger housing, incorporating sustainable principles in design and construction, will be an ongoing challenge for protected area managers in Australia. By adopting sustainable design principles when building or renovating park housing, park management agencies are in an ideal position to promote sustainability while providing comfortable, affordable, and environmentally sustainable living conditions for rangers.

- use of non-toxic materials (e.g. zinc) where possible (Case Study 10.1).

Earthcheck™ has developed a Design and Construct Standard and Benchmarking Indicators, to ensure effective design is implemented at the initial phase of construction, maximising the environmental and economic benefits. The new standard will complement existing international assessment systems such as the Green Building Council assessment tool. Benchmarking performance criteria for design and construction include: an environmental and social policy, positioning (orientation on the north-south axis and impact on the ground), energy efficiency and conservation (use of solar panels, insulation, ventilation), building materials and processes (items with low volatile emissions, recycled materials, sourcing from local suppliers, reduce on-site waste) and protection of air, earth, and water (water recycling, evaporation of wastewater, construction processes, employing locals) (Hyde & Law 2001).

Reducing energy consumption and greenhouse gas production

It is imperative that energy consumption is reduced, especially as burning fossil fuels contributes to global warming. A major consumer of fossil fuels is transport. In Australia transport accounts for approximately 40% of our total energy use (Dunphy & Griffiths 1998). Numerous researchers (including Foran & Poldy 2002, Mason 2003) have predicted that between 2015 and 2030 there will be a significant shortage in liquid fossil fuel (such as petroleum) supplies, which implies that fuel prices are set to soar.

Protected area agencies may reduce energy consumption through practices such as:

- vehicle fleet purchasing or leasing policies, which achieve energy use reduction targets (Snapshot 10.3)
- human resource management policies, which achieve energy conservation through transport services, improved office and workplace environments, and workplace training and induction programs
- operational systems, which utilise low energy options for achieving outcomes (Chapter 11)
- use of alternative renewable energy sources.

For all other non-vehicular, fuel-consumptive machinery there are a number of renewable energy alternatives, including wind and solar power. Wind and solar power are the fastest-growing energy sources in the world. To date renewable resources only provide a small fraction of global energy production, with Germany currently the world's leader in renewable energy use (Sawin 2003). Europe is currently the centre of the wind power industry, but Australia has the research and development capability, and the climate, to lead the way in solar power. According to Dunphy & Griffiths (1998), the amount of solar energy that could be harnessed, if every house in Australia had solar panels on the roof, would supply more electricity than is currently used.

Locations that receive steady strong winds may be able to use wind turbines, subject to other

---

Snapshot 10.3

Government department introduces hybrid vehicles

The Queensland Environmental Protection Agency (EPA) is currently working on a program to introduce electric hybrid vehicles to their fleet. EPA staff drive more than 15 million kilometres for work and use almost two million litres of fuel per year. Their vehicles generate about 4700 tonnes of carbon dioxide a year (EPA 2003). The agency has been trialling the Toyota Primus electric hybrid vehicle, which utilises half the fuel and generates only half the CO_2 as a conventional car. The agency currently has 11 of these vehicles, which should prevent approximately 26 tonnes of CO_2 being emitted each year, reducing the impact that EPA activities have on the environment.

---

environmental considerations. The Australian Antarctic base at Mawson installed two wind turbines in the 2002–03 summer and they are currently supplying 65% of the station's power needs. Although Mawson receives higher than average wind speeds, wind farms in Australia typically operate at wind speeds of approximately 5–9 m/sec (Pyper 2003). Case Study 10.2 discusses sustainable energy initiatives being introduced by the Queensland Parks and Wildlife Service and Environmental Protection Agency.

Case Study 10.2

Sustainable park management

Robert Hughes and Guy Thomas, QPWS

The Queensland Environmental Protection Agency and Parks and Wildlife Service launched a sustainable parks initiative in December 2002. This initiative had the combined benefits of reducing impacts of recreational use and management, while generating cost savings in park operations. This case study looks at what has been achieved and what is being implemented, as part of the sustainable parks initiative at one park.

Moreton Island National Park is a large sand island, off Brisbane in south-east Queensland (Figure 10.3). Most of the 17,000-hectare island is national park, highly valued for nature conservation, recreation, education, and as a place of cultural significance for Aboriginal people. The island is a popular destination for four-wheel driving, weekend camping, and recreational fishing.

The Sustainable Industries Division of the EPA audited the service's sustainable practices on Moreton Island in September 2002. The report indicated significant progress had been made towards sustainability,

Figure 10.3   Moreton Island National Park

North Point

Cape Moreton

Bulwer Township

Moreton Island National Park

Park Headquarters

The Wrecks Campsite

Tangalooma

Moreton

Bay

Kooringal Township

BRISBANE

Stradbroke
Island

0   5   10   15   20
Kilometres

Produced by T. Eeles, EPA.

in several aspects of protected area management. Solar panels were installed to generate electricity for the ranger residences and workshops. These replaced the existing diesel generator, which was expensive to operate, especially when true fuel, maintenance costs, and greenhouse gas emissions were factored into the equation. Other initiatives included new hybrid toilets, rubbish transfer stations, and the use of a solar-powered water pump.

However, the measures had gone only part of the way towards achieving sustainability. The audit team recommended replacing electric kettles with stovetop whistling kettles to take advantage of the higher efficiency of gas appliances, replacing incandescent light bulbs with low wattage fluorescent tubes, replacing existing whitegoods with high energy rated equipment and operating power tools only during generator run time.

Photo voltaic array, Qld (Paul Edwards)

At recreation facilities the team recommended introducing water-wise showerheads on public showers, installing a low wattage solar-powered LED lighting system for the toilets, and converting the remaining septic toilets to hybrid systems.

Much of what has been achieved on Moreton Island has been related to retrofitting, but the EPA/QPWS recognises that major achievements will be made only when the principles of environmentally sustainable development are incorporated into the fundamental planning and development of park facilities. Managing protected areas in a sustainable way provides park management agencies with an opportunity to promote sustainability more broadly in the community. Independent audits are an opportunity to review performance and to give a sound opinion on what has been achieved and what is possible.

Emissions trading

Protected areas can benefit from emissions trading and carbon sequestration. Carbon sequestration involves increasing the amount of carbon that is being stored in soils and vegetation, often by large-scale tree planting. Under the Kyoto Protocol (Backgrounder 2.1) activities that remove carbon dioxide from the atmosphere, or 'carbon sinks', can be used to meet target commitments, and credits gained from carbon sinks can be traded. Landcare Research New Zealand has been involved in the development of a nationwide vegetation and soil carbon monitoring system (CMS) to track changes in New Zealand's forest and soil carbon (Landcare Research 2003). The carbon monitoring system was designed to monitor and assess the extent to which Indigenous forests, shrublands, and soils offset greenhouse gas emissions. Soils are the largest active carbon reservoir in New Zealand, so even small changes in land use can impact on the national CO_2 budget. The carbon monitoring system was adopted as one component of New Zealand meeting its obligations under the Kyoto Protocol.

Table 10.5	Estimates of water consumption relevant to protected areas (American Water Works Association 1996)	
User	Unit	Typical consumption (litres/unit/day)
Hotel	Guest	190
	Employee	38
Public Lavatory	User	19
Restaurant (including toilet)	Customer	34.2
Conventional	Customer	30.4
Short-order		
Camp		
Pioneer type	Person	95
With toilet and bath	Person	171
Day, with meals	Person	57
Day, without meals	Person	49
Campground	Person	114
Picnic park with flush toilets	Visitor	22.8
Visitor Centre	Visitor	19

Reducing water consumption

Water is an increasingly scarce resource and the UN predicts that by 2032 half of the world's population will be living in severely water-stressed areas (IHEI 2003). Reducing water consumption is not only about reducing intake of freshwater, it is also about limiting the amount of wastewater that is finally discharged from operations.

In the past century more than 50% of the world's wetlands have been lost (IUCN 2000b). The IUCN (2000b) predicted that by the year 2025 water consumption would increase by 50% in developing countries and 18% in developed countries. This will have enormous implications for natural ecosystems and human communities because of the finite availability of water.

Australia has the greatest percentage of land mass that is classified as arid/semi-arid (69%); it only has 1% of the world's share of freshwater and drought is a common feature of the environment. It is important that water is well managed and that this is done on a catchment basis. Protected areas are often a significant component of water catchments providing services to maintain water quality and availability. Water consumption (as exemplified for America, Table 10.5), and the amount of effluent and grey water being discharged to the sewer, needs to be reduced. Kavanagh (2002) reports on water/wastewater management and makes the following recommendations towards achieving sustainable water management.

- Wastewater should be treated and reused for all non-potable purposes (grey water and black water should be treated separately).
- Rainwater should be collected as an alternative to sourcing from the mains supply.
- Water-saving devices should be installed wherever possible.
- Dry composting toilets should be used.

These recommendations need to be utilised where possible by protected area managers and staff, but it is also crucial that all visitors to protected areas are made aware of the importance of water conservation and the benefits of utilising such systems.

Sydney Olympics recycling (Graeme Worboys)

Recycling, Point Pelee National Park, Canada (Graeme Worboys)

Solid waste production and minimisation

Australians generate an average of 1.1 tonnes of solid waste per year, which places Australia amongst the top 10 waste generators in the world (Commonwealth of Australia 2001a). Recycling and reduction campaigns can have a significant impact on resource consumption; all it takes are simple changes to work practices, such as recycling or reusing paper. A study conducted by the Centre for Design (2003) compared the life cycle impacts of using landfill or recycling for domestic waste. The results indicated that recycling resulted in significant savings in greenhouse gas emissions, smog precursors, water use, and solid waste. Other options for waste reduction include utilising worm farms to decompose organic matter, reusing packaging and containers, refusing packaging from suppliers, and eliminating waste products from operations. Ultimately, the best approach is to design products and facilities with the notion that there are no wastes and everything is utilised on-site.

Social and cultural commitment

Protected areas cannot operate in isolation. They require the support of adjacent local communities to be able to carry out best environmental practice. In addressing management issues such as visitor and fire management the park neighbours are important stakeholders. Barriers between protected areas and other properties are arbitrary and do not delineate between environmental issues. Water, fire, vegetation, and fauna can move between protected areas and adjacent land and the management of these issues on adjacent land will assist protected area managers. For example, the construction of dams and the use of fertilisers and chemicals on adjacent properties will affect the water quality in protected areas. Likewise surrounding properties need to be involved in hazard reduction burning, to assist with issues of safety, air pollution, and soil erosion. Park visitors impact on neighbours. Conversely neighbouring properties can potentially benefit from tourism businesses.

The public have an expectation that they will be adequately informed and involved in park activities and planning. Reporting to these groups is the responsibility of protected area managers (Chapters 7 and 17). Many of the protected areas have significant cultural values that need to be recognised and protected (Chapters 13 and 18). Social sustainability also includes employing local staff and purchasing (local) products from the local community wherever feasible and ensuring local businesses are given priority for contracts.

Performance reporting

A critical component of achieving sustainability is reporting on progress (De Lacy et al. 2002). Protected area managers have a responsibility to maintain transparent reporting procedures. Environmental performance reporting provides information necessary to assess whether the overall quality of the environment is improving or deteriorating. Forms of reporting include state of the environment reporting, at state, regional, and even local scale, and corporate environmental performance reporting, suitable for associate business activities. Environmental reporting has become a common requirement across all sectors of society, particularly for government agencies (Hockings et al. 2000).

Government environmental reporting

In line with Agenda 21 and the concepts of sustainable development, State of the Environment (SoE) reports are a requirement of governments at the federal, state, and, increasingly, local level. SoE reports generally follow the pressure-state-response model and report more on the condition of the area rather than evaluating processes or delivery (Moore et al. 2003). The process of SoE reporting is continually evolving, but it is now common in many countries and endorsed by organisations such as the Organisation for Economic Cooperation and Development. National SoE reports provide a mechanism for reporting on environmental issues that transcend state boundaries. Local SoE reports should involve extensive consultation with the local community to ensure buy-in by important stakeholders.

Corporate environmental reporting

Corporate Environmental Reports are being much more widely used for businesses to demonstrate their triple bottom line credentials. Corporate Environmental Reports utilise indicators and

Snapshot 10.4

Landcare Research NZ

Landcare Research NZ is an environmental research organisation specialising in sustainable management of resources for production, conservation, business, and community (Landcare Research 2003). Following its philosophy of sustaining the environment, respecting cultures, and conducting business in an ethical manner, Landcare Research NZ provides transparent reporting of its performance. Its annual report includes: our vision; guiding philosophy; performance against statement of corporate intent; science making a difference; our environmental performance; our people; verification and accountability; and financial statements.

Landcare's annual reports are driven by key performance areas that relate to the company values, stakeholder expectations, governance systems, and global issues. Each has interlinked environmental, social, and economic dimensions. The key performance areas are measured using key performance indicators. For more information see the Landcare Research Annual Report 2003: www.landcareresearch.co.nz.

company's performance. Traditionally, companies reported on their financial accounts in order to demonstrate their success to stakeholders. Now, many stakeholders expect to see figures on the company's performance in regard to social and environmental factors as well. De Lacy et al. (2002) describe the contents of triple bottom line reporting as including measures to minimise: resource, water, and energy consumption; the volume and toxicity of wastes generated; and damage to plant and animal species and habitat; and to enhance conservation measures such as land offsets and contributions to environment groups, projects, and research. Landcare Research NZ has been practising triple bottom line reporting for many years (Snapshot 10.4).

Protected area environmental reporting

Protected area managers have a responsibility to work with the local community and stakeholders and to regularly inform them of progress that is made in meeting environmental, social, and economic performance targets. The principles of triple bottom line reporting could be utilised in State of the Parks reports provided for government and public consideration. Information on environmental performance assessment, monitoring, improvement of energy, and water consumption and waste production and social commitments need to be included in these reports. Publicly reporting this information provides an example to other organisations and informs the community of the benefits of adopting sustainable practices.

enable reporting on appropriateness, efficiency, and effectiveness of program delivery (Moore et al. 2003). It is important for organisations, investors, government, and the community to have reliable reporting on a company's environmental, social, and economic performance through such triple bottom line reporting.

The triple bottom line refers to accounting for environmental, social, and economic factors, each of which is to be addressed when assessing a

10.3 Management lessons and principles

Lessons

1 Global environmental problems are directly impacting the viability and effectiveness of protected areas.
2 Protected area conservation management can directly contribute to global environmental management problems.
3 Protected area managers, with their environmental stewardship concerns, can take a leadership

role in global environmental management while undertaking their conservation practices.
4 Environmental management performance standards are critical, especially baseline standards.
5 Protected area management leadership in sustainable environmental management provides a basis for seeking equivalent performance standards from others working with parks such as tourism operators; utility companies; and road construction and maintenance authorities.

Principles

1 As well as conserving biodiversity, geodiversity and cultural heritage protected areas must ensure that they put in place absolute best practice environmental management.

2 The key for doing this is to address the major issues of energy, water, and other resource use and consequent waste production in all park operations.

3 Sustainable management needs to be incorporated in key protected area strategies and planning.

4 The focus should be on outcomes: measuring, monitoring, and improving performance.

5 Key indicators are required for whole of protected area and individual operators in protected areas, which regularly measure, benchmark, and improve upon performance.

6 Protected areas can be an important demonstration to the local community on how to implement environmental best practice.

11

Operations Management

Battles should be won before the barrage…

Sir John Monash, quoted in Palmer, National Portraits

11.1 Planning for operations 290
11.2 Operations implementation 300
11.3 Management lessons and principles 307

A large wild boar thrashing in its capture cage; the repetitive thump of a weed spray unit; the percussion of a jack-hammer on a mountain walking track; and dust rising from road maintenance works are all sights and sounds from protected area management operations. Protected areas need active and effective management to retain the values for which they were reserved. Threatening processes such as weed and feral animal invasions (Chapter 14) are undermining the conservation integrity of protected areas and these threats must be dealt with. Intervention management is often required to conserve biodiversity on small reserves through actions such as disturbance simulation, translocation of species, and species population management (Chapter 12). Nature can no longer take care of its own for many small reserves in fragmented landscapes.

Across the globe, human actions have directly and indirectly undercut the self-sustaining and naturally regenerative capacity of many ecological systems (Meffe & Carroll 1997).

Law enforcement may be needed to deal with human pressures such as unsustainable harvesting, poaching, and vandalism. Poverty contributes to these problems, and civil disorder and war can impact directly on protected areas. Protected area operations management is at the forefront of managing such issues.

Operations are essential activities and tasks that underpin the conservation management of protected areas. Managed correctly, operations directly help in achieving conservation outcomes. They are the major difference between so called 'paper parks' (legally reserved areas with no active management) and parks that are effectively managed and contributing to conservation outcomes.

Conservation management, while critically important, is only a set of tools and approaches whose usefulness and appropriateness are measured by the extent they contribute to long-term conservation of natural patterns and processes (Meffe & Carroll 1997).

Operations management is defined as the management of the productive processes that convert inputs into goods and services (Slack et al. 2001, Bartol et al. 1998). It is considered to be part of the 'Controlling' function of management (Chapter 4) because much of the emphasis is on regulating the productive processes that are critical to reaching organisational goals (Bartol et al. 1998). Protected

Road upgrading works, Ben Boyd National Park, NSW (Graeme Worboys)

area management operations are those inputs, processes, and systems that directly contribute to the achievement of conservation outcomes. There are many and varied operational activities undertaken in protected areas.

> *Management that is logically linked to long-term solutions, to stewardship of the environment can provide the critical intervention needed to conserve biodiversity (Meffe & Carroll 1997).*

Operational management must always be conducted professionally, effectively, and always in the context of the status of land or sea as a protected area. Otherwise operations can themselves become threatening processes. There have been too many instances of 'cowboys' at work in parks. Lazy research, poor planning, poor execution of works, and damaging operations should never happen. Protected areas are fast becoming the last natural lands and seas on Earth and their special status demands suitable respect and caution.

There is an extraordinarily wide scope of protected area management operations. This book describes many of these throughout its chapters on protected area management practice. Many operational case studies are provided. This chapter focuses on generic operations-planning and management, and the principles and practices described here apply wholly or in part to most operations. It has been designed to directly assist effective and responsible operations management. It describes important operational planning considerations; discusses the types of planning approvals that may be needed; encourages the development of effective operational policies and procedures; and describes a range of practical considerations that may assist with the implementation of operations. Monitoring and reporting of operations is also discussed. Prescription burning, a fire management operation, is described in detail to illustrate some operational management considerations, including safety issues. The complexity of multiple, simultaneous operations and incidents that need attention from operations staff are illustrated through a lightly fictionalised 'day in the life of a ranger'.

11.1 Planning for operations

Operations management is guided by a plan. Usually this document is for a 12-month to three-year period and is directly linked to an organisation's strategic plan, the plan of management for a protected area or areas (Chapter 7), the annual budgeting process (Chapter 9), and management effectiveness evaluation system (Chapter 21). It will include priority operations for a protected area organisation for a defined geographic area. Operations are typically action events. They are not the place to establish 'policy or procedures on the run'. The policies and procedures need to be in place. There is a clear role for organisational field management policies that are developed and made available publicly. However, operations management also needs an adaptive capability so that effective responses can be mounted to changes in circumstances. For some operations, this adaptive response may have to be very rapid.

Procedural statements for operations are also essential. Safety is one of the principal drivers for the development of procedural statements. The training courses that accompany them are just as important. The successful implementation of an operational plan will also be influenced by an institution's structure and its governance systems and procedures. Typically the operations plan will deal with many essential but individual protected area operations. Each individual operation will have its own specific plan and will have been influenced by a range of critical planning considerations. These operational planning considerations are described here. Some of the considerations are exemplified by Operation Bounceback (Case Study 11.1).

Operational planning guidance. The Australian Natural Heritage Charter (Chapter 12) and the Burra Charter (Chapter 13) provide clear guidance for protected area management operations. They are considered essential background research reading and inputs for all protected area management operations managers.

Purpose of an operation. The purpose of an operation needs to be clear and an identified strategic priority. In the case of weed and pest animal operations the purpose is protected area threat abatement with the most strategically important introduced species to be targeted first. A rationale for the threat response and the commitment of resources will normally be provided.

Operational context knowledge. An operation needs to be well researched. If it is an introduced species control program, then the biology of weeds or pest animals needs to be known and control techniques

Case Study 11.1

Operation Bounceback, South Australia

Damien Pierce, Department for Environment and Heritage, South Australia

Declaring a patch of land a conservation reserve doesn't guarantee you any conservation outcomes, and if you stood in the Flinders Ranges National Park and Gammon Ranges National Park pre-Bounceback you'd have encountered the usual suspects. Big numbers of feral goats, more than enough Euros (*Macropus robustus*), and rabbit infestation approaching 150 warrens per square kilometre in some areas all contributed to excessive total grazing pressure. Bounceback has developed from an initial feral goat-control effort in the early 1990s, to an evolving program looking at long-term ecological recovery with the following broad aims.

- Link efforts to conserve and enhance biodiversity across the Northern Flinders region.
- Restore the natural ecological processes, with particular focus on core areas of the Flinders Ranges National Park and Gammon Ranges National Park.
- Remove major threats to biodiversity and ecological integrity in the region.
- Develop and demonstrate a best-practice model of integrated ecological management.

There are four key components to Bounceback operations.

Soil erosion, Flinders Ranges National Park/Gammon Ranges National Park, SA (Damien Pierce)

Monitoring and research. Monitoring is conducted for system response and threatening processes. The majority of our monitoring is long-term based on indicators of response, such as vegetation condition, indicators of habitat quality, or the status of populations, such as the viability of Yellow-footed Rock-wallaby (*Petrogale xanthopus*) colonies. We also monitor the threats or pests themselves to gain quicker feedback such as rabbit, fox, or goat populations.

Removal of threats. Our threat abatement activities aren't necessarily radical, innovative, or spectacular. They are based on good science and delivered as an integrated package. The basics address total grazing pressure, introduced predation, and pest plants. The activities include:

- fox control (ground and aerial baiting with dried-meat 1080 baits)
- feral goat control (ground and aerial mustering and culling)
- rabbit control (warren destruction with bulldozer and explosives for follow-up)
- feral cat control (culling and trapping in targeted areas)
- kangaroo management (targeted areas where over-abundant)
- weed control (such as Wheel Cactus (*Opuntia robusta*), Prickly Pear (*Opuntia* spp.)).

Active and direct recovery. Active recovery is activity that doesn't knock out the bad but directly promotes the good. With revegetation, we are reintroducing plants or a seed source through direct seeding. Obviously this is linked heavily with our threat abatement activities to allow any recruitment to survive. It can also include reintroduction of a fauna species, and Bounceback will continue to investigate this option.

District involvement/community partnerships. The northern Flinders Ranges contains many stakeholders with differing land uses, primarily pastoral production, conservation, and tourism. Many Bounceback activities are found off-reserve with the support of landholders. Strong partnerships have been formed with members of the community and various volunteer groups.

Is it working? The short answer is yes. We're seeing signs of recovery in the form of habitat recovery, vegetation recruitment, fauna recovery, and we've arrested the decline of some key species. A major focus of ours is the Yellow-footed Rock-wallaby, and with the help of Bounceback, populations have increased throughout much of the region.

Rabbit warren, Flinders Ranges National Park/Gammon Ranges National Park, SA (Damien Pierce)

Planting saltbush, Flinders Ranges National Park/Gammon Ranges National Park, SA (Damien Pierce)

designed to match these characteristics. If it is a fire management operation such as prescription burning, for example, then fire behaviour characteristics for the treatment area need to be known, as should the climatic 'burning envelope' that best suits the nature of the fire behaviour and fuel reduction sought. If it is the provision of a hardened walking track surface, then research into suitable materials types needs to have been completed.

Too often this background research has either not been done or has been poorly completed. It is a major mistake, and has led to the wasting of precious dollars available for conservation operations. Even worse, it may compound the problems that originated the conservation action response. Operations managers need to obtain the best information possible in designing an operation. This may mean involving researchers, technical experts, and local community experts. It may mean commissioning special reports that will aid the design of an operation.

Previous operations knowledge. Repeating operational mistakes of the past (or 'reinventing the wheel') is a common problem for protected area management and a lot of time and money is wasted because of this. The history of operations needs to be researched and understood. Given that operations have been adequately planned and documented, this information should be available. Unwritten history and local knowledge can assist with gaps in knowledge. Interviewing, recording, and involving previous managers, researchers, practitioners, local experts, user groups, and stakeholders in providing background information and insights is very valuable. In addition, such a group will usually be very interested in maintaining contact and involvement with an operation. It can be a good way of positively involving the community with continued operations within the protected area.

Environmental impact assessment. An assessment of environmental impact of an operation is important and may be a legal requirement. The Commonwealth *Environmental Protection and Biodiversity Conservation Act 1999* has requirements, as does state and territory legislation. An assessment may be brief given a simple operation, but it needs to be completed consistent with standard internal procedures. For repetitive or maintenance operations, it may only be needed in the first instance. Social considerations form an important part of this planning.

Environmental management systems (EMS). Operational systems such as an EMS may provide a formal systems framework for managers managing for operations. This may be recognised in an environmental impact assessment (Thompson 2002). One well-known system, ISO 14001, the international standard for environmental management systems, defines an EMS as 'the part of the overall management system which includes organisational structure, planning, activities, responsibilities, procedures, processes and resources for developing, implementing, achieving, reviewing and maintaining the environmental policy' (Thompson 2002). The ISO 14001 framework may not be suitable for all operations, but it may influence how individual operation plans are developed. Planning, training, procedures, systems, and monitoring are important aspects of an EMS. Monitoring and performance evaluation (Chapter 21) are critical parts of the 'plan-do-monitor-revise-act' continuous improvement cycle an EMS promotes. NSW P&W has introduced an EMS for the Perisher Range Resorts in Kosciuszko National Park.

An EMS is a way of organising our behaviour to make sure that our activities harm the environment as little as possible. As such, it's not actually about managing the environment, rather it is about managing our activities in such a way as to protect the environment from the unintended effects of our actions, and to minimise the negative effects where we do intend to impact the environment (NSW NPWS 2004b).

Planning approvals. There may be a range of approvals required by law, by organisational policy, and by procedures prior to an operation proceeding. Managers will need to account for the 'time out' needed for planning approvals in their operational timetable (Table 11.1).

Sustainability assessment. Protected area management consumes non-renewable resources and causes environmental impacts in its pursuit of conservation outcomes. A holistic approach to environmental management of protected areas is both responsible and desired (Chapter 10). Protected areas are being impacted by the effects of global warming and other global environmental issues. Protected area management operations need to minimise any potential contributions to these global problems.

In the future, internal protected area organisation procedures are anticipated to include sustainability assessment standards for operations. This is most likely to be a benchmarking system

Table 11.1	Examples of operational planning approvals	
Protected area operation	Example of approval required	Approving institution
Major construction or development	Satisfactory environmental impact statement (EIS)	Typically approved by a state or territory government department
(Large) building construction	Management plan EIS or equivalent Building approval	Determining authority Local government
Visitor access improvements	Management plan	Protected area agency
Visitor facilities	Management plan	Protected area agency
Maintenance operations	Operations plan Review of Environmental Factors	Protected area agency
Fire operations	Fire plan Review of Environmental Factors	Protected area agency
Pest animal operations	Pest Animal Plan; Review of Environmental Factors	Protected area organisation

based on quantitative performance. With such a system and use of non-renewable energy (for example) operation approvals may be contingent upon a proposed operation meeting benchmarked baseline sustainability performance standards. Administration systems supporting operations would track actual environmental operational performance against predetermined 'best practice' standards. This information may also be required for future triple bottom line reporting requirements or strategic environmental performance targets set by protected area organisations (Chapter 10). Sustainability approvals for protected area operations may be required for:

- sustainable design of new structures and built facilities for protected areas
- operations in which considerable non-renewable fossil fuels will be used and where substantial greenhouse gases will be generated
- operations that will consume large volumes of water
- operations that will generate solid and liquid waste.

Budget approval. A budget is needed to undertake an operation. Securing budget approval through institutional systems takes time and human energy and this time factor and workload needs to be planned for. Sometimes the budget is approved out of synchrony with planning processes and money is

available but the design is unfinished or the environmental impact assessment needs to be completed. This is where pragmatism can readily replace responsible planning systems. A philosophy of 'the money is here, let's get on with the job' can place enormous pressure on operations managers to proceed despite unfinished or rushed planning. Operations managers need to have the discipline to correctly follow procedures. They also need to anticipate a potential lack of coordination between planning requirements and the budget process from time to time.

Operations schedule. Time and event planning is a critical part of operations management. Given that all operational approvals have been achieved, an operation now needs to be organised so that it is implemented efficiently. Most operations use what is described as a Gant Chart to help achieve this. Gant Charts are used to monitor progress through identification of the range of actions for an operation and the sequencing of those actions (Slack et al. 2001).

Staff competency and capacity. Staff need to be competent to manage an operation (Chapter 4) and there needs to be sufficient staff to meet the forecast work loads and the number of shifts needed (capacity). Specialist operational skills may be required for certain operations (Table 11.2) and such skills need to be recognised in job descriptions, interviews and in

Table 11.2	Selected competency qualifications required for protected area operations staff	
Operational Task	Staff Position	Competency/Proficiency
Chainsaw work	Field Officer	Chain Saw Operation Competency Certificate
Earth-moving	Plant Operator	Plant Operator Certificate
Large truck use	Driver/Operator	Class 5 Licence: Large Trucks
Use of explosives	Field Officer/Powder Monkey	Licence to handle explosives
Air operations	Pilot	Pilot's licence
Air transport	Crewman	Crewman's Certificate
Marine (boat) operations	Field Officer	Coxswain's Certificate
Underwater operations	Diver	Diver's Certificate
Fauna transfer	Veterinary	Veterinary Degree
First aid	First Aider	First Aid Certificate

statements of accountability. On-the-job induction and training are critical for achieving effective and efficient operations.

Operation logistics planning

Operations planning requires, among other things, consideration of a number of important logistical matters.

Transport. Access to and from a workplace will be required. Protected areas may require work in remote locations that could involve aircraft, boats, four-wheel-drive vehicles, or even horses or other animals as access transport.

Accommodation. It may be cost-effective or more practical for operations staff to be accommodated at the actual operations site during a working week. Through workplace agreements of mutual benefit to employer and staff, longer day shifts and shorter working weeks are sometimes negotiated for working at remote and difficult-to-access sites. Accommodation at the work site that meets workplace standards; sanitation; catering; communications; entertainment; first aid; safety; and emergency evacuation are some logistical considerations that will need to be addressed.

Equipment. Operations staff will need to be equipped with the correct personal safety equipment. They will need gloves, masks, and eye and other protective gear for dealing with poisons; they will need ear protection gear for noisy machinery; they will need hard hats, gloves, boots, and safety glasses for construction sites, and other safety equipment. All of the correct safety gear will need to be in place prior to an operation starting. Communications equipment and a working communications system will be essential for safety and logistics.

Evacuation. An emergency evacuation plan should be prepared for any potential medical or other emergency that could occur at an operations site. Such a plan would have endorsement from local medical and emergency authorities. Rehearsing 'mock emergencies' with the multiple authorities involved in such operations is recommended.

Shelter. Poor weather, poor visibility, extreme winds, extreme temperatures, fires, and heavy rain and snow may be encountered during operations in protected areas. The provision of suitable on-site shelter facilities is an important planning consideration.

Environmental impacts. There are many potential operational environmental impacts that need to be managed. Many relate to the quality of supervision, training, and leadership. Training of operational staff and leadership including setting the required standards and demonstrating the required procedures on-site are a critical part of an operation. Training and supervision is especially important if the work is delegated to non-protected area contractual staff with a non-environmental management background. A number of questions have been prepared to provide guidance for operational managers.

- How will the environmental guidelines for the operation be supervised and monitored?
- How will solid, liquid, and gaseous waste impacts generated by operations be minimised?
- What measures will be used by the operation to minimise fossil fuel use?
- What types of fossil fuels or other energy sources will be used?
- What amount of greenhouse gases will be generated?
- Will there be impacts from dust and noise and how will these be dealt with?
- Is there any potential for operational impacts to streams and water-bodies and how will these be dealt with (for example, pH impacts from concreting; petroleum impacts from refueling spills)?
- How will the application of herbicide and pesticide chemicals be kept from impacting streams, water bodies, karst systems, and other non-target effects?
- How will noise impacts be minimised for wildlife, for staff, and for visitors?
- Has all equipment been effectively cleaned and sterilised? This includes transport, heavy plant, and earth-moving equipment. Soil transfer (with potential soil pathogens) and seed transfer on equipment is a major potential problem.
- How will the actual disturbance area of an operation be minimised?

Safety considerations. Safety is paramount, both for operational staff and for those who may be affected by an operation. Operational safety planning questions that may need to considered by the operations manager include the following.

- Do key operational personnel have the correct qualifications and the right amount of suitable experience to undertake the proposed operations?
- Is the correct operational equipment available, and is it in a satisfactory maintenance state to be used?
- Are there any hazards on site that need to be considered (such as unsafe trees; unstable geological features; or dangerous animals)?
- Are there hazards associated with an operation that need to be managed (such as helicopter operations; use of explosives; use of welding equipment; use of cutting and grinding equipment)?
- Are any weapons to be used for animal culling or tranquillising?
- How is public safety to be managed during operations? What public information will be provided?
- Is there a chance that an operation will cause an incident (such as a fire or a pollution event)? What precautionary action is required to deal with this?

Risk management

Risk is defined as 'something happening that will have an impact upon objectives' by Standards Australia, and risk management is defined as the 'culture, processes and structures that are directed towards the effective management of potential opportunities and adverse risks' (AS/NZS 4360:1999 cited in Yates 2003). Risk can be quantified as a function of two interlocking parameters, likelihood; the probability or frequency of an occurrence; and consequence; the outcome or impact of that occurrence. Risk management planning is often closely linked with environmental management systems. A risk profile may be developed for an organisation after completing a risk management evaluation process (Yates 2003). The risk management process typically involves the following steps.

1 *Establish the risk review context*. Set objectives and goals of the risk review process; establish a structured approach including an overall review plan with roles, responsibilities, and deliverables; establish a communications strategy/plan for the process; set criteria against which risk will be evaluated; and document the entire process thereby leaving an audit trail.

2 *Identify risks*. Define all potential areas of risk, their source and impact, including people/processes impacted upon; use a team workshop approach involving 'what if' scenarios to identify potential risks and consequences.

3 *Analyse risks*. Build a risk profile (register); determine existing controls and analyse risks in the context of these controls. Prepare a risk matrix to both rate and prioritise all identified risks, where Risk = function of (Likelihood + Consequence).

4 *Evaluate risks*. Evaluate risks against management's risk acceptance criteria (political, financial, legal, environmental, and social). From this assessment, each risk is either accepted or rejected and prioritised mitigation plans can be established.

5 *Treat risks*. Review each unacceptably high risk and identify potential treatment options (avoidance, reduction, transfer, retention) per risk. Prepare risk mitigation plans for each risk and implement prioritised action plans (actions, responsibility and resourcing, timing, priority). Review, report, and follow up action plans regularly.

6 *Monitor and review*. Monitor all steps in the risk management process and continuously review and improve the process, drawing upon stakeholder feedback and results of action plans.

7 *Communicate and consult*. Continually communicate and consult with both internal and external stakeholders as appropriate at each stage of the risk management process.

Design and materials

Sustainability and environmental impact considerations should provide guidance to the selection and purchase of materials and the design of structures and facilities. Immediate cost may be important, but planning considerations should have regard to both short-term and long-term considerations.

Environmental considerations. An operations manager must pay attention to detail when materials are being purchased. Has the purchasing officer read the Environmental Impact Assessment (EIA)? Is the EIA or equivalent document detailed enough to provide material specifications? Some materials contain chemicals toxic to the environment. Zinc in galvanising; copper and arsenic (and/or other chemicals) in treated timber may be toxic to native plants and aquatic organisms. They are just some examples of materials that will need to be used with care, or in some circumstances not used at all.

Sustainability considerations. Recycled timber and building materials; plantation timbers; weed-free gravel from environmentally approved sources; materials that have been assessed for their life cycle suitability (Thompson 2002) are all examples of sustainability-influenced purchasing. Preferential suppliers for these goods and services would be locals with environmental credentials. Operations managers will need to monitor material purchases to ensure that sustainability requirements are implemented.

Maintenance considerations. Many protected area operations are very costly. They may be remote, require intensive human involvement, and may require costly transport. High capital cost structures and facilities with low long-term maintenance costs may be compelling material choices for many sites. The steel mesh walkway constructed between Thredbo Top Station and Rawson's Pass in Kosciuszko National Park was built based on such a rationale (Case Study 11.2). It has proved to be a success. Long-term repetitive maintenance costs should always form part of the initial costings and materials type decision-making.

Communication and liaison

People are interested in operations in protected areas. This may be for professional or just plain personal or neighbourly interest. At times, for operational logistic reasons, it is essential for people to know exactly what is happening. A range of consultative arrange-

Case Study 11.2

Walking track materials research, Kosciuszko National Park

Catherine Pickering and Wendy Hill, Griffith University

The managers of Kosciuszko National Park face the challenge of selecting the most economic, socially and environmentally appropriate type of walking track, particularly for the increasing number of summer tourists visiting the alpine region. When selecting track types, managers often use the direct cost of track installation, as it is often easy to calculate bases for comparing alternatives. However, if environmental impacts of track surface and the social preferences of users are also considered, then different decisions may be made.

Two options that have been used in the Kosciuszko alpine area are hardened surfaces such as gravel (often used in conjunction with plastic webbing), and a raised walkway constructed from ungalvanised 'cut-corrugated grip' steel mesh. Although the construction costs (per metre) of the raised steel mesh walkway may be greater than the gravel surface (Harrigan 2001), the ongoing costs of maintenance and rehabilitation are much lower, as the steel mesh walkway has little (if any) impact on the flora, fauna, and soils. The raised steel mesh walkway does not require the construction of drainage systems that can add dramatically to the cost of a gravel track. A raised steel-mesh walkway was constructed in the Kosciuszko alpine area in the late 1980s to alleviate multiple tracking and severe erosion, and to prevent further damage to the vegetation associated with the previous unformed track (Worboys & Pickering 2002). It is the most popular track in the alpine area, with approximately 51,500 people per year using sections of this route that leads to the summit of Mt Kosciuszko.

The raised walkway has few negative impacts on vegetation (Pickering et al. 2003). While there was limited disturbance to vegetation, soils, and drainage during its construction (Worboys & Pickering 2002) there is now complete native vegetation cover under the track and on the verge because of the light penetrating properties of steel-mesh, and because its construction did not require removal of vegetation. The raised walkway does not alter drainage patterns, nor provide habitat for weeds either under the track or on the edge of the track. If the track ever needs to be removed, this would simply mean removal of the metal sheets. The steel mesh surface has some limitations including: damage by snow loading that requires the replacement of short track sections, and some walker-caused surface polishing of the steel making it potentially slippery. In some steep sections stone steps have been used instead (Harrigan 2001).

Gravel is the other main track type in the alpine area. Some of the gravel tracks replaced existing bridle trails or roads, while, in other areas, gravel with plastic webbing is being used to replace informal tracks. Large amounts of new surface material are imported during construction, and this, together with the use of heavy machinery, damages vegetation, and alters soil and drainage conditions. In addition to the absence of vegetation on the track surface, gravel tracks have a distinct verge, 20–40% of which is covered by weeds. The cost of controlling weeds along tracks and roads in the Australian Alps is high, and has limited success. The lack of vegetation on the surface of the gravel tracks together with their extensive verge can result in an 11- to 20-fold reduction in native vegetation covered area compared with existing unhardened tracks.

Removal of gravel tracks is expensive, and the restoration success of vegetation and soils is likely

Dr Catherine Pickering inspecting gravel walking track, Kosciuszko National Park, NSW (Graeme Worboys)

to be very limited. The cost of just replanting with native species is over \$120 per metre^2 (Johnston 1998). The success of previous rehabilitation work on an informal/unformed track in the Kosciuszko alpine area has been found to be very limited (Scherrer & Pickering 2002).

Visitors also have opinions about track surfaces. For some, a raised steel-mesh walkway may seem inappropriate in a natural environment. However, among track-users surveyed in the Kosciuszko alpine area (over 3000 respondents in the summer of 2000) the steel-mesh walkway was ranked as the most popular track type (average 4.8 out of possible score of 5), compared to gravel with plastic webbing (2.5) (Johnston & Pickering 2001). Also, visitors observed during the survey very rarely left the elevated steel mesh walkway while walkers on hard gravel surfaces very often moved onto the softer vegetated verges. This potentially contributed to the bare areas and weed mats found on gravel track verges.

When environmental impacts and social preferences are taken into account, it is clear that the raised steel-mesh walkway has many advantages over gravel tracks. It also highlights that the decisions protected area managers make should be based on the full costing of the track over its lifetime, including costs of track maintenance, track removal, and ultimately complete rehabilitation of the landscape.

ments can be put in place depending on the nature of the operation. It is basic courtesy (and good publicity) to advise neighbours, the local community, and stakeholders of operations. It is also essential for cooperative ventures, or for operations such as prescription burning. At the site of an operation, a simple temporary sign is all that is needed. A cheap, plastic-coated (weather-proof) sheet of thick paper with the nature, purpose, and timing of an operation described and securely attached to a temporary display stake works well. People are always interested in what is happening. It is so easy to inform them. Other useful methods include:

*   press releases and other media announcements
*   a newsletter for neighbours and the community that describes the nature and the purpose of the operation
*   a web site describing operations for a protected area that are underway
*   formal liaison meetings with local communities, users, authorities, unions, contractors, and staff.

11.2 Operations implementation

Operations, once commenced, can gain a momentum all of their own. Leadership is required during implementation and this leadership should have regard for the personal safety, needs, and circumstances of the operations team, the project timetable and the scheduling of events to achieve an efficient and effective operation, media management, budget management, meeting reporting requirements, political needs, and local community needs. Effective rostering will ensure that staff needed for long-term tasks are adequately and regularly rested. The operation manager would also have a relieving officer. 'Burn out' of operations managers through lack of rest leave or unwillingness to delegate is unnecessary and constitutes bad organisational management. It also potentially threatens an operation.

Reporting on the progress of operations is both routine and important. Finance reports ensure that the expenditures match estimates and that the project is on target financially. Operational progress reports track progress of an operation against the planned project milestones and outcomes. Sustainability reporting provides environmental management performance information. Staff reporting ensures that payments, leave entitlements, and rostering arrangements are all current. Such reporting also allows a measure of the efficiency and effectiveness of an operation (Chapter 21). There is also a need to adequately document an operation for the historical record at its completion. Backgrounders 11.1 and 11.2 exemplify many aspects of operations planning and implementation with respect to prescription burning. Backgrounder 11.3 gives a typical 'day in the life' of a protected area ranger, including many and various operational matters that must be addressed.

Backgrounder 11.1   Managing a prescription burning operation

---

Prescription burning is also known as burning off, fuel reduction burning, broad area burning, control burning, property protection burning, or hazard reduction burning. The primary aim of prescription burning is to reduce the fuel available to fires prior to the summer fire season. It is undertaken as one of a number of precautions to deal with wildfire incidents and is usually carried out near property. It is also undertaken for ecological reasons.

Before the fire season begins, operation managers may seek to reduce forest (or other natural vegetation) fuel levels for designated areas. Reduced fuel levels reduce the intensity of fire (for a range of fire weather behaviour) and provide greater opportunity to control wildfires. While this does not solve all wildfire control situations, it does provide an advantage in many instances. In undertaking fuel reduction programs by prescription burning, operation managers typically work with local volunteers, councils, and rural fire authorities. There are other effective ways to reduce fuel prior to the fire season. These include tractor-slashing of long grass, mechanical mulching of plant growth (using a tractor and mechanical 'tritter'), and regular clean-ups of fuel such as fallen branches, in cooperation with neighbours or volunteer groups.

Prescription burning is not a panacea for all types of fires. There are many cases, and not only in exceptional years, where wildfires have burnt through areas that had been fuel-reduced only a few months before, or where there had been an intense wildfire the previous fire season.

There is a certain window of weather patterns and conditions within which prescription burning gives firefighters an advantage. Yet there are extremes of temperature and wind-speed where fires are almost uncontrollable, even when there is little fuel on the ground. Forest crown fires are a prime example. Fires may also sweep across closely grazed lands when flammable wind-blown matter acts like a blast furnace of embers spreading the fire.

The ecological effects of prescription burning also need to be considered. Good (1982) studied the effects of prescription burning on subalpine vegetation communities in Kosciuszko National Park. He found that fuel loads actually fell in a fire-free environment as vegetation recovered from the regular burning and grazing imposed by past land-use practices. The shrubby communities with high fuel loads were selectively favoured by regular prescription burns. Left alone, the original natural grassland and herbfield communities, which have lower fuel loads, will in time displace the shrubby communities. A regime of no prescription burning meant a reduction of fuel loads with time. Regular burning may also favour fire-tolerant species, some of which may in turn encourage fires to spread because of their flammable nature. Within a forest, an understorey of fire-tolerant shrubs, for example, may provide light, aerated, free-standing fuel that produces a quick-spreading fire, even if not very hot. Managers must decide if this is desirable.

Alan McArthur pioneered the concept of prescription burning in Australia. His leaflet (McArthur 1962) became an important guide. He advised that if fuel loads were generally below 12 tonnes per hectare, this would benefit firefighters. He stated:

> The aim of area control burning is to reduce fuel accumulation to a level where wildfires will not attain excessive rates of spread or intensities and where, if a fire occurs, suppression forces can attack and control the fires with reasonable ease. One of the essential aims of control burning is to keep possible damage to plants and soil values to an absolute minimum. This can be readily achieved by burning under prescribed conditions which ensures that an unburnt or incompletely consumed layer of fuel remains after the burn (McArthur 1962).

Many debates have followed McArthur's work. Scientists in the 1960s, for example, realised that such burning might leave water catchments stripped of protection. Gilmore (1965) argued that a mosaic pattern of burning should be undertaken to protect soils from erosion. He recommended that burns should leave average fuel loads of 8 to 10 tonnes per hectare (and higher in some plant communities) in the catchments they studied. Today such burns are (or should be) carefully planned and managed. Fuel loads are measured in advance, and computer models simulate likely fire behaviour for given sets of climatic conditions.

Undertaking a prescription burn operation

The operational steps involved in planning and undertaking a typical prescription burn include considerations provided below. Reference should also be made to additional details recommended by the Coroner who conducted the 2001 Ku-ring-gai Chase National Park hearings (Backgrounder 11.2).

Initial planning. A prescription burning plan is devised for an area that reconciles protection of property with conservation needs and cultural heritage protection. This is completed well in advance. Local communities, brigades, and stakeholders are involved in the planning work. The plan, including its perimeters, safety considerations, lighting up strategies, and burning prescriptions is formally approved by a senior, authorised officer.

Preparation. Boundaries are prepared for the fire (fire trails, walking tracks, streams that will hold a fire line etc). Where these do not exist, 'rakehoe' or hand tools are used to construct fire control lines. Fire trails may be upgraded. Boundaries are made more secure. For instance, fuel is cleared from around trees in case they catch fire and fall across the control line or drop burning embers across it. This work can be done some time before the burn. Safety will be a prime concern. Escape routes will be pre-planned, investigated on the ground, and provided accurately on maps as information for fire fighters. Ecologically sensitive areas may need to be identified and excluded by control lines. Fuel loads are assessed in detail for the entire burn area.

Operational planning. The fire's likely behaviour is modelled using computer simulations for that precise area ('the burning block'); or at worst, the 'McArthur fire meters' for forest and grassland are used (Luke & McArthur 1978, Cheney & Sullivan 1997). The operation managers will know how they want the fire to behave, and from this they can deduce the window of weather conditions they will need.

Setting a date. A band of possible dates is set in advance for which it is predicted that suitable weather conditions will prevail. Usually such burns involve local fire brigades and neighbours, so everyone's timetable has to be considered. This will also allow time for equipment to be tested and maintenance undertaken.

Division of tasks. Typically, the operation is divided up—different groups have different duties and work in different sectors. All fire fighting personnel will have received basic training and will know incident control systems. One team may be dropping incendiaries from a helicopter while a ground crew (supported by a firetanker) sets fires elsewhere at precise points along the established control lines. The operation may also be subdivided into time stages. The organisation that initiates the overall plan will probably be responsible for mopping up after the burn-off and also for patrolling the area for the following days until it is declared safe. Written briefing information (including maps and the lighting-up plan) is prepared, and a series of briefing sessions are completed for personnel. Pre-burn on-site familiarisation inspections will be completed.

The lighting-up plan. This involves deciding exactly where and in what order fires will be lit. The correct sequence matters; it determines what burns first, and in what direction fires can move. Fires can be volatile when moving up steep slopes, so these areas are often lit from above. The fires then burn downwards to control lines (where possible) at the base of hills, or into pre-burnt areas. Humidity also affects the behaviour of fires (Cheney & Sullivan 1997), so 'time-of-day' lighting up can help prevent hot burns in sensitive areas, especially if incendiary capsules dropped from a helicopter are precisely placed. This is very useful for getting a 'black edge' for rainforest gullies and stream-side vegetation communities, but one has to calculate precisely. The same technique is often used in remote areas for 'steering' fires.

Informing the public. Neighbours should be advised, and the burn pre-publicised through media such as local papers and on radio so that people are not alarmed when it occurs.

Property protection burn, Blue Mountains National Park, NSW (Graeme Worboys)

Parks field officer backburning, Blue Mountains National Park, NSW (Graeme Worboys)

Arrangements to suppress fire. There have to be enough suitable persons and equipment on standby in case the weather changes without warning. Helicopters that can undertake water bucketing may be needed. They can help crews to hold the more difficult fire-control lines.

Forward control arrangements. A command structure and an agreed system of communications among all parties needs to be in place. So too do arrangements for rescue or evacuation of persons who are injured or in danger. The AIIMS Incident Control System would normally be used, making this excellent training for a wildfire (Chapter 15). A forward control base would be established, with normal incident control procedures for logging communications and for monitoring crews despatch and placement.

Helicopter planning. Aerial incendiary work needs to be precisely managed, with the navigator and bombardier operating as a highly planned and organised team. Incendiary capsules must be placed at intervals that minimise adverse fire behaviour.

Positioning crews. Trained and equipped fire control crews need to be in position to help prevent fires from getting past control lines. Written briefing information will have been provided and briefings conducted so that all personnel are aware of the operation to take place, their role, and safety arrangements including emergency exit routes.

Surveillance. Some crews need to be on standby to patrol the fire overnight. The next day aerial surveillance will be needed to survey the burn, and perhaps to water-bomb flames that persist. Operation managers must make sure the burn is successful, safe, and complete.

Report. A report of the fire operation and the estimated fuel reduction will be prepared and circulated. Media releases of the burn may be developed and released. Computer fire records for the area are updated.

Backgrounder 11.2	Prescription burning tragedy, Ku-ring-gai Chase National Park, NSW

On Thursday 8 June 2000, a routine hazard reduction burn in Ku-ring-gai Chase National Park, north of Sydney, led to the tragic deaths of four NSW NPWS staff and the serious injury of three others. Their aim was to reduce the risk of bushfire damage to nearby residential properties. A full coronial investigation was completed to determine the cause of the incident, and the Coroner gave her finding on 16 July 2001 (NSW NPWS 2004a). As a mark of respect to the officers who lost their lives and for those who were injured, we have reproduced the Coroner's recommendations (NSW NPWS 2004c) here to help ensure they are communicated to future operations personnel who may undertake prescription burns.

Recommendation 1. That hazard reduction burns/prescribed burns on NPWS lands not be undertaken before the plans for such are reviewed and approved by persons qualified in such burns.

Recommendation 2. That no hazard reduction be undertaken in any area where ground crews are to be utilised without such an area being inspected and ground truthed to ascertain safety areas, exits, potential hazards etc.

Recommendation 3. Persons undertaking duties at prescribed burns should be totally familiar with the incident control system and the relevant duties ascribed to positions under that system.

Recommendation 4. That usage of titles or terms not identified under the NPWS Incident Control System no longer be used within the NPWS in relation to fires.

Recommendation 5. That all persons assigned to attend a prescribed burn be notified well in advance of such burn to permit their attending the fireground with suitable safety equipment and sustenance.

Recommendation 6. That a full briefing be carried out with all persons who are to undertake a prescribed burn. Such briefing is to include topographical features, safety areas, exit points, and other relevant features in accordance with the NPWS Incident Control System. All members of the crew are to be given a relevant map and encouraged to seek any information from those carrying out the briefing that may be relevant to their own safety.

Recommendation 7. A senior officer should check and verify that occupational health and safety issues have been assessed prior to prescribed burns being undertaken.

Recommendation 8. At every prescribed burn, an effective control centre is to be established which is staffed by a person who has an advanced First Aid qualification and relevant medical equipment; has the capacity to communicate with the fireground and senior officers; has been briefed as to the burn; and has been supplied with a list of all personnel involved in the burn and their ascribed roles.

Recommendation 9. That the NPWS develop or assist in the development of a suitable method of testing drip torches to ensure that they are manufactured to such a standard to withstand such heat that they do not become dangerous during a fire.

Recommendation 10. That the NPWS has available to the Incident Controller for any burn, information as to the level of training undertaken by crew members who are to undertake a burn and to allocate to each person a more qualified officer to act as their mentor.

Recommendation 11. That no person be permitted to enter a fireground unless suitably attired.

Recommendation 12. To this end, all recruits should be issued and trained as to the care and maintenance of new and appropriately sized fire clothing including two-piece proban treated suits and undershirts, fire resistant footwear, goggles, masks, gloves, and helmet.

Recommendation 13. Consideration be given to personnel on the fireground to be issued with a personal and portable fire protection blanket.

Recommendation 14. That the NPWS stresses to staff that safety of personnel is paramount at all times. Should any one person undertaking prescribed burns or any burn be concerned as to any aspects of safety, they are to be encouraged to bring this to the attention of those who are in authority. No burn is to be undertaken until the concern raised has been considered or addressed at the highest relevant level.

Recommendation 15. That a review be undertaken of the NPWS communication equipment and the effectiveness of such equipment on firegrounds, including the efficacy of a fire relay base, be considered.

Recommendation 16. That there be kept at all times during the prescribed burn or hazard reduction a log of radio and other communications to indicate actions taken on the fireground so that they can later be considered and assessed.

Recommendation 17. That the NPWS assist other fire fighting bodies in formulating a burn guide for the Sydney basin.

Backgrounder 11.3 A day in the life of a ranger

The many demands placed each day on a ranger are vividly suggested in this lightly fictionalised account by former NSW NPWS rangers Rosemary Black (now Charles Sturt University), and Janet Mackay.

My working day generally starts around 8 a.m. Today, apart from a short budget meeting, I have nothing scheduled until 2 p.m. so I tackle my in-tray, which is piled high with papers and correspondence. My priority is commenting on the amendments to the park's plan of management. As I am responsible for a specific part of the park, I focus on those report sections that relate to that area. Most of the amendments are minor, but it's still important that I check it for accuracy and consistency, and ensure that everything proposed is achievable and realistic. I type up a short report of my comments for the district manager.

I then check the rest of my mail. I am pleased to see a letter from the local Rotary club thanking me for the talk I presented last week. It is also nice to see that the local Landcare meeting is still planned for next month and that they will be taking the opportunity to inspect our weed control program.

I'm interrupted by a call from John, one of our law-enforcement officers at head office, letting me know that a case is coming to court at the end of the month. In this instance it involves a man who had brought a dog into the park. I'd warned him on a couple of occasions, but he kept bringing his dog, so a few months ago I booked him. He has refused to pay the fine and so it's going to court, which will require me to attend.

The next job on my list is the eco-tourism grant application, which has to be submitted by the end of the month. We are applying for $150,000 to upgrade and extend a walking track and develop a small picnic area. I haven't done one of these applications before, so I ring the National Office of Tourism to check a few details. The grant application is fairly lengthy, requiring details such as budget, program aims and objectives, benefits of the proposal, and how it links into other regional tourism facilities. I start drafting out some of the sections, but am interrupted by a phone call from Jenny at the ministerial liaison office. She tells me that the minister is planning to come down to the town for a public forum on local nature-based tourism. I knew that the tourism committee had planned the forum, but it is a surprise to learn that the minister is planning to attend it. Jenny explains that she will need a draft itinerary for the minister's visit by the end of the week. Briefing notes for this will be required later.

The phone rings again, this time transferred from the visitor centre. It concerns my section of the park. It's an irate visitor who is camping in the park, and has been disturbed by trail bikes during the night. Apparently the bikes had been in the camping area, despite the fact that they're not allowed there. I take some details from the caller and tell her that I'll send out one of the field staff to the camping area this afternoon, to meet her and check out the situation.

Left over from yesterday is an application for a business operator wishing to get a licence to carry out white-water rafting on the Murray River separating NSW from Victoria. I know from past experience that to deal with this type of licence will require a considerable amount of time. I'll need to deal with head office and the regional office, and also organise a meeting with the operator, and with my Victorian counterpart. Revenue generation from commercial licences has become an important issue as it enables the district to supplement the funds to construct visitor facilities. While the district has a system for processing licences, I have been the region's representative on a state-wide forum reviewing licensing across the state. The aim is to make it more consistent and less confusing for commercial operators. This has meant going to three-monthly meetings in Sydney where staff from across the state meet with head office staff, as well as representatives of the tourism industry, to identify and resolve some of the issues.

I notice from the application that the operator has been conducting his business elsewhere in the state, so I'll need to talk to other districts to see if his operation has conformed with previous licence conditions or if there have been any problems. As he has other licences, I may be able to streamline the process, so only one licence is needed for the operator's activities in all parks in the state. My role is to ensure the operation of the licence is acceptable. The development of the standard licence has been undertaken by concessions and leasing specialists in the region and in head office. I will need to determine the specific licence conditions for this area, such as disposal of garbage and human waste, safety equipment, and access.

The process we use in the district means that, among other things, it is necessary for me to consider the qualifications and experience of the guides, as well as the proposed activities, group sizes, and potential environmental impact. In this case the group sizes proposed do not comply with our standard ratios, which means I will need to meet with the applicant to discuss the issue. This is the third application this year for rafting on this stretch of river, and I am concerned that we may need to limit future licences or offer a system of tendering. This needs to be raised at the next district meeting.

Annette, our secretary, transfers another call to me. This time it's a woman interested in camping in the park next weekend. I describe the area and give her an idea of the facilities available, camping fees, and road conditions. I also let her know that there is a seasonal ranger program of activities in the park that she and her family might like to attend. I put some park pamphlets in the mail for her.

My 2 p.m. appointment today is an on-site meeting with another contractor, who has won the tender to construct a camping area that was proposed in the plan of management many years ago. Now we finally have the money through to be able to start stage 1 of the project. Planning for this project started months ago when I did a review of environmental factors for the proposal. In the course of assessing the site we discovered a small Aboriginal scatter site with stone flakes. As soon as we discovered this, we contacted the regional archaeologist and the local Aboriginal land council to discuss the issue. An archaeological report was done and, after consultation and two on-site meetings with everyone involved, it was decided that the scatter site was not of major significance and the Aboriginal community agreed to us continuing with the project. The review of environmental factors has now been approved.

This project has involved a lot of discussion and liaison with various interested groups. The site has been used by campers on an informal basis over many years, particularly by people fishing, as it is adjacent to the river. Naturally, when this group heard about the plans to formalise the camping area they were concerned, and thought that we might close down the area and introduce fees. Realistically, with revenue generation an imperative these days, we may in fact have to introduce fees. These and other issues were discussed at a community information meeting I organised one evening a few months ago. People interested in the project were invited to come along and discuss their concerns. We also set up a small planning advisory group of interested people, who have assisted me with the planning and development of the camping area. This approach has been really useful as we've all learnt a lot about planning and designing a camping ground. The bonus is that they have also learnt a bit about how our organisation works, while we have learnt more about our community.

This project has involved some long-term planning, with staff collecting information on the number of campers, activities they're involved in, length of stay, and we also did a visitor survey last year to find out what people wanted for the proposed camping site. As well as the small planning team I've had assisting me, I've also visited some campsites in other districts and talked to other rangers about their camping areas. This has given me some good ideas and indications of the pitfalls to avoid. Of course, each site is unique and as this one is adjacent to the river we have to consider carefully issues like erosion, visitor safety, boat access, toilets, and water supply. Once the camping ground has been established we'll need to have a community education program on minimal impact behaviour like disposing of grey water.

This afternoon's meeting is with the contractor and some of the members of the planning team. I have copies of the concept plan, and detailed site plans that were drawn up by the region's engineer. The planning group and I have decided on the type of materials we'd like used, and I've done a project budget that includes the purchase of materials and the contractor's fees. We estimate that once the contractor starts on the job in a couple of weeks, the camping area will be completed in about a month. Of course, this assumes that everything runs according to plan, and you can never predict the weather or supply of materials. We really need to try and stick to this time frame because the minister has decided that she wants to open the new camping ground, and we've already arranged the opening date with the ministerial liaison office. So the pressure's on.

Today's been really busy and I'm running late—I stop and grab a sandwich for lunch on the way to the meeting with the contractor. I arrive at the meeting ten minutes late and everyone else is already there. After introductions, I briefly describe the proposal and we inspect the site, showing the contractor the area and how it relates to the plans. He's a bit concerned about the tight time frame, but seems to understand the situation, and suggests putting on extra staff to speed things up. We discuss the environmental management, and occupational health and safety requirements, which contractors must fulfil when they're working on-site in the park. We cover issues like providing a portable toilet for his staff, and safety issues relating to leaving machinery on-site overnight. Everyone seems happy with the plans, so the planning group leaves. I continue my discussions with the contractor regarding the purchase and supply of materials like timber, cement, hardware, and so on. We also discuss in more detail the exact requirements of the contract, and time frames, and agree on some dates for regular site meetings. I'll give him a ring in a week's time to confirm everything, and let him know how the ordering of the materials is going.

On the way back to the office I notice a kangaroo dead on the road. It must have only just been knocked down, as it wasn't there when I went to the meeting. I feel the animal and it's still warm. It's a female, so I check its pouch, only to find a small joey tucked inside. I put it into a wool sock that I carry in my vehicle and drag the mother off the road into the bush. I then put the joey under my jumper, and head back to the office. As soon as I get in I call Cheryl, one of the volunteers in the wildlife group, who specialises in kangaroos and wallabies. She says she'll come in later and pick up the joey, which seems to be very lively under my jumper!

Just on 4 p.m., with a few issues still to be resolved for the day, the phone rings again. This time it is a local landholder who says he can see smoke to the south of his property, almost certainly within the park. He agrees to go to a high point and get a grid reference on it.

Quickly, I ring the depot and ask four of the staff who are there to stay at the workshop, until we have confirmation or otherwise of a pending fire operation. They know from experience that they may be out for the night, and follow the normal routine of preparing their equipment and fuelling the vehicles. Meanwhile, the district manager and the local fire control officer (FCO) must be advised that we may have a fire. The FCO will put local crews on standby to assist us. I arrange for another ranger and one of the office staff to take a firefighting unit to the suspected fire.

My role is to coordinate the initial response. As the fire increases in size it is likely that I will take on a functional role, and one of the more senior staff will assume control. At this stage though, I alert the rest of the staff in the office that they may be needed, and make a copy of the list of people I will need to call as soon as a fire is confirmed. But before I do anything else I must ring and ask a neighbour to pick up my children, take them home, feed them dinner, and, almost certainly, put them to bed until I can get there. Second, the talk I was going to give to a school group tonight will need to be cancelled or re-scheduled. I'll ring and ask the education officer if he can help out.

A radio call confirms a fire, and it is spreading fast. Everyone swings into action. Two administration staff are deployed to call as many staff as possible, and coordinate their transport towards the fire. One person is required to arrange logistics, including catering, as no one will have had dinner and they will have a long night ahead. The technical officer has been despatched to copy adequate maps, and then to obtain current weather information from the Bureau of Meteorology. This will be used to predict the likely scenario for the fire, and to plan strategies to contain it before morning. I contact head office flight section and advise them that we may need a helicopter for water bombing early tomorrow morning, but we will confirm this during the night. A light plane for reconnaissance would also be useful. With a team of six pursuing these tasks, I have time to ring and let the FCO and the district manager know where we are up to. The district manager indicates he will return shortly to take on the role of incident controller.

Meanwhile, there is a number of forms to fill in and send to head office. With this complete, I receive a radio report, the first from the fireground. The fire is approximately 20 hectares, running uphill and is burning quite hot, particularly for late in the day. The ground crews will need to work hard if we are to contain the fire. It becomes clear the crews will be working through the night. We will need additional resources to be available for the day shift. It is unlikely that there are more than ten staff available from our area, so we will need to get more people from other districts. The crews from the coast will be required to drive up this evening, sleep in town, and start on a day shift in the morning. By this time it is 7 p.m., and I spare a thought for the weary staff settling down for dinner with their families who will have to pack their kit, leave for a three-hour drive, sleep in a strange bed, then be at the fire for anything up to five days. This means the day-to-day duties are left on hold. The dinner you had booked will happen without you—it might be someone's birthday, but it's your job.

The district manager arrives and I brief him on the situation so far, and then take on the role of fire planning officer. The fire operations officer is already in the field. I will need to plan the strategy for the rest of the night, and the next day, on the basis of the information supplied to me from the fireground. As the information comes in, with regularly updated weather information, I attempt to predict how the fire will behave and how best to control it. My role also involves planning the additional resource requirements, and ensuring we keep track of our resources. I have done as much as I can by midnight. The district manager will continue for the night, with the assistance of two support staff. It is time for me to head for home, pick up the sleeping kids, tuck them into their own beds, and plan for tomorrow. It will be another big day.

11.3 Management lessons and principles

Lessons

Senior protected area operations managers are in a privileged and trusted position. They are typically highly qualified, trained, and experienced individuals, and have the responsibility to complete planned operational actions. They have been charged by the community (through their organisation) to under-take conservation actions in protected areas. An operations manager may only be responsible for an operational area for a brief period of time such as three to five years. At the most, a career of 30 years may be committed to an operational area, but it still pales into insignificance when protected areas are being managed not for 100 years, but for 1000 years and longer. Operations managers need to be clear that their role of custodian is essential but successional, just one of many highly competent professionals who will contribute over many years. Arrogance and rigidity do not have a place in this role of custodian. It is not the responsibility of operations managers to be the source and inspiration for all decisions. In fact this is a dangerous position. It is their responsibility however to ensure that the best information is available for decisions, that investments of essential new research are made, and that the operation is adaptive as new and better information comes to light. This is a managerial and inclusive approach. It would include community, stakeholders, and experienced professionals in an exchange of information. Success for a retiring operations manager custodian would be judged by the improved environmental condition of 'their patch' and a supportive community. Key lessons for operational practice are as follows.

1 Protected areas require active, effective, and continuous management if the purposes for which they were reserved are to be retained.
2 Attention to detail by operations managers is critical. Whether it is an initial operation design, the location for an operation, the manner in which it is to be conducted, the materials to be used, the types of expertise to be employed, or other aspects. There is no point in conducting an operation to improve the environment if the net end result is worse. Pre-planning for an operation must be high quality.
3 Leadership during an operation is critical. There will be constant chances for impacts to protected areas through accidents, lack of discipline, or poor planning and implementation techniques. There may even be deliberate acts of vandalism or illegal actions. Eternal vigilance during operations will be critical. Pre-planning will anticipate the accidents, and there will be an on-site capacity to deal with potential incidents.
4 Use of local knowledge and involvement of the community will enhance an operation.
5 Institutional obstacles to learning and action can result in environmental heritage impacts caused by inappropriate works, delays in rehabilitation, or an absence of remedial works. Paralysis of action is a threat to heritage conservation.
6 Adaptive management (Chapter 7) based on quality research and information is essential for the long-term conservation of protected areas.

Principles

1 Effective protected area management operations are an essential and integral part of the conservation of natural and cultural heritage.
2 Operational standards, best practice systems, staff competencies, operational procedures, on-site leadership, and operations team discipline are integral and essential parts of effective protected area operational management.
3 The safety and welfare of operations staff is the primary concern of operations managers.
4 Operational management leadership, inclusiveness, and attention to operational detail are essential parts of successful operational management.
5 Research, operational performance monitoring, and adaptive management are essential parts of successful operational management.
6 The use of local knowledge and involvement of the local community in operations is a fundamental part of an operation.
7 The provision of adequate and timely public information about operations is an essential part of operational management.

Natural Heritage Management

> ... *every race owes certain duties to its descendants, chief among which is to preserve, develop and hand down the great heritage which has been given it. In what sense, then, can a people have performed its duty, if in years to come one of its descendants can say, 'Oh, yes, but where are those forests you talk of, those animals which were the wonder of the world, those birds that scientists came from the four corners of the world to see, that scenery which you say had no peer?' Great that nation whose men can put their fingers on the map and say, 'Here, and here, and here you can see Nature undisturbed. Nature in the same guise as when your forefathers came'* ...

Extract of a letter sent by Romeo Lahey to Beaudesert and Tambourine Shire Councils in 1913 as part of his campaign to establish the Queensland Lamington National Park, quoted in Jannott (1990).

12.1 A national perspective	310	12.6 Managing soils and geology	335
12.2 Conserving fauna	314	12.7 Conserving scenic quality	342
12.3 Conserving flora	328	12.8 Managing fire	343
12.4 Fungi	331	12.9 Management lessons	
12.5 Managing water	332	and principles	349

One might think that protected areas need only to be left alone—unmanaged—for their original species and communities to thrive. Experience shows otherwise. It is not simply that animals, plants, geology, and fungi need to be protected from people. There are many other issues. Feral cats and the introduced fox (*Vulpes vulpes*) may destroy the rare fauna a park was designed to protect; or introduced weeds such as Common Lantana (*Lantana camara*) and Rubber Vine (*Cryptostegia grandiflora*) may overwhelm the original plant communities. There is also the challenge of protected areas becoming, in effect, islands—remnants of native land surrounded by cleared land or dissected by highways. In the past, a given animal or plant might have been eliminated from an area by disease or an intense fire, yet it would have recolonised the area from refuges elsewhere. Today, once eliminated, it may be gone forever. Vegetation communities such as rainforest, brigalow, heathland, or mangroves might once have gradually shifted their borders, perhaps by hundreds of kilometres, as climates, fire regimes, or ocean levels changed. Today they usually have nowhere to move, hence the paradox: humans must intervene to maintain the integrity of natural systems and processes.

From the perspective of many international visitors, Australia is typified as a wide brown land, with kangaroos, eucalypt trees, a magnificent coral reef, the Sydney Harbour Bridge, the Opera House and our brightly coloured birdlife. Many of these images are natural and, thankfully, the difference between the images and reality are still not too far-fetched. However, the natural lands that support these images are fast disappearing. In much of Australia, such images are restricted to our protected areas.

As further changes to our remaining natural areas occur with time, protected areas are becoming increasingly important. Protected areas are fast becoming or have become the last natural lands in some regions. They need active management to maintain their integrity. This chapter deals with the implementation of conservation management practice. We describe some of the methods used by managers to conserve Australia's natural heritage through a wide range of techniques. There is a remarkable diversity of tasks undertaken in managing our natural heritage resources and natural phenomena. Staff require a broad range of skills and expertise to deal with all kinds of issues, including the management of fauna, flora, water, soils, geology, scenery, habitat, fungi and fire. We provide accounts of most of these specialties. Space does not permit us to go into great detail, though for one topic, karst and caves, we have provided additional detail to indicate some of the more specific matters that a manager needs to consider.

No one individual would be expected to possess the whole range of skills and expertise needed to deal with all of these complex areas, but managers are expected to ensure that the right skills are allocated to the right tasks. They do need, therefore, to have some appreciation of the subject areas they are managing.

12.1 A national perspective

The national scope and content of natural heritage management is evident from national and state strategies. For example, one of the three core objectives of Australia's National Strategy for Ecologically Sustainable Development (Commonwealth of Australia 1992a) is 'to protect biological diversity and maintain essential ecological processes and life-support systems'. This approach has also b een promulgated nationally through the Inter-Governmental Agreement on the Environment. The Commonwealth, states, and territories have also agreed on a framework for the conservation of biological diversity, called the National Strategy for the Conservation of Australia's Biological Diversity. The strategy commits protected area staff to an enormous amount of work. Some of the recommended actions for our managers are to:

- manage for biological diversity on a regional basis
- conserve biodiversity in an integrated manner with the community and with other organisations
- establish a comprehensive, adequate, and representative system of protected areas

- strengthen off-reserve conservation of biological diversity
- continue the conservation of Australia's wildlife and threatened species
- involve Aboriginal communities in the conservation management of biodiversity
- integrate biodiversity conservation and natural resource management through ecologically sustainable use of biological resources in pastoralism, fisheries, forestry, water management, tourism and recreation, wildlife, and genetic resource management
- deal with threatening processes
- improve our knowledge of biodiversity resources
- involve the community
- meet the requirements of international conventions.

The Australian Government has also commissioned a series of action plan documents that are aimed at coordinating efforts to conserve animal species. These documents include action plans for rodents, marsupials and monotremes, freshwater fishes, birds, reptiles, and cetaceans and are important reference texts for practising managers (http://www.deh.gov.au/parks/index.html). Progress on conserving Australia's natural heritage has been problematic for the period 1991 to 2001. Experts on Australia's State of the Environment Committee believe that biodiversity significantly declined during this decade (ABS 2004). This view is reflected by the rise in the number of birds and mammals listed as extinct, endangered, or vulnerable from 118 to 165

between 1993 and 2003, an increase of 40% (ABS 2004). On the other hand, the area reserved as protected areas continues to grow, with 10% of Australia reserved as of June 2002. Conservation and protection on private lands is becoming more important, and by the late 1990s more than 1300 conservation covenants have helped protect 774,000 hectares of private land (ABS 2004).

Natural heritage management cannot be effective unless the best scientists, managers, and local experts are working together to achieve agreed objectives. Adaptive management practice (Chapter 7) is an integral part of this approach. This is particularly relevant where scientific understanding of natural changes at a place is only just keeping pace with the dynamic of natural systems and potential external modifiers such as global warming.

The Australian Natural Heritage Charter

In undertaking conservation practice, guidelines have been developed that assist managers. The most important of these at a national level is the Australian Natural Heritage Charter (ACIUCN 2002). The charter provides standards and principles for conserving places of natural heritage significance in Australia, and is administered by the Australian Committee for IUCN. The charter includes a set of procedures for designing a management plan to conserve a given area (Case Study 12.1).

A guide to the Australian Natural Heritage Charter, including case studies of how it has been implemented, has been produced to provide guidance for practitioners (ACIUCN 2003).

Case Study 12.1

The Australian Natural Heritage Charter: A tool for protected area managers

Lorraine Cairnes, Fathom Consulting Pty Ltd

The Australian Natural Heritage Charter (second edition) provides a basis for natural heritage conservation that can be applied throughout Australia. The purpose of the Australian Natural Heritage Charter is to assist with the conservation of natural heritage in terrestrial and aquatic ecosystems. It can be applied to public and privately owned places, to the land of Indigenous owners, to very large or very small areas, to protected and unprotected areas, to areas of international, national, or local significance, and to farms and mining leases. It can be used by non-government and government organisations, land-owners, land managers, decision-makers, voluntary groups, professional practitioners, in fact, everyone with a role in natural heritage conservation.

In 1996, the IUCN acknowledged the leadership being provided by Australia in this area, and distributed copies of the original charter (ACIUCN 1996) to all of its member states.

Development of the charter

The charter was developed with funding from the Australian Heritage Commission. A national steering committee provided perspectives of the Australian Committee for IUCN, the Australian Heritage Commission, the Australian Local Government Association, the Australian Nature Conservation Agency, the Environment Institute of Australia, and Indigenous people. The charter was developed over a two-year period in consultation with key people and organisations around Australia. An initial round of consultation during 1995 resulted in the Interim Australian Natural Heritage Charter (January 1996). A second round of national consultation during 1996 further refined the interim charter, and the Australian Natural Heritage Charter was launched in December 1996. The Charter was revised and updated in 2002 following the planned five-yearly review by users and expert advisors (ACIUCN 2002).

The initial national consultation revealed that:

- there was no consistent natural heritage conservation management process across Australia, but that this was desired by managers and other practitioners
- natural heritage conservation jargon varied around Australia, with practitioners often speaking at cross-purposes due to the use of different terminology; so consistent definitions of terms were required
- because cultural and natural heritage often occurs at the same place, there was a need for a process that could be related to cultural heritage conservation as set out in the Burra Charter (Chapter 13).

The Australian Natural Heritage Charter encompasses the full scope of natural heritage (ecosystems, biodiversity, and geodiversity). It states that natural heritage incorporates a spectrum of values, ranging from existence value at one end through to socially based values at the other. The fundamental concept of natural heritage, which most clearly differentiates it from cultural heritage, is that of dynamic ecological processes, ongoing natural evolution, and the ability of ecosystems to be self-perpetuating. At the cultural end of the spectrum, it can be difficult to clearly separate cultural and natural values, and more than one layer of values may apply to the same place. Central to conservation is the assessment of the natural values of a place and preparation of a Statement of Significance. Without this, conservation management can be seriously misdirected, and can even lead to the loss of the natural values of a protected area. The process outlined in the upgraded Charter (ACIUCN 2002) includes:

- identification of relevant information and stakeholders
- identification of significant natural values and management issues
- development of a conservation policy
- establishment of objectives expressed as desired future conditions
- development of conservation processes and strategies
- development of a conservation plan that details responsibilities, actions, and monitoring
- implementation of the conservation plan
- execution of monitoring that feeds back into assessment of significance, management issues, and establishment of objectives.

Conservation principles

The conservation principles identified in the Charter as being fundamental to natural heritage conservation (Articles 2-7, ACIUCN 2002) are as follows:

Principle 1

The basis for conservation is the assessment of the natural significance of a place, usually presented as a statement of significance.

Principle 2

The aim of conservation is to retain, restore, or reinstate the natural significance of the place.

Principle 3

A self-sustaining condition is preferable to an outcome that requires a high level of ongoing management intervention.

Principle 4

Conservation is based on respect for biodiversity and geodiversity. It should involve the least possible intervention to ecological processes, evolutionary processes, and earth processes.

Principle 5

Conservation should make use of all the disciplines and experience that can contribute to the study and safeguarding of a place. Techniques employed should have a firm scientific basis or be supported by relevant experience.

Principle 6

Conservation of a place should take into consideration all aspects of its natural significance, and respect aspects of cultural significance that occur there.

Conservation processes

The conservation processes outlined in the Charter are as follows.

Regeneration, which involves recovery of natural integrity following disturbance or degradation, with minimal human intervention.

Restoration, which requires returning existing habitats to a known past state or to an approximation of the natural condition by repairing degradation, by removing introduced species, or by reinstatement.

Reinstatement, which means reintroducing to a place one or more species or elements of habitat or geodiversity that are known to have existed there naturally at a previous time but that can no longer be found there.

Enhancement, which involves introduction to a place of additional individuals of one or more organisms, species, or elements of habitat or geodiversity that naturally exist there.

Preservation, which means maintaining the biodiversity and/or an ecosystem of a place at the existing stage of succession, or maintaining existing geodiversity.

Modification, which involves altering a place to suit proposed uses that are compatible with the natural significance of the place.

Protection, which requires taking care of a place by maintenance and by managing impacts to ensure that natural significance is retained.

Maintenance, which involves continuous protective care of the biological diversity and geodiversity of a place.

Conservation plans

Following the processes outlined in the Charter will lead to a conservation plan that consists of:

- identification of the natural values
- a statement of natural significance
- assessment of other management issues and their relevance to management
- a statement of the objectives of conservation, expressed as the desired future condition
- the conservation processes to be used
- responsibilities for decisions and actions
- a monitoring program.

12.2 Conserving fauna

Australia's unique native fauna is known throughout the world. Here, marsupials mainly hold ecological niches that elsewhere are commonly held by placental mammals. There is a high degree of endemicity, and many species have special conservation requirements (Chapter 1). Many species have become extinct since Europeans arrived, and many others are now endangered or threatened. Managing threatening processes is therefore a major part of a manager's work (Chapter 14). Those species that have so far survived must be conserved with the greatest of professional care. A range of management actions is required, many of which are described in this chapter. We begin with a discussion on research.

Facilitating fauna research

Protected area staff may be actively involved in facilitating research. It is a major part of their ongoing and adaptive response to conservation management of species. A classic response of research and adaptive management practices has been described by David Priddel for the Gould's Petrel (*Pterodroma leucoptera*), an endangered species found nesting on small coastal islands off NSW and has been presented here as Case Study 12.2.

Case Study 12.2

Recovery of an endangered species—the Gould's Petrel

David Priddel, NSW Parks and Wildlife

The Gould's Petrel is Australia's rarest endemic seabird. The only place it breeds in the world is on two small islands at the entrance to Port Stephens, NSW. The vast majority of the birds nest in rock cavities on the rugged rainforest-covered slopes of Cabbage Tree Island, and a dozen or so pairs nest on nearby Boondelbah Island.

Banding Gould's Petrel, Cabbage Tree Island, NSW (Don Butcher)

The first comprehensive census of the Gould's Petrel was undertaken in 1982. The initial survey revealed some disturbing facts. Fewer than 300 pairs nested, and breeding success was drastically low (less than 20%). Surveys repeated in each of the subsequent two years yielded similar results. It was also found that the population had declined by more than 25% during the past two decades. The causes of the species' demise were poorly understood.

A research project was initiated to identify the causes responsible for reproductive failure. As expected, the rate of nest failure was exceptionally high. More alarmingly, however, was the discovery that nesting adults were dying in relatively large numbers. Many petrels perished after becoming entangled in the sticky fruits of the Birdlime Tree (*Pisonia umbellifera*). The most prevalent cause of mortality, however, was predation by Pied Currawongs (*Strepera graculina*) and, occasionally, Australian Ravens (*Corvus coronoides*). These predators would kill both chicks and nesting adults to feed their own developing young.

Experimental recovery actions were implemented immediately before the 1993–94 breeding season. Poisoning destroyed Birdlime Trees within the breeding grounds of the Gould's Petrel. Follow-up measures prevented new plants establishing from seed. Shooting reduced Pied Currawong numbers. Their nests were located and destroyed, along with any eggs and young. Ongoing monitoring of the petrel population revealed an immediate rise in the number of petrels' incubating eggs and a marked improvement in breeding success. The culmination of these factors was a four-fold increase in fledgling production. Breeding success now regularly exceeds 55%, and in most years more than 300 young fledge (Figure 12.1).

Clearly, the threats posed by the Birdlime Tree, Pied Currawong, and Australian Raven were able to be ameliorated by appropriate management intervention. The question remained, however, as to why these unusual threats arose, particularly on an island essentially remote from the influences of people. The answer lay in the changes to the vegetation wrought by rabbits since their introduction to Cabbage Tree Island in 1906.

Gould's Petrel breeds in two deep rainforest-covered gullies on the western slopes of Cabbage Tree Island. Rabbits had destroyed much of the rainforest understorey. Without adequate concealment, nesting petrels have been exposed to predators. The sparseness of vegetative cover also makes the petrels more vulnerable to entanglement in the fruits of the Birdlime Tree. An intact understorey captures many of these fruits before they fall to the ground. Fruits caught up in vegetation pose little or no threat to petrels moving about the forest floor.

It was considered that the requirement for long-term control of Pied Currawongs could be eliminated if rabbits were eradicated and the understorey given the opportunity to re-establish. Rabbits were successfully eradicated. The procedure involved the sequential use of three mortality agents: myxomatosis, rabbit calicivirus, and poisoning. Significant changes in the vegetation of Cabbage Tree Island were evident

Figure 12.1   Number of Gould's Petrel fledglings produced on Cabbage Tree Island for ten seasons, 1989 to 1998
(years refer to commencement of breeding season)

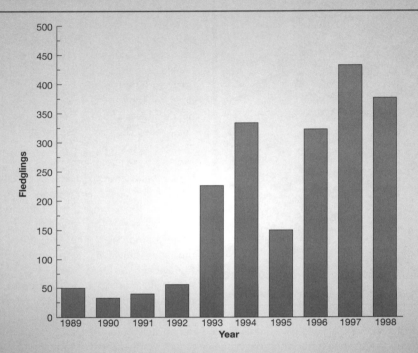

within just weeks of the last rabbit being removed. Before implementation of recovery actions, fewer than 50 young fledged each year. In the late 1990s, reproductive output had risen to 300 young per annum. Following this success, an attempt to establish a second viable colony on Boondelbah Island was initiated and 100 nest boxes were established there between 1999 and 2000. Two hundred chicks were transferred and all but five successfully fledged with some returning to breed. In 2002–2003, 12 nest boxes were occupied and five eggs laid. Surveys in 2001–2002 found that in total there were 1000 birds breeding with 450 young being produced per annum (NSW NPWS 2003).

Undertaking fauna survey and inventory work

Fauna surveys are organised for a number of reasons. They may be undertaken as an initial survey of the biodiversity of an area, as part of a monitoring program, or they may be used to determine the influence of threatening processes over time. A range of field survey techniques are used following the careful preparation of a research design plan. Some of these field survey techniques are described in Table 12.1. Sutherland (1996) provides useful guidance for those seeking more in-depth information on survey techniques.

Identifying species type. Correctly identifying fauna species is a critical part of undertaking any fauna inventory. Diagnostic aids in the form of expert references and keys are available to assist field researchers. In addition, where there is concern over the identity of species, specialist staff and organisations such as the Australian Museum, CSIRO, and contract scientists are available for contractual help.

Table 12.1	Fauna inventory: Some field survey techniques used

Fauna field survey technique	Notes
Direct identification—observation and listening	Skilled observers are invaluable for enhancing information about wildlife. Bird, frog, and some mammal species have distinguishing calls or sounds from which they can be identified. Standard fauna inventory forms have been produced by many organisations to facilitate the recording of observations.
Observation—fauna tracks and diggings	While often difficult to discern, signs of fauna such as footprints and scratchings are an invaluable aid to fauna observers and researchers. Triggs (1996) is a very valuable guide.
Collection and analysis of fauna scats	Predator scats, particularly those of Dingo and the introduced fox, are valuable for rapid inventory of native fauna populations. Researchers have found that the scats from such animals are deposited close to the food source. After *carefully* collecting the scat (given the chance of contracting a disease such as hydatids), it can be dried and analysed for hair and bone content. Scats from native animals themselves are important inventory diagnostics.
Collection and analysis of bird pellets	Some birds regurgitate bone, feather, fur, and other fragments of their meals that they are not able to digest. Owl pellets, for example, contain a wealth of small mammal bones in stratified deposits at some cave sites. They have provided valuable contemporary and historical records of small mammal populations used as prey by the birds.
Fauna signs in their habitat	Animal runways in heath and native grasslands; burrows; nesting hollows; incisions in trees that mark nesting sites; claw marks on trees; and litter and damage to trees and shrubs from animal feeding are all signs that indicate the presence of fauna.
Trapping and collection of insects	Water traps, flight interception traps, light traps and bait traps are methods used for collecting insects.
Spotlighting	Many species are only active during the night. The use of a portable light will reflect the retina colour of animals' eyes. The colour, shape, and size will help in the identification of species.
Call playback	Many animals have distinctive calls, and when these are recorded and played back through a loud speaker, species can be prompted to respond. This technique is effective for locating Koala (*Phascolarctos cinereus*) at certain times of the year.
Use of pit-traps	This is a technique used by zoologists to capture small mammals, reptiles, and invertebrates. Use is made of a barrier and a small container that is sunk into the ground. Animals are directed to the container by the barrier and are captured in the pit as they try to pass through the 'opening' in the barrier.
Reptile searches	This technique is usually undertaken for a small area and during the middle of the day. Favoured habitats for these species (under logs and rocks, in leaf litter, in hollows, and so on) are searched.
Use of hair tubes	A hair tube is a length of plastic pipe (about 90 mm in diameter for small species) that has a bait sealed at one end, and double-sided sticky tape on the side of the pipe. When feeding, the small mammal leaves some hair on the tape, which is subsequently analysed to determine the species.

Table 12.1	(continued)
Fauna field survey technique	Notes
Use of small mammal traps	Collapsible aluminium traps (Elliot traps), which capture their specimens live using a bait, pressure pad, and spring rear-door trap are a common tool of scientists undertaking fauna inventories. Typically, specimens are captured, identified, weighed, measured, and released on-site. Larger live traps (cage traps) are used for the capture of larger specimens.
Use of nets, including harp nets and 'fish' nets.	Harp nets (vertical filaments of nylon organised to form a barrier to bats) are generally placed on bat flight paths. They are designed to minimise their detection from bat sonar signals and to minimise any impact on the bats. Nets are commonly used for the capture of birds.
Use of specialised traps	Large traps are often used for the capture of bigger animals such as Salt-Water Crocodile (*Crocodylus porosus*) of northern Australia. This technique is used when a 'problem' animal needs to be relocated.

Hairs collected from hair tubes are analysed to assess mammal species attracted by the bait inside (Andrew Claridge)

Volunteers and fauna surveys. Enthusiastic volunteers around Australia help enormously with fauna surveys. The Royal Australian Ornithological Union (RAOU, now Birds Australia), for instance, uses volunteers and staff members to carry out regular Australia-wide surveys to monitor bird species. This critical work involves regular bird-banding field days, which are undertaken in locations throughout Australia. Birds Australia contributed to the 2002 Australian Terrestrial Biodiversity Assessment through two 'Birds Atlas' surveys undertaken in

1977–81 and 1998–2001. Six million records of 774 bird species were collected. Differences between the two surveys contributed to describing trends in bird populations for the National Land and Water Resources Audit (NLWRA 2002). This type of monitoring is important for tracking impacts on our birds both nationally and internationally. For example, the Great Knot (*Calidris tenuirostris*), a summer migratory species to Australia from Siberia, is

Elliot traps are used in small mammal studies (Mark Hallam)

Snapshot **12.1**

The Mountain Pygmy-possum
Linda Broome, NSW P&W

Mountain Pygmy-possum (Linda Broome)

The adult Mountain Pygmy-possum fits neatly into the palm of your hand. Until relatively recently, it was only known from the fossil record; however discoveries at Mt Hotham in Victoria, near Guthega, and later at Mt Kosciuszko, confirmed its current wider distribution. Little was known of its ecology until pressures for ski resort expansion and intensive ski slope management created the need for intensive research on the animal, both in Victoria and NSW. The NSW work found that intensive modification of the ski slopes was directly affecting the summer movements of the possum, and compensatory arrangements needed to be put in place. Cross-slope boulder-filled tunnels for the animals have been installed on Mt Blue Cow to allow them to continue their wide-ranging movements safely across the cleared ski slopes. The population numbers are still being actively monitored.

Snapshot **12.2**

Keeping invertebrates on the conservation agenda
Allan Spessa, Environment Australia

An adult female of *Amphylaeus morosus* (sub-family Hylaeinae) (Allan Spessa)

The Commonwealth's *Environment Protection and Biodiversity Conservation Act 1999* requires preparation of five-yearly national State of the Environment (SoE) reports. The first SoE report was published in 1996 prior to the legislation, with the second completed in December 2001. Environmental indicators for SoE reporting have been developed by Environment Australia. Peer-reviewed reports recommending indicators, and appropriate monitoring and reporting methods for the seven environment themes, including biodiversity, have been published. The 2001 SoE report provided assessments of the conservation status of invertebrate taxa (and their habitats) listed under Commonwealth, state, and territory threatened species legislation, as well as the management of threatening processes (ASEC 2001a). Detailed studies and monitoring of populations are unfortunately lacking for the vast majority of Australian invertebrates. Native bees are emblematic of Australia's rich but comparatively little-known invertebrate biodiversity.

Snapshot **12.3**

The Richmond Birdwing Butterfly (*Ornithoptera richmondia*)
Bob Moffatt, NSW Parks and Wildlife

The Richmond Birdwing Butterfly is the largest butterfly to be found in NSW, with a wing span up to 15 centimetres. These brightly coloured butterflies originally extended from Grafton in NSW, to Maryborough in Queensland. The principal larval food source is the Richmond Birdwing Vine (*Pararistolochia praevenosa*), which is found in lowland rainforests. The large-scale clearing of subtropical rainforests during the late nineteenth and early twentieth century resulted in the near elimination of the vines. As a consequence, the population of the butterflies has disappeared from two-thirds of its former range and numbers have

decreased dramatically. Further threats are presented by the imported Dutchman's Pipe Vine (*Aristolochia elegans*), which attracts egg-laying female butterflies. After the eggs have hatched, the larvae then eat the toxic leaves, which kills them.

Urban plantings of the Richmond Birdwing Vine have proved successful in attracting egg-laying females. These plantings, combined with removal of the exotic Dutchman's Pipe Vine, have resulted in schools, community groups, and private landholders reporting increasing numbers of butterfly larvae and adults on their planted vines.

potentially impacted by developments to the Saemangeum tidal flats on the west coast of Korea. The tidal flats are a resting and feeding ground for the birds on their long migration path (Brown 2004). The Birds Australia surveys will help show any changes to such bird populations. Rangers are often involved with facilitating such work. This work is especially important for migratory bird species such as those that are recognised in international treaties such as the Japan–Australia and China–Australia migratory bird agreements.

Radio tracking. This technique enables researchers to gain a much greater understanding of the mobility and the environments that are of importance to species. Radio-tracking work at Blue Cow in Kosciuszko National Park helped researchers to understand the mobility of the endangered Mountain Pygmy-possum (*Burramys parvus*). The analysis warned that unless ski resort management better accommodated the needs of this species, visitor use of the park would contribute to the demise of the species at this site (Snapshot 12.1). Advances in technology have enabled the tracking of the migratory Short-tailed Shearwater from nesting sites on Montague Island on the south coast of NSW to feeding grounds in Antarctica.

Monitoring populations using aerial surveys. Annual aerial surveys are organised to estimate the populations of both native and introduced fauna, such as kangaroos, rabbits, and camels. This effective technique was first introduced for NSW in the 1970s in response to a strong lobby from graziers to cull kangaroo populations in the west of NSW. The CSIRO developed techniques for determining kangaroo populations from aerial surveys (Caughley et al. 1977). Agency staff are directly involved and the yearly surveys (Caughley & Bayliss 1982) provide the benchmark from which they can monitor seasonal fluctuations in the number of kangaroos. The resulting data provides managers with an understanding of the naturally fluctuating kangaroo population numbers as well as the effects of commercial and non-commercial culling practice.

Conserving invertebrates. Invertebrates are a very large and diverse group of fauna that we know very little about. Agencies are increasingly seeing the need to

Snapshot 12.4

Monitoring cave fauna in Tasmania
Michael Driessen, Parks and Wildlife Service, Tasmania

Tasmanian caves contain a rich assemblage of invertebrates, including glow worms, crickets, beetles, and Tasmania's largest spider (*Hickmania troglodytes*). Many cave species are very rare and often restricted to particular cave systems. Caves are increasingly popular places for people to visit and there is concern about visitor impacts on cave fauna, as well as cave formations. The Tasmanian Parks and Wildlife Service has started a program to monitor fauna in two caves in south-east Tasmania. These caves, Exit and Mystery Creek Cave, contain wonderful glow worm displays.

Tasmanian Cave Spider (Stephen Eberhard)

Cave fauna numbers were monitored on a monthly basis from August 1998. Cave fauna are counted by eye in designated areas of the caves. The most common species recorded in the cave are Glow Worm (*Arachnocampa tasmaniensis*), Cave Crickets (*Micropathus tasmaniensis*), Cave Spiders (*Hickmania troglodytes*), Cave Beetles (*Idacarabus troglodytes*) and Cave Harvestmen (*Hickmanoxyomma cavaticum*). This survey provided baseline information for these species against which future impacts may be compared.

acquire more information about this group (Snapshot 12.2) and are actively involved in research and conservation of invertebrates. Insects, for example, play a critical role in maintaining the health and well-being of our ecosystems. They are often underestimated in the role they play, whether it is contributing to natural pollination processes, aiding the decomposition of organic matter, or as a food source for a range of native birds and mammals. Rangers may be directly involved in the conservation of native invertebrates (Snapshots 12.3 and 12.4)

Snapshot 12.5

Conserving the Green and Golden Bell Frog
Modified from NSW NPWS 1999a

The Green and Golden Bell Frog is listed as an endangered species and is the subject of a draft recovery plan. It is a relatively large frog, averaging 60 to 80 mm in size. It is pea-green in colour, with variable amounts of golden brown markings. Formerly, the species was distributed between Brunswick Heads (NSW) to East Gippsland (Victoria) and inland to the tableland areas of NSW. In the last five years, there has been a failure to find the frogs on any tableland sites, and many of the coastal populations have disappeared or have dramatically declined. There are only four large populations in NSW, none of which is protected in conservation reserves, although it is still known from protected areas in Victoria. The frog inhabits marshes, dams, and streamsides, particularly those containing bulrushes (*Typha* sp.) or spike rushes (*Eleocharis* sp.). Conservation of this species requires managers to be actively working with local councils and other authorities to ensure that habitats are not disturbed, that predation from feral cat and fox is minimised, and that any potential for change to drainage patterns or for pollution is minimised. One prominent conservation site for this species is a wetland within the former Sydney 2000 Olympics site at Homebush Bay.

Conserving amphibians. In recent times and particularly during the 1990s, many species of amphibians have experienced major declines in their distribution and abundance. The spectacular black and yellow striped Corroboree Frog of the Australian Alps, an endangered species, has rapidly declined in numbers. This has occurred to such an extent that scientists and managers have established a captive breeding and reintroduction program to help conserve the species while more detailed research is conducted to determine the reasons for the decline. Regrettably, the widespread January 2003 fires in the Australian Alps have burnt many of the Corroboree Frog's subalpine sphagnum bog and wetland habitat areas, making the conservation task more difficult. The Green and Golden Bell Frog (*Litoria aurea*) is similarly receiving active management (Snapshot 12.5).

Conserving reptiles. Australia has a rich diversity of reptiles: crocodiles and sea-turtles of our northern

Snapshot 12.6

Looking after miyapunu: Indigenous management of marine turtles
Rod Kennett, Centre for Indigenous Natural and Cultural Resource Management, Northern Territory University and Nanikiya Munungurritj, Dhimurru Land Management Aboriginal Corporation

Sea turtles (miyapunu) are an important cultural and natural resource to the Indigenous people (Yolngu) of north-east Arnhem Land, Northern Territory. Six of the seven species in the world occur in the region, including nationally and internationally significant rookeries for Green Turtle (*Chelonia mydas*) and Hawksbill Turtle (*Eretmochelys imbricata*). Acting on Yolngu concerns at declining numbers of turtles, the Dhimurru Land Management Aboriginal Corporation in collaboration with the Northern Territory University and the Parks and Wildlife Commission, are engaged in a sea turtle (miyapunu) conservation and research project that combines traditional knowledge and skills with contemporary research and management methods. Activities include the mapping of nesting and feeding habitat, recording traditional knowledge, quantifying the harvest of turtles and eggs, rescuing stranded sea turtles entangled in discarded fishing nets, analysing heavy metal accumulation in turtle tissues, and studying temperature-dependent sex determination. More recently, Yolngu have been using satellite telemetry to track the migrations of nesting turtles back to their feeding grounds to determine which other coastal communities share this migratory resource. In response to findings of the project, traditional owners have closed access to some nesting beaches to reduce egg harvest and four-wheel-drive damage. Dhimurru rangers have made exchange visits with other coastal Aboriginal groups to assist them to establish their own turtle projects and to discuss cooperative management of sea turtles.

Radio tracking sea turtles (Courtesy WWF/WTO, Carol Palmer)

shores, turtles and tortoises of the inland swamps and river systems, vivid green rainforest tree snakes, blind snakes, bearded dragons of our inland deserts, legless lizards, and a wide range of goannas, skinks, and other lizards and snakes. Active conservation work is undertaken on our northern shores to protect large migratory marine turtles (Snapshot 12.6). Similarly, education programs are undertaken to better inform Australians about the role of reptiles in ecosystems. The ACT Government has conserved many habitat areas of the legless lizards found in the territory. Illegal bush rock collection, which was threatening the Broad-headed Snake (*Hoplocephalus bungaroides*), has been actively suppressed in NSW. The vulnerable Western Blue-Tongued Lizard (*Tiliqua occipitalis*) of western NSW, which is threatened by habitat clearing and disturbance, is the subject of an attempt to conserve its habitat.

Conserving birds. Managers spend considerable time helping to conserve the habitat of birds all over Australia. Conservation debates are heard across Australia on issues such as the minimum water flows needed to conserve the wetland habitats of inland streams, the protection of remnant coastal heaths to conserve parrot species, the conservation of wader habitats along coastlines, and the protection of nesting hollows and home range of large predators in forests. The areas at issue include the very heart of Australia's largest and busiest city, Sydney, where an endangered colony of Little Penguin has been conserved.

Conserving mammals. Australia's native mammals, in particular its marsupials and monotremes, need constant conservation action to keep them extant. Besides the loss of habitat, predators such as introduced foxes and cats are having a major impact on our native mammals, and many small to medium-sized marsupials are facing extinction. Considerable innovative work is being done in Western Australia to achieve the exclusion of such predators (through poison baiting and fencing), and this has in turn seen the return of the Numbat (*Myrmecobius fasciatus*) and other species to their former habitats. Rangers may need to use an interim plan or undertake emergency measures to conserve a species until a more adequate recovery plan can be prepared to meet the species' needs. During the early 1990s in the Kangaroo Valley area of southern NSW, colonies of the Brush-tailed Rock-wallaby (*Petrogale penicillata*) were declining rapidly. Following a survey, managers immediately set up a program to control foxes. Most of the fox colonies were actually on private land, but there was enthusiastic support from local landowners. The control program was successful and, as fox numbers fell, the size of the wallaby colony stabilised and the number of young wallabies increased. Similar work is conserving the Yellow-footed Rock-wallaby (*Petrogale xanthopus*) in far western NSW. Baiting for foxes has also helped to conserve the endangered Mountain Pygmy-possum.

Captive breeding. Captive breeding is expensive and difficult. This specialised activity is most suited to organisations such as zoos. In Australia, captive-bred populations of animals have been released into the wild, with mixed success. The reintroduction of the Mala or Rufous Hare-wallaby into the Tanami Desert in the Northern Territory within a fenced feral cat and fox exclusion area is one example of a success story.

Inbreeding and genetic diversity. The impacts of habitat loss and fragmentation may cause populations to decline. As numbers are reduced, managers must become more attuned to the long-term genetic needs of species to prevent extinctions. This complex area of conservation biology must be thoroughly understood by staff so that they can work with scientists to help conserve our native wildlife.

Injured or orphaned animals. Animals may have been injured on our roads, suffered from pollution, ingested garbage (dolphins have been made sick as a consequence of ingesting plastic bags), attacked by feral or domestic predators, or have suffered injuries for other reasons. Injured animals, where possible, should be treated by veterinary surgeons. Many animals however, because of the extent of their injuries, may need to be euthanased. Orphaned animals may be cared for by protected area management agencies, zoos, private organisations such as aquariums, and volunteer groups such as the Wildlife Information and Rescue Service (WIRES). For further information we recommend Hand (1990).

Disoriented, sick, or diseased animals. Sarcoptic mange in wombats, botulism in ducks and other waterfowl, the effects of parasites and lumpy jaw in kangaroos are just some of the diseases and parasites that may need to be addressed. Rangers will usually require the assistance of veterinary surgeons or expert organisations such as zoos or agriculture departments for dealing with diseased animals. Whale strandings (Chapter 15) require a special management response, and agencies train their staff and provide public information to help deal with these events. Typically, whale strandings are managed by protected areas agencies as a cooperative venture between volunteer organisations such as the Organisation for Rescue and Research of Cetaceans in Australia (ORRCA), research organisations such as the Australian Museum, zoos, and private organisations such as aquaria.

Relocating fauna. Handling of fauna requires special skills. Animals that may be causing danger to humans, such as snakes (in the backyard), crocodiles near settlements, and hostile male kangaroos at visitor destinations, are likely candidates for relocation. Each of these species requires special capture and relocation techniques as well as an understanding of the potential stress and trauma that may be incurred by the animals. Releasing wild animals in new areas requires research and careful planning given to considerations such as the territoriality of some species, the need to maintain genetic integrity of wildlife populations and the potential to simply relocate a problem.

Population control and culling of fauna. Some animal populations, because of improvement to their habitat by humans, have increased, so much so that they have become a nuisance or are causing damage. Such populations may be culled under licence, although this is a very sensitive issue near urban populations. Sterilising of animals has been advocated but is not a very practical option.

Managing for traditional use of native animals. Culling of wildlife by traditional owners occurs on many protected areas across the country. This practice is expected to increase in Australia as traditional owners manage more reserves. Conservation and sustainable use of wildlife will be a critical part of the management of traditional tucker (Snapshot 12. 7)

Snapshot 12.7

Kuka Kanyini – looking after wildlife preferred by Anangu of the Pitjantjatjara Lands

George R Wilson, Australian Wildlife Services; Alex Knight, Anangu Pitjantjatjara Yankunytjatjara Land Management; Lynette Liddle, Uluru Kata-Tjuta National Park

In the Pitjantjatjara lands of central Australia, Anangu are working with wildlife managers to restore traditional land management practices. The program known as Kuka Kanyini is increasing species preferred as bushtucker. It is motivating people, maintaining culture, and creating employment.

Aboriginal Land owned by Anangu—the Pitjantjatjara-, Yankunytjatjara-, and Ngaanyatjarra-speaking people of northern South Australia, the Northern Territory, and near Warburton in Western Australia—crosses three states and includes the Australian Government-managed Uluru National Park. Anangu move backwards and forwards across these borders just as they have for æons. The lands and potential participants, including the Anangu Pitjantjatjara Yankunytjatjara Land Management, the Central Land Council, and Ngaanyatjarra Council, are shown in Figure 12.2. In addition to broadly classified Aboriginal Land, the land covers national parks and Indigenous Protected Areas (IPAs). The management of IPAs recognises the active role Anangu play in the stewardship of the land and the critical importance of traditional land management in maintaining the biodiversity of these areas. However, they could form the basis of a biosphere reserve and could play a stronger role in addressing wider social and economic problems.

In spite of major investments in programs for the health and welfare of Indigenous peoples, life expectancies remain 20 years lower than other Australians and incidence of diabetes and kidney failure are in epidemic proportions. In common with Indigenous people in North America and New Zealand, other socio-economic indicators are also of major concern. Breakdowns in traditional culture and lack of leadership are often found in communities in which these statistics are worst. Communities that are making progress redressing the situation are often those that have accepted the responsibility to do something about the situation themselves. Land

Snapshot **12.7**

(continued)

and wildlife are central to Aboriginal culture, so Kuka Kanyini is more than just a wildlife management program.

Kuka Kanyini seeks to strengthen traditional Aboriginal authority and provide fresh food. It blends traditional knowledge customary practice—the Tjukurpa (law)—with scientific knowledge to improve wildlife habitat and harvest species on a sustainable basis.

The most important activity in Kuka Kanyini is gathering Aboriginal knowledge and information so that it can be supported by western technology and scientific information on wildlife. Gathering wildlife and land knowledge complements anthropological studies that are underway as part of the process of documenting native title. Western technology, such as aerial surveys, enables Aboriginal participants to share their information from a different perspective and new insights are obtained.

Remote sensing technology is supporting traditional burning in small patches. Controlled burning is being reinstituted to reduce the impact of wildfire and increase spatial heterogeneity and diversity. Places where customary practice restricts access for hunting and gathering are being identified. They are often those recommended for nature conservation

Ginger Wikilyiri aerial spotting and sharing his knowledge of land and resource management (George Wilson)

reserves because of their biodiversity significance. Endangered species reintroduction programs are being developed to increase the numbers of animals that were locally extinct. Doing so has the profound effect of strengthening culture and correcting the loss of self-esteem, which follows the disappearance of totemic animals for which people are responsible. Aboriginal rangers are baiting and trapping unwanted feral predators—cats and foxes, rabbit warren ripping, and mustering of camels. School and family groups are cleaning out rock holes to enable them to catch water. Keeping them full of water and maintaining bores encourages more even distribution of animals and grazing and increases the productive potential of the landscape. Breeding and release programs for emus and possums are being developed and sustainable harvest rates are being calculated for red kangaroos.

Coordinating management activities throughout the Anangu Lands and exchanging information is a key component of Kuka Kanyini and proactive wildlife management. These activities are not only enhancing biodiversity, but also creating employment, replacing processed food imports, and supporting maintenance of culture. They are a new focus for programs to address health and the motivational challenges facing Aboriginal communities.

Figure 12.2   Location of Anangu Lands in three Australian jurisdictions

Minimising human disturbance to fauna. Fauna may need protection from disturbance by people during breeding seasons. Protected area agencies may need to restrict access temporarily to caves that are known to accommodate over-winter bat roosting sites, given that repeated disturbance may lead to the loss of the bats (through the burning-up of their stored energy reserves). Whales may need protection from too many spectators. Regulatory work may also be required to deal with disturbances such as jet-skiers harassing Little Penguin; four-wheel-drive vehicles disturbing or crushing the nests of endangered bird species nesting or resting on beaches; boats and swimmers pursuing whales and dolphins; shooters killing, wounding, or intimidating animals; bird

Table 12.2	An Australian wildlife year: some fauna management implications (adapted from Reader's Digest (1989))	
Month	Fauna activity	Potential management actions or implications
January	Far north, Magpie Geese (*Anseranas semipalmata*) laying eggs, the wet Eastern Grey Kangaroos (*Macropus giganteus*) give birth Young Platypus (*Ornithorhynchus anatinus*) emerge from burrows Snakes give birth to young	Management for young orphaned animals Visitor use managed to minimise potential negative interactions with young wildlife
February	Sulphur-crested Cockatoos (*Cacatua galerita*) and Long-billed Corellas (*Cacatua tenuirostris*) form large flocks to feed Australian Ravens (*Corvus coronoides*) congregate in groups to feed on insects Warm waters bring Dugong (*Dugong dugon*) and sea snakes further south on the east coast	Mass flocks of Australian Raven in the alpine areas are of immense interest to visitors Dugong in southern waters may need enhanced protection, given that this coincides with peak holiday times
March	Young Sugar Gliders (*Petaurus breviceps*) are ejected from nests Many migratory birds are leaving for their breeding grounds in Asia Silvereyes (*Zosterops lateralis*) head north to Queensland from Tasmania and Victoria	Pest animal control programs (for example, for foxes) may be tailored to protect young fauna (when they are at their most vulnerable)
April	Tasmania's parrots (including the Orange-bellied Parrot (*Neophema chrysogaster*)) migrate to the mainland Far north, the wet is over, parrots nest in termite mounds, Brahminy Kites (*Milvus indus*) nest Little Red Flying-foxes (*Pteropus scapulatus*) are being born Australian Salmon (*Sciaena truttaceus*) and Sea Mullet (*Mugil cephalus*) migrate north to spawn	Fuel reduction burning programs, often undertaken in autumn in the south, will need to consider critical habitat needs for species that over-winter in areas such as heathlands

Table 12.2	(continued)	
Month	Fauna activity	Potential management actions or implications
May	Tasmania—Eastern Quoll (*Dasyurus viverrinus*) males fight for the right to mate Mallee fowl (*Leipoa ocellata*) cocks renovate mounds that will hold the next season's eggs Gang-gang Cockatoos (*Callocephalon fimbriatum*) migrate from highland forests to urban areas in the south-east Black Swans (*Cygnus atratus*) begin to breed	Special vigilance may be required to protect breeding birds such as swans on wetlands in or near urban environments Visitor numbers to northern Australia start to increase
June	Superb Lyrebird (*Menura novaehollandiae*) males are at their most vocal Southern Cassowary (*Casuarius casuarius*) court and mate Fairy Penguins (*Eudyptula minor*) return to colonies after long absences at sea Great White Sharks (*Carcharodon carcharias*) travel from the Great Australian Bight to the south-western and south-eastern coasts Common Wombats (*Vombatus ursinus*), nocturnal in summer, now appear by day and may have last year's cub at heel, and this year's cub in the pouch	Speed restrictions for vehicles may be required to help protect animals such as wombats in protected areas Visitor use may need to be managed carefully to protect wildlife from too much disturbance during the dry in northern Australia Tourism to see Little Penguins may be promoted in southern states
July	Short-beaked Echidnas (*Tachyglossus aculeatus*) are mating and laying eggs Bandicoots are bearing young Antechinuses begin intense breeding Wattlebirds (and many other birds) breeding Whales are migrating north along east and west coasts	Areas of massed winter flowering, which provide energy sources for many birds and animals, may need to be carefully managed during southern winter hazard-reduction burning programs Wandering Short-beaked Echidnas need to be protected from vehicles and other disturbances
August	Common Brushtail Possum (*Trichosurus vulpecula*) mothers reject their half-grown offspring before coming into season for the second time in the year Humpback Whales bear calves off Queensland and Western Australian coasts Peregrine Falcons (*Falco peregrinus*) are incubating their eggs Wading birds begin to arrive from Siberia	Whale-watching programs will need to be carefully managed to minimise any potential impacts to breeding whales Endangered species such as Peregrine Falcon may need enhanced surveillance protection during the breeding period Fox control programs may be required to protect migrating birds

Table 12.2	(continued)	
Month	Fauna activity	Potential management actions or implications
September	Swamp Rats (*Rattus lutreolus*) begin to breed Pythons and tree snakes are laying their eggs Satin Bowerbird (*Ptilonorhynchus violaceus*) males rebuild and redecorate their bowers Many ducks, spoonbills, egrets, and native terns are beginning to breed Migratory birds arrive from the northern hemisphere, including shearwaters	Softening snow in the alpine areas may require special management for oversnow machines in ski resorts, since snow compaction destroys subsnow air spaces required by native animals Fox-control programs may be required to protect migrating birds
October	Brush-tailed Rock-wallaby (*Petrogale penicillata*) joeys are beginning to leave the pouch Eastern Horseshoe Bats (*Rhinolophus megaphyllus*) are beginning to bear young Many birds are breeding Shearwaters continue to migrate from the north	Migrating immature shearwaters may die en route and be found in large numbers along east coast beaches—special educational information may be required for visitors
November	Dingo (*Canis familiaris dingo*) pups and Tasmanian Devil cubs make their first excursions from the den Short-tailed Shearwaters (*Puffinus griseus*) return and start laying their eggs On Great Barrier Reef islands, Flatback Turtles (*Natator depressus*) and Loggerhead Turtles come ashore to lay eggs In the south, fur seals are breeding	Active protection may be required for turtle breeding areas
December	In the far north, Salt-water Crocodiles are laying their eggs Koala (*Phascolarctos cinereus*) of the south-east are at the height of their mating season Bent-wing bats are giving birth Cicadas are at their most numerous and shrill Red-back Spiders are breeding On the Great Barrier Reef, coral polyps spawn in the billions	Access to some caves with breeding colonies of bats may need to be managed carefully during the breeding season

smugglers raiding nests; and reptile smugglers collecting snakes and lizards.

Managing for native and introduced diseases of fauna. Outbreaks of an introduced disease affecting the agricultural industry can sometimes spread disastrously to native fauna. Managers are often involved in state or territory disaster planning groups. For instance, in 1999 there was an outbreak of Newcastle disease among large-scale chicken farms at Mangrove Mountain near Gosford, NSW. NSW State Agriculture quarantined the area. The disease spread rapidly and, as it had the potential to be carried by native birds, NPWS staff were called in. Thankfully, no such link was found. However, some native birds had to be sacrificed so that their blood and tissues could be examined.

Native fauna as 'pests' to human settlements and agriculture. In and around towns and cities native animals are sometimes considered pests. Sulphur-crested Cockatoos (*Cacatua galerita*) can demolish timbered balconies; native Bush Rats (*Rattus fuscipes*) can eat electrical cabling; Australian Magpies (*Gymnorhina tibicen*) can attack local residents in spring; flying-fox colonies can take to roosting above community areas; and kangaroos and ducks can make a mess of golf courses. Managers are often expected to help the community deal with these matters. Sometimes this is a matter of managing public awareness—for instance, arranging media and education campaigns to warn the public about the impending 'magpie swooping season'. Native wildlife may need to be managed to help protect agricultural crops as well as to prevent some dissident farmers taking actions into their own hands. There may be a need for licensed culling of some species.

Minimising road kills. Native species have natural traffic or pathways they follow, moving around by day or night to feeding areas or on longer-term migratory movements. Regrettably our native species have little 'road sense', and many suffer heavily on the roads. In addition, roadsides can enhance food sources for many species; spillage from vehicles such as grain transport trucks or even litter from cars can attract fauna. Within protected areas, roads can be designed so that vehicles move more slowly and cause fewer road kills. Managers can use wildlife warning signs and wildlife underpasses.

Native fauna harmed by pollution. Managers need to know how to deal with a range of pollution events for land and sea, and the special techniques for cleaning pollution-contaminated animals.

Australian fauna management— a year-round requirement

The animals of the Australian continent are seen not only to be extraordinarily diverse but also dynamic. Table 12.2 gives a brief account of the wildlife activity that occurs during an annual cycle, and indicates some of the management implications that arise.

12.3 Conserving flora

The wattle is Australia's national floral emblem. Other plant species have been adopted as state emblems, including the Waratah of NSW and the Pink Heath (*Epacris impressa*) of Victoria. Australia's tremendous diversity of plant species poses many challenges for those responsible for managing and conserving them. Many native species have adapted to increased aridity, low nutrient soils, and increased fire over the past 15 million years (Benson & Redpath 1997). Plants are essential to the web of life and also provide direct benefits to human societies. These benefits include aesthetic and inspirational value (such as the massed blooms of wattles in spring).

Native flora contribute towards clean air and clean water. Plants mitigate climate and are a rich resource of genetic material. Plants also feed natural ecosystems—not simply with what we humans recognise as food, but with wood, cellulose, leaf-litter, pollen, and green forage. There is a wide range of tasks that may be undertaken in managing flora.

Active management of plant communities

Across Australia, rangers are actively intervening to conserve plant communities. There are many threatening processes (Chapter 14) that threaten or wipe out plant species. Some 46% to 53% of these processes still go on after a protected area is reserved (Leigh et al. 1984).

To prevent extinctions, managers must urgently identify the endangered species, where they are, and what they need to survive. For instance, staff in Kosciuszko National Park had to relocate a visitor walking-track on the highest slopes of Mt Kosciuszko. This was done to protect one of Australia's rarest plants, an endangered snow patch species, *Colobanthus nivicola*. In the World Heritage rainforests of Queensland, rangers have had to become law-enforcers and prevent illegal collection of rare palm seeds. Government schemes for flora conservation depend heavily on data provided by local managers. In the ACT, for instance, managers have worked closely with urban planners to conserve native endangered grasslands and woodlands (Snapshot 12.8).

Snapshot 12.8

Protecting endangered ecological communities in the ACT

Bernard Morris, ACT Parks & Conservation Service, Environment ACT

Natural Temperate Grassland and Yellow Box (*Eucalyptus melliodora*)/Red Gum Grassy Woodland have been listed as endangered ecological communities under the *Nature Conservation Act 1980* (ACT). These communities provide habitat for several threatened plants and animals, including the legless lizard *Delma impar* and Button Wrinklewort (*Rutidosis leptorrhynchoides*). Knowledge of the status of these communities and species within the ACT has been developed through extensive and ongoing research and monitoring.

It was recognised that new building developments had the potential to threaten grassland and woodland ecosystems. Once this problem had been recognised, the next step was to draw up detailed ecological maps of areas, so town planners could see the consequences of any decision. This meant identifying the different types of grassland and woodland ecosystems in a way that town planners could readily comprehend and apply. It also meant giving each area a conservation rating. The resulting maps have been accepted and used by planners for many kinds of land use and community decisions within the ACT.

Protecting plant species with commercial potential

The public has taken an extraordinary interest in the Wollemi Pine since its discovery in 1994. It had previously been known only as a fossil (Snapshot 1.3). The fact that it was also an attractive species guaranteed commercial interest in it. Yet the fact that it had long been isolated to a tiny area left it vulnerable to diseases and wildfire (Benson 1996). To this was now added the risk of disturbance from illegal seed-collectors. Rangers decided that wild specimens could best be protected by the commercial release of the plant. Clones of the existing trees were propagated through cuttings. The wisdom of this decision was shown by later research in 1999 that showed the pines were extremely vulnerable to botryosphaeria, a common fungus that strikes plant foliage. A proportion of the revenue from sales of the pine will be reinvested for conserving the species. Woodford (2000) provides further reading for those interested in this remarkable species. Native orchids, tree ferns, waratahs, and other commercially favourable species are in demand from plant propagators, and management arrangements may need to be put in place. Questions such as who has commercial ownership of the genetic resource need to be addressed.

Rehabilitation—preserving genetic diversity

Disturbed sites often need to be rehabilitated by rangers. To do this they often establish small on-site nurseries. Seeds and cuttings are collected locally, and provide tube stock for replantings. Such nurseries may be scaled up to supply major re-planting projects, including government employment programs, or those of Greening Australia and Landcare groups (Snapshot 12.9). Staff or skilled contractors will normally see to the collection of seed from the protected areas. Professional nurseries may be contracted to do the further propagation of the seedlings. This can work well—but one must make sure that they use only sterilised soil. (The last thing one wants is to introduce new weeds or soil pathogens to natural areas.) Such re-planting projects have been successful in areas ranging from harsh heathlands to rainforests, in each case using local plant species.

Snapshot **12.9**

The Wee Jasper Grevillea (*Grevillea iaspicula*)
John Briggs, NSW Parks and Wildlife

The Wee Jasper Grevillea is nationally endangered. It is one of some 300 species of grevillea known in Australia. The plant is confined to the Wee Jasper area where it survives at six sites, with the largest colony containing around 100 mature plants in a one-hectare area. Most sites have less than 10 mature plants each. Four of the six sites are on private property. Protection of the species has for several years been the focus of the Wee Jasper Grevillea Recovery Team, which includes government organisations, four property owners, and the local school teacher. The major threats to the species today are smothering by Blackberry and Sweet Briar (*Rosa rubiginosa*), and grazing by domestic stock and feral goats (*Capra hircus*)—stock have a particular taste for this plant. As a consequence, the Wee Jasper Grevillea tends to survive only in the most precipitous locations where stock have been unable to reach the plants.

Reintroducing and relocating plant species

Plants that have become extinct in an area have sometimes been successfully reintroduced. On occasion some specimens of a species may be moved away from an area that is highly vulnerable to disturbance. Spreading the specimens of an endangered species can reduce the risk of its extinction. Rangers need to beware of mixing genetically distinct varieties, and to make sure the new site is suitable. A successful example is the relocation of the Button Wrinklewort (Snapshot 12.8) from sites near Canberra to more secure reserves. Relocating plants to new areas, however, requires careful evaluation. Associations between species are often overlooked.

Fire as a tool for managing vegetation

Many Australian native plant species have adapted to a range of fire regimes over millions of years and in particular in the last 100,000 years. As a result, we have a complex and diverse mix of vegetation communities that, in many places, require the application of a fire regime to retain the floristic features of the community. Fire is used as a tool to manage vegetation in an attempt to mimic past fire regimes.

The use of fire in managing native vegetation is the subject of much debate. From the perspective of community protection and local ecology, managers need to carefully plan and research their decisions to apply particular fire regimes to an area (Section 12.8 and Chapter 15).

Permanent vegetation plots and transects

To manage the long-term variations of a plant community it is important to have baseline information. Researchers often mark out permanent vegetation plots or transects. These can provide data to monitor a range of effects. For instance, a build-up of fuel after fires can be important data for fire management. Such monitoring can reveal the history of changes to plant communities over time. These data can then be useful for tracking changes caused, for example, by global warming. The longer the data have been collected on a given plot or transect, the more valuable they are. The transects that were set up in 1959 in the alpine area of Kosciuszko National Park, and monitored for 20 years (1959, 1961, 1964, 1968, 1971, and 1978) by researchers Dane Wimbush and Dr Alec Costin, are now invaluable. Between 1999 and 2002 Dr Pascal Scherrer revisited the transects to continue the monitoring and to determine the nature of changes for a 40-year period (Scherrer 2003). In the short term, the transects were intended to monitor the recovery of the alpine area following grazing. Now they provide an opportunity to appraise longer-term effects such as climate change (Case Study 6.4). Similar plots throughout the country provide benchmark data from which we can keep track of what is happening.

Planning to recover threatened flora species

A small team of managers and scientists are often given the task of devising a recovery plan for threatened species. Such plans are outcome-oriented: that is, they lay down clear objectives and also set criteria for measuring success and failure, and propose remedies if the initial plans fail. Such plans include a detailed estimate of costs, which is regularly reviewed. For instance, an interim recovery plan was able to preserve *Ptychsperma bleeseri*, an endangered rainforest palm endemic to Darwin. The plan included such steps as control of wildfires, weed-control, and control of introduced animals

Surveying mangrove species at permanent vegetation plots, Tully River, Qld (Todd Kelly)

(Liddle et al. 1996). Where it is too costly or difficult to have a recovery plan for every species that may be at risk, some rangers have adopted an area (rather than a species) recovery plan.

Protecting boundaries between major plant communities

There may be sharp borders between different plant communities—for instance, between a rainforest on basalt soils and an adjacent *Eucalyptus* woodland on poorer sedimentary soils. Such natural edges or ecotones may have a variety of causes and may also change over time. These areas, which are of scientific interest, should not be damaged. For example, an ecotone may not be the best place to put a walking track. Managers should also be aware that their fire regime might affect rainforest boundaries. Rainforests tend to retreat in the face of regular fires burning into their edges, but expand and invade adjacent communities when fire is absent.

12.4 Fungi

Fungi play a fundamental role in ecosystems, in decomposing organic matter, and in driving the turnover of nutrients. They are also a critical food source for animals. Thirty-eight species of native mammals, for example, feed on fungal matter (Claridge et al. 1996). Some species, such as the Long-footed Potoroo, dine exclusively on subterranean fungi and also help to disperse the spores. Fungi play an important role in the forming of tree-hollows in old growth forest, creating habitat for fauna. Some flora species, particularly in low nutrient environments such as heath, rely on a symbiotic relationship between the plant roots and fungi to assist in the absorption of nutrients. Managers need to understand the importance of fungi as food for specialised species, as recyclers of nutrients, and as signs of healthy ecosystems. A useful introductory text on fungi is *Common Australian Fungi, A Bushwalkers Guide* by Dr Tony Young (2000).

Fungi (Graeme Worboys)

12.5 Managing water

Water is a basic resource for life on this planet. Protected areas provide relatively undisturbed catchments. These in turn provide water to streams and rivers, to underground streams and seepages, and to groundwater. Hence managers have a role in water conservation in a number of ways: managing the catchment, managing groundwater and drainage, maintaining water quality, and preventing pollution.

Protecting catchments

Stable catchments are an important conservation goal. The aim is to have minimal erosion, and enough plant-cover to allow natural and abundant production of high quality water. Catchment protection calls for a wide range of expertise. Increasingly, protected areas are not managed in isolation, but as part of a whole-of-catchment plan. Cooperating with community groups, catchment management committees, and other authorities is now routine (Chapter 19). Some management goals include:

- minimising the harm done by intense bushfires in catchments
- rehabilitating disturbed catchments

- dealing with the consequences of altered flow regimes (Snapshot 12.10)
- preventing damage to catchments.

Mary White's publication *Running Down, Water in a Changing Land* (White 2000b) is recommended reading for protected area managers on the subject of water in Australia.

Economic issues

Water, and catchments, have economic value (Chapter 8). Managers nowadays must often enter negotiations on the value of water and the cost of supplying high quality water. This is a complex task, but it is leading to a greater appreciation of the value of water and the need for catchment protection investments.

Minimising salt damage from rising water tables

Salty groundwater underlies most of the flatter parts of Australia. The destruction of deep-rooted vegetation allows the water level to rise, bringing salt towards the surface. This is not confined to irrigated areas. 'Dryland salt' affects large areas of Australia.

Snapshot 12.10

Water in Barmah Forest

Barmah is a River Red Gum (*Eucalyptus camaldulensis*) forest and wetland on the Murray River in northern Victoria. Construction of the Hume and Dartmouth Dams upstream reversed the natural hydrological cycle. The river has been kept high during the summer to supply water to irrigators, and has then been allowed to drop during winter as water is stored for the following summer. These changes have upset the ecosystems of the Barmah State Park and State Forest. Some low-lying areas were inundated for unnaturally long periods, resulting in the death of red gum trees. On the other hand, higher areas were receiving less water than they used to. These changes were causing significant changes to the vegetation in the forest, and also disrupting the breeding of water birds.

The managing agency, the Department of Natural Resources and Environment, together with the Murray Darling Basin Commission, scientists, and other stakeholders, have devised several methods for addressing the problem. Low lying areas are protected from excessive inundation by construction of regulators, which control the summer flow of water from the river into the forest. The managing agency has also used experimental embankments and other structures to redistribute water in the forest to try to revert to natural flood regimes. Most importantly, a special environmental allocation of water has been set aside to flood the wetlands during winter and spring. Although these measures have not recreated the natural flow regimes, they have helped to mitigate the problems caused by river regulation. Work is continuing both on establishing allocations of water for environmental purposes, and exploring engineering solutions to the problem.

Today farmers are much more aware of the problem, and are fighting back.

The rise or fall of salty groundwater tends to affect an entire catchment. Defeating salt requires a whole-of-catchment approach. Managers of protected areas can assist neighbouring farmers by providing a stable natural setting (Chapter 19). Protected areas then become the 'safe havens' or 'core areas' from which repairs to the rest of the catchment can expand. White (1997) is recommended reading for managers who wish to investigate this subject in greater detail.

Protecting water quality

Waters flowing from protected areas should be of the highest quality. At worst they should be no worse than when the waters entered the protected area! To ensure this, managers must minimise artificial contamination, and ensure healthy water and stream ecosystems. Sampling and monitoring water quality is skilled work and is often contracted out to specialists. Preparing the contract specifications requires special skills.

Managers should appreciate the importance of the relatively pristine streams and water bodies that they manage. Among perennial streams of reasonable size in Australia, there are fewer and fewer that are completely undisturbed. In 1998, Parks Victoria and the NSW NPWS, for example, produced a cooperative report on the catchment values of the Nadgee–Howe Wilderness. The report ranked the six undisturbed streams of these two reserves as being so rare as to be nationally significant (Ecology Australia 1998). Pristine streams provide essential benchmark data for managing other stream systems.

Many streams and rivers have a history of disturbance, particularly to the streamside vegetation, which may lead to non-natural downcutting and erosion of banks. Conservation works to address such problems usually commence at the head of catchments and progressively proceed downstream.

Protecting groundwater

Groundwater conservation is an important issue. This is especially so for limestone and other karst areas, where surface and subsurface flows are integrally related. Aquifers (underground water-bearing strata) across a whole region are often recharged from the protected areas. This helps to keep local water tables and wetlands stable. Aquifers near the surface may also be the source of water for the facilities used by visitors and staff, especially in coastal areas. Such local aquifers need to be carefully managed—local vegetation and ecosystems may depend on them. In some areas, human activities can readily pollute or deplete aquifers.

In Australia the Great Artesian Basin supplies water to a large part of Queensland, New South Wales, and South Australia. Its aquifers are recharged mainly from the eastern part of the

Hedley Tarn, Kosciuszko National Park, NSW (Graeme Worboys)

basin—that is, from along the Great Dividing Range of northern New South Wales and Queensland. Water in the Great Artesian Basin is heavily used for mining and agriculture. Such present use, quite apart from the proposed future increase of it, may be unsustainable. The managers who look after the Mound Springs in Witjara National Park, north-west of Lake Eyre in South Australia may be dealing with 'fossil water' that is up to 2 million years old (White 1997). They may also have special problems:

Already extraction by the thousands of bores which have tapped the artesian water since its discovery in the 1880s has resulted in a considerable drop in water pressure. Water no longer gushes to the surface, it has to be pumped. The fall in pressure is starving some mound-springs, and further pressure decline will put the rest in serious decline. Mound-springs are a lifeline in the desert on which animals depend for survival, and they contain a unique biota which is facing extinction (White 1997, p. 243).

Dealing with snow and ice

The changed environments produced each year by snow and ice in the alpine areas of Australia create special considerations. Fauna research may require radio-tracking of animals within the spaces below the snow (Mansergh & Broome 1994). Snow loading and snow creep requires special design standards for structures. The thawing and re-freezing of soil moisture into long needle-like shards cause the phenomenon of 'needle-ice'. These lift any exposed soil vertically, and help cause erosion during the spring thaw. Managers have responded with special techniques, such as immediately mulching any disturbed and exposed soils. Methods are outlined in the Australian Alps Liaison Committee's draft *Alpine Rehabilitation Manual* (Parr-Smith & Polley 1998).

Ice on mountain roads is a persistent hazard for traffic in winter. Sprinkling road ice with salt is a traditional way to make it melt more quickly. In the long term, this practice is likely to impact on native vegetation. Other, perhaps better, methods are to spread grit on the road, and to make chains compulsory on vehicles. A further issue is the use of over-snow transport machines, and also of snow-grooming machines at ski-resorts. These machines may compact the snow and hence destroy the air space on which small mammals depend for their survival during winter.

12.6 Managing soils and geology

Sand environments

Sand dunes are not necessarily stable. If they happen to be slightly mobile, this may be part of a natural geological process, which needs to be considered when any rehabilitation works are proposed. Natural movement of sand dunes often exposes Aboriginal burial sites, and the remains may require reburial by local Aboriginal people. Community organisations such as Dune Care may become involved in partnerships with protected area agencies. Non–natural dune erosion is a common problem in popular coastal areas.

Soils

When soil is disturbed in protected areas, a number of measures can be taken to minimise damage or erosion. Such work sometimes requires the use of

Snapshot 12.11

Elevated walkway to the summit of Mt Kosciuszko

At Thredbo Village in Kosciuszko National Park during the 1960s, a chair lift was built that gave access to the high alpine area. This led to many more people using the walking route from the Thredbo Top Station (the top chair lift) to the summit of Australia's highest mountain, Mt Kosciuszko (2224 m). By 1974 people had started to wear a track in the easily-compacted alpine soils. By 1977 there were multiple tracks, with up to 15 different routes being developed in some wet sites along the way. A planning study of the area was completed in the same year, and confirmed that the popularity of this walking route to the summit would increase.

Rangers conducted a detailed study of how best to provide a more environmentally sensitive track. They tested some types of gravel (which failed), and an elevated steel mesh walkway that was made of a light and water-transmitting mesh to allow plant growth below it. The mesh was an outstanding success (Worboys & Pickering 2002, Case Study 11.2). Though the initial cost of installing it was high, a careful analysis of the costings convinced managers that a very long-term low-maintenance final product offset the high up-front capital cost. The multiple-tracking and damage to plant cover has been repaired and alpine plants readily grow under the mesh. By the late 1990s some 60,000 visitors were walking the elevated mesh walkway each summer, as compared to 20,000 visitors in the late 1970s. The method has since been used for walkway construction in the Glacier National Park, USA.

Kosciuszko walkway under construction, Kosciuszko National Park, NSW (Graeme Worboys)

Table 12.3	Examples of soil management actions
Management goal	Possible management actions
To control soil erosion	Regulate activities that can cause soil erosion, including overuse of visitor destinations, illegal four-wheel-drive activities, excessive use of horses.
	Manage for a minimum suitable natural ground cover.
	Take steps to control soil erosion where necessary, including revegetation of disturbed areas and 'roll-over' drains for management access tracks.
To minimise the impacts of introduced soil pathogens	Clean earth-moving plant and equipment prior to entry into protected areas.
	Use, if necessary, pathogen treatment solutions for plant and equipment prior to their use in protected areas.
	Provide boot-cleaning stations for hikers at trailheads to reduce the artificial spread of soil pathogens such as Cinnamon Fungus—this method is used in protected areas in south Western Australia (Barrett & Gillen 1997).
	Use of phosphite sprays to treat broad areas of Cinnamon Fungus infestations, for example, in the Stirling Range National Park of Western Australia (Gillen et al. 1997).
To minimise the impacts of soil compaction	Confine plant and equipment to defined routes.
	Use alternative transport techniques, such as helicopters, to eliminate the use of vehicles in areas prone to soil disturbance.
	Use elevated walkways for areas of intensive visitor use to prevent soil disturbance (Snapshot 12.11).
To minimise impacts from trace elements	Many trace elements such as zinc are highly toxic to plants and animals when leached into soils—galvanised and similar products must be used with care and knowledge.
To minimise impacts from introduced seeds	Use clean earth-moving equipment in natural areas.
	Use only clean (weed-seed free) soil, gravel or hay mulch.

contractors with special skills. Soil disturbance in the alpine area in particular leads to rapid soil erosion and special measures are often required (Snapshot 12.11). Some soil management actions are summarised in Table 12.3.

Geological features

Most geological heritage is irreplaceable. Geological pressures are often slow but they are active every day, influencing the Earth's crust, the landforms and the landscape. Managers need to understand these natural processes and make sure that their management activities do not interfere with them.

Karst and caves

Limestone, dolomite, and other karst landscapes and caves need special management. Karst sections develop over geological time, and often show the effects of past climates and landform processes. They often contain features or deposits of great fragility: subfossil bone deposits, speleothems (cave formations), and solution features (Finlayson & Hamilton Smith 2003). All these are usually irreplaceable. Australia has relatively few caves and karst features because of the aridity of its interior and the scarcity of carbonate rocks in the older geological strata of much of the continent.

Jenolan Caves, NSW (Alison Ramsay)

Figure 12.3 shows the distribution of carbonate karsts in Australia. About 4% of the continent is known to possess carbonate rocks, although this is considered to be an underestimate (Gillieson & Spate 1998). Palaeokarst features suggest that Australia probably has the oldest caves in the world (Jennings 1975). In a comprehensive text prepared for IUCN, Watson et al. (1997) proposed a number of guidelines for caves and karst. Some of their specific guidelines are listed in Backgrounder 12.1. Karst management actions are described further in Table 12.4. Steven Bourne describes scientific research management within the Naracoorte Caves World Heritage fossil site (Snapshot 12.12).

Figure 12.3   Carbonate karsts in Australia (A.P. Spate)

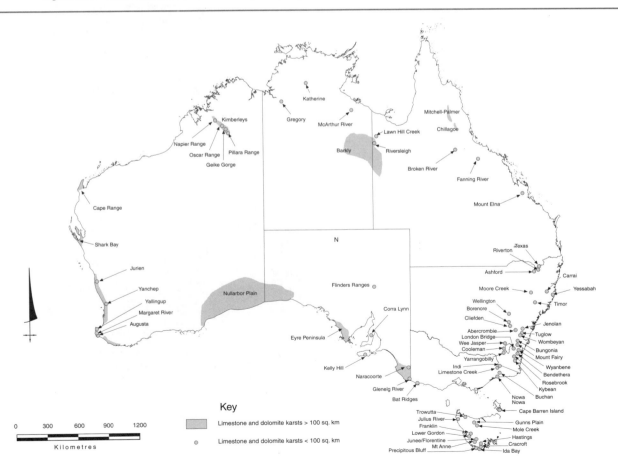

Key

Limestone and dolomite karsts > 100 sq. km

Limestone and dolomite karsts < 100 sq. km

0   300   600   900   1200

Kilometres

Snapshot 12.12

Fossil site management, Naracoorte Caves

Steven Bourne, Department for Environment and Heritage, South Australia

Naracoorte Caves National Park covers 600 hectares in the south-east of South Australia and is a World Heritage fossil site, inscribed in 1994. The research program within the caves is overseen by the Scientific Research Committee, and approved projects are assessed by the park manager to determine methods to minimise impacts to the cave values and to visitor safety, and to deal with logistical problems such as sediment removal and how the site will be managed post excavation. A protocol has been developed that includes a code of conduct for researchers, and the Australian Speleological Federation's Code of Minimum Impact for caving is used if researchers have no previous caving experience. Duplication of research activity is avoided.

Backgrounder 12.1 Guidelines for caves and karst management

Modified from Watson et al. (1997, pp. 4–30)

Two essential characteristics of karst must be taken into account in developing conservation actions. The integrity of karst is closely dependent upon the maintenance of the natural hydrological system, and karst is vulnerable to a distinctive set of environmental influences. Key guidelines for managers are as follows.

- The integrity of any karst system depends on the interactive relationship between land, water, and air. Any interference with this relationship is likely to have undesirable impacts, and should be subjected to thorough environmental assessment. Karst processes are dependent on natural water infiltration processes, natural soil processes, and the natural movement of air within caves. Unnatural soil erosion needs to be managed. A stable vegetation cover needs to be maintained. Groundwaters need to be unpolluted.

- Land managers should identify the total catchment area of any karst lands, and be sensitive to the potential impact of any activities within the catchment, even if not located on the karst itself. Where it is not possible to include the catchment within a protected area, environmental controls, or total catchment management agreements under planning, water management or other legislation should be used to safeguard the quality and quantity of water inputs to the karst system.

- Pollution of groundwater poses special problems in karst and should always be minimised and monitored.

Speleologist, Yarrangobilly Caves, Kosciuszko National Park, NSW (Graeme Worboys)

This monitoring should be event-based rather than at merely regular intervals, as it is during storms and floods that most pollutants are transported through the karst system.

- While recognising the non-renewable nature of many karst features, particularly within caves, good management demands that damaged features be restored as far as is practicable.
- The development of caves for tourism purposes demands careful planning, including consideration of sustainability. Where appropriate, restoration of damaged caves should be undertaken, rather than opening new caves for tourism.
- Imposed fire regimes on karst should, as far as is practicable, mimic those occurring naturally.
- Management planning may need to impose limits on visitor use of caves because of the fragility of the cave system.
- Databases should be prepared listing cave and karst resources within protected areas. Context information for all Australian cave and karst resources is also important.

Table 12.4	Some karst management actions
Management goal	**Management actions**
To protect the cave hydrology	<ul><li>maintaining the natural condition of the catchment as far as possible</li><li>checking any chemicals used for weed-spraying within the catchment to make sure they will not impact the karst system or its plants or animals</li><li>controlling pest animals such as wild pig (*Sus scrofa*) to minimise their impacts on the karst system</li><li>utilising silt traps (that can accommodate peak stream flow conditions) to collect non-natural sediment loads from roads in the karst catchment</li><li>utilising bitumen mixes for road works that do not impact on the cave environments</li><li>building car parks within karst areas so that the water is free-draining through the car park surface to simulate natural water infiltration to cave systems below</li><li>removing any possible sources of pollution, especially petroleum, grey water, sewage, and historical waste dump sources of contaminants—composting toilet systems should be used as an alternative to septic systems for karst areas</li></ul>
To protect caves and cave resources	<ul><li>developing an inventory of all caves and a comprehensive data base</li><li>introducing a cave management classification system as a basis for consistent management of cave environments over time</li><li>managing access to caves classified as 'wild' through a permit system</li><li>utilising management techniques within wild caves to minimise impacts such as closed sections, single-file marked routes, boot-free zones and protective covers such as plastic sheeting</li><li>utilising management techniques within tourist caves to protect cave resources and maximise visitor safety and the opportunity for quality experiences—this could include:<ul><li>access design (stairs, railings, ceiling height)</li><li>safety of in-cave structures</li><li>electrical wiring and equipment design for moist environments</li><li>the placement of lights to show cave formations (when the cave environment and visitor numbers permit)</li></ul></li><li>the colour and duration of cave lighting used to simulate three-dimensional space and to minimise the effects of lampenflora</li><li>the need for cave cleaning as a consequence of lint, body wax, and dirt introduced by visitors, and cave educational features</li></ul>

Table 12.4	(continued)
Management goal	**Management actions**
To protect caves and cave resources (continued)	• the use of gates to manage visitor access to some caves—considerations include the needs of cave fauna (the movement of bats in and out of caves), air flow in caves, the safety of users, and the potential damage to the cave in installing a gate • the need to monitor cave atmosphere conditions to determine whether visitor use or human modifications to caves are influencing natural processes of cave formation/speleotherm formation
To ensure Occupational Health and Safety requirements are implemented for staff and visitors	• the need to monitor caves that are known to contain radon gas—management may need to monitor staff exposure levels • the need to monitor air quality in some caves—excessive carbon dioxide, for example, can build up and be lethal for cavers or even casual visitors
To protect cultural heritage resources within caves	• caves have often been used as burial sites by the Aboriginal community—these sites need to be managed with great sensitivity and in liaison with the Aboriginal community

Geological-type sections and fossil sites

A geological-type section is a scientifically described cross-section that reveals the sequence of the main strata that make up the underlying geology of a region. They are of great scientific value, and are tangible evidence of the Earth's history in that area.

Geological-type sections are found throughout the country and are known to experts in the field. Usually they look after themselves, but they may need to be protected from interference. For example, a misplaced fire-control line put in by a bulldozer could do great damage. An inventory of such sites is needed.

Thredbo landslip stabilisation works (Graeme Worboys)

Table 12.5	Management considerations in utilising geological materials for roads, tracks and other purposes
Proposed management action	Considerations
Use of gravel for a road or walking track through a protected area	Is the gravel material suitable as a road or track base for the environmental conditions the road will be exposed to? Gravel needs to be tested by a suitable laboratory to answer this question. Simple testing could save money for management organisations.
	Is the gravel compatible with the local geology? Granitic rocks vary enormously within a landscape, and subtle variations in rock chemistry can lead to major differences in the physical properties of gravels. In some protected area landscapes, it can be important to ensure that gravels introduced match the local geology. This was particularly relevant in the alpine area of Kosciuszko, where research into evidence of glaciers in the past would be upset if foreign rock-types, including gravels, were introduced.
	Might the gravel introduce unwanted weed seeds or soil pathogens?
	Will the extraction of the gravel cause environmental damage?
	Will the gravel quarry be of value as a geological reference or educational site? Should the site be restored, or left for its educational value?
Use of building stone	Is the building stone to be used compatible with the site geology?
	Is the quarry source an environmentally acceptable location from which to obtain building stone?

Road and track materials, construction stone

Road surface materials and track materials are in constant demand. A number of matters in accessing and utilising such materials, either within protected areas or for protected areas, need to be considered (Table 12.5).

Managing unusual rock types and minerals

Rocks such as serpentinites often contain high levels of trace elements, which can influence vegetation growth and water quality. Such areas may need special management, especially for fire and for any rehabilitation works that may be needed. Serpentinites often contain valuable minerals, and may be the sites of historic mines. Volcanic rocks often contain agates and other collectables of interest to lapidarists. Gold-bearing areas, once mined intensively and now protected areas, often attract the interest of gold prospectors. As protected areas, they must be managed so that

there is no further disturbance. This may require an extensive public education campaign, as well as an on-site presence of rangers.

Managing unusual geological formations

Many geological features tend to be inherently unstable. These include: fault lines; young, rapidly forming erosion valleys in mountainous terrain; and relict unstable slope-deposits, such as the periglacial solifluction deposits of the Thredbo Valley. All these need to be managed carefully. Geological expertise is usually required and managers should employ it as necessary. A landslide in Thredbo Village within Kosciuszko National Park in 1997 tragically led to the loss of life, and the destruction of ski lodges. The subsequent inquest and inquiries were not concluded until 2001, and led to major changes in how the ski resort villages and roads within Kosciuszko National Park were managed (Walker 2001). The stability of geological strata is a critical factor in the management of protected areas.

12.7 Conserving scenic quality

Protected areas form part of a larger landscape. Over much of Australia they provide diversity in a landscape that is substantially agricultural or urban. They are often the visual backdrop to local towns or local properties. Craggy peaks or deeply dissected lands may be appreciated and loved by visitors and local communities. Many scenic features are of great importance to Indigenous communities. Managers can greatly help the cause of protecting these areas by taking into account the scenic values of the landscape and recognising the aesthetic values held by the community.

In conserving scenery, the aesthetic and the visual effect of management actions need to be considered—at the detailed or micro scale as well as at the meso and the macro (or landscape) scale. When designing structures, the choice of colours, materials, shape, and layout should be such as to protect scenic values. Some of the measures to conserve scenic quality are summarised in Table 12.6.

Table 12.6	Some measures for managing scenery
Management goal	**Potential management actions**
To maintain protected areas as an unmodified part of a local landscape	Organisations that manage towers prefer to use local high points on the landscape. Such local high points are often protected areas. During the 1990s, a proliferation of telecommunication towers was introduced into the landscape despite frequent protests by protected area management organisations and local communities. There is a place for such facilities, but they do not need to despoil landscapes. There are many areas in a local landscape that should always be left undisturbed especially when suitable alternative sites are available.
	The location of new structures and facilities within a protected area can be chosen to minimise impacts. Planning by landscape architects can transform the location of new telephone, powerline, road, or building facilities from potential scars to hidden features. This planning is undertaken at micro, meso, and macro scales of scenic protection.
	Organisations that manage utilities or facilities that have created impacts can often be persuaded to become partners in rehabilitating the landscape and restoring its scenic amenity.
To provide opportunities for the community to observe scenery	Rangers may need to trim trees that are blocking the view from scenic lookouts. Occasionally the lookout itself may need to be moved to avoid damaging the habitat of rare or endangered species.
To assist with the rehabilitation of modified landscapes off-park	Rangers may become involved with local community groups such as Total Catchment Management Groups, Landcare, and Rivercare. Their expertise may assist for example with tree-replanting schemes in conservation corridors, in buffer areas, or in the landscape. Satellite imagery may guide the location of landscape-scale conservation corridors.
	Expertise provided by agency staff at bioregional planning level forums might directly influence the conservation of the scenic amenity. The scenic amenity of protected areas is directly affected by development outside their boundaries. Managers need to participate in land-use planning decisions to ensure, as far as possible, that the integrity of the reserve is protected.

12.8 Managing fire

Bushfires have been a natural part of the Australian landscape since at least the Tertiary Period (Chapter 1). Fire has influenced the evolution of our flora and fauna. Humans have altered pre-human fire regimes caused naturally by lightning, or by spontaneous combustion (Bradstock et al. 2002). Based on current knowledge, humans have been present in Australia for more than 40,000 years. Australia has a long history of human use of fire in the landscape. More recently, fire has influenced the landscape through the 200 years of uncontrolled burning associated with European settlement, and nearly 50 years of deliberate hazard reduction burning (Bradstock et al. 2002, Whelan & Baker 1998).

Understanding the historical patterns of fire is integral to efficient fire management planning. The pattern of fire is referred to as a fire regime—that is, how often fires occur (fire frequency), how much heat they produce (fire intensity) and at what time of year they occur (season of fire) (NSW NPWS et al. 1998).

The influence of fire is pervasive on plant and animal communities, so managers need to clearly understand its effects in their local areas. The intended fire regime set by managers should be based on conservation needs, as well as on the need to protect local communities from fire. Changes in fire regimes can transform natural environments. This may be acceptable, provided managers account for the change in clear planning objectives. However, unplanned fires or fire regimes based on inadequate data can produce changes that may threaten biodiversity.

The use of fire as a management tool is a highly debated topic in the community. Flannery (1994) recommended a much more widespread use of fire, his hypothesis being that this was the Aboriginal practice before European arrival. In an article written for the *Sydney Morning Herald*, Benson (1998) warned about the application of hypotheses with limited scientific basis to land management decisions:

> ... *some land management groups have used (information) simplistically to pursue anti-environmental agendas on fire and vegetation management. Flannery's story seems to support the view that at the time of European settlement the Australian bush was composed mostly of grasslands or grassy woodlands ... so some dense vegetation types remaining today are due to a massive 'unnatural regrowth' of trees and shrubs because of a cessation of Aboriginal burning. The logic continues that logging, clearing or frequent burning would restore the pre-European landscape.*

Eucalypt regeneration at Geehi following the January 2003 Australian Alps fires, NSW (Graeme Worboys)

Benson and Redpath (1997) believed the arguments of Flannery (1994) and Ryan et al. (1995) relating to Aboriginal usage of fire were generalised across the Australian landscape, and ignored a complex range of fire regimes that have produced long-standing plant communities in different regions.

Vegetation types such as rainforest, wet sclerophyll eucalypt forest, alpine shrublands and herbfields, and inland chenopod shrublands, along with a range of plant and animal species, would now be rarer or extinct if they had been burnt every few years over the thousands of years of Aboriginal occupation ... Ryan et al. (1995)

and Flannery (1994) ignore much evidence that points to climate as being the main determinant in vegetation change over millions of years ... The adaptation of many plant species to aridity, drought, low nutrient soils and fire does not imply a requirement for them to be frequently burnt. South-eastern Australia's native vegetation is now highly fragmented, after 200 years of clearing, stock grazing and weed infestation. Management of what remains of this vegetation should be based on a scientific understanding of the functioning of ecosystems and the population dynamics of a range of plant and animal species (Benson & Redpath 1997, p. 286).

Backgrounder 12.2 Ecological fire management planning

Roger Good, NSW P&W, and Bruce Leaver, Department of the Environment and Heritage

Fire is very much a part of the Australian natural environment and has over thousands of years contributed to the evolution of most native plant species and vegetation communities to the extent that the dominant sclerophyllous and xeromorphic species have developed mechanisms (adaptive traits) to survive drought and naturally occurring fire regimes.

The majority of native plant species have also developed life cycles that fit within the natural various fire regimes for the regions and landscapes within which they have evolved and established, and many have a degree of dependence on fire to complete their life cycles. Some vegetation communities, particularly the drier sclerophyllous communities produce a mass of dry litterfall and as such provide large amounts of fuel to a fire when an ignition occurs. The resulting promotion of high intensity fire can benefit the community through destruction of disease organisms, grazing insect populations, and parasitic mistletoes. High intensity fire can also clear the understorey and ground litter layers, thereby providing improved seedbed conditions and the reduction of competition for light and soil nutrients.

Under natural conditions the native bushland exists in harmony with the natural fire regimes but unfortunately such scenarios seldom exist today as ever-increasing development within and conflicting demands upon natural bushland areas occurs. Development, whether rural or urban, leads to conflicts between the management of natural areas, including natural fire regimes and that of fire management for the protection of human life and property from the detrimental impacts of fire occurrence.

Plans of management for protected areas seldom address wildfire as an integral part of natural resource management, with fire issues generally being addressed in a separate fire management plan. Unfortunately, this separation of what arguably is the major management issue from the management of protected area values and human use has often led to conflicts between planning objectives and programs, resulting in inappropriate fire regimes or inappropriate levels of impacts on the vegetation and other values from wildfire occurrence.

After every major wildfire occurrence, there is renewed community interest, along with calls for better fire management programs. This exacerbates the problem of integrating fire management into the more general protected area management plans. This situation is no more evident than in upland or mountain protected areas that encompass all or part of a major river system. In such areas, fire management objectives for the protection of human life and property may be in conflict with the management of fire for catchment stability and nature conservation.

The conflict occurs as a result of the perceived need to undertake extensive prescribed burning (hazardous fuel reduction) to reduce the potential impact of any subsequent high intensity wildfire on human life and property, while planning the implementation of fire regimes appropriate to the maintenance of stable natural plant communities. For the high mountains of south-eastern Australia (Snowy Mountains in New South Wales and the High Plains in Victoria) it has been found that in excess of 20 tonnes per hectare (t/ha) of vegetative litter is required to provide protection for some soils and to maintain soil stability on steeper slopes. The objective of most prescribed burning programs is that of reduction of the ground litter layer (fine fire fuel—6 mm diameter litter component) below 10 t/ha; the level that provides for wildfire intensities of less than 3500 Kilowatts per metre (Kwm^{-1}). Fire intensities in the order of 3500 to 4000Kwm^{-1} are low enough to enable short-term direct wildfire suppression operations.

Prescribed burning, if effectively implemented should only remove a proportion of the total fine fuel litter load, so to achieve this, burning is usually undertaken when the weather conditions are mild, (temperatures less than 25°C, winds less than 15 kph, and humidity more than 30%) which in southern Australia is during the autumn and early winter. Similarly, to provide continual benefit in terms of potential wildfire intensities, prescribed burning must be carried out regularly and frequently (between two and seven years) to keep fuel levels below the 10 t/ha threshold. Such burning regimes (low intensity frequent burning) are generally very much different to that which the majority of Australian sclerophyllous and xeric native vegetation has evolved within and to which plant species' life cycles have been adapted. Most native plant species have evolved mechanisms to survive the impacts of drought and natural fire regimes (adaptive traits—epicormic shoots, seed storage capsules, and lignotuber root systems). On the other hand many species benefit from a wildfire event during their life cycle to assist seed release from storage capsules; stratification of soil-borne seed reserves; release of nutrients from the litter layer (ashbed effect); removal of parasitic plants, insect predators, and disease vectors; and the provision of a cleared seedbed in which seed can readily germinate and seedlings establish.

The implementation of prescribed burning is often justified on the basis of these benefits that accrue from a fire event in natural vegetation but a regime of low intensity, high frequency burning can have many detrimental ecological impacts on the vegetation. Leguminous shrub species are common in the Australian bushland and species such as *Bossiaea, Oxylobium, Dillwynia, Hovea, Hardenbergia,* and *Platylobium* produce large amounts of hard seed, much of which is stored in the soil for many years (up to 100 years plus), being scarified by any fire event that occurs. If the fire events are dominated, in terms of frequency by prescribed burning, much of the stored seed is scarified and the understorey can become dominated by these high flammability species, actually increasing the potential for fire ignition and occurrence. In such situations prescribed burning may increase the fire hazard, even though the fine fuel levels have been reduced.

Low frequency high intensity wildfire is an integral factor in the survival and persistence of many species, some having very specific fire requirements in terms of intensity, season, and frequency of occurrence. More frequent low intensity fires, as occurs with prescribed burning, detrimentally impact species having a requirement for high intensity fire once every 50 to 100 years.

Planning for fire management must therefore identify the prime objective for all biophysical areas/vegetation associations within a protected area. Where human life and property issues dominate an area (high-use visitor sites, infrastructure developments, etc) prescribed burning is an appropriate wildfire mitigation and protection program. Such sites seldom cover large areas of a protected area so nature conservation and catchment stability are the prime objectives for the greater part of most protected areas. In these much more extensive areas, which may also encompass wilderness areas, scientific sites, endangered species habitat, and cultural and geoheritage sites of significance, fire regimes must be developed and implemented on an ecological basis to ensure the ecological sustainability and ecosystem functionality of the plant and animal communities of a protected area. Fire regimes that are appropriate to ecological sustainability and functioning are seldom in conflict with catchment management objectives. Such fire regimes do not infer or can include a 'let nature take its course' or a 'natural fire occurrence regime' as human visitation, recreational activities, and the inability of protected area and fire management personnel to guarantee that high intensity wildfires can be contained within planned and defined boundaries, prevail against such management objectives.

Where prescribed burning is recognised as an acceptable and appropriate fire management program, an assessment and quantification of the hazardous nature of the fuel complex must be undertaken. Very few examples exist, where a fuel survey and quantification of the hazardous nature of the fuel has been taken before the implementation of a prescribed burning program. Even where fuel weight alone has been or is the basis of prescribed burning programs, little data is available as evidence of any pre-burn fuel quantification and determination of post-burn fuel accumulation rates. Without such quantification of the fuel loads it is impossible to predict the fire behaviour in terms of intensity and rate of spread of any prescribed burn and hence the actual level of fuel reduction. Similarly, few reports and very little data have been published on the success or otherwise of any prescribed burning program in terms of the level of fuel reduction achieved and the percentage area in any one burn program where acceptable levels of fuel reduction have resulted from a planned burn. This indicates that for the very issue (fuel reduction burning) that dominates protected area management in Australia, research and management have failed to provide the data necessary for effective planning and implementation of major fire management programs. While this situation continues, the development and implementation of prescribed burning regimes, where it is an appropriate management tool, will continue to be a source of conflict in protected area management.

Where fire regimes for a protected area have been developed on sound ecological (vegetation and fuels) data, implementation must still be based on an understanding of the relationship between the fine fuel load, groundcover, slope, aspect, elevation, soil stability, local weather and micro-climate conditions and flammability. Heavy fuel loads (greater than 10 t/ha) do not necessarily infer or contribute to high intensity wildfires. In some high elevation forests fuel loads well in excess of perceived hazardous fuel levels exist, but seldom dry out sufficiently for ignition and subsequent widespread high intensity fire to occur. It is in these forests that some tree species have adapted to very infrequent but very high intensity wildfires, which will continue to occur irrespective of all efforts to prevent such events, as these are a response to very infrequent but extreme fire weather conditions. Such a combination of conditions occurred in 2003 in south-eastern Australia.

The number of days each year when prescribed burning can be effectively carried out must also be considered in the planning of any prescribed burning program. In drier areas or areas where soil moisture deficits regularly occur, even when high rainfall occurs, the number of days when prescribed weather conditions are met may be in the order of 50 to 100 days, but in high elevation areas where the major catchments exist, the number of such days is few. For Kosciuszko National Park which encompasses the headwaters of the major rivers of south-eastern Australia, the average number of days in which prescribed burning can be effectively carried out is only in the order of five to 20 per year. The failure of many prescribed burning programs where the burns have self-extinguished, or alternatively have reached high intensity wildfire proportions, has been as a result of the lack of understanding of the prevailing local weather and micro-climate conditions.

A number of core issues need to be addressed in developing a fire management plan, particularly where a protected area encompasses or is part of a major water catchment area.

- What is the primary focus for fire management?
- Is the total area of the management area part or all of a major catchment(s)?
- What are the land management objectives in terms of catchment management, protection/preservation of natural communities, and use of the land—how will these be effectively addressed and integrated in terms of fire management strategies and actions?
- If catchment stability is a major consideration, what data exists on the soils–slope–elevation–litter weight relationships in terms of maintaining catchment soil stability, particularly if a program of prescribed burning for hazard fuel reduction is proposed? Different soil types on different slope/elevation sites require different levels of litter cover to provide for soil stability. In many situations these levels may actually exceed that desired in terms of hazard fuel reduction.
- What quantitative data for the management area are available on current fuel loads, accumulation rates under different vegetation communities in the range of topographic locations, time since fire, last fire intensities, in terms of impact on the vegetation/vegetative cover, and fuel type?
- What quantitative data are available on the flammability of various fuel types and loads in the management area? Heavy fuel loads do not necessarily equate to high flammability or potential for fire ignition such as in high elevation cool moist forests that produce high fuel loads that only burn in an extreme and prolonged drought situation.
- What quantitative data exists as to fire ignition potential (fire history, lightning ignition history), for the management area and adjoining lands?
- What quantitative data exists as to any paths and sites of regular entry and exit of wildfires to and from the management area?
- What are the real hazards and risks in terms of potential and probability of impact of wildfire on the assets of the area, including both natural values and that of life and property, and have they been adequately assessed?

- If prescribed burning for hazard fuel reduction is planned, have the various fire frequencies (years between implementation) and desired fire intensities been determined for the full range of vegetation types over the full range of toposequences in the management area?
- What levels of fuel are being aimed at in any prescribed burning program (hazard reduction program) and on what basis are these being set?
- Is the prescribed burning to be carried out over all the area or only in quantified 'hazardous' sites/areas, recognising the wide range of elevation, vegetation types, fuel complexes, responses of the vegetation to low intensity fire, and so on that may exist?
- What is the range of prescribed burn intensities being planned for, and is there a capacity to ensure these are effectively met and not exceeded?

If all these questions/considerations cannot be adequately addressed in the identification and development of appropriate fire regimes for a protected area, ecologically sustainable fire management plans and programs cannot be generated. Further planning will be necessary to enable the fire management plan to be effectively integrated with any plan of management for a protected area. In all fire management planning a number of key principles should govern fire management for protected areas. These are as follows.

Principle 1. Fire suppression personnel safety and the protection of life and property are fundamental issues of concern to all fire authorities and land managers and should underpin every fire management strategy.

Principle 2. Fire management strategies must be consistent with the primary land management objectives of the land for which a fire management plan is being developed.

Principle 3. Fire management should be broadly based, involving an integration of fire prevention, preparedness, suppression response capability, and recovery strategies. This will include:

- preparation of an ecologically based fire management plan that is realistic and achievable
- fuel reduction through well developed and implemented prescribed burning in areas of quantified hazardous fuel levels, and mechanical fuel reduction such as slashing and mowing
- early detection and suppression of wildfire ignitions
- strategically located and maintained fire trails and fuel breaks
- a properly trained and equipped fire suppression workforce
- well developed cooperative fire management strategies with adjoining properties/tenures
- well developed community education programs to assist the public in identifying the strategies and actions to be carried out under the fire management plan.

Principle 4. Fire management strategies, including hazardous fuel reduction, should be:

- practical, achievable, and cost-effective
- based on a strategic analysis of risk to the assets (natural, cultural, and physical) that may be impacted by wildfire and any planned burning
- focused on the protection of significant assets and values at risk from wildfire impacts
- based on sound science, particularly that which provides for a clear understanding of the factors that influence fire behaviour (weather, terrain, vegetation type, fuel complex)
- cognisant of the predicted implications of climate change on fire occurrence and behaviour.

Principle 5. A total landscape/catchment approach to fire management, including suppression activities, involving collaboration within and between jurisdictions and across state and territory boundaries.

Principle 6. Community engagement and support in the development, implementation, and review of fire management strategies to ensure that local knowledge, values, and resources are effectively utilised, and to raise awareness, understanding, and appreciation of fire management issues in the community.

Flannery (1998, p. 780) responded with 'In my Introduction I clearly state that "The Future Eaters" is a hypothesis to be tested'. Debates such as these are assisting the ongoing development of our understanding of Australian fire regimes.

Managers need to consider a range of matters when planning for ecological fire management (Backgrounder 12.2). It is clear that the setting of fire management objectives needs to be carefully considered. A fire management plan itself should be based on the best quality data. Fire's effects on plants and animals can vary significantly from one site to another, and a fire prescription that 'works' in one area may not work in another (Bradstock et al. 2002, Whelan & Baker 1998).

In order to conserve environmental values while applying a fire regime, an understanding of responses of flora, fauna, and the ecological community is needed. Finding the best answers demands detailed local knowledge for each area. For instance the Cooperative Research Centre

Fuel measurement, Blue Mountains National Park, NSW (Graeme Worboys)

(CRC) for the Sustainable Development of the Tropical Savannas has mapped burning in northern Australia. They found that in an area of 321,000 km^2 burnt, more than 82% of the fires occurred in the hottest and driest time of the year, which was highly detrimental to fire-sensitive species such as rainforest plants, heath, and cypress-pine.

Research needs to investigate what fire regimes may have been experienced in the past. For example, research in the Lower Snowy area of Kosciuszko clearly identified a much higher fire frequency in the past 200 years. In this case, fire had clearly become much more frequent after European settlement. The wheatfield-like regeneration of native pines (in an area that was previously woodland) was caused by more frequent fires. Banks (1989) did similar research work elsewhere in Kosciuszko. He too found the same pattern of much more frequent fires after European settlement. So at least in this region we have to question the popular theory that fires became less frequent after farmers changed the 'traditional' fire regime.

Air quality also needs to be considered when dealing with wildfires and when planning prescription burns to reduce fuel. On the east coast of Australia, the stable weather of autumn is often very suitable for prescription burns, to reduce fuels prior to the following summer. Yet these same stable conditions can lead to an inversion, which can create smoke pollution and poor air quality for people in cities and towns near the burn. Managers would need to consult local air-quality authorities before burning.

Smoke from fuel-reduction burns near airports and roads can be a hazard to vehicles and aircraft. Typically, managers put notices in newspapers to pre-warn the public. Speed controls and traffic wardens are deployed along nearby roads to minimise risk.

There can also be problems when managers are subject to state-wide or national directives that override their local sense of how a fire is best managed (Snapshot 12.13).

For those with responsibilities for fire management in protected areas, we recommend reading the findings of coroners who have investigated major fire events, the findings of commissions of inquiry into bushfires, as well as Bradstock et al. (2002) and Carey et al. (2003).

Snapshot 12.13

Fire management—Wilsons Promontory National Park

Jim Whelan, Parks Victoria

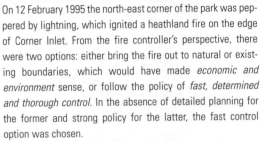

On 12 February 1995 the north-east corner of the park was peppered by lightning, which ignited a heathland fire on the edge of Corner Inlet. From the fire controller's perspective, there were two options: either bring the fire out to natural or existing boundaries, which would have made *economic and environment* sense, or follow the policy of *fast, determined and thorough control*. In the absence of detailed planning for the former and strong policy for the latter, the fast control option was chosen.

Access to the fire was difficult, involving air transport, boat transportation, and four-wheel driving. Water was not available on site, apart from salt water pumped from an adjacent inlet. Aircraft and fire bombing were used extensively. Bulldozers were tracked in through 30 kilometres of wilderness; fire breaks were constructed through woodland, which will take decades to recover; and large quantities of phosphorous-based retardant and salt water were dropped on heathland that is widely understood to suffer from their toxic effects.

This very expensive campaign, involving many thousands of dollars and person hours and producing significant long-term environmental damage, met the *fast and determined control* objective—the fire was contained to 120 hectares and was controlled in the shortest time possible.

Given that burning for ecological purposes was already planned for that sector of the park to restore aged heathlands; the nearest habitation was 25 kilometres to the south; prevailing winds are from the east and west; 90% of the park is surrounded by water; visitor control and evacuation is achievable relatively quickly as there is only one entrance to the park; and the two properties in proximity to the park are eminently defendable—surely the primary objective could have been environmental considerations?

This fire event caused considerable ecological and financial costs to achieve the primary objective of *fast and determined control*. From my perspective I would have preferred to place more emphasis on economic and environmental considerations. The driving force behind our fire management must come from a conservation perspective; if we understand how fire fits into our management of parks we can then look at what further works are required to protect life and property.

12.9 Management lessons and principles

Lessons

1 Management of natural heritage is not easy. It is a complex interdisciplinary process that needs the combined skills of scientists, managers, and local experts.

2 Management must be in tune with the natural dynamic of living things and be adaptive and responsive to emerging threats in achieving conservation outcomes.

3 Managing for the long-term and conservation outcomes is just as important as dealing with short-term objectives.

4 Ecological-based fire management based on good science and understanding is quite critical for the conservation of Australian species for the long-term. There is a need to achieve a capacity to manage fire at a much more sophisticated level.

5 Communicating the reality of natural fire regimes in Australia, and communicating the benefits of biodiversity and ecosystem services to the community, needs to be ongoing and more effective than in the past.

Principles

Management of natural heritage should be based on such conservation principles as those provided in the Australian Natural Heritage Charter (ACIUCN 2002) and in Case Study 12.1. Additional management principles are as follows.

1 Conservation is based on an understanding of ecosystems, biological diversity, and geology. Essential natural and evolutionary processes and biodiversity composition (species and habitats) must be maintained in order to maintain functional integrity.

2   Protected areas should have clear objectives and conservation targets, and be managed according to those objectives.

3   Management should be site-specific and follow an adaptive approach consistent with emerging threats and management needs.

4   Management decisions should be based on sound science and knowledge, including local traditional knowledge.

5   All management interventions should be justifiable in line with the strategic management plan and conservation objectives.

6   Monitoring and research should be based on management needs and prescriptions. All management interventions should be monitored to assess their impact on biodiversity and other protected area objectives.

7   Biodiversity conservation and viability of populations of key species often depends on factors beyond protected area boundaries. Protected areas should be managed within the context of regional land-use planning and development.

8   Protected areas have been established, among other reasons, for the purpose of conserving a representative sample of Australia's natural heritage. Individual protected areas should be managed to enhance their biodiversity contribution to the overall network.

Zoologist and the endangered Mountain Pygmy-possum, Kosciuszko National Park, NSW (Graeme Worboys)

Cultural Heritage Management[1]

> And here at Ubirr the Rainbow Serpent
>
> pressed Herself against rock,
>
> marking and making this place
>
> with self portrait surely
>
> as She made the rest of the land.

From Traditional Owner *by Mark O'Connor*

13.1   An ancient heritage   352
13.2   Types of cultural heritage sites   353
13.3   Conserving cultural heritage   358
13.4   Conserving Aboriginal cultural
       heritage sites   363
13.5   Conserving historic sites   367
13.6   Management lessons
       and principles   370

In Australia the whole landscape is a cultural place, in that it is an artefact of humanity; people have been modifying or giving human meaning to the landscape for at least 40,000 years (Mulvaney & Kaninga 1999). To split natural heritage from cultural heritage often seems tidy and convenient; but conservation staff should remember that this is somewhat artificial. People shape land, and the land in turn shapes people and their culture—a point well made in the National Museum of Australia's permanent 'Land and People' exhibition.

There are deep connections between the physical and biological environment and the rich range of Aboriginal cultures and customs (Chapter 18). As well, since 1788 non-Aboriginal Australians have not only caused major changes to many parts of the landscape and its range of species, but have also established their own strong cultural attachments to the land. Most Australian protected areas (in contrast with English and French national parks, for example, which protect landscapes that include towns and villages) are predominantly natural environments, but include strong cultural links to communities. Cultural heritage plays a vital part in defining group identity; and the value put on certain sites or objects (whether by individuals, groups, or significant proportions of the community) demands that they be actively conserved.

Such significant sites or objects are known generally as 'cultural heritage resources'. These are the physical evidence of past cultural activities. They are valuable, often rare, and could not be replaced if lost. Australia's cultural heritage resources are abundant and diverse. The community has a strong desire to conserve (at least) a representative sample of them for the long term, and this will require active management.

In this chapter we describe some of the ways in which Australia's remarkable array of cultural heritage resources is conserved. It is a brief coverage, but it also offers guidance as to where comprehensive reference material may be found. We focus on cultural heritage that is found within protected areas, and on the practicalities of conserving cultural heritage sites.

13.1 An ancient heritage

A feature of Australia's cultural heritage is the immense antiquity of human occupation. This has led archaeologists to coin the concept of 'Deep Time'. By contrast, most of Europe lay under an ice-sheet during the last ice age, so that its colonisation by modern humans (and by most animals and plants) is relatively recent. In the Americas, there seems to be little evidence of human occupation prior to 14,000 years before present (BP) (Jones 1999, p. 18).

The antiquity of Australian history has only recently been recognised. Nineteenth-century geology and biology destroyed the belief (once held by Sir Isaac Newton and others) that the world was some 6000 years old; yet as late as 1961 Graham Clarke, in the first edition of his *World Prehistory*, considered that there was no convincing evidence for the immigration of people into Australia before post-ice age times. Radiocarbon dating was soon to change this view.

In 1962 Emeritus Professor John Mulvaney (foundation professor of prehistory at the Australian National University) obtained radiocarbon dates of 12,900 BP, and a year later of 16,000 BP, from Kenniff Cave in Queensland. In 1967 Koonalda Cave on the Nullarbor Plain produced a date of 20,000 years. In 1969 a burial in the sand dunes at Lake Mungo was discovered, and carbon-dated at 26,000 to 30,000 BP. It became common to speak of Aboriginal people as having been here for some 30,000 years. By the early 1980s this figure had crept up towards 40,000 BP. Similar dates were also found in Asia to the north. Dates older than 40,000 BP were not easily found.

Yet some archaeologists believed that 40,000 BP was a false limit, produced by the limitations of the carbon-dating method. In the 1990s other methods, including thermo-luminescent dating, proved them right. A much-publicised date of 120,000 to 160,000 BP (obtained at Jinmium in 1996) has been generally discounted; but after excavations in 1988–89, Rhys Jones and others 'put forward two propositions: first, that humans were present on Australia some 53,000 to 60,000 years ago, and second, that they were not there before that time'

Mungo National Park, Willandra Lakes World Heritage Area, NSW (P&W Collection, Gary Bridle)

(Jones 1999, p. 33). Jones noted that if this is correct, people arrived here 'a short time after what most geneticists now believe to have been the explosive expansion of modern *Homo sapiens* out of Africa. The Australian data document one of the key early events of that first global colonisation' (Jones 1999, pp. 33–6). In 1999 one of the Lake Mungo burials ('Mungo 3') was dated to around 60,000 BP. However, Mulvaney and Kaminga (1999) have reinforced the more conservative date of 40,000 BP for the first Aboriginal arrival. Whether the arrival date is 40,000 years BP, 60,000 years BP or older, Australia was a very different place to what we know now. Megafauna were present, the volcanicity, which created Tower Hill in Victoria, hadn't yet occurred, the youngest Pleistocene glaciers to shape Australia's highest peaks hadn't yet formed, and the

lakes at Willandra in western NSW had permanent fresh water (Chapter 1).

Aboriginal elders have sometimes welcomed these older dates as fitting their belief that their ancestors were 'always here'. However, the lack of primates and scarcity of other placental mammals makes Australia an unlikely place for the evolution of *Homo sapiens*. What managers do need to bear in mind is that Aboriginal Australians have deeply influenced the Australian landscape for a period of at least 40,000 years. If we assume there was a new generation every 25 years, this represents some 1600 generations of Australians that have occupied this continent. If we use 60,000 years BP, then it is 2400 generations. If we apply this same arbitrary formula since 1788, only eight generations have been present in Australia since the arrival of the First Fleet.

13.2 Types of cultural heritage sites

Some protected areas are reserved exclusively to conserve cultural heritage resources. Cultural heritage sites have been defined as places that show 'people's interaction with their environment' (Spennemann 1997). As people seek food, tools, other resources or

shelter, they leave marks on the environment. Note, though, that some cultural heritage sites are spiritual, and may not include any physical evidences.

Two major categories of cultural heritage are often distinguished for Australia. These are Aboriginal

Table 13.1	One categorisation of cultural heritage resources (Zerba 1994)		
Environment	Physical Non-moveable	Physical Moveable	Non-physical
Built	buildings, aqueducts	equipment, publications	style, aesthetics, technical knowledge
Modified	mining sites, gardens	machinery, tools	choice of plants, lifestyle
Natural	geology, fossils *in situ*, landforms	fauna, flora	concepts of national park, wilderness
Australian Aboriginal	sacred sites, archaeological sites, art sites	artefacts e.g. stone	language, religion, subsistence lifestyle

and 'historical'. As discussed, Aboriginal sites extend from the first occupation of the continent at least 40,000 years ago to the post European–contact period, and include contemporary sites of significance such as Aboriginal reserves and missions of the late 1800s and early 1900s.

'Historic' heritage includes Macassan (Indonesian) contact sites, and also the extensive signs of settlement and new land-uses since 1788. Different classifications have been developed for Australia's cultural heritage resources, and one of these (Zerba 1994) is presented here (Table 13.1).

Aboriginal cultural heritage

Aboriginal heritage resources are distributed throughout Australia, but vary in type and density according to terrain and local practices. Rock art sites, for example, can occur only where there are suitable rock outcrops. Such sites can be very broadly divided into habitation sites, procurement sites, and spiritual sites (Spennemann 1997). Aboriginal cultural heritage in Australia is rich, varied, and of enormous antiquity (Tables 13.2 and 13.3). It is highly valued by today's Aboriginal community, and it must be very carefully managed. Learning about and experiencing aspects of Aboriginal cultural heritage is becoming increasingly important for visitors to protected areas.

Historic heritage sites

Historical sites include buildings, structures, or remains of general historical interest, and perhaps also of technical, industrial, scientific, or other interest (Flood 1981). They can be divided further into three subclasses: built environments (structures), cultural landscapes (including agricultural and industrial landscapes), and underwater sites (for instance, shipwrecks) (Table 13.4). Built environments include houses, commercial sites, government buildings, power stations and water supplies, churches, temples, cemeteries, bridges, and railway lines (Table 13.5).

Table 13.2	Some Aboriginal sites (Spennemann 1997)		
Standing structures	Surface features	Subsurface features	Natural formations
Mounds	Open campsites	Open campsites	Rock shelters
Fish traps	Isolated artefacts	Isolated artefacts	Scarred trees
Bora bora rings	Dinner-time camps	Dinner-time camps	Carved trees
	Middens	Burials	Rock art sites
	Axe-grinding grooves	Massacre sites	Rock pools

Table 13.3	**Some significant Aboriginal sites and their age** (Flood 1997, 1983; Mulvaney & Kamiga 1999)	
Site	Estimated age	Significance
Mungo burials, NSW	up to about 30,000 BP	Mungo World Heritage Area—oldest known evidence of ritual cremation in the world
Koonalda Cave, South Australia	24,000	Nullarbor Plain, south-west of South Australia—people were obtaining water and flint
Fraser Cave, Tasmania	20,000 BP	South-west Tasmania—occupation of a cave on the Franklin River to survive freezing temperatures
X-ray style art, Ubiri	3000 BP to recent	Kakadu National Park—includes paintings of animals now extinct in the area

Table 13.4	**Some historic cultural heritage resources that may be managed by protected area organisations** (Spennemann 1997, pp. 10, 18)			
	Standing structures	Surface features	Subsurface features	Natural formations
Built environment	Houses and outbuildings	Coastal defence installations	Sewers	
	Shipyards	Mining sites	Cesspits	
	Churches	Airfields	Mine shafts	
	Toilet facilities	Roads	Drains	
	Sawmills	Sports grounds	Utility trenches	
	Bridges	Railway tracks	Tunnels	
	Stores	Aqueducts	Burial grounds	
	Hotels		Water supplies	
	Lighthouses		Cool rooms	
	Shearing sheds			
Cultural landscapes	Streetscapes, Townscapes	Road and rail embankments	Quarries	Beaches
		Public parks	Mines	Recreation areas
		Mine dumps		
Underwater sites	Harbour works (piers, wharves)	Shipwrecks	Pipelines	Natural harbours
		Aircraft wrecks		

Historic wagon, Barrington Tops National Park, NSW (Graeme Worboys)

Old Telegraph Station, Alice Springs, NT (Graeme Worboys)

Table 13.5	Some examples of NSW 'standing structure' (built environment) historic heritage sites and their management			
	Standing structures	Location	Protected area	Management
Built environment	Houses and outbuildings (1836)	Throsby Park, near Moss Vale, southern NSW	Throsby Park Historic Site	Drainage connected, collapsing walls repaired. Managed as a museum
	Shipyards (1936)	Goat Island, Sydney Harbour, NSW	Part of Sydney Harbour National Park	Managed as an operating slip yard
	Churches (1830s)	Hartley Village near the Blue Mountains, NSW	Hartley Historic Site	Used as an active church—extensively repaired
	Toilet facilities (1830s)	Hartley Village near the Blue Mountains, NSW	Hartley Historic Site	Three-hole-to-a-room historic pit toilet—managed as a visitor interest feature
	Sawmills (1860s)	Sawpit Creek, NSW	Kosciuszko National Park	Remnants of an old sawpit used for cutting alpine ash—managed as an interpretive site
	Bridges (1826)	Clares Bridge, Old North Road, NSW	Dharug National Park, Yengo National Park	Old convict road, restored as a walking track
	Stores (1860s)	Yans store, NSW	Kosciuszko National Park	Part of the Kiandra gold fields, managed as a ruin and as a tourist destination, feature of interest
	Hotels (1870s)	Hill End, central NSW	Hill End Historic Site	Used as an operating hotel
	Lighthouses (1881)	Montague Island, southern NSW	Montague Island Nature Reserve	Lighthouse and lighthouse keeper's quarters, used as a visitor destination, and as accommodation for staff and researchers
	Shearing sheds (1875)	Kinchega, western NSW	Kinchega National Park	Woolshed, shearers' quarters, managed as tourist accommodation

13.3 Conserving cultural heritage

Cultural heritage resources need active management because they are essentially non-renewable, and often perishable. They are manifestations of past events, and only a limited number of them were created in the past. Few have survived. Their material fabric also suffers with time, incidents, and disasters. Once destroyed, they cannot be truly recreated. They may be copied or reconstructed, but we cannot renew the spiritual and social and historical moments in which they were created. Each site may be a unique physical manifestation of the activities, ideologies, technologies, and social practices of a particular place and time.

Australians want to protect their cultural heritage and ensure that special sites are conserved for the long term. This may be for a range of reasons, including the following from the Australian Heritage Commission (1985, pp. 17–18).

1   Some sites provide crucial evidence of the emergence of modern humans and their colonisation of this continent at a very early date.
2   Some sites show the complexity of early Aboriginal society in Australia and allow striking comparisons with other centres of early human development, such as the Middle East.
3   Aboriginal rock engravings and paintings supply fascinating evidence of Aboriginal cultural values and creativity. They provide records of Aboriginal perceptions of foreign contact, including Macassan praus (sailing vessels) and European colonists. They record environmental data that are of great interest to science. An example is the paintings of Tasmanian Tiger in Arnhem Land in the Northern Territory, where they have probably been extinct for many thousands of years.
4   Contact sites where Aboriginal and European people first met are of great historical interest. Some, such as massacre sites, have special religious significance to Aboriginal people.
5   Many Aboriginal places where traditional life has persisted have sacred or other symbolic significance to Aboriginal people.
6   Historic structures and places provide us with tangible evidence of modern Australia's past.

They reveal the cultural roots of today's society.
7   Historic places include both the artistic efforts of other eras and examples of structures and settings typical of regional life and work in Australia since 1788. They help us to imagine the past.
8   Historic buildings and remains from other eras provide us with a diversity of building forms that give character and charm to our cities and countryside. Once destroyed they can never be replaced.

Sharon Sullivan provides an account of cultural heritage conservation values and their implications for protected area management (Backgrounder 13.1).

The planning process

The broad principles for conserving cultural heritage are the same both for historic cultural resources and for Aboriginal cultural resources. For both, we need to adopt the planning principles and processes recommended in Chapter 7. We now discuss aspects that are particular to the process of cultural heritage management (Figure 13.1).

The steps of a plan for conserving heritage

Step 1 Identification

First we need to identify accurately the heritage place or object. This means gathering all relevant information about the site, reviewing the literature, consulting archaeologists (or their research papers), and collecting oral histories. Remember to consult cultural heritage registers. State or territory heritage lists, National Trust listings, and the *Register of the National Estate*, the *Commonwealth Heritage List*, and the *National Heritage List* (DEH 2004d) are important sources of information.

The Australian Heritage Council (which replaced the Australian Heritage Commission in 2003) assesses places nominated by the public for the Register of the National Estate. The Register of the National Estate should not be confused with the more recent National Heritage List. Under the system introduced in 2004, the Register of the National Estate is retained as an evolving record of Australia's natural, cultural, and Indigenous heritage places,

Backgrounder 13.1   Cultural heritage in protected areas

Sharon Sullivan, Sullivan Blazejowski and Associates, Nymboida

Cultural Heritage is the value people have given to manifestations of the human past through their associations with them (NSW NPWS 2002a). The criteria for assessing cultural heritage values are based on those used in the *Australian Heritage Commission Act 1975* and the *Burra Charter* of Australia ICOMOS (Marquis-Kyle & Walker 1992) to refer to qualities and attributes possessed by places or items that have aesthetic, historic, scientific, or social value for past, present, and future generations. These values may be seen in places and physical features, but can also be associated with intangible qualities such as people's associations with or feelings for a place or item or in other elements such as cultural practices, knowledge, songs, and stories. When natural elements of the landscape acquire meaning for a particular group, they may become cultural heritage. These may include landforms, flora, fauna, and minerals.

Cultural significance is a concept that assists in estimating the value of cultural heritage manifestations. Those which are likely to be of significance are those that help provide an understanding of the human past, connect us emotionally or spiritually with this past, enrich the present, and will be of value to future generations. Common criteria used in Australia to define cultural significance more closely include the following.

Aesthetic value. This comes from people experiencing the environment and includes all aspects of sensory perception, visual and non-visual, and may include consideration of the form, scale, colour, texture, and material of the fabric or place; the smells and sounds associated with a place or item and its use; emotional response and any other factors having a strong impact on human feelings and attitudes.

Historic value. An item may have historic value because it has influenced, or been influenced by, an historic figure, event, phase, period, or activity. It may also have historic value as the site of an important event. For any given item the significance will be greater where evidence of the association or event survives *in situ*, or where the settings are intact, than where it has been changed or evidence does not survive. However, some events or associations may be so important that the place retains significance regardless of subsequent treatment, such as with massacre or explorers' landing sites.

Scientific value. The scientific or research value of a place or an item will depend upon the importance of the data involved, on its rarity, quality, or representativeness, and on the degree to which it may contribute further substantial information about environmental, cultural, technological, and historical processes.

Social value. This embraces the qualities for which a place or item has become the focus of spiritual, political, national, or other cultural sentiment to a majority or minority group. It is a special meaning important to a community's identity, perhaps through their use of the place or item or association with it. Cultural heritage, which is associated with events that have had a great impact on a community, often has high social value.

Indigenous values are embodied in the cultural, spiritual, religious, social, or other importance an item may have for Indigenous communities. Indigenous items may have other layers of significance as well as those mentioned here; these meanings are defined by the Indigenous communities themselves (Lennon 1999).

This analysis, while using the well-documented term social value as defined above, recognises that in one sense the overarching value of all heritage items is their value to society. The recent NSW National Parks publication, *Social Significance, a Discussion Paper* (Byrne et al. 2001), makes the point that all cultural heritage has a level of social value as an overarching attribute, which includes other attributes such as aesthetic, scientific, or historic value as subsets of this general social value.

Australia's natural and cultural heritage is inextricably entwined. They form a continuum rather than being separate entities. The whole of Australia is a cultural artefact as well as a natural landscape—there is no part of it that has not been profoundly affected by Indigenous and/or settler culture, beliefs, and specific actions, including those of the conservation movement.

Cultural heritage manifestations in protected areas are of particular importance. Many protected areas have cultural values as a major element of their identified significance—for example Carnarvon National Park with its spectacular display of Indigenous art, or Sydney Harbour National Park, containing many of the most important remains of the early history of Sydney. Although this may not be immediately apparent to the manager, many protected areas that have been set aside primarily for their natural values also conserve an important sample of cultural heritage places and landscapes because of their representative element, and their large scale. These cultural heritage manifestations are unlikely to be conserved as well elsewhere.

For example, Kosciuszko National Park was set aside initially for scenic and natural conservation, and for recreation opportunities. Present in Kosciuszko, along with its scenic wonders and its unique alpine environment and directly as a result of the existence of this environment, is an immense range of heritage values not preserved elsewhere. Here the Indigenous people held the great Bogong moth festival and continue to use the Park to this day. Here are their sacred places. Here are displayed the history of pastoralism and mining—huts, yards, mining villages, and remains. Here we see early and lavish visitor accommodation, and the early history of skiing in Australia, the engineering achievements of the Snowy Mountains Scheme, and pioneering evidence of nature conservation. These are all part of the rich tapestry of cultural heritage places in Kosciuszko (Sullivan & Lennon 2003). Equally important are the contemporary memories, associations, traditions, and folkways of all the people who have lived and visited the area over thousands of years. Some of these manifestations—for instance the wild horses that remain from the pastoral era—have an undesirable side in terms of nature conservation but may tell us important stories about human life, achievement, and folly on this continent and in many cases add an important symbolic or sacred element to our understanding of place. The interaction between the natural and cultural heritage values of a protected area adds richness and depth to the story of the place.

This evidence of the past human history has sometimes been seen as essentially in conflict with, or subservient to the conservation of natural values. An earlier style of park management attempted to 'disappear' evidence of past (especially non-Indigenous) use and to exclude or suppress evidence of its continuation. Hence, at Kosciuszko the initial Park approach was to demolish large parts of the historic mining village of Kiandra. Apart from destroying valuable heritage, this approach embittered the previous inhabitants of some protected areas, and created a poor local environment for any form of conservation. Modern managers and certainly the general public are much more aware of the value of cultural heritage and seek to conserve it.

Successful conservation of cultural heritage requires:

- a proper and honest assessment of all the elements of significance, both natural and cultural, of the protected area
- development of policies and priorities, which protect both the natural and cultural heritage and strike a balance in cases of conflict
- close consultation with, and involvement of, the people whose cultural heritage is represented in the protected area
- development among park staff of specialised skills, or access to specialised advice, to effectively protect cultural heritage
- familiarity, on the part of the manager, with best practice methodology for cultural heritage identification and conservation
- perhaps most importantly, an understanding of and commitment to the canons of the Burra Charter.

which are worth keeping for the future. It is a register recognised under the *Australian Heritage Council Act 2003*. The National Heritage List is a list of places with outstanding heritage value to Australia, which is also assessed by the Australian Heritage Council.

This process of collecting information must not be skimped. Simple items such as a brick or a length of barbwire found on a site may give vital information, and should not be undervalued. Consulting experts from a wide range of conservation disciplines

for cultural heritage will ensure the right information is available for decision-making, and texts such as Lennon's (1992) book on historic places on public land in Victoria are a valuable resource. Understanding the social value of a place is critical in determining its significance (Chapter 2, Johnston 1992).

Step 2 Determining cultural significance

This is a simple but critical step. Its aim is to identify and assess the attributes that make a place or

Figure 13.1   Heritage conservation planning framework (Pearson & Sullivan 1995, p. 10)

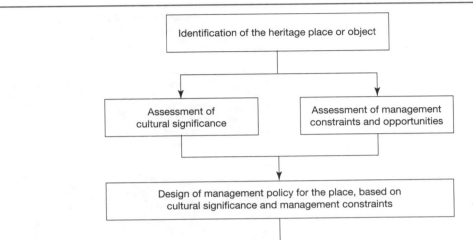

object of value to us and to our society (Kerr 1996). This helps managers set priorities for conservation. There is, however, no single set of criteria for determining significance. Useful guides in this process are Kerr (1996), the Burra Charter, the criteria used for entry onto the Register for the National Estate (Backgrounder 13.2), and Sullivan & Bowdler (1984).

The Burra Charter is the Australian version of the Venice Charter, an international document that deals with the preservation and restoration of historic monuments. There was a particular need for an Australian interpretation of this document. At a meeting at the historic mining town of Burra in South Australia in 1979, the Australian International Council on Monuments and Sites (ICOMOS) Charter for the conservation of places of cultural significance was adopted. ICOMOS is an international organisation under the auspices of UNESCO. It brings together people who are concerned with promoting good practice in caring for culturally important places. The Burra Charter sets standards for heritage conservation in Australia. It defines the basic principles and procedures to be observed. Its principles are adaptable enough to be applied to, for example, a monument, a lawn, a courthouse, an Aboriginal midden, a rock painting or engraving, or even a whole district or a region. It is the critical reference for all managers of cultural heritage in Australia. An easily interpreted step-by-step account of the Burra Charter is given in Marquis-Kyle & Walker (1992).

The Burra Charter includes conservation principles and processes that address the preservation, restoration, reconstruction, and adaptation of cultural heritage. The charter emphasises three essential steps that must be taken, and in the correct order, when managing a heritage site.

1   Assess the cultural significance (gather evidence, analyse evidence, and decide what is significant).
2   Develop a conservation policy and strategy, and gather information; decide conservation policies; and decide a conservation strategy.
3   Carry out the conservation strategy.

The term 'cultural significance' as used in the Burra Charter means aesthetic, historic, scientific, or social value for past, present, or future generations. An issue on which the Burra Charter takes a strong stance is that the total integrity of the site must be preserved, not simply selected features, and not only structures of a particular era. This is partly because we cannot reliably predict the values and interests of future generations. Sometimes what one generation sees as an old ruin, another may see as a historic relic. In general the site's entire history should be documented and protected. The 'jazzing up' of sites to attract tourist dollars is wrong and also counter-productive. In general, authentic history draws more tourists than synthetic or recreated history.

Step 3 Considering management constraints and opportunities

This is the pragmatic part of the assessment process. What is possible and practical for the site? Broadly, there are three possible approaches:

*   *Passive management.* Merely monitor the site (at least until active management is needed).
*   *Active management.* This may mean anything from low-level routine maintenance to a range of conservation strategies sometimes distinguished as: preservation, restoration, reconstruction, and adaptation.
*   *Destruction.* A drastic (and final) option that must sometimes be considered. This could, for example, include salvage archaeology for a site threatened by development.

Step 4 Determining the direction and priority of conservation works

This final planning step includes setting policies, strategies, and methods of monitoring progress. It should include consultation with local communities and stakeholders (Snapshot 13.1).

Snapshot 13.1

Involving the community to assess cultural heritage
Anthony English and Sharon Veale, NSW P&W

Rangers talking with property owner, Culgoa National Park, NSW (Anthony English)

Culgoa National Park is 200 kilometres north of Bourke in NSW, and includes part of the Culgoa River floodplain. In 1996–97 the NSW NPWS made a significant attempt to assess this area's cultural heritage. We made a point of seeking the help of local people, both Aboriginal and non-Aboriginal. We met with them in their homes, on properties, in the main street, and in local museums and community halls. They generously shared with us their memories and ideas—and confronted us with their expectations. Aboriginal people spoke of how they had maintained their traditional life and layered it with elements of European culture, with pastoral work, and with Christianity. The Aboriginal people had strong views on how the park should be managed. Graziers had practical questions for us. How would the new park affect various handshake agreements between them over water, shortcuts, and use of shearing sheds?

Involving the varied local groups in park planning was not trouble-free. Yet we persisted, and succeeded. As we came to share the local people's knowledge and experiences, they felt a sense of inclusion, ownership, and equality. It takes time, and resources, to build trusting and respectful relations with people; but such trust is part of cultural identity and maturity. It helps us say who we are. In short, building such relations is an essential part of managing cultural heritage.

Backgrounder 13.2	Nine criteria used for assessing the national estate significance of a place

Commonwealth of Australia (2003b)

The public can seek the registration of places on the Register of the National Estate. Nine criteria are used for assessing the national estate significance of a place. They include:

1   its importance in the course, or pattern, of Australia's natural or cultural history
2   its possession of uncommon, rare, or endangered aspects of Australia's natural or cultural history
3   its potential to yield information that will contribute to an understanding of Australia's natural or cultural history
4   its importance in demonstrating the principal characteristics of:
    • a class of Australia's natural or cultural places, or
    • a class of Australia's natural or cultural environments
5   its importance in exhibiting particular aesthetic characteristics valued by a community or cultural group
6   its importance in demonstrating a high degree of creative or technical achievement at a particular period
7   its strong or special association with a particular community or cultural group for social, cultural, or spiritual reasons
8   its special association with the life or works of a person or a group of persons of importance in Australia's natural or cultural history
9   its importance as part of Indigenous tradition.

13.4 Conserving Aboriginal cultural heritage sites

We now consider in more detail the ways of conserving cultural heritage sites, beginning with Aboriginal sites. There have been important changes in the way in which Aboriginal sites are managed in Australia. Consistent with increasing Aboriginal self-determination, there has been a shift from sites being assessed for their scientific importance and managed by expert archaeologists, to a situation where Aboriginal communities largely determine a site's future. This often means less emphasis on the scientific value of a site. Many sites may be known only to Aboriginal communities and this needs to be respected by protected area agencies.

Threats to Aboriginal cultural heritage

There are several kinds of threats to heritage (see also Chapter 14). Today we find more and more Aboriginal conservation staff either working for boards of management or within agencies to deal with such issues, usually in consultation with Aboriginal community groups.

Incidents and emergencies. When heavy machinery is used to deal with incidents like bushfires, it needs to be supervised by locally based experts with knowledge of Aboriginal sites. Access to Aboriginal site records is required when planning the use of heavy machinery (Chapter 15). Bulldozer drivers, while fighting a fire, can be relied on to follow directions about avoiding nearby Aboriginal sites, but need clear instructions.

Development. Sites are at risk from property development, from the extraction of resources like timber and minerals, as well as during the installation of utilities and services.. Staff may need to provide special background briefing information for decision-makers (such as local government planning control staff) and developers. Note that conservation agencies may be responsible for Aboriginal sites even on private land.

Tourism. Walking tracks, camping sites, and other facilities must be planned and located with care. Often, the best sites for visitors are the same sites that Aboriginal people used; so placement of visitor facilities needs to be adequately researched. Aboriginal sites may need special protective structures, like boardwalks, to confine and direct visitor use. Sites may also need to be rehabilitated.

Vandalism. Vandalism can mean unforgivable acts of destruction. It can also mean graffiti that defaces without destroying. It can take subtle forms. For instance, someone examining a rock engraving may, without authority, re-etch it with a penknife, the better to view it. In removing spray paint, for example, managers may need the advice of experts to ensure further damage is not done. Damage can also occur through ignorance, such as the touching of art sites by visitors.

Theft. Theft of artefacts does occur. Regular supervision of the site and prosecution are two responses to theft. Education is another means to get the message across. Off-site location of artefacts is an alternative form of protection worth considering.

Impact of animals. Feral pigs may dig up Aboriginal sites and thus confuse the archaeological strata. Inland rock-art sites may be damaged by feral goats or cattle that seek out cool, shaded overhangs. Bird nests, possum nests, rodent nests, and kangaroos using caves formerly occupied by Aboriginal people as shelter sites and rubbing sites are other common problems. Insects such as termites (with their runs), wasps (with their mud nests), and feral bees (with their honeycomb hives) may also disturb sites, especially art sites. Regular maintenance will control most insect problems but with larger ani-

mals, fencing or culling may also be needed. Rangers may face a dilemma when the ripping of rabbit warrens exposes or displaces artefacts.

Impact of plants. Tree roots may invade and enlarge fissures within art sites; lichens and other flora may generate humic acids that stain or degrade sites; and wind-blown shrubs or trees may fall on, or rub against, sites. Blackberry have protected some sites and damaged others. Regular maintenance is the answer.

Erosion. The impacts of wind and water on a site must always be assessed. Where erosion is entirely natural, managers should determine whether they need to intervene; but where it is not, rehabilitation works will usually be needed. Water running along rock surfaces or seeping into art sites through the rock may need to be managed, perhaps with artificial driplines (usually a line of silicone near the entrance to the cave or rock overhang).

Fire. Fires can destroy scarred trees (such as canoe trees) and ceremonial trees. As well, smoke, ash, and soot from fires can affect sites. Campfires beneath rock overhangs can damage rock art.

Managing some types of Aboriginal sites

A site plan should document the values of the site, threats, and the management responses that are necessary, practicable, and affordable. Remoteness or ease of access to visitors may be relevant. Some sites

Engraving site at Bulgandry, NSW, scratched by vandals—before restoration (David Lambert)

Engraving site at Bulgandry, NSW, scratched by vandals—after restoration (David Lambert)

are intrinsically fragile (such as stone arrangements and paintings) and may require special treatment. Other sites are robust and need little or no intervention. Practices common to many sites include rehabilitation of disturbance, salvaging, and regulation. If appropriate, the nature and purpose of site conservation works can be described for visitors using temporary signs.

Open campsites. Open campsites are the most common type of site, and are found throughout Australia (Snapshot 13.2). They can comprise a single artefact or many thousands of artefacts spread over many kilometres. They are usually sites once occupied by Aboriginal people where they may have prepared

Snapshot **13.2**

Cataloguing Aboriginal sites at Namadgi
Trish MacDonald, ACT Parks and Conservation Service

Artefact site, Namadgi National Park, ACT (Trish MacDonald)

Namadgi National Park in the ACT has a large number of campsites and artefact scatters throughout the park. One of these is said to be the highest elevation camping site recorded for Australia. Most sites are protected by being unmarked and inconspicuous to the non-expert. Yet it is important that the rangers know where they are. Hence all sites are located and comprehensively entered into a database Geographic Information System. All rangers have a complete list and map of sites in their area; and this is easily updated. It is quick and easy for field staff to check that work they are carrying out does not affect known sites.

and eaten meals, and worked on stone tools and manufactured artefacts. There may be evidence of cooking fires. The sites are usually located along watercourses, though this is less true at higher altitudes, probably because of the way cold air drains down to valley bottoms on calm winter nights. Erosion is a common threat to these sites. Collection of artefacts is a real problem, and the fencing of a site may cause further damage. Revegetation of a site may be a better means of protection.

Middens. Middens are mounds of discarded items and remains of meals and so on, which may have built up over many generations. Shell middens are common in coastal and riverine areas, and can be mounded or flat. They contain important information about the past, since organic matter is preserved in their alkaline environment. They may contain human burials, as well as the bones of prey animals, plus stone and bone artefacts. Some middens have been protected by fencing or boardwalks or even by sea walls. They are susceptible to erosion and trampling, and may require special structures (such as boardwalks), or stabilisation and revegetation.

Mounds. Mounds are artificially raised habitation sites that are common in riverine areas prone to flooding. They have been built up through regular occupation of the same site over long periods.

Caves or rock shelters. Natural rock formations such as tafoni (a shelter within a rock formed with a cavity and an overhang) gave people shelter against wind and weather. Artefacts and food remains may be found in their floor strata. Both the strata and any art on the walls need vigilant protection. Bushwalkers often use these sites for shelter, and special education and regulations may need to be introduced to protect the sites.

Scarred trees and carved trees. Large old trees may have a number of large scars where bark or wood has been removed from one or more sides of their trunk. This may have been done to make a canoe, a utensil (such as a shield), or a bark shelter. Such trees are common along the inland drainage systems of Australia. Carved trees, with their intricate geolinear designs, have a ritual significance for Aboriginal people. This may be ceremonial; or the carved designs may indicate ownership, or totemic or kinship affiliations. These trees are becoming less common. Determining authenticity is a real issue. Protecting them from wildfire may require special measures. Simple fencing-off does not protect them, since it may attract vandalism, so salvaging may be an option. Carved trees are particularly rare in eastern Australia because of its fire regimes. Managers may choose to avoid advertising their whereabouts.

Quarry sites. Quarry sites occur both in igneous and metamorphic rock outcrops suitable for making stone axes. Aboriginal people usually preferred igneous rocks for making edge-ground tools because of their greater hardness and fine grain. Silica-rich rocks, because of their conchoidal fractures, were favoured for making flaked objects. With careful flaking these rocks give sharp edges, so they were used for knife-edges and spearpoints. Quarries can be difficult to recognise, but normally show evidence of extensive stone working around the outcrops. They are important, not only as sources of Aboriginal tools, but also as the source-point for trade-routes that distributed stone tools throughout Australia. They have to be very carefully protected from collectors and disturbance. A knapping site consists of the flake stone that is left behind at a quarry or a tool-flaking site. Typically it includes discarded flakes and accidentally broken or half-finished products. To protect such sites and their strata from collectors, some managers have fenced off large areas and allowed their revegetation.

Axe-grinding grooves. Axe-grinding grooves are commonly found on flat areas of softer rock, such as sandstone, and near water. The grooves are the result of the final shaping of stone axes and other tools. Water erosion is a problem, as is failure to recognise these features.

Burial sites. Aboriginal peoples practised a range of customs for disposing of their dead, but simple burial or cremation were the two most common. Burials can be found in campsites, caves, middens, and, in many regions, sand dunes. In the last 200 years Aboriginal burials have commonly been linked with Christian rites, in which case their dead were buried in cemeteries. These cemeteries are of great significance to Aboriginal communities today, and agency staff typically work with them to identify and protect such sites. Managers may be involved in helping communities to salvage disturbed sites and to rebury remains.

Recovery and reburial of Aboriginal remains. Regrettably, some Aboriginal burial sites are disturbed by work within the landscape. Also, natural erosion, for instance in sand dunes, can re-expose them. Work is normally stopped when such sites are uncovered, and managers work with the Aboriginal community to ensure a proper reburial. Some managers may also be required to rebury Aboriginal bones that were collected during earlier periods in Australia's history, and that have been retained within collections in museums and scientific institutions. Usually these reburials are respectfully completed in cooperation with Aboriginal communities.

Rock art. Rock art is a spectacular part of the rich Aboriginal cultural heritage, and is found throughout Australia (Snapshot 13.3). It includes drawings, paintings and stencils, engravings, and pickings. The first three of these use a number of colours, including red and yellow ochres, white pipe-clay, and charcoal. Typically they are found in suitable rockshelters and overhangs. The art includes representations of animals, birds, fish, humans, geometric figures, weapons, animal or bird tracks, and Dreamtime figures. A range of designs and colour schemes is used. Such art is easily damaged. Its repair is a specialist area and must involve the Aboriginal communities.

Engravings. Aboriginal rock engravings include carvings and pickings. In particular, the engravings depict, in outline, a range of figures, including animals and birds, their tracks, fish, weapons, humans, and often Dreamtime or mythical figures. These sites, found often on large, flat, open rockfaces, need special protection from horse, foot, and vehicle traffic. Engravings are easily disturbed by wear and tear of rock surfaces as well as by engraving 'enhancement' through photographers and others highlighting the engraving. Education and regulation are two responses to this activity.

Ceremonial grounds. These are commonly known as bora or keeparh grounds, and were used for initiation rituals. A bora ground consists of two circular areas surrounded by low earth banks that have been cleared of all vegetation. The two rings are connected by a path, which may also be marked by earth banks. Stone arrangements were also often built at bora rings or inside them. They vary from simple cairns or piles of stones to elaborate arrangements, circles, or other designs. These sites are easily disturbed and require constant protection. They are regarded as very sacred places.

Natural sacred sites. Many natural features in the landscape have spiritual or historical significance to Aboriginal people and often represent a direct link with their traditional culture. These include rock outcrops, water-holes, caves, and giant tors on mountain sides. Managers need to be sensitive to such sites when managing landscapes.

Snapshot 13.2

Rock art at Namadgi
Trish MacDonald, ACT Parks and Conservation Service

At Namadgi National Park in the ACT the original technique for protecting rock art was 'closing off'—to exclude the public from sites. This is now giving way to 'interpretive management' of sites. In 1989 the University of Canberra ran a Graduate Diploma course in conserving rock art. Namadgi benefited by being the nearest relevant park, and now uses some of the plans for 'best management' that were developed for its sites by the graduate students. Some sites are still kept secret to prevent vandalism; and in one case rangers deliberately re-routed the trail to a remote site. Yet for other sites, like the Yankee Hat art site, the policy has been to install a walking trail and to provide a viewing platform and interpretive signs. This is now one of the most heavily visited sites in the park and yet there has been no damage to it.

13.5 Conserving historic sites

There are many types of historic sites that require active management and conservation work. Examples of a past Australia that protected areas conserve are: old eucalyptus-oil stills deep in the bush, stockyards, shearing sheds, wattle and daub houses, Cornish-style mining cottages, mountain cattlemen's huts, mining races, whaling stations, and lighthouses. Protected area agencies have a special role in Australia since in many parts of the country they are the only organisations able to provide conservation work on rural heritage. The history and heritage of Australia's ordinary citizens could easily be lost.

Threats to historic sites

Theft and vandalism. Regrettably historic structures are vandalised regularly, and require constant supervision. Theft of artefacts occurs from time to time.

Overuse. Managers who observe increasing wear and tear to heritage sites may need to reduce the number of visitors today to conserve the site for the future.

Cape Otway Lighthouse, Vic (David Hill)

Fire. Heritage structures need constant care. Old electricity cabling and tinder-dry timbers are hazards. The activities of visitors and of maintenance staff can add to the risks. Bushfires too should be planned for. There must be adequate fire extinguishers, proper fire hydrants and ample reticulated water, as well as an availability of trained staff.

Water damage. Poor drainage can lead to rising damp, subsidence, and salt damage. These are some of the most serious threats to old buildings, as fixing them is often costly and intricate. The threat of damage or destruction to heritage sites caused by flooding should be considered in plans that are established to conserve these sites.

Storm damage. The Sydney hailstorm of 1999 has been Australia's single most costly natural disaster to date. As this suggests, severe storms are a major threat. Managers may need to urgently re-roof and otherwise protect historic sites after storms.

Animals. Wombats may dig up or destabilise foundations; termites may eat through beams; rats may gnaw electrical wiring; birds may create a fire hazard by bringing nesting materials into the roof. Vigilance and ongoing maintenance are needed.

Plants. Trees can clog drainpipes or fall against buildings during storms. Moss and mould can damage damp interiors, and tree roots can weaken foundations. The routine maintenance plan must deal with all these issues.

Ship's insignia engraved on rock, Quarantine Station, Sydney Harbour National Park, NSW (Graeme Worboys)

Adaptive reuse of historic sites

Managers will normally have a comprehensive plan for each heritage site. Such a plan lists the site's cultural significance and its special management needs. Where there is insufficient money to meet these needs, there may need to be a holding strategy until funds are available. Another option is 'adaptive reuse' of the site as an alternative to letting it fall into neglect. Such 'reuse' must be guided by a conservation plan for the site as well as by a management plan. The Burra Charter provides guidelines for such planning. Adaptations can range from the simple conversion of old shearers' quarters at Kinchega National Park into tourist accommodation, to the extensive refitting of Coolart homestead (Snapshot 13.4).

Snapshot 13.4

Adaptive reuse of Coolart homestead
John Grinpukel, Parks Victoria

Werribee Park and Coolart are two rather similar historic farm sites managed by Parks Victoria. Werribee Park, with its grand mansion and farm buildings, is a major tourist site. At Coolart the surviving buildings are in themselves of less cultural significance, though still an important part of the ensemble. Different conservation plans have led to quite different treatment of these properties; yet both treatments are, we believe, within the rules and spirit of the Burra Charter. At Werribee the emphasis has been on meticulous research leading to preservation, reinstatement, and reproduction of the original detail of the buildings. At Coolart the conservation plan focuses much less on preserving the surviving buildings unchanged. It views them as important, less for the intact detail of their architecture, than as generally indicating the 'grand country house nature' of the ensemble. Hence it allows adaptive reuse, provided this does not diminish the general form and character of the buildings, or of the wider site. The result has effectively built upon the building's significance while enhancing its attractiveness for a range of compatible potential reuses. A detailed survey showed that the original interior decoration and major fittings of Coolart had long been lost, and there was little evidence of their former state. Hence, when refurbishment was necessary, Coolart's conservation plan permitted a redecoration in the general style of the period. We believe this has enhanced Coolart's historic significance.

Conserving cultural landscapes

Increasingly, managers are recognising that their protected area has a role in a wider landscape. Local communities may be vitally concerned about the scenic backdrop that adjacent high mountain peaks provide, for example. They assume that the protected area will continue to create much of the landscape that surrounds their individual homes and properties. Managers need to be aware that protected areas are also part of a cultural landscape.

Staff may also have to manage culturally created landscapes within a protected area. Cultural landscapes have been defined by Lennon & Mathews (1996, p. 4) as:

a physical area with natural features and elements modified by human activity resulting in patterns of evidence layered in the landscape, which give a place its particular character reflecting human relationship with and attachment to that landscape.

Landscapes have historic significance when they, or their components, have strong links or associations with important historic themes, and where the evidence assists in understanding the past (Lennon & Mathews 1996).

For instance, as part of a tourist attraction at Yarrangobilly Caves in Kosciuszko National Park, English-style landscape settings were created in and around historic nineteenth-century buildings. These are still managed as cultural settings today. The coal and oil-shale mines of Newnes, within the Wollemi National Park in NSW, are managed as an old industrial setting. Much the same applies to other old mining sites (Snapshot 13.5), old whaling sites, or to the network of emergency airstrips constructed in the Northern Territory during World War II. They are managed as part of the cultural landscape within a protected area.

There is a range of works for historic structures that require specialist techniques, for example prevention of rising damp, stabilisation of collapsing walls, replacement of decaying materials, application of the correct mortar mixes, and the use of the correct paint colours. Managers need to be alert to the complexity of these works. They also need to make pragmatic decisions at times in order to minimise costs, while still remaining true to the principles of the Burra Charter. Managing underwater archaeological sites requires special expertise. Many wreck sites are actively managed as conservation sites around the coast of Australia.

Snapshot 13.5

Managing mining heritage in Victoria
Roy Speechley, Parks Victoria, and David Banner, Heritage Victoria

Victoria's goldmining history is of national significance. In 1851 the discovery of gold in Victoria started the goldrush. Within a decade 60,000 immigrants had made their way to Australia. By the 1970s there was strong public sentiment in Victoria for conserving the relics of goldmining. This wish was increasingly in conflict with new technologies that made it profitable to re-mine the old fields. Under the *Crown Land (Reserves) Act 1978* many of the nineteenth century goldmining sites were nominated as historic reserves. This antagonised the mining industry, whose leaders pointed to inconsistencies in the ways heritage value was assessed. Yet their protests outraged the vocal heritage lobby. The result was such bitterness and confusion that, we suspect, significant sites were lost and worthwhile mining projects were frustrated.

From 1989, state and federal governments funded a four-year state-wide survey that included a historical survey of mining records for each locality. Once priority sites were identified, fieldwork followed, drawing heavily on the knowledge of local people. A 'typology' was designed to classify the sites. The survey increased the number of recorded sites from a few hundred to over 3000, of which about 2500 were goldmining sites. Goldmining relics were ranked as of Regional, State/National or of Local significance. Only 20% were of Regional, and 6% of State/National significance. This survey transformed the assessment process. Decisions may now be based on reliable historical information. The list of heritage resources is no longer unknown or endless, and the mining industry now supports the process. Thus Victorians can mine their gold—and keep their history.

13.7 Management lessons and principles

Lessons

1 Protected area management in Australia has evolved to recognise the very real importance of social and cultural heritage in the management of protected areas.

2 Conserving cultural heritage can be very costly and time-consuming, and needs to be thoroughly and carefully planned.

Lighthouse, Montague Island Nature Reserve, NSW
(Graeme Worboys)

Principles

1 Protected areas have been reserved, among other things, for the purpose of conserving Australia's cultural heritage.

2 The principles and processes articulated in the Burra Charter provide the basis for managing cultural heritage in Australian protected areas.

3 Assessing conservation significance is a critical part of managing cultural heritage.

4 Aboriginal cultural heritage must be managed as a partnership with the Aboriginal community.

5 Conserving Australia's cultural heritage requires managers to:

- know the type of site
- research it thoroughly
- consult the experts
- plan thoroughly
- consider costs and practicalities
- plan for long-term conservation.

6 Intelligent, practical conservation responses to cultural heritage management are required to achieve responsible conservation outcomes with the limited available funding.

7 Conservation and preservation work should involve the least possible change to the existing structure of the site.

Endnote

1. We recommend (and acknowledge our own debt to) the excellent course notes on Cultural Heritage Management developed separately by two campuses: Charles Sturt University and Gatton College of Queensland University. These are cited as Spennemann (1997) and Zerba (1994). We have been guided by Pearson & Sullivan's (1995) *Looking after heritage places.* All of the above are recommended reading and complement this practical focus. We are also indebted to Sue Feary of NSW P&W for her major contribution to the development and content of this chapter.

Threats to Protected Areas

> *... we find we have a continent degraded by inappropriate use—its soils eroded, salt-affected or destroyed over large areas and its vegetation fast degrading in consequence; its native animals becoming endangered or extinct. We are taking notice at last ...*
>
> From After the Greening: The Browning of Australia *by Mary White (1994)*

14.1	Underlying causes	372	14.4	Minimising environmental threats	392
14.2	Type and nature of threats	377	14.5	Management lessons	
14.3	Preventing environmental threats	386		and principles	400

Australia is a huge country that is seemingly isolated from the rest of the world. Much of it has an appearance of being intact. Yet its clear skies and the vast extent of its seemingly natural landscapes conceal a troubling reality. There is not one bioregion within this remarkable country that has remained unchanged since the arrival of the First Fleet in 1788. Our natural heritage is under perpetual siege: from feral animals, weed invasions, habitat destruction, altered fire regimes, pollution, use by people, and urbanisation. For our rich cultural heritage too, there are many threats—vandalism, fire, water damage, insect pests, and the effects of age threaten historic homesteads and buildings. Globally, Australia is not on its own with such issues. Major impacts and changes to the environment are occurring rapidly and at a global scale. The interconnectedness of the planet's life support system, the biosphere, guarantees that Australia shares any threats to the health of the biosphere with all other nations.

In this chapter, we begin by looking at threats to the environment. These are first put in a world context in discussing the underlying causes of threats. We then describe the enormity of the threats, their types, nature, and causes, and the remedies (or partial remedies) managers have found for them. In doing so we provide a series of snapshots that reflect a wealth of experience in managing natural heritage threats. Following this is a brief discussion of managing threats to cultural heritage.

14.1 Underlying causes

Key underlying causes of threats to the world's environments include:

1 high consumption levels among the richest fifth of the world's population stimulating agro-industrial, tourism, logging, and mining developments that in turn impact on protected areas and on land around them
2 pressure for trade and development with aspirations that downplay or ignore the environmental implications of development policies, which are in turn often driven by high consumption or the need to service debt repayments in many developing countries.
3 poverty among the poorest proportion of the world's population leading to increased pressure on protected areas to supply land and resources (Stolton & Dudley 2003).

We now consider some of the major global concerns that have implications for Australian protected area management and the threats that they face.

Global population growth

Global environmental issues are a result of society, industry, and the increasing number of humans on the planet going about their lives and actions. The world human population in 2003 is estimated to be 6.2 billion, and is growing by 77 million people a year (MacDonald & Nierenberg 2003). Population growth is slowing, but is still forecast by the United Nations to be 9.2 billion people by 2050—an increase of 50% from existing levels (MacDonald & Nierenberg 2003). Climate change, habitat destruction; soil loss; species extinctions; air, water and land pollution; and potable water availability are some of the environmental problems arising from more and more people being present. Australia is not isolated from these problems.

Climate change

Climate change, especially global warming, has substantially developed through the burning of fossil fuels and increasing concentrations of carbon dioxide (a greenhouse gas). Concentrations of carbon dioxide have risen exponentially in the atmosphere since the 1800s. Global mean temperatures have risen approximately 0.7 degrees Celsius since the mid 1800s and Australia has experienced the same rise in temperature (ASEC 2001a, Howden 2003). Regrettably, Australians have contributed a high per capita level of greenhouse gas emissions by world standards, and greenhouse gas emissions increased by 16.9% between 1990 and 1998. Total energy use in Australia has doubled from 1975 to 2000, and usage is increasing at an unsustainable level (ASEC 2001a).

Some sceptics argue that the threat of global warming has been overstated (Lomberg 1998,

Institute of Public Affairs 2003, Lomberg 2003). However, Dr Graeme Pearman, Director of Australia's Commonwealth Scientific and Industrial Research Organisation's climate program, has stated that scientists are trained as sceptics, and that the Intergovernmental Panel on Climate Change climate change reports (IPCC 2001) were put together by 1000 experts from 120 countries, and were reviewed by another 2500 people (Vincent 2003). The report included a forecast that global average temperatures would rise between 1.4 and 5.8 degrees Celsius by the year 2100. Dr Pearman also stated:

> ... the majority of scientists would give their right arm to disprove global warming. It would make their careers (Vincent 2003).

Australia's warmest year on record was 1998, and the 1990s the warmest decade. Globally, effects of these higher temperatures have seen a marked reduction in glaciers, with European glaciers losing 10–20% of their mass since the mid 1980s (Howden 2003). In Australia, snow cover in Kosciuszko National Park has reduced by 30% over 45 years, with the least average snow cover recorded in the last five years (Green 2003). In Antarctica, there has been the retreat and break-up of large ice shelves (Howden 2003).

Other climate change influences include changes in sea-surface temperatures with changes in wave energy and heights; increases and decreases in precipitation; and changes to average sea level with an increase of 10–25 centimetres in the last 100 years being confirmed (Howden 2003). Higher than average sea temperatures have led to mass bleaching of coral reefs. Two recent mass bleaching events on the Great Barrier Reef Marine Park (1998 and 2002) were estimated to have affected over 50% of all its reefs. Scientists advise if temperatures climb to 2 to 3 degrees Celsius above their average, corals may not recover from bleaching. Given that the average tropical sea temperatures are now 0.6 degrees Celsius higher than they were at the beginning of last century, it is estimated that the thermal limits currently seen on coral reefs will be exceeded annually by the years 2030–50 even under the best case climate projections (Hoegh-Guldberg 2003). The potential outcome is a change from coral dominance to algal dominance for affected areas. The ramifications for the Great Barrier Reef Marine Park tourism industry and the fishing industry are clear.

For terrestrial areas, the 2003 bushfires in the Australian Alps occurred in one of the most severe droughts in Australia's history (a one-in-100-years drought):

> Large areas of the country were experiencing serious or severe rainfall deficiencies. Additionally, atmospheric humidity and cloudiness were below normal and daytime temperatures were at record levels. The combination of factors led to an early advanced curing of fuels across most of Eastern Australia. Although many of these factors were also present during previous bushfire events, the high temperatures in the lead up to the 2002/2003 fire season appear to be unprecedented (McLeod 2003, p. 9).

The record high temperatures during the bushfires have been interpreted as an effect of global warming. Global weather changes and extreme weather events are forecast to increase. Changes in global mean temperatures are 'very likely' to affect precipitation patterns, wind velocities, soil moisture and vegetation cover that appear to influence the occurrence of storms, hurricanes, floods, drought, and landslides (UNEP 2002) The Kyoto protocol is a global response to climate change which has the support of the majority of the nations of the Earth. In 2004, Australia was not a signatory to the protocol.

Habitat destruction and species loss

The loss and degradation of habitat is the most important factor causing loss of species (UNEP 2002). Over the past three decades, major losses of virtually every kind of natural habitat on Earth have occurred, and the decline and extinction of species have emerged as major environmental issues (Christ et al. 2003). This trend is not expected to lessen given the steady forecast rise in human population to 9.2 billion people by 2050 (MacDonald & Nierenberg 2003). In recent times, the known rate of extinction among mammals and birds is far higher than the estimated average rate through geological time. All mammals and birds have been assessed for extinction risk, and 24% (1130) of mammal species and 12% (1183) of birds were considered globally threatened in 2000 (UNEP-CBD 2001, Christ et al. 2003). So severe was this problem that the 2002 World Summit on Sustainable Development identified biodiversity conservation as one of its five global priorities (Christ et al. 2003).

For Australia, habitat destruction has accelerated with as much land cleared in the 50 years prior to 2000 as was cleared in the first 150 years of settlement. In 1999, over 400,000 hectares of land were cleared, this being the fifth worst clearing outcome for the planet that year. In 2000, some 500,000 hectares were cleared. This is especially distressing because Australia is a centre of endemism of global significance. Because of its size, age, and geological and evolutionary isolation, over 80% of mammal, reptile, flowering plant, fungi, mollusc, and insect species in Australia are endemic. The number of nationally endangered and vulnerable species increased between 1994 and 2001, by which time Australia had 1451 species and 27 communities listed as either endangered or vulnerable at the national level (ASEC 2001a, 2001b).

There has been an international response to this habitat destruction. The establishment of protected areas has been a key mechanism for maintaining biodiversity and the support systems. This approach is also one of the primary practical responses recognised by the *International Convention on Biodiversity*.

Delegates of the Fourth International Union for the Conservation of Nature (IUCN) World Parks Congress in Caracas, Venezuela in 1992, recognising the urgency of the global trend for habitat and species loss, set an ambitious target of reserving 10% of the terrestrial area of the planet as protected areas by the year 2000. Eleven years later, at the Fifth IUCN World Parks Congress held in Durban South Africa in September 2003, it was announced by IUCN that 11% of the Earth's surface had been reserved as protected area (Worboys 2003c). The international community had risen to the occasion, with many countries such as Madagascar committing to the establishment of major new protected areas before and during the congress. The 2003 congress also concluded that more needed to be done. Very few marine and freshwater protected areas had been established and many of the world's terrestrial ecosystems are poorly conserved.

Potable water

In 2002, about 30% of the population of the Earth lived in countries suffering from moderate to high potable water supply stress. In the mid 1990s, 80 countries with 40% of the world's population were suffering from serious water shortages. By 2020, it is estimated that 66% of the world's people will be living in water-stressed countries and water-use will have increased by 40%, with 17% more water being required for food production (UNEP 2002). Australia also has its water shortage problems, with an increasing level of dispute over water allocations in major catchments such as the Murray-Darling Basin.

Pollution and waste

Pollution of air, land, and water, together with disposal of domestic and industrial waste, are major environmental problems that need managing. The treatment and disposal of sewage is an important issue that must be dealt with in environmentally sensitive areas such as the Great Barrier Reef World Heritage Area, Fraser Island World Heritage Area, and Kosciuszko National Park. Australians generate an average of 1.1 tonnes of solid waste a year, which places Australia in the top 10 waste generators, ranging on a per capita basis from about 800 kilograms/year in the ACT to almost 1400 kilograms/year in Western Australia (ASEC 2001). Waste reduction campaigns and recycling campaigns can be effective and have produced, for the ACT, a 40% reduction in disposable waste and an increase in recycled materials from 118,000 tonnes to 331,000 tonnes in a five-year period between 1993–94 and 1998–99 (ACT Waste 2000).

Australian threats and their causes

Threats to protected areas have many causes. The *National Strategy for the Conservation of Australia's biological diversity* (Commonwealth of Australia 1996b, p. 3) stated:

The loss of biological diversity cannot be slowed effectively unless its underlying causes are directly confronted. These underlying causes are extremely complex; they include the size and distribution of the human population, the level of resource consumption, market factors, and policies that provide incentives for biological diversity depletion, under-valuation of environmental resources, inappropriate institutions and laws, ignorance about the impor-tance and role of biological diversity, underinvest-ment in biological diversity conservation, and inad-equate knowledge of our biological diversity and the rate at which it is being lost.

At a more detailed level there are a number of rea-sons why threats to the environments of protected areas occur. Some of these have been documented to help managers to respond at a local level. A study by Graetz et al. (1995) of satellite images taken between 1990 and 1992 showed that more than half of the potentially arable lands of Australia had been cleared or disturbed. This represents some 39% of the continent. Sixty-five per cent of this area was cleared; and only 7% of the uncleared area was protected. Soil erosion, salinity, pollution of lakes and streams, and urbanisation of coastal areas threaten more species every decade. The Australian National Biodiversity Strategy (Commonwealth of Australia 1996b) lists five main processes that threaten our biodiversity:

1 clearing and modifying native vegetation
2 introducing alien or genetically-modified organisms
3 failing to control pollution
4 altering fire regimes
5 climate change.

Thackway & Cresswell (1995b) collected man-agers' views on the main processes that threatened their protected areas, across Australia's 80 bio-regions. Since only half of these bioregions are adequately represented in the protected area sys-tem, the aim was to work out in which bioregions it was most urgent to acquire land for conservation. They listed 11 threatening processes:

1 agriculture
2 grazing
3 weeds
4 clearing
5 fire
6 feral animals
7 forestry
8 mining
9 salinisation

Snapshot 14.1

Urban pressures on the Dandenong Ranges National Park

Kevin Curran, Parks Victoria

Nearly 80 kilometres of the Dandenong Ranges National Park boundary is abutted by residential suburbs. This causes a num-ber of problems: the removal of native trees and plants, vehicle noise, illegal parking and car dumping, over-the-fence dumping of garden waste and household rubbish, the spread of inap-propriate plant species, water pollution from stormwater runoff and sewage, and the impact of feral cats and dogs. Suburban homes bring an increased risk from bushfires, at a time when Parks Victoria staff are required by law to take all steps to reduce the risk of bushfires. The feeding of native birds is also inappropriate as the non-native diet leads to their dependence and obesity. Local communities and friends groups have supported conservation by helping in revegetation programs and in monitoring Superb Lyrebird populations.

10 tourism
11 urbanisation (Snapshot 14.1).

They found that managers listed the first five of these as the greatest threatening processes for most bioregions in Australia, though managers from bioregions near the heavily populated east coast would clearly have different priorities. Threats to protected areas can also be considered to arise for a number of underlying factors.

Social factors. Australia is a multicultural society with a rich history and diversity of ethnic backgrounds within its communities. However not all commu-nities or generations may be equally sympathetic to or informed about the benefits of Australia's bio-diversity, and this may have led to damage in some protected areas. Given that protected area manage-ment is a relatively new concept in many regional areas, there may also be distrust from local rural communities. Traditional activities may continue despite the fact that they are illegal, including:

• using the parks for illegal hunting or food col-lecting
• scouring public land for resources such as soil, gravel, rock, fence posts, or firewood, long after it has become a reserve

- traditional but illegal annual burning-off by neighbours in what are now protected areas.

Another kind of social problem is when parks close to cities become a place to let off steam for frustrated people. Expensive and distressing vandalism can result. Problems also arise from noisy and intrusive use of vehicles such as trail bikes and jet skis.

Political factors. Australia is a democratic society. Protected areas and their organisations are subject to public opinion and to political processes. They are at the mercy not only of sudden swings in politics but of a series of small changes that can have unintended cumulative effects. Single-issue lobbies can sometimes dominate the debate, without the long-term effects on conservation being properly considered. At other times the community may have real needs that threaten protected areas. The case for putting in a new road, mine or communications tower may seem compelling, yet the cumulative effect of many such individual decisions can destroy the integrity of a heritage area.

Managers, especially when employed by government, are often ill-placed to lobby for conservation. However, a lack of political support during critical parliamentary debates or periods of government decision-making can result in the loss of protected areas or of funding for them. Poor understanding by local communities of what protected area managers believe in and are trying to do may lead to distrust, even sabotage, and political involvement in management decisions. Backlash to some reservation decisions has led to acts of vandalism and even arson. The electorate's concern for the welfare of feral animals, such as brumbies and deer, may mean that rangers are not allowed to cull them, even when they do serious harm to natural environments.

Legal factors. Managers abide by the laws of Australia. Decisions made by a court or an inquiry, though made in the name of justice, may threaten conservation outcomes. Laws may override the power of managers to conserve an area (for instance, by allowing roads and communications towers to be built). Further, there is always the potential that parliament will pass new laws that partly override the charter of a protected area or areas, unless managers are providing timely information to the parliamentary process.

Ecological factors. Human interference can have unintended effects on the ecological balance. Removing a predator like the Dingo may cause kangaroos to proliferate, harming the flora and causing erosion. Unnatural fire regimes (sometimes prescribed by local regulations) may exterminate species that have survived in a region for millennia.

Economic factors. Protected area management needs resourcing. Lack of money often means that an immediate job will not be done well, or that necessary investments in the future (for equipment, infrastructure, and staff) will not be made. A government in financial trouble, or facing economic downturn, may slash available funds, permit pollution-prone developments, or allow increased harvesting of timber, minerals, and river water.

Managerial factors. In a sensitive social, political, or legal environment, managers cannot afford to make mistakes. Protected area agencies have existed in one form or another in all states and the Commonwealth for over 30 years. There is still room for improved management, but there is no place for repeating the trial-and-error methods and the managerial mistakes of the past. Nor can managers afford to work arrogantly and independently of local communities. They must remember that it is a privilege (not a right) to have custodial management of a protected area on behalf of the community. With the pace and degree of change imposed by society, staff no longer have the chance to learn the basic principles of natural and cultural heritage management on the job. The public expects Australian plants, animals, and habitats to be conserved and protected. For this purpose, the public should no more be expected to put up with unqualified (if enthusiastic) managers than with unqualified (if enthusiastic) doctors or dentists. Professional protected area agency staff need prior academic qualifications in appropriate areas of competency (Chapter 9) and should hold at least a Bachelor's degree. Their training should then continue on the job, with experienced managers and experts acting as mentors. The decisions of staff must not be a threat to the very areas they are entrusted to protect.

14.2 Type and nature of threats

A wide range of threat types have been recognised for protected areas. They are summarised in Table 14.1. Many disturbances threaten Australia's bio-diversity and other protected area values. The most common threats are: pollution, fire, visitor use, grazing by domestic stock, edge effects, external effects upon catchments or groundwater, and pest animals. Snapshot 14.2 describes how one small area of Australia suffered a series of invasions and changes to habitat.

Threats to biodiversity and related values

Pollution. Regrettably, pollution in our parks occurs regularly. Garbage may blow in from a nearby tip; or work its way down from upstream or up-slope; or toxic liquid waste may be illegally dumped in bushland streams and creeks. Truck accidents are frequent, causing spills of everything from salt to wheat, fuel oil, and chemicals. Abandoned vehicles, trailers, fuel drums, and other

Table 14.1	Types of threats to protected areas (after Hockings et al. 2000, Ervin 2003, Worboys 2004)
Threat type	Examples of threats
Physical	Fire (arson), severe storm events, geological incidents
Biological	Introduced plants, introduced animals and organisms
Direct human threats	Habitat fragmentation, mining, poaching, hunting and disturbance to fauna, fishing, collecting, grazing and harvesting of flora, trampling, structure development, access development, utility corridors, communications structures, urbanisation, pollution, collecting, managerial damage, vandalism, emergency response damage, arson, squatting, drug cultivation and trafficking, terrorism, and damage from violent conflict
Indirect human threats	Adjoining community and land-use encroachments, impacts to climate, catchments, air and water quality, and poor land use planning
Legal status threats	Absent or inadequate legal protection, lack of clarity of ownership, inadequate legislation
On-ground management threats	Absence of on-ground management, absence of law enforcement, difficulty of monitoring illegal activities
On-ground social threats	Conflict of cultural beliefs and practices with protected area objectives, presence of bribery and corruption, pressures placed on managers to exploit protected area resources, difficulty of recruitment and retention of employees
Socio-political-economic threats	Lack of political support, inadequate funding, inadequate staffing, inadequate resources, absent or unclear policies, and community opposition
Design threats	Adequate geographic size, shape, location, connectivity, and replication of an individual protected area and/or a system of protected areas to achieve effective conservation of biodiversity and other heritage
Managerial threats	Absence of strategic planning, human resource and budget systems, plans of management, effective operations, and effectiveness evaluation systems

Snapshot 14.2

Chronology of changes to habitats and species in the Bega Valley since first European settlement

The Bega Valley today is a rich dairy farming community in southern NSW, bordered by national parks and state forests. As a consequence of habitat clearing and the invasion of pest animals, at least six species of mammals became locally extinct in the 80 years following first European settlement. Lunney & Leary (1988) compiled the following milestones in the changes to local mammal populations.

1830s	First European settlement
1851	Dingo plentiful in the valley
1860s	High number of Koalas in the valley
1880s	Macropods exploited for a thriving fur trade
1886	Brown hares (*Lepus capensis*) introduced to Victoria in 1859, recognised as a major pest in the valley
1890s	'Native cats' (Eastern Quoll) reported as being plentiful
1865–98	Fourteen sawmills operating in the valley
1904	Rabbits, introduced to Victoria in 1859, recognised as a pest in the valley
1908	Noticeable decline in macropod numbers, especially the smaller Red-necked Pademelon (*Thylogale thetis*)
1908	Noticeable decline in the Dingo
1908	Major decline in Koalas—only known from isolated pockets
1909	Foxes, released in Melbourne in 1845, recognised as a pest in the valley
1910	Noticeable decline in 'Tiger Cats' (Spotted-tailed Quoll (*Dasyurus maculatus*)) and 'Native Cats' (coincided with rabbit-poisoning carts)
1910	At least six species of mammal now locally extinct (Parma Wallaby (*Macropus parma*), Red-necked Pademelon, Common Wallaroo (*Macropus robustus*), a rat kangaroo, Eastern Quoll, and Brush-tailed Phascogale (*Phascogale tapoatafa*))
1910	Bega valley essentially cleared farmland

Major new national parks that were established in the late 1990s on lands adjacent to the Bega Valley have helped to conserve the remnant populations of mammals such as Koala and Long-footed Potoroo.

A construction vehicle accident that caused serious diesel pollution to a creek, Kosciuszko National Park, NSW (Graeme Worboys)

rubbish are also commonly found in protected areas and other public land.

Fire. Fire is a natural event in most bioregions, and has helped in their evolution. Aboriginal use of fire shaped the type and structure of vegetation throughout Australia. Today fire regimes are managed to protect human communities. Unless managers know what they are doing, a change in fire frequencies may alter the balance of plants and animals in an area that has been set aside for protection. Careless or cavalier use of fire is a threatening process. Altered fire regimes can threaten plant and animal species. Any change in the frequency and intensity of fires favours some plant species and threatens others, which may become locally extinct. A woodland's grassy understorey (favoured by a given fire regime) will benefit some small mammals but threaten others. As well, regular severe fires may remove the protective plant cover and cause erosion. Arson is an issue of concern, as are deliberately lit fires entering protected areas from neighbouring land.

Illegal grazing. Sheep and cattle grazing have changed even seemingly natural plant communities by selective grazing of one species over another, especially along creeks, rivers, and other watercourses. The opening up and simplification of many habitats also encourages the impact of introduced predators on small native animals.

Edge effects and external impacts on catchments. Many protected areas are islands within an altered landscape. Smaller reserves and those with a large perimeter relative to their area are under special pressure. Compact shapes and uncomplicated boundaries are preferable in a reserve, since the native plants and animals are then less exposed to disturbance from the many threats that occur particularly at the edges. Weed and crop seeds and fertiliser may blow or wash in across boundaries. Predation of native species by cats, dogs, and stock is also prevalent at the edges of native vegetation.

A cat scavenging on King Penguin (*Aptenodytes patagonica*) chicks killed by Southern Giant Petrel (*Macronectes giganteus*), Macquarie Island, Tas (Geof Copson)

Pest animals. Pest animals are those that threaten the agricultural, environmental, and personal resources that humans value (Olsen 1998). There are many introduced species in Australia, including feral populations of at least thirty exotic mammals such as goat, pig, horse (*Equus caballus*), three species of deer,

Pig control program in Namadgi National Park ACT—distributing 'free feed' (Trish MacDonald)

Weed: Monterey Pine (*Pinus radiata*) (Graeme Worboys)

swamp buffalo (*Bubalus bubalis*), fox, cat, black rat, house mouse (*Mus musculus*), and dog (Table 14.2). Five species of amphibian, including the cane toad (*Bufo marinus*), have established themselves in Australia (Recher et al. 1986). As well, house sparrow (*Passer domesticus*), common starling (*Sturnus*

Table 14.2	**Some mammalian pest animals introduced to Australia and some control techniques** (adapted from White 1994, Strahan 1995, Burgman & Lindenmayer 1998, Olsen 1998)			
Mammal	Date of introduction	Location of release, origin	Reason for introduction	Notes
Cat	1788	Introduced in south-eastern Australia by early European settlers	Companionship	The cat is found all over Australia (they are desert-tolerant). Rabbit is the main food-source, but they also take invertebrates, fishes, reptiles, frogs, birds, and mammals. Controls include baiting, trapping, fencing, and shooting.
Goat	1788	NSW, arrived with the First Fleet	Escape from farms	The goat is found in most parts of Australia. They harm native vegetation and the habitats of native animals. Controls include shooting and mustering.
Pig	1788	NSW, descended from European domestic stock; in the north, animals may have arrived from Timor and PNG	Escape from farms	The pig is found in northern WA, NT, Queensland, NSW, and Victoria. They are omnivorous, and prey on small mammals and birds. They cause damage to vegetation and promote soil erosion. Controls include trapping, shooting, and baiting.
Horse	1788?	NSW and elsewhere	Escape from farms	The horse is found across north and inland Australia, and in the Australian Alps. Controls include mustering and some shooting.
Swamp buffalo	1825–43	NT, from Asia	Food	Swamp buffalo are found in NT and small areas of northern WA and Queensland. They harm native flora and wetlands. Control is primarily through shooting.
Deer	1950s	Tasmania, Victoria, from England, India, Ceylon	Escape from farms	Control is primarily through hunting and trapping.
Rabbit	1858	South-eastern Australia, from England	Sport	The rabbit is found in all southern parts of Australia, including Tasmania. They breed quickly, and cause major damage to vegetation, especially in droughts. Controls include poisoning, shooting, and destroying warrens.
Fox	1860s	Released near Melbourne, from Europe	Sport	The fox is found in southern Australia, from deserts to urban regions. They are a predator of native fauna, and also eat insects and fruit. Controls include trapping, shooting and poisoning.

Figure 14.1   Distribution of some pest animals in Australia (SEAC 1996, pp. 4–17)

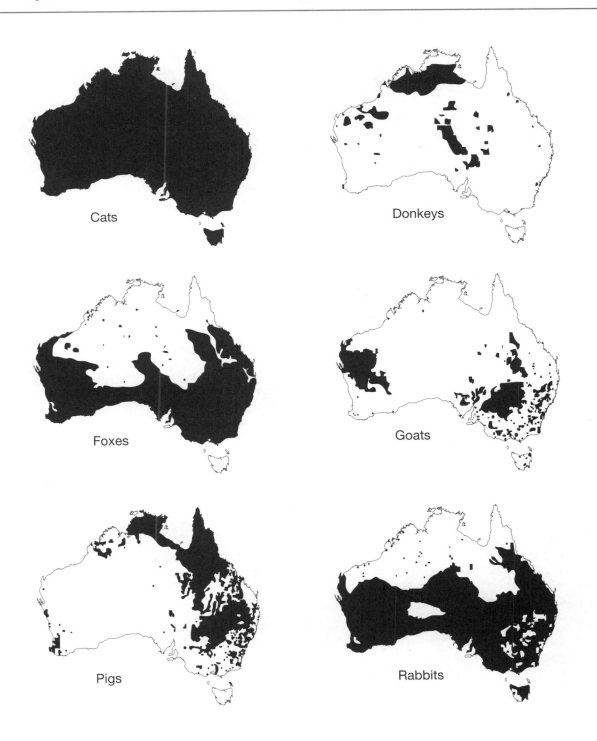

Rockface vandalism, Lamington National Park, Qld
(Graeme Worboys)

Weeds. Australia's vegetation has changed dramatically since 1788. About 10% of today's plant species are introduced (Buchanan 1989). At least 1400 species of plants have established wild breeding populations (Recher et al. 1986). Many weeds have had a devastating impact on Australia. One need only think of the spread of Bitou Bush on our coastlines, Blackberry in the well-watered parts of Victoria and New South Wales, Paterson's Curse in the Riverina, and Rubber-vine in northern Australia. The number of weed species per unit area across Australia is shown in Figure 14.2.

Habitat disturbance. Clearing and fragmentation of natural lands is one of the greatest threatening processes. Even within protected areas, habitat disturbance can occur from many sources: arson, feral animals, clearing for developments, road construction, and the making of utility corridors (for powerlines and for sewerage, water, and fuel pipelines). Bulldozers and emergency vehicles do damage during incident control, as we shall see in Chapter 15.

Visitor use. National parks in Australia are popular destinations. Visitors put pressures upon the environ-

vulgaris), blackbird (*Turdus merula*), and 20 species of introduced fish, including European carp (*Cypinus carpio*), brown trout (*Salmo trutta*), and rainbow trout (*Oncorhynchus mykiss*), all pose threats to habitats and to native animals within protected areas. The distribution of cat, fox, pig, donkey (*Equus asinus*), goat, and rabbit in Australia are shown in Figure 14.1.

The fox, a major pest to Australian fauna (William Logan)

Figure 14.2 Concentration of weeds in Australia
(SEAC 1996, pp. 4–18)

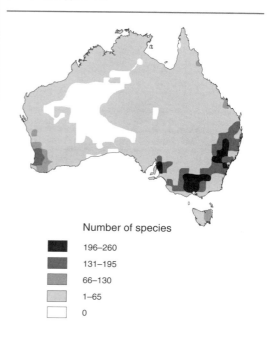

Number of species

■	196–260
▨	131–195
▧	66–130
▫	1–65
□	0

ment. Overuse is a major concern. It may cause soil erosion, habitat disturbance or destruction, disturbance to native animals, as well as pollution from litter and human wastes. Visitors collecting souvenirs, such as rocks, shells, driftwood, and parts of cave formations, can have substantial cumulative impacts. They can also introduce weeds and pathogens.

Vandalism. Vandalism is worst in and near cities. It ranges from the blowing up of toilet blocks (which happened twice at Seven-Mile Beach National Park near Nowra in southern NSW), through to defacing signs, graffiti on buildings or heritage items, and interference with plant and equipment in parks. As well, many roaded but remote areas are used for dumping or burning stolen cars, which can lead to fires.

Illegal actions. Some illegal activities are real threats to park systems. They include the collecting, for sale overseas, of native birds' eggs and chicks or of native reptiles, arson, the release of feral animals such as wild pigs or of domestic pets within parks, the stealing of bush rock, and the destruction of formations in limestone caves. Illegal hunting, both for introduced species such as deer and pig, or for native species such as wallabies, also occurs. Dogs used for hunting can cause significant disturbance to native species. Hunters have also been known to deliberately release prey animals such as pigs in protected areas to establish future hunting opportunities. Illegal use of four-wheel-drive vehicles and trail bikes can cause habitat destruction and soil erosion.

Destruction of park signage by vandals, NSW (Noel Browning)

Introduced agricultural diseases. Measures to contain outbreaks of disease among domestic animals or poultry can threaten wild species that are suspected of being vectors of that disease. In 1999 an outbreak of Newcastle disease within chickens at Mangrove Mountain in NSW led to the testing of some wild birds. If these had been found to be vectors of the disease, there may have been demands for local exterminations. The control of agricultural diseases, such as foot and mouth disease, may have substantial impacts on natural areas.

Threats to soils and geological features

Soil erosion. Non-natural soil erosion is commonly a consequence of overuse, construction activity, or the destruction of vegetation by fires, grazing, or pollutants.

Soil, sand, and gravel extraction. These basic commodities are often illegally extracted from protected areas, though usually on a small scale.

Impacts to stratified deposits. Stratified soil deposits contain priceless evidence of the recent and more distant past. These may occur in lakes or in caves. Humans and feral animals may disturb such sites, jumbling the geological or anthropological record. Jason Bradbury describes (Snapshot 14.3) the work completed to minimise impacts from tourism vessels in the Tasmanian Wilderness World Heritage Area.

Fossil and mineral collections. Fossil sites and geological-type sections record important parts of Australia's past (Chapter 1). Rare and unusual minerals are also protected in many conservation reserves. Illegal collection or disturbance of these sites may destroy what can never be replaced.

People collecting bush-rock. Highly weathered sandstone rocks and boulders, as found in the Sydney Basin, are favoured for ornamental and landscaping uses. In the past many landscapers became used to collecting such rocks from unreserved public land. Many of these areas are now national parks and such collection is illegal. Regrettably, it continues in some areas.

Damage to caves. Caves are vulnerable to overuse and tourism impacts. Commercialised caves are major tourist drawcards in many parts of Australia. Yet the very installation of lights and electrical cables can

Snapshot 14.3

Managing Lower Gordon River erosion—geoconservation of landforms affected by tourism

Jason Bradbury, Department of Primary Industries, Water and Environment, Tasmania

The lower Gordon River is a narrow estuary within the Tasmanian Wilderness World Heritage Area. Its geomorphology is in itself of World Heritage significance. The suite of landforms documents the progressive sedimentary filling of the steep-sided river valley after it was drowned by rising sea level following the last ice age. The lower Gordon River has been a tourism venue for over a century. With more visitors since the 1980s, small and slow cruise boats were replaced by larger and faster vessels and the amount of traffic increased. By 1985 the local Parks and Wildlife Service rangers had noticed that the river banks were starting to collapse. A monitoring program soon found rates of erosion exceeded 1 metre per year in places. In 1987 experiments on a sandy levee bank showed that the rate of erosion increased sharply at speeds above 9 knots (Nanson et al. 1994). This was enough to satisfy the Marine Board and a 9 knot speed limit for vessels greater than 8 metres in length was applied to some three-quarters of the length of the estuary.

After a period of adjustment to the 1989 restrictions the rate of erosion had slowed to 1 to 2 centimetres per year (Figure 14.3). In 1992, and after much consultation, a voluntary users code was produced to show visitors how to reduce damage caused by their vessels. However, people were beginning to question why the Parks and Wildlife Service still regarded the erosion as a matter of serious concern and there were even pushes from some quarters to increase speed limits or re-open closed reaches. A series of radiocarbon dates on bank sediments obtained in 1993 settled the matter. These proved that most banks had been stable for thousands of years, while those nearer the mouth had been very strongly depositional (Bradbury et al. 1995). This combined with further monitoring data supported implementation of a further reduction in commercial vessel speed to 6 knots in 1994.

New 'plume' research after 1997 allowed the determination of true thresholds of erosion, and it became possible to set speed limits for individual cruise vessels based on their wake characteristics. For new cruise boats wave height is now determined by conducting towing basin model tests. This allows a speed limit to be set, a licence issued, and the operator to gain finance before construction of the vessel commences. Further basic research is still required to resolve additional issues. The most general lesson from this work is that it may be much easier to place environmental controls on a new activity at the outset than to try to change established practices after a problem becomes apparent. If it had been normal practice in the early 1980s to apply the precautionary principle then serious degradation of the lower Gordon River may have been prevented.

Figure 14.3   Changes in the measured rate of muddy estuarine bank erosion in response to management controls

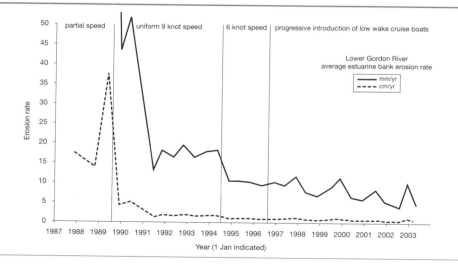

Snapshot 14.4

Management of zinc toxicity
Stuart Johnston and Roger Good, NSW P&W

Domestic stock-grazing in the alpine area of Kosciuszko National Park was withdrawn in the 1950s. The many years of grazing had resulted in severe erosion, for which a rehabilitation and revegetation program commenced in 1959. For the extensive areas of sheet erosion where up to a metre of alpine humus soil had been lost, the areas were sown with a mixture of exotic grasses and legumes, fertilised and covered with a heavy hay mulch. The hay mulch was held in place by galvanised wire netting. Very good ground cover was generally achieved into which native species were planted in subsequent years. After five to seven years, a sudden decline and death of large areas of stable vegetation was noted and initial investigations indicated that the breakdown of the galvanised wire netting was leading to zinc toxicity. A moratorium was placed on the use of galvanised wire netting. Subsequent field trials in 1998 indicated that the organic alpine humus soils have extremely low zinc levels and minute levels of zinc leached into the soil were sufficient to cause death of the recolonising native plant species and consequent renewed erosion in many sites. The use of galvanised products in the alpine area as a result of the trials has now been abandoned.

---

cause damage. In addition, the lights allow so-called *lampenflora* to grow on cave walls. Lint and dirt and debris brought in by people can also alter the low-energy ecology of a cave.

Toxins in soil. Trace elements introduced to soils, such as zinc from galvanised iron or arsenic and copper from treated pine, can harm natural environments. Soil stabilisation in Kosciuszko alpine area, for example, was set back many years by zinc toxicity from galvanised mesh used to help stabilise the very sensitive alpine soils (Snapshot 14.4).

Plant pathogens in soil. Pathogens such as Cinnamon Fungus (*Phytophthoera cinnamomi*) are a threat to native plants. They attack the roots of some plants, causing dieback.

Threats to scenic quality

Poor design and placement of structures and assets. Structures for management purposes within a protected area must be designed to blend in with the natural environment. This means a sensitive choice of colours (not too reflective) and materials, as well as careful positioning and size. In some protected areas, the towers supporting high-voltage electricity lines have been coloured green or grey to minimise their visual impact.

Clearing. Any clearing—even to create a visitor facility such as a lookout—can affect scenic amenity. If it is to be done, great care must be taken.

Inappropriate fire regimes. High-frequency, high-intensity fires may completely transform the structure and composition of the vegetation and therefore the scenic amenity of an area.

Road construction. Access roads in protected areas can be major lines of visual disturbance, or they can blend discreetly into the environment. Managers may need to negotiate a compromise between aesthetics and an engineer's practical requirements (for instance, that the road should have certain grades, widths, and cambers that allow vehicles to use it safely).

Quarries. It is sometimes better to quarry road materials within a protected area than to risk importing alien species and weed seeds. Quarries may need to be managed to minimise soil erosion, visual impacts, and runoff. Today, parts of a quarry are often being actively revegetated while other parts are still in use.

Threats to Aboriginal cultural heritage

Disturbance to sites. Aboriginal cultural heritage sites are easily disturbed by heavy machinery. Whenever it needs to be used, whether for construction, rehabilitation, maintenance, or for managing incidents, managers need to be alert to any Aboriginal sites nearby.

Vandalism. Many rock engravings, art sites, and occupational sites such as middens, are exhibited to the public as examples of Australia's rich Aboriginal cultural heritage. Such sites are vulnerable to vandalism and have often been damaged by spray-painting, graffiti, or other physical damage.

Illegal collection. Certain collectors avidly seek Aboriginal artefacts such as stone axes and tools. Once illegally removed from a site, much of their meaning is lost.

Fire. Some Aboriginal artefacts, such as scarred trees and shield trees, are flammable. Managers should know where they are, and give them special protection during bushfires and burn-offs.

Threats to European cultural heritage

As discussed in Chapter 13, much of the management of historical buildings and structures is concerned with guarding them against threatening processes.

Fire. The timber in historic homes, huts, and other buildings is usually very dry and thus highly susceptible both to bushfires and accidental fires.

Storms. Storms may damage historic buildings; and marine storms may destroy historic breakwaters, piers, and wharves.

Floods, rising damp, and salinity. Floods, even if rare, can severely damage old buildings. So too can rising damp within the walls (caused by poor drainage). Salinity in the Murray–Darling Basin is causing historic buildings in some areas to deteriorate rapidly.

Vandalism. Vandalism is often minor (broken windows, spray paint, graffiti) but it can also extend to arson or total destruction of the historical fabric of a place.

Earthquakes. Some areas of Australia are vulnerable to earthquakes (Chapter 15). Historic structures in these areas may need special protection.

Animals. Rats, Common Wombat, goats, birds nesting in walls, are all potential problems.

Tourism overuse. Historic buildings fascinate the public. Thousands and thousands of people visit such sites each year. This leads to general wear and tear on footpaths, staircases, floorboards, paint work, and so on.

14.3 Preventing environmental threats

Prevention is better than cure. With the multitude of threats facing protected areas today, managers cannot afford to be passive and wait until problems eventuate. They must be proactive, even visionary, and use a mixture of long-term initiatives and simple precautions—such as having the equipment on hand to deal with a polluting spill before it actually occurs. Below we suggest a series of 'visionary' measures, all of which involve more than standard precaution-taking.

Conservation corridors linking public lands

There are exciting possibilities in the linking up of protected areas. Biological studies show there is a close relationship between the size of an 'islanded' reserve and the number of species that it can permanently support. In large reserves, accidents such as fire or disease will often cause the local extinction of a species, after which it recolonises the area. In small isolated reserves, such a species, once lost, is likely to be lost forever. Linking up reserves, and establishing corridors between them, can prevent such fragmentation and the extinctions it promotes (Backgrounder 14.1). The value of corridors linking larger blocks of native vegetation is considered in Chapter 19.

Habitat clearing (Graeme Worboys)

There are also opportunities to link areas of natural environment on a larger scale. When lands are linked up in a north–south direction, or from uplands to lowlands, species may have the ability to survive large-scale effects such as global warming by migrating (over time) to more suitable habitats. Australian governments have some remarkable opportunities to achieve this if they act promptly. For instance, the great coastal escarpment of eastern Australia is largely, for its length through NSW, a series of interconnected public lands (Figure 14.4). These are predominantly state forests and national parks. We envisage that in the future, a whole-of-government policy could see all these lands managed as one entity—a system of remaining natural lands extending the full north–south length of NSW along the Great Escarpment of Eastern Australia and the Australian Alps, and connecting with Victorian and Queensland national parks (Worboys 1996, Pulsford et al. 2003). This would have both conservation and economic benefits, given that the unfragmented conservation corridor could be managed for a range of sympathetic uses. Such a policy could reinforce the conservation of unfragmented landscapes.

Sadly this is the only public land north–south conservation corridor that is still achievable in NSW, but such a large potentially interconnected area also permits many other lands to be linked to it. Similar corridors have, to some extent, already been established in the Australian Alps, the Wet Tropics of north Queensland and in the south-west of Western Australia. With interstate cooperation, a north–south corridor has the potential to be extended to include areas in Victoria such as the Coopracambra and Snowy River National Parks. If this was achieved, the great escarpment conservation corridor could be connected to the Australian Alps (Pulsford et al. 2003, Figure 14.5). This would create an unfragmented conservation corridor of managed public land from Walhalla in Victoria to the Hunter Valley in NSW. Such actions would greatly reduce the potential for losing species in the long term.

Internationally, there are important attempts to achieve continental scale conservation corridors. Hamilton (1997) described a vision for the protection of entire mountain ranges, including initiatives in the Americas (Yellowstone to Yukon) and the Himalaya (the Terai arc) (Figure 14.6). Worldwide there is growing recognition of the importance of conserving as conservation corridors the last remaining unfragmented wild lands.

Transboundary partnerships

There are some exciting opportunities for cooperation across political boundaries, particularly where existing protected areas are located on state or

---

Backgrounder 14.1 Fragmentation of natural habitat and extinctions

Fragmentation of natural habitat occurs when a large expanse of habitat is transformed into a number of smaller patches that are isolated from each other (Wilcove et al. 1986). When the landscape surrounding the fragments is inhospitable to species of the original habitat, and when dispersal is low, remnant patches can be considered 'habitat islands', and local communities will be 'isolates'. Fragmentation can lead to the extinction of species. The principal issue here is that local extinctions, which will inevitably occur from time to time due to chance factors, are no longer reversible. Additional reasons may be that:

- the remaining fragments are smaller than the minimum home range or territories needed by a species
- the fragments lack the diversity of habitats some species need
- predators and pests may build up and invade from the cleared land between the fragmented habitats
- 'edge species' will be unduly favoured
- the fragments may be too small to sustain balanced ecological relationships such as predator–prey, parasite–host, and plant–pollinator
- small populations contain less genetic variation, are more sensitive to chance variations over time, and may be wiped out by maladaptive genetic drift or by natural catastrophes (Soule 1986, Wilcove et al. 1986).

This is discussed further in Chapter 19.

Figure 14.4 **Great escarpment of eastern Australia** (adapted from Ollier 1982)

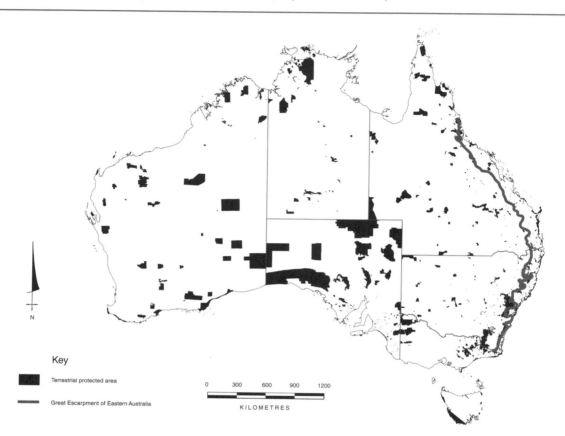

Key

◼ Terrestrial protected area

— Great Escarpment of Eastern Australia

0   300   600   900   1200

KILOMETRES

territory borders. The work of the Australian Alps Liaison Committee provides an internationally recognised model for such cooperation (Thorsell 1990, Crabb 2003). The committee jointly sponsored, with the IUCN, a major international forum on this issue. The resulting publication (Hamilton et al. 1996) has helped promulgate mechanisms for making such partnerships work.

Partnerships with private landowners

Partnerships with private land-owners and government organisations can sometimes produce outstanding initiatives that minimise threats to protected areas. Groups of land-owners may enter into a cooperative agreement that sees the natural heritage values of their properties retained to serve as a conservation corridor between nearby protected lands. An example of this is the Yurangalo Voluntary Conservation Agreement, which was negotiated with a number of property owners on the south coast of NSW. It creates a link between the South East Forest National Park and the Bournda National Park. This natural corridor from protected areas on the coast to those inland at altitudes of about 1000 metres may help save many species during changes induced by global warming.

Education and information

Unless the community is well informed about the nature and status of protected areas, there will be little sympathy for management objectives. The constant provision of information to the community is one way of minimising threats. This may be in the form of a website, or a regular newsletter to all

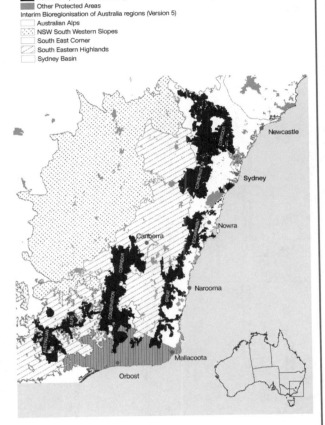

Figure 14.5 Australian Alps–Great Escarpment conservation corridors (Pulsford et al. 2003)

Australia Alps and Great Escarpment Conservation Corridors (2002)
Other Protected Areas
Interim Bioregionisation of Australia regions (Version 5)
Australian Alps
NSW South Western Slopes
South East Corner
South Eastern Highlands
Sydney Basin

relationships. Such cooperation allows 'integrated land-use management' and cooperation on fire management, catchment protection, and pest and weed control. These matters are further discussed in Chapter 17.

Planning and decision-making processes

Many Commonwealth, state and territory acts of parliament (Chapter 2) contain provisions relating to the prevention or mitigation of threatening processes. Planning laws in most states and territories also require some type of assessment of the environmental impacts for any proposed project that affects natural environments. For protected areas, agencies often have their own internal procedures as well as being required to meet formal statutory processes. This formal process is critical. Threats can be minimised by thorough analysis of proposed developments (Backgrounder 14.2).

Managers need to get the legal steps absolutely correct, and should normally take legal advice before signing to approve a development within a protected area. Direct responsibility rests with the officer who signs, but the credibility of the whole organisation is also at stake, since all aspects of a development can be subject to scrutiny through the courts.

Planning of protected areas

Carefully designed zoning schemes can make a great difference to an area's ability to absorb, and delight, visitors. Planners of such schemes must recognise that there are limits to the number of people that can be managed at any given time. This depends on the nature of the site and also the type of experience managers seek to provide (Chapter 16). The zoning plan will distinguish between remote areas that deal with fewer visitors and 'honeypot' areas that cater for larger numbers. The latter may be 'hardened' (for instance by installing boardwalks) to accommodate the numbers, while more sensitive natural areas may be protected by use of permits or in other ways that limit the number of visitors. Park planners may also influence the distribution of visitor loads across a regional landscape. This way, each individual national park does not offer facilities for the whole cross-section of recreational visitors. Instead, a range of recreation

neighbours of a protected area. It may be a whole range of press releases and electronic media announcements about what is happening in protected areas. Involving the community is even better, for instance through schools and education and management planning process. Simple community education programs can target changes in behaviour, and make people aware of why they should not dump lawn clippings over the back fence, allow pets to use bushland, or use the local protected area as a rally circuit for motorbikes.

Working with the community

If protected areas are accepted as a community asset, threats are minimised. Assisting visitors, recreationists, and local communities helps to build good

Figure 14.6 Continental-scale corridors of mountain protected areas (prepared by the World Resources Institute and reproduced in Hamilton 1997)

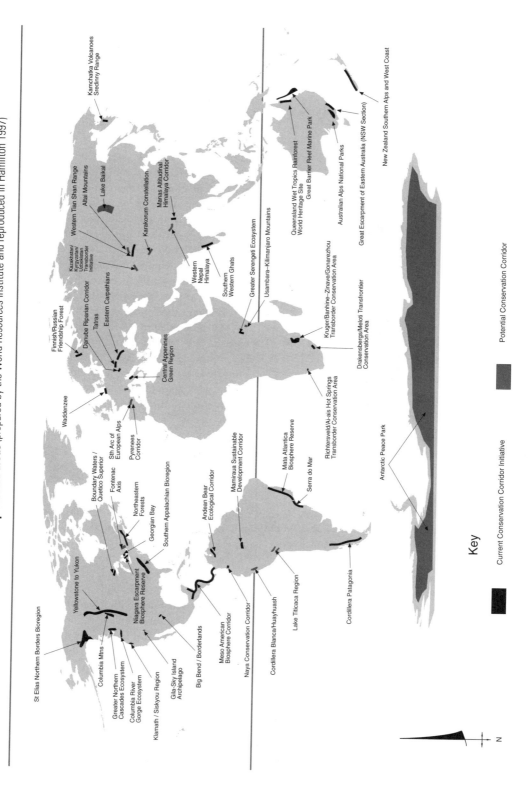

Backgrounder 14.2 Environmental impact assessment

Developments usually involve some disturbance to the land through either the erection of buildings (whether residential, commercial, or industrial) or structures, or by the removal of resources (timber, soil, rock, or minerals). In many cases, the development also may change the land use substantially. While land-use change is often controlled through the planning or zoning processes, it is often also regulated through environmental impact assessment. The environmental impact statement (EIS) has been established as the core of environmental impact assessment. The functions of an EIS are:

- to assist the decision-maker by providing relevant information about the negative and positive impacts of a proposal
- to inform the public so as to allow them to play an effective part in the decision-making process.

An EIS is usually a legal requirement where developments are likely to have a significant impact. They integrate environmental considerations into decision-making processes, concerning both private and public sector developments and activities. Even if an EIS is not required, the NSW NPWS prepares a review of environmental factors (REF) to assess potential impacts from a proposed activity. An REF is similar in nature to an EIS, but much less detailed. Many other agencies prepare an equivalent planning statement under some other name. Most agencies have detailed guidelines for such internal documents. An REF should include the following (NSW NPWS 1994).

1. *Introduction*
    1.1 Location, proponent, name of reserve
    1.2 The need for activity

2. *The activity*
    2.1 Objectives of the activity and justification
    2.2 Description of the activity
    2.3 Associated developments

3. *Alternative proposals* (description of alternative locations and other ways to achieve activity)

4. *The existing environment* (description and significance)
    4.1 Climate
    4.2 Geology, geomorphology
    4.3 Soils
    4.4 Plants (species/communities)
    4.5 Animals (species/habitats)
    4.6 Water catchment
    4.7 Known Aboriginal sites/significance to local Aboriginal Land Council
    4.8 Known historic sites
    4.9 Recreation
    4.10 Landscape

5. *Existing use*

6. *Environmental impacts* (describe impacts of each factor)
    6.1 Impacts on natural environment
        - Climate/greenhouse effects
        - Geology/geomorphology
        - Soils (nutrients/erosion)
        - Plants (species/communities)
        - Animals (species/habitats)
        - Water catchment (quality/drainage)
        - Exotic species
        - Fire
    6.2 Impacts on cultural environment
        - Known Aboriginal sites
        - Significance to local Aboriginal Land Council
        - Known historic sites
        - Likely impact on unrecorded sites
    6.3 Impacts on social environment
        - Traffic/roads
        - Noise
        - Neighbours/local residents
        - Safety
        - Lighting and energy use
        - Education
        - Recreation

7. *Assessment of alternative proposals*

8. *Proposed environmental safeguards*

9. *Consequences of not carrying out the development*

10. *Conclusions and recommendations*

opportunities may be provided across a region, with each park playing its part. This minimises threats, yet maximises the chance for visitors to enjoy themselves (Chapter 16).

Using management systems to deal with threats

Some organisations are developing environmental management systems that can prevent and minimise threats to protected areas. A Parks Victoria system, for example, allowed it to concentrate energies where they are most needed. This Environment Management System (Parks Victoria 1998) has four major components:

1 directions and priorities
2 program development
3 program delivery
4 information and evaluation.

All of these components are interconnected. Part 1 involves analysis of the priorities that need management attention. The natural 'assets' and the threats to them are assessed. In Part 2, ways of avoiding, sharing, minimising, or accepting risks are evaluated. In Part 3, targets and methods for risk reduction are set, and estimates made of the likely level of risk after action is taken, and the likely condition of the 'assets'. Finally, in Part 4, the results are researched and evaluated.

14.4 Minimising environmental threats

Despite the best efforts of managers to research and prevent threats in advance, they must sometimes face direct physical threats to their areas. To be effective, actions involving pest species, for example, may need to be taken at a landscape scale, and involve all stakeholders. A strategic approach is often adopted by management agencies. In relation to pest management this approach involves planning, action, and evaluation. The basic steps are:

- define the problem (in terms of alleviating the damage caused by the pest)
- identify and evaluate the management options and develop the management plan
- implement the management plan
- monitor progress and evaluate results against the stated objectives
- re-evaluate and improve objectives if necessary (Olsen 1998, p. 79).

Dealing with habitat modification and threats to biodiversity

Fire. Techniques for managing fire to minimise threats include:

- ascertaining the natural fire frequency for the plant communities and habitats in an area
- understanding the natural fuel build-ups for each plant community and the kind of fire risk that results

- developing a fire regime that reduces fuel responsibly while otherwise simulating a natural fire regime
- establishing a pattern of mosaic burns that provide diverse and appropriate habitats
- paying special attention to the oldest vegetation communities in the area, and the fire regimes they need (otherwise only those communities that result from more recent or more frequent burns will survive)
- understanding the seasonal pattern of natural lightning-caused fires
- analysing and researching human-caused fires and the degree of risk of arson for any given part of the area.

Construction and developments. When developments are approved in protected areas, there are ways of minimising damage. We need to understand in detail the many possible causes of damage: the parking of construction equipment, the on-site storage of materials, the access to the construction site, the temporary erosion-control measures, even pH controls for construction discharge waters (if substantial concreting is being undertaken). Techniques for minimising their impacts include:

- effective liaison with the construction company and their contractors and subcontractors.

- providing background environmental information and guidelines to the developers, both verbally and in writing
- setting a substantial bond that can be used to repair any environmental damage
- constantly monitoring the development
- having an approved development design that includes a waste treatment system (preferably a completely internalised one) for minimising such effects as stream pollution from sediments, from grease and oil waste, or from the runoff of acid or alkaline waters
- cordoning off of the development site to delimit the area within which disturbance is permitted
- starting rehabilitation while the development is still underway
- offering bonus payments for good environmental results
- having a proviso in the contract that allows managers to halt work for environmental reasons
- contingency plans that deal with potential accidents such as explosions, fire, and so on.

Clearing. Clearing may be necessary for fire breaks, for remote-area helipads, or for temporary staging camps to be used during incident control. Clearings may also be constructed for developments such as telecommunications towers and campgrounds. To minimise the damage, managers should use good quality air photos to select the least harmful spots for clearings.

Introduced plants. There are several techniques for dealing with introduced plants. These are comprehensively described in Buchanan (1989). They include biological control, herbicide sprays, cutting (Snapshot 14.5), slashing, replanting with native species, controlled burns to selectively target weed seedlings (such as Monterey Pine (*Pinus radiata*) invading a eucalypt forest), and hand-clearing. This last method is known in New South Wales as the Bradley Method, after pioneer conservationists the Bradley sisters, who hand-cleared native bushland in Sydney Harbour National Park. An effective combination is to remove weeds and then replant with vigorous native species to prevent reinvasion. Weed control needs to be carefully planned, using precise knowledge of the biology of each weed species

Snapshot 14.5

Willows—what's the worry?

Monica Muranyi and Darren Roso, ACT Parks and Conservation Service

Willows make pleasant shade trees. They were introduced to Australia by early European settlers wishing to 'Europeanise' the landscape. Landholders used them successfully to control stream-bank erosion. However, willows are exotic trees that bring with them some environmental problems. They grow prolifically and block the flow of water around their roots, thereby altering stream ecologies. Their brief annual leaf-fall boosts stream nutrients and exotic algae blooms. Since 1989, staff working in the Murrumbidgee River Corridor have managed willows actively along sections of the Murrumbidgee River. They have conducted surveys of willow species. They have employed control techniques such as steam-injecting the herbicide Roundup through drill or 'frill' holes cut into the willows' trunks, and in some instances have physically removed the trees. The preferred method has been to replant with native trees and then gradually eradicate the willows. In this way they have achieved full control of willows in some areas.

(Snapshot 14.6). It should be part of a strategy (Snapshot 14.7). Parks Victoria, for instance, in their 1999–2000 pest plant management program, identified a total of 351 pest projects in 174 parks and reserves. Thirty-two projects in 26 parks aim to eradicate new infestations. The remaining 319 projects in 99 parks will control existing infestations that harm environmental values. Ninety-three weed species are targeted, the most common being Blackberry, Bridal Creeper (*Myrsiphyllum asparagoides*), Paterson's Curse, African Boxthorn (*Lycium ferocissimum*), Serrated Tussock (*Nassella trichotoma*), Horehound (*Marrubium vulgare*), Ragwort (*Senecio jacobaea*), St John's Wort (*Hypericum perforatum*) and Alligator Weed (*Alternanthera philoxeroides*).

Introduced animals. Pest animals can be controlled by poisoning, shooting, trapping or mustering, by exclusion from areas, by biological control, or by changing the habitat. The key to control is understanding pest biology and pest damage. Controls may, for example, be most effective when populations are stressed or at their lowest, not when the animals are at their most abundant. The control of pest animals is a science in

Snapshot 14.6

Management of Lantana in eastern Australian subtropical rainforests
Daniel Stock, Griffith University

The exotic ornamental weed, Lantana (*Lantana camara* L.), a designated weed of national significance, is considered by some to be one of the greatest threats to south-east Queensland biodiversity due to its capacity to establish in native plant communities. Lantana is widespread in the subtropical rainforest of south-east Queensland and has, for many years, caused concern about its role in successional processes and forest integrity (Fensham et al. 1994, Gentle & Duggin 1997). It is spread by seedling growth, especially along the edges of Lantana patches and in forest clearings. If left unchecked, Lantana becomes the dominant understorey in open forests and tropical tree plantations and can dramatically alter the structure of wet sclerophyll and warm temperate rainforests (Gentle & Duggin 1997). As an invasive weed, its success appears to be due to its reported ability to out-compete native species in frequently disturbed sites and its suspected shade tolerance.

However, in a recent study (Stock & Wild 2002), Lantana was observed in the border ranges area between New South Wales and Queensland to have little capacity to displace forests in the absence of external disturbances. This was researched further (Stock 2004) and Lantana's inability to successfully expand its patch size, under intact rainforest, appears to be due to two main reasons: (i) the increased level of shade provided by the intact canopy of the rainforest compared to the open canopy of the disturbance; and (ii) the particular nature of the intact rainforest. The results suggest that Lantana has little adaptation to shading. By maintaining at least 75% canopy shading the successful encroachment of Lantana will be prevented, and by achieving at least 75% shading in revegetation sites, Lantana

will not thrive in those areas. The management of Lantana patches can be achieved by maintaining the structural integrity of the canopy surrounding a gap and by a combination of physical removal of the Lantana within a gap and the planting of quick growing native species to create shade.

Lantana researcher Daniel Stock, Springbrook National Park, Qld (Daniel Stock)

Snapshot 14.7

Weed management in central Australia
Paul McCluskey, formerly of Parks & Wildlife Commission, Northern Territory

Weed managers from several parks have combined to introduce a Regional Weed Management Strategy, a Weed Control Manual (containing reference data), and an annual Weed Action Plan. Their objectives were: to implement a strategic approach to managing weeds by area and by species; to prevent the introduction and spread of weeds; to maintain weed-free areas; to reduce the impact of existing weeds; to conduct routine weed surveys; to establish suit-

able research procedures; to apply safe control methods; and to educate staff and the public. Vegetation maps, including maps of weed species, were prepared from biophysical data. The strategy was applied successfully to control the weed Mexican Poppy (*Argemone mexicana*) in the Finke River area of the MacDonnell Ranges where seven management zones were established, with different regimes of control and level of priority.

its own right, and needs to be researched carefully. Olsen (1998) is a useful starting point.

Priorities must first be determined. For example, many small marsupials are particularly vulnerable to foxes and cats. Many states have aggressive programs for reducing these species. The work undertaken in Western Australia is outstanding (Snapshot 14.8). The eradication of feral animals from New Zealand islands (Snapshot 14.9), work done by rangers in the ACT on wild pigs (Snapshot 14.10), the control of rabbits and cats on Macquarie Island (Snapshot 14.11), and the management of wild horses in Kosciuszko (Snapshot 14.12) are other excellent examples of pest control projects.

Parks Victoria, in their 1999–2000 pest animal management program, recognised a total of 224 pest animal projects in 68 parks and reserves. Of these, 123 targeted the rabbit, 80 foxes, nine goats,

Snapshot 14.8

Western Shield

Jim Sharp, Department of Conservation and Land Management, Western Australia

Endangered native animals such as the Bilby (*Macrotis lagotis*), Numbat (*Myrmecobius fasciatus*), and Western Quoll (*Dasyurus geoffroii*) are making a comeback in Western Australia due to the groundbreaking wildlife recovery program Western Shield. It is the biggest wildlife conservation program ever undertaken in Australia. Launched in 1996, Western Shield is working to bring at least 13 native fauna species back from the brink of extinction by controlling introduced predators, the fox and cat. The main weapon in the fight against the fox and cat is use of the naturally occurring poison 1080, found in native plants called gastrolobiums or 'poison peas'. While Western Australia's native animals have evolved with these plants and have a high tolerance to the poison, introduced animals do not. Western Shield makes use of this natural advantage. In the south-west forests, scientific research and monitoring has shown that where baiting has reduced fox numbers, there has been a dramatic increase in native animal numbers. Trap success rates for medium-sized mammals in the Jarrah forest of Kingston Block, near Manjimup, reflect a seven-fold increase since baiting began in 1993.

Western Shield involves aerial and hand baiting on almost 3.5 million hectares of Department-managed land. Baiting operations take place four times a year throughout the state from as far north as Karratha to Esperance in the south. Smaller nature reserves are baited more frequently. Around 770,000 1080 baits are dropped from a twin engine Beechcraft Baron each baiting operation—that is more than 3 million baits each year. The plane flies 55,000 kilometres each baiting operation. Monitoring is showing animals once on the brink of extinction in WA are returning and breeding in their natural habitats as a result of fox baiting. Since Western Shield began in 1996, the department has also carried out translocations of animal species. These include the Western Quoll, Dibbler

(*Parantechinus apicalis*), Numbat, Bilby, Southern Brown Bandicoot (*Isodon obesulus*), Western Barred Bandicoot (*Parameles bougainville*), Brush-tailed Bettong (*Bettongia penicillata*), Rufous Hare-wallaby (*Lagorchestes hirsutus*), Tammar Wallaby (*Macropus eugenii*), Shark Bay Mouse (*Pseudomys praeconis*), Noisy Scrub-bird (*Atrichornis clamorosus*), Western Bristlebird (*Dasyornis brachypterus longirostris*), Malleefowl (*Leipoa ocellata*), and Western Swamp Turtle (*Pseudemydura umbrina*). Baiting has been so effective that translocations of between 20 and 40 animals result in the successful establishment of new populations. WA is the only area in the world where three mammals—Tammar Wallaby, Southern Brown Bandicoot, and Brush-tailed Bettong—have been taken off the endangered fauna list because of scientific management action. The small hopping marsupial, the Brush-tailed Bettong, has been relocated to more than 30 places. Just 10 years ago there were three surviving populations. Long-term success is happening at Dryandra Woodland, Tutanning Nature Reserve, Boyagin Nature Reserve, and the south-west forest areas.

The Bilby, which underwent a massive decline in its natural range in the last century, was returned to the Dryandra forest as part of a breeding program. Wild populations of Western Quoll have been re-established in several areas after being restricted to south-west forests and the southern wheatbelt during the 1970s. Western Barred Bandicoots—extinct in the wild on the mainland—have been translocated to field breeding enclosures within Dryandra Woodland, the first time in more than 90 years the bandicoot had existed in the south-west. An important element of the program's success is the cooperation and support of local communities and many private landowners and Land Conservation District Committees have helped with fox baiting by laying baits on their own land where it is next to conservation reserves and state forest.

Snapshot **14.9**

Eradication planning for invasive alien animal species on islands
Pam Cromarty and Ian McFadden, NZ Department of Conservation

New Zealand's Department of Conservation (DoC) is a world leader in the field of invasive alien animal species eradication on islands, particularly rodent eradication. Eradication efforts have focused on Tuhua/Mayor Island (1277 hectares); Raoul and Macauley (2938 and 306 hectares); and Campbell Island (11,216 hectares).

The difference between eradication and control is that control operations manage the impact of invasive alien animal species and are not concerned with removing the 'last animal', while eradication permanently removes the impacts of invasive alien animal species by eliminating the entire population. A number of issues must be dealt with in planning an eradication operation. Failure to consider any one of these can result in an unsatisfactory outcome. Planning for an eradication operation involves research, contingency measures, incorporation of best available techniques, and the flexibility to cope with unexpected difficulties. Planning must ensure each targeted animal species is exposed. Identifying risks, and taking actions to eliminate or minimise them, is mandatory.

One challenge facing DoC as eradication operations become more complex is to ensure that effective communication

and knowledge transfer take place within the organisation. It is vital that the lessons learnt from each operation are recognised and disseminated. DoC has a commitment to learn from all eradication attempts, to reduce the risk of failed operations and to build the capacity to attempt more complex projects. The approach adopted when planning invasive alien animal species eradication programs on islands has several key components:

- a strategic approach considering all eradication programs
- team building including consideration of team dynamics
- skills development for project teams
- peer review to evaluate readiness prior to an operation taking place
- review and debriefings throughout the operation.

The next major challenge is improving the planning and implementation of island quarantine and contingency. Further research needs to be conducted into the long-term effects of eradication, defining long-term restoration goals for islands and island groups, and the improvement of eradication techniques for detecting and managing invasive alien animal species at low numbers.

---

seven cats, seven pigs, six dogs, two introduced bees, and one deer. Parks Victoria calculated that 33 threatened fauna and eight threatened communities will benefit from 72 of the projects in 58 parks. Targeted control of pests may take maximum advantage of when species are at their most vulnerable.

Pollution. Pollution may be as simple as garbage left lying about or soap used in a creek by bushwalkers; or it may be as severe as tankers dumping toxic waste in a national park. Land, air, and water pollution are all problems. Counter-measures include:

- codes of practice for visitors to protected areas
- efficient systems for dealing with wastewater and solid waste (including earthworms or other methods for aerobic decomposition of human waste)
- sewerage works that use advanced aeration treatments and/or chemicals to remove contaminants like phosphates and nitrates

Snapshot **14.10**

Namadgi National Park feral pig control program
Trish MacDonald, ACT Parks and Conservation Service

Scientists from CSIRO Wildlife and Ecology, from ACT Parks and Conservation Service, and key researchers have cooperated to reduce the impact of feral pigs on the ACT's Namadgi National Park. As shooting had proved generally ineffective, they conducted research on the best baits and dosages of poison required. In one test, the use of soaked wheat and warfarin (Rat Sack) proved 100% successful. Each year 12 staff worked 12 days to lay 3500 bait stations for feral pigs. They checked the stations daily for 10 days. They adopted an approach designed to minimise the collateral poisoning of native species: if a feral pig took the bait, the staff replaced it with a poison wheat bait daily over three days. In this way the harm done by feral pigs, through rooting and ripping the grasslands in this 106,000-hectare mountainous national park, has been significantly reduced.

Snapshot **14.11**

Vertebrate pest management in Macquarie Island Nature Reserve
Geof Copson, Parks and Wildlife Service, Tasmania

The control of vertebrate pests on Australia's sub-Antarctic island, Macquarie Island, 1500 kilometres south-east of Hobart, has occurred in several distinct phases. From 1972 to 1978, the fledgling Tasmanian National Parks and Wildlife Service began researching the biology of rabbits on the island. The annual introduction of myxoma virus since 1978 reduced rabbit populations spectacularly and was effective over the whole island by 1985. Following this success, the service began a control/eradication program for feral cats, with spotlighting, trapping, and the testing of poison bait dispensers. (The number of Wekas (a flightless New Zealand rail) had declined rapidly between 1979 and 1986.) The distribution and diets of introduced rodents and ship's rats is currently being studied. While some of Macquarie Island's bird colonies were lost in the early 1980s, the breeding rates of existing bird colonies have increased.

The rabbit has had a major impact on burrow-nesting seabirds on Macquarie Island, Tas (Geof Copson)

- natural wetland filters to reduce nutrients flowing into streams from sewerage plants
- stockpiles of chemicals and other aids to cleaning up oil pollution (on land or sea)
- improved policing to reduce illegal dumping of garbage and toxic wastes.

Illegal activities. Techniques that minimise illegal activities include:

- partnerships with customs agents, the police force, and federal authorities to monitor and check in-coming and out-going wildlife
- monitoring nests of endangered species such as Peregrine Falcon and Superb Parrot (*Polytelis swainsonii*) to minimise the stealing of chicks and eggs
- monitoring protected areas to prevent the theft of rocks, soil, sand, timber, rare plant species, or plants with a natural bonsai form. If theft is occurring, surveillance may be needed to catch those responsible
- not disclosing the location of very rare species such as the Wollemi Pine; or monitoring individual plant specimens to protect the developing seed source

- liaising with the police, where arson occurs regularly.

Climate change. Climate change could bring disproportionately large changes to the seasonal patterns. Sea levels are forecast to increase, which means coastlines will change, and many low-lying coastal environments will be inundated. Protected areas can provide a service for Australian society. The nature of these changes to natural areas could be monitored by carefully selecting representative protected areas nationwide. Permanent transects might be established and monitored long-term to measure how climates and habitats are changing. Thus the protected area system can become an early warning mechanism for climate change. This will also help scientists and managers predict the kind of changes our native species face (Snapshot 14.13). With coastlines and whole bioregions changing, the need for linking up protected areas (as mentioned earlier) will become ever more urgent.

Dealing with threats to soils and geological features

There are several ways to minimise threats to soils and geological sites:

Snapshot **14.12**

Managing a legend – wild horse management in Kosciuszko National Park
Pam Bryant, NSW P&W

Thanks to a poem by Banjo Paterson called *The Man from Snowy River*, the Snowy Mountains and horses are entwined in national folklore with Kosciuszko being a key element of the legend. The reality is, however, that roving herds of feral horses are trampling sensitive alpine plants, polluting high altitude alpine streams, trashing delicate alpine bog communities, and causing increased erosion resulting from the network of brumby tracks crossing the landscape. Horses are severely impacting on the internationally significant natural and cultural heritage values of this unique park. Horses have been in what is now Kosciuszko National Park since the earliest days of European settlement. During the past 20 years the horse population has increased as a result of a lack of active management and in 2002 it was estimated that there were as many as 3000 horses in the park. After the 2002–03 Australian Alps fires, population surveys indicated that horse numbers may have been reduced by up to 50% in many parts of the park. However, without follow-up management their numbers will return to former levels within several years.

In 2000, in response to an increase in environmental impacts resulting from feral horses, the NPWS moved to develop a management plan for them in the fragile alpine area. There was a wide range of conflicting views in the community about the issue and it attracted a high level of media interest. In October 2000 a process of community involvement commenced. A communications plan was developed that identified key stakeholders, including those who could be of the greatest assistance and those who had the potential to have a negative effect on the process. The communication plan established key messages to be promoted throughout the project and communication tools such as information sheets, newsletters, and material for the NPWS web site. A detailed media strategy identified the NPWS spokespeople, detailed media points and included a news release.

Support was sought and gained from the key stakeholders prior to a media announcement about the project. Various stake-holders were then approached to be involved in the Wild Horse Management Steering Committee. The steering committee included representatives from local government and the Snowy Mountains community, the park's advisory committee, horse riders, conservation groups, tourism, scientific experts, animal welfare bodies, and NPWS staff.

The steering committee worked with NPWS to engage the community through a series of public workshops and information sessions. This process highlighted the very wide range of views in the community about the issue. There was, however, general agreement that the alpine area of Kosciuszko needed to be protected from horse impacts, that horses should be managed and that they must be managed humanely. In June 2002 a Draft Wild Horse Management Plan for the alpine area of Kosciuszko National Park was released and placed on public exhibition. This was developed following extensive input from the Wild Horse Management Steering Committee.

While the plan was being finalised volunteer horse riders worked closely with NPWS to trial techniques for trapping horses in the alpine and subalpine area in the Rams Head Range/Dead Horse Gap area. These local horse riders brought a high level of experience and skill that resulted in the successful trapping of 13 horses.

In June 2003 the finalised Horse Management Plan was released and proposed a return to traditional methods of capturing and removing horses from the park. There is still a long way to go before a long-term sustainable removal process is in place. This project does, however, mark the first successful wild horse management strategy in a protected area that has strong community support. We have learnt from this project that staff commitment to the process of community consultation is vital and once made, must be continued. Being open about the process and inviting people to have their say often results in a greater understanding of the issue among the community. It also results in greater ownership of the solution among the key stakeholders.

- advanced soil protection techniques, including spray mulch and seed sprays, and soil-conservation mesh
- hay bales in gullies, and other temporary impoundments to prevent rapid runoff, and

- sophisticated engineering techniques to minimise water erosion on construction sites
- minimising toxic contaminants of soil, such as zinc
- minimising transport of soil pathogens by quarantining techniques, including chemical

baths for heavy equipment and hikers' boots (Snapshot 14.14)

- minimising theft at fossil sites either through the presence of rangers, by security fencing, or by keeping sites secret.

Dealing with threats to cultural heritage sites

Important Aboriginal art sites can be protected by steel mesh security fences. However, the current preference is to use boardwalks or other routes that visitors must stay on, leaving art sites untouched. Walking tracks and roads can also be routed to avoid Aboriginal sites. Animals may be excluded by conventional stock fences. Sites susceptible to fire can be given special protection during fires. Techniques for protecting European cultural heritage sites were discussed in Chapter 13.

Snapshot 14.13

Weeds, tourism, and climate change
Catherine Pickering, Wendy Hill, and Frances Johnston, Griffith University

Weeds are one of the most serious threats to Australia's natural environments and pose a major management issue for protected areas. Weeds have a range of negative impacts, including the displacement of native species and modification of ecosystem functioning. Intact native vegetation is thought to be relatively resistant to weed invasions (Csurches & Edwards 1998, Commonwealth of Australia 1999b). However, once vegetation is disturbed (either from human or natural causes), weeds can become established. For example human disturbance, such as clearing of vegetation for the provision of tourism infrastructure, can alter ecological processes and favour weeds. Problem weeds tend to have high reproductive rates, rapid vegetative growth, and tolerance to a wide range of growing conditions. Tourism use of an area can promote weeds through repeated disturbance that damages vegetation (creating gaps/bare areas), and in some cases through increased nutrification from human waste and other materials that creates weed-favourable habitats (Pickering et al. 2003). Examples of impacts of environmental weeds in the Australian Alps are: alteration of stream ecology (willows); displacement of native species (Broom and Yarrow); alteration of behaviour of native wildlife, such as feeding (willows and blackberries); alteration of the visual appearance of natural areas (Ox-eyed daises, Lupins, Sorrel, Dandelion and Yarrow); and changes in soil chemistry (introduced clovers) (Sainty et al. 1998, Johnston & Pickering 2001a, 2001b). Infrastructure associated with human activity, such as tourism in the Australian Alps national parks, provides habitat for a wide range of non-native plants. A biodiversity survey in Kosciuszko National Park in 2002 of the 27-square-kilometre area from Thredbo Village to the top of Mt Kosciuszko, found that 35% of the total flora was non-Indigenous (Pickering et al. 2002). The vast majority of alien species were associated with tourism

infrastructure. Within the garden areas of Thredbo Village, there were 103 taxa that are not Indigenous to the region, mostly deliberately planted. In the disturbed areas around the resort, again non-Indigenous taxa dominated, with 49 out of the 51 recorded taxa not native to the area. On the ski slopes there was also a relatively high diversity of non-Indigenous taxa—29 alien species or 30% of all species on the slopes.

The presence of so many non-Indigenous taxa within a national park is a concern. Among these taxa are species that could potentially spread from garden beds into adjacent disturbed areas, and even into the natural vegetation, becoming environmental weeds. A high diversity of non-Indigenous flora has also been found for the ski resorts in Victoria, creating similar issues for the management of the adjacent national parks (McDougall & Appleby 2000).

Predicted warmer temperatures and less snow cover from human-induced climate change may result in even greater spread of weeds in the subalpine and alpine areas of the Australian Alps (Good 1998, Buckley et al. 2000, Pickering & Armstrong 2003). Severe climatic conditions are currently a limit on the expansion of alien species into higher altitude habitats. However, the distribution and abundance of alien plants are likely to increase if snow cover declines and average temperatures increase with climate change. Summer tourism and recreational activities are also increasing (Worboys & Pickering 2004) with a commensurate increase in additional support facilities, thereby further exacerbating the weed problem (Pickering et al. 2003). Currently, weed control in the Australian Alps is primarily by herbicide spraying, which is expensive, has limited success for some weeds, and may harm native vegetation (Sainty et al. 1998, Sanecki 1999). More recently, biological control programs have been introduced for weeds such as broom, St John's Wort, Paterson's Curse and

Snapshot 14.13

(continued)

Blackberry. Also, parks agencies are encouraging resorts to use Indigenous plantings rather than exotic plants, both on ski slopes and in the gardens. Active rehabilitation of previously disturbed areas is also critical for weed control. While these strategies tilt the ecological balance in favour of native vegetation, they can be expensive. In addition to these measures, careful selection of tourism infrastructure options is crucial. This should be based on both the environmental and financial cost that can be incurred, in terms of disturbance to native vegetation and the introduction and establishment of weeds.

Snapshot 14.14

Management of Cinnamon Fungus on Kangaroo Island
Robert Furner and Keith Twyford, National Parks and Wildlife Service, South Australia

In 1994, the soil-borne Cinnamon Fungus was first identified on Australia's third largest island, Kangaroo Island, 15 kilometres off the coast of South Australia. Its infestations have been limited to local areas in the west of the island. If left unchecked, Cinnamon Fungus represents the single greatest threat to Kangaroo Island's outstanding biodiversity. A Regional Threat Abatement Planning process was adopted in 1996 to identify protectable areas and to maintain them in a condition that was free of Cinnamon Fungus. Control relied upon three steps: sound evaluation and planning, community awareness and support, and local partnerships. Field and road-side hygiene techniques ensured Cinnamon Fungus was not spread inadvertently in soil and plant tissues adhering to vehicles or hikers. Information sheets and brochures targeted visitors, land managers, and farmers. Signs directed hikers to hygiene stations.

Plant washdown unit for Cinnamon Fungus control, Kangaroo Island, SA (Bob Furner)

14.5 Management lessons and principles

Lessons

1   Threats can arise as a tyranny of small development decisions over time and can change the nature of values of protected areas.
2   Incremental change and gradual erosion of high standards needed for nature conservation is a tactic used by too many pro-development users of natural areas.
3   Managers need to remain vigilant to threats to protected areas. They need to rise above potential personal criticism that can accompany demands for greater utilisation and exploitation of protected areas. Tools such as monitoring and condition reporting are critical for informing the greater public of any changes to conservation standards.
4   Jobs versus the environment will always be a compelling debate, but it is too often a one-sided debate. Experience shows that decisions favouring development and jobs regularly take precedence over the environment, even if it will lead to damage to protected areas. Initiating new commercial activity or use in protected areas needs to be very carefully considered.

NPWS chainsaw crew (Neville Brogan and colleague) 'mopping up', NSW (Graeme Worboys)

5   When managing for threats, the following steps are recommended:

- identify all human-caused threats that require intervention
- quantify each threat individually. Inevitably, there will be more threats than the resources available to address them
- assess the potential for the individual threats for their impact upon the natural resources being managed
- address threats in a holistic manner, ranking each threat on the above assessments
- determine the priority threats in terms of manageability and cost-effectiveness of management actions
- develop strategies and actions to address the threats and to ameliorate/reduce the level of impacts
- establish a legal framework to implement the regulations required to deal effectively with the threats
- while implementing actions, be aware of unintended consequences. Some mitigation may result in exacerbation of the threats
- conduct effective communication with communities for their help in suppressing threats
- conduct community education supplemented with societal benefits, to bring a change in the behaviour that is linked to the threats.

Principles

1   Protected area managers need to intervene so that the impacts of human-caused threatening processes are minimised.

2   Threats to protected areas are inevitable. They need to be anticipated with pre-planned responses, appropriately trained personnel, and readily available logistic support. This pre-planning should take into account the probability that threatening processes will increase over time, especially in response to changing population and use patterns.

3   Dealing with threats may require long-term activity. Interventions need to recognise the long-term budget commitments and stable organisational environment that are necessary to achieve successful conservation outcomes.

4   Immediate, decisive intervention to deal with some threats may save significant future costs.

5   Threatening processes will be diminished where protected areas are managed as part of a community effort to achieve conservation outcomes at a landscape scale.

6   Effective environmental planning processes are necessary to combat threats to protected areas.

7   Regulations may be required to deal effectively with some threats.

8   Effective community education will help minimise threats to protected areas.

15

Incident Management

Lennox: *The night has been unruly. Where we lay.*

Our chimneys were blown down, and, as they say,

Lamentings heard i'th' air, strange screams of death,

And prophesying with accents terrible

Of dire combustion and confused events

New hatched to th' woeful time: the obscure bird

Clamored the livelong night. Some say, the earth

Was feverous and did shake.

Macbeth: *'Twas a rough night.*

From Macbeth, *Act 2, Scene 3, lines 56–64,* **by William Shakespeare**

15.1	Organisations with incident management responsibilities	403
15.2	Managing incident responses	404
15.3	Managing fire incidents	412
15.4	Managing wildlife incidents	417
15.5	Managing incidents arising from natural phenomena	418
15.6	Management lessons and principles	422

If a bushwalker needs rescuing, or a fire is out of control, the media sees this as an emergency, or a disaster. Managers will speak, more technically, of an incident. Emergency Management Australia (1998) defined an incident as an emergency event, or series of events, accidentally or deliberately caused, which requires a response from one or more government agencies.

The media are quick to respond to incidents occurring in Australia, and the general public are well acquainted with rangers and police being involved in incidents such as floods, fires, search and rescues, whale strandings, and major pollution events. Fire incidents include wildfires in natural vegetation on protected areas anywhere on the continent (including alpine and rainforest communities) as well as building, chemical, vehicle, or other fires. Bushfires in particular are part of the folklore and history of Australia. For most of the continent they are the most common incident managed in protected areas. Managers are also involved in dealing with storms and their consequential wind damage, flooding and fires, and their potential to cause beach erosion and the collapse of cliffs, among other natural events. Marine mammal strandings regularly involve managers in cooperation with volunteer whale rescuers such as ORRCA.

People regularly become lost within protected areas, requiring search and rescues to be undertaken, typically with the police in charge. People may create terrestrial incidents through cruelty to animals, road kills, poaching, and illegal collection of flora and fauna species. Senior managers need to consider an organisation's capacity to deal with such incidents as engineering-based problems (including the failure or collapse of buildings or structures such as roads and bridges), vehicle accidents, and traumatic events such as the indiscriminate shooting of visitors at the Port Arthur Historic Site in Tasmania in 1996.

In this chapter, we look at how incident management fits into the wider picture of protected area management. We describe incident management systems used in Australia and identify state initiatives to manage common incidents.

15.1 Organisations with incident management responsibilities

Many organisations and individuals may join forces to manage and resolve an incident. Managers need to be alert to the problems, but also to the opportunities that this cooperation brings. They should also know the legal rights or powers of each group or person, and their likely demands and needs. For instance, when an oil spill threatens sea-birds, the manager should know that the Commonwealth (through the Australian Maritime Safety Authority) has jurisdiction over spills that occur more than 5 kilometres offshore. State law applies closer to land; and these laws give emergency powers to various management bodies during such incidents. Clearly, it is important to have good relations with all relevant bodies and individuals, and these should be established in advance.

Managers must also bear in mind that once human lives and property are at risk, their own control of areas will often be overridden by other authorities with wide emergency powers. Ill-informed use of these emergency powers can bring permanent damage to protected areas. Managers need to ensure that other authorities know the special needs of protected areas (for example, protecting threatened species or culturally significant sites) and that contingency plans take note of these needs. Where possible, the need to protect the environment should be written into the laws that grant such emergency powers. At worst, where this need has not been communicated in advance, timely advice and briefing sessions should be given during the incident.

For instance, when working with local fire authorities during the control of a fire incident in an area, managers may need to negotiate the following understandings:

Search and rescue, Kosciuszko National Park, NSW (Graeme Worboys)

- when firefighting vehicles need access, they will wherever possible follow existing trails
- wherever possible, natural or pre-existing fire-control lines will be used, rather than making new ones
- wherever possible, control lines will be put in without using bulldozers (for example using hand tools, or following natural features such as rivers or rainforest gullies)
- permission should be sought from managers before any use of bulldozers, firefighting chemicals, and aerial burning
- incident managers will be consulted about sensitive plant or animal communities or culturally significant sites in any area where backburning is to be used.

Major incidents can involve a complex mix of Commonwealth, state, and local government authorities as well as volunteer organisations. For example, a marine mammal stranding at Seal Rocks near Newcastle, NSW, in 1992, involved 30 different organisations. When different authorities are collaborating but using different systems, there is a danger that incident administration and communication can become chaotic. Hence, managers need to make certain, well in advance, who has responsibility for what, and which plans and systems are to prevail.

15.2 Managing incident responses

Incidents are a fact of life. We cannot predict exactly when they will occur. Yet for a given region we should have a fair idea of what sort of incidents will happen, in what sort of terrain, and sometimes in what season. Incidents can be anticipated and planned for. Failure to have a suitable plan, suitable logistic support, and trained staff in place could result in serious damage to life, property, and the environment. It could also lead to criticism from local communities and politicians. Organisational policies on how to deal with natural events in protected areas are also an important consideration. Managing for incidents includes:

1   analysis—identification of potential incidents for a protected area

2 research—assessment of local, state, and Commonwealth incident management protocols and systems

3 policy formulation for protected areas—development of organisational responses to incidents

4 incident plans—including incident response systems, heritage protection needs, incident operational techniques, and equipment

5 organisation of logistics—such as plant, equipment, communication

6 training and staff preparation

7 incident response management

8 staff management—potential trauma counselling

9 rehabilitation—potential repair of disturbed areas

10 debriefing—post incident assessment.

Violent events such as cyclones and landslips are natural phenomena. Often, the results of these events can and should be left alone. Managers will sometimes ask of an event, 'Is it an incident?', meaning 'Does it require us to intervene?' Managers also often need to act for the sake of people who live nearby but not in their immediate management area. Good neighbourly relations are very important, even when there is no urgent risk to life or property. Managers need to consider what the wider public expects of them.

In developing a planned response to incidents, a range of questions can be asked, many of which influence the development of policies. Some of these questions are as follows.

1 Does a particular natural event require incident response action? Is it an incident? Does it threaten life, property, or a protected area?

2 What incidents are likely to occur for a protected area?

3 What action is needed to deal with the incident? How is this action to be delivered? How do the control and command systems work?

4 Who needs to be consulted or involved? Who are the stakeholders?

5 For this kind of incident what plans or systems are already in place at national, state, regional, and local level?

6 What management organisations should carry out these plans, and what legal and other powers do their leaders have? What is the correct role of the protected area manager?

7 How can this damage be reduced, both during and after the incident?

8 What safety measures are needed?

9 What skills must staff have to deal with the incidents?

10 What legal and bureaucratic rules or duties must the manager consider?

11 What legal, social, political, economic, and environmental issues will have to be addressed after the incident?

Australian incident plans and management systems

There are national and state plans and systems for the management of incidents. Depending on the nature of the incident, protected area managers can be directly or indirectly involved with them.

When managing an incident and working with other individuals in organisations it is essential to have pre-established rules and guidelines. One such incident management system, the Australian Inter-Service Incident Management Systems (AIIMS) Incident Control System (ICS), has been adopted by many organisations as their standard, and has assisted with cooperative incident management where cross-border cooperative management has been required. The ICS manual (AIIMS 1992) lays down standard procedures that allow different organisations to 'speak one language' and arrange themselves under a single command.

The ICS was derived from the National Interagency Incident Management System, which has been used with considerable success in North America (AIIMS 1992). The North American system was used for example following the terrorist attack on the World Trade Center, New York City on September 11 2001. The ICS is intended to be a national system. It can be used for any incident or emergency, large or small, where various activities or organisations must be coordinated. The ICS is based on a structure of delegation. It ensures that all vital arrangements and communications are properly seen to.

The ICS is based on two key principles.

Management by objectives. The Incident Management Team determines how the incident can and should develop. The desired outcomes or objectives are then communicated to the others, so that everyone is clear about what is to be achieved and how.

Span of control. There is a limit to the number of groups or individuals one person can successfully supervise. Normally no more than five groups or individuals will report to any one person. This way, supervisors can set tasks effectively, and monitor how they are being performed.

The leader of the ICS 'team' is called the Incident Controller. During the initial response to an incident the team may consist of this person only. As an incident grows, he or she delegates more and more functions (as prescribed by the AIIMS manual), while retaining overall control. Shifts are introduced so that team members remain alert and effective to deal with lengthy incidents.

As an incident escalates, the ICS team is divided into four sections: control, operations, planning, and logistics. The leaders of these four sections are known collectively as the Incident Management Team. An example of an ICS for a large incident is shown in Figure 15.1. The Incident Controller has overall responsibility for managing the incident. He or she must prepare the objectives for action and approve the Incident Action Plan (IAP). The other sections have the following functions.

Operations. Operations staff are responsible to the Incident Controller for managing those who combat the incident in the field.

Planning. The planning staff collect and analyse information about the incident: for predicting how it will develop; for maintaining a register that lists where resources and equipment are, and how they are being used; and for preparing alternative strategies to control the incident.

Logistics. The logistics section is responsible for providing facilities, services, and materials to the operations staff.

Part of the beauty of this system is its adaptability. In a large incident, each section might employ scores of people. In a small incident managed by only two people, the second one might be told: 'You'll be in charge both of planning and logistics.' In either case, the same range of duties applies. An Incident Action Plan is developed by the incident

Figure 15.1   The incident control team (AIIMS 1992)

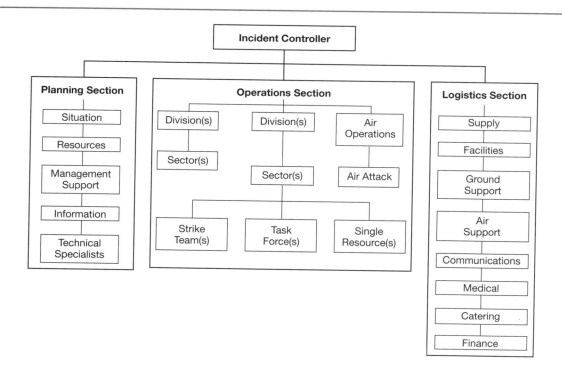

several organisations are cooperating. While some people have criticised the amount of paperwork involved, the system has proven to be successful in dealing with incidents ranging from fires and hail-storms through to outbreaks of livestock disease.

Protected area management agencies have contingency plans for responses to incidents. The Victorian Department of Sustainability and Environment (formerly Department of Natural Resources and Environment—DNRE), for example, has developed contingency plans for oil spills and cetacean strandings. The Wildlife Response Plan for Oil Spills (DNRE 1999b) provides procedures, reporting structures, roles, and guidelines for the rescue and rehabilitation of wildlife affected by an oil spill. The Department suggests that the plan should be implemented in conjunction with the Victorian Marine Pollution Contingency Plan (VICPLAN). The Victorian Cetacean Contingency Plan (DNRE 1999c) outlines appropriate actions to be taken in the event of a whale or dolphin stranding along the Victorian coastline, whether it be a single animal or whole herds.

Operational logistics

Managers need to spend considerable effort to ensure that they have the operational capacity to deal with incidents. This not only includes training

Incident Controller Dave Darlington, 2003 Australian Alps fires, Jindabyne, NSW (Graeme Worboys)

management team for the particular incident, and is reviewed and approved by the Incident Controller. It contains objectives and strategies with specific time frames. The AIIMS ICS system may be used at 'multi-agency' incidents, that is, events where

Multi-agency incident control team, 2003 Australian Alps fires, Jindabyne, NSW (Graeme Worboys)

and organising staff, but also logistic support equipment and cooperative operational arrangements with other organisations (Chapter 11). Incident management often needs the support of equipment such as trucks, tractors, water tankers, light-weight four-wheel-drive firetankers, aircraft, pumps, chainsaws, and mobile catering units. These items all need to be maintained in readiness for an incident. At a minimum, this requires planning and budgetary allocations, and plant replacement systems (Chapter 9). Other practical details such as staff rosters, standby plant and equipment, routines for servicing heavy machines and equipment, catering, first-aid support, communication support, and media support need to be in place. 'Media support' means the organisation's own press or media officers. In the event of an incident, their primary role will be to ensure that the community is very well briefed about what is happening and why.

Incident management skills and training

When television news programs show an incident, they rarely depict the work of incident managers. The camera finds more interest in the heroic acts being played out by crews in the field, such as fighting fires or leading home a rescued child. However, without the work of the people behind the scenes—

the incident managers and planners—the work of the crews can be wasted. The battle against an incident is more likely to be decided by those back at base who plan the strategy. This chapter does not concentrate on the knowledge needed by front-line crews and their immediate leaders. There are many excellent publications that describe first aid, bushcraft, map-reading, practical incident skills such as competence with a four-wheel-drive, and so on.

A manager's greatest resource is the collective skills of his or her staff. Their local knowledge is important; but it is also essential that they have a large suite of skills. These will include some advanced and specialised skills. No one individual can have all these skills; so staff will need to specialise. Teamwork, itself a skill, and leadership are needed (Chapter 4). Training is a fundamental part of preparing for an incident, and managers need to ensure that their staff are trained in the cross-section of skills necessary to deal with the spectrum of events. Some of the roles, and the corresponding skills, needed by an incident response team managing a fire, for example, are listed in Table 15.1.

Occupational health and safety

The safety of all staff involved in an incident is the priority responsibility (Snapshot 15.1). In the first

NSW Rural Fire Service tankers and firefighters near Smiggin Holes, Australian Alps fires January 2003, NSW (Graeme Worboys)

Table 15.1	Skills required for a fire incident (NSW NPWS 1997a)		
	Staff member		
Skill	Incident controller	Incident planner	Incident operations personnel
Leadership	*	*	*
Decision-making	*	*	*
Negotiation	*	*	*
Team work	*	*	*
Media management	*	*	
People management	*		*
Counselling	*		*
Mentoring	*	*	*
Time management	*	*	*
Computing		*	
Weather data analysis	*	*	
Multi-variate analysis		*	
Incident behaviour forecasting	*	*	
Air photo interpretation	*	*	*
Advanced mapping skills	*	*	*
Backburning	*	*	*
Mopping up	*	*	*
Incident control system operation	*	*	*
Chainsaw operation			*
4WD Basic	*	*	*
4WD Advanced			*
Remote area fire training			*
Heavy vehicle licence			*
Tanker operation			*
Motorcycle licence			*
Motor maintenance			*
Pump maintenance			*
Bulldozer supervision		*	*
Bulldozer operation			*
Tractor operation			*
Helicopter safety	*	*	*
Helicopter winching			*
First aid	*	*	*
Advanced first aid			*
Radio procedure	*	*	*
Fire fuel assessment	*	*	*

Snapshot 15.1

Safety and remote-area firefighting

The rough, rugged, and remote sandstone areas of the Blue Mountains and Wollemi National Parks near Sydney are difficult places to fight a fire. In 1984, during a prolonged, dry and volatile summer, a series of fires stretched the capacity of fire-fighting units for weeks. Helicopters were often the only means of getting in; and expert crews were winched down through the forest canopy to cut helipads so that other crews could follow. Fire-control lines were put in by hand, to allow backburning. The incident managers set strict rules for the fire-control campaign in this cliffed landscape.

1   They used only highly trained firefighters.

2   They placed control lines well-back from the wildfires.

3   They made full use of all data about the fires. These included the current rate of spread and flame-length, the likely weather, the types of vegetation and amounts of fuel. Planners used computer simulations of how the fire might behave in given terrain and weather; and they always allowed for worst case scenarios.

4   They set up a forward control point, with reliable radio links.

5   They sent operation controllers to supervise remote helipads to ensure safety near aircraft and to check that all crews were brought in and taken out as planned.

6   There was always an ambulance officer or first aid officer on duty at the forward control point.

7   All who went in by helicopter took an evacuation plan (with maps) so they could escape overland if helicopters were not available. Staff were briefed on evacuation routes.

8   Since the backburning was being done at night, managers sought out helicopters and pilots licensed to operate at night.

9   They meticulously planned the timing of backburns using the data mentioned in item 3, especially humidity levels. On-site weather stations tracked changes in humidity and temperature through the night. The nights were hot and dry, but less so than the days. The strategy of backburning at night worked; and in time all fires were contained.

The only casualty, a firefighter who injured his ankle in the rough terrain, was flown to safety.

Subalpine woodland burning near Dainers Gap, January 2003 Australian Alps fires, Kosciuszko National Park, NSW (Graeme Worboys)

instance, staff who are undertaking physically demanding incident management need to be fit and able to undertake the work. Administration systems need to be in place to manage this requirement. No matter how skilled or experienced staff are, the manager must practise precaution in deploying them to an incident. This is critical when inexperienced staff or external persons such as volunteers are involved. For all personnel, it must be made clear that risk–taking and bravado are unacceptable and may imperil others.

The dangers or special features of the terrain must be fully planned for. Are there likely to be dangerous substances or animals? What if the weather changes abruptly? What else could go wrong? Aspects such as these should not be left to chance and must be routinely considered in organising for incidents.

An efficient system for rotating crews is also needed. Overtired or hungry people can make mistakes that risk lives and threaten incident operations. Managers need to consider how they

Helicopter used for firefighting (Graeme Worboys)

Fire crew awaiting evacuation by helicopter after completing their shift (Graeme Worboys)

will keep track of personnel once they are in the field. How reliable are their systems? The ideal length of shift-time varies. Fire crews usually work 12 hours at a time; however people working with stranded whales in cold water may need to warm up on land again after only 20 minutes, or hypothermia could occur. When an incident goes on for weeks it becomes important to rest people and not keep rotating them back onto duty. Once again, efficient administration systems are needed.

Managers have a range of safety obligations. They need to keep an eye on the reliability of equipment. Has it been maintained as safety standards require, for example? There are strict rules for air safety. Pilots can only work a fixed number of hours per day, with enforced rest days. Managers must insist on a roster system for pilots, and may need to arrange for aircraft to be exchanged, or rested for service, at regular intervals. This is especially important during long and intense campaigns.

Incident administration

A manager may need to set up forward control centres for incidents—in the case of fire, for example. It is crucial to put these in the right places, and to think ahead. What happens, for example, if a fire changes direction? Setting up a forward control centre at the height of an incident is a skill honed by experience. Such centres need to work efficiently from the moment they are set up. Trial and error is unacceptable; the layout and structure of such bases must be carefully pre-planned.

A common serious mistake is to mix crews that are resting with the bustle of crews departing for the incident. Helicopter pads should be set up near the camp, but away from resting personnel. Resting crews need real rest. Food must be adequate, and the site's cooking and ablution facilities free of health risks. Staying healthy in the field is itself a safety issue! Always plan to be ready to treat and evacuate sick or injured personnel. Do not lose sight of the budget. Aircraft, for instance, are not cheap to hire or use. Managers will need to make sure that all flights are justified and cost-effective, and that the right type of aircraft is being used. Managing the budget as the incident proceeds is a fundamental responsibility.

Community consultation during incidents

Managers need to keep the community informed about serious incidents. Early on during a media event they should provide background fact sheets to politicians and the media, and continue to provide regular bulletins. Major incidents may require press conferences, the handling of which requires training and specialist support. Smaller incidents may require regular press releases and media interviews or briefings with individuals. The AIIMS Incident Control System prescribes regular briefings for stakeholders, neighbours, local politicians, and, usually, the media.

Maintaining routines during an incident

Incidents may completely occupy the limelight for a given time, yet managers need to manage the normal routine as well. Visitor facilities must be maintained, accounts and wages paid, and government commitments and deadlines need to be kept. Part of planning for an incident involves planning to keep up with day-to-day routine management.

Post-incident management

Rehabilitation. Preferably, any damage caused to protected areas during an incident should be rehabilitated immediately after the incident, with the cost of this work being a cost against the incident. Some damage can, however, take many years to repair, and may require persistent rehabilitation work.

Debriefing. Incident debriefs are a critical part of managing the total incident operation. Inevitably, in most incidents, there is confusion and conflicting stories about what has happened. Debriefs provide a valuable insight into a whole operation, and how operations can be improved for the next event. Staff counselling may be required immediately following some incidents. Professional assistance is obtained for this type of service (Chapter 9).

Legal issues. Incidents, sadly, can lead to loss of property or life. Formal inquiries (coronial inquests) are often called after the event, and managers may be summoned to give evidence. They will be expected to be able to tell the inquiry exact dates and times, the reasons why certain decisions were made or not

made, and who made the decisions, and how. Clearly this means keeping an accurate account of events during the incident. Like police officers, managers must be trained to record and recall all the legally relevant aspects of an incident. Typically incidents are recorded in an operational 'log', although it is wise also keep a personal log of events. The plans and logs used as part of the AIIMS system assist in this process.

For coronial inquests, parliamentary inquiries, insurance inquiries, and other legal inquiries or actions that can follow an incident, well-organised and accurate information about the incident will be required. Some staff may be subpoenaed and will require special training in how, as witnesses, they should answer questions. It can be an unsettling experience to be 'grilled' by a barrister in an adversarial coronial or court atmosphere. After a major incident and before going into the witness box, managers should always seek legal advice. This is to confirm that they have in fact collected full, accurate, and relevant information, and that they have been briefed on how court proceedings and questioning will take place.

Economic costs of incidents. Managers need to bear in mind that incidents that lead to closure of a protected area may result in serious losses for licensed commercial operators and nearby towns that rely on tourism. Where fault is determined through a legal process, it may lead to compensation payouts. The insurance premiums for protected areas can be influenced by the nature, frequency, and cost of incidents.

15.3 Managing fire incidents

Fire is one of the most common types of incident managed in Australia. While fire is a natural occurrence in the Australian landscape it also needs to be managed to limit the danger it imposes on life, property, and the environment, particularly given that people cause the majority of fires.

Fire mitigation is a huge operation in Australia. Managers respond swiftly, and their pre-planning and preparedness is sophisticated. South-eastern Australia is one of the most fire-prone areas in the world (Figure 15.2). To give some idea of the scale

of effort involved, during the summer of 1997–98, Victorian authorities deployed 1500 trained full-time firefighters, 80 large tankers, 260 lighter four-wheel-drive 'slip ons', 30 heavy bulldozers, 30 light bulldozers and transporters, and 22 aircraft. The Victorian government spent over $50 million on this effort. The statewide target was to immediately attack all fires on public lands, and seek to confine 75% of them to less than 5 hectares. In fact 83% were so confined. The costs of fire-suppression, though large, could seem small to Victorians who

Figure 15.2   Average frequency of large bushfires
(SEAC 1996)

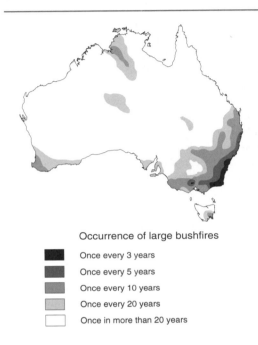

Occurrence of large bushfires

■ Once every 3 years
■ Once every 5 years
■ Once every 10 years
□ Once every 20 years
□ Once in more than 20 years

still remember the 1925–26, 1938–39 and 1982–83 summers notorious for 'Black Thursday', 'Black Friday', and 'Ash Wednesday'. On these occasions 60, 71, and 47 lives, respectively, were lost, and large parts of Victoria were reduced to ash. The 2003 Australian Alps fires, the worst fires experienced in the mountains for over 60 years, stretched the fire-fighting resources of two states and a territory as well as national support resources (Snapshot 15.2).

Most unplanned fires have to be treated as incidents. Fires may be natural (caused by lightning) or due to accident or arson; but under Australian conditions they are usually too volatile, and too dangerous to life and property, to ignore.

Planned fires are a different matter. In parts of northern Australia, large-area prescription burns for management, cultural, agricultural, and ecological purposes are regularly undertaken. So too are the (generally smaller) prescription burns carried out in southern Australia to protect property or for ecological management. When such burns go according to plan they are not classified as an incident. Yet, because even planned burns involve some risk, and because they are part of the basic strategy for dealing with fire, we shall include them in this chapter.

Snapshot 15.2

The 2003 Australian Alps fires

In January 2003, the Australian Alps experienced their largest bushfires in 60 years with an estimated 1.73 million hectares burning (Worboys 2003a). The bushfires burnt across Victoria, New South Wales, and the Australian Capital Territory during a drought that ranks as one of Australia's worst in 103 years of official Australian weather records. The year 2002 had also experienced maximum average temperatures that were one degree above normal, this being the highest departure from normal ever for 100 years of record-taking (Bureau of Meteorology 2003). In January, the bush was tinder dry and day temperatures regularly very hot. Wild electrical storms were all that it would take to start fires, and on 8 January 2003 this is exactly what happened. Easterly moving thunderstorms started 92 fires in the Victorian Alps, 45 (mostly) in Kosciuszko National Park and three in the Australian Capital Territory (Worboys 2003b). By day 10, initial attack had reduced the number of fires from 92 to 10 in Victoria, 45 to 14 in Kosciuszko National Park, and the ACT was fighting their three fires. This proved to be insufficient. The extreme dry and rough, rugged, and remote terrain made it extremely difficult for incident controllers. Despite the presence of over 4000 professional and volunteer fire fighters, helicopters including the Erickson Sky Crane water bomber, fire tankers, bulldozers, fixed wing aircraft, light mobile fire units, and other support equipment on peak days, control lines and suppression efforts were regularly unsuccessful. The area burnt by the fires progressively expanded. On 18 January 2003 during very high wind and extreme high temperatures, fires burnt into the suburbs of Canberra, with the tragic loss of four lives and the loss of 505 houses. It was Canberra's worst ever disaster. The Australian Alps fires expanded and eventually joined to form a single large burnt area of 1.73 million hectares. The fires were finally extinguished on 7 March 2003, some 60 days later. Firefighters had endured eight major weather frontal changes with their accompanying strong winds, high temperatures, and low humidities during this period. Across the Australian Alps, four lives had been lost, 551 houses had been burnt, rural properties impacted, and major water supply catchment areas burnt severely (Worboys 2003b). Five separate inquiries were instigated to investigate the fires and their management.

Planning for fires is a routine task for managers (Snapshot 15.3). They will usually work jointly with community fire brigades. A fire plan means far more than simply arranging to have fire crews and equipment available or on standby. It means clearly understanding the kinds of fire likely to affect a protected area, and how the area's fire-plan fits into that of a larger region (Chapter 12).

Being ready for a fire season involves making sure that staff are adequately skilled and physically fit. It also involves a range of tasks, some of which include ensuring that inter-agency cooperative arrangements are in place, that fire plans are up to date, and that an incident operations plan is complete (including a detailed summary of contacts,

Snapshot 15.3

Fire planning completed by the Southern Regional Fire Association of NSW

In 1994, the executive of the then Southern Regional Fire Association of NSW commissioned a strategy to ensure cooperation among brigades in its region, and to allocate available funds where the priorities were highest. The Association prepared a report titled 'Improving bushfire management for southern NSW' (Dovey 1994). It is a useful example of planning for fire incidents at a regional scale, and has been most influential.

It considered local climate patterns, and evaluated how bushfires behaved and how rapidly they spread in different parts of the region. These things may depend on climate, slope, aspect, and the type and amount of fuel available. The report combined the data on these factors to produce a regional map of likely bushfire behaviour. The report then analysed the historical pattern of causes of bushfires. It found a major variation in the nature and frequency of fires across the region and, in particular, between the more populous coastal areas and the hinterland. The main difference was that arson fires were more common along established fire trails. Lightning strikes were a relatively minor cause of fires.

By combining data on the likely number and likely behaviour of fires with other data on the amount of property at risk in each area, a map was produced of 'bushfire damage potential' across the whole region. The map covered forested and grassland areas, and gave each area a bushfire-threat rating. This helped the Southern Region Bushfire Association give priority to the areas with the highest threat ratings when it allocated its resources within the region.

management systems and procedures). Other considerations include ensuring that computer software for simulating fire-behaviour is operational; heritage data are accessible; a staff roster is in place; a program of prescription burning has been completed; a fire-trail maintenance program has been completed; and the radio communication system is working. Fire observation towers need to be serviceable and accessible; plant and equipment needs to be serviceable; supplies of fuel, incendiaries, and retardants need to be available; and contracts for aviation support need to have been arranged. Finally, contracts for major plant (such as bulldozers) should have been organised and administrative systems put in place. The timing of preparation activities will vary from season to season, and will be influenced by climatic and fuel conditions.

Immediate attack on a fire as soon as it is detected is the best control strategy. Crews may be able to directly attack the fire and suppress it while it is small. In many remote areas this is one of the few options managers have in controlling fires. Remote areas often consist of rugged terrain that is highly unsuitable for the bulldozers and large tankers used to control fires in other areas. Instead, small remote fires are often directly attacked by specialist firefighters using dry firefighting techniques and often supported by water-bombing aircraft. We will now discuss techniques for dealing with fire incidents in more detail.

Rapid detection and response to wildfires. Several methods are used by managers for the early detection of wildfires, including the staffing of fire towers at vantage points during the summer season. These report local weather, plus any fires they sight. Surveillance flights, particularly during afternoons of thunderstorm activity, provide very early warnings for new fires. Mid morning surveillance flights the day after thunderstorms are also used to detect fires (caused by lightning) that have smouldered through the rain or the night-time humidity, and been reactivated by the morning's heat. Commonly, however, local residents and visitors may raise the alarm for fires.

Techniques for rapid response to a wildfire sighting include the following options:

1   Water-bombing can be carried out using aircraft and environmentally acceptable fire retardants. A

second observation aircraft may be used to direct the water-bomber to the target. Often these water-bombers provide a holding strategy until on-ground crews arrive at the scene.

2   The fire can be directly attacked by tankers and crews. Small tankers or 'slip-ons' are useful to reach fires that are inaccessible to larger tankers.

3   Earth-moving equipment can be used to block or encircle the fire with a control line.

4   Helicopter-based fire crews can attack more remote fires. Typically, the first crew is winched down or it rappels (controlled descent utilising ropes and abseiling techniques) from a helicopter and cuts a helipad clearing in the forest for the rest. Direct attack can then begin. Following this, the same helicopter may begin to drop water on the fire using a water bucket serviced by nearby water sources. In volatile conditions, water bucketing helps keep the firefighters safer by reducing the fire behaviour. It may not put out the fire, but it can dampen it enough for fire crews to safely attack its flank.

Tactics will depend on the terrain and the type of fire. When open paddocks or road-systems near a town allow vehicles to move almost at will, direct attack by large tanker units is often very successful. This is especially so if water is available and if control lines (which may be existing roads) are easily set up. For more remote fires, dry firefighting techniques are mainly used.

Dry firefighting techniques. Dry firefighting techniques are the most common approach to fire control in protected areas. Combustion needs three things: fuel, oxygen, and heat. No fuel means no fire. Bare-earth ('mineral-earth') fire-breaks mean no fuel. These fire-breaks can be made by bulldozers, tractors, graders, or by crews using hand tools. They should usually surround the fire, or link to natural fire-barriers such as dams, lakes, rivers, or rock barriers. This linkage forms a control line. On rare occasions, 'wetlines' or water-soaked barriers may be used for an urgent backburn. Typically mineral-earth control lines are put in place well before the fire arrives, and small fires are then set to burn back towards them from the fireward side. This creates a much wider 'blackened' area with little or no fuel. Issues managers should consider in dry firefighting include:

- the Byran-Keetch Drought Index rating, a measure of the relative dryness of forest and other fuels and therefore their potential flammability
- the Fire Danger Rating Index (Luke & McArthur 1978), an index that combines a range of parameters and provides an evaluation of rate of spread or suppression difficulty for specific combinations of fuel, fuel moisture content, and wind speed
- the rate of spread of the fire, for existing and forecast conditions
- the amount and kinds of fuel present near the proposed control line
- how long it will take to construct the control line (by whatever mechanism)
- how wide the control line needs to be in these conditions before one can safely backburn from it
- safety aspects, such as escape routes for personnel
- the use of expert navigators (using air photos) to help place the control lines, particularly where heavy plant is employed
- environmental and heritage assets at risk
- the possible use of natural fire-barriers such as scree slopes, cliffs, and streams (remember that fires sometimes cross these barriers, even with low fuel loads)
- the risk that fire can burn underground and, in very dry conditions, beneath a control line by burning stumps and roots
- the risk from any soils with enough peat or organic matter to nourish fire underground
- the need for support systems such as helicopter water-bombing using small portable reservoirs (buoy walls) set up on ridgetops nearby
- the use of very small portable pumps (in backpacks) for use along creeks and river systems to provide back-up when backburning, or when putting out fire in the numerous log-jams or tangles of vegetation along creek beds
- the use of satellite-based navigation (Global Positioning System (GPS)) to map and place control lines precisely
- the use of aerial thermal-imaging to help locate hotspots when thick smoke is obscuring the fire
- the use of laptop computers to carry data-rich maps of the area into the field and help decision-making, especially if infra-red images and other field-data can be promptly downloaded

- the use of automatic weather stations (temporarily installed) to give precise data on local weather and weather-changes, to be linked to a fire's behaviour
- the use of portable phone systems that can be set up at remote fire-control centres to augment satellite technology
- the use of portable two-way radio transmitter/receiver stations that can keep all teams reliably in touch during the operation.

Essential calculations. Fire incident controllers need the best fire intelligence information possible. Their planning team needs to be constantly providing critical information to service the incident strategic decision-making process. Incident control planners

Fire trail used as a fire control line (Graeme Worboys)

need a range of competencies to achieve this. Skills such as local terrain and heritage knowledge; air photo interpretation; meteorological interpretation; McArthur forest and grassland meter (or equivalent) calculation expertise; and fire behaviour expertise (for variable terrain and variable forest/grassland fuel types) are essential. More advanced skills such as knowledge of computer-based GIS systems, infrared mapping of active fire fronts, use of satellite technology, use of computer fire perimeter forecasting programs, and ability to predict fire behaviour for backburns and optimal burn times using computers are major assets. Fire planners may anticipate a range of questions from incident controllers.

- What is the fire rate of spread for the next 12-hour and 24-hour period?
- Where will the fire perimeter be in 12 hours and 24 hours?
- What is the three-day prognosis for rates of spread of the fire?
- Where is the estimated fire perimeter located in three days?
- For the fire terrain type, how long will it take to construct fire control lines using rakehoes and/or bulldozers?
- Where are the optimum locations for those control lines (for time, heritage protection considerations, and safety considerations)?
- Is there sufficient time to construct the control lines ahead of the approaching fire front?
- What are the optimum weather conditions (for safety and controllability) for a backburn from the control line?
- When are those optimum conditions going to occur during the window of fire weather available?
- How much time will there be to put in place the backburn before the next major frontal system?
- Is it possible to put in place the firefighter resources for the backburn in time?
- Where are the safety routes for firefighters for a remote location?
- How long will it take to evacuate firefighters from a remote location by helicopters?

There will be many more questions. Too often, however, fire-control strategies are put in place with inadequate homework. Repeatedly, the wildfire overruns proposed control lines through poor analysis and planning. Too often damaging bulldozer lines are left half finished. As a consequence, the fire expands to a much larger set of control perimeters.

The role of the fire planner is critical in incident control decision-making process. They need essential competencies to be able to complete this task adequately under extreme time and event deadlines.

15.4 Managing wildlife incidents

There is a range of wildlife incidents that must be dealt with, many of them distressing both to animals and to the people involved. Incidents range from the illegal hunting of native animals, acts of cruelty, to dealing with animals that are injured in road or marine accidents. Rangers may be required to disentangle fishing lines from seals and other marine fauna (the very process being a mini-incident), or relocate kangaroos to more natural environments. The most common marine wildlife incident is the stranding of marine mammals, and in particular, whales. Former ranger Rosemary Black's account of a whale stranding gives an insight into the personal experience and the practical details of managing such an incident (Snapshot 15.4).

Marine pollution events

Marine pollution events are all too common along Australia's coastlines. Typically, protected area managers will deploy crews to retrieve affected wildlife, establish treatment facilities, and contact vets and wildlife volunteer groups to assist. Where large numbers of animals are affected, the zoo and the RSPCA are usually contacted. Sometimes a bird-bander is deployed to record bird statistics and band birds to be released. The *Iron Baron* oil spill of 1995 was particularly devastating. Irynej Skira's account, edited from the 1995 Tasmanian Parks and Wildlife Service annual report, is provided in Snapshot 15.5.

Volunteers often play an important role in marine mammal incidents (courtesy of ORRCA)

Snapshot 15.4

A whale rescue—50 False Killer Whales *(Pseudorca crassidens)* stranded at Seal Rocks, July 1992
Rosemary Black, former ranger with NSW NPWS (now P&W)

In NSW the NPWS coordinates and manages all marine mammal incidents such as the one at Seal Rocks. A marine mammal stranding of any scale calls for excellent organisational and planning skills. It also demands good cooperation between all the individuals and organisations involved. At Seal Rocks it was estimated that 30 different organisations were involved. The district office at Raymond Terrace served as the headquarters for the incident team. The ICS team coordinated and procured the required equipment and answered all the media and public enquiries.

The incident attracted national and international media and public interest. On day one, telephone enquiries jammed the four lines of the office, with an estimated one call every 30 seconds. NPWS regional and head office media officers coordinated the massive interest and produced regular media releases. Such incidents have a huge potential to promote a positive image for conservation agencies, and raise public awareness of marine mammals.

At the scene of the whale stranding there were two sector leaders located at each of the beaches. They were responsible for coordinating the personnel and equipment, as well as implementing the incident controller's decisions for managing the whales. This type of operation requires coordinating large numbers of people and ensuring that their needs are met. In addition to over 100 NPWS staff, there were about 800 volunteers and over 50 army personnel who required portable toilets, food, and equipment such as wet suits. The St John Ambulance was on-site to assist with medical problems that included hypothermia and dehydration, and the Salvation Army supplied meals and hot drinks. Marine mammal experts were brought in from Taronga Zoo, Seaworld, and ORRCA to assist with expert advice to the NPWS staff. Any incident comes at a financial cost. With the Seal Rocks rescue it was estimated to be $85,000.

There have been few whale strandings in Australia of this scale, in which whales were successfully returned to the ocean. Every time this happens we learn a little bit more about coordinating these types of incidents and marine mammal management. To be ready for such incidents, all coastal areas should develop a marine mammal rescue action plan.

15.5 Managing incidents arising from natural phenomena

Extreme weather

Types of violent weather that affect parts of Australia include cyclones, extreme rainfall, freak storms with strong winds and hail, blizzards, prolonged cloud and fog, and dust storms. These become 'incidents' mainly when they affect people. Otherwise it is usually a matter, after the incident, of repairing damage to installations and facilities in the protected area or giving assistance to neighbours in need. Managers may also be responsible for part of a community response, for instance in clearing or rebuilding roads outside their management areas.

Extreme winds. Storms and cyclones do not respect the boundaries of protected areas. Since the main issue is the risk to residents and visitors in the region, managers will usually work closely with community emergency services. A prime concern is the safety of staff members, and especially of those dealing with the incident.

Extreme rain. Extreme rainfall can occur anywhere in Australia, especially in high-rainfall or cyclone-prone zones. Flooding may be confined to low-lying areas, but dangerous landslips and slower mass movement of waterlogged soils can occur anywhere. Flooded rivers may trap bushwalkers in remote areas, or else tempt them into dangerous crossings that can lead to injury or death. Canoeists and white-water rafters may also be trapped. Helicopters provide the ideal means of rescue in such cases, but they have limitations. Flood weather often means driving rain with fog and mist. Clouds can descend to ground level. Typically in such situations, helicopters cannot be used when they are most needed, or else must be used with great caution. Cold, wet weather brings the threat of hypothermia, especially to injured people.

When soils are saturated, four-wheel-drive vehicles can do great damage to tracks in sensitive areas, yet numerous official vehicles may need to

Snapshot 15.5

Wildlife response to the *Iron Baron* oil spill in July 1995

Irynej Skira, Parks and Wildlife Service, Tasmania

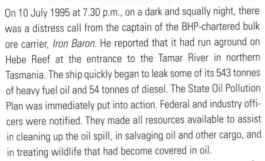

On 10 July 1995 at 7.30 p.m., on a dark and squally night, there was a distress call from the captain of the BHP-chartered bulk ore carrier, *Iron Baron*. He reported that it had run aground on Hebe Reef at the entrance to the Tamar River in northern Tasmania. The ship quickly began to leak some of its 543 tonnes of heavy fuel oil and 54 tonnes of diesel. The State Oil Pollution Plan was immediately put into action. Federal and industry officers were notified. They made all resources available to assist in cleaning up the oil spill, in salvaging oil and other cargo, and in treating wildlife that had become covered in oil.

From a wildlife perspective, the greatest concern was for nearby colonies of Little Penguin. The *Iron Baron* incident was the first oil spill in Australian waters where large numbers of birds were covered in oil. Despite the relatively small amount of oil, the impact on wildlife was extensive. Prevailing winds and currents in the region are predominantly westerly. Over the following month oiled penguins were recovered up to 20 kilometres west and 120 kilometres east of Hebe Reef. Little Penguin made up 98% of the six species of birds and mammals that were taken for treatment or discovered dead due to oiling. Other birds affected included 34 Black-faced Shags (*Leucocarbo fuscescens*), six Australian Pelicans (*Pelecanus conspicillatus*), two Black Swans, one Little Pied Cormorant (*Phalacrocorax melanoleucos*), and two Water Rats (both dead on arrival). Seals and albatrosses were visibly affected by oil but could not be captured for treatment. The rehabilitation program commenced on 11 July and was completed on 29 August, some 50 days later.

The Low Head colony was home to about 1500 Little Penguins. As the oil slick spread to the Furneaux Group, larger penguin colonies were affected. Of the penguins brought in for treatment, 682 were from Low Head, 1119 from Ninth Island, and the remainder from 13 other localities (Holdsworth & Bryant 1995). In total, 2063 penguins were treated of which 1980 were released and 104 died. Only 20 penguins died at Low Head; the remainder were either dead on arrival or euthanased by veterinarians.

A very efficient capture and rehabilitation program was organised and became the most prominent public feature of the oil spill response. The initial assistance of Sydney Taronga Zoo was invaluable. Zoo staff set up the Parks and Wildlife Service oiled sea-bird response, using the oiled wildlife treatment equipment provided by the oil industry. Approximately 50 specialists from around Australia became involved. Some 200 volunteers assisted in night searches for oiled penguins, as well as washing and rehabilitation. An estimated 90% of staff from the nature conservation branch and 70% of Parks and Wildlife Service rangers and land management staff were directly involved in animal care or incident control at some stage.

Rehabilitation of oiled wildlife (Irynej Skira)

Clearing up storm damage (Graeme Worboys)

Snapshot 15.6

Mystery Creek Cave flooding leads to drownings

In 1990, a school excursion to Mystery Creek Cave in south-west Tasmania was underground when a storm in the catchment caused the cave's stream to rise rapidly and turn into a torrent. To escape the fast-rising waters, the school party tried to get back to the entrance; but the extreme conditions brought tragedy. Two students and a teacher were swept to their deaths (Spate 1998).

use them. As well, the media, volunteers, politicians, and other observers often demand access to the incident. Managers may need to plan adroitly to fill empty places in the vehicles and thus limit the number of vehicles used and the damage done. Media helicopters in misty foggy conditions can add to confusion and be hazardous to other aircraft. Afterwards, damage done to tracks needs to be fixed, equipment repaired, and the site cleared of any discarded equipment.

Floods inside cave systems can rise very quickly and be life-threatening for cavers (Snapshot 15.6). Many of Australia's limestone cave systems, especially in the east, have small catchments. Some of these support only temporary streams or seepages. Yet a storm in the catchment can flood underground streams. These can cut off passages on which cavers rely, or flood whole caves. Rescuing cavers is a job for specialists, and is often a race against time to avoid rising waters and the onset of hypothermia. Contingency plans should emphasise both quick response and the need to minimise damage to the caves during the rescue.

Snow. Snowstorms occur in protected areas within Tasmania, Victoria, New South Wales, and occasionally in Queensland and Western Australia. Severe snowstorms or blizzards are more typically confined to the alpine and subalpine areas of Tasmania, Victoria, and NSW. Here they can involve managers in several kinds of incidents. Hikers and skiers often misjudge the severity of an oncoming blizzard or become disorientated in 'whiteout' conditions. Managers need to ensure that snow safety information is provided to visitors. Search and rescues are regular winter events in the Australian high country, and rangers are usually involved in at least a supporting role. Blizzard weather is unforgiving, so rangers who join in rescues in these areas need to be fit and have specialist skills.

Avalanches are less common in Australia than in the steeper and higher snowfields of other continents, but they do occur. Heavy snow falls in some years give rise to overhanging cornices. In a whiteout, skiers may ski over them unaware. Collapsing cornices may also lead to avalanches. Though cornice-avalanches are rare in Australia, they are a real safety issue for managers. Speed is essential in digging out those covered by an avalanche, before they are choked or chilled to death. A major problem is to find out quickly where to dig. For cross-country skiers, prompt rescue is unlikely unless by members of their party, though help may sometimes be summoned in time with the use of mobile phones. Managers should have plans for any populated areas that could be at risk of avalanche after exceptional snowfalls. They should also be aware of overseas developments in avalanche rescue.

Heavy snow, Kosciuszko National Park, NSW (Graeme Worboys)

The alpine parks and resorts are vulnerable in other ways. Deep cold and heavy snow can close roads and bring down powerlines. Cold can also disrupt services such as water supplies and sewerage systems. Service-pumps can break down, and water pipes can freeze or burst. Cold can also play havoc with heavy equipment. For instance, the waxy elements of the crude-oil-based diesel fuel that has been used in Australia may fractionate. Heavy snow-loads can severely damage trees and buildings. Branches that break and fall increase the fuel load for summer fires, and may require an upgraded fire plan. To cope with icy roads and walkways, road authorities and resort managers may apply salt. This affects the environment in ways that must be understood and managed.

Geological incidents

Other than Heard Island, Australia and its administered territories are currently free from volcanic activity. However, geological incidents such as earthquakes, mass movement, and cliff collapse do occur. Some of these are brought on by human activities, while others are natural events.

Earthquakes. Earthquakes of a moderate to severe scale occur irregularly throughout Australia. They are commonly associated with movements along faults and lineaments. On average, Australia has only one earthquake every five years of a magnitude of 6 or more on the Richter scale. This compares with a world average of 140 such events per year (Clarke & Cook 1983). Earthquakes of a magnitude of 6 or more can cause damage if they are shallow and close to populated areas. To date, Australia has had few damaging earthquakes.

Geologists have mapped three broad regions where there is a higher risk of earthquakes: the eastern highlands and coastal regions, the central region, and the western region. Many protected areas are found within these regions. The most active area in the eastern region is the Dalton–Gunning zone, about 50 kilometres north of Canberra. The central seismic zone extends from near Adelaide to the Simpson Desert, while the western seismic region has several distinct zones.

Mass movement. The influence of gravity causes loose material to move down-slope. The movement can seem slow and subtle from day to day, but be large over months or years. It can also be sudden, swift, and devastating, as in a landslide or cliff collapse (Montgomery 1997). The steeper the slope, the more likely it is to slip. As many protected areas have very steep and broken terrain, the potential for such geological incidents is high. Loose substrates, such as unconsolidated rocks and gravel, clay-rich soils, and soil-rock mixes, also invite mass movement. So too do unstable strata structures such as fault planes or joint planes, and steeply dipping structures parallel to the slope. Heavy rain, flooding, earthquakes, or human interference can all trigger mass movement. Most of these natural events occur without warning, such as the collapse of one arch of the natural arch in Port Campbell National Park in 1990 or the Thredbo, Kosciuszko National Park landslide disaster of 1997.

Dangerous terrain and sensitive areas. Steep terrain is also a hazard when dealing with incidents such as bushfires. Managers need to be alert to this. Bulldozers are often used to prepare control lines for fires. In doing so they can encounter small cliffs, unstable blockfields of boulders, scree slopes, steep inclines, and patches of limestone that may be hollow. Bulldozers are best deployed in protected areas with an interpreter/navigator who uses a set of aerial photos and or maps to aid navigation. This officer navigates around potentially dangerous or unstable terrain as well as any sensitive natural or cultural sites. It is their responsibility to know the location of important heritage sites and to protect them from impact. Unstable scree slopes can also be

Snapshot 15.7

Closure of visitor access to Minnamurra Falls

Minnamurra Falls near Kiama in southern NSW was reserved in 1896 when its beauty began to attract visitors. In 1989, cliffs in the narrow canyon of the falls-proper collapsed. Managers commissioned a geotechnical report that found that the canyon was unstable and a potential danger to the public. Managers immediately closed visitor access to the waterfall. After nearly 100 years of continuous visitor use, there were (understandably) many complaints. By 1994, an alternative route was completed for visitors to view the falls from above the canyon. The managers' caution was vindicated when further minor cliff-collapse occurred at the canyon site.

very dangerous when a number of staff are deployed in the same place.

Where terrain may be unstable, managers must plan cautiously for the safety of visitors. A geotechnical report may be required before any proposed structures (including buildings, lookouts, walking tracks, and bridges) are built. Once they are built, a qualified engineer must certify them as safe to use. Managers should check often, and not hesitate to close sites or tracks that may become unstable (Snapshot 15.7).

15.6 Management lessons and principles

Lessons

1 Mutual trust and respect is critical when working with other organisations if effective incident management is to be achieved. Such trust is usually achieved through constant professional and respectful interaction over a long period.
2 Working extensively and professionally with the community is a critical investment for cooperative management. Direct contact with senior local officials and local members of parliament, and constant briefing information about the progress of events is critical.
3 Exchange of information and comprehensive understanding (by all authorities) of the values of

protected areas, and the nature of conservation management is a priceless investment for the future.
4 The constant need for informing new players in incident management requires (at least) annual briefing days at the level of decision-makers as well as field operatives. Out-of-area incident response crews means that more generic state- or territory-wide and national education programs are also required.

Principles

The guiding principles of management for incidents within protected areas are as follows.

1 Sooner or later, incidents will occur.

2 Preparedness is essential. Managers can and should plan in advance, using the 'principles of precaution'.

3 Human life and property take precedence over the environment during an incident.

4 The safety of staff is paramount. Undisciplined or risky behaviour is not acceptable.

5 Staff need extra job-related support during and after an incident.

6 Proven and widely recognised management systems are the best way to protect an area during an incident. It helps if both staff and stakeholders are skilled in the use of Australian standard incident management systems, such as the AIIMS Incident Control System.

7 Having efficient plans, appropriately skilled staff and suitable plant, equipment, and stores accessible, allows for prompt action that can minimise the impacts of incidents.

8 Full and timely data on heritage assets, and on other matters, is crucial. A calm analysis of data, even in the heat of an incident, allows decisions to be made that preserve life and property and limit the impact on the environment.

9 Anticipate official post-incident inquiries, and collect appropriate information during the incident.

10 Staff will respond, and perform better, if given real roles and scope and recognition for their skills.

11 During incidents, expect the unexpected; anticipate the inevitable.

Crocodile warning signs, Daintree River, Qld (Graeme Worboys)

16

Tourism and Recreation

Much of Australia's $70 billion tourism industry is based on our superb and unique natural features and biodiversity and rich cultural landscape. National parks are a key component and (are) popular with domestic and international visitors...

Department of Industry, Tourism and Resources (2003)

16.1 Global tourism and
      environmental performance   427
16.2 Tourism and recreation in Australia's
      protected areas   429

16.3 Managing tourism and recreation   442
16.4 Managing for quality   456
16.5 Management lessons and principles 462

Australia is the type of country that people simply want to see. It is large. It is extraordinarily variable, from tropical landscapes to deserts, snow-capped mountains, and windswept southern ocean islands. Its wildlife is world-famous and mostly unique, and many of its best landscape and heritage areas are inscribed as world heritage areas. It has a living history and cultural heritage dating back to the dreamtime. It is a safe, organised, and reliable place to visit. People simply want to experience and learn more about this country.

We often hear the terms: tourism, nature tourism, ecotourism, visitor use, recreation, and ecologically sustainable use. These terms are defined, for the purposes of this text, in Backgrounder 16.1. The origins of ecotourism are discussed in Backgrounder 16.2.

Australian protected areas receive more visitors per annum than any other outdoor destination type. In 2002, an estimated 84 million visitors were hosted by the combined protected area organisations (Commonwealth of Australia 2003a). Visitors include hikers in the most remote wilderness areas, groups of older Australians arriving by bus at a historic homestead, snorkellers at coral reefs, and volunteers removing weeds from bushland. They all expect the protected area destinations they visit to be well managed. They all expect a quality experience.

However, our most-loved visitor destinations, if left unmanaged, will become degraded. This chapter discusses such management issues. We provide a global context for tourism and protected areas, and discuss Australian tourism and the protected areas values on which it depends. Tourism as a threat to protected areas is discussed, as well as managing for quality tourism, recreation, and visitor experiences. Since visitor management is vitally linked to the sustainability of protected areas, we also present a framework for ecologically sustainable visitor use. For further discussion of these topics we recommend Australian Heritage Commission (2001), Environment Protection and Heritage Council (2003a, 2003b), Eagles & McCool (2002), Eagles et al. (2002), and Newsome et al. (2002).

Backgrounder 16.1 Definitions of tourism and visitor use

Tourism is travel away from home for recreation or pleasure (including the pleasure of taking a look around a new region), and the activities that go with this. It can include visits to friends and spin-offs from business conferences. The term also covers industries and services that aim to satisfy the needs of tourists (GBRMPA 1994, Gee et al. 1997). The tourism system is important, and includes businesses linked to the point of origin, travel, and the destination. Tourism is defined somewhat differently by the World Tourism Organisation and the Australian Bureau of Statistics, as travel for more than 40 kilometres and involving at least one stay overnight (but for less than 12 months). Under this definition, day trippers are not tourists (but can be recreationists).

Visitor use is defined as any use of protected areas by visitors. These include official visitors, volunteers, contractors, park workers, and educational groups, as well as tourists.

Ecologically sustainable use is the use of living things or areas within their capacity to sustain natural processes while maintaining the life-support systems of nature. It is an application of the principle of 'inter-generational equity', since it ensures the benefits of use to present generations do not diminish those available to future generations (GBRMPA 1994).

Recreation is activity voluntarily undertaken, primarily for pleasure and satisfaction, during leisure time (Pigram & Jenkins 1999).

Recreation settings are areas that allow a given activity such as sightseeing, picnicking, camping, rock climbing, or canoeing. They are also sometimes referred to as destinations. The term is also used in relation to the Recreation Opportunity Spectrum (ROS) (Backgrounder 16.3), to indicate the different settings (ranging from wilderness to highly developed) in which recreation takes place.

Backgrounder 16.2   The origins of the concept of ecotourism

Just 21 years ago in 1983, Mexican architect and environmentalist Hector Ceballos-Lascurain coined the word 'ecotourism'. Today ecotourism is a major segment of the tourism industry and a major growth area. As often happens with an emerging phenomenon, it has several similar names: nature tourism, green tourism, adventure tourism, sustainable tourism, appropriate tourism. In describing its evolution, Honey (1999, p. 11) noted that:

> … broadly stated, the concept of ecotourism can be traced to four sources: (1) scientific, conservation, and non-governmental organisation circles; (2) multi-lateral aid institutions; (3) developing countries; and (4) the travel industry and travelling public. Almost simultaneously but for different reasons the principles and practices of ecotourism began taking shape within these four areas and by the early 1990s, the concept had coalesced into the hottest new genre of environmentally and socially responsible travel.

The term *ecotourism* implied a genuine attempt to respect nature and to manage for the future. It linked the tourism industry with the community's concern for the environment, and so was popular both with environmentalists and managers. Common sense dictated that it was simply not sustainable for the tourism industry to degrade its own destinations, soiling its own nest. The industry also realised that a clean, green image was good for business.

Environmentalists helped develop the concept of ecotourism as an alternative to more destructive forms of development like mining, forestry, and agriculture. Local communities and governments often favoured ecotourism as an alternative use of lands. In developing countries, the traditional USA model of national parks (which excluded local peoples from protected areas) was generally not acceptable; but ecotourism offered an alternative economic opportunity without clearing or modifying natural lands or destroying wildlife. It also sat well with the concept of privately owned conservation reserves.

Ecotourism has steadily increased in popularity. Regular newspaper articles, books, societies and conferences on the subject in the late 1990s testify to this. A 1999 survey conducted by Tourism Queensland found that ecotourists represent 30% of the travelling public in Australia (Tourism Queensland 1999).

Swanson (cited in Ceballos-Lascurain 1996) suggested that this is part of a historic change in our culture. In the postwar era, progress and prosperity were considered more important than nature. Ecotourism reflects the emergence of a new breed of traveller who is well educated, values nature for its own sake, plans and acts to control risks of damage to nature, and recognises real limits to growth. Such people seek to foster a new society; one that encourages the participation in tourism of local communities. Travellers today are certainly more discriminating and sensitive to such issues.

The concept of ecologically sustainable development (ESD) in tourism is likely to be an important force over the next few decades. It is a timely emergence, since heritage destinations are receiving more visitors than ever before. Yet Honey (1999, p. 4) warns against over-optimism.

> Around the world, ecotourism has been hailed as a panacea: a way to fund conservation and scientific research, protect fragile and pristine ecosystems, benefit rural communities, promote development in poor countries, enhance ecological and cultural sensitivity, instil environmental awareness and a social conscience in the travel industry, satisfy and educate the discriminating tourist, and, some claim, build world peace. Although 'green' travel is being aggressively marketed as a 'win-win' solution for the Third World, the environment, the tourist, and the travel industry, close examination shows a much more complex reality.

Nature tourism involves travel to unspoiled locations to experience and enjoy nature. It usually involves moderate and safe forms of exercise such as hiking, biking, sailing, and camping. *Wildlife tourism* involves travel to observe animals, birds, and fish in their native habitats. *Adventure tourism* is nature tourism with a kick: it requires physical skill and endurance (rope climbing, deep-sea diving, bicycling, or kayaking) and involves a degree of risk-taking, often in little-charted terrain. Whereas nature, wildlife, and adventure tourism are defined solely by the recreational activities of the tourist, ecotourism is defined by its benefits both to conservation and to people in the host country.

Among definitions, perhaps the best is that of the Australian National Ecotourism Strategy (Commonwealth Department of Tourism 1994): 'Ecotourism is nature-based tourism that involves education and interpretation of the natural environment and is managed to be ecologically sustainable.' This definition should recognise that 'natural environment' includes cultural components and that ecological sustainability involves making an appropriate return to the local community as well as conserving the resource for the future.

Butler (cited in Ceballos-Lascurain 1996) offered a more detailed characterisation. For him, if an activity is to be considered as ecotourism, it must have at least the following nine attributes.

1   It promotes positive environmental ethics and fosters 'preferred' behaviour in its participants.
2   It does not degrade the resource, that is, the natural environment.
3   Facilities and services may support the tourist's encounter with the 'intrinsic resource', but never become attractions in their own right.
4   Ecotourists accept the environment as it is, not expecting it to change or be modified for their convenience.
5   It must benefit the wildlife and environment, contributing to their sustainability and ecological integrity. (This may be through the effects on the local community or economy.)
6   It provides a first-hand encounter with nature. Visitor centres and on-site interpretive slide-shows may be part of an ecotourism activity only if they direct people to a first-hand experience.
7   It actively involves and benefits local communities, thus encouraging them to value their natural resources.
8   It offers gratification through education and/or appreciation rather than through thrill-seeking or physical achievement.
9   It involves considerable preparation, and demands in-depth knowledge on the part of leaders and participants.

16.1 Global tourism and environmental performance

Growth in global tourism has been one of the great phenomena of the late twentieth and early twenty-first centuries. In 2002 there were 715 million international arrivals worldwide—22 million more than in 2001 and 690 million more than in 1950 (Commonwealth of Australia 2003a). The World Travel and Tourism Council (WTTC) has forecast that the number of international arrivals will increase to nearly 1.6 billion by 2020, despite a potential scarcity of petroleum by this time (Commonwealth of Australia 2003a, Mason 2003). Many of the worldwide tourist destinations are protected areas.

In an era of (relatively) cheap petroleum-based fuel, transport systems have delivered visitors quickly and efficiently to visitor destinations around the world. Such tourism is important to the economies of many nations, and brings many benefits to local communities. Managed responsibly, tourism can provide many sustainable benefits to protected areas, including opportunities for education and appreciation of nature and cultural heritage, as well as fostering a conservation constituency (Eagles & McCool 2002).

Tourism has also bought conflicts and caused environmental impacts. Without effective management and responsible action, tourism industry growth can lead to the destruction of environments and destinations and may provide few benefits to local communities (UNEP 2002, Haroon 2002).

The tourism industry, like many other industries, uses resources such as water and energy, contributes to greenhouse gas emissions, and produces solid wastes. Studies in New Zealand showed that:

international tourists travel an average of 23,000 kilometres (return) from their country of origin, and travel an average of 1950 kilometres within New Zealand. Domestic tourists travel an average of 640 kilometres. Transport makes up 69% of an international tourist's energy use, with 85% for domestic tourists. In 2000, New Zealand emitted 30,850 Kilotonne of carbon dioxide, with tourism contributing 6% of these emissions (Becken et al. 2003).

Consumption of resources at a global scale is considerable:

International and national tourists use 80 per-cent of Japan's yearly primary energy supply (5000 million kWh/year), produce the same amount of solid waste as France (35 million tons per year), and consume three times the amount of freshwater contained in Lake Superior, between Canada and the United States, in a year (10 million cubic meters) (Christ et al. 2003).

In Australia, it has been estimated that 40,000 tourism and travel companies will produce 40 mega-tonnes of carbon dioxide by 2008 (De Lacy et al. 2002). Despite the efforts of a few outstanding com-panies, the tourism industry has been slow to achieve substantive environmental performance improve-ments (Worboys & De Lacy 2003).

The tourism industry's peak bodies, the World Tourism Organisation (WTO) and WTTC, have responded to the substantial environmental problems and are aware that growth in tourism is dependent, among other considerations, on the sustainability of destinations. They would also be acutely aware of how global shortages of potable water, higher costs for energy, degraded destinations, and new emissions standards will impact the industry.

The WTO has contributed to international declarations on the environment, environmental codes of ethics, guidelines, and policies that pro-mote sustainable tourism. It sponsored, for example, the first international conference on cli-mate change and tourism and supported the subsequent *Djerba Declaration on Tourism and Climate Change* (WTO 2003) that encouraged the tourism industry to use more energy-efficient and cleaner technologies and urged:

all governments concerned with the contribution of tourism to sustainable development to subscribe to all relevant intergovernmental and multilateral agreements, especially the Kyoto Protocol.

The WTTC has also contributed to environ-mentally sustainable tourism. At the conclusion of the Third Global Travel and Tourism Summit held in May 2003, more than 500 of the world's most influential business and political leaders called on the WTTC to create a new vision and strategy for Travel and Tourism. The resulting strategic docu-ment, *Blueprint for New Tourism*, was launched by the

WTTC, on 7 October 2003 (WTTC 2003). The Blueprint sets as a key goal balancing economics with environment, people and cultures; and indi-cates that 'new tourism' looks beyond short-term considerations to focus:

… on benefits not only for people who travel, but also for people in the communities they visit, and for their respective natural, social and cultural environments.

Similarly, key messages from the non-government organisation Conservation International and its United Nations Environment Program partner rein-force the importance of global sustainable tourism:

Over the past 3 decades, major losses of virtually every kind of natural habitat have occurred, and the decline and extinction of species has emerged as a leading environmental issue. Many of the ecosystems in decline are the very basis for tourism development—coastal and marine areas, coral reefs and mountains, and rainforests—and support a wide range of tourism activities …

Tourism will require careful planning in the future to avoid having further negative impacts on biodiversity. Many of the factors associated with biodiversity loss—land conversion, climate change, pollution—are also linked to tourism development …

At the same time, an increasing number of examples have shown that tourism development guided by the principles associated with eco-tourism—environmental sustainability, protection of nature and supporting the well-being of local peoples—can have a positive impact on biodiver-sity conservation (Christ et al. 2003).

Asia Pacific Economic Co-operation and the Cooperative Research Centre (CRC) for Sustain-able Tourism launched a program for delivering sustainable tourism based on cooperation and quan-tified environmental performance targets (De Lacy et al. 2002). The CRC has been responsible for important research into sustainable tourism. Private hotel chains, ecotourism organisations, magazines devoted to green outcomes, and environmental certification schemes such as Green Globe 21 have all helped. The Rainforest Alliance, a New York-based non-government organisation, has provided leadership for the introduction of a Global Sustainable Tourism Stewardship Council. The Council aims to provide a minimum standard for tourism environmental certification schemes and an accreditation process (Rainforest Alliance 2003).

Despite these initiatives, the extent of on–ground improvement in environmental performance has been disappointing. More effort is needed to implement the principles enunciated in documents such as those cited above.

16.2 Tourism and recreation in Australia's protected areas

In 2003, Australia attracted 4.8 million foreign visitors. This is less than 1% of the world's international tourism market but is critical for Australia's earnings. This number of arrivals was about the same as the previous year. Domestic tourism is also very important for the economy and accounts for about 80% of visitor expenditure in Australia. The number of domestic visitor nights had grown slowly, by about 1% over the previous 3 years. In 2001–02, tourism directly contributed 4.5% to Australia's gross domestic product and was directly responsible for employing 550,000 people, and indirectly a further 397,000 people. Tourism is especially important in regional and rural areas and in 2001–02 employed about 185,000 people or 7% of the rural and regional workforce (Commonwealth of Australia 2003a).

There has been a general increase in domestic tourism, with visitation to some protected areas growing, others steadying, and some have declined. Overall, there has been significant growth in visitor use of protected areas in the last 10 years. In 2001–02, there were 84 million visits to Australia's protected areas (Commonwealth of Australia 2003a). Over 90% of visitors to protected areas are Australian residents, and most (70%) live in cities and access protected areas using private motor vehicles (TTF 2004). We provide a brief insight into visitor use in the desert parks of South Australia (Snapshot 16.1) and Seal Bay on Kangaroo Island (Snapshot 16.2).

Protected areas are a significant tourism attraction for rural and regional Australia and the high proportion (70%) of domestic visitor nights and international visitor nights (23%) spent in these areas reflects, in part, their distribution. World heritage areas such as Uluṟu–Kata Tjuṯa National Park, Kakadu National Park, and the Great Barrier Reef Marine Park are major drawcards for international visitors. In 1994, approximately 3000 licences were

Snapshot 16.1

Four-wheel-drives in the desert
Pearce Dougherty, National Parks and Wildlife Service, South Australia

From the early 1970s the spread of improved and affordable four-wheel-drive vehicles (even among city dwellers) brought an upsurge in recreational travel to remote parts of Australia. This coincided with an opening up of previously inaccessible areas through petroleum and mining exploration. For example, in the early 1960s a French seismic exploration company, searching for hydrocarbons, pushed a seismic line into the Simpson Desert. Known as the French Line, this was the first of an extensive network of access tracks that opened this desert to four-wheel-drive touring. Fuelled by a growing national economy, touring to remote areas has grown to unprecedented proportions. As visitors increased, many unique locations in remote areas began to be degraded. In the late 1980s the South Australian Government attempted to protect remote desert areas by dedicating (and staffing) parks across large areas of the South Australian deserts.

Desert garbage—trailer abandoned on the French Line Track, SA (Pearce Dougherty)

Snapshot 16.2

Balance on the beach: Sustainable tourism at Seal Bay Conservation Park, Kangaroo Island

Keith Twyford and Fraser Vickery, National Parks and Wildlife Service, South Australia; and Graeme Moss, Department for Environment and Heritage, South Australia.

In 1998, Keith Twyford and Fraser Vickery wrote for the first edition of this book:

> Visitors to Seal Bay have increased significantly over the past 10 years ... but may have peaked in 1996 because of fewer day ferries between Adelaide and Kingscote. We still expect around 100,000 visitors a year, up to the year 2000. Over 50% of visitors come on commercial tours, which highlights the park's strong contribution to the region's tourism economy.

Their visitor-use information (Table 16.1) was updated by Dr Graeme Moss in 2003, demonstrating the accuracy of their visitor-use forecasts.

Visitors, Seal Bay Conservation Park, SA (Fraser Vickery)

Table 16.1	Visitation statistics, Seal Bay Conservation Park, 1988–2002				
Year	Total Visitors	% Change	Bus Visitors	% Change	% on Bus Tours
1988	39,623	na	14,505	na	37%
1989	47,227	16%	19,082	24%	40%
1990	63,135	25%	26,950	29%	43%
1991	72,162	13%	30,823	13%	43%
1992	75,827	5%	36,510	16%	48%
1993	82,197	8%	43,349	16%	53%
1994	92,610	11%	52,591	18%	57%
1995	103,334	10%	63,195	17%	61%
1996	112,797	8%	67,520	6%	60%
1997	97,661	−15%	55,822	−21%	57%
1998	96,392	−1%	52,969	−5%	55%
1999	98,693	2%	56,595	6%	57%
2000	102,290	4%	58,338	3%	57%
2001	109,439	7%	61,068	4%	56%
2002	109,389	0%	57,770	−6%	53%

issued to commercial tourism operators for protected areas (TTF 2004).

The tourism industry

The tourism industry and protected areas can potentially enjoy a mutually beneficial relationship. Tourism can provide an economic justification for the estab-lishment of protected areas, as well as opportunities for local people to reduce their dependence on resource extraction. It can build a supportive constituency that promotes biodiversity conservation, and it can provide an impetus for private biodiversity conservation efforts (Christ et al. 2003). However, the health of this relationship depends on the compatibility of their respective needs, and recognition by the industry of all the values afforded by protected areas.

Three broad paradigms for tourism operations conducting their business in protected areas can be identified.

An *eco-tourism operation*:

- has a current licence (or lease) to operate and meets protected area licence/lease requirements
- may be run on a commercial basis by a protected area organisation, with profits being returned to the protected area, or may be run by a private operator, with some profits returned to the protected area
- has a publicly available corporate policy that affirms a commitment to conservation of protected areas
- is certified with an independent environmental certification qualification and usually has full-time environmentally qualified employees
- is professional, with knowledge of environmental management needs, and self-motivated to achieve improved environmental performance within and beyond licence requirements
- provides high quality environmental education for visitors
- provides investments of time and resources to help look after the protected area and works closely with protected area managers.

Eco-tourism operations are becoming more common as protected area agencies seek higher standards of operation in protected areas and some concerned and committed companies provide leadership in environmental management (Case study 16.1). This is the optimum future tourism licence/lease partnership model for protected areas. However, only 0.01% of the Australian tourism industry possess environmental certification qualifications (Worboys & De Lacy 2003) and few of these have working employees with graduate level environmental management qualifications.

A *routine tourism operation* typically:

- has a current licence or lease to operate within the protected area and meets the conditions of this licence or lease
- provides basic services to assist visitors such as access, transport, food, and sometimes accommodation

- is professional in dealings with protected area managers
- runs a profitable and responsible operation
- provides some basic support information for visitors
- may make occasional positive contributions to protected area management
- has no permanent staff with expertise in environmental management.

The routine tourism operator is the dominant type for protected areas and provides many services in supporting visitors. Often routine tourism operators believe that they have little or no environmental impact and do not require regulation (Byrnes & Warnken 2003).

A *development oriented operation* typically:

- has a current licence or lease to operate within the protected area
- has a pro-commercial and profit-centred approach
- provides required licence payments and undertakes mandatory works
- utilises legal and political support to facilitate commercial decisions
- views the protected area as property for the purposes of private commercial gain
- provides tourism-based employment and an efficient service for visitors, but with few or no environmental education services
- has no permanent employee expertise in environmental management.

Concerns about development oriented tourism operations in protected areas are illustrated by the following darkly humorous 'Buy Your Eden' scenario concerning one possible (undesirable) future trajectory for the protected area/tourism relationship.

Economics is the dominant theme in the Buy Your Eden, and the gap between the rich and the poor has widened in 2023. Many protected areas have been privatised and new ecotourism multinationals are running the worldwide system of 'the world's greatest nature' appealing to the prosperous international tourism market. Those fortunate few outstanding protected areas (which were called World Heritage Sites until they were purchased by the consortium of private tourism multinationals) are very well managed for tourism objectives, which often include maintaining biodiversity, especially the charismatic type. But the numerous other protected areas that are not deemed to be of sufficient profit potential are suffering from inadequate investment and many fall prey to the growing number of desperate rural poor (McNeely & Schutyser 2003).

Working relationships with licensed tourism operators can change over time. Changes of attitudes, poor communication, transfer of park managers, and transfer of licences to new operator owners can see operations that commenced as an eco-tourism operation change to a routine operation and even a developer operation. It is critical for protected area managers to be aware of this possibility, especially in the design of initial licence or lease arrangements.

In 2003, the Commonwealth Government's Tourism White Paper (Commonwealth of Australia 2003a) advocated private sector partnerships that will grow tourism in protected areas. Park managers regularly deal with potential tourism operators who 'see the obvious potential' for development of natural lands within protected areas. Such proposals do not, in many cases, meet the long-term interests of the protected area, and can usually be better located outside the park. The tourism industry might self-regulate and become increasingly responsible for protecting the environment. However, in practice it relies on the leadership of conservation agencies to show how to introduce sustainable tourism. It is critical that any partnerships developed are eco-operations. They need to demonstrate they will not compromise conservation outcomes and that they enjoy widespread

community support. In addition, as part of the partnership, tourism operations need to be environmentally competent (qualified) and capable (sufficient resources) to provide the necessary quality of environmental stewardship for protected areas.

Recreation

Recreation, an aspect of park tourism, is an important part of the human experience of protected areas (Pigram & Jenkins 1999). Protected areas have important tourism and recreation values (Chapter 3). Visitors undertake an extraordinary diversity of recreation activities within protected areas. Most activities have a constituency that lobbies in support of its continuation or expansion within the protected area estate. Staff are often required to be involved with facilities supporting bushwalking, skiing, boating, canoeing, caving, four-wheel driving, and a range of other activities. Special arrangements may need to be in place to manage for wilderness areas, where a management presence can detract from the recreational experience of visitors. Adventure recreation activities such as canyoning, white-water rafting, cross-country skiing, abseiling, ice-climbing, and rock climbing may need management attention for safety reasons (response to emergencies in bad weather) and for potential environmental impacts. Dealing with recreational users can be both time-consuming and rewarding for park staff.

The tourism and recreation-use values of protected areas are influenced by a number of geographical, social, managerial, and biophysical factors, including geographical proximity and

Case Study 16.1

Binna Burra Mountain Lodge, Lamington National Park, Queensland

Linus Bagley, Manager, Binna Burra Mountain Lodge

Binna Burra Mountain Lodge, located on Mt Roberts in the McPherson Range, and just inside World Heritage listed Lamington National Park (south-east Queensland) was founded in late 1933. Much of the property today displays natural rainforest regrowth and revegetated areas, while developed sections support the lodge, teahouse café, and a campsite. The property is now listed on the Queensland register of heritage sites.

The original prospectus of the company, dated 10 March 1933, clearly expressed the company's early intentions in respect to environmental protection. Romeo Watkins Lahey, one of the Company's founders, was a great influence in establishing sound environmental principles at the company's inception:

The company is being formed with the objects set out in the Memorandum of Association and in particular to provide tourist facilities and accommodation in beauty spots throughout the State of Queensland and as far as possible to assist in preserving such in their natural state for future generations in accordance with the ideals of the National Parks Association of Queensland.

In recent years Binna Burra Mountain Lodge has adopted a more formalised approach to all aspects of its environmental management. An environmental management plan and a land management plan have been developed to underpin all aspects of the company's operation. Key environmental indicators have been identified and benchmarked through the Green Globe 21 processes. In 1996, the company received the Advanced Accreditation Certification through the Ecotourism Association of Australia's National Ecotourism Accreditation Program.

Binna Burra was one of the initial applicants to the Queensland Government's cleaner production partnership program. This program, delivered through the EPA's Sustainable Industries Division, assists industry to identify areas where environmental performance can be improved and operating costs reduced. The company invested over $40,000 to implement the recommendations of an eco-efficiency assessment. As a result, annual savings of almost $17,000 were achieved. This included savings in diesel fuel ($6500 per annum); water heating ($2700 per annum); ultraviolet water treatment chemical savings ($800 per annum); and sewage-treatment plant chemical savings ($950 per annum). The initiatives achieved electricity savings of 234 MWh and greenhouse gas reductions of 189 tonnes carbon dioxide equivalent per year.

Through the Green Globe 21 program the company's environmental performance is monitored and benchmarked. The major areas of attention include:

- landscaping and land management
- water and wastewater
- solid waste
- cleaning materials
- energy efficiency
- air and noise pollution
- contribution to the local community
- interaction with wildlife
- biodiversity conservation
- safety and emergency procedures
- staff environmental education.

Environmental interpretation is fundamental to the operation. Binna Burra Mountain Lodge attracts visitors who wish to interact with the natural environment and in doing so develop their knowledge, awareness, appreciation, and enjoyment. Interpretation of the natural history and cultural heritage of Lamington National Park is provided by suitably qualified guides. Staff at Binna Burra are dedicated to following best practice methods in regard to environmental conservation, education, and quality of service.

The company's management recognise the importance of working closely with the protected area managers and local rangers. Regular meetings to discuss track maintenance, health and safety, fire control, monitoring outcomes, and visitor activity in the park facilitate a good cooperative working relationship that results in beneficial outcomes for the environmental sustainability of the park.

accessibility to markets, cultural links, availability of services, affordability, peace and stability, positive market image, pro-tourism policies, and availability of attractions (Weaver & Opperman 2000).

Visitor attractions in protected areas may be natural features or destinations with more developed facilities and services such as visitor centres, boardwalks and limestone 'show caves'. Artificial attractions or high impact, derived activities that may diminish the natural or cultural heritage values of protected areas are inconsistent with the concept and purpose of IUCN Category I–IV protected areas. Attributes that contribute to the tourism and recreation values of protected areas are:

- natural attractions, especially scenery, wildlife, and unspoiled nature
- cultural attractions
- access
- diversity of tourism and recreation
- services and facilities
- absence of impacts of use
- affordability
- education opportunities.

The tourism and recreation-use values of protected areas can be described in terms of 'opportunity settings' found within protected areas. Recreation opportunity settings can be defined as

the combination of physical (such as scenery), biological (such as native plants and animals), social (such as family, friends and/or other visitors), and managerial (such as the facilities and regulations imposed at a setting) conditions that give value to a place (Clarke & Stankey 1979). They are what we usually visualise when we think about a destination before a visit.

Tourism and recreation opportunities are means by which a visitor acquires experiences and fulfils aspirations; it has been argued that these experiences fulfil psychological needs and motivations. These needs and aspirations include escape motivation, relaxation and play, strengthening family bonds, prestige, social interaction, educational opportunity, and self-fulfilment (Ryan 1991). Beeton (1989) extends this discussion to protected areas.

Natural areas such as national parks play an important role in both tourist and excursionist satisfaction by providing areas which can potentially offer experiences of challenge, escape, relaxation, self-discovery and spiritual awareness. ... Protected landscapes can provide the matrix for a wide range of tourist experiences involving the utilisation of attraction and facilities in a particular landscape. ... All the elements involved in a traveller's visit and the psychological bene-

Table 16.2	Potential conflicts between recreation groups in protected areas (Eagles et al. 2002)
Conflict	Nature of the conflict
Visitor–visitor (single activity)	Conflicts may occur within one recreational activity. They may occur when there are inappropriate visitor behaviours; different skills and experience levels of visitors; and different expectations of social behaviour.
Visitor–visitor (different activities)	Conflicts may occur between different visitor activity groups. This may occur between motorised and non-motorised recreation; active recreationists (such as cyclists) and passive recreationists (such as nature study); active recreationists (such as downhill skiers) and active recreationists (such as snowboarders); active non-assisted recreationists (such as bushwalkers); and active assisted recreationists (such as horse riders).
Visitor–management operations	Conflicts may occur when a recreational experience is affected by management operations. This may occur when low flying aircraft are required for management duties over remote areas; motor vehicles (over snow, over water, over land) are used in remote areas; controlled burning introduces smoke and creates poor visibility.
Visitor–manager	Managers may be required to regulate visitors closely to protect environmental values. A visitor's experience may be affected by the regulatory action.

fits obtained from these activities constitute the recreation/tourist experience ... The area, including area design, setting, activities engaged in, proximity to and number of other recreationists and their behaviour, the facilities available, and inter-area travel nodes, will all contribute to the on-site experience.

Managing for tourism and recreation opportunity settings is typically achieved through the management planning process (Chapter 7) and the use of tools such as the Recreation Opportunity Spectrum (Backgrounder 16.3) and zoning. A remarkable array of different user groups may want to use a single protected area, and often the same space within that protected area. Overuse or conflicts between groups can arise (Table 16.2).

Characteristics of visitors

People visiting a national park for learning, relaxation, or family recreation have very different needs and expectations from those seeking excitement or challenge. Monitoring the numbers visiting protected areas is important (Backgrounder 16.4) and

so is understanding the characteristics of visitors, their motivations, needs, and expectations.

Tourism and recreation are constantly changing, as are the seasonal patterns of visitor use. A change in either can bring peak loads of visitors that exceed what certain destinations can sustain. Managers should constantly analyse the relationship between the supply (capacity of destinations or settings of a particular sort) and the demand (of visitors who seek or would prefer a given setting). With knowledge of any patterns, they will be in a better position to manage for sustainability. Parks Victoria has developed a visitor management model for Port Campbell National Park to assist it with its operations (Case Study 16.2).

Managers can tailor products, services, and strategies to well-defined visitor groups. This keeps the visitors happier, increases revenue, uses management's time, money, and effort more efficiently, and prevents or reduces damage to sites. Visitors can be characterised in three main ways: demographically (by age, gender, education, life cycle), geographically (by city, region or country of origin), or psychologically (by values, attitudes, motives, and expectations)

Backgrounder 16.3 Recreation Opportunity Spectrum

The Recreation Opportunity Spectrum (ROS) was developed by Clarke & Stankey (1979) of the United States Forest Service. It is an outstanding planning tool for managing natural areas for recreation. Its aim is to help managers give visitors a choice of high quality outdoor recreations by distinguishing and offering a range of recreational settings from remote natural wilderness through to urban and 'developed' settings. This range of settings has been described as a recreation opportunity spectrum. Nature-based tourism experiences range from a short stopover at a scenic lookout, to a guided walk and talk by a ranger, to high-risk adventure activities, or to extended camping tours through remote country lacking in any visitor facilities. The ROS categorises areas by their physical factors such as the naturalness of the area and the presence or absence of roads and visitor facilities; by social factors such as the number of other users; and by managerial factors such as the presence of barriers and signs (NSW NPWS 1997b). It is a fundamental part of a protected area manager's toolkit for managing visitors.

Ideally we should favour those activities that require less modification of the environment. Some activities such as wilderness walking require a much greater area of natural land than those that rely on built infrastructure or intensive interaction. The ROS has been adopted, but not always fully used, by most protected area agencies. A ROS classification system has been developed for use in NSW (Table 16.3). Table 16.3 also gives the different category labels used in the Victorian and Queensland systems. Indicative activities associated with the five ROS classes are given in Table 16.4. A common approach developed for all Australian jurisdictions would be a useful initiative. A similar ROS was used by Nelson & Wearing (1999) to help plan the camping areas of the ACT. They classified existing campgrounds on a spectrum from 'modern' to 'primitive', and identified some gaps in the spectrum that they recommended filling. These included group camping areas, and facilities for campervans and for hut-based camping. In Victoria, a ROS analysis of recreation settings has been completed over 7.2 million hectares—most of the public land in the state. The result (Figure 16.1) illustrates the relatively low proportion of remote areas. With time, such lands will become both rarer and more valuable.

Table 16.3	Recreation Opportunity Spectrum as utilised by the **NSW NPWS** (NSW NPWS 1997b)				
NSW ROS category	Class 1	Class 2	Class 3	Class 4	Class 5
Equivalent Victorian ROS category (DCE n.d.)	Remote	Semi- remote	Roaded-natural	Semi-developed	Developed
Similar Queensland ROS category	Remote	Semi-remote non-motorised	Semi-remote motorised	Natural	Divided into two classes— Intensive and Urban
General description	Essentially unmodified environment of large size	Predominantly unmodified environment of moderate to large size	Predominantly natural environment, generally small development areas	Modified environment in a natural setting— compact development area	Substantially modified environment, natural backdrop
Access	No roads or management tracks—few or no formed walking tracks	No roads— management tracks and formed walking tracks may be present	Dirt roads— management tracks and walking tracks	2WD roads (dirt and sealed), good walking tracks	Sealed roads, walking tracks with sealed surfaces, steps and so on may be present
Modifications and facilities	Modifications generally unnoticeable— no facilities, no structures unless essential for resource protection and made with local materials	Some modifica- tions in isolated locations—basic facilities may be provided to protect the resource (such as pit toilets and BBQs)	Some modifica- tions but generally small-scale and scattered— facilities primarily to protect the resource and public safety— no powered facilities	Substantial modifications, noticeable— facilities may be relatively substantial and provided for visitor convenience (such as amenities blocks), and caravans may be present at times	Substantial modifi- cations that dominate the immediate landscape—many facilities (often including roofed accommodation) designed for large numbers and for visitor convenience
Social interaction	Small number of brief contacts (for example, less than five per day)—high probability of isolation from others—few if any other groups present at campsites	Some contact with others (for example, up to 20 groups), but generally small groups—no more than six groups present at campsites	Moderate contact with others—likely to have other groups present at campsites— families with young children may be present	Large number of contacts likely— variety of groups, protracted contact and sharing of facil- ities common— may have up to 50 sites	Large numbers of people and contacts— groups of all kinds and ages—low likelihood of peace and quiet

Table 16.3	(continued)				
NSW ROS category	Class 1	Class 2	Class 3	Class 4	Class 5
Visitor Regulation	No on-site regulation—off-site control through information and permits may apply	Some subtle on-site regulation such as directional signs and formed tracks	Controls noticeable but harmonised (such as information boards, parking bays)	On-site regulation clearly apparent (such as signs, fences, barriers) but should blend with natural backdrop	Numerous and obvious signs of regulation—rangers likely to be present

Table 16.4	**Indicative tourism and recreation activities undertaken for Recreation Opportunity Classes** (modified from NPWS 2002b)

Activity	Class 1	Class 2	Class 3	Class 4	Class 5
Alpine skiing					■
Snow boarding					■
Cross-country skiing	■	■	■	■	■
Ice climbing	■	■	■	■	■
Picnicking (facility based)			■	■	■
Camping (no facilities)	■				
Camping (facility based)		■	■	■	■
Scenic driving			■	■	■
Four-wheel-driving and registered trail bike riding on road			□	□	
Nature study or cultural awareness	■	■	■	■	■
Horse riding		□	□	□	
Canoeing/ kayaking/white-water rafting	□	□	■	■	■
Boating (motorised)		□	■	■	■
Sailing/sail boarding			■	■	■
Adventure activities	□	□	□	□	
Fishing	■	■	■	■	■
Non-powered flight: hang-gliding, hot air ballooning, paragliding		□	□	□	□
Powered flight: low altitude			□	□	□
Cycling (on existing roads and trails)		□	□	■	■
Bushwalking (on formed tracks, not overnight)	□	■	■	■	■
Bushwalking (remote areas or long distance trails)	■	■	■		
Orienteering/rogaining		□	□		
Cross-country running			□	□	
Caving	□	□	□	□	□
Organised mountain biking				□	□

■ *Activity permitted*
□ *Activity may be permitted subject to certain conditions such as designated sites only*

(Clare 1997). Visitors generally report very high levels of overall satisfaction with their experiences of protected areas:

> *By far the most important factors visitors perceive as affecting the quality of their experiences are the 'natural features' and 'unspoiled nature' of the environment in protected areas. Other factors regarded as being important by Protected Area visitors include the provision of suitable directional signage and maps, and, above all, clean toilets. In some cases, well-planned walking trails and campground facilities are also important (TTF 2004).*

Not all parks, nor all visitor destinations within a park, will appeal to all visitor groups. The type of visitor group will therefore vary, depending on where the park is, the facilities and experiences offered, and other attractions in the region.

An understanding of visitor 'segments' can assist managers to understand the causes of management problems and devise appropriate remedial strategies. Market segmentation can also assist with the design of education and marketing campaigns. Market Solutions (1996) divided visitors into 'market segments' for Parks Victoria, based on their leisure preferences. The approach was used (with permission) by the NSW NPWS as part of their Nature

---

Backgrounder 16.4 Visitor-use data

Catherine Pickering and Yani Grbich, Griffith University

Knowing how many people visit a protected area, how long they stay, and what they do while they are there, is important information for protected area managers. Best practice visitor monitoring procedures for protected area management recommend that two types of visitor data are collected: visitor use and visitor satisfaction. Visitor-use measures include:

- person entry—whenever a person enters a park for any purpose
- person visit—when a person visits the park for the first time or on the first day of their stay to participate in park activities
- person day visit—when a person stays in the park for a day or part of a day, with each day the person stays counting as an additional person day visit
- visitor nights—the count of people staying overnight in a park or protected area for a purpose mandated for the area times the number of nights for each stay (other measures include camper nights, skier nights, bed nights).

Data concerning visitor use can be collected by a variety of means each with its own benefits and limitations.

Vehicle counters. There are a variety of vehicle counters available such as tube, induction loop, and infra beam type counters. Once counters are calibrated, the number of visitors to the area can be estimated by multiplying the calibrated number by the number of cars. For example, Kosciuszko National Park calibrates three people per car and 40 per bus.

Pedestrian counters. Two main types of counters are in use: pressure pad and infrared beam. Laser counters are also in use in some visitor centres.

Ticket sales. Fees and charges can be used to measure visitation when the relationship between revenue and visitor numbers is clear, or sales records are kept.

Ranger and field staff observations. Anecdotal estimates have benefits such as noting overcrowding but cannot provide solid figures.

Concessionaires and leasing. Operators leasing areas within parks and conducting tours are often required to provide data regarding number of tour participants.

Spot counts/sampling studies. For example, seasonal spot counts are undertaken to establish the number of cars in overnight car parks or parked at walking trail heads.

Figure 16.1   Recreation opportunities on public land in Victoria

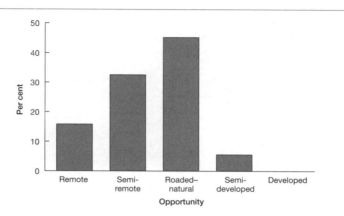

Case Study 16.2

Visitor Management Model for Port Campbell National Park and Bay of Islands Coastal Park

Dino Zanon, Parks Victoria

Port Campbell National Park and the Bay of Islands Coastal Park are located on the Great Ocean Road, Victoria, approximately 250 kilometres west of Melbourne. They are linear parks comprising 65 kilometres of rugged and spectacular coastal scenery and are protected in a strip ranging in width from a few metres to 2 kilometres.

The parks attract large and steadily increasing numbers of visitors (2.1 million visits per year in 1996–97 combined for both parks). In 2001 the annual growth rate of visits was estimated at 3.55%. This increase in visitation suggested that by 2006–07 there would be approximately 3.2 million visits per year to Port Campbell and Bay of Islands combined, causing considerable load on facilities and pressure on the coastal ecosystem.

Managers needed to comprehensively evaluate the cascading effects of the flow of visitors through a sequence of sites along the parks and to estimate the effects of increasing visitor flows through time. They also needed to know if designed capacities for parking, visitor centres, roads, camping areas, and day-use facilities could accommodate projected visitor numbers.

There are many options available to park managers to deal with heavy visitor use. New sites can be opened up; a system of reservations can be implemented; areas can be closed so sites can recover from overuse; facilities can be expanded; or sites can be hardened to accommodate larger numbers of visitors. Each of these strategies will have a different impact on the overall system, and the complex interrelationships between these decisions are almost impossible for a manager to predict. It is in this context where computer simulation of recreation behaviour is of real value.

Recreation Behaviour Simulation, or RBSim (Gimblett & Itami 1997, Itami & Gimblett 2000), is a computer simulation tool integrated with a geographic information system that is designed to be used as a general management evaluation tool for any park. The geographic information systems represent the environment, and autonomous 'intelligent' software agents are used to simulate human behaviour within geographic space. Using RBSim, the park manager can then change a number of variables, including the

Figure 16.2   RBSim output example: predicted average queuing time at car parks, Twelve Apostles, Port Campbell National Park Victoria, for 2001, 2006, 2011

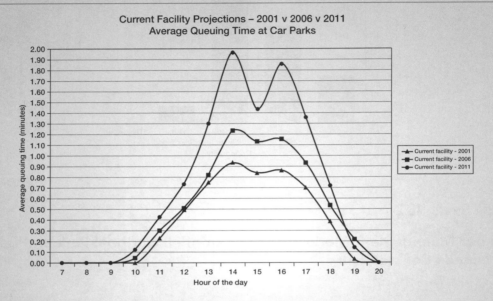

number and kind of vehicles, the number of visitors, the number of parking spaces, road and trail widths, and the total capacity of facilities. In this way, 'what-if' scenarios can be used to explore the consequences of changing variables for quality of visitor experience. The simulation model generates statistical measures of visitor experience to document the performance of any given management scenario. The software provides tables and graphs (Figure 16.2) from the simulation data so park managers can identify points of overcrowding, bottlenecks in circulation systems, and conflicts between different user groups.

RBSim was used to examine the impact of changes in park infrastructure and increasing visitor rates over a 10-year period on a range of sites within the parks. The studies identified likely future problems that needed to be addressed straight away, particularly shortage of car and bus parking at two sites. The RBSim simulation model enabled park managers to collect data using 'real-life' methods, such as minimum, average, or maximum available capacity of car parks at a particular time of day. Simulation makes it possible to identify what visitor management and infrastructure changes are needed and under what conditions, and to predict the effects of those changes on visitor satisfaction.

Tourism and Recreation Strategy. The segments derived by Market Solutions were as follows.

1   Natural adventurers

- Natural adventurers comprised 15% of the total sample and 11% of the NSW sample.
- They are mostly young singles and family groups who enjoy the great outdoors. They

hold very positive attitudes to national parks and display a high level of national park awareness. They enjoy physically demanding and adventurous national park activities.

2   Escape to nature

- These people comprised 12% of the total sample and 12% of the NSW sample.

Bushwalking group, Ben Boyd National Park, NSW (Graeme Worboys)

- They are mostly older families and older couples with higher incomes and education levels. They hold very positive attitudes to national parks and display a high level of national park awareness. Their need for escaping the pace of everyday life and their desire for solitude drives their interest in national parks. 'Non-active' activities appeal the most to this group.

3 Young thrillseekers

- Young thrillseekers comprised 11% of the total sample and 10% of the NSW sample.
- They are mostly young singles and couples, especially males who are highly active in their leisure-time pursuits. They enjoy the adrenalin rush of physically demanding and adventurous activities, but they are not particularly interested in national parks. Their national park awareness is moderate.

4 Out-and-about seniors

- These people comprised 14% of the total sample and 9% of the NSW sample.
- They are mostly mature, socially active older people who are probably retired or nearing retirement. They hold very positive attitudes to national parks and have a high level of awareness. They are most likely to enjoy day tripping for picnics and scenic driving.

5 Nature made easy

- These people comprised 15% of the total sample and 19% of the NSW sample.
- They are mostly family groups and older couples, especially females who are employed in home duties or white-collar occupations. They desire social activities that involve being with the family but do not revolve around the family. Their awareness of national parks is moderate but their attitudes are very positive. Comfort is a key issue for this group.

6 Social relaxers

- These people comprised 11% of the total sample and 11% of the NSW sample.
- They are mostly members of young and middle-aged family groups, especially females. They desire social activities that involve being with family and friends but they do not really enjoy physical activities or the great outdoors. Their awareness of national parks is low and their attitudes to parks are moderate.

7 Family focused

- These people comprised 11% of the total sample and 12% of the NSW sample.
- The main leisure focus for these young and middle-aged families revolves around family activities. It appears that nothing else is of great

importance to this group. They are not very interested in socialising or being active in their leisure time. National park awareness is low and attitudes to national parks are less than positive.

8   Home-based seniors

- These people comprised 8% of the total sample and 10% of the NSW sample.
- These older couples and mature singles are not interested in being involved in physical activities. They are interested in spending time with their family, possibly as observers rather than participants. They have low awareness of national parks.

9   Indifferent youth

- These people comprised 4% of the total sample and 6% of the NSW sample.
- These young singles (mostly males) desire freedom and independence in the company of their friends. They are not interested in physical activities and lack empathy with nature. Their awareness of national parks is low and their attitudes to parks are the least positive of all groups.

The target domestic visitor markets identified for the NSW NPWS based on these segments were: natural adventurers; escape to nature; young thrillseekers; out-and-about seniors; and nature made easy. The NSW NPWS (1997b) also used market segment information to help determine a strategic approach to managing its visitor destinations. For example, many NPWS initiatives in Sydney Harbour National Park such as the 'Discovery Ranger' program are run as commercial operations, so an understanding of the key market segments has been important for their commercial success.

Understanding international visitor use is also important for many protected areas. Close proximity to major metropolitan centres such as Sydney Harbour National Park, or the international status of a protected area as a world heritage area will mean larger numbers of international visitors. The tourism industry regularly undertakes research on overseas tourists, with an emphasis on their country of origin. This information is important for planning to meet the visitor demand.

16.3   Managing tourism and recreation

Unmanaged tourism and recreation is a significant threat to protected areas. Overuse, pollution, vandalism, theft, destruction of habitat, competition for resources (such as potable water), and impacts to wildlife are some of many impacts from use of protected areas by humans (Table 16.5, Snapshots 16.3 and 16.4).

Our goal is minimising and managing such environmental effects to achieve sustainable tourism, recreation and visitor use. Sustainable development principles and practices formally arose as an international initiative from the Rio de Janeiro Earth Summit in 1992 (Chapter 2). Ecologically sustainable use has been defined as:

> *use of living things or areas within their capacity to sustain natural processes while maintaining the life-support systems of nature, and ensuring that the benefits of use do not diminish the potential to meet the needs and aspirations of future generations (GBRMPA 1994, p. 58).*

The travel and tourism industry has developed a formal program of sustainability action (Backgrounder 16.5); and some organisations such as Ecotourism Australia and Green Globe 21 are committed to achieving sustainable tourism through self-regulation of the industry and by promulgating the principles and practices of Agenda 21.

How does one make tourism sustainable? One simple solution, often put forward, is to 'harden' a destination. That is, to change the boggy tracks to boardwalks, to replace the pit toilets with septic or composting toilets, and so on. More and more sites become 'urbanised' under this formula. Furthermore, this is not a general solution, because the very type of site that many visitors wish to see and experience in protected areas (natural destinations) becomes increasingly scarce (Backgrounder 16.6).

An alternative solution is to set an objective such as 'We are going to look after this destination so that in the long term, it is exactly as it is now'.

Table 16.5	**Environmental threats to protected areas from tourism** (Buckley & Pannell 1990, Gee et al. 1997, Green & Higginbottom 2001, Eagles & McCool 2002, Eagles et al. 2002, Newsome et al. 2002, Buckley et al. 2003, Christ et al. 2003)
Element	Examples of threat from tourism and recreation activities
Ecosystems	The construction of accommodation, visitor centres, infrastructure, fences, access roads, walking tracks, and other services has a direct effect on the environment, by vegetation removal, animal disturbance, elimination of habitats, and changes to drainage patterns.
	Wildlife habitat may be significantly changed (travel routes, feeding areas, breeding areas, etc) by tourist development and use.
	Tourism and recreational activities including boating, off-road vehicle use, mountain-bike riding, horse riding, caving, mountaineering, hiking and camping, and loud noise affect natural values.
	Weeds (garden flowers and non-native grasses) and pest animals (cats and dogs) can be introduced by residents accommodated within protected areas.
Soils	Trampling and soil compaction can occur in certain well-used areas. Soil contamination can occur with fertilisers, pesticides, and pollution from vehicles. Soil removal and soil erosion also occurs, and may continue after the disturbance is gone.
Geology	Damage to cave formations and mineral sites can occur from illegal fossil collecting. Sand dunes and reefs are also susceptible to damage.
Vegetation	Concentrated use around facilities has a negative effect on vegetation.
	Transportation may have direct negative effects on the environment (vegetation removal, weed introduction, animal disturbance).
	Fire frequency may change due to tourists and park tourism management.
Water	Visitation increases demands for fresh water.
	Disposal of sewage causes environmental effects even if it is within licence limits.
	Visitation can also lead to solid waste dumped in waterways, erosion of stream banks, and increased turbidity.
Air	Motorised transportation may cause pollution from emissions; smoke from lodge fires can cause pollution in mountain valleys.
	Visitor use can increase energy consumption and cause greenhouse gas emissions.
Wildlife	Major issues include handfeeding, spotlighting, disturbance to nesting birds, disruption of foraging, and loss of energy reserves and local habitat disturbance.
	Fishing may change population dynamics of native species.
	Fishers may demand the introduction of foreign species, and increase populations of target animals.
	Impacts occur on insects and small invertebrates, from effects of transportation and introduced species.
	Disturbance by visitors can occur for all species, including those that are not attracting visitors. Disturbance can be of several kinds: noise, visual, or harassing behaviour.
	Habituation to humans can cause changed wildlife behaviour, such as approaching people for food.
	Vehicle traffic gives rise to wildlife road kills.
Cultural impacts	Theft, vandalism, and overuse can adversely affect cultural sites.

Snapshot 16.3

The problems of overuse in Tasmanian parks
Tim O'Loughlin and Ben Rheinberger, Tasmania Parks and Wildlife Service

In Tasmania during the mid 1980s overuse of parks was severe and worsening. The following are examples of this overuse.

Walking tracks. The major walking tracks had been deepened and degraded until it was not uncommon to have to wade through sections of thigh-deep mud. Even the occasional 'neck-deep dip' was not unknown.

Stomach diseases. Gastroenteritis (diarrhoea and vomiting) among walkers became so common that on the Overland Track (Tasmania's most popular long-distance track) during the summer of 1985–86, about a quarter of all walkers became sick.

Campfires. Bushwalkers' campfires cause two major problems for park managers: fires can escape and turn into bushfires (which then devastate the sensitive areas walkers tend to travel through); and walkers cause local damage by collecting or cutting wood for fuel. For example, in Pine Valley on the Overland Track, the 'campsite' extended for over 1 kilometre due to 'site expansion'.

Rubbish. Many walkers observed the 'Carry In-Carry Out' ethic, but those that didn't left many garbage-bag loads of rubbish to be removed by rangers. Over the 1986–87 summer, rangers carried or helicoptered out three five-tonne truckloads of rubbish from the northern half of the Overland Track alone.

If too many visitors are causing damage then action is taken to change the way in which visitors use the site, when they use it, or how many use it. The advantage of this approach is that it preserves the diversity of our destinations by protecting those most vulnerable to change, the completely natural sites. Those destinations with hardened facilities are easier to protect. In the future, visitors may pay a premium to be able to visit the more natural sites.

Management techniques that can be used to help achieve ecologically sustainable visitor use include:

- determining the recreation setting for a destination
- using ROS to assist with the nature of the facilities for a planned recreation setting
- visitor management practices that include: limits, permits (Snapshot 16.5), dispersal of visitors, concentration of visitors, rules on length of stay, segregating different recreational

Backgrounder 16.5   ESD and the tourism industry

Since the Rio Earth Summit in 1992, three key international organisations, the World Travel and Tourism Council, the World Tourism Organisation, and the Earth Council, have combined to produce *Agenda 21 for the Travel and Tourism Industry: Towards Environmentally Sustainable Development.* This is a program of action for travel and tourism. It lists a number of principles, and recognises ten priority areas for action, all of which are relevant to visitor management. They are:

1 minimising, reusing, and recycling waste
2 managing for energy efficiency and conservation
3 managing freshwater resources
4 managing wastewater
5 controlling hazardous substances
6 efficiency in transport
7 planning and managing land use
8 involving staff, customers, and communities in environmental issues
9 designing for sustainability
10 creating partnerships for sustainable development (Soin 1999, p. 11).

Snapshot 16.4

Sustainable recreational diving in marine protected areas
Derrin Davis, Southern Cross University

Recreational diving is a young, rapidly growing industry. It relies heavily on access to natural areas, and increasingly on marine protected areas (MPAs). The recreational diving 'boom' in Queensland began in about 1979, and the industry quadrupled in size in the period 1979–87 (Centre for Studies in Travel and Tourism 1988). Divers, especially those who are overweighted or unskilled, sometimes bump into corals and break off fragments. They may also fin along the bottom, stirring up clouds of sediment. Recent studies agree that some 70% of the damage is done by only 4% of divers. Some who use cameras or videos are among the offenders. Indirect remedies include improved diver education and pre-dive briefings, particularly about buoyancy control, correct weighting, and reef etiquette.

A dive certificate is a lifetime qualification, requiring neither renewal nor proof of current proficiency. MPA managers might consider requiring that dive operators provide refresher courses unless a certain number of dives are completed within, say, each 12-month period. A diver code of conduct is another possibility. Managers could also introduce a 'no touch' policy, such as exists in many MPAs in the USA, or set limits to the number of divers that use an area per year. Policing such rules is difficult and expensive, especially in extensive MPAs such as the Great Barrier Reef Marine Park. Another tack is to give tour operators certain 'property rights' in defined areas, so that they have a vested interest in maintaining those sites. (A well-maintained dive-site may even have resale value.) Some sites recover faster than others, but in general, a site that attracts more than 5000 divers a year needs careful observation, and one that attracts 20,000 needs urgent management. Benchmarking and monitoring is crucial, so that environmental damage is detected.

activities, seasonal limits, zoning, limits on size of party
- environmental education and interpretation (Snapshot 16.6)
- codes of practice (Snapshot 16.7)
- sponsoring environmental certification schemes for tourism operators, such as the National Ecotourism of Australia (NEAP) scheme or the Green Globe 21 (Backgrounder 10.2) program..

- Determining if any infrastructure is to be provided for a destination and, if it is, how it is to be blended harmoniously with the site.

More site-specific measures might include:

- determining how many visitors a site can handle through planning
- planning the layout of the site for sustainable visitor use, including aesthetic considerations

Backgrounder 16.6   Resisting the obvious

On many occasions the 'obvious' answer to a problem is not the best long-term solution. For a damaged or overused visitor site, an 'obvious' management response is to harden the site. For example, a rough bush track will be changed to a gravelled or paved track. It does not always occur to agency staff that it might have been better to close the track during wet weather, when most of the damage was being done. As soon as there are infrastructure improvements, the nature of the recreation setting changes. So the potential is clear. Incremental, unplanned hardening of sites changes the nature of our destinations towards the developed end of the ROS (Backgrounder 16.3). Some of us may have heard the lament 'I liked it how it used to be … why couldn't they leave it alone?'

Managers receive other criticisms. For many visitors who seek greater comfort, it is 'obvious' that the site will benefit from new facilities such as showers or septic toilets, but these very facilities, if provided, will displace those visitors who are seeking a more natural experience. So, to provide a range of visitor facilities, natural sites need to be actively managed so that use is sustainable without hardening. Planned changes in the nature of recreation settings are acceptable. Incremental and opportunistic/unplanned hardening of sites is poor practice.

Walker impacts, South Coast Track, Southwest National Park, Tas (Ben Rheinberger)

Snapshot 16.5

The Desert Parks Pass
Pearce Dougherty, SA National Parks and Wildlife Service

The Desert Parks Pass was developed as a system of visitor management in the late 1980s. It provides for safe and controlled visits to desert reserves, and earns revenue. The Desert Parks Pass is a mandatory entry permit valid for one year for all of South Australia's Desert Parks. It also gives camping rights. It contains an easy-to-read handbook, full-colour maps, brochures of information and advice, and a conspicuous windscreen sticker. In 1998 it sold for $60 per vehicle (retail). Sales of the Pass bring in about $250,000 per year. Except for production and operating costs ($35 per pass), all funds are spent on visitor services, facilities, protection, and maintenance in the Desert Parks. In some popular locations there is an alternative overnight pass to meet the needs of short-duration visitors.

There are several reasons why this system works. The Pass, with its maps and accessories, is generally seen as good value for money, and its price could even be increased in future without upsetting customers. It is readily available. It is sold over a telephone hotline (1800 number) by credit card, and a system of sales agents has been established along routes to the outback, and in the capital cities of Sydney, Melbourne, and Adelaide from where most visitors come. Tourist agencies are also faxed regular Desert Parks update bulletins, so they can pass on accurate and timely information to potential desert travellers. As well, a Desert Parks web site is now being constructed that will be updated weekly.

Travelling the desert brings risks. The Desert Parks handbook and maps help travellers prepare for their journey and identify safe travel practices. On the ground, all defined routes are marked with low-key colour-coded track-marker posts every 5 kilometres, and a booklet in the pass-kit allows the numbers on each post to be interpreted. The Pass reduces the need for conventional signage, thus reducing maintenance costs in a harsh environment. In promoting the Pass system, the name 'Desert Parks' was designed to be evocative, capturing an image of desert travel. Since 1989 a specific font has been used for this title. Buyers know that funds generated by the Pass go directly back into the parks, and each facility that is funded using income from the Pass displays a small metal sign to say so.

Lessons learned from our experience with the Pass include:

*   ensure the cost is in line with the experience
*   design the system to meet management objectives and visitor needs
*   provide good quality, high value, updated information
*   develop an evocative and creative theme and highlight it
*   provide easy access to information and purchasing points
*   provide a variety of good experiences for the visitor over a range and choice of locations
*   channel income back to the place it was derived, and let people know
*   identify where visitors come from and how to get your message to them.

The pass is not perfect, but if only a fraction of things worked so well!

Snapshot **16.6**

Using interpretation to manage recreation in Lamington National Park
Carolyn J. Littlefair, Griffith University

Increased visitation to natural areas has led protected area managers to seek tools to minimise the impacts of visitors. Interpretation is one such management tool. Interpretation is often preferred by park managers because: it is perceived to be the most cost-effective method (Knudson et al. 1995); it is a light-handed approach and allows visitors the freedom of choice (Brown et al. 1987); and it enhances visitor experiences and satisfaction (Alder 1996, Bramwell & Lane 1993). However, the ability of interpretation to bring about a reduction in the environmental impacts of visitors to natural areas has rarely been quantified. I conducted a research project to assess how well environmental interpretation works as a management tool—specifically, to determine the extent to which interpretation reduced the environmental impacts of visitors on guided walks in national parks.

The research was conducted in Lamington National Park, in the south-east of Queensland, with the assistance of Binna Burra Mountain Lodge. Five different interpretive programs were created and used. The programs were a combination of environmental interpretation, role modelling by the guide of appropriate behaviour, and verbal appeals from the guide for visitors to modify their behaviour. Three common and experimentally convenient environmental impacts were measured: shortcutting of corners, picking up litter already on the track, and noise level of the group. Between 2000 and 2002, 41 walks were studied. The impacts of each group were measured and statistical analysis done on all results.

Results from the measurements of shortcutting found that the interpretive program that consisted of a combination of environmental interpretation, role modelling and verbal appeals, was always the most effective in reducing shortcutting. The programs that had only environmental interpretation or no environmental interpretation were always the least effective in reducing shortcutting. Results of the picking up litter measurements found that verbal appeals from the guide was the only factor that influenced whether litter was picked up. Analysis of the noise results found that no interpretive program significantly reduced the noise level of visitors. There was a small decrease in the amount of shouting and loud talking of visitors in programs with environmental interpretation plus verbal appeals and/or role modelling, but this reduction was only small, and not large enough to be statistically significant.

A key outcome of this study is that it is possible for interpretation to reduce some environmental impacts of visitors. When interpretation did reduce visitor impacts, two significant factors were found relating to its effectiveness. First, environmental interpretation alone did nothing to modify the behaviour of visitors or reduce impacts. Thus, it is important that interpretive programs expressly address an impact or behaviour, to have any chance of reducing it. Second, when the required behaviour is unfamiliar or uncertain to the visitor, role modelling of the correct behaviour by the guide was essential for visitors to behave that way themselves.

Researcher Carolyn Littlefair collecting data, Lamington National Park, Qld (Carolyn Littlefair)

Snapshot 16.7

Australian Alps codes of conduct

The Australian Alps Liaison Committee has provided a leadership role in Australia for transboundary cooperation. Its visitor-use codes of conduct are one of its many success stories. The codes include a Snow Camping Guide; Bushwalking Code; Horse-riding Code; Car-based Camping Code; River Users Code; Mountain Bike Code, and a Huts Code (Figure 16.3). Interpretive signs also assist, such as the famous old sign located for many years on the Lower Snowy River (installed by the Kosciuszko State Park Trust), which stated 'take nothing but photographs, leave nothing but footprints'.

- using renewable energy sources (gas, solar power)
- designing for recycling facilities if appropriate, but also encouraging visitors to take home their garbage
- using recycled and plantation timber and so on.

These and other responses need to be integrated into a coherent planning framework. There are very few sites in protected areas that should not be visited, but there are many that can only accept a few visits a year by a few people and then in conjunction with expert guides. There are many other sites where larger numbers of visitors are ecologically acceptable and actively encouraged and promoted. Planning limits for destinations may vary between hundreds of thousands of visitors per annum to a few visitors per year.

Regional planning should also look at destinations in other public lands, or on private land. The best rock-climbing area may in fact be on a farm or in a state-run pine forest. This may reduce the peak pressure on a park, and so help managers deal with awkard political pressures. Managing sustainably means working within design limits. Integrated planning for tourism would recognise all suitable destinations in an area and would allow for alternative visitor destinations. It is ideal if all involved in such a regional tourism plan use a site classification scheme like ROS so that recreation site information can be exchanged and compared.

Figure 16.3   Visitor codes of conduct produced for the Australian Alps National Parks

Figure 16.4   A planning framework for visitor destinations

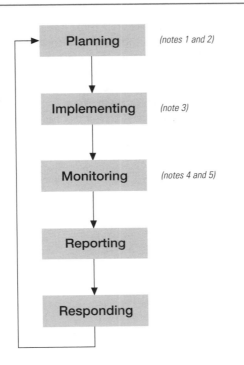

Planning (notes 1 and 2)

Implementing (note 3)

Monitoring (notes 4 and 5)

Reporting

Responding

Note 1. *Any change in the ROS character of a destination is a planned decision. The relative proportion of different recreation settings is managed strategically.*

Note 2. *Planning limits of visitor use for destinations are determined (see steps 1 to 3) so as to achieve sustainable visitor use.*

Note 3. *Sites are managed not to exceed the visitor numbers in the planning limit. Rangers may actively intervene if unsustainable practices occur. Licensed commercial tour operators are allocated a proportion of the site's visitor use planning limit.*

Note 4. *Some selected sites are used to benchmark sustainable use practices for the rest of the state or territory, relative to the objectives set and the criteria established.*

Note 5. *All sites are monitored by rangers, and broad patterns of environmental condition are determined at a state or territory level. Large-scale trends in the sustainable condition of visitor destinations may require responses at a political level.*

Australia has developed national tourism plans, national ecotourism plans, national rural tourism plans, and other major planning documents. All of these emphasise sustainable visitor use. There are also strategic plans at regional level, especially for bioregions, for major water catchments, and for other natural biophysical features.

A planning framework for visitor destinations

Planning, active management of sites, monitoring, and rapid response to unsustainable actions should be the four basic elements of effective visitor management. A planning framework that addresses these elements is shown in Figure 16.4.

This framework is based on the Visitor Impact Management (VIM) model, modified to take account of the cost constraints typically faced by Australian protected area managers. The planning framework incorporates four main steps.

Step 1. Make an inventory of all visitor destinations for a region, state or territory. For this we advocate using the ROS (Backgrounder 16.3).

Step 2. Determine the management objectives for each destination—these objectives may be environmental/ecological, social, economic, or managerial—and how progress towards them will be measured. At a minimum, environmental/ecological objectives need to be established as the primary management objectives for a site. The management objectives should be developed in the context of local, regional, and national planning instruments. This approach is similar to the Limits of Acceptable Change (LAC) and VIM planning frameworks (Backgrounder 16.7).

Step 3. Introduce visitor-use limits for each visitor destination that are consistent with the objectives. The limits will be established by experienced rangers using their theoretical and practical knowledge of a site. Such knowledge includes a working understanding of ecological processes and potential visitor impacts. Rangers will also usually obtain inputs from agency experts and external advisers. Their estimates will err on the side of caution. The preliminary estimates will be confirmed or improved through community consultation usually as part of a public exhibition of a plan for the site. Such planning limits, if they are instituted early in

Backgrounder 16.7	Limits of Acceptable Change, Visitor Impact Management, and other visitor management models

Limits of Acceptable Change (LAC) (Stankey et al. 1985) is based on the premise that human use does cause damage. This is an important point. It sets managers the task of defining (through objectives of management and performance criteria) how they want their destinations to be managed. This approach is quite different from monitoring visitor use to determine if impacts are occurring. The LAC approach sets limits (that can be measured) to the human-induced changes that will be permitted, as well as identifying the remedies managers should provide. The LAC system, whose focus is on wilderness areas, recognises nine steps. It is a relatively costly and complex process but adaptations are possible.

Visitor Impact Management (VIM) was developed by Graefe et al. (1990). The model recognises that managing visitor use is complex, and that impacts are influenced by factors other than use levels. VIM has identified five major sets of considerations that are critical to understanding the nature of recreational impacts and that should be incorporated in any program that models such impacts. They include:

- interrelationships between types of impact
- use–impact relationships
- varying tolerances of impact
- activity-specific influences
- site-specific influences.

The VIM process recognises a number of steps:

- reviewing data
- reviewing objectives of management
- selecting key indicators of impact
- selecting standards for these key impact indicators
- comparing existing conditions to identify probable causes of impact
- identifying management strategies
- implementing the strategies.

Other models of visitor management have been prepared. They include the United States National Parks Service's Visitor Experience and Resource Protection (VERP); Visitor Activity Management Program (VAMP); and the Tourism Optimisation Management Model (TOMM).

---

the life of a protected area, may prove to be well above the current levels of visitor use. Time and experience shows that this is often a temporary situation. Usually these planning limits will be included within a statutory plan and altered after completion of statutory processes.

Step 4. Monitor the sustainable condition of all visitor destinations across a region, state or territory. Individual sites will be monitored relative to the objectives and criteria established in step 2 above. Major trends in their condition would then receive an organisational and strategic response.

This planning framework has the great advantage of not beginning with expensive and

time-consuming reports as the LAC and VIM frameworks require (Backgrounder 16.7). These approaches are probably beyond the resourcing capabilities of most Australian protected area agencies. The planning framework presented here recognises the skills and abilities of rangers and agency staff to establish responsible planning limits for sites. This framework should also be managed so that it is continually appraised. Some sites will be selected as reference sites and will be thoroughly researched so that the planning limits established are evaluated. Adaptive management principles would then apply to improve both management responses and, perhaps, the actual planning limit. This planning framework has been

Hardened site at Murramurang National Park, NSW (Graeme Worboys)

Snapshot 16.8

Sustainable visitor-use limits: Montague Island Nature Reserve

Montague Island Nature Reserve is located offshore from Narooma on the south coast of NSW. The island is rich in wildlife, including an Australian Fur Seal (*Arctocephalus pusillus*) haul-out colony and a major penguin-breeding colony. Previously closed to the public, it was opened for limited eco-tourism consistent with:

- thorough research and understanding of the island's heritage values
- the recreation setting for visitors being linked to the (interesting) cultural heritage precinct including the wharf area, access road, and lightstation complex
- all visitor use being accompanied by a ranger-guide
- a planning limit of up to 90 persons on the island (including management staff) for any 24-hour period
- a detailed research program of wildlife to provide feedback about any effects of the ecotourism so that management may adapt its approach.

The planning limit is defined in the plan of management. Commercial tours operate to the island and stay within the planning numbers. Revenue from ecotourism is reinvested into

Montague Island Nature Reserve, NSW (Graeme Worboys)

management of the island, and pays for the guiding service provided by a ranger. NPWS economic studies (Chapter 8) have demonstrated that the 4300 visitors per annum currently accessing the island help to generate 26 local jobs. In 1999, these efforts won a prestigious international award for ecotourism sponsored by IUCN.

Snapshot **16.9**

Planning visitor use at sub-Antarctic Macquarie Island
Geof Copson, Tasmanian National Parks and Wildlife Service

Macquarie Island is in some ways a model for Antarctica and islands of the Southern Ocean. While tourism had been going on there for around 20 years, there had been little research or monitoring of the effects. Land-based tourism was not considered viable; yet only first-hand experience can provide an appreciation of the isolation, extreme climatic conditions, and grandeur of Antarctica or the bountiful wildlife of these islands. Well-controlled tourist visits can raise the profile of the region, with visitors becoming some of the best advocates for its long-term conservation.

In the late 1980s, limited trials of ship-based tourism commenced at Macquarie Island. Parties were allowed on shore between 7 a.m. and 7 p.m. at two sites, The Isthmus and Sandy Bay. Access was permitted along sections of beaches, and on the formed vehicular tracks at The Isthmus, together with boardwalks and platforms that allowed controlled viewing of penguin colonies, historical sites, and scenic areas.

Four ship visits, totalling about 500 passengers, were allowed each season. The maximum number of visitors ashore at any one time was set at 60 for Sandy Bay and 100 at The Isthmus, in groups of not more than 15 people. In 1995 the limit on the number of ship visits was removed but the other limits remained. This helped smaller vessels with specialised interest groups, such as ornithologists, to visit the island. Tourist guidelines are reviewed after each season and revised if necessary.

It was decided that visitor facilities would be restricted to boardwalks and viewing platforms in sensitive areas, and these were constructed in 1989. High quality interpretation material, in the form of booklets and pamphlets, was prepared for the first season and is still being added to. The costs of provision/maintenance of facilities and interpretation material as well as some field salaries and administrative costs, are funded through a levy on each passenger on board a tourist ship. The environment is carefully monitored for possible side-effects.

The aim is to exclude exotic plants, invertebrates, and pathogens. Visitors are advised of the need to clean their footwear, clothing, and equipment. Baths for washing footwear are provided at the top of ships' gangplanks. It seems to date that these measures are adequate, but the region is fragile and we are only beginning to understand how to manage it.

Construction of boardwalk, Macquarie Island (Geof Copson)

successfully utilised at Montague Island Nature Reserve (Snapshot 16.8).

Planning limits, due to environmental or other sensitivities, have already been applied in a number of protected areas in Australia. Geof Copson (Snapshot 16.9) describes planning limits established for sub-Antarctic Macquarie Island. Alison Ramsay (Snapshot 16.10) describes the work being done by the Jenolan Caves Trust in NSW in applying the VIM model. The detailed work at Jenolan provides a model research approach, and could be used for visitor sites that are recognised as 'benchmark sites' in the planning framework presented by this book. Graeme Moss (Snapshot 16.11) describes research being conducted at the Seal Bay Conservation Park on Kangaroo Island. Finally, Cliff Winfield describes the story of the tingle forest in Western Australia (Snapshot 16.12).

Key destinations

The majority of visitors to protected areas, focus on only a few well-known attractions. These

Snapshot **16.10**

Determining carrying capacity and visitor impacts at Jenolan Caves
Alison Ramsay, NSW P&W

Jenolan Caves is one of Australia's outstanding tourist destinations, boasting a spectacular but fragile natural environment (both above and below the land surface) and a rich cultural heritage. The Jenolan Caves Reserve covers 2500 hectares and has some 350 known caves, 16 of which have been developed for tourist use. It currently attracts about 260,000 visitors a year.

In 1994 the Jenolan Caves Reserve Trust obtained Commonwealth funding under the Sites of National Tourism Significance Program for a study to establish the carrying capacity of the reserve. A three-day workshop of experts found there was not yet enough information to do this. They recommended use of the VIM framework (Graefe et al. 1990). As part of this, several caves are being elaborately monitored for environmental changes. Any such changes are correlated with weather conditions, the times and sizes of cave tours, the number of people using the walking tracks, and the times staff work in the caves.

Preliminary results (after three years) suggest that visitors are not affecting evaporation levels or the number of invertebrates, but heavy metals are accumulating in the sediments, flora, and fauna of the Grand Arch and the first 50 to 100 metres into the caves. Lint, skin, and hair flakes are discolouring cave formations; and the presence of visitors in caves during summer raises carbon dioxide levels significantly. Some accidental spills have also been detected in the streams. Changed practices have resulted, including limits to the size and frequency of tour parties to some caves, and to the amount

Track profile monitoring, McKeown's Valley Track, Jenolan, NSW (Alison Ramsay)

of time staff spend working in the caves. The monitoring process is constantly being refined.

VIM monitoring is costly in staff time and resources; but there is no quick and easy answer to determine how many people an area can sustain, or what impacts are being caused by visitors compared with other factors in the environment. Information about the condition of sites, combined with the classification of the recreation settings, becomes a powerful combination of information for managing these sites.

'honeypot' areas or 'key destinations' can be managed strategically at a whole-of-state or territory level of planning. Some sites will be extremely environmentally sensitive and will need to be managed very carefully. For some others, it may be possible to invest in the more sophisticated facilities such as boardwalks and viewing platforms thus catering for more visitors as well as protecting sites. However, not all potential 'honeypot' areas should have infrastructure developed. There may be strong cultural reasons why a site should be left as a natural destination. Environmental issues may restrict developments. Agencies may simply not be

able to afford the maintenance costs of yet another set of infrastructure.

Detailed planning for such key destinations is critical. Careful selection of the location for key destinations can potentially spread the economic benefits of tourism across a state or territory. Agencies will need to be firm in this planning process as most regions will want key destinations.

Key destinations have the potential to achieve iconic status along with their host protected area, and thereby assist state, territory, and national economies. Tourism marketing campaigns promoting Australia already feature locations such as the

Snapshot 16.11

Balancing seal conservation and sustainable visitor management

Graeme Moss, Department for Environment and Heritage, Kangaroo Island, SA

Australian Sea-lions (*Neophoca cinerea*) have experienced a dramatic decline in their distribution and abundance since the time of European settlement in Australia, primarily because of sealing in the 19th century. The species is Australia's only endemic seal, and is now regarded as rare, with the third most significant breeding colony occurring at Seal Bay on Kangaroo Island, South Australia. This colony is unique because over many years it has become habituated to the presence of humans. The seals do not seem to be adversely affected by regular contact with humans and this has allowed seal watching to become a significant national and international ecotourism attraction and economic venture for Kangaroo Island.

Long-term annual monitoring shows that Seal Bay seal pup population production is hovering somewhere between a very gradual decline to more or less stable (about 160–170 pups per year), despite an increase in visitor numbers from 39,623 in 1988 to 109,389 in 2002. This suggests that the current impact of the careful interpretive officer-guided tourism on seals is sustainable. A collaborative PhD project with the University of NSW is investigating the nature of the impacts of visitors on seals at Seal Bay. Information about seal behaviour and habitat use, and optimal levels of visitor activity are being obtained by both direct observation and an automated time-lapse video system.

Research and monitoring conducted to date by La Trobe University, CSIRO, and the University of California has shed light on the following aspects of Australian Sea-lions' biology.

- Seals will tolerate the physical presence and approach of visitor groups up to about 10 metres, but there are obvious changes in the animal's level of alertness closer than this.
- Some female and male seals have lived at Seal Bay for in excess of 20 years.
- Pup production at Seal Bay is somewhere between declining gradually to stable.
- Traditional breeding areas are used by female seals, and these tend to be areas where visitors have no access.
- The number of non-breeding haulout sites around the island is slowly increasing; however the number of breeding colonies has not increased.
- Marine debris (e.g., fish box plastic tape and liners, rope etc) entanglement-related mortality is a significant threat to seals.
- Both males and females forage out at sea as far as the edge of the continental shelf (100 kilometres), for three to 10 days at a time, while the pups remain on the beach, and learn to forage in nearshore waters.

Inappropriately placed campsite, Zoe Bay, Hinchinbrook Island, Qld (Peter Lavarack)

Snapshot 16.12

Saving the giants
Cliff Winfield, WA Department of Conservation and Land Management

Twenty years ago, no tour of the south-west of Western Australia was complete without a photograph of the car parked in the giant hollowed-out tingle tree in the forest near Nornalup. (Tingle trees are eucalypts, commonly Red Tingle (*Eucalyptus jacksonii*).) In 1989, the annual number of visitors to the Valley had risen to around 100,000 and was still growing. The gravelled car park was now the size of a small football oval. The quaint trail to the other 'giants' had become a labyrinth of goat tracks leading to every big tree in the area. For the huge tree, which had been the main attraction, the end was nigh. Years of people and vehicles trampling around its base had compacted the root zone and strangled its nutrient supply. In 1990, the giant collapsed. The Valley was being visited to death and a visionary approach was needed.

At the time, the 'Valley of the Giants' was part of state forest. The Department of Conservation and Land Management (CALM), in its 1987 regional management plan, recommended that the giants' 'block' (which includes the Valley of the Giants) be included in the adjacent Walpole–Nornalup National Park. The park is located in a unique high-rainfall corner of Western Australia with spectacular landscapes of estuaries, forested hills dissected by rivers, and dramatic coastal scenery.

Work began on a management plan for Walpole–Nornalup National Park in 1990. Meanwhile, the numbers of visitors to the valley continued to increase to around 140,000 a year and urgent action was needed to protect the environment. The then Executive Director, Syd Shea, had seen a treetop walk in Malaysia when attending a forestry conference there. He supported such a concept. Experience in the eastern states had begun to show that such walks were popular, and could not only pay for themselves in visitor fees but help subsidise visitor centres. We drafted a business plan, which was greeted with a mixture of scepticism and enthusiasm, but ultimately it was funded, and project planning commenced.

Valley of the Giants, WA (Courtesy of WA CALM)

Surveys by helicopter located a group of big tingle trees with trunks and crowns intact, and after weeks of assessing the suitability of the new area the planners were confident: this was the site for the treetop walk.

Other visitor facilities had to be planned for as well. More than half of the visitors came in tour coaches, which stopped for a quick look, whereas during peak periods many people sought family walking trails and barbecue facilities. Other features proposed for the 'new-look' valley included a visitor orientation and information area, and access for wheelchair users. Thus, through careful planning and a touch of ingenuity, the Valley of the Giants has been transformed from a moribund curiosity to a vibrant, environmentally sustainable, state-of-the-art, nature-based tourism experience that will delight and inform generations to come.

Great Barrier Reef, the Wet Tropics, South–West Tasmania, and Uluṟu–Kata Tjuṯa National Park. In such places attention to design detail is critical. For example, walks may be designed for brief, medium, and long-stay visitors. Key destinations will also have design limits of visitor use to protect sites and to help the quality of the visitor experience. At all times, key destinations need to be designed so that the integrity of the site is kept intact (Snapshot 16.13).

Snapshot 16.13

Key destination: Minnamurra Rainforest Centre, Budderoo National Park, southern NSW

Budderoo National Park protects an important remnant of the Illawarra rainforest and the scenically attractive escarpment waterfalls of the Minnamurra River in NSW. Initially established as a Crown Reserve in 1896, the rainforest became a national park in 1986. The site was heavily impacted. Funds were obtained to repair the vehicle-damaged areas, weed infestations, damaged walking tracks, and inadequate visitor facilities. A 'clean slate' was available for designers to create the Minnamurra Rainforest Centre. A design with nature principles was adopted and featured:

- boardwalks that replaced eroding walking tracks
- boardwalks that moulded into, and blended with, the landscape
- boardwalk grades that suited all visitors, from older citizens to tots in prams and wheelchair-bound patrons
- loop tracks that suited short-term and long-term day visitors
- a strong emphasis on facilities for education groups, including outdoor classrooms and ample educational sites along the walking routes
- a visitor centre or 'hub' for the site, from which all activities radiated to ensure ease of orientation and cost-effectiveness
- the use of locally propagated rainforest plants for rehabilitation works.

Visitor use of the site grew from 72,000 visitors per annum in 1992 to 130,000 in 1998. Managers have determined a site carrying capacity of about 400,000 visitors. Limited car parking at the site is offset by the use of a shuttle-bus system during peak times. Alternative rainforest destinations are available nearby when the site is full. The site won the prestigious environmental tourism category award at the NSW Tourism Awards for Excellence for the three years that it entered the competition (Worboys et al. 1995).

Tourists at Minnamurra Rainforest Centre, Budderoo National Park, NSW (Ben Wrigley)

16.4 Managing for quality

Tourists and recreationists deserve good service and support. They need care and consideration. They also directly benefit from good planning and the initiative taken by managers early in the establishment of a new national park (Snapshot 16.14). There are several non-invasive ways in which staff can help people have a high quality experience when they visit protected areas.

- Information about destinations should be made readily available in a clear, corporately consistent, concise form, whether in brochures, videos, press releases, articles, or on the Internet.

- Roads to sites should be good, safe, well maintained examples of whatever category they are claimed to be (bitumen road, gravel road, or four-wheel-drive track).
- Tourism operators who transport people to destinations should be encouraged to provide good information that amounts to value adding for nature tourism. Partnerships with such operators could involve managers providing readily usable tour-leader information kits, and brief destination videos that may be shown by coaches en route to destinations. Managers may also provide formal training for tour leaders.

Snapshot 16.14

Rainforests for the people
Harry Creamer, NSW P&W

The campaign to protect NSW rainforests began in 1969 with calls for a national park in the Border Ranges. Over subsequent years several inquiries failed to find a solution to the conflict between rainforest logging and conservation. In 1982 the Wran Labor Government took the historic decision to protect 90,000 hectares of rainforest areas in northern NSW. The decision also involved a restructuring package for the forestry industry. In 1984, another 30,000 hectares of rainforest were added. Many areas were partially logged and included access roads, previously the exclusive domain of logging trucks.

From 1983 the NSW NPWS was involved in managing these new rainforest parks and reserves. For Regional Manager Geoff Martin and his northern region staff it became a priority to provide opportunities for the public to see and enjoy these jewels in nature's crown. Geoff's instinctive leadership pursued a need to quickly provide opportunities for access and appreciation of the rainforest for visitors. It was an alternative economic use of the forests and a means of quickly consolidating a courageous political decision to conserve the forests. The 10-kilometre Washpool Walk, a magnificent rainforest experience, is an example of how creative they were at this time. Old logging roads became new tourism drives. Log dumps became picnic and camping areas, and new walking tracks and innovative bush design furniture cultivated a new clientele. Visitors started to come from traditional coastal tourist towns and a new use of the forests evolved. High quality interpretation and education signs and brochures helped to pioneer these new hinterland tourism destinations of the north coast.

In 1986 seven major rainforest areas of NSW were included on the World Heritage List. The majority were in Geoff Martin's northern region. During the next decade, the Hawke Government's National Rainforest Conservation Program provided funding to make Northern Region rainforests even more accessible, and to increase public understanding, enjoyment, and appreciation of rainforests. Thanks to the pioneering work of Geoff Martin and his staff, these World Heritage rainforests are conserved, and one of the great tourism experiences of northern NSW and Australia. They are of great economic importance to local communities. Regrettably, Geoff passed away soon after his retirement from his national park work. His personal contribution to nature conservation and the establishment and management of protected areas in NSW was officially commemorated on 21 February 2004 when the Premier of NSW, Bob Carr, officially opened the *Geoff Martin Training Centre* at Woody Head, Bundjalung National Park.

Geoff Martin, Chief Ranger, Kosciuszko State Park, NSW (NSW NPWS)

- Destinations should be well and thoughtfully designed or sensitively adapted for the recreation intended. Different sites for a given recreation should offer a range of facilities, from minimal to sophisticated; and visitors should get reliable information as to which they can expect at a given site.
- Design of facilities should be in keeping with the characteristics of the recreation settings in which they are located. Very different types of facilities are appropriate in semi-remote areas, compared with developed settings (Table 16.6). Variation in the level of visitor facilities (from no infrastructure to urban-style facilities) should be planned based on the desired mix of recreation settings.
- For destinations with no facilities, considerable effort will need to be made to ensure that there

Table 16.6	Visitor expectations of services relative to the nature of visitor destination settings based on the ROS (NSW NPWS 1997b)				
ROS category (see Table 16.4)	Class 1	Class 2	Class 3	Class 4	Class 5
Ranger patrol and monitoring	✶	✶	✶	✶	✶
Natural setting, no modifications	✶				
Large expanses of natural scenery	✶				
Natural settings, very basic modifications		✶	✶		
Natural settings, basic modifications with basic road access			✶	✶	
Fire management trails		✶	✶	✶	✶
Walking tracks, basically maintained		✶	✶	✶	✶
Walking tracks developed to a higher standard				✶	✶
Pit toilet facilities		✶	✶	✶	✶
Composting toilets		✶	✶	✶	✶
Septic toilets				✶	✶
Showers				✶	✶
BBQ sites		✶	✶	✶	✶
Parking areas			✶	✶	✶
Formalised camping facilities (basic)		✶	✶	✶	
Formalised camping facilities including furniture				✶	✶
Visitor information signs (basic)		✶	✶	✶	✶
Visitor information signs, display panels			✶	✶	✶
Information maps	✶	✶	✶	✶	✶
Brochures	✶	✶	✶	✶	✶
Booking systems (walking, some areas)	✶	✶	✶	✶	✶
Booking systems (camping, some areas)	✶	✶	✶	✶	✶
Clean facilities		✶	✶	✶	✶
Sites or areas managed sustainably	✶	✶	✶	✶	✶
Facilities and services maintained in a safe, hygienic condition		✶	✶	✶	✶
Vandalism repaired rapidly and basic maintenance achieved for facilities and access		✶	✶	✶	✶

Ranger-guided walk, Border Ranges National Park, NSW
(Graeme Worboys)

ism. Systems need to be in place to ensure that
standards are acceptable.

- Educational information for many key destinations should be available, including teachers'
guides, on-site interpretation signs, and other
support information.

Minimising risk

Risk management is a critical part of managing
for visitors.

Basic first aid. First aid facilities, including staff
trained in first aid, need to be available at destinations. Emergency medical contact procedures
would be pre-determined (Chapter 15).

Communication systems. Staff need to be able to maintain contact with their operational centres for safety
reasons. Communication systems need to be in place
either as permanent fixtures or as temporary, portable
facilities that support staff in more remote areas.

Emergency plans. Risk to staff and visitors can be
minimised through good planning for such potential major incidents as a train accident in a tunnel
within a national park, motor vehicle accidents, or
sickness of people in remote locations. Pre-planning

are no human-generated features such as
garbage. Special arrangements may need to be
made to deal with waste.

- Information provided to visitors prior to their
arrival at a destination should ensure that their
expectations of the facilities and services available match those present. Visitors arriving at a
site managed for its natural (no infrastructure)
condition should not be expecting hot showers.

- This variety of sites and facilities should mean
that licensed commercial tour operators can
select and offer a range of recreation settings.
Visitors who prefer more natural settings will
have that choice. This type of experience may
be increasingly desired, and is already an important niche market.

- Destinations and their facilities should be clean,
tidy, safe, well maintained, and free of vandal-

Parks and Wildlife Rangers Chas Delacoeur (left) and Greg Boehme
assist an injured walker (PWCNT Collection, David Hancock)

Visitors on elevated walkway in the high country of Kosciuszko National Park, NSW (Graeme Worboys)

and training to deal with such incidents helps to minimise risks after events have happened.

Structures and facilities. Infrastructure in protected areas must be safe. Structures and facilities are normally designed and constructed to Australian design standards and independently certified for their safety. Normally there is a time period for which the certificate of compliance relates, and recertification will be required. The tragedy at Cave Creek, Paparoa National Park in New Zealand, is a message to all managers. In April 1995, a cantilevered viewing platform over a limestone hole at Cave Creek collapsed, killing 14 people and injuring four. A subsequent enquiry found that design non-compliance and lack of management systems (as well as other causes) contributed to the collapse (De Bres 1999).

Licensed commercial tour operations. Tour operators run a wide range of visitor services within pro-

tected areas. Their staff need to be trained and accredited to operate a range of services, including the operation of boats, motor vehicles, aircraft, or equipment such as snow skis and scuba-diving gear. They are responsible for the safety of visitors undertaking activities and need systems to manage people. Protected area agencies should ensure that these arrangements are satisfactorily in place.

Insurance. Insurance arrangements need to be in place to deal with injuries or damage arising from visitor use.

Case Study 16.3 describes the asset management system used in New Zealand by the Department of Conservation. As well as addressing provision of quality visitor services and risk management, the system also draws together other aspects of visitor management considered in this chapter: data requirements, market segmentation, ROS, and recreation planning.

Case Study 16.3

National visitor asset management

Mike Edgington, Department of Conservation, New Zealand

In times of increasing competition for scarce government resources, it is absolutely essential that we are able to demonstrably increase the value of conservation assets. The term 'conservation assets' is used here to describe all natural, historic, and cultural values or things on conservation lands. For recreation management there is an obvious suite of assets and values such as facilities and the settings in which they are managed.

In 1987 when the New Zealand Department of Conservation (DoC) was formed there was no inventory of the inheritance. When the inventory was eventually completed the massive asset network included 23,000 structures and buildings, 12,500 kilometres of track, 2200 kilometres of road, 1100 huts, 260 campsites, 50 visitor centres, 15,000 signs, 1500 toilets, and hundreds of picnic areas. In 1987 there was a relatively modest budget to cope with this network and the conflicting and increasing demand for recreation opportunities. At the time there were no standards for facilities. Many had been built for conservation work, for example wild animal control, not recreation. The location and standard of many facilities were not ideal, many were old, maintenance had been deferred, and they required considerable upgrading. DoC attempted to maintain the status quo; however, it was obvious that its resources could not support such a large ageing asset base requiring more input each year to keep it operational. The critical issue was to reduce the maintenance commitment while still providing the widest range of recreation opportunities with the resources available; to implement a sustainable asset management program.

How sustainable is the asset base? The life cycle modelling indicated an annual budget of $80–100 million was required for assets. Analysis of the remaining life of structures and huts showed a need for a higher replacement rate in the short to medium term (next 20 years) and the total capital investment in visitor facilities was $430m. The Department's budget at the time was $30m. With a deteriorating asset base and increasing costs of maintenance and replacement, it was obvious that the system was not sustainable.

Using this information DoC was able to demonstrate the full cost of ownership and prepared a budget case for government to consider. Government agreed to a threefold increase in the Department's recreation funding over a 10-year period (2003–13).

Who wants facilities and at what standard? To identify public demand DoC has consulted with the public, developed asset service standards, and implemented an ongoing process of consultation about the delivery of recreation facilities and opportunities. DoC uses the Recreation Opportunity Spectrum (ROS) to classify the recreation settings and has identified seven visitor (customer) groups, each having different needs in terms of settings, facility standards, and opportunities. The distinguishing characteristics of each group are based on duration of visit, accessibility of sites, activities undertaken, the experience and degree of risk sought, and the facilities and services sought. While each visitor can belong to different groups at different times, at any one time each visitor will be in one of these groups: Short Stop Travellers, Day Visitors, Overnighters, Backcountry Comfort Seekers, Backcountry Adventurers, Remoteness Seekers, and Thrill Seekers.

The distribution of opportunities has been mapped by ROS class and visitor group use. The predominant visitor group at each site is used to specify the level of service provided. Each site has clear and logical boundaries and has been assessed according to its natural/historic values and the use it receives. The assessment criteria are: current visitor numbers, expected future use, the recreational importance of the site and the potential to increase visitors' appreciation of natural and cultural heritage. The higher the value for each criterion, the higher the priority for maintenance and upgrading.

Analysis of use indicates highest demand occurs in the day visitor/front country settings and that there is moderate to low demand for backcountry opportunities. Whether the mix of sites and opportunities meets demand is not well understood and is the subject of research and monitoring. DoC has implemented a national monitoring system and started an ongoing dialogue with the community to determine national, regional, and local demand for recreation opportunities.

Support includes adequate staff appropriately trained in asset management, sufficient financial resources to meet recurring costs, documentation of standards and operating systems, a monitoring process that measures performance against customer needs, business outputs, and strategic outcomes. A key component of DoC's support structure is its information system that links the vision, policies, procedures, standards, and the processes of life-cycle modelling, planning, financial allocations, and performance monitoring.

When DoC's visitor asset management system (VAMS) is assessed against World Commission on Protected Areas framework for evaluating management effectiveness (Chapter 21) it can be concluded that:

- context—VAMS provides an assessment of current standards of recreation opportunities and facilities through baseline inspection processes
- planning—VAMS compares the current standard of facilities against service/legal standards and describes the supply of recreation opportunities
- inputs—VAMS provides a detailed analysis of resources required to maintain facilities at the appropriate standards within an asset life-cycle framework
- processes—VAMS measures through an ongoing (annual) inspection process the standard of facilities against that required for the primary users of the facility/asset
- outputs—VAMS measures the number of facilities/assets provided against the annual business plan predictions.

VAMS has improved efficiency and reduced risk in managing facilities at all levels of the organisation. It has been invaluable in the analysis of needs and risks and has enabled DoC to successfully bid to central government for additional funding.

16.5 Management lessons and principles

Lessons

1 Tourism, visitor use, and recreation in protected areas need to be actively managed. It is a full-time, professional, and scientific-based task, the importance of which must not be underestimated.

2 Of the three licensed tourism operator paradigms, the eco-tourism operator paradigm offers the greatest long-term potential for effective tourism-protected area management cooperative partnerships achieving conservation outcomes for protected areas.

3 The real cost of maintaining all of the tourism and recreation values of protected areas needs to be understood by the public. Constant briefing of the public about achievements is essential.

4 Revenue raised from tourism needs to be reinvested in the protected area system, and the public need to be advised of the investments.

5 Tourists, visitors, and recreationists need to be provided with quality and current information. Modern web-based information systems provide outstanding opportunities to achieve this.

Principles

1　There continues to be a growth in visitation to protected areas, whose sustainability depends on the careful management of visitors.

2　The encouragement and management of appropriate use involves providing a range of opportunities for visitors to interact with the natural and cultural features of protected areas, whenever this is compatible with the goal of conserving natural and cultural heritage values. Effective management demands accurate measurement of visitor impacts.

3　Tourism is the world's largest civil industry, and in Australia the largest tourism operators are protected area agencies. Visitors are welcome; however, the provision of appropriate infrastructure and services should be made in accordance with ecologically sustainable visitor-use principles.

4　Planning tools such as the ROS are essential to the effective management of visitor opportunities. Protected area managers, in cooperation with other land managers, should ensure that within a region, a spectrum of recreation settings is available.

5　The setting of planning limits for visitor destinations is an essential tool in sustainable visitor-use management.

6　Delivering high quality service to visitors is an important component of a protected area manager's job. Management should ensure that visitors, as far as possible, have rewarding and enjoyable experiences.

7　Every natural environment has its own special characteristics that must be taken into account when siting and designing visitor facilities and services.

Boardwalk, Wet Tropics World Heritage Area, Qld (Graeme Worboys)

Department of Conservation information sign, Mount Cook National Park World Heritage Area, NZ (Graeme Worboys)

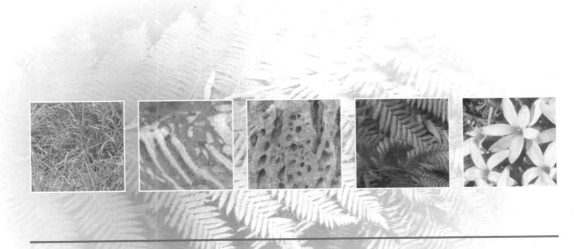

Working with the Community

> *Explicitly or implicitly, every debate on policy is a debate about values. In a democracy, the debate is continuous. It involves everybody.*
>
> From **Discovering Monaro** *by W.K. Hancock*

17.1	A critical working relationship	466
17.2	Making the relationship work	469
17.3	Communicating with stakeholders	474
17.4	Interpretation: Communicating with heart and mind	484
17.5	Management lessons and principles	491

Working with all types of people in the community is one of the most challenging and rewarding aspects of protected area management. There is nothing like shopping at your local supermarket after a hard week's work, when, while purchasing the breakfast cereal, you meet someone with an interest in the park you manage. It may be a neighbouring farmer, a local councillor, or a tourism operator. You are out of uniform and clearly on your day off. Yet the conversation will focus directly on what is happening within the park. Staff working for protected area agencies live and work within a local community and the locals expect you to be their professional contact at all times. Work does not finish when you leave the ranger station and take the uniform off.

On almost any issue you will have to allow for the range of attitudes (and needs) that different people have. This takes skill and common sense.

It also takes patience, strength of character, good humour and, in trying times, the support of colleagues.

This chapter offers both practice and theory on understanding and communicating with stakeholders, as individuals and as groups. Readers will gain insights into:

- how the public may view protected area managers
- why agencies work with the community
- planning for working with the community
- social science as a tool for understanding and interacting with the public—how social science can help
- some theoretical aspects of communication
- some practical mechanisms for working with the community
- the use of 'interpretation' as a management tool.

17.1   A critical working relationship

The community and protected area agencies

The community may perceive modern protected area agencies in many ways. These ways may correspond to the various roles the agencies play in the community and so may change over time. Each role requires a diverse set of skills (Table 17.1); however each of these roles needs to be clearly understood by the community, so that there is no confusion as to what agencies are 'up to'. Communicating these different roles to the community is therefore a critical part of management.

In Chapters 2 and 3 we looked at the social and political context and saw that if the community is unhappy with the concept of protected areas, the concept will die. Likewise, visitation to the protected area will be minimal unless members of the community feel an attachment to the site and share this with friends and relatives. Thus there is real reason for protected area managers to work with the community. But there is more to it than this. The Millennium Development Goals (Chapter 2) highlight the importance of addressing social issues in order to achieve sustainability. The goals include eradicating poverty and hunger and improving access to health services. These goals are now a

major focus of most international programs and protected area organisations have a role in implementing these goals. At the World Parks Congress, in Durban, South Africa 2003, there was a focus on social issues and encouraging community participation in protected area management. Some of the topics included recognition and integration of Indigenous conservation practices and the concept of community conserved areas.

As the population continues to grow, involving the community in protected area management, and the creation of protected areas, becomes increasingly important. The demands of an ever-increasing population for infrastructure and services place pressure on natural and cultural spaces. Protected area managers and community groups need to work together if the values of such spaces are to be maintained.

The community itself is interested in what is happening in protected areas. They want to be involved. In fact, if they perceive that protected areas are being managed in a way that clashes with their understanding or interests, they will be quite vocal. Hence, protected area agencies need to keep the community up to date with contemporary park management principles, as well as with day-to-day happenings at the 'local reserve'.

For protected area managers to work in isolation from the community is neither practical, desirable, nor usual. Apart from legal processes that prescribe formal consultation procedures, managers are interacting with the community every day on what are regarded as routine matters.

Friends of the Great Victoria Desert Parks, SA, during a desert clean-up in the unnamed conservation park (Dene Cordes)

Table 17.1	**Examples of the many 'public faces' of protected area organisations** (with related actions and public interaction skills)	
'Public faces' of protected area management agencies	Related actions	Public contact skills
Partner and community member	Feral animal control; helping each other in emergencies	Being a good and friendly communicator
Conservation innovator and project developer	Harmonising park conservation with off-park conservation efforts; operating at a landscape level	Negotiation skills to create support for conservation strategies
Business manager	Marketing services; promotional image; and working with local tourism organisations	Appropriate targeting to reach audience; media management
Educator	Environmental and cultural education activities	Effective techniques and skills to convey messages
Operator	Operational activities (such as prescribed burning)	Informing community about activities and risks; consulting them on proposals and involving them on site
Government department (for statutory management bodies)	Legislative responsibilities and authority	Law-enforcement; consultation for management plans

Snapshot **17.1**

Tarra–Bulga National Park: a little help from some friends
Craig Campbell and Angie Gutowski, Parks Victoria

'Why is a $350,000 Visitors Centre never open?' A local resident and café owner asked the question of the Victorian Premier on a popular radio talkback program. In 1990 a visitor centre had been opened in Victoria's Tarra–Bulga National Park to interpret and promote the natural values of the park. For the first few years, two rangers were rostered on duty over weekends; one was to run the centre. By early 1993, cutbacks to rostered time meant that the centre was mostly closed on weekends. Such were the circumstances that prompted the Friends of Tarra–Bulga National Park to form in December 1994 to staff the visitor centre.

The group is a mix of retirees, working people, locals, and those from the surrounding district. They staff the centre for six hours each day, approximately 150 days per year. In a typical day a volunteer will operate the slide program, answer the telephone, help set up children with activities, accompany adults who are looking at the displays, sell items such as posters over the counter, keep a tally of numbers of visitors to the centre, and

answer any number of questions from visitors. In the short time that the 'Friends' have been operating, they have achieved much and have greatly helped the management of the park.

School students on a suspension bridge at Tarra-Bulga National Park, Vic. (Craig Campbell)

Cooperative fire fighting, tourism management, pest animal control, weed spraying, road maintenance, school programs and joint interpretive displays are some examples. As well, staff may have amicable dealings with a whole range of user groups, volunteer groups, stakeholders, and business people in their normal management duties. Snapshot 17.1 provides an example of local cooperation.

There is mutual advantage, as well as genuine good-neighbourly behaviour, in most of these actions. There are also many issues that need to be resolved. These may be considered as routine and

Snapshot **17.2**

Parks Victoria and tourism following the 2003 Australian Alps fires
Dianne Smith, Parks Victoria

The January 2003 fires in Victoria's north-east and Gippsland regions affected 64 public land licensed tour operators and restricted access to the parks. Parks Victoria worked closely with the licensed tour operators, the Victorian Tourism Operators Association (VTOA), Tourism Victoria and other government agencies, to provide accurate and timely information and a coordinated recovery strategy. Several workshops were held and VTOA were commissioned to present a report summarising the effect of the fires on the associated licensed tour operators.

To assist licensed tour operators, local tourism, and visitors, the Parks Victoria web site maintained a detailed section dedicated to the latest available information regarding access to fire-affected areas. Parks Victoria promoted reasons to return to the fire-affected areas, via a newspaper lift-out and flyers distributed at Visitor Information Centres, as well as other locations.

Parks Victoria allocated $50,000 to involve licensed tour operators in the fire impact assessment process at specific visitor sites. Five licensed operators were contracted to assist with the assessment process with a minimum of 10 days work guaranteed.

include dealing with illegal activities such as arson, trail bike riding, timber gathering, and so on. In the longer term, managers will usually be respected if they take a tough line but are careful to be fair. Court action over matters such as arson (the illegal burning off of park land) is usually needed only once in a given area. Many locals may actually applaud such a firm stand.

Some neighbours, for whatever reason, may be unwilling to cooperate on projects such as weed and pest control and this just has to be accepted. Others may attack the agency over fire policies or over lack of action in burning off, or in maintaining trails. Some of these criticisms may be fair, but if managers respond sincerely and follow through with action, the criticism will usually diminish with time. For many local communities, the 'new chum' park agencies need to demonstrate that they are bona fide before they are trusted. With sensitive handling, even such contentious issues as pollution from neighbours' pesticides and fertiliser can be resolved.

The community expects protected area agencies to manage well; they expect staff to be competent and trustworthy. They also expect to be kept informed, especially of major projects and sometimes to be involved or consulted (Snapshot 17.2). An agency that fulfils these expectations can expect a positive image in the community over time. This in turn creates a positive climate for action, since the agency's policies and programs are then much more likely to be accepted by local people, government, and business. The Nowra District of NSW P&W 'Good Neighbour Policy' is a useful example of this type of approach. The principles articulated in the policy include the following.

1   The right of neighbours to the quiet enjoyment of their land is recognised and respected.
2   The responsible management and stewardship of protected area lands is recognised.
3   Generally accepted standards of good neighbourly behaviour will be undertaken.
4   The practical resolution of management matters at a local level is a priority (NSW NPWS 1993).

17.2 Making the relationship work

Planning for working with the community

All interactions with the public, from park visitors to government ministers, must be planned and targeted to achieve positive outcomes. Through planning, programs can become proactive rather than defensive. As David Lloyd of the Great Barrier Reef Marine Park Authority puts it, 'We make frequent deposits of information and trust in the good times to ensure understanding and trust when times are tense'. Since communicating with the public is an integral part of protected area management it should be a routine part of the process of developing the annual operational plan.

The term 'public contact' covers a wide range of communication activities. It includes public relations, education, interpretation, providing information, and extension services (services providing technical advice to the community) (Hockings et al. 1995). Each of these may require somewhat different approaches, resources, and activities, but it is useful to plan them together, as part of an overall plan for public contact.

Planning contact with the public involves defining a desired future and then selecting strategies and actions for moving from 'here' to 'there'. Chapter 7 should be referred to for planning principles. However, managers need to consider some points that are specific to public contact planning.

All planning needs a goal. Sometimes this may be as concrete as in other areas of planning: to build a track with volunteers, or to revegetate a site. But with public contact plans, the goal may be personal or conceptual rather than physical. It is the hearts and minds of people that need to be focused on. Thus planners must specify their goals precisely. Just what do we want to communicate to the public? Usually it is something based on the organisation's corporate goals and objectives, the features of the park, or a particular issue.

To plan public contact we need to know with whom we will be working. Who are these people

Volunteer Venturer scouts repainting heritage stairs, Montague Island Nature Reserve, NSW (Graeme Worboys)

Methods for gathering this sort of information are given below. Equipped with knowledge of 'who and what', the planning process should then consider the 'how'. Managers and staff need to be prepared for

Snapshot 17.3

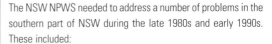

Communication plan—Southern Region of NSW NPWS

The NSW NPWS needed to address a number of problems in the southern part of NSW during the late 1980s and early 1990s. These included:

1   poor information services for visitors
2   limited responses to neighbours, stakeholders and the local community about what was happening within protected areas
3   inability to deal with a range of negative media articles
4   few 'good news' stories.

The resulting community relations/communications plan (reviewed and updated annually) produced the following actions:

- a five-year investment in high quality visitor-information displays at important destinations throughout the region
- a complete upgrading of all visitor information brochures
- a complete upgrading of the agency's high quality protected area destination marketing posters
- employment of a professional media officer
- training for all senior staff and media contact staff in media interview techniques
- developing a database of neighbours
- producing a newsletter for neighbours
- systematic release of positive press articles
- targeting areas of 'anti-park opinion' with regular talkback radio features about 'what is happening'
- developing cooperative visitor information projects (such as visitor touring maps) with local tourism organisations
- providing immediate responses to negative press stories, including deploying spokespersons to major events that have the chance of becoming negative
- proactive engagement of the media in positive conservation stories
- familiarisation tours (by bus) for prominent locals and the media
- open days, openings, and other media launches.

The outcome was very positive for the NSW NPWS and the community.

and groups; what are their needs, concerns, attitudes, socio-economic status, and ways of communication? There are five important questions to answer.

1   Who are the stakeholder groups and what is there about the ways they perceive and behave that may affect the protected area?
2   What community or environmental issues and attitudes may affect the relationship?
3   How are decisions made and power shared in the community?
4   Which communication channels and media can best reach all potential stakeholder groups?
5   What impacts will our management plans have on the local and wider community?

the activities they must undertake when working with the community. Staff need to be briefed about the relevant people and groups and equipped with the skills for the job (Snapshot 17.3).

As with any type of planning, the outcomes of a public contact plan can be used to gauge its success. This may involve collecting data that reveals social changes. For example, whether people's attitudes towards a park have changed may be measured by questionnaires and by other social science methods.

Understanding the community

To solve conservation problems we must first ask, 'What is their cause?' The immediate answer may seem obvious. Yet the underlying causes, which may be related to attitudes, to socio-economic welfare, and so on, will need to be dug out before we can form a strategy for action. To work with people on solutions, we need to know how people in a given community interact and communicate. The 'nil-tenure' approach is one possible method for getting different stakeholders to work together (Case Study 17.1).

Some of these considerations are relatively new to conservation agencies, which have traditionally focused on biogeographical studies. Social sciences on the other hand have amassed a large body of knowledge since the 1960s, which is very relevant (Machlis 1992). Areas of interest include: ethnography and community assessment, psychology of attitudes and behaviour, communication and education, social impact assessment, organisational sociology, media studies, and conflict resolution. Some social science methods that can be applied to protected area management are introduced here.

Identifying stakeholders. Managers often consult the public when making decisions with a wide impact. To ensure that the process is efficient and fair, they must carefully consider who should be involved: is it some local matter relevant to park neighbours who share a risk from fire, or is it a conservation matter of national concern? The people who are to be consulted are often referred to as 'stakeholders'. There is no rigid definition, but 'the term is typically used to define persons or groups who have an interest in or who could be affected by an issue or a situation. It should also include persons or groups

Case Study 17.1

Management of wild dogs and foxes: a nil tenure approach to a landscape issue

Rob Hunt, NSW P&W

The management of wild dogs (including dingoes, feral domestic dogs, and their hybrids) has traditionally been an extremely emotive and, at times, political issue. The relationship between managers of private and public lands has been cautious at best. While public land managers have acknowledged the financial impact of wild dog attacks they have rarely acknowledged the emotional distress endured by landholders or indeed agency field staff.

A meeting of the NSW NPWS South West Slopes Regional Advisory Committee and local landholders in November 2000 identified the need for a local solution to the problem of wild dog predation of livestock. Landholders supported the concept of forming a committee (working group) comprised of local land managers (both public and private), the Yass Rural Lands Protection Board, and the local wild dog control specialist.

The approach adopted by this committee was that of 'nil tenure'—that is, the process does not differentiate action according to the tenure of the land concerned, whether it be park or private. This process, now adopted by the Rural Lands Protection Board in NSW for the management of all pest species, involves the following stages:

- initial consultation (identify level of support for change)
- establishing a local working group
- identifying any hidden issues
- mapping and defining an operational area, making sure that it is not too large relative to available resources
- confirming membership of working group, ensuring all land managers are represented via the working group
- mapping the issue (nil tenure)
- evaluating and costing current best practice control and monitoring methods
- mapping solutions and identifying priority areas (nil tenure)
- identifying land manager costs (using a tenure map overlay)
- securing resources and signing off on commitments to a cooperative plan
- reviewing the program by the working group after a trial period.

The implementation of the Brindabella and Wee Jasper Wild Dog/Fox Control Plan has identified the value of a nil tenure approach when addressing landscape issues such as wild dog management. A significant reduction in stock losses, along with a notable improvement in working relations between public and private land managers, has ensured the commitment of resources. The implementation of a robust monitoring system, which not only monitors stock losses but also the response of native animals and pest species to wild dog and fox control over time, will enable the program to be continually evaluated by the working group.

The nil tenure approach highlights the benefits of focusing on the 'common problem' rather than criticising the efforts of adjoining land managers. The establishment of the working group, and the ability of this group to effectively address the wild dog issue, has amended over 20 years of negative relationships between private and public land managers in the area. More importantly, it has had a positive impact on the well-being of farmers in the area, who now feel that something positive is being done to address the constant financial and emotional impact of wild dogs. Through this consultative process local land managers have taken ownership of the issue and have identified and pursued the resources required to successfully implement a local solution.

who perceive themselves as affected' (Forrest & Mays 1997, p. 32). Examples of stakeholders are:

- neighbours who may be directly impacted by an issue (fire), or are directly responsible for one (intrusion of pets)
- park visitors who may be directly impacted by an issue (restricted access), or are directly responsible for one (track erosion)
- employees of an industry whose practices may be altered or restricted (mining, grazing, agriculture, logging); that is, people whose sources of income are at stake
- representatives of private companies whose practices may be altered or restricted (tour leaders, extractive industries)
- the wider national (and even international) community that is concerned with conservation, or alternatively with economic development
- various community groups with specific concerns (such as a frog conservation society, bird watchers, the Country Women's Association)
- government agencies with varying agendas and responsibilities, for example the department that controls land use around the protected area
- local businesses
- non-government organisations, particularly those with a conservation focus
- international conservation organisations (World Heritage Committee, IUCN, WWF)
- any person or group who expresses an interest.

Stakeholders may approach the agency voluntarily, but managers should attempt as comprehensive an outreach program as possible. This may include contacting major groups, placing notices in the mass media or on-line, mail-outs to local people and visiting neighbours and local institutions. Sometimes a core set of stakeholders may already exist in the form of a regional joint committee that has been formed on a previous occasion.

Social and cultural impact assessment. Social and cultural impact assessment (SIA) is a means of assessing and predicting the probable effects of a development project or a policy change on individuals and communities. Issues include:

- physical and psychological health
- material well-being and welfare
- traditions, lifestyles, and institutions
- interpersonal relationships (Furze et al. 1996).

Surveys and questionnaires. Agencies often use surveys and questionnaires to understand visitor activity, attitudes, preferences, and characteristics (see also Chapter 16). Methods are derived from the long history of survey research in the social sciences, as well as from modern market research techniques. A brief overview of designing a survey is given here, but further reading is recommended. The basic steps in developing a survey are as follows (Furze et al. 1996).

1 *Identify* the information required.
2 *Sampling.* Specify the number of people to be contacted (sample size) and the process for selecting them (for example, random selection from a population stratified into groups).
3 *Choose the type of survey.* Surveys can be conducted on a group or on a one-to-one basis. They can be done in-person, by mail, or over the phone. They can involve in-depth interviews, or responses to multiple-choice questions. They can be administered by the researcher or self-administered by the respondent.
4 *Design the survey instrument.* The instrument itself can be in the form of a booklet for a mail survey, or a script for a telephone or personal interview. Visual material such as maps and photographs may be required. Questions can be open-ended and qualitative ('How … ?'), or closed-ended and quantitative ('How many … ?'). Focus groups of stakeholders are often used to help refine the survey instrument and to make sure respondents will answer the questions in the terms intended.
5 *Pre-test.* Pre-tests are trial runs of the survey, using a small group of respondents. Even after extensive use of focus groups, surveys can contain errors or problems that can be sorted out at this stage. It is important to make sure that the 'delivery' and administration procedures will be efficient and effective.
6 *Data analysis and interpretation.* Analysis and interpretation of quantitative data typically requires the use of statistical techniques. Statistical packages also exist for the analysis of qualitative responses, and these usually involve identifying key phrases, or detecting underlying meanings and themes.

An example of a community attitude survey conducted on behalf of the NSW NPWS (Frank Small and Associates 1993) illustrates the type of feedback that a survey can provide for an agency (Snapshot 17.4).

Rapid rural appraisal (Furze et al. 1996). Rapid rural appraisal is a technique used to gain a quick yet effective understanding of the social, economic, and political processes in any community. Typically a team of two or three researchers spend a week in a community, gathering a range of different data, including:

- *secondary data review*—a review of published and unpublished data
- *direct observation*—personal visits and observations aided by an observational checklist
- *key indicators*—shortcuts to insights about community social conditions and change
- *semi-structured interviews*—interviews that follow a set of points, which permit probing and allow the interviewer to follow up on unexpected responses, without the requirement that all the checklist points must be covered in any one interview
- *key informants*—identifying those best able to give information on particular topics or to give special points of view.

Snapshot 17.4

Community surveys

During 1993 Frank Small and Associates conducted a number of surveys for the NSW NPWS. This provided benchmarking information that helped the NPWS in working with the community. Here are some of its findings.

- The public endorsed the work of the NPWS and recognised its contribution to protecting and conserving plants, animals, and habitats.
- The community expected the NPWS to be an environmental watchdog.
- There was little understanding of its role in managing cultural heritage.
- The public considered that the NPWS manages its protection and conservation responsibilities well, but its environmental watchdog role less well.
- Almost all residents (95%) supported and valued the NPWS.
- Staff of the NPWS are considered one of its greatest strengths.
- There was a view that the NPWS was under-resourced.
- There was high endorsement (81%) for greater environmental education for the public.
- There was support for guided tours.
- There was general support for user-pays.

A stakeholder survey and summer and winter visitor surveys were also completed. They provided further valuable background information for plans and actions.

Rapid rural appraisal is becoming much more widespread with the increasing demand for protected area managers to work with their local community.

Community assessment. Community assessment can be a valuable tool for improving the focus and effectiveness of interaction and communication with the local community. Such an assessment looks at several issues, as described below (Forrest & Mays 1997).

- *Community history.* What historic events and experiences may affect their perceptions of the protected area and of conservation?
- *Social and political climate and the dynamics of decision-making.* Begin by identifying formal and informal community leaders. These are the people who act as spokespersons, or to whom others look for advice. Consider the relationships among community members and groups—are they all included in the political process and how do they interact? Do not neglect more formal channels of communication, such as committees and other organisational networks.
- *Channels of communication and preferred forms of interaction.* This involves examining and ranking the available media or channels of communication: local newspapers, radio, leaflet drops, public meetings, and so on.

17.3 Communicating with stakeholders

A basic explanation of communication is given in Backgrounder 17.1. Good communication skills are essential for working with the community.

There are various ways that staff can communicate with their various stakeholders and we will describe some of these in more detail (Table 17.2). In Case Study 17.2 Stuart Cohen describes how the NSW NPWS communicated information about the 2003 Australian Alps bushfires to stakeholders.

Being fluent in a language is not enough to make one a good communicator. This requires a range of further skills, such as effective listening, critical thinking, and persuasive presentation (Scott 1998). In their professional and private lives, managers will use these skills to communicate with staff and stakeholders as well as to family and friends. Selecting the right channel of communication is often the key to having one's message understood. For different purposes conservation staff will use different channels: brochures, signs, displays, mass media and so on. Snapshot 17.5 shows the importance of using channels and media that suit the audience.

Backgrounder 17.1 — What is communication?

Communication involves a communicator who chooses and sends a message to a receiver. The receiver interprets that message. Feedback from the receiver lets the communicator know if the message was correctly understood (Queensland Department of Environment and Heritage 1996). The chances of success depend on such things as the technique used to transmit the message and the 'noise' produced by differences in cultural backgrounds, personal style, and attitudes. For example, if you are feeling shy another person may mistake this for unfriendliness. There are three main forms of human communication: verbal, non-verbal, and visual.

Verbal communication refers to words that are written or spoken. Technically speaking, it is an exchange of linguistic signs (via conversations, speeches, letters, memos, and so on).

Non-verbal communication often accompanies and adds to verbal communication. As actors know, the same statement may be made with quite different tones or emotions. It provides cues to the workings of the subconscious and to the attitudes and value systems that underlie verbal statements. Dress, facial expression, and posture are all non-verbal elements in spoken communication. Fonts, letter size, and page-layout are non-verbal elements of written communication.

Visual communication is an important part of literacy. It is the understanding of graphic signs. For example, a red circle with a red line through its middle over an outline of flames is used to warn us not to light fires. Understanding such signs is called visual literacy. Such graphics can be used as warnings, or to highlight certain points, sell commodities, or persuade (Scott 1998).

Table 17.2	Communication methods
Communication method	Notes
Face-to-face	This is always a useful method, whether one is resolving issues, having a good neighbourly yarn, or conducting a guided tour. Regrettably, lack of time often prevents this approach.
Formal lectures, talks, presentations	These can vary from talks for Rotary to conference presentations or information sessions at schools.
Newsletters, web pages, advertisements	These should provide regular information to stakeholders on 'what is happening'.
Using the media	The rural media welcome regular press releases on the routine work of the agency. Electronic media interviews and live talkback sessions are especially useful. The media is a crucial communications tool during incidents.
Openings, launches, and other events	The unveiling of a new 'product', or of a completed project, is a chance for a 'launch' to generate media interest.
Marketing information, background information	Television stations will use a suitable and well-made video about natural and cultural heritage destinations. Brief (15-second to 30-second) segments may be used as community service announcements.
Sausage sizzles and 'open days'	'Getting to know you' gatherings, such as a sausage sizzle with the local fire brigade or volunteer groups, are a marvellous mechanism, as are 'open days' at parks.

Case Study 17.2

Managing public information—Australian Alps bushfires, 2003

Stuart Cohen, NSW NPWS

Bushfires are always frightening events but under extreme conditions they can be utterly terrifying, even for veteran firefighters. Managing public information in this environment is about providing concise, accurate, and timely information that can help individuals and communities overcome their fears and allow them to make good decisions in the midst of chaos. Public information during an emergency should neither panic people nor make them complacent, but should seek to achieve a balance that creates vigilance and compels individuals to make preparations.

On 8 January 2003, under appalling weather conditions, lightning ignited 45 separate bushfires across the 690,000-hectare Kosciuszko National Park in southern NSW, as well as over 100 other fires across NSW and Victoria (Snapshot 15.2). Firefighting authorities were faced with an unprecedented demand for information from the multitude of small communities, townships, and landholders surrounding the park.

Typically, an individual or a small team of people manages public information. However, this was not a normal situation and so the multi-agency teams managing the fires established a Public Information Unit (PIU), which at its peak included 16 people. The primary objective of the PIU was to ensure that the local communities took the appropriate preparations and considered all contingencies in the event they came under attack from wildfire.

The challenges facing the PIU were many and varied. National media attention was not focused on the fires in Kosciuszko National Park, but on other major fires in the ACT and Victoria, and so the adjacent communities did not fully appreciate the gravity of the situation and hence did not fully appreciate the need to prepare against the threat of fire.

While the media remains a traditional tool for getting critical information out quickly it has obvious limitations. Stories are often rewritten, reinterpreted, and homogenised in order to streamline the product. Control over broadcast times and frequency is also limited. In response the PIU adopted other tools to ensure everyone in the community had access to detailed information.

The cornerstone of the PIU's information campaign was the 'Fire Facts Summary', a document that contained facts on the fires, their status, strategies employed, weather forecasts, road closures, emergency service updates, information about livestock management, health warnings, and any other details that would assist the local communities. A map, updated daily to provide a visual aid in understanding the progress of the fires, accompanied this. Importantly these were delivered by both fax and email to a constantly evolving list of stakeholders.

The summary and map, which was updated twice daily during the peak of the emergency, was posted on a number of web sites. Emails allowed information to be directed specifically at a very large but targeted audience. They contained detail that the media could not convey with the same frequency or to such a specific audience and they were easily forwarded on to others. This allowed up-to-date information to be distributed, virtually instantaneously, to a large number of people with specific interest in the emergency. Estimates put the number of people receiving the summary at somewhere between 10,000 and 40,000 people daily and it could have easily been more.

Each day a hard copy of the Fire Facts Summary and map were enlarged and put on display in 75 prominent locations throughout the region including post offices, general stores, and pubs. The PIU also delivered a total of 21,000 newsletters to outlying communities, held 18 public meetings for 3000 people, conducted regular briefings with key local stakeholders, and kept up a constant flow of community service announcements on local radio.

Another effective means of providing the public with information was via a heavily advertised, 24-hour public information telephone service. The operators worked from the fire facts summary and map and were updated as developments occurred. More than 20,000 callers used this service during the peak of the crisis.

PIU field liaison officers were sent to outlying areas under threat, while other unit staff telephoned several hundred landholders on properties adjacent to the park to ascertain each property owner's vulnerability and ability to defend their property. In many instances the PIU was able to allay immediate fears, offer advice, and assess the relative risk to different people.

In the aftermath of the Kosciuszko bushfires a detailed analysis of the PIU performance and outputs found that the community was overwhelmingly satisfied with the way in which public information had been provided during the fires. The Fire Facts Summary proved to be the most popular information tool while email, web sites, and the 24-hour public information line and radio were popular means of accessing the information.

The PIU ensured that the community was well informed, and prepared against the threat of fire. This was achieved through a multifaceted approach to information management that was able to make the most of Internet technology.

Snapshot 17.5

The right channels

David Lloyd, Great Barrier Reef Marine Park Authority (GBRMPA)

Direct liaison has been GBRMPA's secret in achieving good relations with the commercial fishing industry and with the farming and grazing industries. The two main secrets of success in this area are:

- Choose for your liaison officer a suitable person from that same industry, so that he or she will bring background knowledge and grassroots contacts.
- Speak to these sectors on their own terms, using their own language and familiar communication channels. For fishers this includes regular visits to the fishing wharves to talk (or yarn) and using video rather than printed information—the 'Deckhand' video series is highly successful.

Working with print and electronic media

The strongest advice we can offer is to receive professional training before dealing with the media. Adversarial interviews are common and staff need to be adequately prepared. The following tips have been adapted from Kennedy (1999) and Dawson & Cohen (1999).

The interview

- It is your interview.
- Pick the right location—preferably outside.
- Dress sensibly. Look professional. Create the image that suits the message and the audience. (For instance: Policy issue—collar and tie; fire issue— open-necked field-type shirt and safety helmet; rural issue and audience—open-necked shirt).
- Prepare 'A' points (points you will make at any possible opportunity during the interview).
- Prepare 'B' points (points you may need to respond to if the journalist raises them, but not otherwise).
- Prepare 'C' points (points that you need to be aware of, but will not respond to, or only very briefly).
- Do a dummy run.
- 'Add value' to everything you say.
- Turn all negatives into positives.
- Beware of making admissions that may be used out of context.
- Say 'no' nicely—sometimes it is not possible to answer a question.
- Never repeat a negative phrase used by the interviewer.
- Re-focus questions by choosing key words from the question.

Venturer scout Bernadette Dadday describing for television the heritage of Montague Island Nature Reserve, NSW (Graeme Worboys)

- Steer the interview your way—be concise—do not use jargon.
- Control the interview by returning to your 'A' points.
- Silence the interjecting interviewer.
- Summarise the 'A' points.

NSW NPWS communications team Stuart Cohen and Penny Spoelder (Graeme Worboys)

Fact sheets

These information sheets can be handed out at an interview, along with a press release. They make it easier for the interviewer. Commonly, for adversarial interviews, they address the key issues and provide statistics and other evidence supporting your agency's position. These sheets commonly take the sting out of what may otherwise be a threatening news story and can even change the direction of an interview. For interviews where the agency is taking the initiative, fact sheets are essential allies.

Media releases

These are usually brief. They should carry that day's date and be on official letterhead, with a catchy headline. The lead paragraph encapsulates the newsworthy issue. The media release should cover the basic questions of who, what, when, where, how, and why. The release should include quotes and factual information, be simple, and be interesting. It should list a person and a phone number to contact for further information. Note that journalists are trained to ignore media releases with the previous day's date and also that a release may be ignored if it arrives too late in the day.

Recognising items of interest to the media

Too often very interesting stories are not told because staff have seen them as un–newsworthy or 'more of the same' in the daily routine. Journalists might have a different view. They see news value in stories that are linked to:

- proximity—'the fire was 3 kilometres from town …'
- prominence—'the minister said today …'
- timeliness—'reports just to hand reveal …'

Snapshot 17.6

The Cooma necklace
Stuart Cohen, NSW P&W

In August 1991, a farmer located Aboriginal bones on the side of an eroding creek near Cooma in south-east NSW. A necklace consisting of 326 pierced kangaroo teeth mixed with large quantities of red ochre was buried with the skeletons. The necklace was dated at 7000 years—an extraordinary antiquity and the oldest site known for the tableland area near Cooma. With the close cooperation of the Merrimans people, the Aboriginal remains were reburied in a traditional manner, but at their request the necklace was not reburied. Instead it was retained in safe-keeping as permanent evidence of the richness of Aboriginal culture. The news was greeted by the media with appropriate amazement. By any journalist's standard it was a great story that would do well on page 1. And so it went … 'Amazing discovery', 'Necklace found in burial site', '7000 year-old burial found'.

In the weeks that followed we learnt how the story had appeared in newspapers around the world. What was clear to us, but not so clear to the media, was that the story was about cooperation and mutual respect between the Aboriginal people, whose ancestors were accidentally unearthed, and the NPWS, which had helped manage the process with sensitivity. At least, in the Australian media, the Aboriginal community were able to spread their message: that they are happy to cooperate with landholders who made these types of finds. While a great story of discovery was told, it was the process of getting there that we felt best about. Today, people may consciously remember the necklace and the burial, but they will unconsciously remember the Aboriginal Land Council chairman standing shoulder to shoulder with the NPWS archaeologist and the direct involvement of the Aboriginal community in managing their heritage. Changing people's attitudes is a subliminal process, not a frontal assault.

- impact—'over 70 whales have stranded …'
- extremes—the longest, the biggest, the shortest, the oldest (Snapshot 17.6)
- threats to wildlife (Snapshot 17.7)
- conflict—'the Premier today attacked the agency over …'
- oddity—'the male Antechinus dies after mating'.

Snapshot 17.7

Doing a good turn for Little Terns (*Sterna albifrons*)
Peter Reed, NSW P&W

Every year around November, Little Terns nest at Lake Wollumboola. This is a shallow saline lake south of Crookhaven Heads near Nowra on the NSW south coast. The lake is a significant habitat for thousands of water birds, including many species of migratory waders protected under international treaties. I have been looking after the Little Tern population at Lake Wollumboola ever since 1994 when Nowra District received a memo from head office. The memo requested that we attend a meeting to discuss a decline in breeding pairs of Little Terns in NSW. In the three years since the recovery team was formed, Lake Wollumboola has become one of the most intensively managed colonies in NSW and the number of chicks has increased steadily. The threats that used to worry us most have almost disappeared. Human disturbance to the colony was due to ignorance and disinterest. This has been greatly reduced through a very active public awareness campaign about the plight of this cute little bird. The campaign included radio interviews, a community service announcement on television, numerous media releases, interpretive signs, and a general spreading of information throughout the community. The decision to tell the public the locations of the colonies within NSW was not taken lightly, since there is always the risk that it could result in deliberate destruction of the eggs and chicks. In this case it turned out to be the best decision. It has produced an educated and sympathetic public that has helped the service to protect colonies throughout NSW.

After the first season, we realised that as well as a warden there must be local volunteers. These could respond quickly to a threat such as tidal flooding, which requires lots of willing hands to erect sandbag walls that keep the nests above the highest tides. Volunteers could be asked to monitor the site in the warden's absence. To solicit support at the local level, a public meeting was held at the nearby township of Culburra. This attracted a small group of interested local people who now form the nucleus of an emergency-response team.

Working with the media

Staff can establish an effective working relationship with local media people and can assist by providing ideas for stories and by helping them meet tight media deadlines. They can also help by providing supporting factual information, photos, and other aids and props. One tip in working with the media: there is never a situation where comments can safely be made 'off the record' and the camera is 'always rolling'—never assume that filming has stopped at the end of the formal questions.

Working with stakeholders

To work with other organisations or with the community towards common goals one needs a range of techniques, from conflict resolution to group decision-making. In Chapter 7 we described forms of management and planning that invite the com-

Snapshot 17.8

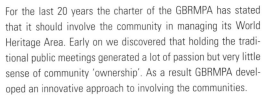

Community involvement in the Great Barrier Reef
David Lloyd, Great Barrier Reef Marine Park Authority (GBRMPA)

For the last 20 years the charter of the GBRMPA has stated that it should involve the community in managing its World Heritage Area. Early on we discovered that holding the traditional public meetings generated a lot of passion but very little sense of community 'ownership'. As a result GBRMPA developed an innovative approach to involving the communities.

The reef is too large to police, so we have to rely on the people who use the marine park to care for the reef and behave appropriately. Public education is a key management tool. In the past we have used a standard range of methods to communicate and educate:

- producing information about the reef and how we want people to behave there, and making it available in a variety of formats—printed and audio visual
- conducting public education programs, including advertising in the mass media
- arranging for marine and reef issues and examples to be included in school curricula and in teacher support materials
- involving our staff in face-to-face meetings with the public and encouraging people to comment or make written representations on our management proposals.

This was a good foundation to build on, but as issues emerged we found we needed to do better. People feel suspicious, uncertain, and threatened when they don't understand what is happening and suspect that government authorities are making decisions that threaten their interests. They fear for their livelihoods, their futures, and their children's futures. They can then become very vocal, negative, and extreme. In response, GBRMPA has now integrated three things—communications, community education, and participation—into all our processes across the whole agency. In effect, we make sure we have considered the economic, social, political, and recreational needs of the coastal communities and that all our programs include:

- education of the interest groups so they understand enough of the issues and options to have effective input
- education of GBRMPA's staff about the needs and ideas of interest groups
- conflict resolution between GBRMPA and interest groups to achieve mutual respect, understanding, and acceptance of management decisions
- conflict resolution between interest groups to achieve mutual understanding of each other's views.

We have found other methods of increasing the community's involvement and support. In some we communicate directly with competing interest groups and in others we help them to resolve their conflicts. These methods include:

- developing a 25 Year Strategic Plan
- establishing community-based Regional Marine Resources Advisory Committees
- creating Councils of Elders and memoranda of understanding with Indigenous communities
- setting up issue-based workshops and fostering negotiations or conflict resolution between different stakeholders
- keeping in touch with all affected groups
- creating joint projects between users and the GBRMPA, such as the Cooktown Offshore Management Plan
- producing training materials
- encouraging interest groups to make representations to the planning processes.

munity to participate. This is good human relations and may be required by law or by the agency's policy (Snapshot 17.8). A further advantage is that the community, being involved in the management

Snapshot **17.9**

Involving local government in conservation—dog and cat control

Samantha Bradley, Parks Victoria

We spent many hard years developing ways to control domestic cats and dogs in urban areas around Dandenong Ranges National Park in Victoria. Yet the community kept rejecting our recommendations, often considering them as 'extremist'. Acceptable solutions were found only after we set up what we think is a world first: a working group whose members made up a broad cross-section of the community, including conservation and animal welfare groups, pet-breeding organisations, and vets. The working group relied on pet-owners complying, so without wide community support the recommendations could never have worked. As a result, domestic pets now do less harm to the area's native fauna.

Snapshot **17.10**

Managing community participation and consultation programs

Lyn Webber, NSW P&W

The service invested in training its media staff in techniques of community consultation. The training included action-learning workshops. The media staff were then actively involved in community consultation programs such as planning forums or 'visioning' processes. Important planning insights gained from these processes included the following.

1　Define the purpose of community involvement.
2　Define exactly what the nature of the consultation is for the stakeholders.
3　Understand the audience.
4　Research any previous work.
5　Define how the consultation fits into the agency's overall management process.
6　Focus on quality decisions.
7　Provide timely feedback.
8　Undertake formal evaluation of the consultation process.

process, comes to better understand the issues and their complexities (Snapshot 17.9).

Community participation work requires many skills, which improve with experience. At the heart of the process are human interactions—individuals and groups working together with understanding, integrity, and commitment. Lyn Webber (Snapshot 17.10) describes some key lessons learnt in community participation.

Working in an organisation

The internal 'community' of the organisation is as important as the wider community. Protected area staff need to be able to communicate with each other to achieve their goals. Possible barriers to good communication include power relations, gender issues, and ideological differences. Chapter 9 gives more detail on methods for positive 'people management' in terms of organisational cultures, internal workshops, techniques that support the conservation philosophy, and practical help.

Working with schools

Partnerships with the education system can directly assist conservation in the long term. Partnerships could include:

- involving staff in helping develop the educational curriculum on conservation
- involving staff in training teachers
- developing teaching-guides for sites and destinations as part of curriculum development
- making sites available for educational groups
- organising tour guides and local heritage experts for educational groups
- operating specialist educational facilities such as the 'Earth-Education' program at Kosciuszko National Park and the Bournda Field Studies Centre at Bournda National Park in southern NSW.

An important part of a ranger's job is working with children to educate them about conservation. Schools regularly have excursions to parks and historic sites and it is important to make these enjoyable as well as informative. Effective planning of school programs is vital. This includes visiting local schools, developing information packs, preparing displays or audiovisuals that suit the precise age group, and designing some fun activities for the time they are in the park. Ham (1992) gives useful suggestions for interesting activities.

Marketing and promotion

Protected areas deliver a whole range of services to the community. This requires a sophisticated understanding of consumers and matching delivery of services to consumer wants. Conducting market research is the first component, followed by tailoring services to demand and then communicating and advertising these services to target groups. Specialist contractors are normally used to carry out market research and provide major advertising services for release on radio and television. Local districts can undertake local advertising through newsletters, local media outlets, community networks, and so on. In some protected areas, visitors are charged fees for entry or for the use of services and facilities (Chapter 8). Optimising the income from visitor charges requires an effective marketing strategy. Informing the community about where visitor fees are being spent can be a great form of promotion, especially if all visitor fees are going towards improving visitor facilities. Marketing and promotion may be directed towards enhancing the profile of an agency, or to reinforce policy. Case Study 17.3 outlines an example of how marketing can be used to increase visitor awareness and alter how protected areas are represented.

Case Study 17.3

The use of marketing as a tool in protected area management

Russell Watkinson and Doon McColl, Wet Tropics Management Authority

The Wet Tropics World Heritage Area (WTWHA) is the oldest continually surviving tropical rainforest on Earth. It is home to more than 3000 plant species and at least 25 vertebrate animals regarded as rare, threatened, or endangered.

Tropical North Queensland is a significant destination for international visitors to Australia with an estimated two million people visiting the WTWHA each year. Some 100 commercial operators access the WTWHA. Presenting the WTWHA to visitors is one of the key responsibilities of the Wet Tropics Management Authority. The tourism industry has a pivotal role in promoting and presenting the WTWHA to visitors. The marketing of the WTWHA by tourism organisations helps establish visitor expectations before they arrive in North Queensland. The Wet Tropics Management Authority has found that harnessing marketing concepts and working with the tourism industry has been the way forward for more effective visitor management in the WTWHA.

The Wet Tropics Management Authority developed a Nature Based Tourism Strategy to ensure a strategic approach to the management and development of tourism activities within the WTWHA. It is based upon a partnership approach among management agencies, the tourism industry, local government, conservation groups, Aboriginal people, and the broader community. It sets a strategic framework for the development and management of tourism in the WTWHA.

The strategy identified the need for a marketing plan to provide a framework for a coordinated approach between the tourism industry and management agencies towards marketing the WTWHA, and to ensure the region is able to benefit from appropriate tourism development while minimising conflicts between stakeholders.

A Visitor Awareness and Images Survey conducted at the Cairns airport as part of developing the marketing plan indicated that while the existence of the rainforest and its World Heritage status had high recognition, the existing logo for the WTWHA had very low visitor penetration and had not been adopted by the tourism industry. The name 'Wet Tropics' also had negative connotations and did not convey the desired image of the WTWHA to visitors. A separate research project undertaken by the Rainforest Cooperative Research Centre assessed how printed materials developed by the tourism industry identified and represented the WTWHA. This research indicated that 'World Heritage' was not being effectively used in marketing materials and that some commercial publications represented inaccurate, and indeed misleading, images and information on the WTWHA. Inappropriate behaviour was shown in some promotional images, such as people touching and feeding animals. The study suggested that representations of the WTWHA have powerful 'impacts' on visitor perceptions and expectations prior to arrival, and ultimate environmental behaviour on-site.

It was clear that a new branding was required for the WTWHA to more accurately reflect World Heritage values and connect with visitors. Following extensive consultation including input from Aboriginal people, the new branding for the WTWHA was developed incorporating a frog and a leaf logo.

The implementation of the World Heritage branding is seen as a significant milestone in the development of an effective partnership between the Wet Tropics Management Authority, land management agencies, the tourism industry, and Aboriginal people.

In order to oversee implementation of the marketing plan, and reinforce the new branding, a Tourism Partnership Committee was formed and undertook a number of initiatives including:

- free CDs of the logo and approved images
- branded World Heritage displays available to regional tourism operators
- placement of logo on road signs
- rebranding of promotional material and signage
- upgraded Visitor Centre displays and products with new branding
- information packs for visiting journalists.

A style manual outlines how the new branding should be used. This will encourage a consistent and appropriate representation of the area to visitors. Photographs have been provided in an image library that incorporate accurate descriptive text to be used with the images. Following consultation with Aboriginal people, images are not provided of sites that have cultural sensitivity. Management messages are incorporated to make visitors better aware of appropriate behaviour within the WTWHA.

Having overseen implementation of the branding, the Tourism Partnership Committee will now monitor marketing practices. This latter role will be particularly important in ensuring that marketing of sites is consistent with site-use designation and availability, as outlined in the Nature Based Tourism Strategy. Compliance with the new branding and marketing guidelines is voluntary; however, an accreditation system for tour guides is being developed and there is the potential that compliance with marketing guidelines will become a requirement of accreditation.

By working in partnership with the tourism industry, the Wet Tropics Management Authority hopes not only to build visitor awareness and support for the WTWHA, but to influence the way the WTWHA is marketed, and how operators present their World Heritage-based products to customers. The program won a marketing award at the Regional Tourism Awards in 2002.

17.4 Interpretation: Communicating with heart and mind

The best of Australia's heritage is preserved within the protected area system. This system holds a wealth of interest. Interpretation communicates what is special about protected areas and, by creating a greater understanding, makes them better appreciated. It helps visitors have more positive and meaningful experiences. It is also important to long-term conservation. 'Interpretation' is a technical term that means 'expert guidance'. For instance, a sign explaining how and why a historic house was built would be 'an interpretive sign'. Interpretation has long been at the heart of managing protected areas.

Interpretation first arose within the United States National Park Service around the end of the nineteenth century. Early interpreters had specific roles such as providing education about the environment and (later) enhancing visitors' enjoyment or helping to minimise visitor impacts (Beaumont 1999). It was not until the 1950s that a theoretical examination of interpretation began. At this time, Freeman Tilden suggested the 'principles of interpretation'. These are still relevant today, as is his definition:

Interpretation is an educational activity which aims to reveal meanings and relationships through the use of original objects, by first hand experience and by illustrative media, rather than simply to communicate factual information (Tilden 1982, p. 8).

The way rangers explain sites can help them to educate the community, to foster management objectives, and to give visitors an enjoyable and profound experience. As interpretation is one of the major ways staff interact with the community, we devote the last part of this chapter to it.

Outstanding interpretive experiences

The best interpretive experiences are long remembered. Their impact may provide dinner-table talk for many years and influence visitors' friends as well as themselves.

How do we help visitors to have such outstanding experiences? What works? Snapshot 17.11, a fictional visitor's experience, offers us some

Information shelter, Bungle Bungle National Park, WA (Graeme Worboys)

Entrance sign, Keep River National Park, NT (Graeme Worboys)

basic human guidelines about how interpretation can help a visitor to a national park before we look at more technical details.

Role of interpretation and interpreters

Interpretation can have many roles and benefits for protected area agencies. It is often referred to in their mission and vision statements and is a major activity of Australian agencies (DNRE 1999d). When planned and implemented effectively, interpretation is a valuable and central management tool. It is also a potential money-spinner—an important consideration in the modern corporate atmosphere. The four roles of interpretation—promotion, enjoyment, management, conservation (Beaumont 1999)—can be described as follows.

Interpretive sign, Keep River National Park, NT (Graeme Worboys)

Snapshot **17.11**

Interpretation and marketing—as experienced by a visitor

Here, in point form, is a report a visitor might give of a thoroughly satisfactory experience.

- We easily found information on guided walks and tours in the local newspaper and in the 'What's Happening' guide. Our caravan park showed a short video on the national park.
- The pamphlet provided detailed information on what to bring for the guided tour, what footwear and clothing were suitable, how long it would take, and how much it would cost. The pamphlet also mentioned that our fees would help to pay for the conservation of the site.
- Access to the area was signposted clearly and we were helped by the rangers' free access-touring map, which we picked up at our accommodation.
- When we got there, well-sited information signs made it clear where to park the vehicle and where to find the toilets and the start of the guided walk.
- The design of the national park's entrance sign and its welcoming message were basic but impressive.
- The temporary sign explaining the new work in the car park area was helpful.
- The toilet block was basic but clean.
- There was a clear sign to mark the rendezvous point and an information display explained both the route of the walk we were taking and some other activities. It matched the map on the information brochure.
- The display board provided some very interesting and easy-to-read information about the park and its history.
- The ranger arrived on time. She was courteous and relaxed and looked very neat but casual in her uniform.

- There was a round of introductions for the group; and then she explained what would happen on the tour. She collected the tour fee and gave us our tickets in the shape of an attractive souvenir postcard of our destination for the walk.
- Her presentation was impressive. First, the group size of 15 was small enough to allow us all to hear what was being said. Second, she waited until the group was comfortably gathered before giving her presentation at each spot. It is so hard to hear what is being said on a 'talk while you walk tour'. This was just perfect. What she said was really interesting. Her enthusiasm for what she was describing was infectious.
- Even more impressive was that she really knew her facts about each spot. She covered everything from the rocks, to the history, to the animals and plants.
- She was not pretentious. There were no big words used and she answered all the questions honestly. There were one or two things she didn't know and she said so honestly.
- We asked her some tough questions about how the park was managed. Her answers were not too bad, but she also encouraged us to talk to the senior ranger about matters of detail. She mixed some of this serious stuff with humour and a great smile.
- Her guided tour was perfect, but there were some useful display signs on the path that helped us understand the place for ourselves.
- The complimentary orange juice at the end of the walk rounded things off very well.

Promotion

A protected area management agency can:

- promote public understanding of its goals and objectives
- disseminate information about the managed area
- advertise recreational programs.

While there are marketing aspects to this role, integrity is required to avoid 'propaganda' that can undermine community relations.

Visitor enjoyment

- This is a major objective of interpretation and is especially important in ecotourism.
- Interpretation helps develop a keener awareness, appreciation, and understanding of the area visited and enriches the visitor's experience.
- It helps orientate visitors, allowing them to find the recreation they prefer and to do so safely and with enjoyment.

Interpreter Pat Hall, leading a group of visitors at Fitzroy Falls, Moreton National Park, NSW (Stuart Cohen)

Management

- It can persuade visitors to treat sites respectfully, without need for regulations and policing.
- It can be used to subtly direct most visitors' attention towards less fragile sites.

Who is listening and reading?

Good interpretation will always use a mix of methods and media. However, the appropriate mix will depend on the audience. Are they adults or children? What do they want from the experience? Are they first-time or repeat visitors? What are their preferred activities? Where do they come from? What is their cultural context? How do they like to learn (Backgrounder 17.2)?

Visitor surveys and other social science methods described earlier can be used to classify an audience. As well, some agencies now attempt to divide an audience into 'market segments' to identify its different requirements (DNRE 1999d). Skilled presenters can 'feel out' their audience by asking questions. They can then adapt or personalise their presentations accordingly.

Clear and simple English is especially important when dealing with visitors. Inappropriate jargon can easily alienate the general public, or cause dangerous misunderstandings. Hence a good staff-member needs to be 'multilingual'—able to adapt their language to their hearers. For instance, if talking to a group of tourists he or she might say *Many of the animals in this region are found nowhere else on Earth.* To a group of zoology students the same talk might run *Many of the mammals are endemic to this region*; and to a group of scientists or fellow-managers it might run *Endemism is high among the mammals.* The last version is the most concise, but the least clear to non-experts.

Staff can vary their register, both in speech and writing (including public signage). Register is a term for the tone and style that makes speech or writing suitable for a given audience. Different registers may be required for writing, conversation, and formal speech. For instance, *Hi folks, I'm here today to tell you about the very interesting sand goannas we have round here* might be the right register for an informal talk to a group of adults and older children. *Monitor habitat is declining locally by an estimated 10% a year* might be acceptable only if writing for an audience that has some expertise in biology. In a pamphlet for Year 5 science students the correct register might be: *Hi, I'm Gertie the Sand Goanna. Here's what I look like. Can you see in this picture the sort of places I like to sleep and where I like to feed? These places are called my habitat. There used to be lots of it around here. Lately there isn't so much …* Note that the two things that most readers find off-putting are long sentences and long abstract words. When preparing publications, staff may seek advice on the appropriate register; but they will need their own sense of register when they meet with or give talks to the public.

Many conservation staff are biology graduates who may have been trained to suppress all emotion from their writing and to state conclusions in the most abstract way possible. This may be a good principle when writing up a scientific experiment, but it is a most serious mistake when writing for the public. To write or speak in a lively way and to enthuse the public (or a team of one's co-workers) means activating their emotions.

The many faces of the interpreter

Guides lead many of the interpretive activities (tours, talks, group activities, and so on). To meet

Backgrounder 17.2   Theories of learning and communication

To understand interpretation and how it can be used to achieve conservation goals, it is useful to consider learning and communication theory. Studies in these fields look at how people take information 'on board'. This can help staff choose the best ways to communicate their messages. Theory (and observation) tells us that people require certain conditions to learn and that different people prefer different styles of learning. They may prefer communication that provokes mental processes (cognitive learning), or emotional states (affective learning), or that requires action (kinaesthetic learning). Interpretation programs can often be improved if they are based in theory. For example, observational learning theory would suggest that people will learn and repeat actions they have observed and which they have perceived as having had positive outcomes. This mode of learning opens the doors for interpreters and resource managers to demonstrate practical skills, best-practice environmental actions—recycling, minimal impact camping and so on—and positive conservation attitudes.

the broad goals of interpretation a guide needs to don many caps. The qualities and roles of a good guide–interpreter are sometimes indefinable and may vary from place to place and depend on the audience. It is only recently that standards and measures for interpretation have been formalised— for example the EcoGuide Australia Certification Program developed by Ecotourism Australia specifically for nature and ecotourism guides. A number of aspects of the guide's role have now been formally defined and studied, including:

1 *a leader and organiser* of group activities
2 *an educator* communicating messages and skills to the group
3 *a public relations representative*—the interpreter will be judged as a representative of their organisation and at the same time their account of the organisation and its goals will be noted
4 *a host* for visitors, ensuring their needs, desires and safety are attended to
5 *a motivator* for the group to act in environmentally sensitive ways
6 *an entertainer* providing fun and comradeship and breaking the ice
7 *a conduit*—through whom the experiences (both conscious and subliminal) of the site are facilitated (Pond 1993).

Hence the personal skills and qualities of a good environmental interpreter may include:

- enthusiasm, which sparks the audience's enthusiasm
- knowledge of the subject

- group management skills
- enjoying being with people and caring about their well-being
- a commitment to protecting the site/object being interpreted
- leading by example, demonstrating appropriate visitor behaviour
- a sense of humour and perspective
- credibility
- being honest about what s/he knows about a particular topic
- research skills
- organising and planning skills
- ability to evaluate and refine interpretive skills (Howard 1999).

Some good and not-so-good types of interpretive guides are described in Table 17.3. But do not despair if you do not think you are the right person for the job—many skills can be learnt and people of many types can make good interpreters.

While 'in-house' staff often deliver interpretation, there is an emerging tendency to outsource this work or to use seasonal staff. The costs and benefits of doing so need to be weighed and care must be taken that the contracted interpreters maintain their skills and continue to fulfil the management's objectives.

Types of interpretation and media used

Interpretation occurs in many settings:

- visitor centres with information displays, brochures, and multimedia
- museums for cultural and natural history displays

Table 17.3	Types of interpretive guides (adapted from Ham 1992)
Types of guides	**Traits**
Cops	• Perceive visitors as threats to the environment • Tolerate audiences by issuing many rules for visitor behaviour
Machines	• Regurgitate the same performance without modification • No spontaneity, personal input, or adaptation to different audiences • Disapprove of questions or requests to change their format
Know-it-alls	• Focus on imparting information to suggest superiority • Cannot admit lack of knowledge, prefer to pretend
Hosts	• Perceive audience as guests • Offer all clients the opportunity to speak and contribute to discussions • Happily take questions, chat, and joke • Respond to audience needs even if it means deviating from planned interpretation

- guided tours using non-motorised or motorised transport such as bicycles, horses, rafts, canoes, kayaking, buses, powered boats
- signage on trails.

Different interpretation types and media are required for all these cases. They can be broadly classified as:

- *personal*—attending services such as information centres, conducted activities, talks, live interpretations, and cultural demonstrations

- *non-personal or 'static' interpretation*—printed materials, signs, exhibits, self-guided walks, pre-recorded tour commentaries on cassettes or videos, virtual tours, and other electronic media.

The positive and negative aspects of some interpretive media are given in Table 17.4. Other examples of interpretation activities are:

- *theatrical performances*—messages delivered through plays, pantomimes, puppet shows, and other performance arts

Interpretation sign using international symbols, Mimosa Rocks National Park, NSW (Graeme Worboys)

Table 17.4	**Positive and negative aspects of some interpretive media** (Beaumont 1999)	
Method	Positive aspect	Negative aspect
Personal contact	Messages can be more powerful, adaptable, spontaneous, and up to date.	May reach only a small proportion of visitors.
Signs	Low cost; always there; visitors set own pace; visitors read what interests them.	Passive; one-way communication; can be intrusive on the landscape.
Brochures	Low cost; offer more depth of information than signs; can be taken home to read; not intrusive on landscape.	Require a certain level of reading skills; one-way communication.
Self-guided trails	Low cost; visitors can set own time and pace; suitable for remote areas without personnel; large number of visitors can experience interpretation in small groups.	Interpretive story not flexible; no personalised version or clarification of complex points.

- *role-plays*—often are fun and highly participatory; by acting out parts in a thematic story, people (especially children) may gain a more personal understanding
- *club-based activities*—programs consisting of regular participatory activities, informative talks, and so on for kids and adults, which can help to develop skills and understanding relating to nature and conservation while providing opportunities for outdoor recreation (Brisbane Forest Park's Junior Rangers program is a successful example).

Principles for interpretation planning

Finally, there are five widely accepted principles that should provide the basis for interpretation planning (Ham 1992).

Interpretation is neither teaching nor 'instruction' in the academic sense. Clients of interpretation are not a 'captive' audience like students in a classroom. They will only pay attention to interpretation if they choose. Thus, interpretation methods must capture attention.

Interpretation must be enjoyable for visitors. Entertainment is not the main goal in using interpretation, but it is one of its essential qualities. While 'fun' differs between groups of people, informality is a common element. Adopting a conversational tone, playful competitions, or role-plays are all ways to achieve a less academic feel. Likewise, exhibits should avoid the appearance of formal education tools: game-like, 3-D, participatory, moving exhibits are best. Talks and tours that include humour, audience participation, discussion and music, will hold attention for longer. However, interpreters should not go too far towards pure entertainment.

Interpretation must be relevant for visitors. When information or concepts are related to the visitors' interests, personalities, and experiences, it becomes clearer and more interesting for them. Ideas should be presented in a way that is meaningful—related to something the audience already knows and cares about. Common techniques include analogies, metaphors, and using common language; for example, certain geological processes can be compared to a layered onion skin, or forest succession to the steps in building a house.

Interpretation must be well organised so that visitors can easily follow it. Well structured, easy-to-follow interpretive messages are more likely to be understood and to keep people's attention. This may mean reducing the number of concepts introduced. Each person's ability to deal with a number of categories of information, or main ideas, is limited by their prior experience or interest in the subject being presented. Ham (1992) suggested that interpretation organised around five or less ideas is more likely to sustain the interest of most people in a variable audience. This can be shown by simple observation, or by asking visitors what they remember from the information provided.

Interpretation should have a theme, not simply a topic. The messages (related to heritage resources and other ideas called for in the public contact plan) can be organised into topics that form the main items to communicate and interpret. The point to be made about each topic—the idea or moral of the story—may be called the theme. For example, in a presentation about birds in the area (the topic) you could choose any of several themes: birds are an integral part of the local ecosystem; people should work together to preserve bird habitat; different bird species use different levels of the vegetation.

People may forget isolated details, but still remember and perhaps take to heart the overall moral or theme. They are more likely to understand the interpretation if it is structured around a theme. Thus it is not enough to be entertaining and provide interesting facts; themes are needed to bind it all together. The theme should be made clear early on, whether in a talk, sign, or brochure.

Over-interpretation

When a site is over-explained this takes away the visitor's personal sense of adventure and discovery. A presentation must not go on for so long or be so crammed with facts and themes that people become tired or feel they have been left no space in which to evolve their own response to the site. Guides should not hurry people away from an area as soon as they have finished speaking. Interpreters are facilitators, not instructors. They need to respect the creativity and individuality of their audience.

17.5 Management lessons and principles

Lessons

The success or failure of protected areas as a land use will be dependent on public support. Although protected areas bring a rich array of benefits, experience shows that the task of engaging support among some communities is not easy. Investment in communicating and involving the community in the benefits of parks, their management and, activities is an ongoing priority for agencies.

1   Consistent information needs to be made available to the community about what is happening within a protected area. A monthly newsletter

Local advisory committees at the South East Forests National Park launch, NSW (Graeme Worboys)

posted on a web site, and emailed and/or mailed to park neighbours is a good start.

2   Planned information programs can be very effective when they include a range of methods and media such as information signs, regular radio and television interviews, launches and official events, open days and field days, attendance at local community meetings, and briefing local members and senior local officials.

Principles

1   Effective protected area management can only occur with the support of the community. Human values will always drive management goals.

2   The needs and desires of people must be considered from the outset and throughout the management process.

3   Agencies, in working with the community to achieve conservation outcomes, must understand the community and be part of it.

4   To communicate effectively, agencies need to understand the community's needs, attitudes, values, and behaviour.

5   Communicating with the community is an integral component of protected area management planning.

6   Interpretation is a means of communicating to the community the exceptional heritage values of protected areas. It thereby facilitates conservation outcomes by promoting understanding and appreciation of protected areas.

7   Interpretation and working with the community are effective mechanisms for reinforcing the importance and benefits of conservation.

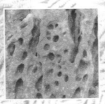

18

Indigenous People and Protected Areas

Julie Collins, Armidale, NSW[1]

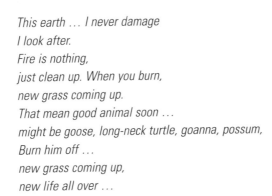

"

This earth … I never damage
I look after.
Fire is nothing,
just clean up. When you burn,
new grass coming up.
That mean good animal soon …
might be goose, long-neck turtle, goanna, possum,
Burn him off …
new grass coming up,
new life all over …

All these places for us …
all belong Gagadju.

We use them all the time.
Old people used to move around,
camp different place.
Wet season, dry season …
always, camp different place …

We come from earth … bones.
We go to earth … ashes.

Keep going …
hang on like I done.

From Gagadju Ways by Bill Neidjie, Elder of the Gagadju (Kakadu) tribe

18.1	Who are Indigenous people?	494
18.2	'Caring for country'	495
18.3	Land rights	498
18.4	Indigenous people and protected areas	499
18.5	Implications for Aboriginal land management	503
18.6	Management lessons and principles	507

The first peoples of Australia may well be the world's oldest continuous culture, spanning at least 40,000 years and possibly much longer (Roberts & Jones 1994). Aboriginal and Torres Strait Islander people have suffered the effects of dispossession as a result of the legal fiction of *terra nullius*. At present, native title issues are proving to be contentious at a national level, reflecting the paramount importance of environmental and land issues to Aboriginal people.

Indigenous people hold land tenure over approximately 15% of the land mass of Australia (Thackway et al. 1996). In the future, a portion of the conservation estate may also be owned by Aboriginal people, facilitated by the *Native Title Act 1993* (amended 1998) (Cth) and other legislation. It is vitally important, therefore, that conservation practices are linked to the interests of Aboriginal people.

In this chapter, we describe some of the concepts and background information that can help Indigenous and non-Indigenous land managers to work together, including:

- the nature of Indigenous people's relationship with the land and sea

Aboriginal ranger at Nitmiluk National Park, NT, lights speargrass for an early dry season burn (PWCNT Collection, David Hancock)

- the nature of their resource use practices
- various models of Indigenous protected area management
- skills required for Indigenous and non-Indigenous land managers to work together.

18.1   Who are Indigenous people?

Indigenous people have the right to maintain their distinctive and profound relationship with their lands ... and resources ... (UN 1993).

Indigenous or 'first peoples' are 'the "original" or oldest surviving inhabitants of an area, who have usually lived in a traditional homeland for many centuries' (Stevens 1997, pp. 19-20). Their subsistence practices (now or until relatively recently at least) rely on the use of local resources and ecosystems. The actual number of Indigenous people surviving today is a matter of definition (Kempf 1993). It is estimated that between 200 and 600 million of the 5.5 billion people living on Earth are Indigenous (UN 1993, Stevens 1997). Comprising only 5 to 10% of the world's population, Indigenous groups contribute as much as 90 to 95% of the world's cultural diversity (Gray cited in Stevens 1997). They inhabit more than 70 countries, in habitats ranging from the Arctic to the Amazon, claiming as traditional homelands 20 to 30% of the Earth's surface: four to six times more territory than is encompassed within the entire global protected area system. Many of these environments are fragile or under threat from development and are characterised by high levels of biodiversity; they are therefore significant to global conservation. Typically, Indigenous groups have suffered from the colonisation of their land by others, with their populations decimated by violence and disease (Kempf 1993, Furze et al. 1996).

However, although Indigenous rights are far from secure, Indigenous people are increasingly active on the world stage, fighting for rights to land and self-determination, and the preservation of the environment (Burger 1990, p. 135). A growing literature extols the virtues of Indigenous 'earth honouring' (Jacobs 1994) and the 'Wisdom of the Elders' (Knudtson & Suzuki 1992). Traditional knowledge and wisdom of Indigenous peoples can help us to develop more sustainable relationships between people and resources. It can also help us to understand that cultural diversity itself serves as a form of insurance, which can expand the capacity of our species to change (McNeely 1995).

A number of international resolutions describe general principles on the rights of Indigenous people and how these relate to environmental protection. Most notably:

Indigenous people and their communities, and other local communities, have a vital role in environmental management and development because of their knowledge and traditional practices. States should recognise and duly support their identity, culture and interests and enable their effective participation in the achievement of sustainable development (United Nations Rio Declaration on Government and Development, Principle 2.2).

18.2 'Caring for country'

Aboriginal perceptions of land

It is part of our responsibility (to be) looking after our country. If you don't look after country, country won't look after you (Bright 1994, p. 59).

Although there is much variability, certain themes are common to all Aboriginal belief systems. Aboriginal law was established during the Dreaming when ancestral beings created the landscape and all living species, and brought order and meaning to the world (Bennett 1983). Land is a topographical record of the journeys of the ancestral beings, a physical manifestation of the truth of the moral system. The active ingredient is the spiritual essence that animates all things (Berndt & Berndt 1983). Aboriginal people believe that to maintain social order it is necessary to maintain the physical environment, and this includes the preservation of all species. In return, the country will look after them, providing food, water, and other necessities. 'Country is a place that gives and receives life', it is a 'nourishing terrain' (Rose 1996, p. 7). The law of the land is also the law of the sea and sky; all are 'country'.

Individual men and women hold particular relationships with land inherited from parents and arising from their own conception and birth sites (Goodall 1996). Yet these relationships do not allow automatic rights, but confer obligations and responsibilities to care for the land through ritual and land management. 'Caring for country' usually means that custodians will learn, perform, and teach the ceremonies belonging to the Dreamings for that land; they will protect the land, its species and

people from unauthorised use, and they will husband the land, use it, harvest it, and manage it in such a way as to maintain its productivity, for example, through the use of fire. Country cared for in the proper way is 'quiet', in contrast to 'wild', uncared for country, land 'without its songs and ceremonies' (Aboriginal and Torres Strait Islander Commission 1995, Head & Hughes 1996).

Traditional land management practices and traditional ecological knowledge

Fire, grass, kangaroos and human inhabitants, seem all dependent on each other for existence in Australia, for any one of these being wanting, the others could no longer continue (Major Mitchell cited in Rolls 1981, p. 249).

Land-use patterns

Before the Europeans arrived, land for Indigenous Australians was inalienable: not a commodity to be bought or sold. The basic social unit was a local exogamous 'clan', composed of around 25 people. They would use a small core 'estate', containing Dreaming sites, which they would protect and care for (Stanner 1965). The clans' economic range was much larger, overlapping the estates of other clans to which they were connected through ties of marriage and language. Clans would split into family groups when resources were scarce, and would also merge to form a large group to exploit seasonally abundant resources. Base camps were rotated so as not to locally deplete resources. The 'tribe' denoted the largest area a group felt free to travel in, and which usually had a common language and a collective name. Tribe territory size was highly variable, and inversely proportional to the productivity of the ecosystem (Jones 1990).

Land and its products were the basic economic resource of Aboriginal societies. Harvesting of this resource was not passive; many techniques were used to increase the productivity of game and plants (Goodall 1996).

Use of fire

Big fires come when that country is sick from nobody looking after with proper burning (Rose 1995, p. 89).

We burn off the grass to give new growth to plants and food for animals, we give them space to breathe and live … (Peter Danaja cited in Collins 1998, p. 48).

Habitat firing was and is the most ecologically significant practice of Aboriginal people (Lewis 1989), although the extent of its impact on Australian vegetation is disputed. Aboriginal firing, sometimes described as 'fire-stick farming' (Jones 1969) created a mosaic of vegetation in different stages of regeneration (Kimber 1983). These microhabitats favoured different assemblages of plants and animals, thus supporting high species diversity. Aboriginal fire regimes could also protect pockets of fire-sensitive plants, such as those in rainforest, from hot burns. Fires were set at different times of year in different habitats. Burning not only kept the vegetation open to facilitate hunting and travelling, but it promoted the growth of plant foods and extended the habitat of prey. Around camps, as well as providing warmth and light and controlling snakes and mosquitos, fires could be used for signalling purposes, allowing groups to assess how many people were in the area, so that over-exploitation did not occur (Jones 1969, Myers 1980, Head & Hughes 1996). Most importantly, apart from economic objectives and ecological outcomes, in some cases fire was used to fulfil responsibilities to country: to protect sacred sites and to 'clean up country', 'to imprint a human signature' (Rose 1992). Resource management through fire contin-

Snapshot 18.1

Conservation: A family matter
David Major, Cultural Heritage Division, NSW P&W, Aboriginal person, south coast of NSW

Uncle Thirriwirri is the ironbark tree; he teaches me patience and tolerance.

Grandfather Sun lets me know I am alive each morning, teaches me constancy, and Grandmother Moon brings me the dreaming time where I learn each night.

My songlines teach my past and give me the key to my future. The dreaming trails that criss-cross the land teach me the ways to go; and the forbidden places, they teach the ways to survive in this land we belong to.

The stone artefacts are there to teach us how to make our tools and where the best camping spots and good stone can be found.

The middens teach us how to properly harvest from Gadu, the sea, so as not to eat too much of one shell, to let it replenish.

The sacred places and story places protect places of ceremony and lore, and also those plants and animals that are at risk of disappearing. Nowadays in the valleys of the long-footed potoroo, that place would be sacred and forbidden for hunting and burning to allow our little brothers and sisters to grow strong and numerous. As their numbers grew and they spread out from their sacred places, we would know that it was alright to collect their skins for warmth and meat for sustenance, because their sacred places would ensure survival.

Where Uncle Mingarl, the grass tree, was disappearing that place would be forbidden, to allow that special tree that give us spears and glue, to return.

All these types of flora and fauna are our cultural heritage; they carry our lores and our memories and they are the key not only to the future of Koori people but also that of all Australians. Learning to live with the environment in such a harsh country is not just a matter of conservation but of survival—mentally, physically, and spiritually.

Note: The cultural information in this Snapshot is held by the Aboriginal people of south-east NSW as custodians and guardians of their heritage.

ues today in the great deserts and the tropical savannas and forests of Australia (Langton 1998).

Resource use

> *Everything comes up out of ground-language, people, emu, kangaroo, grass. That's law (Rose 1996, p. 9).*

> *Our life is like a circle, we follow the seasonal cycle … Each season has its own way of telling us the best place and time to hunt. Everything has a role telling us when it is the season to hunt or not to hunt a particular species (Peter Danaja cited in Collins 1998, p. 46).*

Rose (1996) claimed that resource use by Aboriginal people is traditionally part of a holistic system, a unified 'Dreaming ecology' where the spiritual, social, and ecological are interwoven in relationships of cause and effect (Snapshot 18.1). The shared creation of Aboriginal law established a bond between humans and non-human entities that can best be described as 'totemism', establishing symbolically the spiritual relationships between human beings and their environment (Berndt & Berndt 1970). Sacred sites often provide refuge areas for wildlife, as they are frequently of high resource value and totemic animals cannot be killed in their vicinity (Newsome 1980, Bennett 1983, Chapter 2). Totems protect food resources, and also species that do not have 'economic value', for example flies and leeches, but which are important for ecosystem functioning (Bennett 1983).

Traditional ecological knowledge

> *Aboriginal and western systems of knowledge are parallel, co-existing, but different, ways of knowing. Scientific descriptions of nature and precepts of the natural world cannot subsume traditional ways of knowledge (Langton 1998, p. 8).*

A realisation that Western belief systems have promoted the unsustainable use of resources and environmental degradation has led to the validation of Traditional Ecological Knowledge (TEK). Respect for TEK can be empowering for Indigenous people (Williams & Baines 1993); however, it can be problematic. Knowledge is often

presented as an abstract system— fragmented and decontextualised (Pannell 1996). This ignores the localised nature of such knowledge and implies an ecological knowledge base that is distinct from other cultural elements. More correctly, TEK is the product of holistic sociocultural practices. Thus the term Indigenous Knowledge System (IKS) is probably a more apt term as it reflects the totality of information, practice, belief, and philosophy that is unique to each Indigenous culture. Indigenous knowledge exists wherever there are living Aboriginal customary systems (Langton 1996).

Snapshot 18.2

Traditional Aboriginal land management and biological survey on the Anangu Pitjantjatjara Lands, South Australia
Peter Copley, South Australian Department for Environment, Heritage and Aboriginal Affairs, and Lynn Baker and Bradley Nesbitt, Wallambia Consultants

The Biological Survey of South Australia is an ambitious program aiming to systematically sample the variety of broad-scale ecological patterns across the state. The Anangu Pitjantjatjara Lands in the state's far north-west are currently being jointly surveyed with Anangu Pitjantjatjara.

A male and a female consultant, both with biological and cross-cultural expertise, were contracted to:

- liaise between Anangu and the scientists
- translate and interpret both English–Pitjantjatjara/ Yankunytjatjara and Pitjantjatjara/Yankunytjatjara–English
- systematically record traditional ecological information obtained during each survey trip.

Documentation of TEK has identified food, medicinal, and other cultural uses of many plants and animals and has assisted in the recognition of several easily confused or overlooked plant species (such as *Solanum* spp. and some grasses). Knowledge held by traditional owners has assisted with the location of several rare species of mammals, birds, and reptiles. It has also highlighted the relatively recent regional extinction (mostly about 40 to 60 years ago) of all critical weight-range mammal species (Burbidge & McKenzie 1989) between the size of the Spinifex Hopping-mouse (Tarkawarra or *Notomys alexis*) and the Black-footed Rock-wallaby (Waru or *Petrogale lateralis*).

Such knowledge is based on generations of accumulated knowledge, and often on a lifetime's experience. Even the longest conventional biological field study is brief in comparison. There is now a recognised need to integrate the knowledge of Indigenous peoples into national strategies for environmental and conservation management (Snapshot 18.2). Partnerships between 'science' and Indigenous knowledge form a strong basis for environmental protection.

18.3 Land rights

They give us rations, a little bit of rations when the managers were here, but we are hungry for our own ground (Milli Boyd cited in Goodall 1996, p. 335).

A settlement that involves self-government and self-determination in the fullest sense is what is needed, not just land and sea rights (Mick Dodson cited in Cordell 1993a, p. 112).

Aboriginal participation in protected areas is fundamentally a land rights issue (Woenne-Green et al. 1994). Since the European invasion, Aboriginal people have consistently asserted claim to their land and their law, claims that are as much about power, justice, and self-determination as they are about land and territory. The onslaught of invasion took a terrible toll on the Indigenous Australian population, through introduced disease, violence, poverty, and repression. Much knowledge of Dreaming and ceremonies was lost in this way. Yet contemporary Aboriginal people still retain a cultural perspective that holds land to be a central organising principle (Cordell 1993a, Goodall 1996). The majority of contemporary Aboriginal people know to which land they belong, even if they do not have access to that land or to all the stories and ceremonies. They expect to identify themselves by and with the land and to authorise their political standing with reference to it. Land continues to be a primary goal and desire.

Although governments have been returning tracts of land to traditional owners over the past 25 years, especially in the Northern Territory where the *Aboriginal Land Rights Act 1976* (Cth) has given many Aboriginal people title to their traditional land, it was not until 1992 that Aboriginal land rights became a major issue of public concern in Australia.

The landmark 1992 Mabo High Court decision brought Australia into line with other nations that had inherited English common law (Nettheim 1998). The court decided that the pre-existing land rights of Indigenous Australians had survived the assertion of British sovereignty and may still survive today, provided that people still maintained their connection to land under their own laws and that 'native title' had not been extinguished by acts of governments. The High Court has also recognised that native title may be recognised not only over unallocated crown land, but also over areas where use is consistent with concurrent native title, for example, national parks (Wootten 1994). The Wik decision of the High Court in 1996 held that Queensland pastoral leases were not wholly inconsistent with native title, which is relevant as many national parks in Queensland are situated on former pastoral leases. These high-profile court decisions have led to the Commonwealth and then the state governments legislating to incorporate Indigenous land management into Australian land ownership and management systems.

The *Native Title Act 1993* (Cth) (NTA) defines 'native title' as the rights and interests that are possessed under the traditional laws and customs of Aboriginal peoples and Torres Strait Islanders. The *Native Title Amendment Act 1998* (Cth) was passed in reference to the Wik discussion and clarified leaseholders' rights in respect to pastoral and mining use. With particular reference to national parks, the amendments allow the states and territories to replace the right to negotiate (RTN), contained in the 1993 Act, with their own processes, and claimants have to satisfy a stringent—and retrospective—registration test before they can qualify for an RTN.

18.4 Indigenous people and protected areas

Protected area paradigm— from 'wilderness' to 'cultural landscape'

A national park must remain a primordial wilderness to be effective. No men [sic], not even native ones, should live inside its borders (Grzimek & Grzimek 1977, p. 177).

We didn't even know the parks existed until the authorities started sending our people to prison for hunting ... I am a traditional healer in this region ... but now I can't get some drugs because I'm not allowed to gather medicine in the park. When people decided that we should not get anything from the park, did they not know that we do not have a hospital? ... If you were in my place, would you let the person die, or would you go to that park and gather the medicine? (Njiforti & Tchamba 1993, p. 173).

There are fundamental differences between Indigenous and non-Indigenous conceptions of nature (Morrison 1997). While Indigenous societies place people at the axis of the natural world, integral to the whole, one thread of the conservation movement is profoundly romantic in that it sees the human species as an intruder to the natural world. Protected area management has tended, in part, to reflect this philosophy. Globally, protected areas have focused on preserving endangered species, habitats, and ecosystems (Stevens 1997), while Indigenous people have been evicted from their homelands in the name of these causes. In Australia, parts of the landscape have been classified as 'wilderness', even though such areas were inhabited and managed by Indigenous Australians for millennia (Langton 1996).

In recent decades, however, Indigenous 'involvement' in conservation and protected area management has emerged as a much lauded, but highly charged domain of policy and practice (Birckhead et al. 2000). Such changes have been facilitated by a growing international and national recognition of the rights of Indigenous peoples, the realisation that the conservation of biodiversity is unlikely to succeed without the support of local and Indigenous communities, and that denying their resource rights eliminates any incentive to conserve these resources (Ghimire & Pimbert 1997). In contrast to 'wilderness', with its inherent denial of human agency in ecosystems, 'cultural landscapes' has emerged as a more appropriate protected area category for the vast majority of socioecological contexts that inform and renew nature's diversity (Ghimire & Pimbert 1997, Australian Conservation Foundation 1998). Furthermore, the IUCN (1994) advocated that taking into account the needs of Indigenous peoples, including subsistence resource use, is a specific objective of the creation of national parks.

Models for cooperation—the Australian experience

Aboriginal land which is just national park is like a table with one leg or like a bird. It's not very stable. Shove it and it will fall over. Just one leg is not enough for Aboriginal land. It has to have other legs there: the leg that Aboriginal law and ownership provides; that Aboriginal involvement in running the park provides; that an Aboriginal majority on the board of management provides (Tjamiwa 1992).

The Royal Commission into Aboriginal Deaths in Custody (1991) identified Aboriginal involvement in protected area management as a significant measure for the reinforcement of Indigenous culture. It could create links to traditional lands and empower aspects of traditional knowledge as a major contributor to conservation management (Toyne in Woenne-Green et al. 1994). Uluru–Kata Tjuta, Kakadu, and Booderee in particular have been commended as models for Aboriginal ownership and joint management of national parks. These parks are all leased back to the Commonwealth agency, Environment Australia, and managed by a board, the majority of whom are Aboriginal. In addition, most states and territories are developing their own models of Indigenous involvement in national park management with varying degrees of empowerment for Indigenous participants. Models for Indigenous involvement in protected area management are illustrated in Snapshot 18.3.

Giving protected area status to Indigenous homelands, and involving Indigenous peoples in the management of existing protected areas is a mechanism to accommodate Aboriginal interests in the management of a pre-determined conservation framework (Snapshot 18.4). For non-Aboriginal protected area managers, co-management translates into greater access to spheres of traditional knowledge for managing country, Aboriginal assistance in conducting environmental research and in interpreting cultural and natural history information, and revenue-sharing from tourism (Cordell 1993a). But it also causes conflict, confusion, and can challenge long-held attitudes regarding ecosystem management. For Aboriginal owners, co-management arrangements may include funding for

community projects such as housing, income from tourism, control of cultural sites, and some support for the continuity of traditional resource management practices. However, national park status precludes certain development avenues, and Aboriginal owners do not have the option to close or degazette the national park or exercise substantial control over tourist numbers. On the other hand, some conservationists fear that 'unscientific' management of protected areas by Indigenous people may result in degradation arising from actions such as hunting and the introduction of animals such as dogs (see, for example, Martin 1992).

Snapshot **18.4**

Bama involvement in the management of the Wet Tropics of Queensland World Heritage Area (WTWHA)

Bruce Lawson, Manager, Aboriginal Resource Management Program, Wet Tropics Management Authority

Bama is the generic name used by the majority of Rainforest Aboriginal people of Far North Queensland to describe themselves as an Indigenous collective. The realisation that the 1992 High Court *Mabo* decision had significant implications for Aboriginal land and resource rights and consequently for the overall management of the WTWHA prompted the Wet Tropics Ministerial Council to agree to requests by peak Rainforest Aboriginal lobby groups to undertake a comprehensive review of management.

Begun in 1996, the review was overseen by an all-Aboriginal steering committee made up of representatives from various Aboriginal groups. An interdepartmental technical reference group was also established to provide policy advice and to comment on proposals arising from the review. Not wishing to be associated with the production of yet another set of recommendations that only gathered dust on some government bookshelf, both groups were committed to identifying workable strategies that provided real opportunities for change. One hundred and sixty-seven recommendations were provided to inform the development of the proposed regional agreement. In Queensland, where there is currently much discussion relating to the use of Indigenous Land Use Agreements (under the amended *Native Title Act 1993*), the review has the advantage of having moved beyond much of the rhetoric, with the provision of a practical strategy for the resolution of competing land-use interests.

Snapshot **18.3**

Joint management at Uluru–Kata Tjuta National Park

In 1985, Uluru–Kata Tjuta National Park was handed back to its traditional owners. The joint management arrangement that has evolved since between the Indigenous Anangu people and the federal protected area agency is now lauded around the world as the Kakadu/Uluru model of joint management. Prior to handing back Uluru–Kata Tjuta to its traditional owners, several conditions were made, including:

- Anangu would have freehold title under Commonwealth legislation
- there would be immediate lease back of the land to the Commonwealth Government under the authority of the then ANPWS, to maintain community access
- a board of management with a clear Aboriginal majority would be established for the national park
- a lease would detail the obligations of ANPWS to the Aboriginal owners, along with details of Anangu rights of use and residence in the park
- an annual rental payment and a portion of park entrance fees would to be paid to Anangu by ANPWS.

For a summary of Uluru–Kata Tjuta joint management, see Furze et al.(1996).

While opportunities for Aboriginal participation in the conservation estate are increasing, they are still subject to the vagaries of government policies, often determined without consultation with Aboriginal people (Woenne-Green et al. 1994). We believe that jointly managed protected areas have achieved, and can continue to achieve, much for both Indigenous people and for conservation. But success requires people with goodwill, flexibility, and much dedication.

Indigenous protected areas

An alternative model for Aboriginal involvement in protected area management is the recently developed Indigenous protected areas concept, as described in Snapshots 18.5, 18.6 and 18.7. This model gives Aboriginal people greater control over the declaration of protected areas on their land, as well as subsequent management.

Management of sea country

Just as for the terrestrial environment, the lives of coastal Aboriginal people and Torres Strait Islanders are intrinsically interwoven with the marine environment—or sea country. Snapshot 18.8 describes the ongoing process of recognition for Aboriginal rights in the case of marine protection.

Aboriginal 'reserves' outside the government system

The fact that Indigenous lands have remained areas of high conservation value is not simply a historical legacy (Dale 1997). Indigenous peoples have continued to manage their lands sustainably outside the protected area system. Indeed, Indigenous communities are themselves seizing the initiative and declaring their own protected areas without waiting for official recognition (Cordell 1993b).

Snapshot **18.5**

Indigenous peoples managing Indigenous Protected Areas— Nantawarrina, South Australia
Steve Szabo and Richard Thackway, Environment Australia

Through extensive consultation and discussion between the relevant stakeholders the following definition has emerged:

- An Indigenous Protected Area is governed by the continuing responsibilities of Aboriginal and Torres Strait Islander peoples to care for and protect lands and waters for present and future generations.
- It includes land and waters over which Aboriginal and Torres Strait Islanders are custodians, and which shall be managed for cultural biodiversity and conservation, permitting customary sustainable resource use and sharing of benefits. It also includes areas within the existing conservation estate, that are or have the ability to be cooperatively managed by the current management agency and the traditional custodians (Environment Australia 1996).

On the ground, individual IPAs are characterised by the following elements:

- the control, rights and responsibilities of the local Indigenous community are recognised and maintained
- application of the IUCN Guidelines in establishing an Indigenous Protected Area (IPA) aims to involve and empower the local Indigenous community rather than reduce their participation and role in the process

- the establishment and management of IPAs fosters new partnerships
- natural and cultural values are managed equally—by protecting one, the other is protected.

Twelve pilot IPA projects were implemented in 1996 to test the on-ground feasibility of the concept. A sign of the success of the concept was that eight projects have indicated that they will proceed to declaration as IPAs. The first IPA in Australia, Nantawarrina in the Gammon Ranges within the upper northern Flinders Ranges in South Australia, was proclaimed in mid-1998.

The success of the IPA concept to date can be largely attributed to a mutual partnership between Indigenous peoples and government nature conservation agencies. Indigenous peoples have shown themselves willing to engage in the process provided they are equitable partners with government for the purposes of cultural management and biodiversity conservation. Indigenous landholders themselves need to control the pace and direction of the process and to seek the involvement of outside agencies when they are ready. IPA proposals need to be developed and managed by local people to suit local aspirations. Government assistance to local communities should not be overly prescriptive on what can or cannot be funded, but be used to meet mutually beneficial goals.

Snapshot **18.6**

A traditional owner's perspective on Indigenous protected areas

In 1998, Anangu Pitjantjatjara Land Management (APLM) received a Commonwealth grant to initiate discussions with traditional owners across the Anangu Pitjantjatjara (AP) Lands about how such a concept might have relevance to them and, if appropriate, how it might be implemented.

Frank Young, Director of APLM, says that the IPA idea:

> … has been brought to the Anangu people so that they can think about it and decide if they want it to happen here. After thinking about it, some people might say that they want to look after their own country. They can look after the animals, put up fences and kill foxes and get rid of all the cats as well. Perhaps tourists might come and visit as well, like they do in the national parks.

But as Frank points out, a pilot IPA project may be the best way to establish a working model suitable for the AP Lands: 'We have already started on a plan with IPA for Walalkara and

Wataru to see how that will go.' Both will involve patchburning to make the country come back green (nyaru) after rain. This will provide food for animals and bush food for Anangu. It will protect other important plants like figs, mulgas, and minyera from fire and protect nganamara (Malleefowl) as well. Perhaps a small area could be set aside for reintroductions of special animals like wayurta (Common Brushtail Possum) and waru (Black-footed Rock-wallaby). 'Maybe it will succeed and from this we will know how IPAs will work in our area. Perhaps it will work or maybe not. So we will watch this first to decide.'

The important point for Anangu is that IPAs present an opportunity 'for Aboriginal people to care for their own land and look after the animals as well, and also the trees and the grass in all the lands'. Training will be necessary in some non-traditional land management methods, but this will be important to ensure that Anangu can run and manage the IPAs themselves.

Snapshot **18.7**

Learning about issues for Indigenous people in protected area management
Jocelyn Davies

Australia's protected area system is finally recognising Indigenous peoples as protected area managers. This is particularly the case in the 15 IPAs declared by Indigenous land owners in areas of significance for biodiversity and cultural conservation, totalling 13.5 million hectares. In these parts of the National Reserve System, Indigenous people are the protected area managers, accessing funding incentives from the Commonwealth government, policy direction from IUCN (World Conservation Union) frameworks and professional advice from regional Indigenous organisations and land management support services.

Walalkara IPA in the Pitjantjatjara Yankunytjatjara lands, the north-western region of South Australia, is one area where IPA declaration and management programs have been the catalyst for traditional owner families to spend time in the bush away from the communities where 'sitting down' has become the way of life for many, with drastic health consequences. Caring for country at Walalkara involves various adaptations of traditional land management techniques such as patchburning, now with drip torches, to make mosaic habitats and fuel reduced zones that might break the path of summer wildfires; observing the activity and breeding success of endangered mallee fowl; and cleaning rockholes, water sources that were once critical for human occupation and

continue to be important for cultural teaching and for animals. Here, teaching and learning about country is now not just an activity for the younger generation of traditional owners. It also involves scientists, students, volunteers, and tourists who come to the area to learn through biodiversity research that involves traditional owners, or just to enjoy and learn from the experience of being on country with traditional owners.

Most of my own visits to Walalkara have been leading classes of natural resource management university students to learn from Robin and Angela Kankanpakantja and their family members about contemporary management of country for cultural and biodiversity values that draws on traditional knowledge and skills and on the toolkits of scientists. The experience changed many of the students' lives, leading them to explore and question their own values and ways of relating to nature and people, as well as consolidating their interest in Indigenous ways of relating to country. This learning will help to ensure they will not readily accept protected area management systems that continue to marginalise Indigenous people. I expect many of them will contribute positively in their own careers to bringing the Australian protected area system to maturity through full recognition of Indigenous people in protected area management.

Snapshot 18.8

Indigenous peoples and marine protected areas
Dermot Smyth, Smyth and Bahrdt Consultants

In Australia, recognition of Indigenous people's rights and interests in the management of protected areas in the sea has lagged behind such recognition on land. In the mid 1970s, when Kakadu and Gurig National Parks in the Northern Territory were being established as Australia's first Aboriginal-owned and jointly managed national parks, the Great Barrier Reef Marine Park (GBRMP) was established under Commonwealth legislation that contained no recognition of Indigenous interests. Consequently, the initial management arrangements established for the GBRMP provided no meaningful involvement of Aboriginal traditional owners over whose sea country the marine park had been established.

Coastal Aboriginal groups viewed much, if not all, of the GBRMP as comprising their traditional clan estates for which they have both customary ownership and management responsibility. The vastly differing concepts of ownership and management responsibility for the sea presented, and still presents, a major hurdle to the incorporation of Aboriginal interests in the management of GBRMP. For the first 10 or 15 years, GBRMPA dealt with Aboriginal groups only as 'stakeholders' (Chapter 14) whose primary interest in the marine park was restricted to traditional fishing and hunting.

Since the late 1980s, improved measures to involve Aboriginal people in the management of the GBRMP have taken place. Following the recognition by the High Court in 1992 of the continued existence of native title, and the subsequent passage of the *Native Title Act 1993*, native title claims have been made over significant areas of the marine park, challenging the government's assertion of ownership of the sea.

In 1998 the Federal Court in Darwin made Australia's first determination in respect of a marine native title claim. The Court found that the Aboriginal traditional owners of Melville Island, north-east of Darwin, continue to hold native title rights to the sea surrounding the island, and that these rights include subsistence hunting and fishing, access to their marine clan estates and protection of their cultural sites in the sea. However, the Court also found that marine native title is not exclusive, does not include commercial rights to marine resources, does not give native title holders the right to control access by others and that native title must yield to other legal rights in the sea (for example commercial and recreational fishing rights). While this decision was upheld in an appeal to the full bench of the Federal Court in 1999, the full implications of marine native title in Australia will remain unclear until a determination is made by the High Court.

The experience of the GBRMP holds many lessons for establishment, planning, and management of marine protected areas elsewhere in Australia.

18.5 Implications for Aboriginal land management

Recognition of Indigenous rights and sites in protected areas

Following on from the recognition of Indigenous rights concerning land, sea, and natural resources as well as a valuation of IKS, shifts in the philosophy of natural resource management are occurring, such as those described below.

Resource use. Respect for Indigenous resource use within protected areas is one of the most fundamental and controversial issues for policy-makers. One of the key aspects is definition of 'traditional' use.

Use of fire. While Aboriginal traditional burning regimes are being replicated in some protected areas, tension exists over the way in which these are melded with Western science and associated views. Use of fire by Aboriginal people is controlled through legislation and policy. Under the *Native Title Amendment Act 1998* consultation with Aboriginal people is mandatory during the preparation of fire management plans by protected area managers. The Act has also made prescribed burning a notifiable act, except in the event of an emergency. Thus the relevant Aboriginal people must be notified prior to a prescribed burning operation.

Rights to access and control areas of cultural and spiritual significance. While non-Indigenous resource management agencies make the distinction between

Snapshot 18.9

A walk through time

Walking tracks may have cultural heritage significance (Pentecost 1999). They can have historical, economic, ceremonial, and recreational value for traditional owners. During 1994, a cultural heritage survey project relocated and recorded part of the routes of several Aboriginal tracks on Girramay traditional land in the closed forests of the Cardwell and Kirrama Ranges. The project employed Indigenous people and archaeological consultants and was funded by the Wet Tropics Management Authority and ATSIC.

> Aboriginal walking tracks tell us things about what people did together; things which often cannot be determined from other cultural heritage sites. They demonstrate links with other tribes; for instance, suggesting that social connections through marriage, trade and shared ceremonies were maintained. While a midden may well tell us what people ate, a walking track can show where they went to obtain food (Pentecost 1999).

Parks and Wildlife Ranger Lance Spain patrols the Katherine River in Nitmiluk National Park, NT (David Hancock)

'cultural' and 'natural' heritage, Indigenous people make no such distinction. For Aboriginal people, all is part of a holistic system, a 'cultural landscape'. While sacred sites are sites where life and law continue to be brought into being, the whole continent is criss-crossed with the tracks of Dreamings (Rose 1996, Snapshot 18.9), and many places retain contemporary significance to Aboriginal people. With increasing tourism interest in culturally significant sites, traditional owners naturally have a strong interest in involvement in site management.

Opportunities and issues for Aboriginal managers

> *Too many times Indigenous Rangers are not taken seriously. It is about time we step forward and be recognised as the true preservers of nature's splendour (Brim 1998).*

For Aboriginal people, participating in protected area management represents an opportunity to maintain connections to country. In recent years there has been a growing emphasis on funding programs for Aboriginal participation in all aspects of conservation and land management. Various pieces of legislation have provided for Aboriginal ownership and joint management of protected areas. However, many Indigenous people continue to remain alienated from a controlling interest in the management of protected areas that overlie their traditional estates (Lawson 1999).

For Aboriginal communities involved in negotiations with protected area management authorities, it is necessary to understand that such organisations are generally hierarchical and bureaucratic. While agreement may be achieved at a local level, administrative and legislative responses may be very slow. Therefore, incremental as opposed to radical change is more likely to achieve desired management outcomes for Indigenous people (Lawson 1999). 'Big picture' goals can be achieved through the cumulative effect of small concessions to Indigenous interests. For Aboriginal communities involved in negotiations with protected area management authorities, a lack of funding for the negotiation process can be a problem, particularly if traditional owners are dispersed over a wide geographic area. Funding for the negotiation process has been provided by the IPA scheme in some instances, for example Mutawintji (Mark Sutton, Mutawintji Land Council, pers. comm.) and at Byron Bay (Yvonne Stewart, Arakwal Aboriginal Corporation, pers. comm.). Many Aboriginal people see involvement in protected area management as a means of achieving economic independence.

An Aboriginal majority on the board of management has become an accepted part of the joint management regime. For Indigenous people seeking a more active role in protected area management, involvement may include the day-to-day running of the park as rangers and field officers, interpretive tours, and artefact production for tourists. However, protected area management bodies usually have limited funding available, particularly in the case of remote area parks, and this is unlikely to change substantially with Aboriginal ownership. Therefore, despite the rhetoric, employment opportunities may still be limited. Attempts to accommodate Aboriginal traditional owners within existing agency career paths are not always successful. It may be difficult for Aboriginal people to fulfil obligations to country and family while undertaking full-time employment and, in areas with high tourist visitation, it can be exhausting and confronting to be constantly 'on show' to a public seeking an 'authentic' Indigenous experience.

Flexibility in training and employment conditions is therefore essential. Contract employment has been introduced, for example as interpretive tour guides by Parks Australia and the NSW Aboriginal Discovery Ranger program. Yet there is some resistance within management authorities to anything that may be perceived as a 'dropping of standards'. Furthermore, even when 'community ranger' positions are created, which do not require formal tertiary qualifications, Aboriginal people

Snapshot 18.10

Empowerment (Djabugay Ranger and Land Management Agency)

Community rangers in remote areas of Cape York often feel frustrated by their lack of law enforcement powers. To combat this, the Queensland Parks and Wildlife Service has introduced a 'Law Enforcement' course. Several Djabugay Rangers have taken the course and are now able to charge people breaking the law in protected areas in Djabugay country (Barron Gorge National Park and the Jum Rum Creek Conservation Park/Kuranda).

Snapshot 18.11

Go for it!
Modified from Rainforest Aboriginal News *(1998)*

Advice from participants in the Cairns TAFE Certificate in Aboriginal and Torres Strait Islander Natural and Cultural Management to any Indigenous people thinking of enrolling in the course:

> Go for it. Because we must keep our culture strong, teach our young and teach non-indigenous people the importance of how our people survived and looked over what many people take for granted. Don't give up on your dreams (Wesley Epong).

> … it is a deadly course. You will like it because it gives you respect to yourself and others, knowledge of culture, and how it is important for other indigenous people to get involved in something like this course (Ronald Johnson).

may themselves feel dissatisfied with what they consider to be an inferior position (Snapshot 18.10) with a lower salary (Mark Sutton, pers. comm.).

Aboriginal ranger traineeships and courses in cultural heritage management are offered by some institutions (Snapshot 18.11). Such courses strive to be culturally appropriate by providing tuition from Aboriginal elders, as well as the provision for external or short-term study.

During their career, protected area managers are frequently required to move to another place. For Aboriginal people this may not be suitable as many wish only to be involved in the management of their own country. On the other hand, some may prefer to work away from their own country as problems can arise if their professional position conflicts with their role within their kin and tribal network (Barry Cain, Aboriginal Sites Office, NSW NPWS, pers. comm.).

Working with Aboriginal communities

One of the most challenging tasks for non-Aboriginal protected area managers is the means by which they incorporate Aboriginal interests and input within existing agency structures. It would seem imperative that more respect for Indigenous knowledge and skills be acknowledged as relevant 'qualifications' become a part of the agency culture. Personnel working in protected area management need to be able to understand Indigenous perspectives and aspirations. Aboriginal traditional owners and park managers need to develop a consensus about the values that are to be managed within a protected area (Case Study 18.1).

Case Study 18.1

Looking After the Munda (land)—joint management of Witjira National Park

Stephen Arnold, National Parks and Wildlife, South Australia

Chances are you've at least heard of Witjira National Park, located in central northern South Australia on the Northern Territory border. Apart from its biological diversity, Witjira is unique in other ways. It holds great cultural significance for the Lower Southern Aranda, Lurtitja, and Wangkangurru people. Witjira National Park is managed according to a joint management model, although in its own special way.

A new era in joint management began for me recently when I travelled to Witjira to survey a route across untracked country. The route was to be built as a response to lengthy delays in visitor movement to and from the Simpson Desert as a result of flooding along the major access track. I arrived at Dalhousie Springs and met Dean Ah Chee, ranger and law man, and two old men, Binjy Lowe and Harry Taylor. Both of these men had spent years on the land and are now the custodians of the area's traditional culture. Together, we left Dalhousie and began our survey, armed with a map and a GPS.

NPW SA and Irrwanyere staff at the end of work to relocate the camp ground at Dalhousie Springs, Witjira National Park, SA (Stephen Arnold)

Having just completed a road and track construction and maintenance course, I was feeling pretty confident about my ability to find and mark a route across country that had not felt the lash of our civilisation since seismic lines were rolled 30 years ago. My confidence evaporated shortly after entering a few waypoints on the GPS. Backtracking to move a marker, I spotted a black object ahead of the vehicle. Even as my mind screamed 'don't run over it' the wheels crunched my new GPS into bits of black plastic and glass. Dean, Binjy and Harry were quite amused as I sheepishly admitted that I must have left it on the bonnet at the last checkpoint.

So there we were, with no GPS. I furrowed my brow over the map even as Binjy and Dean discussed the route. Despairing of getting the job done, I appealed to them to help. 'We've got to go that way,' Binjy declared, 'His way no good. Too many creek, no good country.' Righto, I thought, let's do it. That was the beginning of a very successful trip in which Binjy's fantastic knowledge of the country enabled us to survey the entire route with no navigational assistance at all. I relegated myself to follower, recorder of landmarks, and basher of star droppers.

At night, we camped together—the old men warming up by the fire as Dean and I prepared a feed of meat and vegetables. This was the first time I had spent a decent amount of time with these people, or they with me. As the trip progressed, there was a sense of mutual respect growing all the time. I saw how Dean looked after the old men, and how he deferred to their knowledge and expertise. I saw how the old men told Dean what they knew, so that he could become the custodian of that precious knowledge. They saw how I was willing to participate in cooking and organising, and how I was happy to defer to their knowledge.

18.6 Management lesson and principles

Lesson

1   Establishing understanding and good working relationships between non-Aboriginal managers and Indigenous communities is a work in progress. Cultural differences can mean it can be tough for both parties. The history of European land-use in Australia doesn't always help either, but more and more successful cooperative protected area projects reflect the rewards that mutual respect and mutual understanding is bringing.

Principles

1   Recognition must be given to the special situation of Indigenous people in relation to land management in Australia.
2   Traditional knowledge of natural resource management should be respected and incorporated into protected area management systems.
3   Protected area managers must understand and respect Indigenous beliefs and management systems.

4   Co-management of protected areas has proved to be one effective means of benefiting Indigenous people and conservation of the environment.
5   Protected area managers must be committed to negotiate on equal terms with Indigenous people.
6   Protected area managers must ensure that they are working with those Indigenous people who have a right to speak for their 'country'.
7   Managers need to recognise different systems of land tenure in different cultures that need to be taken into account in planning and management.

Endnote

1.   This chapter was written at the request of the authors by Julie Collins, and subsequently edited by the authors.

Petroglyphs, Mootwingee Historic Site, NSW (Graeme Worboys)

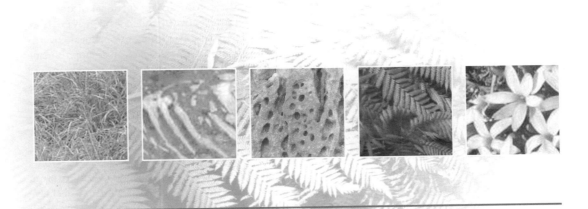

Linking the Landscape

The sand with its slow-motion breakers
is the inland sea, it spills
over half a continent, its tides swirling in millennial time
through archipelagos of red-earth hills
with here and there blue shadows of forest
floating like seawrack
… in galleries far ahead
the for-sale culture meets the forever one.

From Dot Paintings *by Mark O'Connor*

19.1 Importance of linkages   511
19.2 Conservation at a regional scale   515
19.3 A menu of policy instruments   525
19.4 Management lessons and principles   530

Even a perfectly functioning protected area system on public land will not ensure that Australia's biodiversity is adequately conserved. The natural diversity of Australia is evident across all the landscapes and land tenures. Achieving a comprehensive, adequate, and representative reserve system requires the targeted and strategic establishment of protected areas on both public and private land (Chapter 5). In addition, an exclusive focus on public land fails to recognise the interactions between and surrounding land uses; and does not take account of the fact that many public reserves are too small to sustain viable populations and ecological processes; nor do they allow for movement of animals across their natural range (Bennett 2003).

Furthermore, conservation on public land alone will not provide the degree of ecosystem integrity required to support Australia's rural industries. Biodiversity conservation, together with problems such as salinity, soil loss, and reduced water quality, demand an integrated landscape-scale approach to land management. Establishing conservation linkages at a landscape scale is necessary to achieve both conservation and rural sustainability objectives. It is also essential that the majority of the Australian landscape that is outside the reserve system be managed in an environmentally sustainable way.

Landscapes arise from the interaction of people with their environment over time (Phillips 2002). Landscape therefore includes biophysical, social, cultural and economic aspects of the environment. With landscape-scale management, there is reduced emphasis on jurisdictional distinctions, such as those between public and private land; national park and state forest; or one local government area and another. Instead, an attempt is made to establish networks of protected areas within and between catchments and bioregions. Attention is also given to sustainable management of the matrix of other land uses in which protected areas are embedded. This means that landscape management must integrate social, cultural, economic, and institutional concerns with biodiversity outcomes. Protected area managers need to work with a wide range of people in order to achieve these goals. As far as possible, management needs to be carried out with community support at the national, regional and local levels. Landscape management demands a coordinated effort from

Granite landscape: South Bald Rock with tors, Girraween National Park, Qld (Brett Roberts)

national, state, and local governments, non-government organisations, regional communities, and private land managers.

Delegates at the 2003 IUCN World Parks Congress in South Africa were concerned with how integrated landscape management can support protected areas, and recommended that governments, non-government organisations, and communities:

- adopt and promote protected area design principles that reflect those inherent in the world network of biosphere reserves
- adopt design principles for protected areas that emphasise linkages to surrounding ecosystems and ensure that the surrounding landscapes are managed for biodiversity conservation
- recognise the need to restore ecological processes in degraded areas both within protected areas and in their surrounding landscapes to ensure the ecological integrity of protected areas
- recognise that the presence and needs of human populations consistent with biodiversity conservation within and in the vicinity of protected areas should be reflected in the overall design and management of protected areas and the surrounding landscapes

- recognise the importance of participatory processes that link a diverse array of stakeholders in stewardship of the landscape linkages
- ensure that principles of adaptive management are applied to protected areas
- adopt and promote a policy framework and incentives that encourage active involvement of local communities in biodiversity stewardship (WPC Recommendation 09).

This chapter outlines some of the major features of conservation management at a landscape scale. We begin by summarising those characteristics of ecosystems that require us to establish linkages and adopt an integrated landscape approach to biodiversity conservation. We then examine the 'non-jurisdictional' scales at which such efforts are currently being mounted—bioregions, biosphere reserves, catchments, and regional natural resource management (NRM). Many of these efforts require the application of a range of policy instruments and processes directed towards achieving sustainable land management on private property. We review the current 'menu' of such instruments and processes, and indicate the way forward for landscape management in Australia.

19.1 Importance of linkages[1]

Since 1788, forest ecosystems in Australia have been reduced from 9% of the land area down to about 5% (Commonwealth of Australia 1996a). Over the same period, the area of woodland has decreased from 21% to 14% of the continent. About 75% of rainforests, more than 60% of coastal wetlands in southern and eastern Australia, nearly 90% of temperate woodlands and mallee, and more than 99% of temperate lowland grasslands in south-eastern Australia have been lost (Industry Commission 1997). Although clearing has slowed since the 1970s, the average annual amount of land clearing between 1991 and 1995 was still estimated to be over 471,828 hectares (Krockenberger 1998). In addition to clearing, the remaining native vegetation on private property is threatened with incremental degradation from rising water tables, grazing, weeds, and insect attack. Loss of native vegetation is one of

rural Australia's major environmental problems. As well as the biodiversity implications, impacts associated with this loss include dryland salinity, weed invasion, and soil erosion.

Native vegetation in rural areas provides nature conservation and aesthetic benefits, as well as contributing to the productive capacity of land and catchment health. It is 'important for a proportion of all farmland to have some native vegetation for ecological stability, to regulate hydrological processes and for long-term sustainability of farm production' (Jenkins 1996, p. 1).

Until recently, the effort to conserve biodiversity has largely concentrated on reserving areas of public land. While these protected areas on public land are clearly the keystone of any conservation effort, native vegetation on private property is also of importance, particularly for those vegetation types that are not well represented on public land. Over two-thirds of Australia is managed by private landholders (Commonwealth of Australia 1996b).

Many Australian ecosystems are fragmented, with relatively undisturbed areas located within an extensive matrix of moderately and severely disturbed environments. Human activities such as urban, agricultural, and industrial development have isolated populations of native plants and animals into islands surrounded by environments that are often hostile to them.

> *Animal species vary greatly in their level of habitat specialisation and their tolerance to habitat disturbance and change. These attributes are important influences on how they perceive a particular landscape and the level of connectivity that it affords. Such species are tolerant of human land use and are able to live in, and freely move through, a patchwork of degraded natural habitats … In contrast, there are many organisms that are sensitive to habitat change and degradation … For these species, survival and maintenance of connectivity in disturbed landscapes depends on the provision of suitable habitat (Bennett 2003, p. 49).*

While extensive areas of some ecosystems, such as the foothill and mountain eucalypt forests of south-eastern Australia, remain relatively intact, others such as the White Box (*Eucalyptus albens*) woodlands in the central and south-west slopes of NSW or the Salmon Gum (*Eucalyptus salmonophloia*) woodlands of the Western Australian wheatbelt, exist only in a small fraction of their original extent. In some bioregions such as the South West Slopes of NSW, there has been extensive clearing of native vegetation and a considerable proportion of the remaining native vegetation is located on private property. In some bioregions, only a small fraction of the original vegetation is located within protected areas. These severely depleted systems are the most urgent focus for a landscape approach to conservation management. Even where relatively large areas of natural environment remain, a landscape approach can add to the protection of native species, as well as addressing problems such as salinisation of agricultural land downstream.

With the fragmentation of a habitat, the size of populations of particular species may be reduced considerably, whereas other species are able to colonise the area once it has become fragmented. For example, Temple (1991) distinguished three types of bird populations that differ in their sensitivity to habitat fragmentation:

- area-sensitive birds, which have large spatial requirements that cannot be met in fragments of their habitat below a critical minimum size
- isolation-sensitive birds, which have difficulty dispersing between isolated fragments of their habitat
- edge-sensitive birds, which originated in extensive and contiguous ecosystems that featured few ecological edges, where different systems abut each other (such as forest and farmland).

One of the most damaging landscape-level effects of fragmentation is the loss of a keystone species—that is, a species that provides a key link between a number of species such as a pollinator, seed disperser, or important prey species. The introduction of non-Indigenous species may also have a major effect on small remnants. The reduced interior to edge ratio that accompanies fragmentation results in increased pressure from predators, competitors, parasites, and disease. For example, when an aggressive species such as the Noisy Miner (*Manorina melanocephala*) invades a small remnant of native vegetation or a linear reserve such as a roadside, they drive out smaller birds. Edge effects can penetrate far into a habitat. In a severely fragmented landscape, virtually all of the remaining habitat may be so close to edges that almost no interior habitat remains. Edge-sensitive species are particularly vulnerable to population decline.

A patchy landscape can be characterised by the size and type of patches (their internal quality), as well as how those patches are arranged in space and time (that is, the connectivity of the ecosystem). An understanding of ecosystem connectivity across the landscape can inform effective biodiversity management at a landscape scale. Connectivity concerns how patches of relatively undisturbed environment are connected spatially, temporally, genetically, and ecologically. Links between patches can be made through physical connections such as corridors of native vegetation, or through dynamic processes such as dispersal mechanisms (for example, a tree growing within a forest can transmit and

Community planting at the opening of the Burnett Bushcare Support Centre, Qld (Environment Australia Collection)

receive pollen from any other tree that lies within the range of bees or other pollinating insects—pollination provides a connection among the trees).

Conservation linkages attempt to build and connect areas of natural habitat, thereby reducing fragmentation and the extent of environments 'hostile' to native plants and animals. There is considerable evidence that linkages can enhance the viability of populations (Bennett 2003). Landscape linkages can involve:

- linear strips of suitable vegetation or habitat that provide a pathway or corridor between two or more larger areas of habitat
- a series of 'stepping stones' that enable movement of native biota between two or more larger areas of habitat
- a habitat mosaic in which boundaries between suitable and hostile environments are not clearly defined, but which occur as gradients, such that species can make some use of a range of habitats (Bennett 2003).

The degree of connectivity of patches in a landscape can be characterised by three phases: disconnected (the landscape is broken into many isolated sites), critical (a single large section may be connected but the remainder of the landscape remains as isolated patches), and connected (most of the landscape is connected, with only a few isolated patches). When spatial connectivity is critical, the dynamics of systems are inherently unpredictable, and changes in landscape connectivity can result in rapid fragmentation of habitats. Landscape management should aim to provide for, and where necessary recreate, as much connectivity as possible between patches of natural vegetation. Disconnected landscapes need to be modified so that they move through the critical phase and, where possible, return to a connected state. Modifications typically involve recreation of connections through establishment of corridors, and consolidation of patches so that their edge area is minimised.

The theory of meta-population dynamics deals with colonisation and extinctions in patchy environments. At a landscape scale, there may be a relationship between populations in several fragments as individuals move between them. The dynamics of the population (that is, how numbers change over time, and how they are spatially distributed) may be considered in terms of semi-isolated subpopulations composing a 'meta-population'.

Subpopulations are considered to belong to a single meta-population if there is some level of dispersal flow between them. The long-term survival of meta-populations and hence species depends on the persistence of subpopulations. Subpopulation extinction and recolonisation is a natural process, but the meta-population will only continue if there remain sufficient source subpopulations. The probability that at least one of the subpopulations is stable generally requires that a large minimally disturbed area is present—though the definition of 'large' varies with the area requirements of particular species.

It is not only the total size of a population that influences its vulnerability. Other factors include the geometrical character of this distribution and the dynamics of how individuals within the population move among the habitable portions of the landscape. Competition, predation, herbivory, and other interactions among species are also important types of connectivity within an ecosystem. The degree of connectivity is crucial, because unless isolation is complete, the interchange between populations may be sufficient to recolonise a patch after a local extinction has occurred and thereby prevent a regional extinction in the entire group of populations.

Meta-population theory has influenced the emerging science of landscape ecology, which is concerned with survival of species, communities, and ecosystems in fragmented landscapes. Whereas meta-population research and traditional conservation biology are concerned with single species, landscape ecology concentrates more on studies of flows and processes and considers these to be the crucial factors in maintaining biodiversity at a landscape scale. Landscape ecology also examines the degree of difference between habitat patches and their surroundings, the original size and shape of the patches, and the role of habitat corridors in facilitating dispersal and maintaining viable meta-populations. A combination of landscape ecology and meta-population dynamics is proving to be fruitful in studies of the processes that are taking place in fragmented landscapes.

However, our science remains limited in its capacity to explain the complexity of environmental processes at a landscape scale, and therefore to provide guidance on the management and conservation of ecosystems. The value of corridors, for example, is still a controversial topic. Though studies have shown that animals are present in corridors, they have not conclusively demonstrated that corridors play a major role in connecting populations through movement and dispersal. Since they have a high proportion of edge, they may attract large populations of edge-inhabiting predators. In this way, they may become traps or sinks that actually deplete populations of some species. Narrow corridors are likely to be all edge, and may present a hazard to dispersing individuals. Nonetheless, and importantly, corridors still:

- provide habitat for some native species
- facilitate the movement of some native plants and animals
- contribute to mitigation of land degradation arising from soil loss, rising water tables, and salinity
- contribute to the aesthetic values of the landscape.

Some native species, particularly those that use overstorey structural elements, may persist even in moderately sparse remnants of the original vegetation. For example, Arnold et al. (1987) found that those birds of the Western Australian Wandoo Woodland that foraged above ground were also found in pastoral land where there remained a moderate canopy cover of trees. It is possible for some native fauna to persist in modified landscapes, as long as some scattered trees remain.

Although, as we have noted, there is still much to learn, the message from landscape ecology seems to be to:

- retain large, minimally disturbed areas that contain stable source populations
- reconstruct and restore ecosystems where there are no remaining large areas of intact environment to provide a stable source population
- build upon isolated remnants as the core of a restoration effort
- retain and protect small bushland remnants
- establish and maintain connections between these remnants
- retain and regenerate isolated trees, which can be of some value for certain fauna when restoration is not possible.

The following sections discuss how we might attempt to achieve these outcomes, both at a strategic level and with respect to on-ground works.

19.2 Conservation at a regional scale

Many issues confronting conservation managers occur at scales that do not match the familiar national, state, and local tiers of government. Many environmental problems are best addressed at the scale of the region, ecosystem, or catchment. This has meant that our public institutions have not been well placed to mount effective responses to these issues. In Chapter 2, we considered various modes of governance that can be used to establish protected areas. This broadening of governance possibilities beyond traditional government–managed protected areas is being manifest in an emerging emphasis on working at a regional scale. Regions have become the focus of environmental governance, particularly through integrated catchment and NRM approaches. Potential advantages of regional-scale conservation planning include:

- capacity to engage all stakeholders
- opportunity to build on activity at the property and local levels
- capacity to integrate social, economic, and environmental dimensions
- appropriateness of this scale for negotiating trade-offs, determining priorities, and investment sharing (Meadowcroft 1997, Read & Bessen 2003).

Regional landscape-scale natural resource management poses challenges for all land managers. There is much to learn, for example, about how to best manage low-intensity grazing in grassy woodland environments, or how to take timber products from forests in a manner that sustains other forest values. An adaptive approach such as that advocated in Chapter 7 would be an intelligent way for managers to respond to the complexity and lack of knowledge confronting them. Different management options can be explored, with care taken to ensure that they do not result in irreversible impacts. Over time, those options that best enable satisfaction of management objectives can be identified and adopted. It is a process of learning through experience, rather than being constrained by a fear of making mistakes. Traditional management solutions may not address at a landscape scale the crucial natural resource management issues—salinity, wetland degradation, soil loss, weed invasion, predation by feral animals on native species, and biodiversity loss.

It is not only biological aspects of landscape management that are amenable to an adaptive approach. Exploration of the most workable institutional arrangements, the most effective mix of policy instruments, and the most democratic means of facilitating engagement of all stakeholders in planning and decision-making also demand an adaptive approach.

People live in landscapes. Many people derive their income from using these landscapes. Protected area agencies are part of regional communities, and as such, they need to encourage and facilitate sustainable development. Protected areas are important for the economic prosperity of some regions (Chapter 8). Protected landscapes (IUCN Category V, Backgrounder 3.1), for example, support sustainable use of natural resources across landscapes that have been significantly shaped by people. Most Category V areas are in Europe, but examples can also be found in other parts of the world. Many more areas, particularly in the developing world, have the potential to be recognised as protected landscapes (Phillips 2002). Australia has 172 terrestrial protected areas recognised as Category V, covering 788,779 hectares. The largest are in South Australia, where two Category V Indigenous Protected Areas cover 506,000 hectares.

Activities outside protected areas also need to be managed to foster the vitality of local communities, while maintaining the long-term health and viability of catchments and ecosystems. Protected area managers can provide expertise and technical support, and in turn can benefit from local knowledge and experience. Agencies must develop partnerships with local groups and individuals. Groups that have an important part to play in helping achieve sustainable landscape management at the regional level include:

- local government
- farmer organisations
- Indigenous organisations

- Landcare groups
- catchment authorities
- local tourist authorities

Snapshot 19.1

WildCountry partnerships
Wilderness Society (2004)

WildCountry is a landscape-scale conservation initiative involving several non-government, government and community organisations, including the Wilderness Society and the Australian Bush Heritage Fund. There are currently WildCountry projects in Western Australia, South Australia, Cape York, and Northern Australia. It is envisaged that WildCountry will assist establishment of large conservation areas through reservation, management, or agreement with land-users and landholders.

The Gondwana Link project, for example, seeks to connect the ecosystems of inland Western Australia with the southwest corner: from the woodlands of WA's Goldfields to the karri and jarrah forests of the Margaret River area, a distance of almost 1000 kilometres. This will require a cooperative effort from a broad range of government, community, and non-government organisations. Initially, a partnership has been established between the Wilderness Society, Greening Australia, Fitzgerald Biosphere Group (Case study 19.1), Friends of the Fitzgerald, Mallee Fowl Preservation Group, and the Australian Bush Heritage Fund. A full-time Project Coordinator has been employed.

In South Australia, a WildCountry project aims to provide conservation connectivity from the Eyre Peninsula to the WA border region. In the eastern end, due to the relative isolation of existing reserves within a cleared agricultural region, developing and restoring links is a priority.

The project is a long-term venture that requires involvement of landholders, Indigenous communities, local communities, and government. The South Australian Government has a policy commitment to support the project, and is implementing this through its 'NatureLinks' program. NatureLinks aims to integrate biodiversity protection and management over an area of 21 million hectares. The 'Ark on Eyre' project has 170 landholders working with local authorities to reduce feral animal numbers around Venus Bay Conservation Park. In October 2004, the Hambidge, Hincks, and Memory Cove Wilderness Protection Areas were established under the *Wilderness Protection Act 1992* (SA). These 136,372 hectares of new protected areas are part of the South Australian Government's contribution to the Naturelink program.

- state and Commonwealth government agencies
- conservation groups
- educational institutions.

Regional environmental management is increasingly being delivered through newly established community-based organisations. However there are also concerns that longer-term biodiversity conservation may not be adequately managed through such regional organisations due to a range of factors.

- Regional boundaries are often defined according to catchments, which while having some advantages (see below), do not match with the bioregional framework being used for the National Reserve System (Chapter 5).
- Regional groups typically focus 'off-reserve', whereas landscape biodiversity conservation should be undertaken according to regional priorities irrespective of land tenure.
- Regional groups may not have the necessary resource capacity, specialist knowledge, or statutory authority to successfully plan and deliver biodiversity outcomes.
- Regional groups may focus more on local and regional priorities, perhaps at the expense of state and national priorities (Curtis & Lockwood 2000, Read & Bessen 2003).

Many of these concerns can be addressed by the establishment of robust partnerships between national and state government agencies, local government, regional bodies, and civil society institutions such as conservation groups. Examples of the emergence of such partnership arrangements at a 'meta-regional' scale are given in Snapshot 19.1. More regionalised examples are discussed in relation to bioregions, biosphere reserves, catchments, and NRM plans.

Bioregions

Bioregional planning provides a framework that allows for all land tenures within a bioregion to be managed in a complementary way to achieve long-term nature conservation objectives (Bridgewater et al. 1996). Bioregional planning is based on the IBRA ecological classification of the landscape (Backgrounder 1.3). Bioregions can be used as a framework for integrated planning in the achievement of conservation, social, and economic objectives. The bioregional approach has considerable support in policy:

> There is an increasing body of opinion that advocates that individual conservation initiatives and conservation actions that deal with subsets of biodiversity and/or small land areas need to be integrated better into a larger planning framework based, not just at the species and genetic level, but at the regional scale to address wider landscape and ecological processes. This is a prerequisite for meeting the objectives of ecologically sustainable development. Both the National Strategy for the Conservation of Australia's Biodiversity and the Draft NSW Biodiversity Strategy advocate bioregional approaches to conservation planning (NSW NPWS 1998, pp. 69–70).

Conservation strategies need to shift to a landscape scale, utilising a broad range of conservation initiatives in an integrated way, to achieve biodiversity conservation goals. However, to date there have been no successful examples of integrated bioregional planning in Australia. While the NSW NPWS, for example, has carried out bioregion-based assessment in the Riverina, and attempted to prepare a bioregional plan for the Cobar Peneplain bioregion, such efforts fall short of true landscape-scale planning because, among other things:

- a single protected area agency cannot hope to provide all the necessary skills nor make available sufficient funds
- protected area agencies in Australia do not have overall accountability for bioregional planning
- the very nature of the task requires a partnership between government agencies, local councils, community groups, and individual landholders.

As Miller (cited in NSW NPWS 1996, p. 72) observed: 'the stupendous scales adopted for these forays into bioregional planning call for creative capacities equal to the task ... and [are] better described as a "leap" rather than as just another step in our continuing experiment with Nature'. Clearly, coordination at a landscape scale poses significant challenges for land managers. Each manager must be willing to work closely with other managers who might have goals that are quite different from their own. The following questions with which the NSW NPWS is grappling are probably typical of the concerns facing protected area management agencies across Australia.

1. Will conservation move beyond core protected areas or will the integration of conservation and economic imperatives see a watering-down of protection to allow for greater commercial exploitation?
2. Will the involvement of local stakeholders, some of whom are not sympathetic to conservation, weaken the forms of protection applied? Is this necessary, however, to gain community support for new initiatives?
3. Should private land be regulated or are voluntary mechanisms all that are required to achieve nature conservation on private land?
4. What should a protected area agency's role be in off-reserve conservation?
5. Given a limited amount of funding for nature conservation, what proportion should go to off-park versus on-park conservation?
6. Of the funding that goes to off-park conservation, what proportion should go to voluntary versus involuntary measures?
7. What focus should a protected area agency place on developing economic instruments for biodiversity conservation?

The following extract from a workshop paper (Pressey 1995) describes the importance, from a nature conservation perspective, of meeting these challenges.

> Protected areas ... exist in a matrix composed, not only of natural and semi-natural areas without protected area status, but of highly altered areas such as cities and agricultural fields. These areas outside protected areas but with some level of conservation management, together with many others in natural, semi-natural and highly altered states, make up the 'unreserved matrix' ... The basis of bioregional management is ... that protected areas are part of a range of management approaches for maintaining biodiversity in a region and that the right approaches should be applied to the right places. Part of the task is to establish protected areas in places where they are

School children involved in rainforest rehabilitation, Norfolk Island Botanic Gardens (Paul Stevenson)

needed, part is to 'soften' the unreserved matrix or make it less hostile to dispersal and more complementary to the more natural, reserved parts … The broad goal is therefore not to represent a sample of every species, community or ecosystem in a system of reserves but to maximise the persistence of biodiversity in the landscape, whether it is reserved or not.

Biosphere reserves

The biosphere reserve concept, discussed in Chapter 3, is one approach to management at a landscape scale. Biosphere reserves aim to preserve genetic resources, species, ecosystems, and landscapes; foster sustainable economic and human development; and act as a demonstration of what can be done in relation to local, national, and global issues of conservation and sustainable development. A biosphere reserve has three roles:

(i) a 'conservation role' (providing protection of genetic resources, species, and ecosystems, on a world-wide basis), (ii) a 'logistic role' (providing interconnected facilities for research and monitoring in the framework of an internationally coordinated scientific program), and (iii) a 'devel-

opment role' (searching for rational and sustainable use of ecosystem resources and hence for close cooperation with the human populations concerned) … Naturally, the relative importance of the three roles will vary from case to case, depending on the great diversity of situations the world over: but it is the combined presence of these three roles which characterises the Biosphere Reserve concept. In other words, a 'Biosphere Reserve' that does not have a protected core area is not a true Biosphere Reserve. A National Park that does not care for the sustainable development of surrounding regions and for the basic needs of their population, is not a true Biosphere Reserve either (Batisse 1990, p. 111).

Each biosphere reserve should contain three elements: one or more *core areas* devoted to long-term conservation of nature; a clearly identified *buffer zone* in which activities compatible with the conservation objectives may occur; and an outer *transition area* that is devoted to the promotion and practice of sustainable development (Cresswell & Thomas 1997) and may contain a variety of agricultural activities, settlements, or other activities. The strength of biosphere reserves is the emphasis

Lower Snowy River, Kosciuszko National Park, NSW—a World Biosphere Reserve (Graeme Worboys)

of their management objectives on the integration of human and natural systems.

In practice, the application of the biosphere reserve concept in Australia has been limited. There are currently 12 Australian biosphere reserves. However, nine of these (Uluṟu–Kata Tjuṯa, Croajingolong, Hattah–Kulkyne and Murray–Kulkyne, Wilsons Promontory, Kosciuszko, Yathong, Unnamed, Prince Regent River, and Macquarie Island) are managed as protected areas, and do not contain buffer or transition zones. This probably reflects the fact that the original motivation to have these areas declared biosphere reserves was more about seeking international recognition than addressing landscape-scale management issues.

One of the major problems is that the agencies managing these areas do not have jurisdiction over the zoning or planning of areas outside the protected area boundaries (Parker 1993). This makes it difficult to establish buffer and transition zones on the land around the existing reserves. In some cases, the adjacent land is managed by an agency not previously engaged in the biosphere reserve program. In the case of Macquarie Island, the extension is simply not feasible. In all cases, there is no established framework under which the necessary negotiations can take place. The perception, perhaps less widely held today than it was a few years ago, that the aim of biodiversity conservation is to include as much land as possible in national parks, has also inhibited the establishment of true biosphere reserves in Australia (Parker 1993).

There have been few successful initiatives to broaden the management base of these ten biosphere reserves to include other stakeholders and managers of adjacent land. The exceptions to this are Uluṟu–Kata Tjuṯa and Prince Regent River, where the biosphere reserve status may have encouraged enfranchising of local people in the management of these areas (Parker 1993). In the other seven reserves, the application of the biosphere reserve concept has generally been limited to preservation and monitoring functions that could have been accomplished simply through their status as protected areas.

However, there are three biosphere reserves (Fitzgerald River in Western Australia, Bookmark in South Australia, and Mornington Peninsula in Victoria) that are in the spirit of the concept.

In November 2002, UNESCO approved the Mornington Peninsula Biosphere Reserve—the

Case Study 19.1

Stepping outside—a bioregional and landscape approach to nature conservation

John Watson, WA Department of Conservation and Land Management

The Fitzgerald River National Park (FRNP) in Western Australia is relatively large (about 330,000 hectares) and is considered to be mostly in 'pristine' condition—a true benchmark in a world network of protected areas (Watson & Sanders 1997). Because it has a large central wilderness zone (about 70,000 hectares), is located on a remote section of Australia's coastline, and has low visitor levels in most of its areas, it should therefore surely be able to 'look after itself'. Wrong!

The FRNP may be more secure than many smaller areas but it is not immune from either internal or external influences that, over time, will lead to a deterioration of its true 'representativeness' and its nature conservation value. In December 1989 three lightning strikes 40 kilometres apart on a day of extreme fire weather resulted in a wildfire of some 149,000 hectares—almost 50% of the park by area—and most of this burnt within eight hours!

Lesson 1—even large protected areas can be significantly impacted by natural 'catastrophic' events that may be quite normal and indeed beneficial in a pristine landscape but which have severe ramifications in what is now an 'island' reserve in a 'sea' of cleared agricultural land.

The FRNP is mainly comprised of a low plateau and a series of river basins with significant upper catchments outside the park in the cleared farmlands. Changes in groundwater and surface hydrology have resulted in rising water tables, increased salinity, more rapid surface runoff, and greater silt levels in waterways, all of which then impact on the downstream FRNP, in particular its riparian systems, its inlets and estuaries, and its adjacent marine values.

Lesson 2—undesirable impacts can occur even on large protected areas through inward drainage processes, and reach right into central core areas. Protected areas with predominantly outward drainage, such as mountains, are clearly less vulnerable in this context (although they do have their own suite of other threatening processes).

The FRNP has not escaped the Fox that is believed to have had a major impact on Western Australia's native fauna species. Whereas baiting programs (using the naturally occurring 1080 (sodium monofluoroacetate) poison) have been outstandingly successful in combating this predator, Foxes continue to move in from outside reserve boundaries.

Lesson 3—you may be able to control feral animals within your protected area but there is typically a further population waiting to move in again from outside and replace them when the opportunity arises.

For all of the above reasons—and many other examples that could have been used—it is clear that we must take a *landscape* approach to protected area management and must 'network' outwards—both physically and socially.

Physical networks and linkages

Over the past 22 years since its designation as a World Biosphere Reserve, the Fitzgerald has evolved from the original gazetted 'core' area of 278,000 hectares to a 'model' biosphere reserve of some 1.3 million hectares, including a 'buffer and corridor zone' around the national park core, and a 'zone of cooperation' that incorporates the farming areas and towns beyond the buffer zone.

We are now working together with the local community at three levels within the 1.3-million-hectare 'Fitzgerald Biosphere' landscape to address physical linkages and networks.

1   In addition to the buffer zone around the outside of the national park we also recognise major corridor linkages of uncleared vegetation, sometimes up to 10 kilometres wide. These extend in both directions along the coast from the national park core area, and inland along several river foreshore reserves and through a broad linkage via the Pallinup River area and Ravensthorpe Range (Figure 19.1). These buffer and major corridor linkages are mainly crown land—essentially existing or proposed protected areas, whether formally part of the conservation estate, or shire reserves with other primary purposes.

2   During 1996, Environment Australia provided funding to produce an integrated vegetation management plan for the zone of cooperation. This was completed in March 1997 and it identified important remnant vegetation patches, poorly conserved vegetation types, and rare vegetation communities (Robinson 1997). A review of all catchments was carried out and priority actions were identified. Salinity prediction maps and vegetation change maps, produced by CSIRO, were used to help identify suitable areas where corridors could be re-established to provide interconnected east–west and north–south linkages between large remnant patches of vegetation. These strategic plantings have now been fine-tuned to include prescriptions for species selection and placement.

3   Within the framework described in points 1 and 2, individual farmers and local catchment groups are continuing to develop revegetation and cropping strategies to further combat rising groundwater salinity and provide more localised vegetation corridors and protection of on-farm remnant vegetation.

Social networks and linkages

It is often said that the most important resource in an organisation is its people—the same is true for protected areas. However, in most protected areas such as national parks and conservation reserves, there are very few people, sometimes only the park ranger and family plus visitors to the area, who could be regarded as temporary or itinerant residents.

Again, in the social context protected areas must look outward and encompass ownership and support from the broader landscape around them. The biosphere reserve model provides an identifiable framework to achieve this—the challenges are to increase community awareness of protected area values, to extend that awareness to the expanded physical networks and linkages discussed earlier, and to encourage pride and support for working and living together in the total biosphere landscape.

There are various ways in which this can be achieved, for example through 'friends groups' and public involvement in planning, but I will give two different examples here.

Figure 19.1   Fitzgerald Biosphere Reserve showing national park core area and surrounding buffer-corridor zone

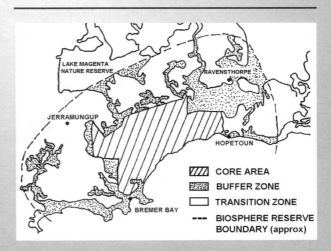

1   Around 1994, as part of a major biological study of the buffer/corridor zone and zone of cooperation around the FRNP (Sanders 1996), we established a schools monitoring program. Every school within the biosphere reserve was provided with basic equipment, and with the assistance of staff, encouraged to set up ongoing monitoring of 'their bit of bush'—typically an area of remnant vegetation relatively close to the school. The methods used included direct observation, fauna trapping, and vegetation/plant identification in transects and quadrats. The studies were incorporated into the school curriculum. This program provided a valuable educational opportunity for the students. It also provided useful data and further consolidated community support for the biosphere reserve. Approximately 25% of the entire community population assisted or visited at least one of the sites—imagine extrapolating that proportion to a 'hands on' conservation project in a city! The project ran over a three-year period and has continued in those schools with appropriately skilled teachers.

2   The Mallee Fowl Preservation Group is a voluntary organisation based around the west of the Fitzgerald Biosphere Reserve. Two of the group's main study sites, where the threatened Malleefowl still survives, are located in the zone of cooperation in the Cocanarup Nature Reserve and Peniup Reserve. The group produced a community action plan for Malleefowl in its area, and has been successful in promoting the bird as a 'flagship' species. This has provided a focus for on-farm conservation of wildlife habitats. As the Malleefowl lives throughout the Fitzgerald Biosphere, this local community action has made a significant contribution towards nature conservation at the landscape level.

Bringing it all together

Clearly cooperation is required between people at all levels to ensure a landscape-scale approach to nature conservation and protected areas—local residents, school children and teachers, agency personnel, and funding bodies at the local, state, national, and even international level. At the physical/social interface, we are currently using an even broader landscape approach, namely the expanded implementation of these principles from the 1.3-million-hectare biosphere reserve to the whole of the South Coast Region. The 'flagship' for this is not a species but our 'macro-corridor' concept. This is a vision of an unbroken network of wide corridors along the coast from Esperance to Albany and inland to other major protected areas—some 500 kilometres long and 100 kilometres or so inland—with the Fitzgerald Biosphere Reserve as the central 'hub' (Figure 19.2). Hopefully this initiative will galvanise broad community awareness in our entire region and benefit not only the Fitzgerald Biosphere but also all of the other 150 or so protected areas within the regional landscape.

Figure 19.2 Macro-corridor network in the South Coast Region of Western Australia

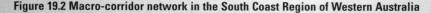

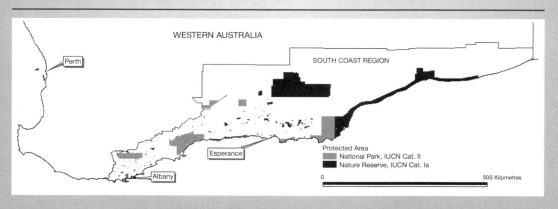

first biosphere reserve in the world to encompass substantial urban areas. The reserve has a core area of 9300 hectares; 63,600 hectares of buffer zone (including 50,100 hectares of marine environment); and a transitional zone of 141,300 hectares (18,500 hectares marine), for a total area of 214,200 hectares. It includes the whole of the Mornington Peninsula Shire and parts of the municipalities of Frankston, Casey, Cardinia, and Bass Coast, including Phillip and French Islands. The Mornington Peninsula National Park is part of the core area, and other land uses in the buffer and transition zones include viticulture, horticulture, dairying, cattle grazing, fishing, aquaculture, and residential.

Bookmark Biosphere Reserve is located on the Murray River in semi-arid mallee county near the Victorian border. The reserve comprises Dengalli Conservation Park, Murray River National Park, Chowilla Regional Reserve, and the old Calperum and Taylorville stations (Snapshot 3.1). Case Study 19.1 describes the management approach taken for the Fitzgerald River Biosphere Reserve, which includes the Fitzgerald River National Park as a core area.

Catchment management

Regional environmental management in Australia has often been undertaken by institutions based on catchment plans. This has been in part due to the importance of physical factors, such as soil and water, to agricultural industries. Catchments are also often convenient in terms of size and ability to clearly identify boundaries. Catchment boundaries are essentially determined by topography. They provide a natural unit for integrated planning. Catchment-scale institutions can also provide the regional perspective necessary for effective landscape management.

Catchment management reforms are a national process. Throughout NSW for example, water and native vegetation management regulatory plans are currently being developed to improve catchment protection, provide river systems with environmental flows, and help conserve biodiversity and cultural heritage. State Forests, through audited ecologically sustainable forest management and the NSW NPWS through voluntary conservation agreements partnerships with landholders, are con-

tributing to this approach (Pulsford et al. 2003). As part of the recent reforms in natural resource management in NSW, Catchment Management Authorities (CMAs) have been established under the *Catchment Management Authorities Act 2003*, with boards reporting to the Minister for Infrastructure and Planning and Minister for Natural Resources. Thirteen CMAs have been established across the state. The CMA's responsibilities include:

- preparing catchment plans and associated investment strategies, in consultation with local government and the catchment's communities
- recommending and managing incentive programs to implement plans
- providing landholders with access to data needed to prepare Property Vegetation Plans (PVPs) and implement catchment plans
- allocating funds to support development of PVPs and for associated incentive programs
- monitoring performance against catchment plans and certified PVPs
- developing transparent procedures for the resolution of local disputes (DIPNR 2003).

In Victoria, nine catchment management authorities have been established to implement regional catchment strategies. These strategies were developed by the previous catchment and land protection boards, in consultation with regional communities. The strategies attempt to provide the basis for delivery of catchment management and sustainable agriculture programs. Catchment management authorities combined the roles of the previous authorities including catchment and land protection boards, river management authorities, salinity implementation groups, water quality groups, and sustainable regional development committees. Other roles of the authorities include identifying priority activities and work programs under the regional catchment strategies, and providing advice to the state and federal governments on resourcing priorities. The establishment of these catchment management structures is an attempt to provide an integrated approach to sustainable development and conservation of land and water resources.

Catchment-scale institutions are providing the regional perspective necessary for effective landscape management. Such organisations make good use of local knowledge, and are particularly suited to dealing with issues related to soil and water management. Empowering regional communities makes for more flexible, locally relevant land-use decisions, and increases landholder support for the outcomes of planning processes.

However, they do have limitations. Their members are generally appointed by a government minister rather than elected. This means they may not adequately represent the views of the communities that they serve. Tensions also exist between providing clear direction through state and Commonwealth legislative frameworks, and maintaining flexibility to accommodate regional differences (ANZECC 2000). Unless legitimate state and national interests are adequately considered, there is a danger that local and regional interests will largely determine land management outcomes. Concentrating decision-making power at the regional level may result in continued erosion of public good values such as biodiversity conservation (Curtis & Lockwood 2000). To gain the benefits of devolution to regions, while still maintaining the wider public interest, priorities or standards need to be set in relation to key natural resource management issues. Examples of standards include catchment-based caps on water allocations and salt exports, and requirements that an ecologically sustainable proportion of each regional vegetation type is conserved. Such priorities and standards need to be developed though cooperative partnerships between state agencies and regional stakeholders (Snapshot 9.2).

Regional natural resource management

Integrated NRM is being furthered by a number of recent policy frameworks promulgated by the NRM Ministerial Council (Chapter 3). These include the *National Action Plan for Salinity and Water Quality* (NAP), *Framework for Extension of the NHT*, *National Framework for NRM Standards and Targets*, *National NRM Monitoring and Evaluation Framework*, and the *NRM Action and Implementation Plans for Reconciliation* (Jaireth 2003). The NAP is a joint $1.4 billion program between the Commonwealth, states, and territories. The NHT extension involves $1.032 billion from the Commonwealth for 2002–03 to 2006–07, to be partly matched by states/territories. Delivery of these funds is to be on a regional basis, with 21 NAP regions and 60 NHT regions having been established.

New arrangements between the Commonwealth and the states for distribution of NHT and NAP funds require the preparation of regional plans devised by a regional body, based on targets, which form the basis for regional NRM investment and identify the contributions of all parties. Plans will be accredited based on whether they have:

- comprehensive coverage of issues
- decisions based on sound scientific analysis
- effective key stakeholder involvement
- consistency with other planning processes
- specific targets at the regional scale
- strategic, prioritised actions
- provision to evaluation, continuous development, and review.

In Tasmania, for example, the *Natural Resource Management Act 2002*:

- established the mechanism for devolving NRM decision-making to regions

Snapshot 19.2

Blackwood Biodiversity Program, south-west Western Australia

Read & Bessen (2003)

The aim of the Blackwood Biodiversity Program is to protect and enhance as many bush areas as possible within the Blackwood catchment that are of high biodiversity value. A model was developed by the WA Department of Agriculture to rank the 24,000 patches of bush over 1 hectare that are located within the Blackwood catchment. Computer-based modelling was used for selecting priorities, based on remnant area; area to boundary ratio; proportion of original vegetation remaining on public land; proportion of remnant vegetation potentially exposed to salinity; distance to a publicly managed protected area; and number of remnants potentially forming links.

Priority sites identified through this process were then 'ground-truthed' using local knowledge and networks to identify any sites not containing high biodiversity values or not privately owned. More than 6000 hectares of high value remnants have been protected, through fencing and revegetation for buffering or connecting purposes, and protective covenants have been placed over 30% of the priority sites.

- established membership and functions of an NRM Council
- established membership and functions of regional NRM committees
- identified processes for regional plan development and accreditation.

The Tasmanian NRM Council provides advice to the Tasmanian Government on NRM priorities; implementation and administration of national and state NRM programs; accreditation for regional NRM plans; and the effectiveness and efficiency of activities carried out under regional plans. There are three Regional NRM Committees, with boundaries matching the Cradle Coast Authority, Northern Tasmanian Municipal Organisation, and Southern Tasmanian Councils. Functions of the Regional NRM Committees include:

- identification of regional priorities
- preparation of a regional plan that must identify NRM issues; develop targets; and identify actions to achieve these targets
- facilitating plan implementation, including obtaining and allocating the necessary funds

- monitoring and evaluating plan implementation.

Lowe et al. (2003) concluded that the effectiveness of integrating biodiversity conservation into regional NRM plans is dependent on:

- active involvement of the range of stakeholders over an extended period to build trust, understanding, and decision-making abilities, as well as driving attitude and behavioural change
- knowledge of the biodiversity assets of region, threats to the services they deliver, and ways to mitigate these threats
- clear statements of desired biodiversity outcomes with realistic targets
- clarity in responsibilities and cooperation between the three tiers of government and other institutions
- explicit mechanisms to translate national, state, and bioregional biodiversity outcomes to the level of individual properties
- willingness to use an appropriate mix of policy instruments.

19.3 A menu of policy instruments

Historically, landholders have enjoyed extensive rights, and had few responsibilities with respect to natural resource management on their properties. Over the past few decades, in some Australian jurisdictions, there have been stricter limits placed on landholders' rights, especially with respect to the clearing of native vegetation. There has also been an increasing emphasis on cost-sharing and providing landholders with incentives to adopt best practice land management.

The justification for government intervention with respect to landholders' property rights rests on two foundations (MacKay et al. 2000). First, land management practices on one property can impose costs on other properties and infrastructure (damage to buildings, roads, pipelines, and so on). These costs arise from the effects land-clearing and other activities have on down-catchment salinisation, and on water quality and quantity. Second, clearing native vegetation gives rise to a loss of public

benefits associated with biodiversity and amenity values. Governments have a role in protecting the public interest with respect to biodiversity and amenity values. They can also represent the interests of down-catchment communities where these communities do not already have the institutional structures in place to represent themselves.

A growing number of landholders appreciate the value of maintaining native vegetation on their properties. However, these pro-conservation attitudes may not be translated into conservation-oriented decisions and activities. There are a number of factors that may prevent or discourage a landholder from looking after their native vegetation. Barriers to conservation and management include economic pressures to earn a living, the difficulty that some landholders have in breaking with their traditional style of land management, and the social pressure in some rural communities of not being seen to be a 'greenie' (Campbell 1994,

Chamala 1992, Vanclay 1992). Furthermore, the causes of biodiversity loss on private land can be traced back to a number of social, economic, and institutional factors, including 'the failure of markets to value all biodiversity considerations, incomplete specification of property rights, poor institutional arrangements, failure to distribute information, inadequate resources allocated for biodiversity conservation, and a general lack of awareness of the value of biodiversity' (Young et al. 1996, pp. 11–12). If policy instruments are to address effectively the continued loss and degradation of native vegetation, they need to take these barriers and factors into account.

Given this complexity, it is unrealistic to expect a single policy instrument to be effective. A range of complementary instruments must be employed. There is now a 'menu' of policy instruments available for achieving sustainable landscapes. Broadly, these can be classified into five categories: regulations, economic (market based) instruments, education, framework strategies, institutional reform, and partnerships. The challenge for governments and management agencies is to develop integrated packages that may incorporate:

- legislation or regulations that can be used to create an institutional framework for management, set aside areas of land, and enforce standards and prohibitions
- formal agreements between agencies and landholders
- economic measures such as subsidies and tradeable permits to encourage landholders to provide conservation benefits to the community, assist efficient allocation of resources, and allow for the equitable distribution of costs and benefits of conservation activities
- education to convince people of the need to change behaviour, gain support for policy instruments, and ensure the ability to apply instruments (Dovers 1995).

Regulations

Regulatory instruments are often referred to as 'command and control' measures, and have traditionally been favoured by governments to carry out environmental policy. Environmental regulations are often established through legislation. Regulations may be directed towards a range of purposes.

Preventing action. A major use of legislation is to legally prevent certain actions from occurring. For example, land-clearing is not permitted in Australian mainland states and territories, unless a permit is applied for and obtained. Victoria, for example, has clearing regulations in place for land above 0.4 hectares. Applications to clear native vegetation take into account habitat values, genetic diversity, carbon storage, land degradation, groundwater, water resources, sustainable use of land, and the aesthetic values of the area. Local councils in Victoria determine applications where areas are less than 10 hectares. Advice must be sought from the Department of Sustainability and Environment for all other applications. Before the 1989 amendments, between 1972 and 1987, native vegetation was being cleared at an average rate of 15,000 hectares per year (DNRE 1996b). Between 1990 and 1992, clearing was reduced from approximately 6160 hectares to 3150 hectares per year (DNRE 2000a). More recent estimates have suggested that Victorian vegetation is being cleared at a rate of 2500 hectares per year (DNRE 2000b).

Requiring action. Legislation can identify when action must be undertaken. For example, the *Environmental Management and Pollution Control Act 1994* (Tas) defines and deals with three main classifications of activities that may cause environmental harm, and for a 'Level 2' activity a formal Environmental Impact Assessment must be undertaken.

Establishing institutions and process. Planning and management institutions and processes are often established through legislation, such as the NRM processes established in Tasmania under the *Natural Resource Management Act 2002* mentioned above.

Technology bans. Technology bans may also provide an effective method of reducing the impacts of some agricultural practices, for example, such as particularly inefficient irrigation methods. The advantage of this form of regulation is that it focuses on eliminating the technology that is causing the problem. Such regulations are also relatively easy to enforce.

Self-regulation. Some industries have adopted a self-regulatory approach for achieving professional, safety, legal, social, or environmental objectives. A common method of self-regulation is the use of 'Codes of Practice', through which an industry strives to achieve best practice. Such codes are most effective when they incorporate incentives to encourage compliance. The code of practice developed by the sugar industry, for example, addresses a range of sustainability issues pertaining to that industry. Another approach is the use of environmental management systems. One widely adopted EMS is the International Standards Organisation's ISO 14000 suite of standards for environmental management. ISO 14000 requires a management system aimed at: setting environmental policy and defining environmental goals; establishing a program to meet the goals and implementing that program in day-to-day operations and emergency situations; measuring performance in achieving those goals and taking action when the targets are not met; and progressively improving the system by repeating the cycle. Management systems are verified by audit and the result is an official certificate. Components can include an environmental performance evaluation to measure, analyse, assess, and describe an organisation's environmental performance against agreed criteria for appropriate management purposes; or a life cycle assessment (Backgrounder 10.1).

Well-constructed regulations create a clear understanding of what can and cannot be done, and enable the community, through its parliamentary representatives, to protect natural resources. While regulatory approaches can lead to positive outcomes, they also have limitations. Regulations can create animosity and drive local people away from developing productive partnerships with other land managers. They generally do not give landholders any incentive to improve management of their properties. It is also very difficult to effectively enforce restrictions, for example, on clearing native vegetation when the areas concerned are remote, enforcement staff are few, and the regulations are not supported by the local community (Young 1996). Clearing regulations do not prevent the incremental degradation of remnant native vegetation from threats such as rising water tables, the impact of grazing, fire or fertiliser, invasion of exotic weeds, and insect attack. Other instruments that can be used to address the shortcomings of a purely regulatory approach to native vegetation conservation include economic incentives, voluntary programs, management agreements, and education programs. On the other hand, as mentioned above, regulatory approaches can be used to set and enforce standards, and have slowed the rate of native vegetation-clearing. Regulations are a necessary component of enabling governments to create markets in natural resources such as water. Regulatory mechanisms may also be needed if other methods are likely to be relatively costly to implement or ineffective.

Economic instruments

Economic instruments tend to be more flexible and provide greater individual choice than regulations, and may offer more cost-effective ways for achieving environmental objectives (James 1997). Economic measures can reduce environmental destruction and promote conservation either by increasing the cost of activities that damage ecosystems, or increasing the returns gained from conservation activities.

Increasing costs of damaging activities can be achieved by imposing an environmental tax. Attaching a price to an activity that has an adverse environmental effect may influence the behaviour of individuals simply because it can make environmental best practice the most cost-effective alternative. This approach may help address, for example, the damage that one landholder's clearing activities may have on the productivity of downstream properties.

Subsidies can be paid to landholders for providing nature conservation values that benefit the rest of the community. This means, in effect, paying landholders to manage areas of native vegetation for their conservation values, rather than putting them to some other environmentally damaging use that may generate an economic return. But who should pay for such subsidies? Here economists generally make use of the 'beneficiary-pays principle'. Two aspects of the beneficiary-pays principle are:

- user-pays, whereby supply costs are paid for by those who derive a direct benefit such as increased farm productivity

- beneficiary-compensates, whereby supply costs are paid for by those who derive an indirect or intangible benefit, such as maintenance of bio-diversity or aesthetic values (Donaldson 1996).

In the case of user-pays, the owner who enjoys improved productivity on their property as a result of conservation works should fund at least part of the cost of such works. On the other hand, works designed to maintain or enhance biodiversity values benefit many people throughout the community, so the beneficiary-compensates principle suggests that such works should be funded by governments from taxation revenue.

In general, a variety of economic instruments can be employed, including incentives, levies, market creation, and full-cost pricing.

Incentives

Transition incentives are one-off payments that can assist landholders to meet new regulations. They are directed towards minimising the difficulties for landholders in adjusting to a new policy regime, and to encourage landholder support for the new management standards. The incentive payments available in South Australia between 1985 and 1991 for protecting vegetation following refusal of a clearance application could be considered an example of a transition incentive.

Catalytic incentives are relatively small payments that are designed to secure landholder participation in best practices, but not to fully compensate them for the net costs of adoption. These incentives attempt to capture the goodwill that landholders have to undertake on-ground works in the public interest. Small payments can minimise the public investment required to achieve on-ground out-comes. However, they are unlikely to be sufficient to change the land management practices of those landholders who do not already have both a strong commitment to achieving environmental outcomes, and the financial resources to cover those costs of adopting best practice that are not covered by the incentive payment. Examples of catalytic incentives include Landcare grants, and Greening Australia fencing assistance programs (ANZECC 2000).

Full cost-sharing incentives are based on a comprehensive identification of the costs and benefits of a best practice, as well as who bears the costs and who enjoys the benefits. Cost-sharing is then done according to the beneficiary-pays or polluter-pays principles. Cost–benefit analyses typically recommend much higher payments than would be made under a catalytic approach. Formal cost-sharing frameworks need to be linked to formal management agreements or covenants that secure the public investment. Full cost-sharing incentives are likely to attract more landholders to adopt best practices than other forms of incentive, and are the most equitable approach. However, it has been estimated that public funding for vegetation management would have to rise by at least an order of magnitude in order to support full cost-sharing (ANZECC 2000).

Environmental levies

Environmental levies are used to raise funds for environmental management. The South Australian Government, for example, passed legislation to allow regional catchment boards to levy people in their region to provide funds for programs addressing land and water degradation. The use of environmental levies is becoming more popular with local councils. A number of councils in NSW such as Coffs Harbour and Warringah have received ministerial support to increase the 'environmental levy'. Several councils in Western Australia (Katanning Shire Council, Cunderdin Shire Council, and Tammin Shire Council are examples) have raised their environmental levies to secure the employment of Landcare coordinators.

Market creation

In some cases, scarce environmental assets can be efficiently managed by creating tradeable property rights. For example, in the case of water management, environmental flows are first secured through regulation, and the market is then used to determine the price of the remaining water. Quotas are allocated to individual water users who may trade them. Establishment of markets for water was a major part of the Council of Australian Governments water reform agenda. Water markets have been established in Victoria, NSW, and South Australia, but further work is required to improve their effectiveness.

Full-cost pricing

In the past, users of natural resources such as water and timber from native forests have not provided the full costs associated with their production. Although this situation is changing, many natural resources are still under-priced. Governments have often not attempted to recover all the costs they face in producing the resource. For example, irrigation subsidies have primarily involved water prices that do not provide an adequate rate of return for investments made in public infrastructure such as dams, weirs, pipelines, and channels. In some cases, prices have not been set high enough to cover the government's operating costs. For example, in rural NSW it has been estimated that in 1992–93 there was a subsidy to irrigators of $403 million. In addition, the external costs associated with water or timber use are generally not included in the price. Charging full price encourages resource users to reduce their consumption, or to make more effective use of their current allocations.

Education

There has been considerable investment of resources over the two decades in awareness raising and education programs. There is evidence that these activities have contributed to increased awareness and understanding of land management issues, and that these changes enhance landholder capacity to adopt best practice. There is a link between concern about salinity and adoption of best practices, suggesting an important role for community education activities that raise awareness of biodiversity values and increase understanding of the complex processes that contribute to land and water degradation. Environmental education can encompass a variety of activities (field days, workshops, printed material, films, and so on) spanning all levels of government. Governments are providing resources for activities such as property management planning and community education and awareness programs about biodiversity.

Framework strategies and plans

A number of strategies in Australia address resource management issues. We have already considered many of these elsewhere in the book—examples include the *National Strategy for the Conservation of Australia's Biological Diversity*, *National Strategy for Ecologically Sustainable Development*, *National Forests Policy Statement*, and so on.

Institutional reform and partnerships

Reforming existing institutions. Organisations are continually evolving. A current trend is to expose public sector agencies to what can generally be considered 'market-based' mechanisms (including the contracting out of services, corporatisation of agencies, and so on) (Backgrounder 4.1). The main potential advantages are better coordination of efforts, more effective use of scarce resources, and potential to overcome existing inter-agency conflicts and 'turf-wars'. The disadvantage is that it may simply internalise conflict and prevent examination of different perspectives in resource management.

Establishing new institutions. Effective management of environmental problems may be perceived to require the establishment of new institutions. Landcare groups, for example, emerged in response to the need for a local-organisational structure with which to address rural land degradation. A more recent example is the establishment in 2002 of the Tasmanian NRM Council and Regional NRM Committees mentioned above.

Partnerships and agreements. It is widely recognised that increased complexity and diversity in governance arrangements as well as environmental, economic, and social activity means that individual organisations have limited capacity to achieve their objectives relying solely on their own resources.

Management agreements. Management agreements can be used to achieve environmental outcomes. Such agreements generally have low administrative costs, high levels of community involvement, and community and political acceptability. For landholders with genuine concern for the protection of natural resources, voluntary agreements can be effective. Under the Land for Wildlife program in Victoria, landholders are encouraged to register properties that are being managed for conservation purposes. Landholders become part of a network and receive recognition for their efforts, as well as support and advice. In 2000, the scheme involved over 5000 landholders, covering an area of 520,000 hectares. However, for many landholders, concern is not enough. For this reason, voluntary agreements are more successful when combined with legislation or financial incentives to encourage landholders to enter into them.

An example of the use of a range of instruments and processes to achieve conservation outcomes on private property is given in Snapshot 19.3.

Snapshot 19.3

Woodland Watch—community conservation of eucalypt woodlands, Western Australia
Read & Bessen (2003)

Woodland Watch is a World Wide Fund for Nature (WWF) project, based in the wheatbelt of Western Australia. Eucalypt woodlands in this region have been extensively cleared, with only 3% remaining of their original extent. The project targeted landholders with good quality remnant woodland, especially those community types severely under-represented in public protected areas. The locations of such areas were identified and assessed for their conservation significance; field days were held to raise awareness of their importance, management requirements, and the diversity of woodland types; and various instruments were used to improve their conservation status. As of 2003:

• working partnerships have been developed between WWF and the Western Australian Herbarium, the WA Department for Conservation and Land Management, the Avon Catchment Council, Greening Australia (WA), National Trust (WA), North Eastern Wheatbelt Regional Organisation of Councils, Alcoa, and community groups

• 85 landholders with high quality woodlands were actively participating in the project

• 87 high quality woodland sites had been identified and assessed.

• 87 flora surveys had been conducted, with 12 new populations of rare or priority flora species located, range extensions established for numerous other species, and several potentially new species identified

• significant woodland areas were protected by either conservation covenants (2135 hectares, 20 landholders), Land For Wildlife Agreements (3143 hectares, 25 landholders), or other voluntary conservation agreements.

19.4 Management lesson and principles

Lesson

Unless protected areas are very large, or are interconnected with natural areas within a land or seascape, the diversity of fauna and flora conserved will diminish or become more vulnerable with time as protected areas become islands within a modified landscape. Establishing linkages in the landscape is vital for achieving nature conservation goals as well as for the health and sustainability of rural communities.

Principles

We believe the elements required for a successful landscape-scale approach to natural resource management in Australia are as follows.

1   Protected areas are only one of a number of land-use types within a region but an essential component of any bioregional planning. Integrated planning and management across all land tenures is essential for achieving natural

Tree nursery, NSW (Graeme Worboys)

resource management goals, including biodiversity conservation.

2 Protected areas should be integrated into a regional context to ensure that there is connectivity of all components of biodiversity as well as ecosystem processes within the bioregional landscape.

3 Protected area agencies are part of the social and economic fabric of regions, and must develop partnerships with other agencies and individuals to further both conservation and regional development outcomes.

4 Protected area managers should engage with adjacent land managers to ensure that activities outside protected areas are compatible with protected area objectives, while fostering the vitality of local communities and maintaining the long-term health and viability of catchments and ecosystems.

5 The biosphere reserve model should be considered an effective institutional instrument for integrated decision-making involving protected areas across jurisdictions within bioregions.

6 Protected area managers should engage with other institutions to promote more effective coordination and avoid overlapping or competing jurisdictions and mandates. Protected area agencies should lobby for and create opportunities for improved coordination among government, community, and private sector agencies at all levels to facilitate integrated funding, planning, implementation, and monitoring for landscape-scale conservation.

7 Protected area agencies should assist with building capacity among other sectoral agencies to enable them to contribute more effectively towards biodiversity conservation goals.

8 Maintaining and enhancing nature conservation values on private land can be achieved through a mix of policy instruments, including regulations, economic incentives, partnerships, management agreements, and conservation purchases.

9 Protected area managers should participate in and facilitate environmental education and training for achieving conservation goals. Such programs are a vital mechanism for gaining an understanding of regional communities and issues, informing the community about conservation, encouraging behavioural change, and building institutional capacity to deliver improved environmental outcomes.

Endnote

1 This section draws on a workshop presentation by Green & Klomp (1996).

20

Marine Protected Areas

Karen Edyvane and Michael Lockwood, University of Tasmania

20.1	Importance of marine protected areas	533
20.2	Australian marine environments	534
20.3	Marine management issues	539
20.4	Marine park management	541
20.5	Management lessons and principles	552

While Australia has established reserves for terrestrial ecosystems for many decades, the formal conservation of Australia's marine environments and biodiversity is a relatively recent phenomenon. In 1990, the Prime Minister, Bob Hawke, announced at the meeting of the General Assembly of the International Union for the Conservation of Nature (IUCN) in Perth that Australia would establish a national, representative system of Marine Parks by the year 2000. This was facilitated by the establishment of a National MPA Taskforce and funding under the *Ocean Rescue 2000 Program* (Muldoon & Gillies 1996). The National Representative System of Marine Protected Areas (NRMSPA) is being developed cooperatively by Commonwealth, state, and territory agencies responsible for the conservation, protection, and management of marine environments (ANZECC 1998, 1999; Section 5.2).

Though there have been significant advances in the establishment of MPAs in Commonwealth waters, and in some State jurisdictions (Western Australia, Victoria, Queensland), progress has been very slow in other jurisdictions (Tasmania, Northern Territory). Victoria has recently established a network of MPAs that fully protect 5.2% of Victoria's waters (Snapshot 5.1). The recent re-zoning of the Great Barrier Reef Marine Park has resulted in 33% of the park area reserved in highly protected `green zones' (Case Study 20.2). Within tropical Australia, marine tourism has largely driven the establishment of large, multiple use MPAs, such as the Great Barrier Reef Marine Park and Ningaloo Reef Marine Park.

This has been assisted by a high degree of research and marine conservation management efforts, and the need to manage marine-based tourism, coastal development and fishing on an integrated ecosystem basis. In contrast, fisheries management has largely driven marine conservation management in temperate Australia. As a consequence, MPAs have historically been typically small, and focused principally on fisheries habitat management objectives, such as the protection of critical nursery, breeding, spawning, and feeding habitats of commercial species (Edyvane 1996).

MPAs in Australia can range from small, highly protected no-take marine reserves, to large, multiple-use marine parks (such as the Great Barrier Reef Marine Park) that permit activities such as fishing (commercial and recreational), tourism, and recreation. MPAs can be established for a variety of purposes and can provide for a range of activities while still protecting the environment. For example, they may be reserved for conservation, fisheries management, research, education, social and historical importance, tourism or recreational use, or a combination of any of these, and may also include neighbouring coastal lands and islands.

In this chapter we indicate the importance of MPAs, summarise some of the main biophysical features of Australian marine environments, the major issues faced by MPA managers, and outline specific management concerns related to marine biodiversity conservation; fishing and aquaculture; tourism and recreation; and research and monitoring.

20.1 Importance of marine protected areas

Increasingly, MPAs are being recognised as a vital tool in the protection and wise use of Australia's oceans. As a signatory to international agreements such as the *Convention on Biological Diversity*, Australia has a special responsibility for conserving and managing its marine and coastal environments and their resources. Marine Protected Areas (MPAs) are one element of an integrated marine management framework. While MPAs set aside areas for the specific conservation of biodiversity, the management of sustainable use (protection from overuse or over-exploitation) and ecosystem health (protection from pollution or habi-

tat degradation) are also key elements in maintaining the overall integrity of the ecosystem, conserving marine biodiversity and ecological processes (Kenchington 1990, Ray & McCormick-Ray 1992). Approximately 80 international agreements relate to the management of Australia's oceans. About half of them relate to managing the marine environment, including fisheries. These include the *UN Convention on the Law of the Sea 1980*, the *Convention on the Conservation of Antarctic Marine Living Resources 1980* and the *World Heritage Convention*. Despite these obligations, the 2001 State of the Environment Report for Australia noted that 'there has been a very slow improvement in the state of marine resource management' (ASEC 2001a).

Australia's Oceans Policy was released by the Australian Government in 1998; the International Year of the Ocean. This policy includes support for some innovative approaches to integrated oceans management, including the concept of regional marine plans (Case Study 5.2). These plans acknowledge the need to take an ecosystem approach to natural resource management, striking a balance between environmental, economic, and social objectives (Commonwealth of Australia 1998a).

MPAs play a major role in the conservation and management of marine environments by:

- protecting representative, rare, and unique marine ecosystems, habitats, and species
- protecting areas of cultural, educational, and recreational importance
- assisting fisheries management by acting as propagation areas supporting wild fisheries, providing 'insurance' against fisheries misman-agement in the rest of the marine environment, and by helping scientific research to understand natural fisheries and ecosystem dynamics
- supporting recreational diving and tourism.

Fully protected ('no-take') MPAs are now viewed as a key tool to help reverse widespread overfishing and habitat disturbance (Murray et al. 1999, Lubchenco et al. 2003). The declining state of the oceans and the collapse of many fisheries create a critical need for more effective management of marine biodiversity, populations of exploited species, and the overall health of the oceans. There is now widespread, international scientific consensus that the establishment of highly protected MPAs are essential in sustainable fisheries management through protection of sensitive habitats and species, the provision of reference sites, and assistance with stock management (Murray et al. 1999, Halpern 2003, Gell & Roberts 2003). Their important role is magnified by the fact that very little is known about most marine species and importantly, their ecological interactions; marine species undergo large seasonal and environmental fluctuations; and information on the possible effects of large-scale climatic and oceanographic changes is poor.

MPAs are increasingly being recognised as a vital tool in implementing an ecosystem-based approach to fisheries management (Ward & Hegerl 2003). However, because there are gaps of knowledge about how reserves work and because they are perceived to be taking something else away from dwindling fisheries, they are often vigorously resisted (Lubchenco et al. 2003). At present, fully protected marine reserves encompass only 1% of the world's oceans. At the World Summit on Sustainable Development in Johannesburg in 2002, countries agreed on ambitious targets for creating national networks of MPAs by 2012 and rebuilding overexploited fisheries by 2015. Similarly, the 2003 World Parks Congress in Durban has recommended a target of 20–30% of each bioregion for strict protection. Recent models estimate that a global MPA network meeting the World Parks Congress target might cost between $5 billion and $19 billion annually to run and would probably create around one million jobs (Balmford et al. 2004). Although substantial, gross network costs are less than current government expenditure on harmful subsidies to industrial fisheries. They also ignore potential private gains from improved fisheries and tourism and are dwarfed by likely social gains from increasing the sustainability of fisheries and securing vital ecosystem services.

20.2 Australian marine environments

Australia's marine environment extends from the coast to the boundary of its 200 nautical mile Exclusive Economic Zone (EEZ) and includes external territories in the Indian Ocean, South Pacific, Southern Ocean, and Antarctica. Australia's ocean territory extends from the northern tropical waters of the Arafura Sea and Torres Strait to Prydz Bay—the southernmost element of the jurisdiction in the Antarctic, and extends from the isolated islands of Cocos (Keeling) and Christmas in the west to Norfolk in the east (Figure 5.3). As the world's largest island, continental Australia has a wide range

Quicksilver pontoon tourist destination, Great Barrier Reef Marine Park, Qld (Graeme Worboys)

of coastal and marine environments, which stretch approximately 32,000 kilometres, from the tropical northern regions to temperate southern latitudes. Along this extensive, 61,700-kilometre (including nearby islands) coastline, there are a wide range of habitats and biological communities including rocky shores, sandy beaches, algal reefs, kelp forests, which dominate the temperate south, and coral reefs, estuaries, bays, seagrass beds, mangrove forests, and coastal saltmarshes, which dominate the tropical north. All major groups of marine organisms are represented, and many of the highly diverse groups and species are endemic– particularly in the southern temperate Australian waters. Australia has the world's highest species diversity and largest areas of tropical and temperate seagrasses, largest area of coral reefs, highest mangrove species diversity, and third-largest area of mangroves (Zann 1995).

Australia's marine environment also include less understood mid-water, outer shelf, and deepwater habitats, which include features such as the seamounts of the Southern Ocean, the extensive and massive Lord Howe Ridge system off eastern Australia, a long chain of seamounts arising from the Tasman abyssal plain and stretching from Tasmania up to the deep waters off the Great Barrier Reef, numerous deep-sea canyons and the astounding biologically active mid-ocean ridge systems such as the Macquarie Ridge

and South Tasman Rise. Intensive surveys have recorded only 5% of the Australian ocean's physical terrain, and less than 2% of its life and habitats.

Coastal and marine habitats

The coastal habitats of Australia are largely influenced by the interplay of physical factors, particularly oceanography, climate, and coastal geology (Zann 1995). The inshore environments of Australia range from low-energy muddy, tidal deltas that dominate the tropical north, to the swell-dominated rocky coasts and sandy shores of the temperate south. Muddy sediments dominate the tropical environments of Australia, either under mangrove forests or as wide, near-flat shores. Rocky shores are more widespread in temperate marine environments (particularly in the Great Australian Bight and Tasmania), occupying approximately 30% of the coastline. While sandy beaches are common all around Australia, they occupy approximately 50 to 70% of the temperate coastline, particularly along the eastern and western coast—regions swept by the Eastern Australian Current and the Leeuwin Current, respectively (Fairweather & Quinn 1995). Coral reefs occupy 14% of Australia's coastal habitat and are essentially restricted to tropical regions.

Australian estuaries occur over a wide range of geological and climatic conditions and consequently display a great variety of form (Saenger 1995). Australia has 783 major estuaries; 415 in the tropics, 170 in the subtropics and 198 in temperate areas (Bucher & Saenger 1991). In the semi-arid and arid regions of Australia, low rainfall and variable climate result in irregular freshwater inputs and prevalence of 'inverse estuaries'—estuaries that generally flow only after local rains have fallen. Gulf St Vincent and Spencer Gulf represent the largest temperate inverse estuaries in Australia, while Shark Bay and Exmouth Gulf (in Western Australia) are the largest tropical estuaries. Naturally rich in nutrients, estuaries are ecologically highly productive and are important habitats for adult marine fauna as well as critical nursery habitats for the juveniles of many species.

Estuaries were the preferred site for European settlement. In a comprehensive assessment of 972 estuaries, the National Land and Water Resources Audit found that almost half are degraded in some way, usually owing to land-use practices or human settlement pressure. For the other estuaries, pressure is assumed to be light owing to low levels of human use.

Coastal saltmarshes

Coastal saltmarshes in Australia are intertidal plant communities dominated by herbs and low shrubs, which are often associated with estuaries. When viewed in a world context, Australian saltmarshes are not as distinctly 'Australian' as are the various terrestrial communities of the continent (Adam 1995). Although there is a high degree of endemism at the species level in the saltmarsh flora, at generic and family level there is a strong similarity (in structure and composition) between Australian saltmarshes and those elsewhere (Adam 1990). Australia has 13,595 km^2 of coastal saltmarshes, with most extensive areas occurring in the tropics (Bucher & Saenger 1991). Saltmarshes provide habitat for numerous organisms of both terrestrial and marine origin. The most important fish habitats in intertidal wetlands are seagrasses and mangroves.

Mangrove forests

Australia has the third largest area of mangroves in the world (approximately 11,500 km^2) and some of the most diverse communities (Robertson & Alongi 1995). The mangrove flora of Australasia

(the area including New Guinea, New Caledonia, Australia, and New Zealand) has approximately five times the species richness of all other regions excepting the neighbouring region of Indo-Malesia (Duke 1992). A total of 43 species are found in Australia, which represents 58% of the world's mangrove diversity. One species, the newly discovered *Avicennia integra*, appears to be endemic to Australia. Mangrove forests show their greatest development on humid, tropical shorelines where there are extensive intertidal zones and abundant fine-grained sediment, high rainfall and freshwater runoff from catchments.

Mangroves are of critical ecological importance to estuarine and coastal food chains, and a vital habitat, nursery, and feeding area for juvenile fish, crabs, and prawns. Crab larvae from mangrove forests are a major food source of juvenile fish in north Queensland estuaries, while juveniles of some prawns feed directly on mangrove detritus or on other fauna that is mangrove-dependent. Some important commercial and recreational fish species are directly linked to mangroves during at least part of their life cycles, such as the Banana Prawn (*Penaeus merguiensis*), Bream (*Acanthopagrus australis*), and Grunter (*Pomadasys kakaan*) (Robertson & Alongi 1995). Other commercial fisheries have more indirect but equally important connections to mangroves. In the wet tropics of north-eastern Australia, the seagrass meadows on which juvenile Tiger Prawns (*Penaeus esculentus*) depend often occur immediately seaward of mangrove forests (Robertson & Lee Long 1991). Mangroves in Australia are also important nursery grounds for a variety of small non-commercial fish species that are an important food source for carnivorous estuarine fish such as barramundi and trevallies and jacks (*Carangidae*), and offshore fish such as mackerel.

Seagrass beds

Seagrasses are productive, widespread, and ecologically significant features of Australia's tropical and temperate near shore environments. Australia has the highest biodiversity of seagrasses in the world, the largest areas of temperate seagrass, and one of the largest areas of tropical seagrass. Australian seagrasses are characterised by high speciation and also endemism, especially in temperate regions.

Australia has over 30 species of seagrass. Of the 55,500 km² of seagrass meadows in Australia, the largest meadows occur in tropical Australia, in Torres Strait (17,500 km² or 31.5%) and Shark Bay (13,000 km² or 23.4%).

Seagrass communities are of considerable importance in the processes of coastal ecosystems because of their high rates of primary production and their ability to trap and stabilise coastal sediments and organic nutrients. Seagrasses also provide important feeding grounds for birds and nursery habitat for juvenile fish and prawns and are a critical habitat for many commercial fish and invertebrates (Cappo et al. 1998, Butler & Jernakoff 1999). In tropical areas seagrasses serve as the staple diet of Dugongs and adult Green Turtles. Dugongs' specialist requirement for seagrass habitat makes the species vulnerable to changes in seagrass location and extent. In temperate Australia, large areas (approximately 450 km²) of seagrass meadows (dominated by *Posidonia*) have been lost or degraded due to nutrient pollution in areas such as Westernport Bay and Gulf St Vincent or sand-mining in Cockburn Sound (Kirkman 1997).

Rocky reefs

Rocky shores are the main hard substrata of headlands along the open coasts and sometimes within estuaries. Rocky reefs occur along the entire length of the Australian coastline, mostly within shallow ocean waters. They are widespread in temperate marine environments (particularly in the Great Australian Bight and Tasmania). Prominent areas of reef occur seaward of most headlands and rock platforms. Significant areas of subtidal rocky reef are also found in deeper waters offshore, around islands, and in estuaries that include drowned river valleys.

Species diversity is particularly high in temperate rocky reef habitats (Keough & Butler 1995). The southern coastline has the world's highest diversity of red and brown algae (around 1155 species), bryozoans (lace corals), crustaceans, and ascidians (sea squirts). Diverse assemblages of brown, red, and green macroalgae, along with sponges, ascidians, and other invertebrates enhance habitat complexity and provide many opportunities for specialisation (Andrew 1999). Furthermore, the large macroalgae (such as kelp) that partially cover most rocky reefs enhance overall species diversity by providing patches of shaded habitat favoured by distinct assemblages of organisms (Kennelly 1995).

Rocky reef provides refuge and feeding opportunities for a wide variety of fish and mobile invertebrates. Small fish and invertebrates can escape predators by hiding in cracks and crevices. Larger fish, along with octopus and cuttlefish, appear to use rocky reef as cover from which they can ambush passing prey. Pelagic fish are also attracted to rocky reef areas by aggregations of small baitfish. Many species of invertebrates and algae live on or within rocky habitat and provide a diverse range of foods. Some fish along with abalone and sea urchins eat algae associated with rocky reefs (Jones & Andrew 1990). Many commercially and/or recreationally important fish and invertebrates such as rock lobster and abalone depend on rocky reef habitat for some or all of their life.

Sandy and soft-sediment shores

Sandy and soft-sediment shores and their biotic assemblages are unlike any other marine benthic (sea bed) habitat. Sandy and soft-sediment habitats are characterised by their three-dimensional nature (depth is an important variable); range of grain size, depth, and chemistry of the sediments (which exerts a profound influence on the types of organisms living within them); large size range of organisms that include some that ingest the sediment matrix, as well as living on or within it; and the lack of large attached plants and dominance of microscopic primary producers (Fairweather & Quinn 1995). Importantly, soft-sediment habitats provide the contiguity of habitat with other types, such as seagrasses, mangroves, saltmarsh, and open ocean that facilitates the movement of organisms among them—both between and within different stages of their life cycles. As elsewhere in the world, unvegetated, soft-sediment shores remain one of the most under-researched marine benthic habitats in Australia.

Coral reefs

The central Indo-Pacific is the world's centre of reef coral diversity, with at least 50% of the world's coral reefs, covering an area greater than 300,000 km² (Vernon 1995). Australia has the largest area of coral reefs of any nation and the world's largest single coral reef complex, the Great Barrier Reef (GBR). There are also large areas of coral reefs in Torres Strait, the Coral Sea Territories, and central and northern WA. The Tasman Sea reefs (Elizabeth and Middleton Reefs and Lord Howe Island fringing reef) are the highest-latitude coral reefs in the world, thriving in conditions that are marginal for coral growth elsewhere. Australia's coral reefs form seven distinctive groups: high-latitude reefs of eastern Australia (Solitary Islands, Lord Howe Island, Elizabeth and Middleton Reefs); the GBR; reefs of the Coral Sea (e.g. Ashmore Reef, south to Cato and Wreck reefs); reefs of northern Australia and shallow, turbid waters of the eastern Arafura Sea; Cocos (Keeling) Atoll and Christmas Island; reefs of the North-west Shelf; reefs of coastal WA such as Ningaloo (Snapshot 20.1) and Houtman Abrolhos.

The central Indo-Pacific region of northern Australia and South-East Asia supports by far the richest invertebrate fauna associated with coral reefs in the world, with more than 100 coral genera and some 500 coral species (compared with only 25 genera in the Western Atlantic and about 60 species in the Caribbean). Australia, because of its geographic position within the world's centre of marine biodiversity, is critical to the conservation of corals because of the level of degradation in the region's equatorial reefs. Approximately 70% of all central Indo-Pacific coral reefs have been significantly degraded (Vernon 1995). This degradation has occurred primarily through overfishing (which has effectively removed the top of the food pyramid in most South-East Asian and Japanese areas of reef), eutrophication, and increased sedimentation (from urban outfall, deforestation, agricultural runoff, and coastal development) and direct intrusive activities (principally through subsistence food gathering—particularly mining practices, shell collecting, and unregulated tourism).

Australia's reefs are less affected by human activities than those in other countries, mainly due to low to moderate levels of use and their remoteness, but elevated nutrients and sediments resulting from inland soil erosion are a threat in our non-arid regions. Threats to Australia's coral reefs include elevated nutrients in the inner GBR, coral bleaching events (Backgrounder 20.1), outbreaks of

Quicksilver tourist glass bottom boat observation area, Great Barrier Reef Marine Park, Qld (Graeme Worboys)

crown-of-thorns (*Acanthaster*) starfish in the outer central and northern GBR and Tasman reefs, and damage from tropical cyclones.

Offshore habitats

Australia's marine environments also include lesser known mid-water, outer shelf, and offshore, deep-water habitats. Australia's continental shelf covers about 2.5 million km², half of which is less than 50 metres deep. Its width varies from 15 kilometres off the coast of New South Wales to 400 kilometres in the Timor Sea (ASEC 2001a). The inner continental shelf encompasses the waters and seabed from the shore to the midshelf (about 50 metres deep). Beyond this is the outer shelf extending to the 'break' (typically 150 metres). The continental slope encompasses the area beyond the shelf break into water depths of 4000 metres or more.

The continental shelf consists mainly of soft sediments but, in places, may comprise gravel and pavement rock. These form substrate for a large variety of invertebrate species, including sponges and bryozoans. These long-lived communities are of critical importance as habitat and food for fish. Most of the slope environment is too deep for plants because of the reduced light conditions, and the main living organisms are the animals that live on and in the sediments. The dominant large animals are marine worms, crustaceans, echinoderms (e.g. sea urchins), and shellfish. The epifauna include hydroids, sea-pens, small bryozoans, and sponges. Beyond the continental slope and shelf break, the deep offshore environments include features such as seamounts, deep-sea canyons, and biologically active mid-ocean ridge systems. The fauna of the Tasmanian seamounts (Snapshot 20.4) are remarkably different from those of seamounts in the northern Tasman Sea (on the Lord Howe Rise and Norfolk Ridge) within the EEZ of New Caledonia (de Forges et al. 2000). The biodiversity of other Australian seamounts is still poorly understood and there is little or no information on Antarctic seamounts.

Intensive surveys have recorded only 5% of the Australian ocean's physical terrain, and less than 2% of its life and habitats. Some studies of this environment have only been undertaken in the last five years. Surveys of benthic habitat in the Twofold Shelf region off Victoria yielded over 60,000 individuals from 803 species (Coleman et al. 1997). About half of these species were previously undescribed.

Snapshot 20.1

Ningaloo Reef, WA

Ningaloo Reef is Australia's largest fringing reef, stretching 230 kilometres along a very lightly populated coastline. The reef has an estimated 300 coral species, 500 fish species, and 600 mollusc species. It is a marine park under both WA legislation (for the inshore waters) and Commonwealth legislation (for offshore waters). The MPA lies on the migration path of the Humpback Whale (*Megaptera novaeangliae*), is home to turtles and Dugong (*Dugong dugon*), and regularly hosts migratory birds. It is the only place in the world known to be visited regularly and in significant numbers by the Whale Shark (*Rhincodon typus*).

Recent surveys of the Ningaloo Marine Park show that, apart from a few localised areas, the benthic (bottom-dwelling) communities in the park, including corals and macroalgae, are in excellent condition. This suggests that to date the impact of human activity has been minimal. Most evidence of human activity is litter associated with recreational fishing activities, particularly in the vicinity of popular moorings such as Bundegi and Coral Bay, and fishing areas. Coral damage from boat moorings, boats, and divers was observed at several locations in the marine park. Coral bleaching (Backgrounder 20.1) has not been extensive and has been observed on relatively few individual colonies.

Backgrounder 20.1 Coral bleaching

Coral bleaching is a stress condition in reef corals that involves a breakdown of the symbiotic relationship between corals and unicellular algae (zooxanthellae). These microscopic plants live within the coral tissue and provide the coral with food for growth and their normal healthy colour. The symptoms of bleaching include a gradual loss of colour as zooxanthellae are expelled from the coral tissue, sometimes leaving corals bone white. Bleaching stress is also exhibited by other reef animals that have a symbiotic relationship with zooxanthellae such as soft corals, clams, and sponges. The stress factor most commonly associated with bleaching is elevated sea temperature, but additional stresses such as high light intensity, low salinity, and pollutants are known to exacerbate coral bleaching. If the causal stress is too great or for too long, corals can die.

20.3 Marine management issues

MPAs face a number of issues that, while similar to those facing terrestrial protected areas, have particular characteristics unique to the marine environment.

Lack of information. Less than 5% of Australia's marine waters have been mapped, and only about 2% of the marine biota has been surveyed. Many marine species remain undescribed.

Historically, marine research priorities have focused on economically important species and their critical habitat (mangroves, seagrasses) and species of high conservation significance (i.e. cetaceans, seals, seabirds, turtles). Very little is known about marine flora and marine invertebrates, apart from the few species that are harvested commercially or recreationally (ASEC 2001a).

Low level of public awareness. Public awareness of Australia's marine environments, habitats, and species is poor, compared to terrestrial ecosystems, due to the general uninhabitable, inaccessible nature of the marine environment. This is particularly the case for the cool, temperate marine ecosystems of southern Australia.

Openness and connectivity. The connectivity and 'openness' of marine ecosystems (the extent and rate of dispersal of nutrients, plankton, reproductive propagules, and organisms) has a marked influence on the structure and dynamics of marine populations, communities, and ecosystems. These characteristics give rise to special reserve design and management requirements for marine environments (Halpern & Warner 2003). In managing human activities within MPAs, the idea of boundaries to protect critical areas is purely nominal. Because water has the potential ability to transport not only nutrients and marine organisms, but also pollutants, into or out of protected areas, MPAs need to provide for buffer and transition zones, if specific areas are to be protected. This principle of a buffer zone protecting a core site from impact was originally developed under UNESCO's Biosphere Reserve system (Section 3.3), and although this is reasonably well established for terrestrial environments, its application to marine environments is very recent. While reserves and closed areas work well across a size range spanning <1 km^2 to >5000 km^2, the key to success is matching reserve size to the scales of managements of the organisms that they are designed to protect (Roberts et al. 2003). For sedentary animals living on coral reefs, reserves of <1 km across can maintain populations (and augment local fisheries), especially where established in networks. For more mobile species, larger reserves are required to sustain populations.

Multiple use. Large multiple use MPAs are an ideal tool for establishing buffer and transition zones to protect core areas. Small, high protection areas can be encompassed and protected within a broader integrated management regime afforded by an MPA. This concept has been encapsulated in the phrase: 'islands of protection in a sea of management' and is widely acknowledged, both nationally and internationally, as providing the best protection for marine environments, while allowing sustainable resource use (Kelleher & Kenchington 1992).

MPAs as fisheries management tools. While there is strong scientific evidence and consensus of the benefits of MPAs as a fisheries management tool (Murray et al. 1999, Halpern 2003, Lubchenco et al. 2003), there is strong resistance in Australia (and globally) to establishing fully protected MPAs for fisheries management purposes (Ward & Hegerl 2003). Critics argue that most commercial species are too mobile to benefit, that marine reserves are only appropriate in very special cases (usually small-scale tropical fisheries), and that it is too risky to implement them on a larger scale until there is more and stronger experimental proof of their efficacy. Advocates outline the results of widespread MPA research, and call for a precautionary approach, arguing that existing fisheries management approaches have failed (Gell & Roberts 2003).

Adjacent land management. The condition of near-shore marine environments is heavily influenced by what happens on the adjacent land. Integrated coastal and catchment policies and plans are required to provide an integrated framework that enables control of such influences, and minimisation of their deleterious effects. Implementation requires community involvement, cooperation of industries and governments, and alignment of regulatory regimes between Commonwealth, state, and local governments. There is still no nationally applicable coastal zone policy, and delivery of effective catchment management across all jurisdictions is still some way off.

Position of commercial stakeholders. Commercial stakeholders, particularly those related to fishing, gas, and oil extraction, have significant power in relation to MPA establishment and management (Roberts et al. 2003). Their influence is even more substantial than equivalent resource interests related to terrestrial environments (forestry, mining, agriculture). In part this is because of the relative absence of public stakeholders, particularly those that support biodiversity conservation. In this regard, scientists have a key role to play.

Effective Indigenous participation. Rights of Indigenous peoples to the use of marine resources have been acknowledged in several court decisions relating to harvesting of traditional food species, and in legislation in the case of Torres Strait Islander use of the marine environment. The High Court decision establishes that the *Native Title Act 1993* recognised native title rights in relation to the territorial sea. The decision also establishes the primacy of the public rights to fish and navigate.

Community participation. Public participation is one of the most important elements in the management of MPAs. Unless users are persuaded that the restrictions are reasonable and likely to achieve a useful purpose, measures to protect the environment are likely to be costly and ineffective. This is particularly true in marine environments where the environment and its resources are common property (Section 2.2) and where it is often very difficult to define and close boundaries or to establish and check entry points. Hence, without public and user commitment to the planning process and its outcomes, any marine or maritime management strategy will probably fail. Essential issues of marine conservation strategy and the role of protected areas within it cannot be addressed without involvement of all users. If it is to be effective, the involvement must not be dominated or directed by sectoral interests.

20.4 Marine park management

MPA management tasks encompass:

- controlling or eradicating introduced marine organisms
- enforcing 'no-take' zones
- dealing with pollution events such as oil spills
- supervising recreational fisheries
- licensing and supervision of boating and related activities and facilities (Snapshot 20.2)
- supervising recreation and tourism activities
- undertaking education about MPA establishment, values, and management
- undertaking environmental impact assessment (Snapshot 20.3)
- undertaking and supervising research and monitoring.

Some of the day-to-day management considerations in Australia's largest MPA, the Great Barrier Reef Marine Park (GBRMP), and World Heritage Area are described in Case Study 20.1. A major planning and management tool for most protected areas, but one that has particular utility in the case of MPAs, is zoning (Section 7.3). A new zoning plan for the GBRMP is described in Case Study 20.2. This plan applies comprehensive scientific

principles in the identification and selection of highly protected zones, as part of the implementation of ecologically representative MPA systems (Section 5.2). In offshore deepwater ecosystems, such as the Tasmanian Seamounts Reserve, horizontal zoning has been used to protect sea-floor habitats (Snapshot 20.4).

Governor Island Marine Reserve, Bicheno, Tas (Graeme Worboys)

Snapshot **20.2**

Mooring policy for WA Marine Conservation Reserves
CALM (2004)

Use of WA marine environments is increasing as more people enjoy recreational swimming, fishing, or diving, or undertake a tour with a commercial operator. With this increase of use, there is a need to manage locations where vessels drop an anchor or install a mooring so as to protect the environment. Marine conservation reserves in WA are vested in the Marine Parks and Reserves Authority. The Authority's function includes the development of marine reserve policies and management plans, and oversees their implementation by the Department of Conservation and Land Management. A mooring policy has been developed to:

- maintain the ecological and social values of marine conservation reserves by minimising impacts of uncontrolled mooring and anchoring activities
- enhance user safety, access, and equity in relation to moorings in marine conservation reserves
- provide a framework to accommodate present and future mooring usage patterns in marine conservation reserves.

Moorings can, where appropriately designed and sited, play an important role in protecting areas of high conservation values (such as coral reefs and seagrass) by minimising the need for anchoring, thus reducing potential anchor damage. Moorings also facilitate better access to locations of interest such as dive sites and provide a level of security with regard to safety for vessels. Private ownership of moorings in multiple-use areas can create a perception of exclusive use and displacement of existing users. Therefore, a system of public and private moorings needs to be provided for, where recreational and commercial users have the ability to access areas with minimal impact to the environment and in an equitable manner. Because of the range of biophysical characteristics and patterns of use from reserve to reserve, there is a need for a policy that provides a number of management options, thereby allowing an open, transparent, accountable, and flexible approach.

Snapshot **20.3**

Environmental impact assessment on the Great Barrier Reef
GBRMPA (2004)

Environmental impact assessments (EIAs) are undertaken by the Great Barrier Reef Marine Park Authority (GBRMPA) for developments such as pontoons, jetties, pipelines, dredging, and marinas. All proposals for 'permitted uses' within the marine park undergo an EIA. Approximately 700 assessments are conducted and decisions made each year. This includes up to 30 major projects.

The GBRMPA has a transparent, rigorous, and risk-based process for EIA. Proposed developments undergo thorough consultation, planning, evaluation of alternatives, and negotiation. Information is compiled to evaluate whether an approval (permit) or refusal will be in the best interests of the community. Assessment criteria include ecological, social, economic, and Indigenous interests, as well as current and future use of the proposed location. Potential developments that may have a significant impact on the Great Barrier Reef World Heritage Area may be referred to the Minister for the Environment and Heritage to seek advice under the *Environment Protection and Biodiversity Conservation Act 1999*. There are four main stages of an EIA: scoping, assessment, implementation, and audit. Approvals are only granted to proposals that are ecologically sustainable. Applications classed as low-risk undergo a streamlined assessment process.

There are fixed assessment costs for an application to the GBRMPA. These vary from several hundred dollars to over $70,000. For complex developments consultants may be employed by the developer for specialist studies. If a permit is granted there may be additional costs associated with monitoring and supervising permit conditions. The GBRMPA has a policy of recovering these costs from the developer. A major project will generally take between six months to three years for planning, assessment, construction and management. In some cases very large projects may continue to develop over ten years or more and require ongoing management by the GBRMPA.

Biodiversity conservation

MPAs are increasingly being used as a tool to protect rare and threatened species. The recent establishment of a system of Dugong Protection Areas in the Queensland to protect dugongs from

mesh netting and protect their seagrass feeding habitat is the world's first system of protected areas specifically for the conservation of Dugong. More recently, in NSW, the feeding, mating, and pupping sites of the endangered Grey Nurse Shark

Case Study 20.1

Day-to-day management of the Great Barrier Reef Marine Park

GBRMPA (2004)

The Great Barrier Reef Marine Park Authority and the Queensland Parks and Wildlife Service (QPWS) are jointly responsible for the day-to-day management of the Great Barrier Reef Marine Park and World Heritage Area (GBRMP). The Day-to-day Management Program guides the field operations and routine activities. This program is primarily delivered by QPWS rangers and conservation staff. Protection of park values against illegal activities is also achieved in cooperation with the Queensland Boating and Fisheries Patrol, Queensland Water Police, Coastwatch, and the Australian Maritime Safety Authority. There are approximately 100 QPWS marine park officers employed under the Day-to-day Management Program working out of 14 centres between Cooktown and Gladstone. The QPWS Marine Parks Officers manage the GBRMP through:

- resource protection
- visitor education and services
- monitoring
- surveillance and enforcement.

Activities such as tourist operations, commercial and recreational camping, and traditional hunting in reef areas require a permit. QPWS Marine Parks staff assess and approve such activity permits. Resource protection activities in the park include the installation of public moorings and 'no anchor' markers to reduce the impact of anchoring on coral reefs. Facilities are provided to ensure that visitors gain maximum enjoyment from the use of the World Heritage Area, with minimal damage to the delicate environment. In addition to public moorings and 'no anchor' markers, infrastructure includes marine parks signage, camping grounds, day-use facilities, and walking tracks. Over $2.6 million is allocated each year toward infrastructure management and maintenance.

The GBR has suffered two mass coral bleaching events (Backgrounder 20.1) in the last five years: in the summers of 1998 and 2002. The mass bleaching event that occurred in the summer of 2002 affected between 60% and 95% of reefs in the GBRMP. While most reefs that were surveyed survived with relatively low levels of coral death, some locations suffered severe damage with up to 90% of corals killed. Up to 5% of reefs on the GBR have been severely damaged during each of the last two major bleaching events. Full recovery of these badly damaged reefs will take many years to decades. Climate change is a global phenomenon that is beyond the scope of the GBRMPA to manage directly. Nevertheless, the GBRMPA is involved in collaborative research projects to monitor sea temperatures and coral bleaching. Importantly, pressure from terrestrial runoff, overfishing, and losses in biodiversity will affect the ability of coral reefs to cope with stress from coral bleaching. By minimising such pressures the GBRMPA aims to ensure that coral reefs are relieved from other pressures when coping with coral bleaching, giving them a better chance of surviving and recovering from these events.

Education is the most effective strategy to encourage compliance with marine park management principles. Though enforcement action and prosecution are two of the most important tools that managers have available, the dedication to user education indicates that these are not necessarily the tools of first opportunity, nor tools of last resort. Marine parks inspectors have the discretionary power to decide a course of action on a case-by-case basis.

Monitoring programs measure the condition of natural values and evaluate the effectiveness of management. Resource assessment and monitoring activities identify critical sites that will need to be specially managed; assess threats to park values; assess the effectiveness of current and proposed management of sites; and monitor experiences and attitudes of visitors to assist future communication strategies.

Policing the GBRMP costs around $1.7 million a year. Boat and aircraft patrols operate in the GBRMP on a daily basis, checking on activities and monitoring ecological conditions. The QPWS uses 10 primary vessels to patrol the park and also makes about 140 charter surveillance flights each year. The Queensland Boating and Fisheries Patrol carries out surveillance of remote offshore areas where illegal fishing has been identified as a major problem, as well as targeting inshore closed-area trawling and netting. Marine park inspectors appointed by the GBRMPA have the power to stop and search a vessel or aircraft; order a person from the park; seize any vessel, aircraft, or article the inspector reasonably believes to have been used or involved in an offence; give general direction to ensure compliance with the *Great Barrier Reef Marine Park Act 1975*.

The *Great Barrier Reef Marine Park Act 1975* provides for penalties of up to $22,000 for an individual who enters or uses a zone for a purpose other than that allowed for in a zoning plan. In certain circumstances, the owner of the vessel may also be liable and face penalties of up to $220,000, or, where the owner is a company, $1.1 million. Offences connected with the Environmental Management Charge (Case Study 8.2) range from $4000 to $8000.

(*Carcharias taurus*), have been protected in 10 Critical Habitat Areas. Within these areas, diving and fishing rules apply within a 200-metre aggregation zone, and a 800-metre buffer zone.

In the past five years, there has also been an increasing adoption of MPAs as a means of conserving marine invertebrates and protecting them from extractive industries. In South Australia, intertidal rocky shores (down to one-metre depth) are protected from harvesting and collecting on a statewide basis. In NSW, intertidal ecosystems have been protected through the establishment of 13 Intertidal Protected Areas. These are temporary fishing closures and complement the NSW MPA system by protecting rocky shore habitats.

Tourism and recreation

Marine tourism is a major focus of many MPAs and is also a significant part of Australia's economy (ASEC 2001a). There is a tremendous range of activities including boat cruising, whale watching, recreational fishing, recreational scuba diving and snorkelling, and exploring historic shipwrecks. Environmental aspects of marine tourism in MPAs are dealt with by regulations controlling discharges from vessels, island resorts, and tourist pontoons, and the establishment of moorings for tourist vessels to minimise anchor damage and improve safety. These regulations are managed by the states and the Northern Territory with varying degrees of effectiveness.

Sustainable tourism and recreational management requires the following types of information:

1   levels of use; interactions, perceptions and motivations of these users, particularly tourists, recreational users
2   limits of acceptable change (Chapter 16); performance/environmental indicators; trigger mechanisms/thresholds
3   information and interpretation needs of tourism operators and other recreation users.

Marine tourism has been a strong driver for the establishment of MPAs, particularly in Queensland and WA. This includes Ningaloo Marine Park (which protects Ningaloo Reef) and Shark Bay Marine Park in WA. Ningaloo Reef Marine Park (declared in 1987) includes the 260-kilometre-long Ningaloo Reef. It is the only large reef in the world found so close to a continental

Case Study 20.2

Re-zoning of the Great Barrier Reef Marine Park

On 1 July 2004, the new Great Barrier Reef Representative Areas Program (RAP) Zoning Plan for the GBRMP was enacted. The zoning plan was developed in response to concerns that current levels of protection were inadequate to maintain the biological diversity and ecological integrity of the GBRMP.

The objective of the RAP was to increase the protection of biodiversity within the park through increasing the extent of Marine National Park Zones, also called Green Zones or 'no-take' zones. This initiative is designed to:

- maintain biological diversity of the ecosystem, habitat, species, population, and genes
- allow species to evolve and function undisturbed
- provide an ecological safety margin against human-induced disasters
- provide a solid ecological base from which threatened species or habitats can recover or repair themselves
- maintain ecological processes and systems.

The zoning plan will result in a six-fold increase in highly protected areas Green Zones to around 33% of the park. Under former management arrangements only 4.5% of the park was protected from extractive practices, such as fishing and collecting. The park now features the world's largest network of no-take zones, representing all 70 bioregions within the GBRMP.

The RAP had several phases:

- classification—mapping the marine diversity in the GBRMP into bioregions
- review—determining the extent to which the existing zoning protects the biodiversity shown by the bioregions
- identification—identifying networks of candidate areas that will achieve the biological objectives of RAP
- selection—selecting from among the options of candidate areas a network to maximise benefits and minimise detrimental effects while considering social, economic, cultural, and management implications.

To assist the RAP, two independent steering committees were formed to provide advice to the GBRMPA. The preparation of the plan was guided by four 'Biophysical Operational Principles' and 11 'Social, Economic, Cultural and Management Feasibility Operational Principles' developed by the independent steering committees. Local knowledge was essential to implementing these principles.

Summary of Biophysical Operational Principles

1. Have no-take areas the minimum size of which is 20 kilometres along the smallest dimension (except for coastal bioregions, refer to Principle 6).

2. Have larger (versus smaller) no-take areas. This principle must be implemented in conjunction with principle 3.

3. Have sufficient no-take areas to insure against negative impacts on some part of a bioregion—'sufficient' refers to the amount and configuration of no-take areas and may be different for each bioregion depending on its characteristics. For most bioregions, three to four no-take areas are recommended to spread the risk against negative human impacts affecting all Green Zones within a bioregion.

4. Where a reef is incorporated into no-take zones, the whole reef should be included.

5. In each reef bioregion, protect at least three reefs with at least 20% of reef area and reef perimeter included in no-take areas.

6. In each non-reef bioregion, protect at least 20% of area. Two coastal bioregions, which contain finer scale patterns of diversity due to bays, adjacent terrestrial habitat, and rivers require special provisions (not reproduced here).

7. Represent cross-shelf and latitudinal diversity in the network of no-take areas.

8. Represent a minimum amount of each community type and physical environment type in the overall network taking into account Principle 7. This principle is to ensure that all known communities and habitats that exist within bioregions are included in the network of no-take areas.

9. Maximise use of environmental information to determine the configuration of no-take areas to form viable networks. The network of areas should accommodate what is known about migration patterns, currents, and connectivity among habitats. The spatial configurations required to accommodate these processes are not well known and expert review of candidate networks of areas will be required to implement this principle.

10. Include biophysically special/unique places. These places might not otherwise be included in the network but will help ensure the network is comprehensive and adequate to protect biodiversity and the known special or unique areas in the GBRMP.

11. Include consideration of sea and adjacent land uses in determining no-take areas. For example, existing no-take areas and areas adjacent to terrestrial National Parks are likely to have greater biological integrity than areas that have been used heavily for resource exploitation.

Summary of Social, Economic, Cultural, and Management Feasibility Operational Principles

1. Maximise complementarity of no-take areas with human values, activities, and opportunities. This is achieved by placing Green Zones (or no-take areas) in locations that:

    *   have been identified through a consultative process that is participatory, balanced, open, and transparent
    *   Traditional Owners have identified as important and in need of high levels of protection
    *   minimise conflict with Indigenous people's aspirations for their sea country
    *   protect areas that the community identifies as special or unique, such as places of biological, cultural, aesthetic, historic, physical, social, or scientific value
    *   minimise conflict with non-commercial extractive users such as recreational fishers
    *   minimise conflict with commercial extractive users
    *   minimise conflict with all non-extractive users.

2. Ensure that final selection of no-take areas recognises social costs and benefits.

3. Maximise placement of no-take areas in locations that complement and include present and future management and tenure arrangements. These arrangements include:

    *   existing or proposed zoning plans, management plans, or other related management strategies for marine areas by federal, state, or local government authorities

- existing or proposed tenure and management strategies for coastal areas (mainland and islands) in the region
- Native Title claim areas and issues.

4. Maximise public understanding and acceptance of no-take areas, and facilitate enforcement of no-take areas. This is achieved by:

- having Green Zones that are simple shapes
- having Green Zones with boundaries that are easily identified
- having fewer and larger Green Zones rather than more and smaller Green Zones.

The Draft Zoning Plan was developed using the best available information from scientists, government agencies, local communities, industry, and other organisations, together with information from the 10,190 submissions received during the formal community participation phase in 2002. The second formal phase of community participation involved more than 10,000 packages of information, more than 50,000 submission forms, and more than 76,000 draft zoning maps distributed throughout Australia. GBRMPA received more than 21,000 submissions on the draft zoning plan.

In recognition that there could be some impacts from the new plan, independent socio-economic assessments were undertaken by Bureau of Resource Sciences and the Bureau of Tourism Research, in conjunction with GBRMPA, to quantify the range of benefits and costs for all affected parties. These assessments formed the basis of the Regulatory Impact Statement that was submitted to parliament with the revised plan. The Australian Government has agreed in principle to a structural adjustment package for commercial fishers and others that may be adversely impacted by the implementation of the zoning plan.

Snapshot 20.4

Tasmanian Seamounts Marine Reserve
DEH (2004g)

The Tasmanian seamounts are located 170 kilometres south of Hobart in Australia's Exclusive Economic Zone. There are approximately 70 seamounts, each situated in unusually close proximity to one another. Seamounts are remnants of extinct volcanoes found in Australia's deep marine environment. They are typically cone-shaped, 200 to 500 metres high and several kilometres across at the base with slope gradients of 20–30°. They rise sharply from the ocean floor at depths of 1000–2000 metres beneath the sea surface, and peak at depths of 660–1940 metres below the sea surface.

Seamounts are frequently places of high biodiversity and the Tasmanian seamounts are important as places of aggregation of spawning Orange Roughy (*Hoplostethus atlanticus*). A CSIRO study of southern seamounts (Koslow & Gowlett-Holmes 1998) collected a number of rare and previously undescribed fish species, and between 25% and 50% of the invertebrate species collected were believed to be new to science and between 31% and 48% known only from this region.

The Commonwealth declared the Tasmanian Seamounts Marine Reserve in May 1999, comprising an area of 370 km_, to protect a sample of the unique bottom-dwelling communities associated with the seamount region. This is the first marine reserve of its type in the world. The major pressure on seamounts is commercial fishing. Past trawling on the shallow seamounts south of Tasmania damaged this habitat through scraping of the substrate and removing the reef aggregate and associated flora and fauna. Under the Tasmanian Seamounts Marine Reserve Management Plan commercial activity is prohibited below a depth of 500 metres, while in the upper 500 metres there is a multiple use zone, where limited commercial fishing may occur.

land mass; about 100 metres offshore at its nearest point and less than seven kilometres at its furthest. From mid March to mid May each year visitors from all around the world converge on Ningaloo Reef for the experience of a lifetime—diving with the giant whale shark (Snapshot 20.5, see also the photo in the colour section).

In Western Australia, a total of 109 whale-watch boat-tour operators are licensed (CALM 2003). In 2003, a total of 86 boat-based dolphin interactions licences, three in-water dolphin interaction licences, and three boat-based Dugong interaction licences were issued in WA. Nationally, whale watching is rising in popularity and is subject to state controls and national guidelines. The aim is for people to view cetaceans and learn about them without interfering with their migration, feeding, and breeding. Several coastal towns (such as Merimbula and Eden in NSW) have improved their economies by fostering whale and dolphin watching.

Fisheries and aquaculture

Fishing is a major activity undertaken in multiple-use MPAs and is one of Australia's major marine industries. Fishing activity is generally regulated within the overall objectives of a MPA through management prescriptions outlined in individual MPA management plans. However, the management of fishing within the MPA is largely undertaken by sectoral interests, under various fisheries legislation and responsibilities, and enforced through regulations. These sectoral management tasks include: licensing and supervising equipment and maximising the sustainable efficiency of commercial fisheries; allocating sustainable yield between commercial and recreational fishing inter-

Snapshot 20.5

Management of whale shark watching, Ningaloo Marine Park
CALM (2004)

Whale sharks are the world's biggest species of fish, reaching more than 12 metres long and weighing up to 12 tonnes. They are a protected species in WA waters. Whale Sharks visit Ningaloo Reef, off the north-west coast of WA, between March/April and June/July each year. Their appearance has resulted in the development of an increasingly popular, seasonal ecotourism industry. Western Australia is the only easily accessible place in the world to be visited by Whale Sharks on a regular basis.

Commercial Whale Shark watching in Ningaloo Reef Marine Park is one of the fastest-growing and most important nature-based tourism attractions in WA. Over the past decade, the industry has grown from an estimated 1000 passengers in 1993, to more than 2500 in 1998, 4195 in 2001–02 and 4975 in 2002–03 (CALM 2003). Approximately three-quarters of the visitors are from overseas. In 1998, Whale Shark tourism was estimated to contribute approximately $9 million a year to the WA economy.

In 1998, the Department of Conservation and Land Management (CALM) released a management plan to ensure the ecological sustainability of whale shark watching. The overall objectives were to improve the management of whale shark interactions and provide the scientific basis for modifying management controls to minimise any impacts. The management

program details the research needed to gain a better understanding of the species and to determine the possible links between environmental factors and whale shark numbers. This includes instigating long-term monitoring programs to assess how and why numbers of Whale Sharks at Ningaloo varied naturally as well as to determine if any impacts are occurring as a result of increasing tourism pressure.

Whale Shark watching is managed by the CALM under a licensing system. Only persons who are licensed by CALM are permitted to conduct commercial Whale Shark interaction tours within Ningaloo Marine Park. Currently, 15 charter boat operators are licensed. Groups of up to 10 people at a time can swim with the animals under strict guidelines prepared by CALM in consultation with the tour operators. Two codes of conduct have also been developed that apply to people swimming with Whale Sharks and to vessels (both private and commercial) operating in the vicinity of a Whale Shark. CALM has also developed log books for tourism operators who take divers out to swim with the Whale Sharks. The data from these logbooks is used to build up profiles of individual sharks and migratory habits. Demand for interaction licences is increasing. Revenue from Whale Shark licence fees (approximately $37,000 a year) is used to fund research and monitoring programs.

ests; licensing and supervision of recreational boating and related activities; and supervising the use of sites protected for the purposes of research, education, recreation, and tourism.

Fisheries managers use a combination of instruments to achieve the desired level of usage within the MPA: establishing area boundaries for specific activities; enforcing closure during parts of the year critical to life histories of species or for longer periods; setting size limits, maximum permitted catches, and harvest limits; prohibiting or limiting use of unacceptable equipment; licensing or issue of permits to provide specific controls or to limit the number of participants in a form of use; or limiting access by setting a carrying capacity which may not be exceeded (Kenchington 1990).

The MPAs created for fisheries purposes in Australia range from large closures to eliminate specific gear types (trawling) to smaller areas designed for specific habitat protection and protection of fishery nursery grounds. These MPAs contribute to the local, and possibly regional, conservation of biodiversity, although this has been only poorly documented (Ward & Hegerl 2003). The objectives and management of these fishery closures vary from regulatory closures designed to allocate resources across individual sections of the fishing industry, to voluntary and non-regulatory community-based closures designed to rebuild fish stocks across a region.

Many Australian fisheries are fully or overexploited (ASEC 2001a). While fisheries management is largely undertaken by sectoral interests, there is increasing recognition of the role of fully protected MPAs as a critical fisheries management tool and framework for the ecosystem-based management of fisheries (Ward & Hegerl 2003). The extent to which MPAs will achieve fisheries and biodiversity conservation depends on the objectives set by the stakeholders and MPA managers. Many fisheries, including tropical and temperate reef fisheries, are inherently multispecies, multigear fisheries and are difficult to manage by traditional methods. While these fisheries should benefit from the establishment of no-take marine reserves, there remains strong resistance in Australia to establishing MPAs for fisheries management goals.

In February 2001, the American Association for the Advancement of Science meeting in San Francisco released a Scientific Consensus Statement on Marine Reserves and Marine Protected Areas signed by 161 of the world's leading marine scientists, arguing that sanctuaries result in long-lasting and often rapid increases in the abundance, richness, and productivity of marine organisms. The statement declared that there is now compelling scientific evidence that marine reserves conserve both biodiversity and fisheries, and urged the immediate application of fully protected marine reserves as a central oceans management tool.

By protecting marine ecosystems and their populations, reserve networks can reduce risk by providing important insurance for fishery managers against overexploitation of individual populations (Dugan & Davis 1993, Murray et al. 1999, Halpern 2003). Fully protected reserves provide both fisheries and biodiversity conservation benefits through:

- protecting coastal ecosystem structure and, functioning
- protecting exploited populations and fisheries, enhancing production of offspring which help restock fishing grounds
- supplementing fisheries through spill-over of adults and juveniles into fishing grounds
- providing a refuge from fishing for vulnerable species
- preventing habitat damage and promoting habitat recovery
- provide enriched opportunities for non-extractive human activities
- improve scientific understanding of marine ecosystems
- maintaining biodiversity by promoting development of natural, 'unfished', biological communities
- facilitating ecosystem recovery after major human or natural disturbances.

Government regulators and the industry have recognised the need to accelerate sustainability into all aspects of the seafood industry, with the development of environmental, economic, and social indicators into fisheries management, and the need to address the effect of fishing on the environment beyond the immediate target species. As Commonwealth fisheries develop management plans they have to do a strategic assessment under

Snapshot 20.6

Management of the Great Barrier Reef trawl fishery

GBRMPA have working on an East Coast Trawl Plan with the Queensland Government and industry. The introduction of the still disputed plan follows years of challenging negotiations. Significant outcomes of the negotiations on the plan to date include:

- a capped trawl effort at 1996 levels, with an immediate 15% reduction
- mandatory turtle exclusion devices and bycatch reduction devices
- the closure of lightly trawled and untrawled areas (approximately an additional 20% of the GBRMP)
- strict monitoring and recording of bycatch species, which may result in further protective measures.

Increasingly, in the last four years, environmental management systems have been voluntarily adopted by the industry. The most effective of these so far has been the global market-driven ecolabelling program of the Marine Stewardship Council.

Bycatch Action Plans for two Commonwealth-managed fisheries have been prepared. Turtle exclusion devices are compulsory for the entire Great Barrier Reef Marine Park and Hervey Bay for all trawling. Large areas of the Park are closed to trawling, and the number of boats in the trawl fishery has been reduced.

Education sign, Bicheno, Tas (Graeme Worboys)

the *EPBC Act 1999*. There has been progress in implementing sustainable principles in some areas such as the Great Barrier Reef Marine Park where trawling has been reduced and by-catch reduction devices are now mandatory (Snapshot 20.6). In addition, Indigenous councils have implemented management plans to control their take of Dugongs and turtles at several locations.

Recreational fishing is a popular sport in Australia and is increasingly being undertaken in areas in multiple-use MPAs or adjacent to MPAs. The best estimate indicates that 30,000 tonnes of seafood is taken per year by about five million people. About 73% of recreational fishing activity is in saltwater. Within zoned or multiple-use MPAs, recreational fishing is principally controlled through gear restrictions (usually under fisheries regulations). Controls on recreational fishing are usually through size limits, gear or bag limits, and closures by season

or area. Several states have introduced angling licences to include marine recreational fishing and a national survey was instigated in 2000 to gather information on the extent of recreational fishing and fishing by Indigenous peoples. Given the high level of recreational fishing activity within and adjacent to MPAs, there is a clear need for more research on the recreational catch rates (and target species), and also, data on possible 'spill-over' benefits of fully protected MPAs.

Few aquaculture developments occur within MPAs; however, inappropriate siting of sea-based aquaculture has the potential to affect MPAs through impacts on water quality and wildlife interactions (seals, dolphins, cetaceans, seabirds). The value of aquaculture production has been growing at 14% per year since 1989. Until the 1990s, most commercial aquaculture was for oysters (edible and pearl), and a limited amount of fish and prawn culture. Recently the cage culture of fish (Atlantic Salmon, Southern Bluefin Tuna) has grown rapidly. The growth of aquaculture has brought new environmental management issues under scrutiny, particularly adjacent to and within MPAs. For example, GBRMPA introduced water quality standards for prawn farm discharges into the GBRMP.

Education

Effective education and promotion programs are a key element of managing MPAs. Educational measures ensure that those user groups that are affected by a proposed MPA are aware of their rights and responsibilities under a management plan and that the community at large supports the park's objectives. Few countries can afford the cost of effective enforcement in the presence of a generally hostile public. Conversely, where public support exists, costs of enforcement can be very low (Kelleher & Kenchington 1992). A well designed education and public involvement program can generate political and public enthusiasm for a proposed MPA and its management. Local ownership can generate support and pride and commitment to the MPA's objectives.

In Australia, marine education and interpretation programs in many tropical MPAs are undertaken by a range of tourism operators (e.g. Great Barrier Reef Marine Park, Ningaloo Marine Park, Shark Bay Marine Park), in addition to programs undertaken by park rangers and interpretive centres. For temperate MPAs, marine education programs have generally been very limited. However, underwater interpretation and education trails have been successfully established at Port Noarlunga Aquatic Reserve in South Australia and Tinderbox Marine Reserve in Tasmania. While the former trail is supported by educational groups (Port Noarlunga Aquatic Centre) and dive groups, resources are very limited. There is a critical need to extend formal marine education programs and support facilities for temperate MPAs generally.

Research and monitoring

As noted in Section 2.3, the relative paucity of information about marine processes, habits, and organisms is a major impediment to environmentally sustainable marine management in general, and effective MPA selection and management in particular. While efforts continue to remedy this situation (excellent work is being done by CSIRO Marine Research; Cooperative Research Centres for the Great Barrier Reef World Heritage Area, Coastal Zone, Estuary and Waterway Management, and Antarctica and the Southern Ocean; and National Marine Science Centre and the Australian Institute of Marine Science) there is still a long way to go.

The ecological integrity and protection of marine resources in MPAs depend on regular

Kelp, Bicheno, Tas (Graeme Worboys)

monitoring and research into the effects of human usage. Three of the key elements of a management program for MPAs are essentially monitoring tasks:

- surveillance to assess how people are using the area
- obtaining information on the condition of the MPA and the impacts on it from human use and other factors
- assessment of likely impact of new or altered uses such as facility development, and establishment of management conditions for that use (Kenchington 1990).

Monitoring should be broadly based and able to assess impacts and also detect expected as well as unexpected outcomes of protection. Many recreational, educational, and scientific uses of reserves are compatible with full protection. However, it is important to monitor reserves to ensure such uses are not causing harm. Regular monitoring can not only test the effectiveness of certain management regimes to provide protection for species and habitats, but can also assess the effectiveness of reservation, by examining areas before and after establishment as MPAs. In temperate Australasia, long-term monitoring of Marine Reserves in Tasmania (Edgar & Barrett 1997, 1999) and New Zealand (Babcock et al. 1999, Kelly et al. 2000), has demonstrated strong conservation and fisheries benefits.

20.5 **Management lessons and principles**

Lessons

1 Sustainable multiple use MPAs should provide areas of high protection 'no-take' zones over large areas, as well as providing for levels of sustainable use and access. Zoning should be used to establish buffers to protect core areas of protection.

2 Linkages between marine environments should be recognised—marine organisms, their food and pollutants can travel considerable distances.

3 Linkages between marine and terrestrial environments should be recognised—provision must be made for control of activities occurring outside a MPA which may adversely affect features, natural resources, or activities within the area.

4 A collaborative and interactive approach between the governments, agencies, industry, and other stakeholders is essential.

5 Science-based operational principles and targets are essential for effective preservation and management of marine biodiversity, and should be developed by scientific stakeholders.

6 Protecting a significant fraction of the sea from fishing will help sustain biological diversity, ecosystem functioning, and provide resilience against human and natural catastrophes.

Principles

1 MPAs should contain a representative sample of marine ecosystems in the IMCRA planning framework for the Australian Exclusive Economic Zone, including state, territory and Commonwealth waters.

2 MPAs should be designed to incorporate zones ranging from core conservation zones, affording high levels of protection, to sustainable multiple-use zones, accommodating a spectrum of human activities.

3 MPAs should, where possible, facilitate, integrate, and assist the sustainable management of economically important species.

4 MPAs should, where appropriate, provide economic benefits.

5 MPAs should, where appropriate, protect the rights of non-extractive uses (recreation, education, tourism, navigation, and so on).

6 Public education about the marine environment, and the role and benefits of MPAs should be provided to ensure MPAs are successfully established and community-owned.

7 Research and monitoring of marine life, habitats, ecosystems, and human activities within MPAs and adjacent areas is essential to ensure effective management.

8 Effective compliance promotion and law enforcement should be provided for each MPA.

Evaluating Management Effectiveness

Marc Hockings, University of Queensland
Fiona Leverington, Queensland Parks and Wildlife Service
Robyn James, Queensland Parks and Wildlife Service[1]

An effective leader knows the way, shows the way, and goes the way.

Anonymous

21.1 Purposes of management effectiveness evaluation 554

21.2 Developing evaluation systems 558

21.3 Guidelines for evaluating management effectiveness 563

21.4 Management lessons and principles 567

How well are protected areas meeting their conservation objectives and protecting their values? Are we able to manage them to cope with increasing threats such as weeds, feral animals, and overuse? How do we measure this and adapt management so that protected areas will be maintained for now and the future?

All over the world, huge investments of money, land, and human effort are being put into protected area acquisition and management. However, declaration alone does not guarantee the conservation of values. Now that more than 10% of the world's land surface is in some form of protected area, attention is turning from increasing the quantity of protected areas to improving the quality of their management. In most cases we have little idea of whether management of protected areas is effective. And more importantly, what little we do know suggests that many protected areas are being seriously degraded by a variety of threats. Many seem to be in danger of losing the very values for which they were originally protected.

Internationally, attention is focusing on the extent of threats such as poaching and encroachment on protected areas. Carey et al. (2000)

reported that most protected areas face multiple, serious threats and that their values have been significantly degraded. And here in Australia there are also serious threats to many protected areas. These include weeds, feral animals, excessive visitor use, and impacts of surrounding land-use (Chapter 14). As with other countries we also appear to lack the capacity to effectively manage many protected areas. For example, some studies have reported that funding for protected areas may be as low as 30% of the amount required (Beeton 2000).

We clearly need to understand what is happening and then invest often scarce resources into managing existing areas to cope with escalating threats and pressures. Therefore, methods of evaluating management effectiveness are growing in importance. An increasing number of people have been developing ways to monitor and evaluate the effectiveness of protected areas (Hockings 2003). This has led to a growing awareness that evaluation of management effectiveness and application of findings is at the core of good protected area management. Essentially, evaluation enables managers to reflect on experience, allocate resources efficiently, and, assess and plan for potential threats and opportunities.

21.1 Purposes of management effectiveness evaluation

Management effectiveness evaluation measures the degree to which a protected area is protecting its values and achieving its goals and objectives. It enables managers to improve conservation and management of protected areas. Evaluation enables managers to allocate resources efficiently and plan for potential threats and opportunities. Because evaluation involves judging management, some people see it as negative or threatening. However, management effectiveness evaluation should be a positive process, which allows us to correct and learn from our mistakes and build on success. The purpose of evaluation should be made clear at the beginning of the process. Four broad purposes for evaluation are outlined below.

Promoting better protected area management, including a more reflective and adaptive approach. By comparing evaluations over time, changes and threats may be

noticed as well and the effects of changes to management. For example, an individual park evaluation may indicate that the condition of visitor facilities and visitor satisfaction at a particular national park is declining. Sometimes a significant outcome of evaluation is to demonstrate effective management practices and to provide justification for their continued support (Case Study 21.1).

Guiding project planning, resource allocation, and priority setting. Some conservation organisations are now developing models to set priorities and allocate resources. Evaluation plays a key role in these models, which generally establish a minimum acceptable standard for different criteria and then assess protected areas against these standards. The conservation importance of protected areas, their suitability for particular uses (such as tourism), and their current threats are usually taken into account.

Findings of evaluation can also influence resource allocation by indicating which programs are most effective in achieving objectives.

Providing accountability and transparency. Evaluation can provide reliable information to the public, donors, and other stakeholders about how resources are being used and how well an area is being managed. For example, the public often want concrete evidence that funding is benefiting conservation or that a particular project is achieving its goals. Where protected areas are managed by more than one party, through joint management arrange-

ments, regular and impartial evaluations provide a basis for ensuring that obligations are met.

Increasing community awareness, involvement and support. As mentioned above, a chronic resource shortage is a common feature of protected area systems, and public support—sometimes serious public concern—is needed to convince governments to provide better resourcing. Evaluation processes can alert the community to threats and can demonstrate the need for better support for or resourcing of protected areas. Results, especially from independent evaluators, can spur public action on park management issues.

Case Study 21.1

Is the management plan achieving its objectives?

Glenys Jones, Parks and Wildlife Service Tasmania

When the first management plan for the Tasmanian Wilderness World Heritage Area was being developed, an insistent question kept emerging: 'How would we know if management under the plan was actually achieving its objectives?' To address this question, the Parks and Wildlife Service set out to develop a practical system for monitoring and evaluating the effectiveness of the management plan. The result was a management evaluation system that integrates monitoring, evaluation, and reporting into the overall management cycle for the area. This system is operating successfully in the Tasmanian Wilderness World Heritage Area (TWWHA) and provides informed feedback that enables managers and stakeholders to see how management is progressing in relation to the area's objectives. This feedback helps guide adaptive management and continuous improvement in management performance, consistent with international environmental management system standards such as ISO 14004. The management evaluation system is simple and flexible, and can be scaled up or down to suit a broad range of management contexts.

When evaluation of management effectiveness was first recognised as an essential—but missing— component of the Parks and Wildlife Service's management process, a consultant with expertise in evaluation (Dr Helen Dunn) was engaged to work with departmental staff to develop an evaluation framework for the 1992 TWWHA management plan. An important associated aim of the project was to enhance the capacity of the managing agency to undertake effectiveness evaluation to allow progress in management of the area to be reported. The experience and lessons learnt in establishing an evaluative approach to management are described in Jones and Dunn (2000).

By the time the first management plan for the TWWHA was due for revision, planning staff within the managing agency were well positioned to integrate a structured approach to evaluation into the new (1999) management plan (Parks and Wildlife Service Tasmania 1999). This management plan was awarded the Planning Institute of Australia's State and National Awards of Excellence in the category for Environmental Planning or Conservation, and the Planning Minister's Award for overall winner across all categories.

The management evaluation system for the Tasmanian Wilderness World Heritage Area is illustrated in Figure 21.1. The integration of performance monitoring, evaluation, and reporting into the management cycle for the protected area generates informed feedback that enables managers to learn from and improve on past management approaches, and so progressively improve management performance.

Figure 21.1   Management cycle for the Tasmanian Wilderness World Heritage Area (Parks and Wildlife Service Tasmania, 2004)

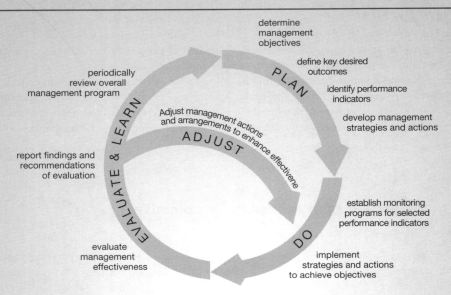

Two key documents support the management evaluation system: the management plan and a linked 'State of the Park Report', which evaluates the effectiveness of management under the plan. The contents of the management plan include:

- management objectives
- clear statements of key desired outcomes from each objective (that is, statements of the on-ground results that would be expected if the objective were fully realised)
- prescriptions for management strategies and actions to achieve the objectives
- requirements for performance monitoring, evaluation, and reporting
- requirements for review of the management plan.

The contents of the 'State of the Park Report' include evidence of management effectiveness; stakeholders' assessments of management performance; and proposed actions for enhancing management performance.

A number of key messages emerge from our experience with the evaluation process.

Integrate evaluation into the management system

- Ensure that evaluation of management effectiveness is integrated into the overall management plan or program.
- Minimise problems associated with 'shifting goal posts' by aligning management and evaluation programs to stable long-term mandates such as the obligations of legislation, the World Heritage Convention, or long-term funding arrangements.
- Facilitate continuity and consistency in evaluation programs through the use of long-term tenured positions for key staffing roles.

Monitoring performance indicators

- For each management objective ask 'How would we know if management was working well?' and just as importantly 'How would we know if management was failing?' The answers to these questions assist in developing meaningful statements of key desired outcomes, and suggest the types of performance indicators that should be monitored for evidence of management effectiveness.
- Use the in-depth knowledge of those with management responsibility and/or expertise in particular fields to assist in identifying appropriate and practical performance indicators and monitoring methodologies.
- Make sure your indicators are monitoring effectiveness in achieving the key desired outcomes, not just activities or processes.
- Remember, you can't monitor everything! Prioritise monitoring needs so that they will compete realistically alongside other demands on the total management budget.
- Start monitoring programs simply, with a core set of essential performance indicators. Expand the program as time and experience dictate.
- Where possible, integrate monitoring programs for performance indicators into the relevant operational management program.
- Ensure that data used in the evaluation are scientifically valid and/or from reliable sources. Identify all sources of data for the evaluation.
- Get baseline or reference data on performance indicators as early as possible so that changes over the management period can be detected.

Assessments of performance and feedback into ongoing management

- In addition to measured performance data, invite assessment and critical comment on management performance from those who can best provide legitimate and credible assessments for each objective or area of management responsibility. The inclusion of external sources for assessments increases the credibility of the evaluation.
- Ask assessors to identify the key factors that contributed positively to, and that hindered or threatened management performance over the management period.
- Ensure that the findings of evaluation feed back into and inform ongoing management decisions and budget processes.
- Celebrate successes! Use the findings of evaluation to give recognition to management programs that have been demonstrated to be effective, and to acknowledge the people behind those programs.

Fostering adoption of evaluation

- Encourage agency adoption of an evaluative approach to management through the influence of appropriate stakeholders, advisory forums, or the establishment of formal requirements for performance reporting.
- 'Sell' the advantages of evaluation as a means of providing sound information that equips managers and other decision-makers to make the best use of resources; increasing transparency in management; improving on-ground management results; and reducing community conflicts.
- Foster agency learning in evaluation by working with staff who are receptive to new ideas and who can take the lead in establishing evaluation programs and so become role models for others to follow.
- Foster internal and external support networks for the evaluation program. Stakeholder support can provide a protective buffer for the program during periods of potentially destabilising change such as agency restructures, personnel changes, or changes in government.
- Develop community expectations for transparency and accountability in management through performance evaluation and reporting in regular 'State of the Parks Reports'.

21.2 Developing evaluation systems

The Fourth World Parks Congress in Caracas in 1992 recommended that IUCN develop a system for monitoring management effectiveness of protected areas. To address the issue, IUCN convened an international task force within its World Commission on Protected Areas (WCPA). The work of the task force resulted in a publication, *Evaluating effectiveness: a framework for assessing management of protected areas* (Hockings et al. 2000), which provides a framework and principles for evaluation of management effectiveness. The framework and examples of its application were presented at the Fifth World Parks Congress held in Durban, South Africa 2003.

The WCPA task force developed a framework rather than a standard global methodology, because different situations require different types of assessment. In particular, there are major differences in the time and resources available for assessment in different parts of the world. Issues of scale and differences in (i) the nature of management objectives, (ii) threats and impacts, and (iii) available resources, all affect the choice of evaluation methodology. The framework helps in the design of assessment systems by providing a structure and process for developing an evaluation system, together with a checklist of issues that need to be measured, suggesting some useful indicators and encouraging basic standards for assessment and reporting.

The WCPA framework is based on the premise that the process of management starts with establishing a vision (within the context of existing status and pressures), progresses through planning and allocation of resources and, as a result of management actions, produces results that (hopefully) lead to the desired outcomes. Monitoring and evaluation of these stages provide the link that enables planners and managers to learn from experience. Figure 21.2 presents a common framework within which evaluation and monitoring programs can be established, combining context, planning, inputs, processes, outputs, and outcomes. The elements to be measured are summarised in Table 21.1.

Ideally, assessments should cover each of these elements, which are complementary rather than alternative approaches to evaluating management effectiveness. However, assessments are driven by particular needs and resources and a partial evaluation can still provide very useful information (Snapshot 21.1).

Snapshot 21.1

Evaluating the dingo education campaign, Fraser Island

There have been serious concerns about human safety and dingoes on Fraser Island World Heritage Area, especially after a child was fatally attacked in 2001. Queensland Parks and Wildlife Service (QPWS) commissioned an external evaluation by Environmetrics Pty Ltd and Beckmann and Associates to assess the effectiveness of dingo education strategies on the island. The evaluation was able to investigate the topic in detail with a comprehensive literature review, followed by stakeholder consultation, visitor surveys, and a sign audit and analysis of Fraser's dingo-related interpretive material.

The results showed that QPWS was communicating key messages very well to almost all audiences but highlighted some areas where communication approaches could be enhanced. Certain recommendations have already been implemented. For example, improved liaison with commercial operators at island entry points ensures that all tourists receive the dingo-awareness message upon arrival, commitment to face-to-face interpretation at campsites has been guaranteed, and a new communication plan has been developed.

Figure 21.2   Evaluation in the management cycle

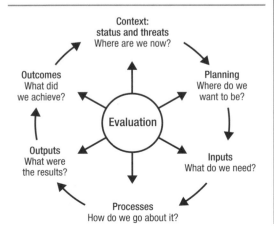

Table 21.1	Summary of the WCPA framework		
Elements of evaluation	Explanation	Criteria that are assessed	Focus of evaluation
Context	What is the current situation? Assessment of importance, threats, and policy environment	Significance Threats Vulnerability National context Partners	Status
Planning	Are the design of the area, planning systems, and plans adequate? Assessment of protected area design and planning	Protected area legislation and policy Protected area system design Reserve design Management planning	Appropriateness
Inputs	Are resources for management adequate? Assessment of resources needed to carry out management	Resourcing of agency Resourcing of site	Adequacy
Processes	How is management carried out and does it meet relevant standards? Assessment of the way in which management is conducted	Suitability of management processes	Efficiency and appropriateness
Outputs	What were the results? Assessment of the implementation of management programs and actions; delivery of products and services	Results of management actions Services and products	Effectiveness
Outcomes	What has been achieved? Assessment of the outcomes and the extent to which they achieved objectives	Impacts: effects of management in relation to objectives	Effectiveness and appropriateness

The framework provides a structure for designing an evaluation system (Backgrounder 21.1). It can be used to develop a system for evaluation at any level, from a whole protected area system (or for sites across nations such as World Heritage areas) down to a single protected area or part of it. It can also be used to look at specific issues of management such as community involvement. For each of the framework elements above, specific indicators and questions will be framed, so that the evaluation meets its stated objectives. The types of indicators for each element are discussed below.

Context: what is the current situation? This review element aims to answer the following questions.

- Why is the area/system important: what are its values on a local, regional, and global scale?
- What are the stresses and threats facing the protected area/system?
- What is its broad policy and managerial environment?
- What is the role and effect of stakeholders on management?

This information helps put management decisions in context and is also critical for management planning. If the area has a management plan, much of this information may already be compiled. This may be the main assessment used to identify priorities within a protected area network, or to decide on the time and resources to devote to a special project.

Planning: are the design of the area, planning systems and plans adequate? The planning element of evaluation asks questions such as:

- How adequate is protected area legislation and policy? Is the legal status and tenure of the site clear?
- How do site characteristics, such as size and shape, influence management?
- Is there an up-to-date management planning process?

The selected indicators for evaluation will depend on the purpose of assessment and its scale. For whole protected area systems, ecological representativeness and connectivity will be particularly important. Assessment of individual protected areas will focus on the shape, size, location, and detailed management objectives and plans.

Inputs—are resources for management adequate? This element considers the adequacy of available resources—staff, funds, equipment, and facilities—in relation to the management needs of an area. The evaluation will assess:

- Does the site have the resources needed to meet its management objectives?
- Are resources used in the best way?

Process—how is management carried out and does it meet relevant standards? Assessment looks at how well management is being carried out. Indicators may include policy development, enforcement, maintenance, community involvement, and systems for natural and cultural resource management. These basic questions are being asked:

- Are the best systems and processes for management being used, given the context and constraints under which managers are operating?
- Are established policies and procedures being followed?
- What areas of management need attention to improve the capacity of managers to undertake their work (more resources, staff training, and so on)?

Outputs—what are the results? Output monitoring focuses on whether targets (set through management plans or a process of annual work programming) have been met as scheduled and what progress is being made in implementing long-term management plans. The questions are:

- Has the management plan and/or work program been implemented? If not, why not?
- What are the results or outputs from the management process?

Outcomes—what has been achieved? This question evaluates whether objectives of a protected area have been achieved: principally whether values have been conserved and whether threats to these values are being effectively addressed. Outcome evaluation is most meaningful where concrete objectives for management have been specified either in national legislation, policies, or site-specific management plans. The main questions are:

- Has management maintained the values of the site and achieved the other site management objectives?
- Are threats to these values being adequately addressed?

Outcome evaluation is the most important test of management effectiveness. To be accurate, outcome evaluation will often require long-term monitoring of the condition of the biological and cultural resources of the system or site, socio-economic aspects of use, and the impacts of management on local communities. The selection of

| Backgrounder 21.1 | Process for developing a monitoring and evaluation system |

Developing a system for monitoring and evaluating management effectiveness using the WCPA Framework involves making a number of decisions about the purpose and scale of the assessment process. Following this process can help in selecting an appropriate methodology from the range of existing systems available, adapting these as necessary to meet particular needs, or in developing a methodology to meet local needs and circumstances. A process for establishing an evaluation system is set out in Figure 21.3.

Figure 21.3   Process for establishing an evaluation system

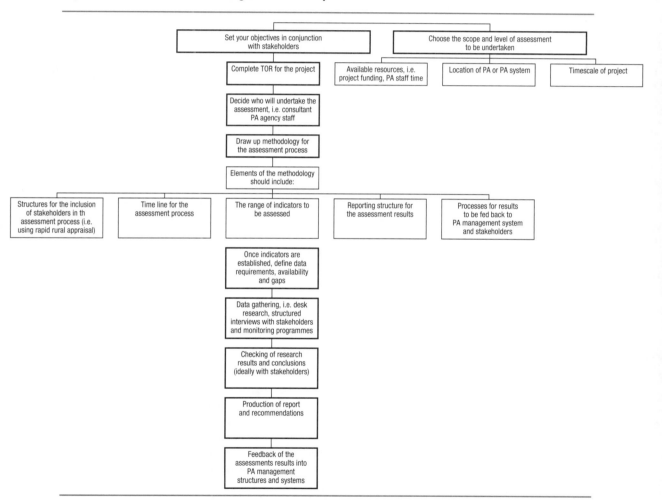

indicators to be monitored is critical so that resources are not wasted on monitoring features that cannot help manage the most critical issues. In the absence of monitoring results, qualitative assessments and expert opinion may still yield useful conclusions.

Applying the framework

Several methodologies for evaluating management effectiveness are now being applied all over the world (Backgrounder 21.2). Many of these are based on the WCPA framework. Depending on available time and resources and the objectives of evaluation, processes range from complex to simple and cheap. While most of the developmental work on management effectiveness evaluation was carried out in terrestrial (especially forest) protected areas, work in marine protected areas is now accelerating and additional guidance on relevant indicators and monitoring programs for marine environments has been prepared (Pomeroy et al. 2004).

| **Backgrounder 21.2** | Internationally applied systems for evaluating management effectiveness |

Major studies and guidebooks that provide the basis for a number of evaluation studies include:

- World Bank WWF Forest Alliance tracking tool (McKinnon 2003)
- RAPPAM methodology (Ervin 2001)
- Marine protected areas evaluation guidebook (Pomeroy et al. 2003)
- Enhancing our heritage toolkit (Hockings et al. 2001)
- ProArca CAPAS (Courrau 1999)
- WWF/CATIE (Cifuentes et al. 2000)
- 5S Threat Analysis (TNC 2002)
- Adaptive management-based evaluation (Margoluis & Salafsky 1998, 2001).

Three major evaluation systems that have been developed and applied in a number of countries around the world are outlined below.

Rapid assessment and prioritisation of protected areas management (RAPPAM). WWF International has developed and field-tested a tool for assessing the management effectiveness of protected area systems at a national level, collecting and comparing information on all or most of the sites within a system (Ervin 2003). The methodology has been applied in countries such as China, South Africa, Bhutan, Georgia, Lao PDR, and Russia. Evaluation consists of review of available information and a workshop-based assessment using the Rapid Assessment Questionnaire (available on the WWF web site, www.panda.org), analysing findings and making recommendations. The process involved park staff, local communities, scientists and NGOs. The results allow comparisons to be made across sites with the intention of:

- identifying management strengths and weaknesses
- analysing threats and pressures
- identifying areas of high ecological and social importance and vulnerability
- indicating the urgency and conservation priority for individual protected areas
- helping to improve management effectiveness at the site and system level.

Enhancing our heritage: monitoring and managing for success in natural world heritage sites. This methodology was developed as part of a four-year United Nations Foundation, IUCN, and UNESCO project to improve monitoring and evaluation in natural World Heritage sites. A monitoring and assessment toolkit based on the WCPA framework has been developed to help managers and stakeholders assess current activities, identify gaps, and discuss how problems might be addressed (Hockings et al. 2001). The methodology involves field monitoring, workshops, and stakeholder interviews. It uses a participatory process involving field staff, local communities, other stakeholders, and both local and international NGOs. The project web site (www.enhancingheritage.net) provides full details of the methodology and examples of its application. The methodology is being applied at 10 pilot sites including Bwindi Impenetrable National Park.

Bwindi Impenetrable National Park is managed by the Uganda Wildlife Authority (UWA) primarily to conserve its rare afromontane vegetation, and associated wildlife including mountain gorillas. Gorilla-based tourism provides a major source of income for the area and has been the main focus of conservation. However, the need for knowledge and assessment of a far wider range of species and interactions was recognised. The evaluation project is thus helping update natural resource information including vegetation maps and endemic species, especially lesser known plants and animals. The project is also researching the sustainability of local non-timber forest product harvesting and assessing systems that monitor such harvesting. Methods for minimising crop raiding by wild animals, including the research and testing of new methods and deterrents, are also being examined.

In Bwindi, UWA has developed good relationships with the local stakeholders around the park. In particular, 20% of the park's revenue from entrance fees is directed towards meeting the basic social and economic needs of the local people. The local people took an active part in park management assessment, which was a new and challenging step for the park managers. The results have led to an increased awareness of management issues and conservation objectives. One positive outcome has been the handing over of 4.2 km^2 of land by local people for gorilla conservation (Stolton & Chong-Seng 2003).

Reporting progress in protected areas: a site-level management effectiveness tracking tool. The World Bank/WWF Alliance for Forest Conservation and Sustainable Use has developed a simple, site-level questionnaire-based assessment system for tracking progress in management effectiveness (Stolton et al. 2003). The Tracking Tool has been applied in 122 protected areas in 23 countries. Early findings have revealed that the major threats to these areas are encroachment, poaching, and logging. This rapid method is not, however, designed to replace more thorough methods of monitoring and assessment for purposes of adaptive management.

21.3 Guidelines for evaluating management effectiveness

Based on experience in management effectiveness evaluation over the past decade, a number of general guidelines have been developed. These guidelines are derived from discussion held at an international workshop on management held in Melbourne in February 2003 (Braun 2003). The guidelines are listed in brief, grouped according to relevant aspects of evaluation.

Support and participation

Effective evaluation needs a high level of support and commitment from protected area management agencies as well as from other parties involved. This is essential both for the smooth running of the evaluation process and for making sure the evaluation results in the changes needed in management. Evaluation should be integrated into management so that it becomes an accepted, integral part of doing business (Case Study 21.1).

Good communication is essential from the beginning of the evaluation and at all stages throughout. Evaluation always involves a group of people, including, at a minimum, the evaluators and management agency staff, and usually a range of other stakeholders. Building a team with the evaluators and the participants is also important. In most cases, the evaluation process should be regarded as a team effort to obtain positive change, rather than as a potentially threatening and punitive process.

Purpose, objectives, and scope

The evaluation system should have clearly defined objectives and an implementation plan. It is important

Penguin researcher Amy Jorgensen (right) and volunteer monitoring Fairy Penguin population numbers, Montague Island Nature Reserve, NSW (Graeme Worboys)

at the beginning to also understand the levels of resourcing and support that can be expected. All parties need to agree on the project expectations. The different purposes of management effectiveness evaluation influence how the evaluation process is designed and implemented. Often an evaluation process can be designed to fulfil several purposes.

The scope and scale of the evaluation need to be established at the outset. The scope of evaluation can be very broad—the evaluation of all aspects of management—or specific—for example, looking at how effective a particular education program or weed-control initiative has been. The scope should also specify whether this is a one-off evaluation, a time-bound evaluation (for example, over the life of a short-term project), or the establishment of a continuing program.

The scale can vary from system-wide (or even embracing a number of national systems) to a specific protected area or a location. Evaluations of broad scale and scope are likely to be relatively superficial but can provide vital information for management at high levels, such as system-wide resource prioritisation, advocacy, and policy directions (Snapshot 21.2). Localised or issue-specific evaluations are useful for improving management at a practical on-ground level. In practice an evaluation program for a large or complex protected area will consist of a number of monitoring and evaluation projects operating at differing scales and with different objectives (Case Study 21.2). Integrating these into a coherent program is frequently a challenge.

Methods

Once the objectives, scope and scale have been clarified, the primary aspects of developing an evaluation methodology are:

- selecting the evaluation team
- defining the questions or indicators (usually in several 'layers' where lower-level questions or indicators help to answer higher-level questions)
- choosing the ways in which information should be obtained (literature studies, interviews, questionnaires, observations, scientific studies, and so on)
- deciding how the information will be analysed and reported

Snapshot 21.2

Evaluation of effectiveness of protected area management in India
Singh (2003)

An evaluation was conducted by the Indian Institute of Public Administration and the Centre of Equity Studies (commissioned by the Indian Ministry of Environment and Forests). The aim was to survey the status of protected areas in India, including the legal and administrative status, socio-economic pressures, management planning and implementation, staffing, research, monitoring, and tourism.

The previous evaluation done (1984–87) identified gaps in management and resources and led to significant increases in the investments on the protected area network. Other benefits included amendments to the laws governing wildlife and protected areas, and to government acceptance of recommendations regarding ecodevelopment activities around protected areas. The current study will also help the Government of India to:

- assess the effectiveness of remedial measures in response to the last evaluation
- highlight other issues needing attention
- recommend legal and policy changes
- prioritise protected areas for special attention and investment
- help the government take stock of its performance.

- considering how the information can be applied to meet the evaluation objectives.

Designing a methodology for evaluation can be a daunting task for managers. However, there has been a great deal of thought put into existing methodologies, and the use or adaptation of these can save considerable resources as well as allow comparability of results between projects or sites. For example, a guidebook for evaluating marine protected areas has been developed, based on the WCPA framework (IUCN 2003e).

Adopting a methodology does not mean all the indicators, survey methods, or reporting proformas of a previous project need to be used. These can and should be tailored to fit specific needs. Flexibility should be retained and methodologies can be improved over time, though changes will

Case Study 21.2

Evaluating management effectiveness in the Great Barrier Reef Marine Park

Jon Day, Great Barrier Reef Marine Park Authority

Around 50 monitoring and management programs are currently being undertaken for physical, biophysical, biological, social, and governance aspects (see examples in Table 21.2) of the Great Barrier Reef Marine Park. Although assisting in evaluation these projects have been undertaken as 'stand-alone' tasks and collectively do not constitute a systematic evaluation of management effectiveness across the entire Marine Park.

Table 21.2	Examples of evaluation assessments undertaken in the Great Barrier Reef	
Type of evaluation	Comments	Reference
State–Pressure–Response model	Summarised in the *State of the Reef Report 1998*	Wachenfeld et al. (1998)
Day-to-day management reports	Reporting quarterly and annually against targets set for such aspects as vessel patrols	DDM (2002)
Reactive Monitoring Report for Great Barrier Reef World Heritage Area	Report to World Heritage Committee assessed against five priority action areas; updated annually 2000–02	GBRMPA (1999)
Effects of sea dumping on nearby fringing reefs and seagrasses	A reactive monitoring program with decision thresholds developed to manage effects of port developments (dredging and dumping) on nearby corals and seagrasses	Benson et al. (1994)
Environmental effects of prawn trawling in the Great Barrier Reef	A five-year study into the effects of trawling on seabed communities in the Far Northern Section of the Great Barrier Reef	Poiner et al. (1998)
Long-term monitoring of key organisms across the Great Barrier Reef	Annual monitoring of status and natural variability of populations of corals, algae, and reef fishes from 48 reefs and crown of thorns starfish from 100 reefs to assist with management decisions	Sweatman et al. (2000)
Effects of line fishing	Monitoring recovery of exploited stocks following baseline surveys and manipulations of fishing closure strategies	Mapstone et al. (2003)
Audit of performance of East Coast Trawl Plan	Audit of the East Coast Trawl Management Plan to examine how well the trawl fishery is managed against the ESD objectives of Queensland fisheries legislation	Huber (2003)
Assessment of a new network of no-take areas against biophysical principles	Sets measurable objectives for 11 biophysical operating principles against which the new 'no-take' zoning network can be assessed	Day et al. (2000)

In a move toward a more holistic park-wide evaluation, the Great Barrier Reef Marine Park Authority (GBRMPA) has developed seven Key Performance Indicators (KPIs) derived from the Authority's main objectives. These KPIs do not replace the more detailed assessments, but provide a 'broad-brush' evaluation at a park-wide scale and are included in the agency's Annual Report.

The *State of the Reef Report* is a dynamic web-based product that allows the GBRMPA to continuously update information and increase user flexibility and access to information. It is available on the GBRMPA web site http://www.gbrmpa.gov.au/corp_site/info_services/publications/sotr/. Some of the challenges regarding evaluating management effectiveness on the Great Barrier Reef Marine Park include:

- the problems of 'shifting baselines'
- the difficulties of monitoring a dynamic marine ecosystem
- the setting of appropriate targets
- the maintenance of ecological integrity and natural processes within limits or ranges of variation that are considered 'acceptable' or 'desirable'.

Establishing an effective and appropriate evaluation system has not been without its problems. Some monitoring methodologies require destructive sampling/ killing of individual species, often at a questionable level of appropriateness in a World Heritage Area. New non-destructive sampling practices (such as Baited Remote Underwater Videos) have been developed by the Australian Institute of Marine Science that assist monitoring while minimising the damaging side-effects. There is also a need for complementary monitoring outside the GBRMP to understand the wider context; and more social/economic evaluation and monitoring to enable sound decision-making (the 'triple-bottom line' approach to monitoring and reporting).

The GBRMP is not a typical marine protected area because of its enormous size and complexity. Nevertheless like most marine protected areas, the evaluation of management effectiveness still has a way to go before it becomes totally integrated into the adaptive management cycle.

lessen the comparability of results and should be kept to those essential for better evaluation. The methodology used should be simple, repeatable, and transparent. Limitations to the process, including knowledge gaps, should always be identified. There is a danger that evaluations can over-simplify reality by interpreting indicators to mean more than they really do. Methods should be:

- cost-effective—if they are too expensive they will not be adopted
- replicable—to allow comparability across sites and times
- robust and statistically valid—they must be able to withstand scrutiny
- simple—very complex tools can alienate field staff and stakeholders
- field-tested—pilot studies before major projects are essential

- documented in manuals or other formats—so that they can be reviewed
- credible, honest and non-corrupt—the results need to be shown to be genuine
- able to yield unambiguous results—or to have the greatest explanatory power possible
- congruent with management and community expectations
- rapid—the evaluation process should draw on and review longer-term monitoring where possible, but should not be overly time-consuming.

Analysis and reporting

Results of evaluations can be simply tabulated, can be compared across time or areas, or can be analysed more thoroughly, looking for answers to complex questions. Management of protected areas is a very complex process and it can be very difficult to attribute causes to results, whether good or bad (for example, is the death of the forest patch due to inappropriate burning regimes, an unknown pathogen, a natural cycle, or a combination of all these factors?). Generally, it is more important to understand what is happening within a program and what might be contributing to success or failure.

Evaluation that assesses all the elements discussed above with clearly framed questions, carefully chosen indicators, good monitoring, and sound methodology is most likely to reveal a useful understanding of the effectiveness of management and provide a solid basis for improving management over time.

The way that evaluation findings are reported must suit the intended audiences. Methods of presentation, language, and terminology used in collecting and reporting evaluations should be commonly understandable, though language that is more technical will be appropriate for selected audiences. Possible methods of communication include reports in hard copy and on the Internet, attractive publications and brochures to increase public interest, presentations to managers, decision-makers, interest groups, and other stakeholders, field days and special events, media coverage, and displays. Assessment reports should identify strengths and weaknesses of management. Changes should be recommended or support current good practice.

Applying results

The evaluation process itself is a vital learning experience, which enhances and transforms management. Evaluation often has impacts on management well before a formal report is prepared. Getting people together to talk about management and focus on issues of effectiveness provides a valuable opportunity for increased understanding, improved learning, and the exchange of different viewpoints.

The findings and recommendations of evaluation also need to feed back formally into management systems to influence future planning, resource allocations, and management actions. Evaluations that are integrated into the managing agency's culture and processes are more successful and effective in improving management performance in the long term.

21.4 Management lessons and principles

Lessons

1   There is a central role for good science in management effectiveness evaluation. Management effectiveness evaluation provides a vital link between science and management in two ways. It provides a framework to direct resources towards the monitoring of the most critical factors—biophysical, cultural and socio-economic. It puts science into a context that is more easily understood and incorporated into management.

2   Management effectiveness evaluation provides a mechanism for adaptive management (Chapter 7)—feeding the results of research and monitoring into management on the ground and giving a basis for decision-making.

3   Studies such as ecological integrity measurement are a vital link between the past, present, and future, setting benchmarks and providing the information so that future generations understand why management decisions were made.

4   It is important that management effectiveness evaluation involves appropriate stakeholders—including Indigenous and local communities, on-ground park staff, non-government organisations, and experts—in all phases of evaluation, from design to adoption of recommendations. Evaluations must listen to the needs of local people and staff, and suggest appropriate responses to these needs.

5   Two key factors determine whether evaluation findings will 'make a difference': a high level of commitment to the evaluation by managers and owners of the protected areas; and adequate mechanisms, capacity, and resources to address the findings and recommendations.

Principles

1   Evaluation of management effectiveness is a vital component of responsive, proactive protected area management.

2   Evaluation works best where it is an integrated part of an effective management cycle, and where it has support from all levels of management as well as from stakeholders.

3   Good communication, team-building, and stakeholder involvement is essential in all phases of the project.

4   An accepted framework for evaluation is useful: the IUCN WCPA framework provides a flexible model for the development of a wide range of evaluation systems.

5   Evaluation works best with a clear plan: a clear purpose, scope, and objectives are needed.

6   Evaluation that assesses each of the elements of protected area management and the links between them will obtain the most comprehensive picture of management effectiveness.

7   Methods, questions, and indicators need to be carefully chosen to be relevant and useful in assessing effectiveness, and to be as cost-effective as possible. It is desirable for indicators to have some explanatory power, or be able to link with other indicators to explain causes and effects.

8   Evaluation of management effectiveness is best if it is backed up by robust, long-term monitoring.

9   Evaluation findings must be communicated and used positively.

10 Advice from evaluation needs to be clear and specific enough to improve conservation practices and it needs to be realistic, addressing priority topics and feasible solutions.

Endnote

1   This chapter was written at the request of the authors by Marc Hockings, Fiona Leverington, and Robyn James and subsequently edited by the authors.

Futures and Visions

There's a divinity that shapes our ends,
Rough-hew them how we will

From Hamlet, Act 5 Scene 2 *by William Shakespeare*

If the average life of a species is in the vicinity of a million years, we need to start thinking of what is, for humans, the very long term. We cannot of course foresee the dangers, or the technologies, of the distant future. But on the principle that 'well begun is half done', we can begin to plan and prepare for it.

Australia is a fortunate country. It has exceptional natural and cultural heritage and a world-class system of protected areas, albeit that they are still evolving. It has a stable political environment, and the resources to support innovative and creative futures. It also has people with the desire, understanding, skills, and capacity to create new systems. Despite mounting pressures on our remaining natural lands, we have seen established in one generation a reserve system that spans over 10% of the continent, managed by professionals who are respected worldwide.

We should not underestimate this achievement—or what it has cost those who have worked for it. Many conservationists and conservation agency staff have suffered in the struggle to protect our heritage. Even today, some people in the community simply do not agree with conservation and have opposed the concept of protected areas at every step. But there have been many positive and constructive changes with time. Rural opposition is slowly changing to acceptance as people understand the benefits protected areas can deliver to regional communities. Park proponents too are listening and responding more appropriately to local views, needs, and aspirations. There are more cooperative partnerships. Increasingly, local knowledge and local ideas are becoming an essential part of protected area management. This is the beginning of a future where local communities will be the proudest supporters and defenders of 'their' conservation reserves. We believe that protected areas of the future will be national icons, seen as part of what makes Australia a great nation. Many parks such as Kakadu, Great Barrier Reef, Wilsons Promontory, and Cradle Mountain–Lake St Clair have already achieved this status. Others will follow.

Protected areas have had a major influence on Australia's recent land use. No other policy has had the same capacity to stem the tide of development and consequent environmental impact. This influence has been most evident on the highly urbanised

Nourlangie Rock, Kakadu National Park, NT (Graeme Worboys)

east coast of Australia. Sydney has been called the 'city of national parks' since it has the good fortune to be surrounded by them. A giant arc of protected areas starts at our county's oldest park, Royal National Park on the south side of the city. It then extends to include outstanding reserves such as Blue Mountains, Wollemi, Yengo, and Brisbane Water National Parks before finishing on the northern side of the city at Ku Ring Gai Chase National Park. To top it off, right in the heart of the city are the famous waterways of Sydney Harbour National Park. Visitors who fly over the Royal National Park while descending into Sydney's main airport can clearly see the line between Sydney's urban sprawl and the naturalness of this historic reserve. It is a stark contrast. Park status has protected these environments for more than 125 years.

Yet the creation of national parks over 10% of our country is not enough if we are to conserve our heritage. They must be part of a larger national effort for conservation. Activating the necessary coalition of energy and resourcefulness by the community and its political leaders is an essential step. It will be part of Australia's future. The choice of not taking this path is inconceivable, given Australia's pride in its natural and cultural heritage—a legacy that is simply too precious to lose.

There is also a moral responsibility. Australians are the custodians of an assemblage of plants, animals, environments, and cultural heritage that are found nowhere else. We have an obligation to conserve this heritage, recognising the intrinsic values inherent in the natural world, as well as our stewardship on behalf of the world's peoples—current and future generations.

Australia has signed many international conventions and agreements on conservation and the environment; but Australians will need to be more innovative than we have been to meet the core objective of conserving our extant natural and cultural heritage. There are outstanding opportunities for improvements. We need to reserve and manage protected areas for the long term. The implementation of a National Reserve System with respect to both terrestrial and marine protected areas is a major step forward. All political parties accept the concept of protected areas. We should be able to think and act beyond the normal cycle of elections and the rigours of the annual treasury allocation. When we speak of the long term we mean really long term. We do not mean 100 or 200 years—we suggest 1000 years or more; and even that is a short time frame in terms of Aboriginal history or the development of natural systems.

Thinking and planning for 1000 years into the future needs a different mindset from the normal. It could too easily be treated with scepticism, but we see it instead as a challenge to establish the frameworks that can promote conservation for the very long term, and bring benefits to the world over many generations.

So how do we conserve lands for more than 1000 years? There are no clear precedents, since few, if any, ancient civilisations have set out to do this. Many of them took their own environments for granted. Today we live in a very different world. The notion of conserving natural lands, keeping them protected for centuries or millennia is relatively recent, yet it is now a worldwide endeavour. In practice, such thinking should be reflected in long-term objectives for protected areas. These objectives would be guided by knowledge of past trends and realistic future scenarios for perhaps 25 to 50 years at a time, but always with the very long term considered. Performance would be reviewed and monitored. Objectives will need to be refined over time relative to the views of new generations of Australians.

Australia since the 1960s has already seen two generations of career professionals looking after its protected areas. The annual cycle of operations, budget management, downsizing, expanding, restructuring, and meeting the requirements of governments is a consuming reality for protected area organisations. It should, however, never prevent the task of reviewing management performance against such long-term objectives. We believe investments that help to conserve protected areas for the very long term are fundamental to Australia's future.

Long-term thinking poses particular challenges in relation to privately managed protected areas. We are only beginning to establish governance models that can deliver sustained conservation outcomes from such areas. Establishing a land and seascape-scale network of protected areas, located within a matrix of sustainably managed non-park lands and seas, is a long term project for humanity.

Benefits beyond boundaries

Some 3000 delegates from 157 nations attended the Vth World Parks Congress 2003 in Durban South Africa. Under the congress theme 'benefits beyond boundaries', delegates reviewed progress over the previous 10 years and prepared goals and actions (the Durban Action plan) for the next 10 years.

The goals for the next 10 years are as follows.

1   Protected areas' critical role in global biodiversity conservation fulfilled.
2   Protected areas' fundamental role in sustainable development implemented.
3   A global system of protected areas linked to landscapes and seascapes achieved.
4   Improved quality, effectiveness, and reporting of protected areas management in place.
5   The rights of Indigenous peoples, mobile peoples, and local communities recognised and guaranteed in relation to natural resources and biodiversity conservation.
6   Empowerment of younger generations fulfilled.
7   Significantly greater support for protected areas from other constituencies achieved.
8   Improved forms of governance, recognising both traditional forms and innovative approaches of great potential value for conservation, implemented.
9   Greatly increased resources for protected areas, commensurate with their values and needs, secured.
10 Improved communication and education on the role and benefits of protected areas (IUCN 2003b).

These goals provide some telling motivation and direction for Australian protected area management.

Completing the protected area system

We have already noted the task of establishing a reserve system linked to landscapes and seascapes. The protected area system of Australia is incomplete (Chapter 5). Marine and freshwater systems and many Australian bioregions are inadequately conserved. In several places throughout this book we have indicated the significance of establishing

large-scale conservation connections between protected areas. We have also described the shift from reservation of particular areas of public land as national parks, to conserving biodiversity across landscapes and rural land tenures. These two approaches to large-scale conservation can be used as a basis for linking protected areas at a continental scale.

Achieving a capacity to manage

Competency. To achieve protected areas for the long term, we need to have the very best managers. By cooperative agreement between the Commonwealth and the states, a national training program needs to be established to train managers of both public, private, and Indigenous protected areas to the highest standards.

Legal protection. Protected areas need the best legal protection, at federal and at state or territory level. National leadership is needed to achieve both high level protection and long-term conservation objectives. Incremental impacts are a very real threat and there needs to be protection against short-term imperatives and 'short-sightedness'. The status and significance of Australia's protected areas deserves the highest level of protection, as well as recognition in an enhanced Australian Constitution.

Funding. At the Durban Congress it was estimated that an additional US$15 billion per year is needed to adequately fund management of the current global protected area estate. And this estate is continuing to expand. While the figure required to adequately manage Australia's protected areas is unknown, it is certainly more than the combined current public and private parks revenue. This must be remedied. Expanding the constituency of park supporters will be vital in this regard.

Building partnerships

The challenges of implementing the 'new paradigm' for protected areas (Chapter 2) must be met. Much is being done to address the rights and concerns of Indigenous Australians through initiatives such as resource-use agreements and Indigenous Protected Areas. Local communities, in some places, remain disaffected by protected area establishment and management. Agencies must continue to work towards healing old wounds and demonstrating the benefits of parks for local people. There is already a broadly based constituency supporting protected areas. This must be further expanded. Establishing partnerships will be crucial.

Extending our education

A concerted national effort is needed to elevate the level of understanding and debate about ecologically sustainable land and sea management, the values of protected areas, and appropriate ways in which we can use and relate to natural environments. Expanding both formal and informal education programs and activities are central to this.

Getting our house in order

Protected area agencies need to adopt sustainable management practices (Chapter 10), and be accountable for their environmental performance. They must play a leadership role in this regard. Equally important is demonstrating that the expected benefits from protected areas are actually being delivered. More attention must be given to evaluating management effectiveness (Chapter 21).

Leading and learning

Australia has internationally recognised capacity, skills, and knowledge in protected area management. This imparts a responsibility to provide leadership internationally and within our region, while at the same time recognising that we have much to learn from the rest of the world. International linkages need to be fostered and established as part of an Australian effort to support global conservation initiatives.

Managing change

The world we live in is changing rapidly. The global population in 2004 was six billion and it is expected to grow to eight billion by 2030. This is approximately the same period as a working career for a protected area manager. During this time, changes likely to affect protected area management include continued habitat destruction, increasing fuel costs, water shortages, impacts of global warming, and an ageing population. Protected area managers need to be prepared and have the capacity to respond appropriately.

Reflection

Our protected areas feature the very best examples of Australia's natural and cultural heritage. They have worked effectively for at least four generations to ensure that part of the country is kept intact for the long term. We are optimistic that they will endure for millennia. Protected areas are a source of national and international pride, and their status as a central part of Australian life will continue to grow. They benefit from multi-partisan support from Australian political parties. They need a truly national and visionary approach to their management. Australians must establish the greatest possible long-term protection that a democracy can offer. The necessary community support is there, and will grow. The conservation of Australia's great natural and cultural heritage depends upon the skills and expertise of our future managers, and on the partnerships they can establish with the Australian and international communities. Their success will be the inheritance of future generations.

APPENDICES

Appendix 1
Chronology of Major Events

Table A1.1 gives a chronology of some key events related to protected areas. The chronology focuses on Australia, together with international events that serve to provide a wider context. The authors acknowledge the particular contributions of Robert Snedden, Patricia Worboys, and Neville Gare to this compilation. An extended version of this chronology, with citations, can be found in Worboys et al. (2004).

The chronology was compiled with the assistance of the following sources: Helms (1896), Brockman (1959), Cooma-Monaro Express (1959), Carson (1962), Costin (1962), Lewis (1967), Mosley (1968), AGPS (1972), AGPS (1974), Specht et al. (eds) (1974), Rata (1975), Goldstein (1979), AGPS (1983), Lunney & Moon (1987), WCED (1987), Stoddart & Parer (1988), Brown (1989), Richards et al. (1990), Charles (1994), Hill (1994), Commonwealth of Australia (1997a), Commonwealth of Australia (1997b), Aplin (1998), Commonwealth of Australia (1999a), Mulvaney & Kamminga (1999), Anderson (2000), Batisse (2001), Mulligan & Hill (2001), Smyth (2001), Sutherland & Muir (2001), Lawrence et al. (2002), UNEP (2002), Dovers (2003), Finlayson & Hamilton-Smith (2003), Harding & Traynor (2003), McNeely & Schutyser (eds) (2003), Wagner (2003), Logan (2004), and WTMA (2004).

Table A1.1	Chronology of major events
Year	Event
40,000 BP	Confirmed earliest date Australian Aborigines arrived in Australia
252 BP	India: earliest formally recorded protected area when Emperor Asoka passed an edict for the protection of animals, fish, and forests
1641	USA: *Great Ponds Act*, Massachusetts set aside some 90,000 acres forever open to the public for '*fishing and fowling*'
1788	Rabbits in Australia: five rabbits arrive with the first fleet, later to become a major pest from the 1860s onwards
1832	USA: Hot Springs Reservation, Arkansas were reserved when Congress set them aside so that they '*shall not be entered, located, or appropriated for any other purpose whatever*'
1833	USA: Birth of a national park idea and emergence of an awareness of the great aesthetic and cultural qualities of 'primitive' America, and the need for their preservation: '*and what a splendid contemplation too, when one (who has travelled these realms and can duly appreciate them) imagines them as they might in the future be seen (by some protective policy of government) preserved in their pristine beauty and wildness, in a magnificent park ....* '*A nations park, containing man and beast, in all the wild and freshness of their nature's beauty*' (George Catlin)

Table A1.1	(continued)
Year	Event
1840	New Zealand: Mt Egmont reserved
1842	Bunya Pine protection: special decree by Governor Gipps (colony of NSW) that no timber licences were to be granted in any rainforest north of Brisbane containing Bunya Pine (in recognition of its importance to Aboriginal people)
1863	John Gould's *Mammals of Australia* published that urges laws be enacted to protect Australia's larger species and condemns the introduction of exotic species
1863	*Waste Lands Act 1863* (Tasmania) introduced, enabling areas to be preserved for their scenic value
1864	Marsh's book published on land and water degradation
1864	USA: first of a new national-level model of protected areas. Yosemite (California) established by US Congress 1 July 1864. '*to commit them (Yosemite Valley and the Mariposa Grove of Big Trees) to the care of the authorities of that State for their constant preservation, that may be exposed to public view, and that they may be used and preserved for the benefit of mankind*'
1866	Jenolan Caves Reserve (NSW) declared on 2 October 1866
1866	Tower Hill Public Park (Victoria) declared, upgraded to a national park in 1892
1872	USA: world's 'first' national park, Yellowstone, formally reserved in 1872
1879	Australia's first national park— 'The National Park' in NSW near Sydney was formally declared on 26 April 1879 for recreation and use by acclimatisation societies, later to become The Royal National Park (1955)
1880	Field Naturalists Club of Victoria established
1881	*Birds Protection Act 1881* (NSW)— native birds protected, and provision for the gazettal of bird preserves
1882	Fern Tree Gully Reserve (Victoria): 167 hectares reserved to protect tree ferns
1891	South Australia's first national park, Belair, established on 19 December 1891
1892	Victoria's first national park, Tower Hill, established
1893	The Australasian Association for the Advancement of Science formalised the concept of establishing reserves for the protection of flora and fauna
1895	*Parks and Reserves Act 1895* (WA) enacted
1895	First national park in WA at Swan View, known as 'The National Park'
1896	Scientist Richard Helms argues against grazing and burning in Kosciuszko: '*what right has one section of the community to rob the other of the full enjoyment of an unsullied alpine landscape. The husbandman on the farm by the river, the artist and tourist who seek the picturesque, the botanist and zoologist who came in pursuit of plants and animals, are all interfered with. And why? Because some inconsiderate people are allowed to do as they please … A great number of species must become extinct at a very early date unless the burning is prohibited by law, or better still, the highland area proclaimed a reserve for the preservation of its rare fauna and flora*'
1898	Temporary reserve, Wilsons Promontory (36,842 hectares) declared as a site for a national park
1901	Creation of the Commonwealth of Australia: crown lands mostly under state control

Table A1.1	(continued)
Year	Event
1906	First specific Australian national park legislation established following a long advocacy by R.M. Collins: Queensland *State Forests and National Parks Act 1906* provided for the creation of national parks that could only be alienated by parliament
1908	Romeo Lahey begins conservation campaign for the forests of the McPherson Range, Queensland
1908	Snowy Mountains National Chase (NSW) established for the purpose of public recreation and the protection of game in the area around Mt Kosciuszko
1913	Stirling Range National Park Western Australia established
1914	Myles Dunphy forms the Mountain Trails Club— the genesis of the Natural Primitive Areas Council
1915	Lamington National Park, Queensland established, after seven years of campaigning by Romeo Lahey, achieving the vision of R.M. Collins
1915	Tasmania: *Scenery Preservation Act 1915* established the Scenery Preservation Board—the first special central authority for parks and reserves in Australia
1916	Mt Field National Park and Freycinet National Park declared, the first national parks in Tasmania
1916	United States National Parks Service established
1916	Forestry Commission of NSW established
1924	Cobourg Peninsula (Northern Territory) becomes the first flora and fauna reserve in north Australia
1927	Honorary Rangers established (NSW)
1931	Snowy Indi Primitive Reserve proposed for the Snowy Mountains by Myles Dunphy
1932	Myles Dunphy unveils a proposal to the Mountain Trails Club in April 1932 for a Greater Blue Mountains National Park (NSW)
1932	Hinchinbrook Island National Park (Queensland) gazetted
1933	NSW National Parks and Primitive Areas Council formed in October 1933 with Myles Dunphy as Secretary
1937	Green Island, Great Barrier Reef (Queensland) established as a national park
1939	Major fires affect the forests of south-eastern Australia, 79 people killed in Victoria
1944	*Kosciusko State Park Act 1944* passed by NSW parliament, Kosciusko State Park Trust established to maintain the then largest single protected area in Australia (518,220 hectares)
1945	United Nations Educational, Scientific and Cultural Organisation established
1948	International Union for the Conservation of Nature (IUCN) established as a means of promoting conservation worldwide
1950	Myxoma virus (rabbit control) released into the wild near Corowa, Victoria
1952	Victorian National Parks Association formed to lobby for park legislation
1956	*National Parks Act 1956* (Victoria) providing for the control of all national parks in Victoria and establishing the National Parks Authority

Table A1.1	(continued)
Year	**Event**
1957	National Parks Association of NSW formed to lobby for the establishment of improved national parks legislation and the creation of a professional management organisation
1957	Director of National Parks (Victoria) appointed with a staff of 12
1958	Ayers Rock–Mount Olga National Park (Northern Territory) established
1959	Superintendent Kosciuszko State Park (NSW) Neville Gare, a forester, appointed with a staff of four—first professionally qualified national park manager in NSW
1959	Master plan for Kosciuszko National Park announced
1961	World Wildlife Fund (later to become the World Wide Fund for Nature) established as an international non-governmental organisation to mobilise support for conservation
1962	*Silent Spring*: Rachel Carson's influential book published, exposing the problems of indiscriminate use of pesticides especially the impacts of DDT on the food chain
1962	Wildlife Preservation Society of Queensland formed, which led the Great Barrier Reef campaign
1962	A paper entitled *National Parks in Australia* published describing the status of national parks in Australia and calling for better security of tenure, scientific management, financial provisions
1962	First IUCN World Conference on National Parks held in Seattle, Washington USA, 30 June to 7 July 1962—issues discussed include effects of humans on wildlife, species extinction, economic benefits of tourism, park management
1963	The Kosciuszko State Park Trust accepts a recommendation by the Australian Academy of Science for the creation of a Primitive Area over the summit area of Mt Kosciuszko protecting it from development by the Snowy Mountains Scheme
1964	Tidbinbilla Fauna Reserve established— the first formal reservation of land in the ACT for nature conservation purposes (subsequently expanded and renamed Tidbinbilla Nature Reserve)
1964	Wildlife Preservation Society co-founded by poet Judith Wright McKinney, an ardent activist for environment protection of the Great Barrier Reef and Rainforest
1965	Draft Kosciusko State Park master plan— the first protected area park plan of management with zoning in NSW, and possibly the first in Australia launched for public comment
1965	Australian Conservation Foundation formed
1966	Embryonic National Parks Service (NSW) established within the Lands Department in advance of proposed legislation for a new National Parks Service
1966	Len Webb put forward a series of National Park proposals, designed to protect the full range of the remaining habitats of the Queensland Wet Tropics
1966	*South Australia National Parks Act 1966*— national parks come under the control of the National Parks Commission, staff employed to manage the parks, and 86 new reserves established
1967	National Parks and Wildlife Service (NSW) established via the *National Parks and Wildlife Act 1967*
1967	First ranger training school in NSW conducted at Royal National Park in November 1967

Table A1.1	(continued)
Year	Event
1968	10 million hectares reserved as protected areas in Australia
1969	Land Conservation Council (Victoria) established in Victoria to plan for balanced land use
1969	Edgar Report on grazing in Kosciuszko National Park released in April and recommends the removal of all grazing from the park
1969	Future of the Great Barrier Reef (Queensland) Symposium conducted by the ACF and chaired by Sir Garfield Barwick, which provided direct input to State and Federal elections later in 1969; Save the Great Barrier Reef Campaign Committee established in Brisbane in September
1970	Two Royal Commissions established to inquire into petroleum exploration and production drilling in the Great Barrier Reef Region— the Commissions found that there should be no further exploration until the results of further research were known, with drilling on the Reef prohibited indefinitely
1970	Ministerial Council on National Parks agree on a definition of national park: *'a relatively large area set aside for its features of predominantly unspoiled natural landscape, flora and fauna, permanently dedicated for public enjoyment, education and inspiration and protected from all interference other than essential management practices so that its natural attributes are preserved'*
1970	Bouddi National Park, Australia's first marine national park, reserved in September by protecting 283 hectares of the sea floor
1970	December 1970, first shipment of woodchips left Eden (NSW) for Japan— start of the Eden NSW woodchip debate and conflict
1971	Convention on Wetlands of International Importance (Ramsar Convention) adopted on 2 February 1971 in Ramsar, Iran
1972	United Nations Conference on the Human Environment, Stockholm, Sweden held to sound the alarm about the perilous state of the Earth and its resources—concept of sustainable development introduced—United Nations Environment Program established
1972	Second IUCN World Conference on National Parks held in Yellowstone 18–27 September 1972—issues addressed include: the effects of tourism on protected areas; park planning and management; social, scientific, and environmental problems within national parks in wet tropical, arid and mountain regions
1972	UNESCO Man and the Biosphere Program formally endorsed by United Nations Conference on the Human Environment in Stockholm to establish a coordinated world network of Biosphere Reserves
1972	UNESCO Convention Concerning the Protection of the World Cultural and Natural Heritage adopted
1972	Wildlife Conservation Report by Select Committee of the Commonwealth Parliament—recommendations for secure protected areas representing all habitat types, establishment of a Commonwealth wildlife conservation authority, setting the Great Barrier Reef aside as a marine park, migratory species agreements
1973	Convention on International Trade in Endangered Species of Wild Fauna and Flora (CITES) adopted
1974	*Conservation of major communities in Australia and Papua New Guinea* (Specht Report) provided the first national evaluation of the conservation of major plant communities of Australia and highlighted major deficiencies in the protected area system
1974	Lake Pedder flooded

Table A1.1	(continued)
Year	**Event**
1974	Report of the Committee of Inquiry into the National Estate prepared—recommended that: adequate funding for protected areas be provided; new protected areas be selected based on objective scientific criteria; administration and management services for proper protection be allocated
1974	Australia becomes a party to the World Heritage Convention
1974	*National Parks and Wildlife Act 1974* (NSW) passed, rationalising the workings of previous fauna and flora legislation and the 1967 NPW Act
1975	*National Parks and Wildlife Conservation Act 1975* (Cth) passed—established the Australian National Parks and Wildlife Service and established the statutory office of the Director of National Parks and Wildlife Service, to be responsible for national policies for wildlife and nature protection; international agreements for national parks and wildlife; research, survey and inventory; specialist assistance; national statistics; and cooperating with the Territories
1975	*Great Barrier Reef Marine Park Act 1975* (Cth) establishes the Marine Park, the Marine Park Authority, and a Consultative Committee
1975	National Parks and Wildlife Service (Queensland) established by amalgamating the National Parks branch of Forestry with the Fauna Conservation Branch of Primary Industries
1975	Australian Heritage Commission established
1975	First personal computer goes on sale
1976	Great Barrier Reef Marine Park Authority appointed
1977	Uluṟu-Kata Tjuṯa (Ayers Rock-Mount Olga) National Park (Northern Territory) declared—the first protected area to be established under the *NPWC Act 1975*
1978	Fitzgerald River National Park (Western Australia) listed by UNESCO as a World Biosphere Reserve
1978	*Flora and Fauna Guarantee Act 1978* (Victoria) passed by the Victorian Parliament, the legislation designed to conserve species and communities
1979	Kakadu National Park (Stage 1) declared—Stages 2 and 3 follow in 1984 and 1987
1979	Burra Charter adopted by the Australian Branch of ICOMOS at Burra Burra, South Australia
1980	Rangers (NSW) agreement with the establishment of minimum qualifications for rangers as an Associate Diploma in a relevant discipline
1980	World Conservation Strategy prepared by IUCN, WWF and UNEP
1981	Gurig National Park (Northern Territory) becomes the first jointly managed national park in Australia
1981	Associate Diploma in Park Management developed by the Riverina and Goulburn College of Advanced Education (later to become part of Charles Sturt University)
1981	Great Barrier Reef Marine Park (Queensland) and Stage 1 of Kakadu (NT) inscribed as World Heritage Properties

Table A1.1	(continued)
Year	Event
1982	Third IUCN World Congress on National Parks in Bali, Indonesia focused on the role of protected areas in sustaining society; the inadequacy of the existing global network of terrestrial protected areas; the need for more marine, coastal, and freshwater protected areas; improved ecological managerial quality of existing protected areas; a system of consistent protected area categories to balance conservation and development needs; and links with sustainable development
1982/83	Franklin River blockade—Federal Government intervenes to halt the Gordon-below-Franklin dam
1983	16 February, Ash Wednesday—hot and dry conditions intensify fires in South Australia and Victoria, 76 people killed
1983	National Conservation Strategy for Australia developed
1983	Our Common Future and its advocacy for sustainable development generated by a UN commission
1985	Uluṟu-Kata Tjuṯa National Park (Northern Territory) returned to Aboriginal ownership and leased back for joint management
1985	International Conference on the Assessment of the Role of Carbon Dioxide and other Greenhouse Gases, Villach, Austria
1987	*Our Common Future* published—report of a World Commission on Environment and Development Independent Commission chaired by Gro Harlem Bruntland to bring forward strategies for achieving sustainable development
1992	The Joint National Forests Policy Statement launched by Prime Minister Paul Keating—implementation through Regional Forest Agreements
1992	Intergovernmental Agreement on the Environment, including that the concept of ecologically sustainable development should be used in the assessment of natural resources, land-use decisions and approval processes
1992	National Strategy for Ecologically Sustainable Development
1992	Fourth IUCN World Congress on National Parks and Protected Areas, 'Parks for Life' held in Caracas Venezuela—emphasised the relationship between people and protected areas, the need for identifying sites of importance for biodiversity conservation, and a regional approach to land management, aimed to extend the protected area network to cover at least 10% of each major biome by the year 2000
1992	United Nations Conference on Environment and Development (UNCED) held in Rio de Janeiro, Brazil resulting in an action plan for sustainable development (Agenda 21), as well as a Convention on Biological Diversity
1992	UN Framework Convention on Climate Change
1992	Mabo decision handed down by the High Court
1993	*Native Title Act 1993* (Cth) passed
1996	National Strategy for the Conservation of Australia's Biological Diversity promulgated
1996	State of the Environment Report, Australia published by the State of the Environment Advisory Committee
1996	Parks Victoria established 12 December 1996, part of the new Department of Natural Resources and Environment

Table A1.1	(continued)
Year	Event
1997	South East Forests National Park (NSW) launched, creating NSW's 100th national park and reserving 90,000 hectares of forests, effectively ends the anti-woodchip debates of the south-east forests of NSW
1997	Comprehensive, adequate and representative criteria for a forest reserve system
1997	60 million hectares of Australia reserved as protected areas
1998	First Indigenous Protected Area formally proclaimed at Nantawarrina, Flinders Ranges, South Australia
1999	Commonwealth's *Environment Protection and Biodiversity Conservation Act 1999* gazetted
2002	77 million hectares of land reserved as protected areas in Australia (10% of land area)
2003	Fifth IUCN World Congress on National Parks and Protected Areas, 'Benefits beyond boundaries' held in Durban South Africa— emphasised linkages in the landscape and seascape; developing the capacity to manage; building broader support; evaluating management effectiveness; governance; building a secure financial future; building comprehensive protected area systems; marine protected areas; and communities and equity

Appendix 2
Extent of Australian Protected Areas

There are many government agencies and organisations across Australia that have direct management responsibilities for protected areas. These organisations include the three levels of government—Commonwealth, state and territory, and local—as well as private conservation land managers. This section provides a summary of those protected areas managed by the major Commonwealth and state and territory agencies. The locations, sizes, and types and IUCN categories (Backgrounder 3.2) of protected areas in each jurisdiction are summarised in the following figures and tables. The tables have been compiled from the Collaborative Australian Protected Area Database (2002).

Table A2.1	Proportion of land managed as protected area in Australia
	Area (ha)
Total land area of Australia	768,826,956
Total land area of terrestrial protected areas in Australia	77,461,951
Percentage area of terrestrial protected areas	**10.08**

Commonwealth protected areas

Figure A2.1   Protected areas under Commonwealth jurisdiction

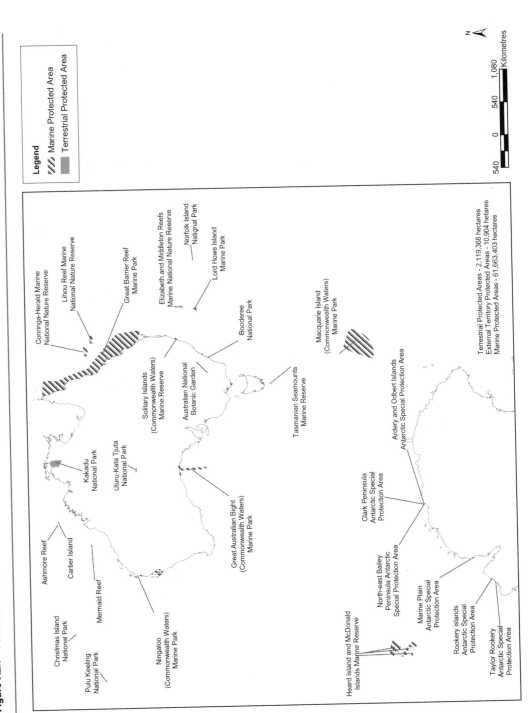

Legend

/// Marine Protected Area

▨ Terrestrial Protected Area

Christmas Island National Park
Pulu Keeling National Park
Ashmore Reef
Cartier Island
Mermaid Reef
Ningaloo (Commonwealth Waters) Marine Park
Kakadu National Park
Uluru-Kata Tjuta National Park
Conringa-Herald Marine National Nature Reserve
Lihou Reef Marine National Nature Reserve
Great Barrier Reef Marine Park
Elizabeth and Middleton Reefs Marine National Nature Reserve
Norfolk Island National Park
Lord Howe Island Marine Park
Solitary Islands (Commonwealth Waters) Marine Reserve
Australian National Botanic Garden
Booderee National Park
Macquarie Island (Commonwealth Waters) Marine Park
Great Australian Bight (Commonwealth Waters) Marine Park
Tasmanian Seamounts Marine Reserve
Ardery and Odbert Islands Antarctic Special Protection Area
Clark Peninsula Antarctic Special Protection Area
Heard Island and McDonald Islands Marine Reserve
North-east Bailey Peninsula Antarctic Special Protection Area
Marine Plain Antarctic Special Protection Area
Rookery Islands Antarctic Special Protection Area
Taylor Rookery Antarctic Special Protection Area

Terrestrial Protected Areas - 2,119,368 hectares
External Territory Protected Areas - 10,904 hetares
Marine Protected Areas - 61,663,403 hectares

N

Kilometres
540   0   540   1,080

Commonwealth protected areas (continued)

Table A2.2	Protected areas by type managed by the Commonwealth Government

TERRESTRIAL PROTECTED AREAS, 2002

Type	Number	Area (ha)
Botanic Garden	1	90
National Park	3	2,119,278
Total	**4**	**2,119,368**

MARINE PROTECTED AREAS, 2002

Type	Number	Area (ha)
Antarctic Special Protection Area	1	2088
Historic Shipwrecks	14	827
Marine National Nature Reserve	2	241,714
Marine Park	5	53,097,234
Marine Reserve	4	6,531,734
National Nature Reserve	3	1,786,500
National Park	2	3307
Total	**31**	**61,663,403**

EXTERNAL TERRITORY PROTECTED AREAS, 2002

Type	Number	Area (ha)
Antarctic Special Protection Area	5	1362
National Park	3	9542
Total	**8**	**10,904**

TOTAL ALL PROTECTED AREAS	**43**	**63,793,675**

Table A2.3	Commonwealth protected areas by IUCN category

TERRESTRIAL PROTECTED AREAS, 2002

IUCN Category	Number	Area (ha)
Category II	3	2,119,278
Category IV	1	90
Total	**4**	**2,119,368**

MARINE PROTECTED AREAS, 2002

IUCN Category	Number	Area (ha)
Category IA	23	14,400,033
Category II	3	1,799,021
Category IV	2	10,958,037
Category VI	3	34,506,312
Total	**31**	**61,663,403**

EXTERNAL TERRITORY PROTECTED AREAS, 2002

IUCN Category	Number	Area (ha)
Category IA	5	1362
Category II	3	9542
Total	**8**	**10,904**

TOTAL ALL PROTECTED AREAS	**43**	**63,793,675**

Australian Capital Territory protected areas

Figure A2.2   Protected areas under ACT jurisdiction

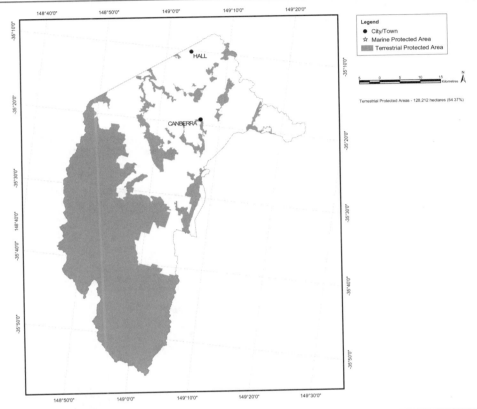

Table A2.4	Protected areas by type managed by the ACT Government		

TERRESTRIAL PROTECTED AREAS, 2002

Type	Number	Area (ha)
National Park	1	106,112
Nature Reserve	8	22,100
TOTAL ALL PROTECTED AREAS	**9**	**128,212**

Table A2.5	ACT protected areas by IUCN category

TERRESTRIAL PROTECTED AREAS, 2002

IUCN Category	Number	Area (ha)
Category II	9	128,212
Total	**9**	**128,212**

Table A2.6	Proportion of land managed as protected area in the ACT

	Area (ha)
Total land area of ACT	235,813
Total area of terrestrial protected areas in the ACT	128,212
Percentage area of terrestrial protected areas	**54.37**

New South Wales protected areas

Figure A2.3   Protected areas under NSW jurisdiction

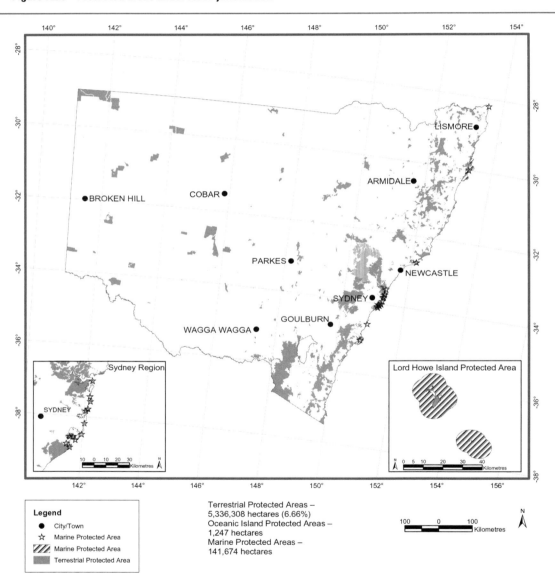

New South Wales protected areas (continued)

Table A2.7	Protected areas by type managed by the NSW Government plus IPAs

TERRESTRIAL PROTECTED AREAS, 2002

Type	Number	Area (ha)
Karst Conservation Reserve	4	4409
National Park	162	4,470,014
Nature Reserve	363	798,037
Reserve	9	21,662
Flora Reserve	103	35, 394
Indigenous Protected Area	1	480
Total	**642**	**5,329,996**

MARINE PROTECTED AREAS, 2002

Type	Number	Area (ha)
Aquatic Reserve	15	2076
Marine Park	3	139,598
Permanent Park Preserve (Lord Howe Island)	1	1247
Total	**19**	**142,921**

TOTAL ALL PROTECTED AREAS	**661**	**5,472,917**

Table A2.8	NSW protected areas by IUCN category

TERRESTRIAL PROTECTED AREAS, 2002

IUCN Category	Number	Area (ha)
Category IA	300	735,056
Category IB	20	1,545,503
Category II	166	2,950,308
Category III	8	521
Category IV	96	48,171
Category V	7	5271
Category VI	10	22,142
To Be Advised	35	23,024
Total	**642**	**5,329,996**

MARINE PROTECTED AREAS, 2002

IUCN Category	Number	Area (ha)
Category IV	18	141,674
Category II	1	1247
Total	19	142,921

TOTAL ALL PROTECTED AREAS	**661**	**5,472,917**

Table A2.9	Proportion of land managed as protected area in NSW

	Area (ha)
Total land area of NSW	80,121,268
Total land area of terrestrial protected areas in NSW	5,329,996
Percentage area of terrestrial protected areas	**6.65**

Northern Territory protected areas

Figure A2.4   Protected areas under Northern Territory jurisdiction

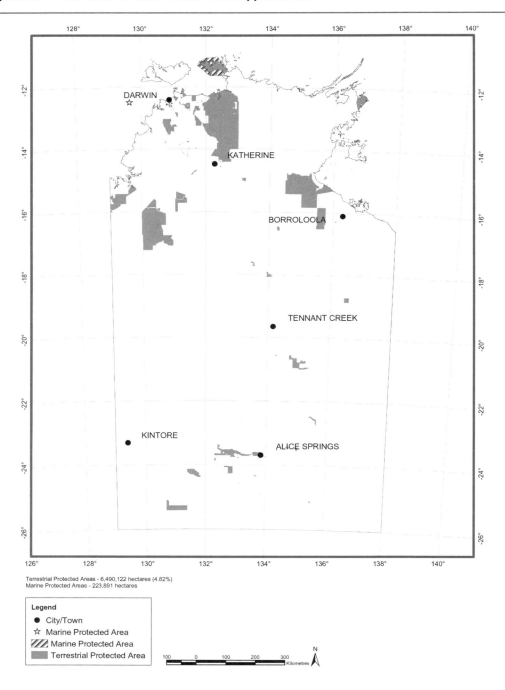

Terrestrial Protected Areas - 6,490,122 hectares (4.82%)
Marine Protected Areas - 223,891 hectares

Legend
● City/Town
☆ Marine Protected Area
▨ Marine Protected Area
▨ Terrestrial Protected Area

Northern Territory protected areas (continued)

Table A2.10	Protected areas by type managed by the Northern Territory Government plus IPAs

TERRESTRIAL PROTECTED AREAS, 2002

Type	Number	Area (ha)
Coastal Reserve	1	12,300
Conservation Reserve	14	48,208
Historical Reserve	15	7916
Hunting Reserve	1	1605
Management Agreement Area	5	26,228
National Park	12	3,391,559
National Park (Aboriginal)	4	575,814
Nature Park	13	24,155
Nature Park (Aboriginal)	1	3038
Other Conservation Area	31	172,630
Protected Area	3	12,710
Indigenous Protected Area	1	100,993
Total	**101**	**4,377,156**

MARINE PROTECTED AREAS, 2002

Type	Number	Area (ha)
Aquatic Life Reserves	2	279
Marine Park	1	223,612
Total	**3**	**223,891**
TOTAL ALL PROTECTED AREAS	**104**	**4,601,047**

Table A2.11	Northern Territory protected areas by IUCN category

TERRESTRIAL PROTECTED AREAS, 2002

IUCN Category	Number	Area (ha)
Category IA	4	44,284
Category II	25	4,050,874
Category III	6	7174
Category V	56	184,783
Category VI	10	90,041
Total	**101**	**4,377,156**

MARINE PROTECTED AREAS, 2002

IUCN Category	Number	Area (ha)
Category II	1	223,612
Category IV	2	279
Total	**3**	**223,891**
TOTAL ALL PROTECTED AREAS	**104**	**4,601,047**

Table A2.12	Proportion of land managed as protected area in the Northern Territory

	Area (ha)
Total land area of the Northern Territory	134,778,762
Total land area of terrestrial protected areas in the Northern Territory (excluding Commonwealth protected areas)	4,377,156
Percentage area of terrestrial protected areas	**3.25**

Queensland protected areas

Figure A2.5   Protected areas under Queensland jurisdiction

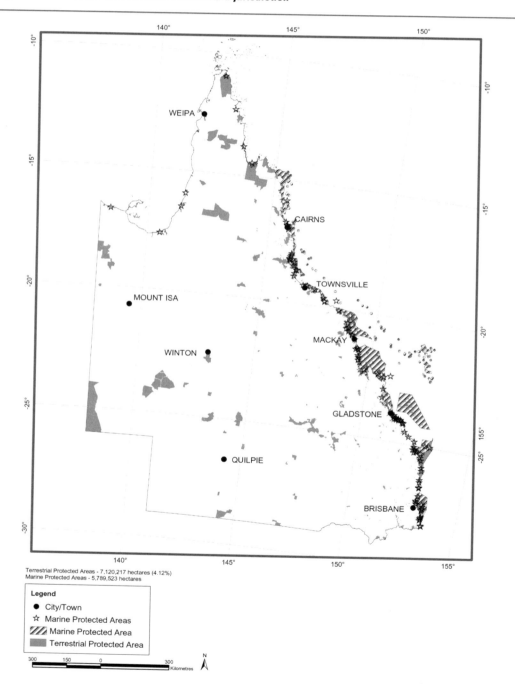

Terrestrial Protected Areas - 7,120,217 hectares (4.12%)
Marine Protected Areas - 5,789,523 hectares

Legend
- ● City/Town
- ☆ Marine Protected Areas
- ▨ Marine Protected Area
- ▨ Terrestrial Protected Area

Queensland protected areas (continued)

Table A2.13	Protected areas by type managed by the Queensland Government plus IPAs

TERRESTRIAL PROTECTED AREAS, 2002

Type	Number	Area (ha)
Conservation Park	174	34,940
Feature Protection Area	25	1253
National Park	213	6,661,888
National Park (Scientific)	7	52,181
Resource Reserve	36	348,678
Scientific Area	51	21,139
Indigenous Protected Area	2	138
Total	**508**	**7,120,217**

MARINE PROTECTED AREAS, 2002

Type	Number	Area (ha)
Dugong Protection Area	16	637,298
Fish Habitat Area	78	729,150
Marine Park (excluding the Great Barrier Reef Marine Park)	7	5,243,546
Total		**5,789,523**[1]

TOTAL ALL PROTECTED AREAS		**12,909,740**

1 *To avoid double-counting due to overlapping areas, total area figure is based on union of Dugong Protection Areas, Fish Habitat areas, and Marine Parks*

Table A2.14	Queensland protected areas by IUCN category

TERRESTRIAL PROTECTED AREAS, 2002

IUCN Category	Number	Area (ha)
Category IA	51	21,139
Category II	215	6,662,825
Category III	174	34,940
Category IV	9	52,412
Category V	25	1253
Category VI	34	347,648
Total	**508**	**7,120,217**

MARINE PROTECTED AREAS, 2002

IUCN Category	Number	Area (ha)
Category IA		5799
Category II		258,698
Category IV	101	5,514,675
Category VI		10,352
Total		**5,789,523**[1]

TOTAL ALL PROTECTED AREAS		**12,909,740**

1 *To avoid double-counting due to overlapping areas, total area figure is based on union of Dugong Protection Areas, Fish Habitat areas, and Marine Parks*

Table A2.15	Proportion of land managed as protected area in Queensland

	Area (ha)
Total land area of Queensland	172,973,671
Total land area of terrestrial protected areas in Queensland	7,120,217
Percentage area of terrestrial protected areas	**4.12**

South Australian protected areas

Figure A2.6   Protected areas under South Australian jurisdiction

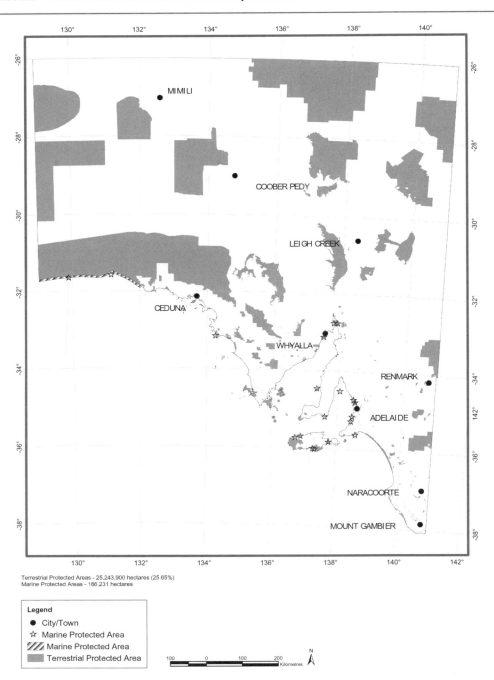

Terrestrial Protected Areas - 25,243,900 hectares (25.65%)
Marine Protected Areas - 186,231 hectares

Legend
● City/Town
☆ Marine Protected Area
▨ Marine Protected Area
▓ Terrestrial Protected Area

South Australian protected areas (continued)

Table A2.16	Protected areas by type managed by the South Australian Government, plus IPAs

TERRESTRIAL PROTECTED AREAS, 2002

Type	Number	Area (ha)
Conservation Park	217	5,848,009
Conservation Reserve	52	257,600
Game Reserve	10	25,799
National Park	19	4,416,196
Native Forest Reserve	58	14,187
Recreation Park	13	3152
Regional Reserve	7	10,601,706
Wilderness Protection Area	5	70,069
Indigenous Protected Area	5	3,442,500
Total	**386**	**24,679,218**

MARINE PROTECTED AREAS, 2002

Type	Number	Area (ha)
Aquatic Reserve	14	15,790
Conservation Park	2	202
Historic Shipwreck Protection Zone	1	78
Marine National Park	1	126,292
Marine Park	1	43,869
Total	**19**	**186,231**

TOTAL ALL PROTECTED AREAS	**405**	**24,865,449**

Table A2.17	South Australian protected areas by IUCN category

TERRESTRIAL PROTECTED AREAS, 2002

IUCN Category	Number	Area (ha)
Category IA	93	6,247,172
Category IB	9	2,215,803
Category II	16	2,645,201
Category III	184	186,541
Category IV	73	1,890,124
Category V	2	506,000
Category VI	9	10,988,377
Total	**386**	**24,679,218**

MARINE PROTECTED AREAS, 2002

IUCN Category	Number	Area (ha)
Category IA	1	78
Category IB	2	202
Category II	15	59,659
Category VI	1	126,292
Total	**19**	**186,231**

TOTAL ALL PROTECTED AREAS	**405**	**24,865,449**

Table A2.18	Proportion of land managed as protected area in South Australia

	Area (ha)
Total land area of South Australia	98,422,137
Total land area of terrestrial protected areas in South Australia	24,679,218
Percentage area of terrestrial protected areas	**25.07**

Tasmanian protected areas

Figure A2.7   Protected areas under Tasmanian jurisdiction

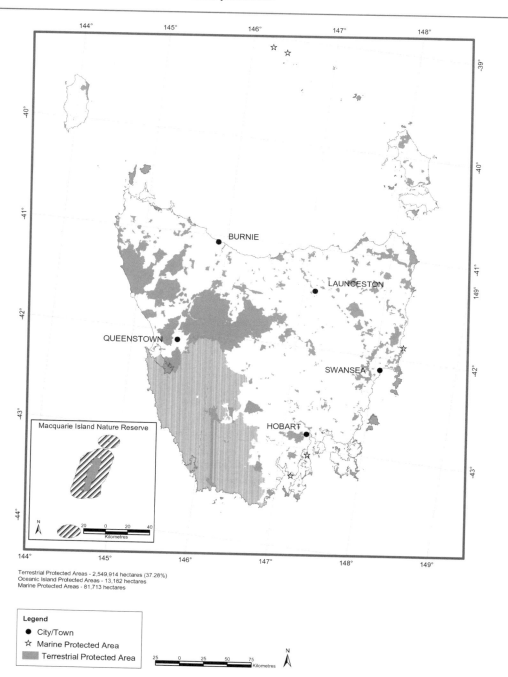

Terrestrial Protected Areas - 2,549,914 hectares (37.28%)
Oceanic Island Protected Areas - 13,182 hectares
Marine Protected Areas - 81,713 hectares

Legend
● City/Town
☆ Marine Protected Area
▨ Terrestrial Protected Area

Tasmanian protected areas (continued)

Table A2.19	Protected areas by type managed by the Tasmanian Government, plus IPAs

TERRESTRIAL PROTECTED AREAS, 2002[1]

Type	Number	Area (ha)
Conservation Area	160	531,282
Conservation Covenant	67	9343
Forest Reserve	190	174,724
Game Reserve	11	11,690
Historic Site	15	16,057
National Park	19	1,431,126
Nature Recreation Area	18	59,223
Nature Reserve	61	23,190
Other Conservation Area	1	18,000
Private Nature Reserve	1	120
Regional Reserve	21	228,654
State Reserve	60	44,302
Indigenous Protected Area	5	2203
Total	**629**	**2,549,914**

MARINE PROTECTED AREAS, 2002

Type	Number	Area (ha)
Marine Nature Reserve	3	176
Nature Reserve	1	81,538
Nature Reserve (Macquarie Island)	1	13,182
Total	**5**	**94,895**
TOTAL ALL PROTECTED AREAS	**634**	**2,644,809**

1   *Figures for State Forest Reserves not included.*

Table A2.20	Tasmanian protected areas by IUCN category

TERRESTRIAL PROTECTED AREAS, 2002

IUCN Category	Number	Area (ha)
Category IA	52	21,015
Category II	33	1,502,856
Category III	55	17,645
Category IV	259	173,863
Category V	73	90,677
Category VI	157	743,858
Total	**629**	**2,549,914**

MARINE PROTECTED AREAS, 2002

IUCN Category	Number	Area (ha)
Category IA	2	94,720
Category IV	3	176
Total	**5**	**94,895**
TOTAL ALL PROTECTED AREAS	**634**	**2,644,809**

Table A2.21	Proportion of land managed as protected area in Tasmania

	Area (ha)
Total land area of Tasmania	6,840,133
Total land area of terrestrial protected areas in Tasmania	2,549,914
Percentage area of terrestrial protected areas	**37.28**

Victorian protected areas

Figure A2.8 Protected areas under Victorian jurisdiction

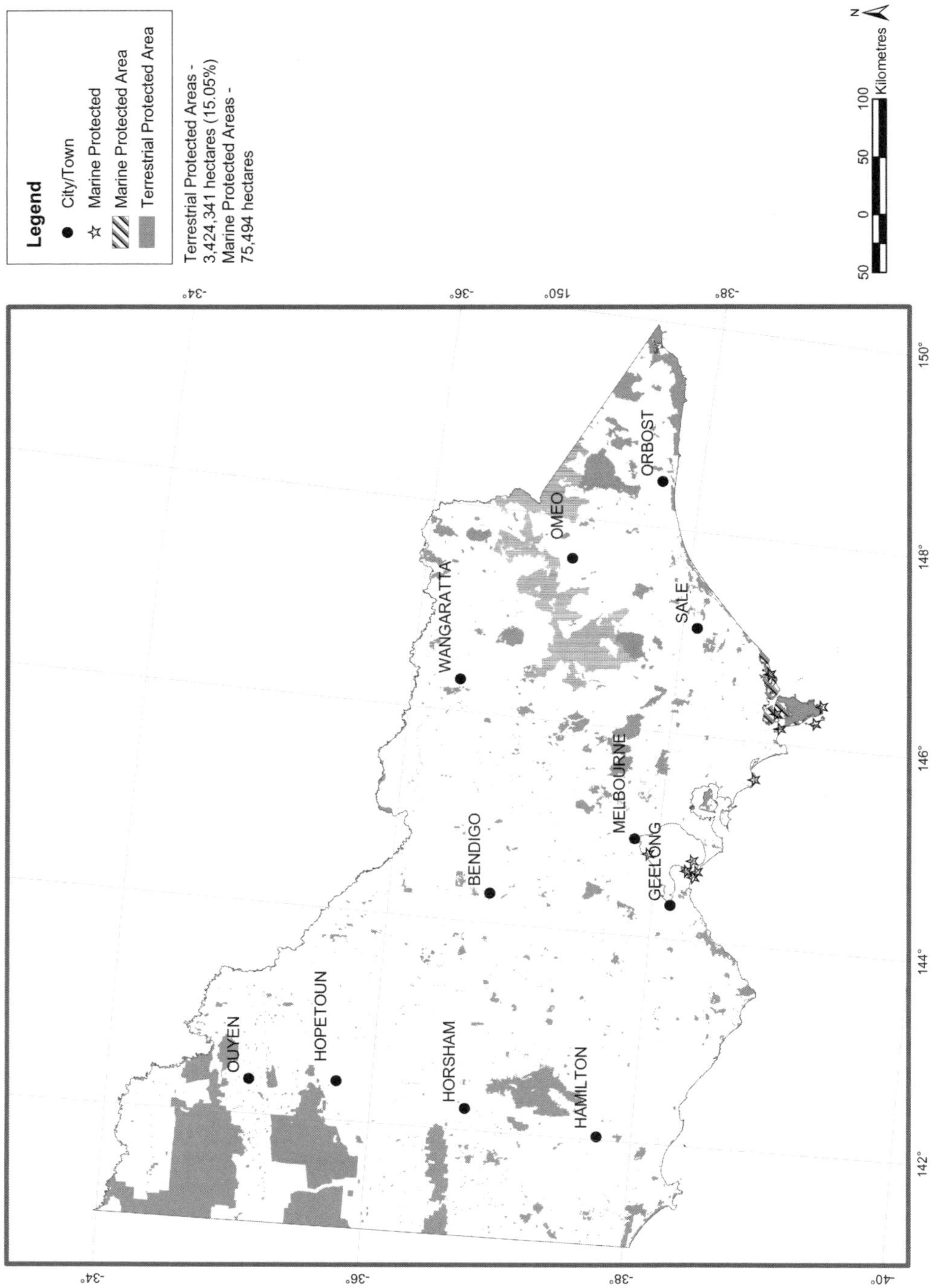
Legend
City/Town
Marine Protected
Marine Protected Area
Terrestrial Protected Area
Terrestrial Protected Areas - 3,424,341 hectares (15.05%)
Marine Protected Areas - 75,494 hectares
50 0 50 100 Kilometres
N
OUYEN
HOPETOUN
HORSHAM
HAMILTON
BENDIGO
GEELONG
MELBOURNE
WANGARATTA
SALE
OMEO
ORBOST
142°
144°
146°
148°
150°
-34°
-36°
-38°
-40°

Victorian protected areas (continued)

Table A2.22	Protected areas by type managed by the Victorian Government

TERRESTRIAL PROTECTED AREAS, 2002

Type	Number	Area (ha)
National Park	36	2,590,586
Natural Features Reserves	1518	168,877
Nature Conservation Reserves	341	203,623
Other Park	9	52,479
Phillip Island Nature Park	1	2118
Reference Area	35	20,459
State Park	31	183,696
Wilderness Park	3	202,050
Indigenous Protected Area	1	453
Total	**1975**	**3,424,341**

MARINE PROTECTED AREAS, 2002

Type	Number	Area (ha)
Fisheries Reserve	5	3357
Marine & Coastal Park	3	62,143
Marine Park	2	8980
Marine Reserve	1	1014
Total	**11**	**75,494**

TOTAL ALL PROTECTED AREAS	**1986**	**3,499,835**

Table A2.23	Victorian protected areas by IUCN category

TERRESTRIAL PROTECTED AREAS, 2002

IUCN Category	Number	Area (ha)
Category IA	377	224,296
Category IB	3	202,050
Category II	59	2,778,310
Category III	263	69,891
Category IV	1075	45,700
Category VI	198	104,094
Total	**1975**	**3,424,341**

MARINE PROTECTED AREAS, 2002

IUCN Category	Number	Area (ha)
Category II	1	1014
Category VI	10	74,481
Total	**11**	**75,494**

TOTAL ALL PROTECTED AREAS	**1986**	**3,499,835**

Table A2.24	Proportion of land managed as protected area in Victoria

	Area (ha)
Total land area of Victoria	22,754,364
Total land area of terrestrial protected areas in Victoria	3,424,341
Percentage area of terrestrial protected areas	**15.05**

Western Australian protected areas

Figure A2.9   Protected areas under Western Australian jurisdiction

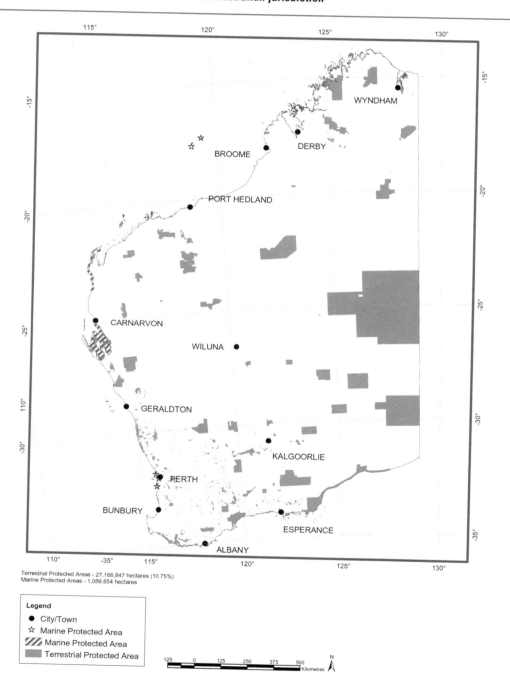

Terrestrial Protected Areas - 27,168,847 hectares (10.75%)
Marine Protected Areas - 1,089,654 hectares

Legend

● City/Town
☆ Marine Protected Area
▨ Marine Protected Area
▨ Terrestrial Protected Area

Western Australian protected areas (continued)

Table A2.25	Protected areas by type managed by the Western Australian Government, plus IPAs

TERRESTRIAL PROTECTED AREAS, 2002

Type	Number	Area (ha)
Conservation Park	26	704,202
Miscellaneous Conservation Reserve	74	299,923
National Park	71	5,092,172
Nature Reserve	1135	10,825,062
Indigenous Protected Area	2	10,247,488
Total	1308	27,168,847

MARINE PROTECTED AREAS, 2002

Type	Number	Area (ha)
Marine Nature Reserve	1	114,540
Marine Park	6	975,114
Total	**7**	**1,089,654**

TOTAL ALL PROTECTED AREAS	**1315**	**28,258,501**

Table A2.26	Western Australian protected areas by IUCN category

TERRESTRIAL PROTECTED AREAS, 2002

IUCN Category	Number	Area (ha)
Category IA	1129	10,810,293
Category II	115	5,929,043
Category III	6	74,236
Category IV	15	14,848
Category VI	34	10,339,632
Total	**1299**	**27,168,052**

MARINE PROTECTED AREAS, 2002

IUCN Category	Number[1]	Area (ha)
Category IA		711,863
Category II		877
Category VI		376,913
Total	**7**	**1,089,654**

TOTAL ALL PROTECTED AREAS	**1315**	**28,258,501**

1   *IUCN Categories determined by management zones within protected areas*

Table A2.27	Proportion of land managed as protected area in Western Australia

	Area (ha)
Total land area of Western Australia	252,700,808
Total land area of terrestrial protected areas in Western Australia	27,168,847
Percentage area of terrestrial protected areas	**10.75**

International designations

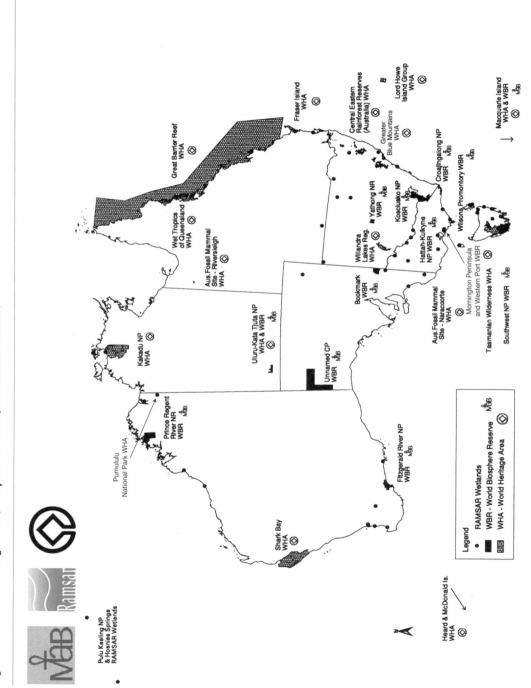

Figure A2.10   World Heritage Areas, Biosphere Reserves, and Ramsar Wetlands

References

AALC n.d., *Weed Management Manual for the Australian Alps National Parks*, Australian Alps Liaison Committee, Canberra.

Abelson, P. 1979, *Cost Benefit Analysis and Environmental Problems*, Saxon House, Farnborough.

Aboriginal and Torres Strait Islander Commission 1995, *Recognition, Rights and Reform: A Report to Government on Native Title Social Justice Measures*, Commonwealth of Australia, Canberra.

ABS 1995, *1995 Year Book Australia*, Australian Bureau of Statistics Catalogue No 1301, Commonwealth of Australia, Canberra.

ABS 1996, *Australians and the Environment*, Australian Bureau of Statistics, Commonwealth of Australia, Canberra.

ABS 1997, *Environmental Issues: People's Views and Practices*, Australian Bureau of Statistics, Commonwealth of Australia, Canberra.

ABS 1999, *Australian Demographic Statistics, June Quarter, 1999*, Australian Bureau of Statistics Catalogue No 3101, Commonwealth of Australia, Canberra.

ABS 2003a, *Australian Demographic Statistics*, URL: http://www.abs.gov.au/Ausstats/abs%40.nsf/b0666059243 0724fca2568b5007b8619/6949409dc8b8fb92ca256b c60001b3d1!OpenDocument.

ABS 2003b, *Year Book Australia: environment − environmental views and behaviour*, URL: http://www.abs.gov.au/Ausstats/abs@.nsf/Lookup/B188426A2452BF4BC A256CAE001599BD.

ABS 2004, *Measures of Australia's Progress, 2004*, Australian Bureau of Statistics Catalogue No 1370, Commonwealth of Australia, Canberra.

ACF 1998, *Draft Policy: Wilderness and Indigenous Cultural Landscapes in Australia*, Australian Conservation Foundation, Melbourne.

ACIUCN 1996, *Australian Natural Heritage Charter*, Australian Committee for the International Union for the Conservation of Nature, Australian Heritage Commission, Canberra.

ACIUCN 2002, *Australian Natural Heritage Charter (2nd edn)*, Australian Committee for the International Union for the Conservation of Nature, Australian Heritage Commission, Canberra.

ACIUCN 2003, *Protecting Natural Heritage: Using the Australian Natural Heritage Charter (2nd edn)*, Australian Committee for the International Union for the Conservation of Nature, Australian Heritage Commission, Canberra.

ACT Waste 2000, *No Waste by 2010: A Waste Management Strategy for Canberra*, URL: http://www.act.gov.au/nowaste.

Adam, P. 1990, *Saltmarsh Ecology*, Cambridge University Press, Cambridge.

Adam, P. 1995, 'Saltmarsh', in *The State of the Marine Environment Report for Australia Technical Annex 1: The Marine Environment*, Zann, L.P. & Kailola, P. (eds), GBRMPA, Townsville.

Adams, P. 1999, 'Central Eastern Rainforest Reserves World Heritage Listing', in *Compendium of Case Studies: Protected Area Management Principles and Practice*, NSW NPWS, Queanbeyan.

AGPS 1972, *Wildlife Conservation*, Report from the House of Representatives Select Committee, AGPS, Canberra.

AGPS 1974, *Report of the National Estate*, Report of the Committee of Inquiry into the National Estate, AGPS, Canberra.

AGPS 1983, *National Conservation Strategy for Australia*, Department of Home Affairs and Environment, AGPS, Canberra.

AIIMS 1992, *Incident Control System: The Operating System of Australian Inter-service Incident Management System*, Australian Association of Rural Fire Authorities, Melbourne.

Ajzen, I. & Fishbein, M. 1977, 'Attitude-behaviour Relations: A Theoretical Analysis and Review of Empirical Research', *Psychological Bulletin*, 84: 888–918.

Ajzen, I. & Peterson, G.L. 1988, 'Contingent Value Measurement: The Price of Everything and the Value of Nothing?', in *Amenity Resource Evaluation: Integrating Economics with Other Disciplines*, Peterson, G.L., Driver, B.L. & Gregory, R. (eds), Venture, State College.

Alder, J. 1996, 'Costs and Effectiveness of Education and Enforcement: Cairns section of the Great Barrier Reef Marine Park', *Environmental Management*, 20(4): 541–51.

Allan, R. 2003, 'El Niño', in *Ecology: An Australian Perspective*, Attiwill, P. & Wilson, B. (eds), Oxford University Press, Melbourne.

Allen, C. 2002, *An Adaptive Management Background Paper*, Presented to the workshop Adaptive Management in Australian Catchments: Learning from Experience, Albury.

Allen, G.R., Midgley, M. & Allen, M. 2002, *Field Guide to the Freshwater Fishes of Australia*, CSIRO Publishing, Melbourne.

Allmendinger, P. 2002, 'The Post-positivist Landscape of Planning Theory', in *Planning Futures: New Directions for Planning Theory*, Allmendinger, P. & Tewdwr-Jones, M. (eds), Routledge, London.

American Water Works Association 1996, *Federal Water Use Indices*, URL: http://www.eere.energy.gov/femp/technologies/water_useindices.cfm.

ANCA 1993, *Biosphere Reserves in Australia: A Strategy for the Future*, Commonwealth of Australia, Canberra.

ANCA 1996, Application of IUCN Protected Area Management Categories: Draft Australian Handbook, Australian Nature Conservation Agency, Canberra.

Ander, G.D. 2003, *Daylighting Performance and Design*, John Wiley and Sons, New York.

Anderson, E. 2000, *Victoria's National Parks: A Centenary History*, State Library of Victoria and Parks Victoria, Melbourne.

Andrew, N. (ed.). 1999, *Under Southern Seas: The Ecology of Australia's Rocky Reefs*, University of New South Wales Press, Sydney.

ANPWS 1992, *The Action Plan for Australian Birds*, Commonwealth of Australia, Canberra.

ANZECC 1996, *Benchmarking and Best Practice Program: Best Practice in Staff Training Processes*, National Parks Service, Melbourne.

ANZECC 1998, *Guidelines for Establishing the National Representative System of Marine Protected Areas*, Environment Australia, Canberra.

ANZECC 1999, *Strategic Plan of Action for the National Representative System of Marine Protected Areas: A Guide for Action by Australian Governments*, Environment Australia, Canberra.

ANZECC 2000, *National Framework for the Management and Monitoring of Australia's Native Vegetation*, Environment Australia, Canberra.

Aplin, G. 1998, *Australians and Their Environment: An Introduction to Environmental Studies*, Oxford University Press, Melbourne.

Archer, M., Hand, S.J. & Godthelp, H. 1994, *Riversleigh: The Story of Animals in Ancient Rainforests of Inland Australia*, Reed, Sydney.

Arnold G.W., Maller, R.A. & Litchfield, R. 1987, 'Comparison of Bird Populations in Remnants of Wandoo Woodland and in Adjacent Farmland', *Australian Wildlife Research*, 14: 331–41.

Arnstein, S. 1969, 'A Ladder of Citizen Participation', *Journal of the American Institute of Planners*, 4: 216–24.

ASEC 2001a, *Australia State of the Environment 2001*, CSIRO Publishing, Collingwood.

ASEC 2001b, *Human Settlements: Australia State of the Environment 2001*, CSIRO Publishing, Collingwood.

Ashe, G. 1972, *The Art of Writing Made Simple*, W.H. Allen, London.

AS/NZS 1999, *Risk Management*, Standards Australia International Ltd, Sydney & Standards New Zealand, Wellington.

Atkinson, I.A.E. & Taylor, R.H. 1992, *Distribution of Alien Mammals On New Zealand Islands (2nd edn)*, New Zealand Department of Conservation, Wellington.

Attiwill, P. & Wilson, B. (eds) 2003, *Ecology: An Australian Perspective*, Oxford University Press, Melbourne.

Auditor-General of Victoria 1995, *Managing Parks for Life: The National Parks Service*, Special Report No. 34, Victorian Auditor-General's Office, Melbourne.

AUSLIG 1990, *Atlas of Australian Resources: Vegetation*, Commonwealth of Australia, Canberra.

Australian Conservation Foundation 1998, *Draft Policy: Wilderness and Indigenous Cultural Landscapes in Australia*, Australian Conservation Foundation, Melbourne.

Australian Greenhouse Office 2002, *The Kyoto Protocol and Greenhouse Gas Sinks*, URL: http://www.greenhouse.gov.au/international/kyoto/fs-sinks.html.

Australian Heritage Commission 1985, *Australia's National Estate: The Role of the Commonwealth*, AGPS Canberra.

Australian Heritage Commission 2001, *Successful Tourism at Heritage Places*, AGPS, Canberra.

Babcock, R.C., Kelly, S., Shears, N.T., Walker, J.W. & Willis, T.J. 1999, 'Changes in Community Structure in Temperate Marine Reserves', *Marine Ecology-Progress Series*, 189: 125–34.

Balmford, A., Gravestock, P., Hockley, N., McClean, C.J. & Roberts, C.M. 2004, 'The Worldwide Costs of Marine Protected Areas', *Proceedings of the US National Academy of Sciences*, 101(26): 9694–7.

Banks, J.C.G. 1989, 'A History of Forest Fire in the Australian Alps', in *The Scientific Significance of the Australian Alps: Proceedings of the First Fenner Conference on the Environment*, Good, R. (ed.), Australian Alps Liaison Committee & Australian Academy of Science, Canberra.

Barham, E. 2001, 'Ecological Boundaries as Community Boundaries: The Politics of Watersheds', *Society and Natural Resources*, 14: 181–91.

Barlow, B.A. 1981, 'The Australian Flora: Its Origin and Evolution', in *Flora of Australia, Vol. I: Introduction*, AGPS, Canberra.

Barrett, S. & Gillen, K. 1997, 'Mountain Protected Areas of South Western Australia', *Parks*, 7(1).

Bartol, K., Martin, D., Tein, M. & Matthews, G. 1998, *Management: A Pacific Rim Focus* (2nd edn), McGraw-Hill, Sydney.

Bates, G. 1995, *Environmental Law in Australia* (4th edn), Butterworths, Sydney.

Batisse, M. 1990, 'Development and Implementation of the Biosphere Reserve Concept and its Applicability to Coastal Regions', *Environmental Conservation*, 17: 111–16.

Batisse, M. 2001, 'World Heritage and Biosphere Reserves: Complementary Instruments', *Parks*, 11(1): 38–43.

Bayet, F. 1994, 'Overturning the Doctrine: Indigenous People and Wilderness—Being Aboriginal in the Environmental Movement', *Social Alternatives*, 13(2): 27–32.

Bazerman, M. 1998, *Judgement in Managerial Decision Making* (4th edn), Wiley, New York.

Beal, D.J. 1995, 'A Travel Cost Analysis of the Value of Carnarvon Gorge for Recreational Use, *Review of Marketing and Agricultural Economics*, 63: 292–303.

Beal, D.J. & Harrison, S.R. 1997, *Efficient Pricing of Recreation in National Parks: A Queensland Case Study*, Presented to the AARES Conference, Gold Coast.

Beard, J.S. & Webb, M.J. 1974, *Vegetation Survey of Western Australia*, University of Western Australia Press, Perth.

Beaumont, N. 1999, 'Ecotourism: The Contribution of Educational Nature Experiences to Environmental Knowledge, Attitudes and Behaviours', PhD Thesis, Australian School of Environmental Studies, Faculty of Environmental Sciences, Griffith University, Brisbane.

Becken, S., Simmons, D. & Hart, P. 2003, *Tourism and Climate Change: New Zealand Response*, Presentation to the 1st International Conference on Climate Change and Tourism, World Tourism Conference, URL: http://www.world-tourism.org/sustainable/climate/brochure.htm.

Beckmann, R. 1996, 'Portrait of Australia', in *Australia: State of the Environment 1996*, CSIRO, Melbourne.

Bedward, M., Pressey, R.L. & Keith, D.A. 1992, 'A New Approach to Selecting Fully Representative Protected Area Networks: Addressing Efficiency, Reserve Design and Land Suitability with an Iterative Analysis', *Biological Conservation*, 62: 115–25.

Beeton, R.G.C. 2000, *LGAQ National Parks Enquiry: Final Report*, Local Government Report of Queensland Inc., Brisbane.

Beeton, R.J.S. 1989, 'Tourism and Protected Areas', in *National Parks and Tourism*, Bakson, P., Nyman, S. & Sheppard, D. (eds), NSW NPWS, Sydney.

Bell, S. & Morse, S. 2003, Measuring Sustainability: Learning from Doing, Earthscan, London.

Beltrán, J. 2000, *Indigenous and Traditional Peoples and Protected Areas: Principles, Guidelines and Case Studies*, Best Practice Protected Area Guidelines Series No. 4, ed. A Phillips, IUCN WCPA, Gland.

Bennett, A.F. 2003, *Linkages in the Landscape: the Role of Corridors and Connectivity in Wildlife Conservation* (2nd edn), IUCN, Gland.

Bennett, D.H. 1983, 'Some Aspects of Aboriginal and Non-Aboriginal Notions of Responsibility to Non-human Animals', *Australian Aboriginal Studies*, 2: 19–24.

Bennett, J. 1984, 'Using Direct Questioning to Value the Existence Benefits of Preserved Natural Areas', *Australian Journal of Agricultural Economics*, 28: 136–52.

Bennett, J. 1995, The Economic Value of Recreation Use of Gibralter Range and Dorrigo National Parks, NSW NPWS, Sydney.

Bennett, J., Blamey, R. & Morrison, M. 1997, *Valuing Damage to South Australian Wetlands Using the Contingent Valuation Method*, LWRRDC Occasional Paper 13/97, LWRRDC, Canberra.

Benson, J. 1996, 'Threatened by Discovery: Research and Management of the Wollemi Pine *Wollemia nobilis*', in *Back from the Brink: Refining the Threatened Species Recovery Process*, Stephens, S. & Maxwell, S. (eds), Surrey Beatty and Sons, Sydney.

Benson, J. 1998, 'Spinning Science into a Yarn', *Sydney Morning Herald*, 2 October.

Benson, J.S. & Redpath, P.A. 1997, 'The Nature of pre-European Native Vegetation in South-eastern Australia: A Critique of Ryan, D.G., Ryan, J.R. and Starr, B.J. (1995) The Australian Landscape—Observations of Explorers and Early Settlers', *Cunninghamia*, 5(2).

Benson, L.J. 1994, 'Introduction and Overview', in *Townsville Port Authority Capital Dredging Works 1993: Environmental Monitoring Program*, Benson, L.J., Goldsworthy, P.M., Butker, I.R. & Oliver, J. (eds), Townsville Port Authority, Townsville.

Benyus, J.M. 1998, *Biomimicry: Innovation Inspired by Nature*, Quill, New York.

Berndt, R.M. & Berndt, C.H. 1970, *Man, Land and Myth in North Australia: The Gunwinggu People*, Ure Smith, Sydney.

Berndt, C.H. & Berndt, R.M. 1983, *The Aboriginal Australians: The First Pioneers*, Pitman Publishing Pty Ltd, Carlton.

Bigg, T. 2003, *The World Summit on Sustainable Development: was it worthwhile?*, URL: http://www.iied.organisation/wssd/wssdreview.pdf.

Binning, C. 1998, *Beyond Roads, Rates and Rubbish: Interim Report*, CSIRO Wildlife and Ecology Division, Canberra.

Birckhead, J., Collins, J., Sakulas, H. & Bauer, J. 2000, 'Caring for Country Today? Towards an Ecopolitics of Sustainable Indigenous Land Management Regimes', in *Nature Conservation in Production Environments: Managing the Matrix*, Saunders, D., Craig, J. & Mitchell, N. (eds), Surrey Beatty and Sons, Sydney.

Bradbury, J., Cullen, P., Dixon, G. & Pemberton, M. 1995, 'Streambank Erosion, Revegetation and Management of the Lower Gordon River, Tasmanian Wilderness World Heritage Area', *Environmental Management*, 19: 259–72.

Bradstock, R.A., Williams, J.E. & Gill, A.M. (eds) 2002, *Flammable Australia: The Fire Regimes and Biodiversity of a Continent*, Cambridge University Press, Oakleigh, Victoria.

Bramwell, B. & Lane, B. 1993, 'Interpretation and Sustainable Tourism: The Potential and the Pitfalls', *Journal of Sustainable Tourism*, 1(2): 71–80.

Brandon, K. 1996, *Ecotourism and Conservation: A Review of Key Issues*, Environment Department Paper No. 33, The World Bank, Washington, DC.

Brandon, K., Redford, K.H. & Sanderson, S.E. 1998, *Parks in Peril: People, Politics and Protected Areas*, The Nature Conservancy, Island Press, Washington.

Braun, A. (ed.) 2003, *Preparing for the 2003 World Parks Congress—Managing Effectively in the Face of Change: What Lessons Have We Learned?*, World Commission on Protected Areas, URL: http://www.iucn.org/themes/wcpa/theme/mgteffect.html.

Briassoulis, H. 1989, 'Theoretical Orientations in Environmental Planning: An Inquiry into Alternative Approaches', *Environmental Management*, 13: 381–92.

Bridgewater, P.B., Cresswell, I.D. & Thackway, R. 1996, 'A Bioregional Framework for Planning a National System of Protected Areas', in *Approaches to Bioregional Planning*, Brickwoldt, R. (ed.), DEST, Canberra.

Briggs, J.D. & Leigh, J.H. 1995, *Rare or Threatened Australian Plants*, CSIRO Publishing, Melbourne.

Bright, A. 1994, 'Burn grass', in *Country in Flames: Proceedings of the 1994 Symposium on Biodiversity and Fire in North Australia*, D. Rose (ed.), DEST and the North Australia Research Unit, Canberra & Darwin.

Brim, L. 1998, *Rainforest Aboriginal News*, 2: 7.

Brockman, C.F. 1959, *Recreational Use of Wild Lands*, McGraw Hill, New York.

Broome, L.S., Blackley, S. & Tennant, P. 1996, 'Long-footed Potoroo *Potorous longipes* Research Plan in South-eastern NSW', Unpublished Internal Report, NSW NPWS, Sydney.

Brown, A.J. 1989, *The South East Forests Information Kit*, Conservation Council of the South East Region, Canberra.

Brown, R. 2004, *Memo for a Saner World*, Penguin Books, Camberwell.

Brown, P.J., McCool, S.F. & Manfredo, M.J. 1987, 'Evolving Concepts and Tools for Recreation User Management in Wilderness: A State-of-knowledge Review', in *Proceedings—National Wilderness Research Conference: Issues, State-of-knowledge, Future Directions*, Lucas, R.C. (ed.), USDA Forest Service, Intermountain Research Station, Ogden, UT.

Bryce, W.J., Day, R. & Olney, T.J. 1997, 'Commitment Approach to Community Recycling: New Zealand Curbside Trial', *Journal of Consumer Affairs*, 31: 27–69.

BTR 1999, *Tourism's Economic Contribution 1996–97*, Bureau of Tourism Research Paper No. 5, BTR, Canberra.

Buchanan, R.A. 1989, *Bush Regeneration: Recovering Australian Landscapes*, TAFE, Sydney.

Bucher, D. & Saenger, P. 1991, 'An Inventory of Australian Estuaries and Enclosed Marine Waters: An Overview of Results', *Australian Geographical Studies*, 29: 370–81.

Buckley, R. & Pannell, J. 1990, 'Environmental Impacts of Tourism and Recreation in National Parks and Conservation Reserves', *Journal of Tourism Studies*, 1(1).

Buckley, R.C., Pickering, C.M. & Warnken, J. 2000, 'Environmental Management for Alpine Tourism and Resorts in Australia', in *Tourism and Development in Mountain Regions*, Goode, P.M., Price, M.F. & Zimmermann, F.M. (eds), CABI Publishing, New York.

Buckley, R., Pickering, C.M. & Weaver, D.B. (eds) 2003, *Nature-based Tourism, Environment and Land Management*, CABI Publishing, New York.

Bulbeck, C. 1993, *Social Sciences in Australia: An Introduction*, Harcourt Brace, Sydney.

Burbidge, A.A. & McKenzie, N.L. 1989, 'Patterns in the Modern Decline of Western Australia's Vertebrate

Fauna: Causes and Conservation Implications', *Biological Conservation*, 50: 143–98.

Bureau of Meteorology 2003, *Bureau of Meteorology February 2003 media release*, URL: http://www.bom.gov.au/announcements/media_releases/vic/20030228.html.

Burenhult, G. (ed.) 2003, *People of the Past—the Illustrated History of Humankind: The Epic Story of Human Origins and Development*, Fog City Press, San Francisco.

Burger, J. 1990, *The Gaia Atlas of First Peoples: A Future for the Indigenous World*, Penguin, London.

Burgman, M.A. & Lindenmayer, D.B. 1998, *Conservation Biology for the Australian Environment*, Surrey Beatty and Sons, Sydney.

Butler, A. & Jernakoff, P. 1999, *Seagrass in Australia*, CSIRO, Melbourne.

Byrne, D., Brayshaw, H. & Ireland, T. 2001, *Social Significance: A Discussion Paper*. NSW NPWS, Sydney.

Byrnes, T. & Warnken, J. 2003, 'Establishing Best-practice Environmental Management: Lessons from the Australian Tour-boat Industry', in *Nature-based Tourism, Environment and Land Management*, Buckley, R., Pickering, C. & Weaver, D.B. (eds), CABI Publishing, New York.

Cain, B. 2000, Aboriginal Sites Office, NSW NPWS, personal communication.

Callicott, J.B. 1986, 'On the Intrinsic Value of Nonhuman Species', in *The Preservation of Species: The Value of Biological Diversity*, Norton, B.G. (ed.), Princeton University Press, Princeton.

CALM 1996, *Shark Bay Marine Reserves Management Plan, 1996–2006*, Department of Conservation and Land Management, Perth.

CALM 2003, *Annual Report 2002–2003* Department of Conservation and Land Management, URL: http://www.naturebase.net/anrep_02-03.html.

CALM 2004, Mooring Policy for the Management of Moorings in WA's Marine Conservation Reserves URL: http://www.calm.wa.gov.au/national_parks/marine_mooring_policy.html.

Campbell, A. 1994, *Landcare: Communities Shaping the Land and the Future*, Allen and Unwin, Sydney.

Campbell, S. & Fainstein, S. 2003, 'Introduction: The Structure and Debates of Planning Theory', in *Readings in Planning Theory* (2nd edn), Campbell, S. & Fainstein, S. (eds), Blackwell, Oxford.

Cape Woolamai Steering Committee 1989, *Cape Woolamai State Faunal Reserve Environment Effects Statement: Visitor Facilities*, CWSC, Cowes.

Cappo, M., Alongi, D.M., Williams, D.M. & Duke, N. 1998, *A Review and Synthesis of Australian Fisheries Habitat Research: Major Threats, Issues and Gaps in Knowledge of Coastal and Marine Fisheries Habitats Volume 1: A Prospectus of Opportunities for the FRDC Ecosystem Protection Program*, Australian Institute of Marine Science, Townsville.

Carey, C., Dudley N. & Stolton S. 2000, *Squandering Paradise: The Importance and Vulnerability of the World's Protected Areas*, WWF International, Gland.

Carey, G., Lindenmayer, D. & Dovers, S. 2003, *Australia Burning: Fire Ecology, Policy and Management Issues*, CSIRO, Melbourne.

Carson, R. 1962, *Silent Spring*, Houghton Mifflin, Boston.

Carter, B. 1996, 'Private Sector Involvement in Recreation and Nature Conservation in Australia', in *National Parks: Private Sector's Role*, Charters, T., Gabriel, M. & Prasser, S. (eds), USQ Press, Toowoomba.

Carter, M., Vanclay, F.M. & Hundloe, T.J. 1989, *Economic and Socio-economic Impacts of the Crown of Thorns Starfish on the Great Barrier Reef*, Report to the Great Barrier Reef Marine Park Authority, Townsville, Institute of Applied Environmental Research, Griffith University, Brisbane.

Caughley, G., Sinclair, R.G. & Wilson, G.R. 1977, 'Numbers, Distribution and Harvesting Rate of Kangaroos on the Inland Plains of New South Wales, *Australian Wildlife Research*, 4: 99–108.

Caughley, J. & Bayliss, P. 1982, 'Kangaroo Populations in Inland New South Wales', in *Parks and Wildlife: Kangaroos*, Haigh, C. (ed.), NSW NPWS, Sydney.

CBD 2003, *Sustaining Life on Earth: How the Convention on Biological Diversity Promotes Nature and Human Well-being*, URL: http://www.biodiv.organisation/doc/publications/guide.asp.

Ceballos-Lascurain, H. 1996, *Tourism, Ecotourism and Protected Areas*, IUCN, Gland.

Centre for Design 2003, *Paper and Packaging LCA*, URL: http://www.cfd.rmit.edu.au/life_cycle_assessment/paper_packaging_lca.

Centre for Studies in Travel and Tourism 1988, *The Queensland Recreational Scuba Diving Industry Study Report*, Centre for Studies in Travel and Tourism, James Cook University of North Queensland, Townsville.

Chamala, S. 1992, 'Factors Influencing the Effectiveness of Group Extension in Soil Conservation', in *Proceedings of the 5th Australian Soil Conservation Conference: Volume 8 — Group Extension*, Hamilton, G.J., Howes, K.M. & Attwater, R. (eds), Department of Agriculture, Perth.

Chape, S., Spalding, M. & Sheppard, D. 2003, 'Chapter 1: Global Overview', in *State of the World's Protected Areas* (first draft for comment), Spalding, M. & Chape, S. (eds), UNEP-WCMC, Cambridge.

Charles, I. 1994, *Parks, People and Preservation: Recollections of the Establishment and First 25 Years of the NSW National Parks and Wildlife Service*, NSW NPWS, Sydney.

Cheney, P. & Sullivan, A. 1997, *Grassfires Fuel, Weather and Fire Behaviour*, CSIRO Publishing, Melbourne.

Christ, C., Hillel, O., Matus, S. & Sweeting, J. 2003 *Tourism and Biodiversity: Mapping Tourism's Global Footprint*, Conservation International, Washington, DC.

Cifuentes, A.M., Izurieta V.A. & De Faria H.H. 2000, *Measuring Protected Area Management Effectiveness*, WWF Centroamerica, Turrialba.

Clare, V. 1997, 'Assessment of Supply and Demand: Background Report to the NSW NPWS Draft Nature Tourism and Recreation Strategy', Unpublished Internal Report, NSW NPWS, Sydney.

Claridge, A.W., Castellano, M.A. & Trappe, J.M. 1996, 'Fungi as a Food Resource for Mammals in Australia', in *Fungi of Australia Vol. 1B: Introduction—Fungi in the Environment*, Mallett K. & Grgurinovic C. (eds), Australian Biological Resources Study, Canberra.

Clarke, I.F. & Cook, B.J. 1983, *Geological Science: Perspectives of the Earth*, Australian Academy of Science, Canberra.

Clarke, R.N. & Stankey, G.H. 1979, *The Recreation Opportunity Spectrum: A Framework for Planning, Management and Research*, General Technical Report PNW-98, USDA Forest Service, Fort Collins.

Cogger, H.G. 1994, *Reptiles and Amphibians of Australia*, Reed, Sydney.

Coleman, N, Gason, A.S.H. & Poore, G.C.B. 1997, 'High Species Richness in the Shallow Marine Waters of South East Australia', *Marine Ecology*, 154: 17–26.

Collaborative Australian Protected Areas Database 1997, *Terrestrial and Marine Protected Areas*, Natural Resources System Program, Environment Australia, Canberra.

Collaborative Australian Protected Areas Database 1999, *Terrestrial and Marine Protected Areas*, Natural Resources System Program, Environment Australia, Canberra.

Collaborative Australian Protected Areas Database 2002, *Terrestrial and Marine Protected Areas*, Department of Environment and Heritage, Canberra.

Collins, J. 1998, 'Contemporary Hunter-gatherers in North-central Arnhem Land', Honours thesis, School of Environmental and Information Sciences, Charles Sturt University, Albury.

Commonwealth Department of Tourism 1994, *National Ecotourism Strategy*, AGPS, Canberra.

Commonwealth of Australia 1990, *Ecologically Sustainable Development: A Commonwealth Discussion Paper*, AGPS, Canberra.

Commonwealth of Australia 1992a, *The National Strategy for Ecologically Sustainable Development*, AGPS, Canberra.

Commonwealth of Australia 1992b, *Intergovernmental Agreement on the Environment*, DEST, Canberra.

Commonwealth of Australia 1996a, *Investing in our Natural Heritage*, statement by senator the Honourable Robert Hill, Minister for the Environment, AGPS, Canberra.

Commonwealth of Australia 1996b, *The National Strategy for the Conservation of Australia's Biological Diversity*, DEST, Canberra.

Commonwealth of Australia 1997a, *Natural Heritage Trust: A Better Environment for Australia in the 21st Century*, Environment Australia and Department of Primary Industries and Energy, Canberra.

Commonwealth of Australia 1997b, *Hazards, Disasters and Survival*, Emergency Management Australia, Canberra.

Commonwealth of Australia 1998a, *Australia's Oceans Policy*, Commonwealth of Australia, Canberra.

Commonwealth of Australia 1998b, *Comprehensive Regional Assessments and Regional Forest Agreements*, Commonwealth of Australia, Canberra, URL: http://www.rfa.gov.au.

Commonwealth of Australia 1999a, *Threat Abatement Plan for Competition and Land Degradation by Feral Rabbits: National Feral Animal Control Plan*, Natural Heritage Trust, Environment Australia, Canberra.

Commonwealth of Australia 1999b, *The National Weeds Strategy: A Strategic Approach to Weed Problems of National Significance*, Commonwealth of Australia, Canberra.

Commonwealth of Australia 2001a, *Human Settlements: Australia State of the Environment 2001*, CSIRO Publishing, Canberra.

Commonwealth of Australia 2001b, *Inland Waters: Australia State of the Environment 2001*, CSIRO Publishing, Canberra.

Commonwealth of Australia 2003a, *Tourism White Paper: A Medium to Long Term Strategy for Tourism*, Commonwealth of Australia, Canberra.

Commonwealth of Australia 2003b, *Australian Heritage Council Act 2003, No 85*, AGPS, Canberra.

Commonwealth of Australia 2004, *South-east Regional Marine Plan Implementing Australia's Oceans Policy in South-east Marine Region*, National Oceans Office, Hobart.

Collaborative Australian Protected Areas Database 2002, *Terrestrial and marine protected areas*, Department of Environment and Heritage, Canberra.

Conner, N. & Gilligan, B. 2003, *Socio-economic Benefits of Protected Areas: Concepts and Assessment Techniques as Applied in New South Wales, Australia*, NSW NPWS, Sydney.

Cooma Monaro Express 1959, 'State Park Supplement: Outlines Plans for Development', *Cooma Monaro Express*, 12 August 1959.

Corbett, D. 1996, *Australian Public Sector Management* (2nd edn), Allen and Unwin, Sydney.

Cordell, J. 1993a, 'Who Owns the Land? Indigenous Involvement in Protected Areas', in *Indigenous Peoples* and *Protected Areas: The Law of the Mother*, Kempf, E. (ed.), Earthscan, London.

Cordell, J. 1993b, 'Boundaries and Bloodlines: Tenure of Indigenous Homelands and Protected Areas', in *Indigenous Peoples and Protected Areas: The Law of the Mother*, Kempf, E. (ed.), Earthscan, London.

Costin, A. 1962, 'National Parks in Australia', in *National Parks: A World View*, Cahalane, V.H. (ed.), American Committee for International Wildlife Protection, New York.

Cotgrove, R. 2004, University of Tasmania, Hobart, personal communication.

Courrau, J. 1999, *Strategy for Monitoring the Management of Protected Areas in Central America*, PROARCA CAPAS Program, The Nature Conservancy, Arlington.

Crabb, P. 2003, *Managing the Australian Alps: A History of Cooperative Management of the Australian Alps National Parks*, Australian Alps Liaison Committee and the Centre for Resource and Environmental Studies, Australian National University, Canberra.

Creighton, J.L. 1981, 'The Use of Values: Public Participation in the Planning Process', in *Public Involvement and Social Impact Assessment*, Daneke, G.A., Garcia, M.W. & Priscoli, J.D. (eds), Westview, Boulder.

Cresswell, I.D. & Thomas, G.M. 1997, *Terrestrial and Marine Protected Areas in Australia*, Environment Australia, Canberra.

CSIRO 1986, *Weather and Climate*, CSIRO, Canberra.

Csurches, S. & Edwards, R. 1998, *National Weeds Program: Potential Environmental Weeds in Australia—Candidate Species for Preventative Control, Biodiversity Group*, Environment Australia, Canberra.

Curtis, A. & Lockwood, M. 1998, 'Natural Resource Policy for Rural Australia', in *Agriculture and the Environmental Imperative*, Pratley, J. & Robertson, A. (eds), CSIRO, Melbourne.

Curtis, A. & Lockwood, M. 2000, 'Landcare and Catchment Management in Australia: Lessons for State-sponsored Community Participation', *Society and Natural Resources*, 13: 61–73.

Dale, A. 1997, 'Principles for Negotiating Conservation on Indigenous Lands', in *Conservation Outside Nature Reserves*, Hale, P. & Lamb, D. (eds), University of Queensland, Brisbane.

Daneke, A. 1983, 'Public Involvement: What, Why and How', in *Public Involvement and Social Impact Assessment*, Daneke, G.A., Garcia, M.W. & Priscoli, J.D. (eds), Westview, Boulder.

Davey, A.G. 1998, *National System Planning for Protected Areas*, World Commission on Protected Areas Best Practice Protected Area Guidelines Series No. 1, ed. A. Phillips, IUCN, Gland.

David, F.R. 2001, *Strategic Management: Concepts* (8th edition), Prentice-Hall, New Jersey.

Dawson, A. & Cohen, S. 1999, *Media Training Manual*, NSW NPWS, Sydney.

Day, J.C., Fernandes, L., Lewis, A., De'ath, G., Slegers, S., Barnett, B., Kerrigan, B., Breen, D., Innes, J., Oliver, J., Ward, T.J. & Lowe, D. 2000, 'The Representative Areas Program for Protecting Biodiversity in the Great Barrier Reef World Heritage Area', *Proceedings of the 9th International Coral Reef Symposium, Bali, Indonesia, October 2000, Vol 2*.

DCE n.d., *An Inventory of Recreation Settings on Major Areas of Public Land in Victoria*, Occasional Paper Series No. 4, DCE, Melbourne.

DDM 2002, 'Day-to-day Management: Six Month Report for Great Barrier Reef Marine Park and Related Areas', Unpublished Report to GBRMPA, Townsville.

De Bres, J. 1999, *Tragedy at Cave Creek: What Should a Chief Executive do?*, Department of Conservation, Wellington.

DEC 2004, *Who Cares about the Environment in 2003?* Department of Environment and Conservation, Sydney.

De Forges, B.R., Koslow, J. & Poore, G.C.B. 2000, 'Diversity and Endemism of the Benthic Seamount Fauna in the Southwest Pacific', *Nature*, 405: 944–7.

De Lacy, T., Battig, M., Moore, S. & Noakes, S. 2002, *Public/Private Partnerships for Sustainable Tourism: Delivering a Sustainability Strategy for Tourism Destinations*, Sustainable Tourism CRC, Gold Coast.

DEH 2002, *Department of Environment and Heritage Annual Report 2001-02*, Commonwealth of Australia, Canberra.

DEH 2004a, National List of Extinct, Extinct in the Wild, Critically Endangered, Endangered, Vulnerable, and Conservation Dependent Species, URL: http://www.deh.gov.au.

DEH 2004b, *Wetlands and Migratory Birds Website*, URL: http://www.deh.gov.au/water/wetlands.

DEH 2004c, *Best Practice in Park Management*, URL: http://www.deh.gov.au/parks/best-practice.

DEH 2004d, *New National Heritage System*, URL: http://www.deh.gov.au/heritage/publications/factsheets/general.html.

DEH 2004e, Parks and Reserves, URL: http://www.deh.gov.au/parks/index.html.

DEH 2004f, Calperum and Taylorville Stations, URL: http://www.deh.gov.au/parks/biosphere/bookmark/index.html.

DEH 2004g, Tasmanian Seamounts Marine Reserve, URL: http://www.deh.gov.au/coasts/mpa/seamounts.

Department of Industry, Tourism and Resources 2003, *Pursuing Common Goals: Opportunities for Tourism and Conservation*, Commonwealth of Australia, Canberra.

DEST 1994, *Australia's Biodiversity, An Overview of Selected Significant Components*, Biodiversity Series Paper No. 2, Commonwealth of Australia, Canberra.

De Villiers, M. 2001, *Water: The Fate of our Most Precious Resource*, Phoenix Press, London.

DIPNR 2003, *Annual Report 2002–2003*, Department of Infrastructure, Planning and Natural Resources, Sydney.

DNRE 1996a, *National Data Standards on Protected Areas*, Report to the ANZECC Working Group on National Parks and Protected Areas, URL: http://www.ea.gov.au/Parks/best-practice/reports/data-standards/index.html.

DNRE 1996b, *Planning Guidelines for Native Vegetation Retention Controls*, DNRE, Melbourne.

DNRE 1999a, *Draft Forest Management Plan, North East Forest Management Area*, DNRE, Melbourne.

DNRE 1999b, *Wildlife Response Plan for Oil Spills*, DNRE, Melbourne.

DNRE 1999c, *Victorian Cetacean Contingency Plan*, DNRE, Melbourne.

DNRE 1999d, *Best Practice in Park Interpretation and Education*, Report to the ANZECC Working Group on National Park Management Benchmarking and Best Practice Program, DNRE, Melbourne.

DNRE 2000a, *Native Vegetation: Planning Guidelines for Native Vegetation Retention Controls*, URL: http://www.nre.vic.gov.au/4A2567D0024CB2.

DNRE 2000b, 'Restoring our Catchments: Victoria's Draft Native Vegetation Management Framework', DNRE, Melbourne.

Donaldson, S. 1996, 'A Community Perspective', in *Sustainable Management of Natural Resources: Who Benefits and Who Should Pay?*, Price, R. (ed.), Land and Water Resources Research and Development Corporation, Canberra.

Dovers, S.R. 1995, 'Information, Sustainability and Policy', *Australian Journal of Environmental Management*, 2: 142–56.

Dovers, S.R. 1997, 'Sustainability: Demands on Policy', *Journal of Public Policy*, 16: 303–18.

Dovers, S.R. 1998, *Public Policy and Institutional R & D for Natural Resource Management: Issues and Directions for LWRRDC*, Report to the Land and Water Resources Research and Development Corporation, Centre for Resource and Environmental Studies, ANU, Canberra.

Dovers, S.R. & Mobbs, C.D. 1997, 'An Alluring Prospect? Ecology and the Requirements of Adaptive Management', in *Frontiers in Ecology: Building the Links*, Klomp, N. & Lunt, I. (eds), Elsevier, Oxford.

Dovers, S. 2003, 'Discrete, Consultative Policy Processes: Lessons from the National Conservation Strategy for Australia and National Strategy for Ecologically Sustainable Development', in *Managing Australia's Environment*, Dovers, S. & Wild River, S. (eds), The Federation Press, Sydney.

Dovey, S. 1994, *Improving Bushfire Management for Southern NSW*, NSW NPWS, Sydney.

Dryzek, J.S. 1997, *The Politics of the Earth*, Oxford University Press, Oxford.

Dryzek, J.S. & Braithwaite, V. 2000, 'On the Prospects for Democratic Deliberation: Values Analysis Applied to Australian Politics', *Political Psychology*, 21: 241–66.

DSE 2004, Marine National Parks for Victoria, URL: http://www.dse.vic.gov.au/dse/nrencm.nsf.

Dudley, N., Hockings, M. & Stolton, S. 1999, 'Measuring the Effectiveness of Protected Areas Management', in *Partnerships for Protection—New Strategies for Planning and Management for Protected Areas*, Stolton, S. & Dudley, N. (eds), Earthscan, London.

Dudley, N., Gujja, B., Jackson, B., Jeanrenaud, J., Oviedo, G., Phillips, A., Rosabel, P., Stolton, S. and Wells, S. 1999, 'Challenges for Protected Areas in the 21st Century', in *Partnerships for Protection: New Strategies for Planning and Management of Protected Areas*, Stolton, S. & Dudley, N. (eds), WWF International, IUCN and Earthscan, London.

Dugan, J.E. & Davis, G.E. 1993, 'Applications of Marine Refugia to Coastal Fisheries Management', *Canadian Journal of Fisheries and Aquatic Sciences*, 50(9): 2029–42.

Duke, N.C. 1992, 'Mangrove Floristics and Biogeography', in *Tropical Mangrove Ecosystems*, Robertson, A.I. & Alongi, D.M. (eds), AGU Press, Washington.

Dunphy, D. & Griffiths, A. 1998, *The Sustainable Corporation: Organisational Renewal in Australia*, Allen and Unwin, Sydney.

Du Toit, J.T., Rogers, K.H. & Biggs, H.C. 2003, *The Kruger Experience: Ecology and Management of Savanna Heterogeneity*, Island Press, Washington.

Eagles, P.F.J. & McCool, S.F. 2002, *Tourism in National Parks and Protected Areas, Planning and Management*, CABI Publishing, New York.

Eagles, P.F.J., McCool, S.F. & Haynes, C. 2002, *Sustainable Tourism in Protected Areas: Guidelines for Planning and Management*, Best Practice Protected Area Guidelines Series No. 8, ed. Phillips, A., IUCN WCPA, Gland.

Ecology Australia 1998, *The Intrinsic Values of the Nadgee/Howe Area of South Eastern Australia*, Report prepared for Parks Victoria and the NSW National Parks and Wildlife Service, NSW NPWS, Sydney.

Edgar, G.J. & Barrett, N.S. 1997, 'Short Term Monitoring of Biotic Change in Tasmanian Marine Reserves', *Journal of Experimental Marine Biology and Ecology*, 213(2): 261–79.

Edgar, G.J. & Barrett, N.S. 1999, 'Effects of the Declaration of Marine Reserves on Tasmanian Reef Fishes, Invertebrates and Plants', *Journal of Experimental Marine Biology and Ecology*, 242(1): 107–44.

Edyvane, K.S. 1996, 'The Role of Marine Protected Areas in Temperate Ecosystem Management', in *Developing Consistent National Criteria for the Identification and Selection of a Representative System of Marine Protected Areas*, South Australian Aquatic Sciences Centre, Adelaide, 22–23 April 1996.

Edyvane, K.S. 2000, 'The Great Australian Bight', in *Seas at the Millennium: An Environmental Evaluation*, Sheppard, C. (ed.), Elsevier, Amsterdam.

Edyvane K.S. & Andrews, G.A. 1995, *Draft Management Plan for the Great Australian Bight Marine Park*, Prepared for Primary Industries South Australia, South Australian Research & Development Institute, West Beach.

Emergency Management Australia 1998, *Australian Emergency Management Glossary*, Australian Emergency Manual Series, Tasmanian State Emergency Service, Hobart.

Environment Australia 1996, 'Records of the 2nd Meeting of the Working Group on Iindigenous Protected Areas', Biodiversity Group, Environment Australia, Canberra.

Environment Australia 1998, *National Strategy for Ecologically Sustainable Development*, Environment Australia, Canberra.

Environment Australia 2001, *Directory of Important Wetlands in Australia* (3rd edn), Environment Australia, Canberra.

Environment Australia 2003a, *The Commonwealth Marine Protected Areas Program*, Commonwealth of Australia, Canberra.

Environment Australia 2003b, *Commonwealth Marine Protected Areas*, URL: http://www.ea.gov.au/coasts/mpa.

Environment Australia, CSIRO & NOO 2003, *Australia's South-east Marine Region: A User's Guide to Identifying Candidate Areas for a Regional Representative System of Marine Protected Areas*, Commonwealth of Australia, Canberra.

Environment Defenders Office (Tas) 2001, *The Environmental Law Handbook* (2nd edn), EDO, Hobart.

Environment Protection and Heritage Council 2003a, *Going Places: Developing Natural and Cultural Heritage Tourism in Australia: Issues Paper*, AGPS, Canberra.

Environment Protection and Heritage Council 2003b, *Going Places: Developing Natural and Cultural Heritage Tourism in Australia: Key Opportunities*, AGPS, Canberra.

Environs Australia 2003, *About Us*, URL: http://www.environs.org.au/about.

EPA 1997, *Who Cares About the Environment in 1997?*, NSW EPA, Sydney.

EPA 2003, 'Fleet changes reduce emissions', *EQ: News*, 21: 5.

Ervin, J. 2001, *Improving Protected Area Management: WWF's Rapid Assessment and Prioritisation Methodology*, WWF, Gland.

Ervin, J. 2003, *WWF: Rapid Assessment and Prioritization of Protected Area Management (RAPPAM) Methodology*, WWF, Gland.

Fairweather, P.G. & Quinn, G.P. 1995, 'Marine Ecosystems: Hard and Soft Shores', in *The State of the Marine Environment Report for Australia Technical Annex 1: The Marine Environment*, Zann, L.P. & Kailola, P. (eds), GBRMPA, Townsville.

Fallding, M. 2000, 'What Makes a Good Natural Resource Management Plan?' *Ecological Management and Restoration*, 1(3): 185–93.

Fenna, A. 1998, *Introduction to Australian Public Policy*, Addison Wesley Longman, Melbourne.

Fensham, R.J., Fairfax, R.J. & Cannell, R.J. 1994, 'The Invasion of *Lantana camara L.* in Forty Mile Scrub National Park, North Queensland', *Australian Journal of Ecology*, 19: 297–305.

Ferguson, I.S. & Greig, P.J. 1973, 'What Price Recreation?', *Australian Forestry*, 36: 80–90.

Ferrier, S., Pressey, R.L. & Barrett, T.W. 2000, 'A New Predictor of the Irreplaceability of Areas for Achieving a Conservation Goal, Its Application to Real-world Planning and a Research Agenda for Further Refinement', *Biological Conservation*, 93: 303–25.

Figgis, P. 1999, *Australia's National Parks and Protected Areas: Future Directions*, Australian Committee for IUCN, Sydney.

Figgis, P. 2003, *Private Lands Conservation: The Australian Experience*, Presented to the 5th World Parks Congress, Durban, South Africa.

Finlayson, B. & Hamilton-Smith, E. 2003, *Beneath the Surface: A Natural History of Australian Caves*, University of New South Wales Press, Sydney.

Flannery, T.F. 1994, *The Future Eaters*, Reed, Sydney.

Flannery, T.F. 1998, 'A Reply to Benson and Redpath (1997)', *Cunninghamia*, 5(4): 779–82.

Flood, J. 1981, *Identification and Recording of Historic Sites: Industrial and Historical Archaeology*, National Trust of Australia, Sydney.

Flood, J. 1983, *Archaeology of the Dreamtime: The Story of Prehistoric Australia and her People*, Collins, Sydney.

Flood, J. 1997, *Rock Art of the Dreamtime: Images of Ancient Australia*, Angus and Robertson, Melbourne.

Flyvberg, B. & Richardson, T. 2002, 'Planning and Foucault: In Search of the Dark Side of Planning Theory', in *Planning Futures: New Directions for Planning Theory*, Allmendinger, P. & Tewdwr-Jones, M. (eds), Routledge, London.

Foran, B. & Poldy, F. 2002, 'Between a Rock and a Hard Place: Australia's Population Options to 2050 and Beyond', *People and Place*, 11(1).

Forrest, C.J. & Mays, R.H. 1997, *The Practical Guide to Environmental Community Relations*, John Wiley & Sons, New York.

Fowler, H.W. 1926, *Modern English Usage*, Oxford University Press, Oxford.

Fox, A. 1979, 'Reflections', in *Australia's 100 Years of National Parks*, Goldstein, W. (ed.), NSW NPWS, Sydney.

Frank Small and Associates 1993, *Northern NSW Community Attitudes Towards the National Parks & Wildlife Service*, NSW NPWS, Sydney.

Friedmann, J. 1987, *Planning in the Public Domain: From Knowledge to Action*, Princeton University Press, New Jersey.

Furze, B., De Lacy, T. & Birckhead, J. 1996, *Culture, Conservation and Biodiversity: The Social Dimension of Linking Local Development and Conservation through Protected Areas*, John Wiley & Sons, London.

GBRMPA 1994, *The Great Barrier Reef: Keeping it Great— a 25 Year Strategic Plan for the Great Barrier Reef World Heritage Area 1994–2019*, Great Barrier Reef Marine Park Authority, Townsville.

GBRMPA 1999, *The Great Barrier Reef World Heritage Area: Framework for Management*, Report to World Heritage Committee, GBRMPA, URL: http://www.gbrmpa.gov.au/corp_site/info_services/publications/brochures/protecting_biodiversity/gbrwha_management_framework.pdf.

GBRMPA 2004, *Environmental Impact Assessment*, http://www.gbrmpa.gov.au/corp_site/management/eim/eia/index.html.

Gee, C.Y., Makens, J.C. & Choy, D.J.L. 1997, *The Travel Industry* (3rd edn), Van Nostrand Reinhold, New York.

Gell, F.R. & Roberts, C.M. 2003, 'Benefits Beyond Boundaries: The Fishery Effects of Marine Reserves', *TRENDS in Ecology and Evolution*, 18(9): 448–55.

Gentle, C.B. & Duggin, J.A. 1997, 'Allelopathy as a Competitive Strategy in Persistent Thickets of *Lantana camara* L. in Three Australian Forest Communities', *Plant Ecology*, 132: 85–95.

Ghimire, K.B. & Pimbert, M.P. 1997, 'Social Change and Conservation: An Overview of Issues and Concepts', in *Social Change and Conservation: Environmental Politics and Impacts of National Parks and Protected Areas*, Ghimire, K.B. & Pimbert, M.P. (eds), Earthscan Publications, London.

Gifford, R. 2003, 'Global Climate Change', in *Ecology: An Australian Perspective*, Attiwill, P. & Wilson, B. (eds), Oxford University Press, Melbourne.

Gillen, K., Danks, A., Courtenay, J. & Hickman, E. 1997, 'Threatened Species Management on the South Coast of Western Australia', *Parks*, 7(1).

Gillespie, R. 1997, *Economic Value and Regional Economic Impact: Minnamurra Rainforest Centre, Budderoo National Park*, NSW NPWS, Sydney.

Gillieson, D. & Spate, A. 1998, 'Karst and Caves in Australia and New Guinea', in *Global Karst Correlation*, Daoxian, Y. & Zaihua, L. (eds), Science Press, Sydney.

Gilmore, D.A. 1965, 'Hydrological Investigations of Soil Loss and Vegetation Types in the Lower Cotter Catchment', MSc thesis, Botany Department, ANU, Canberra.

Gimblett, H.R. & Itami, R.M. 1997, 'Modeling the Spatial Dynamics and Social Interaction of Human Recreationists Using GIS and Intelligent Agents', *MODSIM 97—International Congress on Modeling and Simulation*, Hobart, Tasmania, 8–11 December.

Goldstein, W. (ed.) 1979, *Australia's 100 Years of National Parks*, NSW NPWS, Sydney.

Gommon, M.F., Glover, J.C.M. & Kuiter, R.H. 1994, *The Fishes of Australia's South Coast*, State Print, Adelaide.

Good, R. 1982, 'Effects of Prescribed Burning on Sub-alpine Vegetation', MSc thesis, University of NSW, Sydney.

Good, R.B. 1998, 'The Impacts of Snow Regimes on the Distribution of Alpine Vegetation, in *Snow: A Natural History; An Uncertain Future*, Green, K. (ed.), Australian Alps Liaison Committee, Canberra.

Goodall, H. 1996, *From Invasion to Embassy: Land in Aboriginal Politics in New South Wales*, Allen and Unwin, Sydney.

Gore, R. 1997, 'The Dawn of Humans: The First Steps', *National Geographic*, 191: 72–99.

Government of South Australia 1998, *Great Australian Bight Marine Park Management Plan*, Department of Environment, Heritage and Aboriginal Affairs, Adelaide.

Graefe, A.R., Kuss, F. & Vaske, J. 1990, *Visitor Impact Management Volume 2: The Planning Framework*, National Parks and Conservation Association, Washington.

Graham, J., Amos, B. & Plumptre, T. 2003, *Governance Principles for Protected Areas in the 21st Century*, prepared for the 5th World Parks Congress, Durban, South Africa.

Graetz, R.D., Wilson, M.A. & Campbell, S.K. 1995, *Landcover Disturbance over the Australian Continent: A Contemporary Assessment*, Commonwealth of Australia, Canberra.

Green, D.G. & Klomp, N.I. 1996, *Ecosystem Connectivity and its Implications for Managing Biodiversity*, presented to Setting Conservation Priorities Workshop, La Trobe University, Wodonga.

Green, K. 2003, 'Impacts of Global Warming on the Snowy Mountains', in *Climate Change Impacts on Biodiversity in Australia*, Howden, M., Hughes, L., Dunlop, M., Zethoven, I., Hilbert, D. & Chilcott, C. (eds), Commonwealth of Australia, Canberra.

Green, R. & Higginbottom, K. 2001, *Negative Effects of Wildlife Tourism on Wildlife*, Wildlife Tourism Research Report No 5, CRC for Sustainable Tourism, Gold Coast.

Green Globe Asia Pacific 2003, *Performance Level Benchmarks*, URL: http://www.greenglobe21.com/ Documents/General/Performance%20Level%20 Benchmarks.xls.

Greig, P.J. 1977, 'Cost–benefit Analysis in Recreation Planning and Policy-making', in *Leisure and Recreation in Australia*, Mercer, D. (ed.), Sorrett, Melbourne.

Groombridge, B. & Jenkins, M.D. 2000, *Global Biodiversity: Earth's Living Resources in the 21st Century*, UNEP-WCMC, World Conservation Press, Cambridge.

Groombridge, B. & Jenkins, M.D. 2002, *World Atlas of Biodiversity: Earth's Living Resources in the 21st Century*, UNEP-WCMC, University of California Press, Cambridge.

Grubb, M. 1999, *The Kyoto Protocol: A Guide and Assessment*, Royal Institute of International Affairs, Washington.

Grzimek, B. & Grzimek, M. 1977, *Serengeti Shall Not Die*, Fontana, London.

Guzowski, M. 1999, *Daylighting for Sustainable Design*, McGraw-Hill Professional, New York.

Hajkowicz, S.A. 2002, *Regional Priority Setting in Queensland: A Multi-criteria Evaluation Framework*, Report for the Queensland Department of Natural Resources and Mines, CSIRO Division of Land & Water, Canberra.

Halpern, B.S. 2003, 'The Impact of Marine Reserves: Do Reserves Work and Does Reserve Size Matter?' *Ecological Applications*, 13(1): S117–S137.

Halpern, B.S. & Warner, R.R. 2002, 'Marine Reserves Have Rapid and Lasting Effects', *Ecology Letters*, 5(3): 361–6.

Halpern, B.S. & Warner, R.R. 2003, 'Matching Marine Reserve Design to Reserve Objectives', *Proceedings of the Royal Society of London Series B—Biological Sciences*, 270: 1871–8.

Ham, S. 1992, *Environmental Interpretation: A Practical Guide for People with Big Ideas and Small Budgets*, North American Press, Golden.

Hamilton, L.S. 1997, 'Maintaining Ecoregions in Mountain Conservation Corridors', *Wild Earth*, 7(3): 63–6.

Hamilton, L.S., Mackay, J.C., Worboys, G.L., Jones, R.A. & Manson, G.B. 1996, *Transborder Protected Area Cooperation*, Australian Alps Liaison Committee, Canberra.

Hamilton-Smith, E. 1998, 'Changing Assumptions Underlying National Park Systems', in *Celebrating the Parks: Proceedings of the First Australian Symposium on Parks History*, Hamilton-Smith, E. (ed.), Rethink Consulting, Melbourne.

Hammitt, W.E. & Cole, D.N. 1998, *Wildland Recreation* (2nd edn), John Wiley and Sons Inc., New York.

Hancock, W.K. 1972, *Discovering Monaro: A Study of Man's Impact on his Environment*, Cambridge University Press, Cambridge.

Hand, S.J. (ed.) 1990, *Care and Handling of Australian Native Animals: Emergency Care and Captive Management*, Surrey Beatty and Sons, Sydney.

Harden, G.J. 1990, *Flora of New South Wales Volume 1*, Royal Botanic Gardens, Sydney.

Harden, G.J. 1991, *Flora of New South Wales Volume 2*, Royal Botanic Gardens, Sydney.

Harden, G.J. 1992, *Flora of New South Wales Volume 3*, Royal Botanic Gardens, Sydney.

Harden, G.J. 1993, *Flora of New South Wales Volume 4*, Royal Botanic Gardens, Sydney.

Harding, R. & Traynor, D. 2003, 'Informing ESD: State of the Environment Reporting', in *Managing Australia's Environment*, Dovers, S. & Wild River, S. (eds), The Federation Press, Sydney.

Harmon, D. & Putney, A.D. (eds) 2003, *The Full Value of Parks: From Economics to the Intangible*, Rowman and Littlefield, Lanham.

Haroon, A.I. 2002, 'Ecotourism in Pakistan: A Myth?', *Mountain Research and Development*, 22(2): 110–12.

Harrigan, A. 2001, 'Kosciuszko Summit Area Track Techniques', in *Mountain Walking Track Management: An Australian Alps Best Practice Field Forum*, McKay, J. (ed.), Australian Alps Liaison Committee, Canberra.

Head, L. & Hughes, C. 1996, 'One Land, Which Law? Fire in the Northern Territory', in *Resource, Nations and Indigenous Peoples*, Howitt, R., Connell, J. & Hirsch, P. (eds), Oxford University Press, Melbourne.

Healey, P. 1997, *Collaborative Planning: Shaping Places in Fragmented Societies*, Macmillan Press, London.

Heatwole, H. 1987, 'Major Components and Distributions of the Terrestrial Fauna', in *Fauna of Australia*, Vol. 1A, AGPS, Canberra.

Helms, R. 1896, 'The Australian Alps, or Snowy Mountains', *Journal of Royal Geographical Society of Australasia*, VI(4).

Herford, I., Lloyd, M. & Hammond, R. 1995, *West Cape Howe National Park Management Plan 1995–2005*, CALM, Perth.

Higgins, A. & Thompson, D. 2002, 'Life Cycle Assessment', in *Tools for Environmental Management: A Practical Introduction and Guide*, Thompson, D. (ed.), New Society Publishers, British Columbia.

Hill, M. 1994, *A Brief History of the Australian National Parks and Wildlife Service (ANPWS) and its Successor the Australian Nature Conservation Agency, 1973–1994*, Australian Nature Conservation Agency, Canberra.

Hillary, R. 1997, *Environmental Management Systems and Cleaner Production*, Wiley, Chichester.

Hockings, M. 2003, 'Systems for Assessing the Effectiveness of Management in Protected Areas', *Bioscience*, 53(9): 823–32.

Hockings, M., Leverington, F. & Carter, B. 1995, *Public Contact Planning for Conservation Management*, Department of Management Studies Occasional Paper 3(2), University of Queensland, Gatton.

Hockings, M., Stolton, S. & Dudley N. 2000, *Evaluating Effectiveness: A Framework for Assessing the Management of Protected Areas*, IUCN WCPA, Gland.

Hockings, M., Stolton, S., Dudley, N. & Parrish J. 2001, *The Enhancing our Heritage Toolkit: Book 1*, UNESCO and IUCN, URL: http://www.enhancingheritage.net/docs_public.asp.

Hoegh-Guldberg, O. 2003, 'Coral Reefs, Thermal Limits and Climate Change', in *Climate Change Impacts on Biodiversity in Australia*, Howden, M., Hughes, L., Dunlop, M., Zethoven, I., Hilbert, D. & Chilcott, C. (eds), Commonwealth of Australia, Canberra.

Holdsworth, M. & Bryant, S. 1995, 'Rescue and Rehabilitation of Wildlife from the Iron Baron Oil Spill in Northern Tasmania', *Tasmanian Naturalist*, 117: 39–43.

Honey, M. 1999, *Ecotourism and Sustainable Development: Who Owns Paradise?*, Island Press, Washington.

Howard, J. 1999, *Interpretive Guiding Management in Ecotourism*, Open Learning Institute, Charles Sturt University, Albury.

Howden, M. 2003, 'Climate Trends and Climate Change Scenarios', in *Climate Change Impacts on Biodiversity in Australia*, Howden, M., Hughes, L., Dunlop, M., Zethoven, I., Hilbert, D. & Chilcott, C. (eds), Commonwealth of Australia, Canberra.

Howden, M., Hughes, L., Dunlop, M., Zethoven, I., Hilbert, D. & Chilcott, C. 2003, *Climate Change Impacts on Biodiversity in Australia*, Outcomes of a workshop sponsored by the Biological Diversity Advisory Committee, 1–2 October 2002, CSIRO Sustainable Ecosystems, Australia.

Hubbard, L., James, D. & Smart, J. 2001, *A Guide to the State Project*, NSW NPWS, Sydney.

Huber, D. 2003, 'Audit of the Management of the Queensland East Coast Trawl Fishery in the Great Barrier Reef Marine Park', Unpublished Interim Report, GBRMPA, Townsville.

Hughes, O.E. 1998, *Australian Politics* (3rd edn), Macmillan, Melbourne.

Hundloe, T., McDonald, G.T., Blamey, R.K., Wilson, B.A. & Carter, M. 1990, *Non-extractive Natural Resource Uses in the Great Sandy Region*, Institute of Applied Environmental Research, Griffith University, Brisbane.

Hunger, J.D. & Wheelan, T.L. 2001, *Essentials of Strategic Management*, Prentice-Hall, New Jersey.

Hutton, D. & Connors, L. 1999, *A History of the Australian Environment Movement*, Cambridge University Press, Melbourne.

Hyde, R. & Law, J. 2001, *Green Globe 21—Designing Tourism Infrastructure: Steps to Sustainable Design*, CRC for Sustainable Tourism, Gold Coast.

ICAC 1999, *Best Practice, Best Person: Integrity in Public Sector Recruitment and Selection*, Independent Commission Against Corruption, Sydney.

IHEI 2003, 'Water Saving: Bathrooms and Toilets', *GreenHotelier*, (29): insert.

IISD 2003, 'A Summary Report of the 5th IUCN World Parks Congress', *Sustainable Developments*, 89(9): 1–15.

Imber, D., Stevenson, G. & Wilks, L. 1991, *A Contingent Valuation Survey of the Kakadu Conservation Zone*, Resource Assessment Commission Research Paper No. 3, AGPS, Canberra.

IMCRA 1998, *Interim Marine and Coastal Regionalisation for Australia: An Ecosystem-based Classification for Marine and Coastal Environments (Version 3.3)*, Environment Australia, Canberra.

Industry Commission 1997, *A Full Repairing Lease: Inquiry into Ecologically Sustainable Land Management*, Industry Commission, Canberra.

Institute of Public Affairs 2003, Institute of Public Affairs website, URL: http://www.ipa.org.au.

IPCC 2001, *Climate Change 2001: The Scientific Basis—Contribution of Working Group 1 to the Third Assessment Report of the Intergovernmental Panel on Climate Change*, Cambridge University Press, Cambridge.

Itami, R.M. & Gimblett, H.R. 2000, 'Intelligent Recreation Agents in a Virtual GIS World', *Proceedings of Complex Systems 2000 Conference*, University of Otago, Dunedin, New Zealand, 19–22 November.

IUCN 1980, *World Conservation Strategy: Living Resource Conservation and Sustainable Development*, IUCN, Morges.

IUCN 1994, *Guidelines for Protected Area Management Categories*, IUCN, Gland.

IUCN 1998, *Economic Values of Protected Areas: Guidelines for Protected Area Managers*, IUCN, Gland.

IUCN 2000a, *Financing Protected Areas*, IUCN, Gland.

IUCN 2000b, *Vision for Water and Nature: A World Strategy for Conservation and Sustainable Management of Water Resources in the 21st Century—Compilation of all Project Documents*, IUCN, Gland and Cambridge.

IUCN 2003a, *The Durban Accord*, 5th IUCN World Parks Congress Durban, South Africa, 8-17 September, 2003, IUCN WCPA, Gland, Switzerland.

IUCN 2003b, *The Durban Action Plan*, 5th IUCN World Parks Congress Durban, South Africa, 8-17 September, 2003, IUCN WCPA, Gland, Switzerland.

IUCN 2003c, *Recommendations of the 5th IUCN World Parks Congress*, Durban, South Africa, 8-17 September, 2003, IUCN WCPA, Gland, Switzerland.

IUCN 2003d, *Message of the 5th IUCN World Parks Congress to the Convention on Biological Diversity*, IUCN WCPA, Gland, Switzerland.

Ivancevich, J., Olekalns, M. & Matteson, M. 1997, *Organisation behaviour and management*, Irwin, Sydney.

Jacobs, J.M. 1994, 'Earth Honouring: Western Desires and Indigenous Knowledge', *Meanjin*, 53: 304–14.

Jacobs, M. 1995, 'Sustainability and "the Market": A Typology of Environmental Economics', in *Markets, the State and the Environment: Towards Integration*, Eckersley, R. (ed.), Macmillan, Melbourne.

Jaireth, H. 2003, 'Integrated Natural Resource Management', in *Innovative Governance: Indigenous peoples, Local Communities and Protected Areas*, Jaireth H. & Smyth, D. (eds), Ane Books, New Delhi.

James, D. 1997, *Environmental Incentives: Australia's Experience with Economic Instruments for Environmental Management*, Environment Australia, Canberra.

Jarrott, J.K. 1990, *History of Lamington National Park*, National Parks Association of Queensland, Brisbane.

Jarvis, T.D. 2000, 'The Responsibility of National Parks in Rural Development', in *National Parks and Rural Development: Practice and Policy in the United States*, Machlis, G.E. & Field, D.R. (eds), Island Press, Washington.

Jenkins, S. 1996, 'Native Vegetation on Farms Survey: A Survey of Farmers' Attitudes to Native Vegetation and Landcare in the Wheatbelt of Western Australia', *Research Management Technical Report 164*, Western Australian Department of Agriculture, Perth.

Jennings, J.N. 1975, 'How Well Off is Australia for Caves and Karst? A Brief Geomorphic Estimate', *Proceedings of the 10th Australian Speleological Federation Conference*.

Johnston, C. 1992, *What is Social Value?*, AGPS, Canberra.

Johnston, F.M. & Pickering, C.M. 2001a, 'Alien Plants in the Australian Alps', *Mountain Research and Development*, 21(3): 284–91.

Johnston, F.M. & Pickering, C.M. 2001b, 'Yarrow, *Achillea millefolium L.*: A Weed Threat to the Flora of the Australian Alps', *Victorian Naturalist*, 118: 230–3.

Johnston, S.W. 1998, *Marritz Precinct Plan: Perisher Village Kosciuszko National Park—Principles and Procedures of Soil Restoration and Native Revegetation*, NSW NPWS, Jindabyne.

Johnston, S.W. & Pickering, C.M. 2001, 'Visitor Monitoring Including Social Expectations for Track Planning', in *Mountain Walking Track Management—An Australian Alps Best Practice Field Forum*, McKay, J. (ed.), Australian Alps Liaison Committee, Canberra.

Jones, G. & Dunn, H. 2000, 'Experience in Outcomes-based Evaluation of Management for the Tasmanian Wilderness World Heritage Area, Australia: Case Study 1', in *Evaluating Effectiveness: A Framework for Assessing the Management of Protected Areas*, Hockings, M., Stolton, S. & Dudley, N. (eds), IUCN, Gland and Cambridge.

Jones, G.P. & Andrew, N.L. 1990, 'Herbivory and Patch Dynamics on Rocky Reefs in Temperate Australasia: The Roles of Fish and Sea Urchins', *Australian Journal of Ecology*, 15, 505–20.

Jones, R. 1969, 'Firestick Farming', *Australian Natural History*, 16: 224–8.

Jones, R. 1990, 'Hunters of the Dreaming: Some Ideational, Economic and Ecological Parameters of the Australian Aboriginal Productive System', in *Pacific Production Systems: Approaches to Economic Prehistory*, Yen, D.E. & Mummery, J.M.J. (eds), ANU, Canberra.

Jones, R. 1999, 'Folsom and Talgai', in *Approaching Australia: Papers from the Harvard Australian Studies Symposium*, Bolitho, H. & Wallace-Crabbe, C. (eds), Harvard University Press, Massachusetts.

Kaplan, S. & Kaplan, R. 1989, 'The Visual Environment: Public Participation in Design and Planning', *Journal of Social Issues*, 45: 59–86.

Karolides, A. 2002, 'An Introduction to Green Building: Part 1: Resource Efficiency', *RMI Solutions*, 18(3): 7–8.

Karolides, A. 2003, 'An Introduction to Green Building: Part 2: Environmental Sensitivity with Building Materials', *RMI Solutions*, 19(1): 5–7.

Kaufman, J.L. & Jacobs, H.M. 1996, 'A Public Planning Perspective on Strategic Planning', in *Readings in Planning Theory*, Campbell, S. & Fainstein, S. (eds), Blackwell, Oxford.

Kavanagh, L.J. 2002, *Water Management and Sustainability at Queensland Tourist Resorts*, Research Report Series, CRC for Sustainable Tourism, Gold Coast.

Kelleher, G. & Kenchington, R. 1992, *Guidelines for Establishing Marine Protected Areas*, IUCN, Gland.

Kelly, S., Scott, D., MacDiarmid, A.B. & Babcock, R.C. 2000, 'Spiny Lobster, *Jasus edwardsii*, Recovery in New Zealand Marine Reserves', *Biological Conservation*, 92(3): 359–69.

Kempf, E. (ed.) 1993, *Indigenous Peoples and Protected Areas: The Law of the Mother*, Earthscan, London.

Kenchington, R. 1990, *Managing Marine Environments*, Taylor and Francis, New York.

Kennedy, P. 1999, *Working with Media: Internal Training Notes*, NSW NPWS, Sydney.

Kennelly, S.J. 1995, 'Kelp Beds', in *Coastal Marine Ecology of Temperate Australia*, Underwood, A.J. & Chapman, M.G. (eds), University of New South Wales Press, Sydney.

Keough, M.J. & Butler, A.J. 1995, 'Temperate Subtidal Hard Substrata', in *The State of the Marine Environment Report for Australia Technical Annex 1: The Marine Environment*, Zann, L.P. & Kailola, P. (eds), GBRMPA, Townsville.

Kerr, J.S. 1996, *The Conservation Plan: A Guide to the Preparation of Conservation Plans for Places of European Cultural Significance*, National Trust of Australia, Sydney.

Kiernan, K. 1996, *Conserving Geodiversity and Geoheritage: The Conservation of Glacial Landforms*, Forest Practices Board, Hobart.

Kimber, R. 1983, 'Black Lightning: Aborigines and Fire in Central Australia and the Western Desert', *Archaeology in Oceania*, 18: 38–44.

Kirkman H. 1997, *Seagrasses of Australia*, Department of the Environment, Canberra.

Kirkpatrick, J. 1983, 'An Iterative Method for Establishing Priorities for the Selection of Nature Reserves: An Example from Tasmania', *Biological Conservation*, 25: 127–34.

Knapman, B. & Stanley, O. 1991, *A Travel Cost Analysis of the Recreational Use Value of Kakadu National Park*, Resource Assessment Commission, Canberra.

Knudson, D.M., Cable, T.T. & Beck, L. 1995, *Interpretation of Cultural and Natural Resources*, Venture Publishing, State College.

Knudtson, P. & Suzuki, D. 1992, *Wisdom of the Elders: Honouring Sacred Native Visions of Nature*, Allen and Unwin, Sydney.

Koslow, J.A. & Gowlett-Holmes, K. 1998, *The Seamount Fauna off Southern Tasmania: Benthic Communities, Their Conservation and Impacts of Trawling*, CSIRO Division of Marine Research, Hobart.

Koteen, J. 1997, *Strategic Management in Public and Nonprofit Organizations* (2nd edn), Praeger, Westport.

Krockenberger, M. 1998, 'Falling Down: Land Clearing in Australia', *Habitat Australia Supplement*, February: 1–8.

Land Conservation Council 1991, *Wilderness Special Investigation Final Recommendations*, LCC, Melbourne.

Landcare Research 2003, *Annual Report 2003*, Landcare Research, Lincoln, New Zealand.

Ladson, T. & Argent, R. 2002, 'Adaptive Management of Environmental Flows: Lessons for the Murray Darling Basin from Three Large North American Rivers', *Australian Journal of Water Resources*, 5: 89–102.

Lahdelma, R., Salminen, P. & Hokkanen, J. 2000, 'Using Multi-criteria Methods in Environmental Planning and Management', *Environmental Management*, 26(6): 595–605.

Landre, B.K. & Knuth, B.A. 1993, 'Success of Citizen Advisory Committees in Consensus-based Water Resources Planning in the Great Lakes Basin', *Society and Natural Resources*, 6: 229–57.

Lane, M. & McDonald, G. 1997, 'Not all World Heritage Areas are Created Equal: World Heritage Area Management in Australia', in *National Parks and Protected Areas: Selection, Delimitation and Management*, Pigram, J.J. & Sundell, R.C. (eds), University of New England, Armidale.

Langton, M. 1996, 'Art, Wilderness and *Terra Nullius*', in *Ecopolitics IX Conference Papers and Resolutions*, Northern Land Council, Darwin.

Langton, M. 1998, *Burning Questions: Emerging Environmental Issues for Indigenous Peoples in Northern Australia*, Centre for Indigenous Natural and Cultural Resource Management, Northern Territory University, Darwin.

Lawrence, D., Kenchington, R. & Woodley, S. 2002, *The Great Barrier Reef, Finding the Right Balance*, Melbourne University Press, Melbourne.

Lawson, B. 1999, 'Towards Culturally Appropriate and Culturally Sustainable Nature-based Recreation and Ecotourism Management: An Introduction to the "Indigenous Cultural Landscape Model"', Masters thesis, School of Environmental and Information Sciences, Charles Sturt University, Albury.

Leaver, B., 1999, 'Land Use Allocation in the State of Tasmania, Australia', in *Compendium of Case Studies: Protected Area Management Principles and Practice*, NSW NPWS, Queanbeyan.

Leigh, J.H., Boden, R. & Briggs, J. 1984, *Extinct and Endangered Plants of Australia*, Macmillan, Australia.

Lennon, J. 1992, *Our Inheritance: Historic Places on Public Land in Victoria*, Department of Conservation and Environment, Melbourne.

Lennon, J. 1999, *Conserving the Cultural Values of Natural Areas: A Discussion Paper for Australia*, ICOMOS, Developed in association with the University of Canberra, Canberra.

Lennon, J., Egloff, B., Davey, A. & Taylor, K. 1999, *Conserving the Cultural Values of Natural Areas: A*

Discussion Paper, Jane Lennon and Associates and University of Canberra, Canberra.

Lennon, J. & Mathews, S. 1996, *Cultural Landscape Management: Guidelines for Identifying, Assessing and Managing Cultural Landscapes in the Australian Alps National Parks*, Jane Lennon & Associates, Melbourne.

Leopold, A. 1949, *A Sand County Almanac*, Oxford University Press, New York.

Lewis, H.T. 1989, 'Ecological and Technological Knowledge of Fire: Aborigines versus Park Rangers in Northern Australia', *American Anthropologist*, 91: 940–61.

Lewis, T.L. 1967, 'Kosciusko State Park Master Planning', *Australian Ski Year Book*.

Liddle, D.T., Taylor, S.M. & Larcombe, D.R. 1996, 'Population Changes from 1990 to 1995 and Management of the Endangered Rainforest Palm *Ptychosperma bleeseri* Burret (Arecaceae)', in *Back from the Brink: Refining the Threatened Species Recovery Process*, Stephens, S. & Maxwell, S. (eds), Surrey Beatty and Sons, Sydney.

Lindberg, K. 1998, 'Economic Aspects of Ecotourism', in *Ecotourism: A Guide for Planners and Managers, Vol. 2*, Lindberg, K., Epler-Wood, M. & Engeldrum, D. (eds), Ecotourism Society, Bennington.

Lindblom, C.E. 1979, 'Still Muddling, Not Yet Through', *Public Administration Review*, 39: 317–26.

Lockwood, M. 1998, 'Contribution of Contingent Valuation and Other Stated Preference Methods to Evaluation of Environmental Policy', *Australian Economic Papers*, 37: 292–311.

Lockwood, M. 1999, 'Humans Valuing Nature', *Environmental Values*, 8: 381–401.

Lockwood, M. & Tracy, K. 1995, 'Nonmarket Economic Valuation of an Urban Recreation Park', *Journal of Leisure Research*, 27(2): 155–67.

Lockwood, M., Loomis, J. & De Lacy, T. 1993, 'A Contingent Valuation Survey and Benefit Cost Analysis of Forest Preservation in East Gippsland, Australia', *Journal of Environmental Management*, 38: 233–43.

Lockwood, M. & Spennemann, D.H.R. 1999, *Value Conflicts Between Natural and Cultural Heritage Conservation: Australian Experience and the Contribution of Economics*, Presented to the workshop Heritage Economics: Challenges for Heritage Conservation and Sustainable Development in the 21st Century, Australian Heritage Commission, Canberra.

Logan, W. 2004, 'ACT Parks and Reserves (Protected Areas) Significant Milestones', Unpublished Report, ACT Parks and Conservation Service, Canberra.

Lomberg, B. 1998, *The Skeptical Environmentalist: Measuring the Real State of the World*, Cambridge University Press, Cambridge.

Lomberg, B. 2003, Decision regarding complaints against Bjorn Lomberg, Danish Committees on Scientific Dishonesty, URL: http://www.lomberg.com.

Loomis, J.B. & Walsh, R.G. 1997, *Recreation Economic Decisions: Comparing Benefits and Costs* (2nd edn), Venture, State College.

Lothian, J. A. 1997, 'Attitudes of Australians Towards the Environment: 1975 to 1994', *Australian Journal of Environmental Management*, 1: 79–99.

Lowe, K., Fitzsimons, J., Straker, A. & Gleeson, T. 2003, *Mechanisms for Improved Integration of Biodiversity Conservation in Regional Natural Resource Management Planning within Australia*, Synapse Research & Consulting & Associates for Land & Water, Brisbane.

Lubchenco, J., Palumbi, S.R., Gaines, S.D. & Andelman, S. (eds) 2003, 'The Science of Marine Reserves', *Ecological Applications* 13(1): S3–S228.

Luke, R.H. & McArthur, A.G. 1978, *Bushfires in Australia*, AGPS, Canberra.

Lunney, D. & Leary, T. 1988, 'The Impact on Native Animals of Landuse Changes and Exotic Animals in the Bega District Since Settlement', *Australian Journal of Ecology*, 13: 67–92.

Lunney, D. & Moon, C. 1987, 'The Eden Woodchip Debate (1969-1986)', *Search*, 18(1).

Lyden, J.F., Twight, B.W. & Tuchman, E. 1990, 'Citizen Participation in Long-range Planning: The RPA Experience', *Natural Resources Journal*, 30: 23–138.

MacDonald, M. & Nierenberg, D. 2003, 'Linking Population, Women, and Biodiversity, in Worldwatch Institute', in *State of the World: Progress Towards a Sustainable Society*, Earthscan, London.

Machlis, G.E. 1992, 'The Contribution of Sociology to Biodiversity Research and Management', *Biological Conservation*, 62: 161–70.

Machlis, G.E. & Field, D.R. (eds) 2000a, *National Parks and Rural Development: Practice and Policy in the United States*, Island Press, Washington.

Machlis, G.E. & Field, D.R. 2000b, 'Conclusion', in *National Parks and Rural Development: Practice and Policy in the United States*, Machlis, G.E. & Field, D.R. (eds), Island Press, Washington.

MacKay, J., Lockwood, M. & Curtis, A. 2000, *A Review of Policy Options to Enhance the Management of Dryland Salinity*, Johnstone Centre, Albury.

MacKinnon, J., MacKinnon, K., Child, G. & Thorsell, J. 1986, *Managing Protected Areas in the Tropics*, IUCN WCPA, Gland.

Mainieri, T., Barnett, E.G., Valdero, T.R., Unipan, J.B. & Oskamp, S. 1997, 'Green Buying: The Influence of Environmental Concern on Consumer Behaviour', *Journal of Social Psychology*, 137: 189–204.

Maltby, E. 1997, 'Ecosystem Management: The Concept and the Strategy', *World Conservation*, 3: 3–4.

Manins, P., Holper, P., Suppiah, R., Allan, R., Walsh, K., Fraser, P. & Beer, T. 2001, *Atmosphere: Australia State of the Environment 2001*, Commonwealth of Australia, Canberra.

Mansergh, I. & Broome, L. 1994, *The Mountain Pygmy Possum of the Australian Alps*, New South Wales University Press, Sydney.

Mapstone, B.D., Campbell, R.A. & Smith, A.D. 1996, *Design of Experimental Investigations of the Effects of Line and Spear Fishing on the Great Barrier Reef*, CRC Reef Research Centre Technical Report No. 7, CRC Reef, Townsville.

Mapstone, B.D., Davies, C.R., Little, L.R., Punt, A.E., Smith, A.D.M., Pantus, F., Lou, D.C., Williams, A.J., Jones, A., Russ G. R. & MacDonald, A.D. 2003, *The Effects of Line Fishing on the Great Barrier Reef and Evaluations of Alternative Potential Management Strategies*, CRC Reef Research Centre, Townsville.

Margoluis, R. & Salafsky, N. 1998, *Measures of Success: Designing, Managing, and Monitoring Conservation and Development Projects*, Island Press, Washington.

Margoluis, R. & Salafsky, N. 2001, *Is Our Project Succeeding? A Guide to Threat Reduction Assessment for Conservation*, Biodiversity Support Program, Washington, DC.

Marguglio, B. 1991, *Environmental Management Systems*, ASQC Quality Press, New York.

Margules, C. 1989, 'Introduction to Some Australian Developments in Conservation Evaluation', *Biological Conservation*, 50: 1–11.

Margules, C.R., Nicholls, A. & Pressey, R.L. 1988, 'Selecting Networks of Reserves to Maximise Biological Diversity', *Biological Conservation*, 43: 63–76.

Marion, J.L. & Leung, Y. 1997, *An Assessment of Campsite Conditions in Great Smoky Mountains National Park: Research/Resources Management Report*, USDI National Park Service, Atlanta, GA.

Market Solutions 1996, *National Parks Visitor Segmentation Study*, Market Solutions (Australia) Pty Ltd, Melbourne.

Marquis-Kyle, P. & Walker, M. 1992, *The Illustrated Burra Charter*, ICOMOS, Brisbane.

Martin, R. 1992, 'Of Koalas, Tree-kangaroos and Men', *Australian Natural History*, 24(3): 22–31.

Mason, C. 2003, *The 2030 Spike: Countdown to Global Catastrophe*, Earthscan, London.

McArthur, A.G. 1962, *Control Burning in Eucalypt Forests*, Forest and Timber Bureau, Canberra.

McDougall, K.L. & Appleby, M.L. 2000, 'Plant Invasions in the High Mountains of North-eastern Victoria', *Victorian Naturalist*, 117: 52-59.

McDowell, D. 1995, *IUCN and the World Heritage: A Review of Policies and Procedures*, IUCN, Gland.

McHarg, I.L. 1992, *Design with Nature*, John Wiley and Sons, New York.

McKinnon, K. 2003, *Reporting Progress at Protected Area Sites: A Simple Site-level Tracking Tool Developed for the World Bank and WWF*, World Bank/WWF Alliance for Forest Conservation and Sustainable Use, Washington.

McLain, R.J. & Lee, R.G. 1996, 'Adaptive Management: Promises and Pitfalls', *Environmental Management*, 20: 437–48.

McLeod, R. 2003, *Inquiry into the Operational Response to the January 2003 bushfires in the ACT*, ACT Government, Canberra.

McNeely, J.A. 1995, *Expanding Partnerships in Conservation*, IUCN and Island Press, Washington.

McNeely, J.A. & Schutyser, F. (eds) 2003, *Protected Areas in 2023: Scenarios for an Uncertain Future*, IUCN, Gland.

Meadowcroft, J. 1997, 'Planning for Sustainable Development: Insights from the Literatures of Political Science', *European Journal of Political Research*, 31(4): 427–54.

Meffe, G.K. & Carroll, C.R. (eds) 1997, *Principles of Conservation Biology* (2nd edn), Sinauer Associates, Sunderland.

Menkhorst, P.W. 1995, *Mammals of Victoria*, Oxford University Press, Melbourne.

Mercer, D. 2000, *A Question of Balance: Natural Resource Conflict Issues in Australia* (3rd edn), The Federation Press, Sydney.

Mitchell, R.C. & Carson, R.T. 1989, *Using Surveys to Value Public Goods: The Contingent Valuation Method*, Resources for the Future, Washington.

Montgomery, C.W. 1997, *Environmental Geology* (5th edn), McGraw Hill, New York.

Moore, S.A., Smith, A.J. & Newsome, D.N. 2003, 'Environmental Performance Reporting for Natural Area Tourism: Contributions by Visitor Impact Management Frameworks and their Indicators', *Journal of Sustainable Tourism*, 11(4): 348–75.

Moote, M.A., McCalaran, M.P. & Chickering, D.K. 1997, 'Research Theory on Practice: Applying Participatory Democracy Theory to Public Land Planning', *Environment Management*, 21(6): 877–89.

Morrison, J. 1997, 'Protected Areas, Conservationists and Aboriginal Interests in Canada', in *Social Change and*

Conservation: Environmental Politics and Impacts of National Parks and Protected Areas, Ghimire K.B. & Pimbert M.P. (eds), Earthscan, London.

Morrison, M.D., Blamey, R.K., Bennett, J.W. & Louviere, J.J. 1996, A Comparison of Stated Preference Techniques for Estimating Environmental Values, Choice Modelling Research Report No. 1, University of New South Wales, Canberra.

Mosely, J.G. 1968, National Parks and Equivalent Reserves in Australia: Guide to Legislation, Administration, and Areas, Special Publication No 2, ACF, Canberra.

Muldoon, J. & Gillies, J. 1996, 'The Ocean Rescue 2000 Marine Protected Area program'; in The State of the Marine Environment Report for Australia Technical Annex 1: The Marine Environment, Zann, L.P. & Kailola, P. (eds), GBRMPA, Townsville.

Mulligan, M. and Hill, S. 2001, Ecological Pioneers: A Social History of Australian Ecological Thought and Action, Cambridge University Press, Cambridge.

Mulvaney, J. & Kamminga, J. 1999, Prehistory of Australia, Allen and Unwin, Sydney.

Murray, S.N., Ambrose, R.F., Bohnsack, J.A., Botsford, L.W., Carr. M.H., Davis, G.E., Dayton, P.K., Gotshall, D., Gunderson, D.R., Hixon, M.A., Lubchenco, J., Mangel, M., MacCall, A., McArdle, D.A., Ogden, J.C., Roughgarden, J., Starr, R.M., Tegner, M.J. & Yoklavich, M.M. 1999, 'No-take Reserve Networks: Sustaining Fishery Populations and Marine Ecosystems', Fisheries, 24(11): 11–25.

Myers, F.R. 1980, 'Always Ask: Resource Use and Land Ownership Among Pintupi Aborigines of the Australian Western Desert', in Resource Managers: North American and Australian Hunter-gatherers, Williams, N.M. & Hunn, E.S. (eds), Westview Press, San Francisco.

Nanson, G.C., von Krusenstierna, A., Bryant, E.A. & Renilson, M.R. 1994, 'Experimental Measurements of River Bank Erosion Caused by Boat Generated Waves on the Gordon River, Tasmania', Regulated Rivers: Research and Management, 9: 1–14.

Neal, B., Erlanger, P., Evans, R., Kollmorgen., Ball, J., Shirley, M., Donnelley, L. & Hack, B. 2001, Inland Waters: Australia State of the Environment 2001, Commonwealth of Australia, Canberra.

Nelson, H. & Wearing S. 1999, 'Campground Strategy for ACT Parks and Forests', Unpublished report to NSW NPWS, Sydney.

Nettheim, G. 1998, 'Native Title: The 1998 Amendments and International Law', Unpublished report.

Newsome, A.E. 1980, 'The Eco-mythology of the Red Kangaroo in Central Australia', Mankind, 12(4): 327–33.

Newsome, D., Moore, S.A. & Dowling, R.K. 2002, Natural Area Tourism: Ecology, Impacts and Management, Channel View Publications, Sydney.

Njiforti, H.L. & Tchamba, N.M. 1993, 'Conflict in Cameroon: Parks For or Against People?', in Indigenous Peoples and Protected Areas: The Law of the Mother, Kempf, E. (ed.), Earthscan, London.

NLWRA 2002, Australian Terrestrial Biodiversity Assessment, Land and Water Australia, Commonwealth of Australia, Canberra.

NRMMC 2004, Directions for the National Reserve System: A Partnership Approach, Department of the Environment and Heritage, Canberra.

NSW NPWS 1993, Nowra District Neighbour Policy, NSW NPWS, Sydney.

NSW NPWS 1994, Environmental Planning and Assessment Manual, NSW NPWS, Sydney.

NSW NPWS 1995, Annual Report 1994–1995, NSW NPWS, Sydney.

NSW NPWS 1996, Heritage Highlights 1996, NSW NPWS, Sydney.

NSW NPWS 1997a, Eden District Incident Action Plan, NSW NPWS, Sydney.

NSW NPWS 1997b, Draft Nature Tourism and Recreation Strategy, NSW NPWS, Sydney.

NSW NPWS 1998, National Parks Visions for the New Millennium, NSW NPWS, Sydney.

NSW NPWS 1999a, The Green and Golden Bell Frog, Rare and endangered species information sheet, NSW NPWS, Sydney.

NSW NPWS 1999b, 'C-Plan Conservation Planning Software User Manual for C-Plan (Version 2.2)', Unpublished report by NSW NPWS, Hurstville.

NSW NPWS 2001, Plan of Management Manual 2001, NSW NPWS, Sydney.

NSW NPWS 2002a, Cultural Heritage Conservation Policy, NSW NPWS, Sydney.

NSW NPWS 2002b, Recreation planning framework for NSW parks, NSW NPWS, Sydney.

NSW NPWS 2003, 'NSW Saving Our Threatened Native Plants and Animals: Recovery and Threat Abatement in Action', 2003 Update, NSW NPWS, Sydney.

NSW NPWS 2004a, Mt Ku-ring-gai Coronial Inquiry, URL: http://www.nationalparks.nsw.gov.au.

NSW NPWS 2004b, Perisher Range Resorts Environmental Management System, URL: http://www.nationalparks.nsw.gov.au.

NSW NPWS 2004c, Summary of Coroners' Findings, URL: http://www.nationalparks.nsw.gov.au.

NSW NPWS 2004d, Corporate Policies and Guidelines, URL: http://www.nationalparks.nsw.gov.au.

NSW NPWS, Rural Fire Service, State Forests of NSW, and Department of Land and Water Conservation 1998, *Living with Fire: Bushfire Management, the Environment and the Community*, AGPS, Canberra.

O'Riorden, T. & O'Riorden, J. 1993, 'On Evaluating Public Examination of Controversial Projects', in *Advances in Resource Management*, Foster, H.D. (ed.). Belhaven Press, London.

Ostrom, E. 1990, *Governing the Commons: The Evolution of Institutions for Collective Action*, Cambridge University Press, Cambridge.

Ollier, C.D. 1982, 'Great Escarpment of Eastern Australia: Tectonic and Geomorphological Significance', *Journal of Geological Society of Australia*, 29: 13–23.

Olsen, P. 1998, *Australia's Pest Animals: New Solutions to Old Problems*, Bureau of Resource Sciences, Canberra.

Orams, M.B. & Hill, G.J.E. 1998, 'Controlling the Ecotourist in a Wild Dolphin Feeding Program: Is Education the Answer?', *The Journal of Environmental Education*, 29(3): 33–8.

Orna, E. & Pettitt, C. 1998, *Information Management in Museums* (2nd edn), Gower Publishing Co. Ltd, Hampshire.

Pakulski, J., Tranter, B. & Crook, S. 1998, 'The Dynamics of Environmental Issues in Australia: Concerns, Clusters and Carriers', *Australian Journal of Political Science*, 33: 235–52.

Pannell, S. 1996, '*Homo Nullius* or "Where Have All the People Gone?" Refiguring Marine Management and Conservation Approaches', *Australian Journal of Anthropology*, 7(1): 21–42.

Parker, P. 1993, *Biosphere Reserves in Australia: A Strategy for the Future*, ANCA, Canberra.

Parks Australia 2002, *Policy Development Guide*, Department of Environment and Heritage, Canberra.

Parks and Wildlife Commission of the Northern Territory 2002, *Public Participation in Protected Area Management Best Practice*, prepared for the Committee on National Parks and Protected Area Management Benchmarking and Best Practice Program, Parks and Wildlife Commission of the Northern Territory, Darwin.

Parks and Wildlife Service Tasmania 1991, *Tasmanian Wilderness World Heritage Area Management Plan*, PWST, Hobart.

Parks and Wildlife Service Tasmania 2000, *Best Practice in Protected Area Management Planning*, Parks and Wildlife Service Tasmania, ANZECC, Tasmania.

Parks and Wildlife Service Tasmania 2004a, *A History of the Parks and Wildlife Service*, URL: http://www.parks.tas.gov.au/manage/about/history.html.

Parks and Wildlife Service Tasmania 2004b, *State of the Tasmanian Wilderness World Heritage Area—An Evaluation of Management Effectiveness*, Report No. 1, Department of Tourism Parks Heritage and the Arts, Hobart.

Parks Victoria 1998, *Directions in Environmental Management*, Parks Victoria, Melbourne.

Parks Victoria 2000, *State of the Parks 2000*, Parks Victoria, Melbourne.

Parr-Smith, G. & Polley, V. 1998, *Draft Alpine Rehabilitation Manual for Alpine and sub-Alpine Environments in the Australian Alps*, Australian Alps Liaison Committee, Shannon Books, Melbourne.

Parsons, W.T. & Cuthbertson, E.G. 1992, *Noxious Weeds of Australia*, Inkata, Melbourne.

Pearson, M. & Sullivan, S. 1995, *Looking After Heritage Places: The Basics of Heritage Planning for Managers, Landowners and Administrators*, Melbourne University Press, Melbourne.

Pentecost, P. 1999, *Rainforest Aboriginal News*, 3.

Pettigrew, C. & Lyons, M. 1979, 'Royal National Park: A History', in *Australia's 100 years of National Parks*, Goldstein, W. (ed.), NSW NPWS, Sydney.

Phillips, A. 2002, *Management Guidelines for IUCN Category V Protected Areas: Protected Landscapes/Seascapes*, Best Practice Protected Area Guidelines Series No. 9, ed. Adrian Phillips, IUCN WCPA, Gland.

Phillips, A. 2003, 'Turning Ideas on their Head: The New Paradigm for Protected Areas', in *Innovative Governance: Indigenous Peoples, Local Communities and Protected Areas*, Jaireth H. & Smyth, D. (eds), Ane Books, New Delhi.

Pickering, C., Appleby, M., Good, R., Hill, W., McDougall, K., Wimbush, D. & Woods, D. 2002, 'Plant Diversity in Subalpine and Alpine Vegetation Recorded in the Thredbo 2002 Biodiversity Blitz', in *Biodiversity in the Snowy Mountains*, Green, K. (ed.), Australian Institute of Alpine Studies, Jindabyne.

Pickering, C.M. & Armstrong, T. 2003, 'Potential Impact of Climate Change on Plant Communities in the Kosciuszko Alpine Zone', *Victorian Naturalist*, 120: 263–72.

Pickering, C.M., Hill, W. & Johnston, F. 2003, 'Ecology of Disturbance: The Effect of Tourism Infrastructure on Weeds in the Australian Alps', *Celebrating Mountains: Proceedings of an International Year of the Mountains Conference*, Australian Alps Liaison Committee, Canberra, November 2002.

Pigram, J.J. & Jenkins, J.M. 1999, *Outdoor Recreation Management*, Routledge, London.

Pimbert, M.P. & Pretty, J.N. 1997, 'Parks, People and Professionals: Putting "Participation" into Protected Area Management', in *Social Change and Conservation: Environmental Politics and Impacts of National Parks and Protected Areas*, Ghimire, K.B. & Pimbert, M.P. (eds), Earthscan, London.

Pitt, M.W. 1992, *The Value of Coastal Land: An Application of Travel Cost Methodology on the North Coast of NSW*, presented to the 36th Annual Conference of the Australian Agricultural Economics Society, Canberra.

Poiner, I., Glaister, J., Pitcher, R., Burridge, C., Wassenberg, T., Gribble, N., Hill, B., Blaber, S., Milton, D., Brewer, D. & Ellis, N. 1998, *Final Report on Effects of Trawling in the Far Northern Section of the Great Barrier Reef: 1991–1996*, CSIRO Division of Marine Research, Cleveland.

Pomeroy, R., Parkes, J. & Watson, L. 2003, 'How Is your MPA Doing? A Guidebook of Natural and Social Indicators for Evaluating Marine Protected Area Management Effectiveness', IUCN WCPA-Marine/ WWF MPA Management Effectiveness Initiative, URL: http:// effectivempa.noaa.gov/docs/ dguidebk.pdf.

Pomeroy, R.J., Parks, J.E. & Watson, L.M. 2004, *How is Your MPA Doing? A Guidebook of Natural and Social Indicators for Evaluating Marine Protected Area Management Effectiveness*, IUCN, Gland.

Pond, K.L. 1993, 'Interpretation and the Role of the Guide', in *The Professional Guide: Dynamics of Tour Guiding*, Van Nostrand Reinhold, New York.

Potter, C. 1998, *International Forest Conservation Protected Areas and Beyond*, Environment Australia, Canberra.

Pouliquen-Young, O. 1997, 'Evolution of the System of Protected Areas in Western Australia', *Environmental Conservation*, 24(2): 168–81.

Powell, J.M. 1976, *Environmental Management in Australia 1788–1914*, Oxford University Press, Melbourne.

Powell R. & Chalmers, L. 1995, *Regional Economic Impact: Gibraltar Range and Dorrigo National Parks*, NSW NPWS, Sydney.

Pressey, R.L. 1995, 'Information for Bioregional Management', in *Approaches to Bioregional Planning: Proceedings of the Conference*, Breckwoldt, R. (ed.), DEST, Canberra.

Pressey, R.L. 1998, 'Systematic Conservation Planning for the Real World', *Parks*, 9(1): 1—6.

Pressey, R.L., Bedward, M. & Keith, D.A. 1994a, 'New Procedures for Reserve Selection in New South Wales: Maximising the Chances of Achieving a Representative Network', in *Systematics and Conservation Evaluation*, Forey, P.L., Humphries, C.J. & Vane-Wright, R.L. (eds), Oxford University Press, Oxford.

Pressey, R.L., Johnson, I.R. & Wilson, P.D. 1994b, 'Shades of Irreplaceability: Towards a Measure of the Contribution of Sites to a Reservation Goal', *Biodiversity and Conservation*, 3: 242–62.

Pressey, R. L. & Logan, V. S. 1995, 'Reserve Coverage and Requirements in Relation to Partitioning and Generalisation of Land Classes: Analysis for Western New South Wales', *Conservation Biology*, 9(6): 1506–17.

Pressey, R.L., Hager, T.C., Ryan, K.M., Schwarz, J., Wall, S., Ferrier, S. & Creaser, P.M. 2000, 'Using Abiotic Data for Conservation Assessments Over Extensive Regions: Quantitative Methods Applied Across New South Wales, Australia', *Biological Conservation*, 96: 55–82.

Pulsford, I., Worboys, G.L., Gough, J. & Shepherd, T. 2003, 'A Potential New Continental Scale Corridor for Australia', *Mountain Research and Development*, 23(3).

Pyper, W. 2003, 'Extreme Power', *ECOS*, 116: 8–9.

QPWS 2000, *User-pays Revenue*, Benchmarking and Best Practice Program, ANZECC, Canberra.

Quarmby, D. 1998, 'The Early Tasmanian National Parks Movement', in *Celebrating the Parks: Proceedings of the First Australian Symposium on Parks History*, Hamilton-Smith, E. (ed.), Rethink Consulting, Melbourne.

Queensland Department of Environment and Heritage 1996, *The Art of Interpretation Workbook*, Queensland Department of Environment and Heritage, Brisbane.

Qureshi, M.E. & Harrison, S.R. 2001, 'A Decision Support Process to Compare Riparian Revegetation Options in Scheu Creek Catchment in North Queensland', *Journal of Environmental Management*, 62: 101–12.

Rainforest Alliance 2003, *Rainforest Alliance News*, 17 October 2003, URL: http://www.ra.org/programs/ sv/index.html.

Rata, M. 1975, 'Preserving the Natural and Cultural Heritage', *Proceedings of the South Pacific Conference on National Parks and Reserves*, Wellington New Zealand, 24–27 February.

Ray, G.C. 1976, *Critical Marine Habitats: Definition, Description, Criteria and Guidelines for Identification and Management*, International Union for the Conservation of Nature and Natural Resources, Morges.

Ray, G.C. & McCormick-Ray, M.G. 1992, *Marine and Estuarine Protected Areas: A Strategy for a National Representative System within Australian Coastal and Marine Environments*, Consultancy report, Australian National Parks and Wildlife Service, Canberra.

Read V. & Bessen, B. 2003, 'Mechanisms for Improved Integration of Biodiversity Conservation', in *Regional NRM Planning Report*, prepared for Environment Australia, Canberra.

Read Sturgess and Associates 1994, *The Economic Significance of Grampians National Park*, DCNR, Melbourne.

Read Sturgess and Associates 1999, *Economic Assessment of Recreational Values of Victorian Parks*, Read Sturgess and Associates, Melbourne.

Reader's Digest 1989, *The Australian Wildlife Year: A Month-by-month Guide to Nature*, Reader's Digest Services Sydney.

Recher, H., Lunney D. & Dunn, I. 1986, *A Natural Legacy: Ecology in Australia* (2nd edn), A.S. Wilson, Sydney.

Reeve, I., Marshall, G. & Musgrave, W. 2004, *Resource Governance and Integrated Catchment Management*, Institute for Rural Futures, University of New England, Armidale.

Resource Assessment Commission 1992, *Forest and Timber Inquiry Final Report*, Volume Vol. 2B, AGPS, Canberra.

Rich, T.H. & Vickers-Rich, P. 2000, *Dinosaurs of Darkness*, Allen and Unwin, Sydney.

Richards, B.N., Bridges, R.G., Curtain, R.A., Nix, H.A., Shepherd, K.R. & Turner, J. 1990, *Biological Conservation of the South-East Forests*, AGPS, Canberra.

Robbins, S.P., Bergman, R., Stagg, I. & Coulter, M. 2003, *Foundations of Management*, Prentice Hall Australia, Sydney.

Roberts, C.M., Branch, G., Bustamante, R.H., Castilla, J.C., Dugan, J., Halpern, B.S., Lafferty, K.D., Leslie, H., Lubchenco, J., McArdle, D., Ruckelshaus, M. & Warner, R.R. 2003, 'Application of Ecological Criteria in Selecting Marine Reserves and Developing Reserve Networks', *Ecological Applications*, 13(1): S215–S228.

Roberts, R.G. & Jones, R. 1994, 'Luminescence Dating of Sediments: New Light on Human Colonisation of Australia,' *Australian Aboriginal Studies*, 2: 2–17.

Robertson, A.I. & Alongi, D.M. 1995, 'Mangrove Systems in Australia: Structure, Function and Status', in *The State of the Marine Environment Report for Australia Technical Annex 1: The Marine Environment*, Zann, L.P. & Kailola, P. (eds), GBRMPA, Townsville.

Robertson, A.I. & Lee Long, W.J. 1991, 'The Influence of Nutrient and Sediment Loads on Tropical Mangrove and Seagrass Ecosystems', in *Proceedings of the Workshop on Land Use Patterns and Nutrient Loading of the Great Barrier Reef Region*, Yellowlees, D. (ed.), James Cook University, Townsville.

Robinson, C.J. 1997, *Integrated Vegetation Management Plan for Fitzgerald River Biosphere Reserve Zone of Co-operation*, CALM, Albany.

Roggenbuck, J. 1987, 'Park Interpretation as a Visitor Management Strategy', in *Metropolitan Perspectives in Parks and Recreation*, Proceedings of the 60th National Conference of the Royal Australian Institute of Parks and Recreation, Royal Australian Institute of Parks and Recreation, Canberra, Australia, 25–29 October.

Rolls, E. 1981, *A Million Wild Acres*, Penguin, Melbourne.

Rolston, H. 1983, 'Values Gone Wild', *Inquiry*, 26: 181–207.

Rolston, H. 2003, 'Life and the Nature of Life in Parks', in *The Full Value of Parks: From Economics to the Intangible*, Harmon, D. & Putney, A.D. (eds), Rowman and Littlefield, Lanham.

Rose, D.B.,1988, 'Exploring an Aboriginal Land Ethic', *Meanjin*, 47(3): 378–87.

Rose, D.B. 1992, *Dingo Makes Us Human: Life and Land in an Aboriginal Australian Culture*, Cambridge University Press, Melbourne.

Rose, D.B. 1995, *Aboriginal Land Management Issues in Central Australia*, Central Land Council, Alice Springs.

Rose, D.B. 1996, *Nourishing Terrains: Australian Aboriginal Views of Landscape and Wilderness*, Australian Heritage Commission, Canberra.

Royal Commission into Aboriginal Deaths in Custody 1991, *National Report: Overview and Recommendations*, AGPS, Canberra.

Ryan, C. 1991, *Recreational Tourism: A Social Science Perspective*, Routledge, London.

Ryan, D.G., Ryan, J.R. & Starr, B.J. 1995, *The Australian Landscape: Observations of Explorers and Early Settlers*, Murrumbidgee Catchment Management Committee, Wagga Wagga.

Rydin, Y. 2003, *Conflict, Consensus and Rationality in Environmental Planning*, Oxford University Press, Oxford.

Saddler, H., Bennett, J., Reynolds, I. & Smith, B. 1980, *Public Choice in Tasmania*, ANU, Canberra.

SA DEH 2001, Great Australian Bight Marine Park, URL: http://www.environment.sa.gov.au/coasts/pdfs/gab_mp.pdf.

Saenger, P. 1995, 'The Status of Australian Estuaries and Enclosed Marine Waters', in *The State of the Marine Environment Report for Australia Technical Annex 1: The Marine Environment*, Zann, L.P. & Kailola, P. (eds), GBRMPA, Townsville.

Sainty, G., Hosking, J. & Jacobs, S. 1998, *Alps Invaders: Weeds of the Australian High Country*, Australian Alps Liaison Committee, Canberra.

Salafsky, N., Margoulis, R. & Redford, K. 2001, *Adaptive Management: A Tool for Conservation Practitioners*, WWF, Washington.

Sanders, A. 1996, *Conservation Value of Fitzgerald Biosphere Reserve Buffer/Transition Zone Phases I–IV*, CALM, Albany.

Sanecki, G. 1999, 'Yarrow (milfoil) *Achillea millefolium* L.: Results for the Monitoring of Current Control Techniques and Testing of Alternative Chemical Control Techniques', Internal Report, NSW NPWS, Jindabyne.

Sanecki, G.M., Sanecki, K.L., Wright, G.T. & Johnston, F.M. 2003, 'The Control of *Achillea millefolium* in the Snowy Mountains, Australia', *Weed Research*, 43: 357–61.

Saunders, D., Margules, C. & Hill, B. 1998, *Environmental Indicators for National State of the Environment*

Reporting—Biodiversity, Department of the Environment, Canberra.

Sawin, J. 2003, 'Charting a New Energy Future', in *State of the World 2003: A Worldwatch Institute Report on Progress Towards a Sustainable Society*, Starke, L. (ed.), Earthscan, London.

Scherrer, P. 2003, 'Monitoring Vegetation Change in the Kosciuszko Alpine Zone, Australia', PhD thesis, Faculty of Environmental Sciences, School of Environmental and Applied Sciences, Griffith University, Gold Coast.

Scherrer, P. & Pickering, C.M. 2002, 'Restoration of Alpine Herbfield Vegetation on a Closed Walking Track in the Australian Alps', *Ecological and Earth Sciences in Mountain Areas*, Banff, Canada 6–10 September 2002.

Schmitt, K. & Sallee, K. 2002, *Information and Knowledge Management in Nature Conservation*, URL: http://www.uwa.or.ug/Mist.pdf.

Schuh, C. & Thompson, D. 2002, 'Environmental Indicators', in *Tools for Environmental Management: A Practical Introduction and Guide*, Thompson, D. (ed.), New Society Publishers, British Columbia.

Scott, S.M. 1998, *The Fundamentals of Communication*, Prentice Hall, Sydney.

SEAC 1996, *Australia: State of the Environment 1996*, CSIRO, Melbourne.

SECARC 1998, *Access to Heritage: User Charges in Museums, Art Galleries and National Parks*, Commonwealth of Australia, Canberra.

Selin, S.W., Schuett, M.A. & Carr, D. 2000, 'Modeling Stakeholder Perceptions of Collaborative Initiative Effectiveness', *Society and Natural Resources*, 13: 735–45.

Selman, P.H. 2000, *Environmental Planning: The Conservation and Development of Biophysical Resources* (2nd edn), Chapman, London.

Senn J.A. 1990, *Information Systems Management* (4th edn), Wadsworth Publishing Co, Belmont.

SERMP 2004, *South-east Regional Marine Plan: Implementing Australia's Oceans Policy in the South-east Marine Region*, National Oceans Office, Hobart.

Serventy, V. 1999, *An Australian life: Memoirs of a Naturalist, Conservationist, Traveller and Writer*, Fremantle Arts Centre Press, Fremantle.

Shindler, B. & Brunson, M. 1999, 'Changing Natural Resource Paradigms in the United States: Finding Political Reality in Academic Theory', in *Handbook of Global Environmental Policy and Administration*, Soden, D. & Steel, B. (eds), Marcel Dekker, New York.

Shultis, J. 2003, 'Recreational Values of Protected Areas', in *The Full Value of Parks: From Economics to the Intangible*, Harmon, D. & Putney, A.D. (eds), Rowman and Littlefield, Lanham.

Simpson, K. & Day, N. 1989, *Field Guide to the Birds of Australia*, Viking O'Neil, Melbourne.

Sinden, J.A. 1978, 'The Application of Benefit–Cost Principles to Recreation Management: Case Study of Colo Shire', in *Australian Project Evaluation*, McMaster, J.C. & Webb, G.R. (eds), Australia and New Zealand Book Company, Sydney.

Sinden, J.A. 1987, 'Community Support for Soil Conservation', *Search*, 18(4): 188–94.

Sinden, J.A. 1990, *Valuation of Unpriced Benefits and Costs of River Management*, DCE, Melbourne.

Sinden, J.A. & Worrell, A.C. 1979, *Unpriced Values: Decisions Without Market Prices*, John Wiley & Sons, New York.

Sinden, J.A., Koczanowski, A. & Sniekers, P.P. 1982, 'Public Attitudes to Eucalypt Dieback and Dieback Research in the New England Region of New South Wales', *Australian Forestry*, 45: 107–16.

Singer, P. 1998, 'New Ideas for the Evolutionary Left', *The Australian*, 17 June: 48–9.

Singh, S. 2003, 'Evaluating the Management Effectiveness of Protected Areas in India: Second National Evaluation of Indian Protected Areas', in *Securing Protected Areas and Ecosystem Services in the Face of Global Change—Managing in the Face of Global Change: The Role of Evaluating Management Effectiveness*, Leverington, F. & Hockings, M. (eds), IUCN/WRI Ecosystems, Parks and People Project, IUCN, Gland.

Singleton, G., Aitken, D., Jinks, B. & Warhurst, J. 1996, *Australian Political Institutions* (5th edn), Addison Wesley Longman, Melbourne.

Skeat, A. & Skeat, H. 2003, *Financial Issues and Tourism Systems to Make Tourism and Others Contribute to Protected Areas—the Great Barrier Reef*, presented to the 5th World Parks Congress Sustainable Finance Stream, Durban, South Africa.

Slack, N., Chambers, S. & Johnston, R. 2001, *Operations management* (3rd edn), Prentice Hall, London.

Smart, J.M., Knight, A.T. & Robinson, M. 2000, *A Conservation Assessment for the Cobar Peneplain Biogeographic Region: Methods and Opportunities*, NSW NPWS, Sydney.

Smyth, D. 2001, 'Joint Management of National Parks', in *Working on Country: Contemporary Indigenous Management of Australia's Lands and Coastal Regions*, Baker, R., Davies, J. & Young, E. (eds), Oxford University Press, Melbourne.

Smyth, D. 2003, *Joint Management of National Parks in Australia*, presented to the 5th World Parks Congress, Durban, South Africa.

Soin, M.S. (ed.) 1999, *Ecotourism and Environment Handbook: A Practical Guide for the Tourism Industry*, Environment and Ecotourism Committee, New Delhi.

Soule, M.E. (ed.) 1986, *Conservation Biology: The Science of Scarcity and Diversity*, Sinauer, Sunderland.

Sorensen, T. & Auster, M. 1999, 'Theory and Practice in Planning: Further Apart Than Ever?', *Australian Planner*, 36(3): 146–9.

Spate, A. 1998, NSW NPWS Karst Investigations Officer, personal communication.

Spearritt, P. 2003, *Green Space Audit of South East Queensland*, URL: http://www.brisinst.organisation.au/resources/spearritt_peter_greenspaceaudit.html.

Specht, R.L. 1970, 'Vegetation', in *The Australian Environment* (4th edn), Leeper, G.W. (ed.), CSIRO, Melbourne.

Specht, R.L., Roe, E.M. & Boughton, V.H. 1974, *Conservation of Major Plant Communities in Australia and New Guinea*, CSIRO, Melbourne.

Spennemann, D. 1997, *Cultural Heritage Management*, Open Learning Institute, Charles Sturt University, Albury.

Stankey, G.H., Cole, D.N. & Lucas, R.C. 1985, *The Limits of Acceptable Change (LAC) System for Wilderness Planning*, Forest Service, US Department of Agriculture, Ogden.

Stanner, W.E.H. 1965, 'Aboriginal Territorial Organisation: Estate, Range, Domain and Regime,' *Oceania*, 36: 1–26.

Stewart, Y. 1999, Arakwal Aboriginal Corporation, personal communication.

Stevens, S. 1997, *Conservation Through Cultural Survival: Indigenous Peoples and Protected Areas*, Island Press, Washington.

Stevenson, P. 2004, Senior Planning Officer, Parks Strategic Development Section, Parks Australia, Commonwealth Department of Environment and Heritage, personal communication, 10 February.

Stock, D. H. 2004, 'The Dynamics of *Lantana camara* L. Invasion of Subtropical Rainforest in Southeast Queensland', PhD thesis, School of Environmental and Applied Sciences, Griffith University, Gold Coast.

Stock, D.H. & Wild, C.H. 2002, 'The Capacity of Lantana (*Lantana camara* L.) to Displace Native Vegetation', in *Papers and Proceedings of the 13th Australian Weeds Conference*, Spafford Jacob, H. & Moore, J.H. (eds), Plant Protection Society of Western Australia, Perth.

Stoddart, E. & Parer, I. 1988, *Colonisation of Australia by the Rabbit*, CSIRO Australia.

Stoekl, N. 1994, 'A Travel Cost Analysis of Hinchinbrook Island National Park', in *Tourism Research and Education in Australia*, Faulkner, B. (ed.), Bureau of Tourism Research, Canberra.

Stolton, S. & Chong-Seng, L. 2003, *Enhancing our Heritage—Monitoring and Managing for Success in*

Natural World Heritage Sites, presented at the 5th World Parks Congress, Durban, South Africa.

Stolton, S. & Dudley, N. 2003, 'Threats to the System', in *State of the World's Protected Areas* (draft for comment), Spalding, M. & Chape, S. (eds), UNEP-WCMC, Cambridge and IUCN WCPA, Gland, URL: http://www.iucn.org/themes/wcpa/wpc2003/english/outputs/intro.htm.

Stolton, S., Hockings, M., Dudley, N., McKinnon, K. & Whitten, T. 2003, *Reporting Progress in Protected Areas: A Site-level Management Effectiveness Tracking Tool*, World Bank/WWF Alliance for Forest Conservation and Sustainable Use.

Strahan, R. (ed.) 1995, *The Mammals of Australia*, Reed, Sydney.

Streeting, M. & Hamilton, C. 1991, *An Economic Analysis of the Forests of South-eastern Australia*, Resource Assessment Commission Research Paper No. 5, AGPS, Canberra.

Strom, A.A. 1979a, 'Impressions of a Developing Conservation Ethic', in *Australia's 100 Years of National Parks*, Goldstein, W. (ed.), NSW NPWS, Sydney.

Strom, A.A. 1979b, 'Some Events in Nature Conservation over the Last Forty Years', in *Australia's 100 Years of National Parks*, Goldstein, W. (ed.), NSW NPWS, Sydney.

Stunden, G. 2002, *What is Sustainable Development?*, URL: http://www.jeanlambertmep.org.uk/downloads/briefings/0204sustaindevelopment.doc.

Sugden, R. & Williams, A. 1978, *The Principles of Practical Cost–Benefit Analysis*, Oxford University Press, Oxford.

Sullivan, S. & Bowdler, S. (eds) 1984, 'Site Surveys and Significance Assessment in Australian Archaeology', *Proceedings of the 1981 Conference on Australian Prehistory*, ANU, Canberra.

Sullivan, S. & Lennon, J. 2003, 'Cultural Values', in *An Assessment of the Values of Kosciuszko National Park*, Independent Scientific Committee Report, NSW NPWS, Queanbeyan.

Sutherland, J. & Muir, K. 2001, 'Managing Country: A Legal Overview', in *Working on Country: Contemporary Indigenous Management of Australia's Lands and Coastal Regions*, Baker, R., Davies, J. & Young, E. (eds), Oxford University Press, Melbourne.

Sutherland, W.J. (ed.) 1996, *Ecological Census Techniques: A Handbook*, Cambridge University Press, Cambridge.

Sutton, M. 1999, Mutawintji Land Council, personal communication.

Sweatman, H., Cheal, A., Coleman, G., Fitzpatrick, B., Miller, I., Ninio, R., Osborne, K., Page, C., Ryan, D., Thompson, A. & Tompkins, P. 2000, *Long-term Monitoring of the Great Barrier Reef*, Status Report No. 4, Australian Institute of Marine Science, Townsville.

Taylor, N. 1998, *Urban Planning Theory since 1945*, Sage, London.

Temple, S.A. 1991, 'The Role of Dispersal in the Maintenance of Bird Populations in a Fragmented Landscape', *Acta XX Congressus Internationalis Ornithologici*, Vol. IV, New Zealand Ornithological Congress Trust Board, Wellington.

Thackway, R. & Cresswell, I.D. (eds) 1995a, 'Towards a Systematic Approach for Identifying Gaps in the Australian System of Protected Areas', in *Ecosystem Monitoring and Protected Areas*, Herman, T.B., Bondrup-Nieslen, S., Wilison, J.H.M. & Munro, N.W.P. (eds), Proceedings of 2nd International Conference on the Science and the Management of Protected Areas, Halifax.

Thackway, R. & Cresswell, I.D. (eds) 1995b, *An Interim Biogeographic Regionalisation for Australia: A Framework for Establishing the National System of Reserves*, ANCA, Canberra.

Thackway, R., Szabo, S. & Smyth, S. 1996, *Indigenous Protected Areas: A New Concept in Biodiversity Conservation: Biodiversity—Broadening the Debate*, ANCA, Canberra.

Thomas, J.F. 1982, 'Recreation Value', in *On Rational Grounds: Systems Analysis in Catchment Land Use Planning: Developments in Landscape Management and Urban Planning 4*, Bennett, D. & Thomas, J.F. (eds), Elsevier, Amsterdam.

Thompson, D. 2002, *Tools for Environmental Management: A Practical Introduction and Guide*, New Society Publishers, Gabriola Island.

Thorsell, J. (ed.) 1990, *Parks on the Borderline: Experience in Transfrontier Conservation*, IUCN, Gland.

Thorsell, J. 1997, 'Nature's Hall of Fame: IUCN and the World Heritage Convention', *Parks*, 7: 3–7.

Tilden, F. 1982, *Interpreting Our Heritage*, University of North Carolina Press, Chapel Hill.

Tjamiwa, T. 1992, 'Tjunguringkula Waakaripai: Joint Management of Uluru National Park', in *Aboriginal Involvement in Parks and Protected Areas*, Birckhead, J., De Lacy, T. & Smith, L.J. (eds), Aboriginal Studies Press, Canberra.

TNC 2002, *The Five-S Framework for Site Conservation: A Practitioner's Handbook for Site Conservation Planning and Measuring Conservation Success*, The Nature Conservancy, Arlington, URL: http://www.consci.org/scp.

Tourism Queensland 1999, *Ecotrends*, Tourism Queensland, Brisbane.

Tranter, D. (ed.) 1996, *Ecotourism: Incorporating the Fitzroy Falls Charter*, Nowra District Advisory Committee of the NSW NPWS, Nowra.

Triggs, B. 1996, *Tracks, Scats and Other Traces: A Field Guide to Australian Mammals*, Oxford University Press, Melbourne.

Trusty, W. 2003, 'Issues in Materials Selection', *RMI Solutions*, 19(2): 26–27.

TTF 2004, *Natural Partners: Making Protected Areas a National Tourism Priority*, Tourism and Transport Forum, URL: http://www.ttf.org.au.

Tuler, S. & Webler, T. 1999, 'Voices From the Forest: What Participants Expect of a Public Participation Process', *Society and Natural Resources*, 12: 437–54.

Ulph, A.M. & Reynolds, I.K. 1978, *An Economic Evaluation of National Parks*, CRES Report R/R4 1978, ANU, Canberra.

UN 1992a, *Convention on Biological Diversity*, UNEP-CBD, Montreal.

UN 1992b, *Report of the United Nations Conference on Environment and Development (Agenda 21)*, United Nations, Rio de Janeiro.

UN 1993, *Draft Declaration on the Rights of Indigenous Peoples as Agreed Upon by the Members of the Working Group at its Eleventh Session*, E/CN.4/Sub.2/1993/29.

UN 2003, *United Nations Division for Sustainable Development*, URL: http://www.un.organisation/esa/sustdev/index.html.

UNEP 2002, *Global Environmental Outlook 3: Past, Present and Future Perspectives*, Earthscan, London.

UNEP–Ozone Secretariat 2003, *Public Information*, URL: http://www.unep.org/ozone/Public_Information/index.asp.

UNEP-SCBD 2001, *Global Biodiversity Outlook*, UNEP Secretariat of the Convention on Biological Diversity Montreal, Canada.

UNEP-WCMC 2000, *Protected Area Resource Centre (PARC)*, URL: http://www.unep-wcmc.org/protected_areas/pavl/parc.htm.

UNESCO 1989, *The World Heritage*, UNESCO, Paris.

UNESCO 1999, *World Heritage*, URL:. http://www.unesco.org/whc/index.htm.

Vanclay, F. 1992, 'The Social Context of Farmers' Adoption of Environmentally-sound Farming Practices', in *Agriculture, Environment and Society*, Lawrence, G., Vanclay, F. & Furze, B. (eds), Macmillan, Melbourne.

Vandenbeld, J. 1988, *Nature of Australia: A Portrait of the Island Continent*, Collins, Melbourne.

Van Der Ryn, S. and Cowan, S. 1995, *Ecological Design*, Islander Press, New York.

Van Schaik, C., Terborgh, J., Davenport, L. & Rao, M. 2002, 'Making Parks Work: Past, Present, and Future', in *Making Parks Work: Strategies for Preserving Tropical Nature*, Terbough, J., Van Schaik, C., Davenport, L. & Rao, M. (eds), Island Press, Washington.

Vernon, J.E.N. 1995, 'Coral Reefs: An Overview', in *The State of the Marine Environment Report for Australia*

Technical Annex 1: The Marine Environment, Zann, L.P. & Kailola, P. (eds), GBRMPA, Townsville.

Vickers-Rich, P. & Rich, T.H. 1993, *Wildlife of Gondwana: The 500 Million Year History of Vertebrate Animals from the Ancient Supercontinent*, Reed, Sydney.

Vincent, P. 2003, 'The Debate Hots Up', *Eco: The Green Lifestyle Guide, Sydney Morning Herald*, Sydney, Australia, Thursday 30 October.

Wachenfeld, D., Oliver, J. & Morrissey, J. (eds) 1998, *State of the Great Barrier Reef World Heritage Area 1998*, GBRMPA, URL: http://www.gbrmpa.gov.au/corp_site/info_services/publications/sotr.

Wagner, L. (ed.) 2003, 'A Brief History of the World Parks Congress', in *Sustainable Developments*, 89(9): 1.

Walker, B. 2001, Inquiry into certain matters including NPWS urban communities and road maintenance responsibilities arising from the Coroner's report on the Thredbo Landslide, URL: http://www.nationalparks.nsw.gov.au.

Walsh, F.J. 1996, 'The Relevance of Some Aspects of Aboriginal Subsistence Activities to the Management of National Parks: with Reference to Martu People of the Western Desert', in *Aboriginal Involvement in Parks and Protected Areas*, Birckhead, J., De Lacy, T. & Smith, L.J. (eds), Aboriginal Studies Press, Canberra.

Wanna, J., O'Faircheallaigh, C. & Weller, P. 1992, *Public Sector Management in Australia*, Macmillan, Melbourne.

Ward, T. & Hegerl, E. 2003, *Marine Protected Areas in Ecosystem-based Management of Fisheries*, A report prepared for discussion at the Vth World Parks Congress, DEH, Canberra.

Ward, T.J., Heinemann, D. & Evans, N. 2001, *The Role of Marine Reserves as Fisheries Management Tools: A Review of Concepts, Evidence and International Experience*, Bureau of Rural Sciences, Canberra.

Watson, J.R. 1991, 'The Identification of River Foreshore Corridors for Nature Conservation in the South Coast Region of Western Australia', in *Nature Conservation 2: The Role of Corridors*, Saunders, D.A. & Hobbs, R.J. (eds), Surrey Beatty and Sons, Sydney.

Watson, J., Hamilton-Smith, E., Gillieson, D. & Kiernan, K. (eds) 1997, *Guidelines for Cave and Karst Protection*, IUCN, Gland.

Watson, J. & Sanders, A. 1997, 'Fitzgerald River National Park Biosphere Reserve 1978–97: The Evolution of Integrated Protected Area Management', *Parks*, 7(1): 9–19.

WCED 1987, *Our Common Future*, Oxford University Press, Oxford.

WCMC 1999, *Homepage*, URL:, http://www.wcmc.org.uk.

Weaver, D. & Opperman, M. 2000, *Tourism Management*, John Wiley and Sons, Sydney.

Webler, T. and Renn, O. 1995, 'A Brief Primer on Participation: Philosophy and Practice', in *Fairness and Competence in Citizen Participation: Evaluating Models for Environmental Discourse*, Renn, O., Webler, T. & Wiedemann, P. (eds), Kluwer, Dordrecht.

Wentworth Group of Concerned Scientists, The, 2003, *Blueprint for a National Water Plan*, WWF Australia, Sydney.

Whelan, R.J. & Baker, J.R. 1998, 'Fire in Australia: Coping with Variation in Ecological Effects of Fire', in *Bushfire Management Conference Proceedings: Protecting the Environment, Land, Life and Property*, Sutton, F., Keats, J., Dowling, J. & Doig, C. (eds), Nature Conservation Council of NSW, Sydney.

White, M.E. 1990, *The Nature of Hidden Worlds*, Reed, Sydney.

White, M.E. 1994, *After the Greening: The Browning of Australia*, Kangaroo Press, Sydney.

White, M.E. 1997, *Listen … Our Land is Crying: Australia's Environment—Problems and Solutions*, Kangaroo Press, Sydney.

White, M.E. 1998, *The Greening of Gondwana* (3rd edn), Reed, Sydney.

White, M.E. 1999, *Reading the Rocks*, Kangaroo Press, Sydney.

White, M.E. 2000a, personal communication.

White, M.E. 2000b, *Running Down, Water in a Changing Land*, Kangaroo Press, Sydney.

White, M.E. 2003, *Earth Alive: From Microbes to a Living Planet*, Rosenburg Publishing, Sydney.

Whitehouse, J.F. 1992, 'IVth World Congress on National Parks and Protected Areas, Caracas, Venezuela', *Australian Zoologist*, 28: 39–46.

Whitelock, D. 1985, *Conquest to Conservation: History of Human Impact on the South Australian Environment*, Wakefield Press, Adelaide.

Wilcove, D.S., McLellan, C.H. & Dobson, A.P. 1986, 'Habitat Fragmentation in the Temperate Zone', in *Conservation Biology: The Science of Scarcity and Diversity*, Soule, M.E. (ed.), Sinauer, Sunderland.

Wilderness Society 2004, *Wild Country*, URL: http://www.wilderness.org.au/campaigns/wildcountry.

Wildlife Australia 1996a, *The 1996 Action Plan for Australian Marsupials and Monotremes*, prepared for the IUCN/SSC Australasian Marsupial and Monotreme Specialist Group, Commonwealth of Australia, Canberra.

Wildlife Australia 1996b, *The Action Plan for Australian Cetaceans*, Commonwealth of Australia, Canberra.

Wildlife Australia 1997, *The 1997 Action Plan for Australian Frogs*, Commonwealth of Australia, Canberra.

Wilks, L.C. 1990, *A Survey of the Contingent Valuation Method*, Resource Assessment Commission Research Paper No. 2, AGPS, Canberra.

Williams, J., Read, C., Norton, T., Dovers, S., Burgman, M., Proctor, W. & Anderson, H. 2001, *Biodiversity: Australian State of the Environment 2001*, Department of Environment and Heritage, Canberra.

Williams, N.M. & Baines, G. (eds) 1993, *Traditional Ecological Knowledge: Wisdom for Sustainable Development*, Centre for Resource and Environmental Studies, ANU, Canberra.

Williams S.E. 2003, 'Impacts of Global Climate Change on the Rainforest Vertebrates of the Australian Wet Tropics', in *Climate Change Impacts on Biodiversity in Australia*, Howden, M., Hughes, L., Dunlop, M., Zethoven, I., Hilbert, D. & Chilcott, C. (eds), Commonwealth of Australia, Canberra.

Wilson, E.O. 1992, *The Diversity of Life*, Penguin, London.

Wilson, S. & Swan, G. 2003, *A Complete Guide to Reptiles of Australia*, Reed New Holland, Sydney.

Wilson, G.A. 1997, 'Factors Influencing Farmer Participation in the Environmentally Sensitive Areas Scheme', *Journal of Environmental Management*, 50: 67–93.

Wimbush, D.J. & Costin, A.B. 1979, 'Trends in Vegetation at Kosciusko: III Alpine Range Transects, 1959–1979', *Australian Journal of Botany*, 27: 833–71.

Woenne-Green, S., Johnston, R., Sultan, R. & Wallis, A. (eds) 1994, *Competing interests: Aboriginal Participation in National Parks and Conservation Reserves in Australia: A Review*, Department of Employment, Education and Training, Canberra.

Woodford, J. 2000, *The Wollemi Pine*, The Text Publishing Company, Melbourne.

Wondolleck, J.M. and Yaffe, S.L. 2000, *Making Collaboration Work: Lessons from Innovation in NRM*, Island Press, Washington.

Wootten, H. 1994, 'Mabo and National Parks', in *Competing interests: Aboriginal Participation in National Parks and Conservation Reserves in Australia: A Review*, Woenne-Green, S., Johnston, R., Sultan, R. & Wallis, A. (eds), Department of Employment, Education and Training, Canberra.

Worboys, G. L. 1996, *Conservation Corridors and the NSW Section of the Great Escarpment of Eastern Australia*, presented at the IUCN World Conservation Congress, Montreal.

Worboys, G.L. 2003a, 'A Brief Report on the Australian Alps Bushfires', *Mountain Research and Development*, 23(3): 294–5.

Worboys, G.L. 2003b, 'The 2003 Australian Alps Bushfires: Notes for Research Purposes', Unpublished report.

Worboys, G.L. 2003c, *The 5th IUCN WCPA World Parks Congress*, A Report to the CRC for Sustainable Tourism, CRC for Sustainable Tourism, Gold Coast.

Worboys, G.L. 2004, 'Protected Area Management Effectiveness Evaluation Requirements', PhD Confirmation Report, Canberra, Australia.

Worboys, G.L. & De Lacy, T. 2003, *Tourism and the Environment: It's time!*, Presented at the 11th Ecotourism Australia National Conference, Adelaide, South Australia, 10–12 November 2003.

Worboys, G.L., Hall, P., Kennedy, P., Hahn, P., Collins, D. & Mills, K. 1995, 'Managing Ecotourism: the Minnamurra Rainforest Centre', in *Managing the Place*, Baird, I.A. (ed.), Royal Australian Institute of Parks and Recreation, Sydney.

Worboys, G.L. & Pickering, C.M. 2002, *Managing the Kosciuszko Alpine Area: Conservation Milestones and Future Challenges*, Mountain Tourism Research Report No. 3, CRC for Sustainable Tourism, Gold Coast.

Worboys, G.L. & Pickering, C.M. 2004, 'Tourism and Recreation Values', in *An Assessment of the Values of Kosciuszko National Park: Report of the Independent Scientific Committee*, NSW NPWS, Queanbeyan.

Worboys, G.L., Worboys, P.M., Gare, N. & Snedden, R. 2004, 'Protected area management: a chronology of key events', Unpublished report.

Wright, J. 1971, 'Seven Songs from a Journey', in *Collected Poems 1942–1970*, Angus and Robertson, Sydney.

WTMA 2004, *Chronology of Key Events*, Internal Working Document, Wet Tropics World Heritage Area Management Authority, Cairns.

WTO 2003, *Djerba Declaration on Tourism and Climate Change*, URL: http://www.world-tourism.org/sustainable.htm.

WTTC 1997, *WTTC update—New Environmental Standard Demonstrates Compliance with Agenda 21*, URL: http://www.hospitalitynet.org/news/4000904.html.

WTTC 2003, *BluePrint for New Tourism*, URL: http://www.wttc.org.

Yaffee, S.L. 1997, 'Why Environmental Policy Nightmares Recur', *Conservation Biology*, 11: 328–37.

Yaldwyn, J.C. 1978, 'Chairman's summing up during "General Discussion on Part 3"', in *The Ecology and Control of Rodents in New Zealand Nature Reserves*, Dingwall, P.R., Atkinson, I.A.E. & Hay, C. (eds), New Zealand Department of Conservation, Wellington.

Yates, K. 2003, 'Risk Management: A Best Practice Framework', prepared for the Second Edition of Worboys et al. 2001, *Protected Area Management: Principles and Practice*.

Young, A.M. 2000, *Common Australian Fungi: A Bushwalkers Guide* (revised edn), UNSW Press, Sydney Australia.

Young, M.D. 1992, *Sustainable Investment and Resource Use: Equity, Environmental Integrity and Economic Efficiency*, UNESCO, Paris.

Young, M.D. 1996, 'Mixing Instruments and Institutional Arrangements for Optimal Biodiversity Conservation', in *OECD International Conference on Incentive Measures for Biodiversity Conservation and Sustainable Use*, OECD, Paris.

Young, M.D., Gunningham, N., Elix, J., Lambert, J., Howard, B., Grabosky, P. & McCrone, E. 1996, *Reimbursing the Future: An Evaluation of Motivational, Voluntary, Price-based, Property Right, and Regulatory Incentives for the Conservation of Biodiversity*, DEST, Canberra.

Young, R. & Carter, M. 1990, *The Economic Evaluation of Environmental Research: A Case Study of the South-east Forests*, Presented to the 34th Annual Conference of the Australian Agricultural Economics Society, University of Queensland, Brisbane.

Young, W.J. (ed.) 2001, *Rivers as Ecological Systems: The Murray Darling Basin*, The Murray Darling Basin Commission, Canberra.

Zann, L.P. 1995, *Our Sea, our Future: Major Findings of the State of the Marine Environment Report for Australia*, Great Barrier Reef Marine Park Authority, Commonwealth of Australia, Canberra.

Zerba, A. 1994, *Cultural Resources Study Book*, University of Queensland, Gatton.

Zutshi, A. & Sohal, A. 2001, *Environmental Management Systems Implementation: Some Experiences of Australian Auditors*, Department of Management, Monash University, Caulfield East.

Index

Aboriginal Deaths in Custody   499
Aboriginal Land Rights Act 1976 (Cth)   498
Aboriginal people
   and Australian landscape   353
   axe-grinding grooves   366
   burial sites   366, 479
   caring for the country   495–8
   caves or rock shelters   365
   ceremonial grounds   367
   communities   505–6
   cultural heritage   354, 358, 363–7, 385–6, 399
   culture and customs   352
   development   363
   engravings   366
   and fauna conservation   323–4
   impact of plants   364
   land management   497, 503–7
   land rights   97, 498
   land-use   495–6
   management lesson and principles   507
   managing cultural heritage sites   364–7
   natural sacred sites   367
   open campsites   365
   perceptions of land   495
   middens   365
   mounds   365
   and protected areas   36, 37–8, 82, 97, 98, 138, 323–4, 494–507
   quarry sites   366
   reburial of remains   366, 479
   resource use   497
   rock art   358, 366
   sacred or carved trees   365–6
   threats to cultural heritage   363–4, 385–6
   tourism   363–4
   traditional ecological knowledge   497–8
   use of fire   378, 496–7
ACT   329, 587
adaptive management practices   118
administration
   of assets, standards and systems   262–4
   of finances   257–62
   lessons and principles   267–8
   of people   251–6
   policy and process considerations   264–7
   and protected areas management   251
aerial surveys   320
agencies
   integration of   123
   protected area management   xxviii–xxix, 95, 96, 123, 136
Agenda 21   50, 51–2, 285
agricultural
   diseases, introduced   383
   landscapes   20
Alexander, Karen   42
alpine
   areas   19, 334
   habitat   24
amphibians   30–1, 321
ancient heritage   352–3
animals, introduced   379–82, 393–6, 471–2, 481
annual reports   260
aquaculture   548–50
arid lands   18–19, 24
art and heritage   71
artists and protect areas   81
Asia Pacific Economic Community   48
audits   260
Australian Alps fires   413, 468, 476–7
Australian Alps–Great Escarpment conservation corridors   389

Australian Alps Liaison Committee 98
Australian Alps National Park 448
Australian and New Zealand Environment and
    Conservation Council (ANZECC) 155, 262
Australian Bush Heritage Fund 103
Australian Conservation Foundation 42
Australian continent, distinctive characteristics 12–13
Australian Greens 42
Australian Heritage Council (AHC) 99, 358, 360
Australian Heritage Council Act 2003 99, 360
Australian Inter-Service Incident Management Systems
    (AIIMS) 405, 411
Australian National Biodiversity Strategy 375
Australian National Heritage Charter 163, 290
Australian Natural Heritage Charter 311–14
Australian Sea Lion (*Neophoca cinerea*) 32
Australia's Oceans Policy 534
awards, industrial 252–3
axe-grinding grooves 366

Barmah Forest 333
Bay of Island Coastal Park 439–40
Bega Valley 378
benchmarking 262–3, 277, 279, 294–5
benefit–cost analysis 129, 211, 237–48
    accounting for time 246–7
best practice 262–3
Bilby (*Macrotis lagotis*) 395
Biodiversity Convention 80
biodiversity *see* biological diversity
biodiversity of life on Earth 21
biological diversity (biodiversity) 20–3, 80, 81, 134,
    374, 377, 392–7, 542–3
    see also Convention on Biological Diversity
biological inventory 163, 208
biological survey of South Australia 170
bioregions, Australia's 20–3, 516–18
biosphere 21, 518–23
biosphere reserves 94–5, 98
birds 28–9, 314–16, 318, 322
Birds Australia 318
Black-eared miner (*Manorina melanotis*) 95
Blackwood Biodiversity Program 524
Blue-lined Goatfish (*Upeneichthys lineatus*) 142
Blue Mountains (NSW) 94
Bookmark Biosphere Reserve 95
Briars Park 61
Brisbane Forest Park (BFP) 125–6
Brittle Star (*Ophionthrix spongicola*) 142
Brown, Bob 42
Brutland Report 51
Budderoo National Park 456

budgets 258–60, 261, 295
Bunya Mountains National Park 41
burning, prescription 300–4, 344–7
Burra Charter 163, 290, 361–2
bushfires *see* fire
business planning 189, 222–3
Button Wrinklewort (*Rutidosis leptorrhynchoides*) 329,
    330
Byles, Marie 42

Cambrian Period 9
capital works projects 259
captive breeding 322
carbon monitoring system (CMS) 282
Carboniferous Period 8
Carson, Rachel 38
catchment management 523–4
catchments, water 332
cats 380, 381, 481
cave fauna 320
caves 10, 94, 336–40, 383, 420
central eastern rainforest reserves (Queensland and
    NSW) 94
Christmas Island 97
Cinnamon fungus 400
citizen participation 53–4
clearing 393
climate 13–15
    Australia's 14–15
    changes 14, 271–2, 372–3, 397, 399–400
coastal habitats 535–9
coastal saltmarshes 536
coastlines 18
Cocos (Keeling) Island 97
code of conduct 257
Collaborative Australian Protected Area Database
    (CAPAD) 61, 105
*Commonwealth Environment Protection and Biodiversity
    Conservation Act 1999* 136
communication
    definition 475
    methods 475
    skills 263–4
    and stakeholders 475–83
community and protected area management 466–92
community assessment 474
competitive tendering 121
conservation 38–44, 78, 80, 82–4, 133
    agencies 120
    biodiversity 542–4
    corridors linking public lands 386–7
    cultural heritage 69, 82, 358–63, 399

cultural landscapes 369
of fauna 314–28
of flora 62, 69, 159–60, 328–31
natural heritage 314–31, 342
outcomes and clear English 263–4
planning and species 134
and private property 221, 388
regional scale 515–25
and scenic quality 342, 385
status 134, 135
Conservation and Land Management Act 1984 (WA) 69
Conservation Management Networks (CMN) 62
Conservation-Plan (C-Plan) 146
construction and biodiversity 392–3
construction stone 341
contracting-out 121
controlling in organisations 126
Convention on Biological Diversity 51, 52, 85, 90, 113, 133, 189, 533
conventions 90
Coolart homestead 368
coral cays 18
coral reefs 537–8
corporate environmental reporting 285–6
corporate sponsorship 221
corporatisation 121–2
corruption 260–2
costs, supply 226–7
Council of Australian Governments (COAG) 66
Cretaceous Period 6–7
Culgoa National Park 362
cultural and environment 69–75
cultural commitment and protected areas 285
cultural heritage, Aboriginal 354, 358, 363–7
cultural heritage conservation 69, 82, 358–63, 399
cultural heritage management 352
ancient heritage 352–3
conserving historic sites 367–9, 399
historic heritage sites 354–7, 399
lessons and principles 370
in protected areas 359–60
cultural heritage sites 353–8, 399
cultural impact assessment 473
cultural inventory 164, 208
cultural landscapes 369
culture and conservation 82–4

Dandenong Ranges National Park 375
data
analysis 163, 168–74, 179–84
collection 167–74, 207–9
and land use 168

monitoring 168, 174
operational 168
reference 167–8
decision-making by managers 128–9, 194, 213, 389
deep ecology 80
deer 380
demographics 72–3
Desert Parks Pass 446
development
and Aboriginal cultural heritage sites 363
and biodiversity 392–3
Devonian Period 8
digital information and managing protected areas 179
donkeys 380, 381
Dorrigo National park 244
Dugong (*Dogong Dugong*) 11
Dunphy, Miles 41, 42, 80
Durban Accord 89, 132

Earth Alive 13
Earth Sanctuaries 104
earthcheck 277, 279
earthquakes 386, 421
East Gippsland forests 247
ecological considerations and management 129
ecological factors and protected area management 376
ecological fire management planning 344–7
ecologically sustainable development (ESD) 53, 70, 174
economic considerations and protected areas 129, 135, 208, 228, 527–8
economic factors and protected area management 376
economic impacts on regional areas 80, 84, 208, 244
economic management 248–9
economic rationalism 49
ecosystem services 81–2, 228
ecosystems, Australian 134–5, 155–6, 512
ecotourism 426–7, 431, 442
education 72, 529, 550–1, 573
El Niño–Southern Oscillation (ENSO) 15
emissions trading 282
employee assistance programs 256
energy consumption 279–80
engravings 366
enterprise agreements 252–3
environment and values 69–75, 78–84
Environment Protection Act 1970 (Vic) 68
Environment Protection and Biodiversity Conservation (EPBC) Act 1999 67–8, 97, 293
environmental accountability and protected area management 271
Environmental Assessment Council Act 2001 (Vic) 154

Environmental Effects Act 1978 (Vic)   68
environmental agencies   123
environmental condition   164
environmental considerations and management   129
environmental impact assessment   293, 391
environmental legislation   122–3, 144
   Commonwealth   67–8, 133, 136
   state and territory   68–9, 133, 136
environmental levies   528
environmental management   82
Environmental Management Systems (EMS)   277,
   293–4, 392
environmental performance   275–7
Environmental Planning and Assessment Act 1979 (NSW)
   68
environmental planning and protection   68
environmental protection   68, 525–30
   international   85–91
Environmental Protection Act 1994 (Qld)   68
environmental reporting   285–6
Environmental Research Information Network (ERIN)
   178
environmental threats
   minimising   392–400
   preventing   386–92
environmental valuation   237–48
   travel cost method   238–42
   value measurement   237–8
Environmentally Sustainable Development (ESD)   444
erosion
   and Aboriginal cultural heritage sites   363
   threat to protected areas   383, 384
estuaries   18
ethics   257
ethnicity   72
Eucalyptus   24
European Economic Union   48
evaluation in organisations   126
evolutionary development, Earth's   4–9
evolutionary heritage and protected area management
   3–9
existence values   82
external territory protected areas   106
extinctions
   and habitat fragmentation   387
   invertebrate   33
   mammal   32

facilities and services, pricing   223–37
False Killer Whales (*Pseudorca crassidens*)   418
fauna   17, 159
   cave   320

conservation   314–28, 395
   culling   323
   disease management   328
   evolutionary development   4–9
   invertebrate   28, 33
   management   325–7, 395
   marine   34
   minimising human disturbance   325
   pests to agriculture   328
   population control   323
   relocation   323
   research   314
   road kills   328
   surveys   316–28
   terrestrial   28–33, 134
feral animals   310
Figgis, Penny   42
filing systems, office   263
finances
   administering   257–62
   lessons and principles   248–9
   and protected areas   220–3
fire
   and Aboriginal cultural heritage sites   363, 364, 386
   Aboriginal use of   378, 496–7
   ecological management planning   344–7
   and European cultural heritage   386
   management   343–9, 378–9, 392, 412–17
   prescription burning   300–4, 344–7
   and vegetation management   330
firefighting, safety and remote-area   410
first aid   459
fish
   freshwater   17
   introduced   382
fisheries   548–50
Fitzroy Falls Visitors Centre   223
Flinders Rangers National Park   291–3
floods   420
flora   11, 17
   conservation   62, 69, 159–60, 328–31, 511
   evolutionary development   4–9
   marine   34
   rehabilitation   329
   terrestrial   23–7, 134
   threatened species   330–1
   see also plants; vegetation
Flora and Fauna Guarantee Act 1988 (Vic)   69
Food and Agriculture Organisation (FAO)   85
Forestry Act 1916 (NSW)   69
forests
   closed   25

landscapes   16–17
low closed   26
low open   26
old growth   155
open   25
policy, national   154–6
reserves in North East Victoria   156–60
tall closed   24
tall open   24–5
see also Regional Forest Agreement (RFA)
fossil collection   338
fossil sites   340
Australian mammal   10
fox (*Vulpes vulpes*)   310, 380, 381, 471–2
Franklin River   42
Fraser Island   18, 41, 94, 127
fraud   260–1, 262
frogs   28
fungi   331

Gammon Ranges National Park   291–3
genetic diversity   322, 329
genetic resources   265–7
Geographic Information Systems (GIS)   163, 178, 179
geological features, threats to   383–5, 397–9
geological formations, managing unusual   341
geological incidents   421–2
geological-type sections   340
geology   335–41
Giant Gippsland Earthworm (*Mega scolides australis*)   28
Gibraltar Range National Park   244
Global Environment Facility (GEF)   85
global environment management   270
global protected area network   137
global representative system of marine protected areas   141–2
goats   380, 381
Golden Perch (*Macquaria ambigua*)   17
goodwill   256
Gordon River   42, 384
Gould's Petrel (*Pterodroma leucoptera*)   314–16
governance
and protected area management   56–63, 103
government
action, coordinating   66–7
departments   65–6
environmental reporting   285
organisational model, whole-of-government   123
governments
Commonwealth   64–5, 133, 136, 143–4, 154, 389
federalism   64, 66
and forest protection   154–6

local   67
protected area policy and management   47–50, 64–7, 123, 136, 144, 227
and reserve selection   144
state   65, 133, 136, 154, 389
territory   65, 133, 136, 154, 389
grasslands
hummock   27
landscapes   19
tussocky   27
grazing   84, 379
Great Artesian Basin   333–4
Great Australian Bight Marine Park   140, 215–16
Great Barrier Reef   42, 94, 98, 140, 234–6, 469, 477, 480, 541, 542, 543–4, 545, 550, 565–6
Great Knot (*Calidris tenuirostris*)   318
Green and Golden Bell Frog   321
Green Globe 21   277, 442
Green Turtle (*Chelonia mydas*)   321
greenhouse gas production   279–80
Greening Australia   66, 329
Greenpeace   90

habitat destruction   373–4, 387
habitat modification   392–7
habitats   134, 387
coastal and marine   535–9
offshore   538–9
Heard Island   94, 98
heathlands   19, 26
heavy machinery and Aboriginal cultural heritage   363
Heritage Act 1993 (SA)   69
Hinchinbrook Island National Park   212–13
historic heritage sites   354–7, 367–9
Historical Cultural Heritage Act 1995 (Tas)   69
Holocene Epoch   4
horses (wild)   398
housing, ranger   255–6, 278–9
human behaviour and nature   80
Humpback Whale (*Megaptera novaengliae*)   11
hunting, illegal   383

ice and snow   334
identity and conservation   82–4
illegal action in protected areas   383, 397
impact assessment   189
Incident Control System (ICS)   405, 406, 411
incident management   403
administration   411
community consultation   411
fires   412–17
management lessons and principles   422–3

natural phenomena   418–22
occupational health and safety   408–11
operational logistics   407–8
organisations with responsibilities   403–4
plans and management systems   405–7
post-incident management   412
responses   404–12
skills and training   408
wildlife   417–18
indigenous cultures and protected areas   36, 37–8,
   82–4, 89, 97, 133, 138, 359, 499–503
indigenous protected areas   97
Indonesia   98
industrial awards   252–3
information
analysis and application   179–84
collection and data   167–74, 207
disseminating   180–4
lessons and principles   186–7
management of   163
management systems   184–6
needs, scope of   163–7
office filing systems   263
storage, retrieval and presentation   174–9
strategic use of   165–7
tracking systems   263
infrastructure and facilities   164
initiatives
and protected area management 56–64
institutional reform   529–30
Intergovernmental Agreement on the Environment
   66, 133
Interim Biogeographic Regionalisation of Australia
   (IBRA)   22–3, 134–5, 136, 138, 139, 161
Interim Marine and Coastal Regionalisation for
   Australia (IMCRA)   34, 161
international conventions   90
International Council on Monuments and Sites
   (ICOMOS)   85, 361
international environmental protection   85–91, 602
International Union for the Conservation of Nature
   (IUCN)   78, 87, 88, 91–2, 93, 94, 105, 110, 160,
   161, 164, 374, 584
interpretation   484–91
intrinsic worth of nature   79–80
introduced animals   379–82, 393–6
introduced plants   310, 382, 393–4
invertebrate
conservation   319, 320
terrestrial   28, 33
Iron Barron oil spill   419
islands   18

JANIS report   155
Jenolan Caves   39, 453
Jurassic Period   7
jurisdictions, Australian   xxviii–xxix, 96, 140

Kakadu National Park   94
Kangaroo Island   224–6, 400, 430, 454
karst landscapes   19, 336–40
King Billy Pine (*Athrotaxis selaginoides*)   155
Kings Park (NSW)   39
knowledge   163
Kosciuszko National Park   75, 176–7, 200, 201, 202–4,
   298–9, 329, 335, 360, 385, 398
Ku-ring-gai Chase National Park   303
Kyoto Protocol   51, 52, 282

La Niña   15
Lake Eyre   17
Lake Pedder   42
lakes   17–18
Lamington National Park   41, 432–4, 447
land
management, Aboriginal   497, 503–7
private   221, 388, 510
public   386–7, 510
rights, Aboriginal   97, 498
tenure   97, 103
Land Conservation Council (Victoria)   43, 133, 147–54
Landcare   66
Landcare Research NZ   286
landforms   16
landscapes, Australian   16–20, 134, 147, 509–31
land-use
history   164
planning   68, 147–60, 189, 191
Land Use Planning and Approvals Act 1993 (Tas)   68
Lantana (*Lantana camara*)   310, 394
leading in organisations   124–6
Leafy Seadragon (*Phycodurus eques*)   142
leasing   263
legal considerations and management   130, 208
legal factors and protected area management   376
legislation, protected area   44–7, 67–9, 114, 122–3, 133,
   136, 144, 265–7
Leopold, Aldo   80
licensing   263
lichen   24
life cycle assessments   276
Little Penguin (*Eudyptula minor novaehollandiae*)   172
Little terns (*Sterna albifrons*)   479
Loggerhead Turtle (*Caretta caretta*)   11
Long-footed Potoroo (*Potorous longipes*)   31, 33

Lord Howe Island Group   94
Lucas, John   39

Mabo   97
Macdonald Island   94, 98
Macquarie Island   94, 397, 452
Mallee scrubland   95
mammals   31–3, 322
Man and the Biosphere (MAB) Program   94
management
    adaptive   198
    and capacity building   119, 254–5
    and decision-making   128–9, 194, 213, 389
    effectiveness   553–68
    environment, global   270
    front-line   119, 254
    of information   163
    lessons and principles   160–1, 217–18, 248–9,
        267–8, 286–7, 307–8, 349–50, 370,
        400–1, 422–3, 462–3, 491–2, 507, 530–1,
        552, 567–8
    middle level   119, 254
    organisational   114–15
    planning   217–18
    processes   113–30, 164–5
    strategic   196
    top-level   119, 254
    tourism and recreation   442–63
    see also planning; protected area management
Management by Objectives (MBO)   196
manager, attributes of a competent protected area
    127–8
managerial considerations and protected areas   130
mangrove forests   536
marine
    environments, Australian   534–9
    fauna   34
    flora   34
    habitats   535–9
    landscapes   19
    management issues   539–41
    national parks   142
    national reserve system   138–44
    park management   541–51
    pollution   417
marine protected areas (MPAs)   88–9, 104–5, 106, 110,
        111, 138–44, 445, 503, 532–52, 586
market-based approaches   120–1
markets
    protected area policy and management   47–51
marsupials   32, 33
Masked Owl (*Tyto novaehollandiae*)   159

media
    interviews   477–8
    print and electronic   477–8
    releases   478
Messmate Stringybark (*Eucalyptus obliqua*)   156
middens   365
Millennium Summit   51, 52, 466
mining heritage in Victoria   369
Minnamurra Falls   422
Minnumurra Rainforest Centre   456
monitoring programs   168, 174
monotremes   32
Montague Island Nature Reserve   172, 451
Monterey Pine (*Pinus radiata*)   393
Moreton Island National Park   280–1
Mornington Peninsula   61
mounds   365
Mountain Ash (*Eucalyptus regnans*)   81, 156
Mountain Pygmy-possum   319
Mt Field (Tasmania)   41
multicriteria analysis (MCA)   213–14
Murray Cod (*Maccullochella peelii*)   17
Murray–Darling Basin   17, 272, 374

Naess, Arne   80
Namadgi National Park   365, 396
Nantawarrina (South Australia)   501
Naracoorte Caves   10, 94, 338
National Competition Policy (NCP)   48, 120–1, 236
national estate   99, 358, 363
National Forest Policy   154–6
National Heritage List   99, 358, 360
National Heritage Trust   48, 65, 136
National Landcare Program (NLP)   48
National Oceans Policy   138
national parks   39–41, 77–8, 134, 142, 571
National Parks and Wildlife Act 1974 (NSW)   67, 69, 189
National Parks and Wildlife Act 1972 (SA)   69, 104
National Parks Foundation of South Australia   222
National Representative System of Marine Protected
        Areas (NRSMPA)   138, 140, 141
National Reserve System (NRS) Program   103, 105,
        136
National Strategy for the Conservation of Australia's
        Biological Diversity   67, 133, 310, 374
National Strategy for Ecologically Sustainable
        Development (NSESD)   53, 67, 133, 310
Native Title Act 1993 (Cth)   494, 498
Native Vegetation Act 2003 (NSW)   48
natural heritage
    Australia today   12–13
    conserving fauna   314–28

conserving flora 328–31
conserving scenic quality 342
evolution of Australia's 3–11
fire management 343–9
fungi 331
lessons and principles 349–50
management 309–14
and protected areas 82
soils and geology, managing 335–41
water management 332–4
natural phenomena 418–22
natural resource management (NRM) 511, 524–5
Natural Resource Management Ministerial Council
    (NRMMC) 59, 60, 98, 105, 136
nature
    intrinsic worth 79–80
    values of 78–84
Nature Conservancy 90
Neogene Epoch 5–6
New South Wales 39–40, 41, 146, 147, 588–9
New Zealand 117, 155, 286, 396, 461–2
Ningaloo Marine Park 140, 539, 548
Nitmiluk (Katherine Gorge) National Park 62
non-governmental protected areas 103–4
non-recreational uses 164
non-use values 82, 220, 227
Norfolk Island Boobook Owl (*Ninox novaeseelandiae*)
    96
Norfolk Island National Park and Botanic Garden 96,
    97
Northern Hairy-nosed Wombat (*Lasiorhinus krefftii*) 33
Northern Territory 205–7, 590–1
Numbat (*Myrmecobis fasciatus*) 395

occupation and conservation 73
occupational health and safety 408–11
oceanic islands protected areas 106
offshore habitats 538–9
oil spills 419
open camp sites 365
operations
    implementation 300–7
    logistics planning 296–7
    management 289–90, 307–8
    planning for 290–300
Ordovician Period 9
organisation structures, public sector 120
organisation sustainability 264–5
organisational
    goals 115
    goals and budgets 258–60

goals and leading 124–5
management 114–15, 256–7
principles 256–7
organising
    levels of 119
    and protected areas 116–23
    work 118
    work, vertically and horizontally 119–20

Palaeogene Epoch 6
Papua New Guinea 98
Parks Australia 98, 136
Parks Victoria 164, 165, 181–4, 392, 468
parliamentary processes 265–7
performance
    assessment and monitoring 275–7
    improvement 277–8
    managing for 126–30
    measures 126
    reporting 285
Permian Period 8
personal development 84
pest animals 379–82, 393–7, 471–2, 481
Pheasant Snail (*Phasianella australis*) 142
physical inventory 163
pigs (feral) 380, 381, 396
planning
    adaptive 192–4, 198
    approach to process 195–8
    approaches to 190–5
    business 189, 222–3
    economic 189
    financial 189
    functional 189
    incident 189
    incremental 192
    land use 68, 147–60, 189, 191
    levels 115
    management 189–218
    operational 116, 189
    organisational 189
    participatory 194–5
    processes 195–8
    projects 116
    and protected areas 115, 189, 190–1, 199–217,
        389–92
    rational comprehensive 191–2
    site 189
    species recovery 189
    strategic 115, 196–7
    systems 197–8

tactical　115–16

plant(s)

  and Aboriginal cultural heritage sites　364

  commercial potential　329

  communities, active management　328–9

  communities, classification of　24–7

  communities and boundaries　331

  introduced　310, 382, 393–4, 399–400

  rare or threatened Australian　27

  see also flora, weeds

Pleistocene Epoch　4–5

Plowman, Cathy　42

political considerations and protected areas　130, 144, 154

political factors and protected area management　376, 525–30

pollution　68, 374, 396, 417

population　72

  growth, global　372

Port Campbell National Park　439

postmodern times and protected area management　55–6

poverty and protected areas　89

Powerful Owl (*Ninox strenua*)　159

Precambrian Era　9

prescription burning　300–4, 344–7

pricing services and facilities　223–37

  setting a price　228–37

  supply costs　226–7

pricing, supply and demand　229–31

private property and conservation　221, 388, 510

private protected areas　97, 103, 133

private sector organisations　122

privatisation　121–2

Proteaceae　24

protected area agency organisational structures within Australia　120–3

protected area management

  actions　211–17

  agencies　xxviii–xxix, 95, 96, 123, 136

  attributes of a competent manager　127–8

  and Australia's evolutionary heritage　3–9

  basic principles　113

  and community　466–92

  contemporary influences　47–56

  decision-making　128–9, 194, 213, 389

  departments　95

  digital information and technology　179

  economic considerations　129, 135, 208, 228

  environmental accountability　271

  environmental/ecological considerations　129

financing　220–3

goals　209–11, 222

governance　56–63, 103

identifying and analysing the issues　209

information model　184–5

legal considerations　130, 208

lessons and principles　160–1

management process　113–30, 164–5

managerial considerations　130

marketing and promotion　482

objectives　209–11, 217, 222

planning　115, 189, 190–1, 199–217

political considerations　130, 144, 154, 208

pricing services and facilities　223–37

remote sensing　169

responses and initiatives　56–64

social considerations　129–30, 135, 208

society, culture and values　69–75

staff motivation　125

and treats to Australian protected areas　374–6

values　79–84

zoning　214–16

Protected Area Management Committee　98

protected area paradigm

  emerging　57–8

  traditional　57

protected areas

  biological inventory　163, 208

  cultural heritage　359–60

  cultural inventory　164, 208

  demand for recreation opportunities　227

  economic impacts　80, 84, 129, 135, 208, 244

  and ecosystem services　81–2, 228

  education and research　81

  environmental condition　164

  environmental reporting　286

  establishing　113, 132–61

  global network　137

  and grazing　84, 379

  and illegal actions　383, 397

  and indigenous cultures　36, 37–8, 82–4, 89, 97, 133, 138, 359, 499–503

  international designations　94–5

  international system of　91–3

  non-recreational uses　164

  personal and social benefits　84, 466

  physical inventory　163, 208

  planning　115, 189, 190–1, 199–217, 389–92

  and poverty　89

  social and cultural commitment　285

  social and land-use history　164

socio-economic costs and benefits　164
spirit, culture and identity　82–4
and sustainability　270, 271, 274–86
threats to　371–401
and timer industry　84
types of　91–105
values　78–84
visitor use　164, 165, 208–9, 382–3, 422, 425–7, 429–42, 486–91
see also environmental protection
protected areas, Australia　23, 105–11, 570–3
Aboriginal people　36, 37–8, 82, 97, 98, 133, 138, 323–4, 494–507
artists and writers　81
categories　97
chronology of major events　576–83
and Commonwealth　97–9, 133, 143–4, 585–6
departments and agencies　95
ecological factors　376
economic factors　376
establishing　113, 132–61
extent of　584–602
external territory and oceanic islands　106–7
and historic cultural heritage sites　355–7
history　36–47
human–nature interaction　78
indigenous　97, 133, 138
and interpretation　484–91
jurisdictions　xxviii–xxix, 96
legislation　44–7, 67–9, 114, 122–3, 133, 136, 144, 265–7
management effectiveness　553–68
managerial factors　376
marine　88–9, 104–5, 106–7, 110, 111, 138–44
multiple use in　104–5
non-governmental　103–4
original　36–8
private　97, 103, 133
public　97, 103
and recreation　429–42
re-emergence of　38–47
scientific selection methods　144–7
state and territory　99–102, 133
terrestrial　105–11, 133–8
tourism　164, 165, 208–9, 382–3, 422, 425–7, 429–42
protected areas, threats to
agricultural diseases, introduced　383
Australia　374–6
caves　383
clearing　393
climate change　372–3, 397, 399–400

construction and development　392–3
edge effects　379
fire　378–9, 392
geological features　383–5, 397–9
habitat destruction　373–4
habitat disturbance　382
illegal actions　383, 397
illegal grazing　379
introduced animals　393–7
introduced plants　393–4, 399–400
legal factors　376
management lessons and principles　400–1
minimising environmental threats　392–400
pest animals　379–82, 393–7, 471–2
political factors　376
pollution and waste　374, 377–8, 396–7
population growth　372
potable water　374
preventing environmental threats　386–92
scenic quality　385
social factors　375–6
soils　383–5, 393–7
species loss　373–4
tourism　399–400, 443
type and nature of　377–86
underlying causes　372–7
vandalism　383
visitor use　382–3, 399–400, 443
weeds　382, 393–4, 399–400
public lands and conservation corridors　386–7
public protected areas　97, 98, 137, 510
public sector organisation structures　120
purchaser–provider organisational model　121
Purnululu National Park (WA)　94

quarries
Aboriginal sites　366
threat to protected areas　385
Queensland　592–3

rabbits　380, 381, 397
radio tracking　320
rain, extreme　418–20
rainforests　24, 155, 456, 457
Ramsar Convention　17, 94, 98
Ramsar wetlands　94
rangers　304–7
rapid rural appraisal　473–4
recreation
Australia's protected areas　429–42
management　442–63
marine areas　544–8

opportunities, demand for 227

pricing 229–31

settings 425

supply costs 226–7

see also tourism

Recreation Opportunity Spectrum (ROS) 214, 435–8

recruitment 251–2

Red-fronted Parrot (*Cyanoramphus novaezelandiae cookii*) 96

Red Velvetfish (*Gnathanacanthus goetzeei*) 142

reefs 537–8

reference collections 174–8

regional development 54–5

Regional Forest Agreement (RFA) 136, 147–51, 154–6, 168, 174

regulations, environmental 526–7

religion 73

remote sensing in protected area management 169

reptiles 29–30, 321–2

research 168–74

  capacity 174

  design 173

  risk management 174

reserve selection 144–7

resource allocation 69

Richmond Birdwing Butterfly (*Ornithoptera richmondia*) 319

Rio Earth Summit 88

risk management 174, 267, 297

River Red Gum (*Eucalyptus camaldulensis*) 333

rivers 17–18

Riversleigh (Lawn Hill National Park) 10, 94

road surface materials 341

rock art, Aboriginal 358, 366

rock types, managing unusual 341

rocky reefs 537

Royal National Park 39–40, 134

Rubber Vine (*Cryptostegia grandiflora*) 310

Rufous Hare-wallaby (*Lagorchestes hirsutus*) 33

Russell Falls (Tasmania) 41

salaries 253–4

Salmon Gum (*Eucalyptus salmonophloia*) 512

salt damage from rising water tables 332–3

saltmarshes 536

sand environments 335

Sandhill Dunnart (*Sminthopsis psammophila*) 171

scenic quality, conserving 342, 385

schools 481–2

scientific reserve selection methods 144–7

scleromorphic ('hard-leaved') floras 24

scrub 26

seagrass beds 536–7

Sea-lions (*Neophoca cinerea*) 454

Sea Turtles (*miyapunu*) 321

semi-arid lands 18–19

settlement patterns 72

Shark Bay World Heritage Area 11, 94

Short-tailed Shearwater (*Puffinus tenuirostis*) 172

shrublands 26–7

Silent Spring 38

Silurian Period 9

Sky Lily (*Herpolirion novae-zelandiae*) 81

Slender Blue-tongue Lizard (*Cyclodomorphus venustus*) 171

snow 334, 420–1

social benefits and protected areas 84, 466

social commitment and protected areas 285

social considerations and protected areas 130, 135

social factors and protected area management 375–6

social history 164

social impact assessment 473

social norms, changing 73–5

society and conservation 69–75, 135, 375–6

socio-economic costs and benefits 164

soils 16, 335–41, 383–5, 385, 397–9

Sooty Owl (*Tyto tenebricosa*) 159

Sooty Tern (*Sterna fuscata*) 96

South Australia 38, 168, 170–1, 222, 594–5

South East Forest National Park (NSW) 25

South-East Regional Marine Plan 143–4

Southern Bluefin Tuna (*Thunnus maccoyii*) 104

Southern Regional Fire Association of NSW 414

species diversity, Australian 22, 134

species identification 316

species loss 373–4

spirituality and conservation 82–4

staff

  accountability 253

  and anti-discrimination laws 257

  competency 253, 295

  contract 252

  delegation of authority 253

  development 119, 254–5

  employee assistance programs 256

  and goodwill 256

  health and safety 255

  housing 255–6, 278–9

  induction 252

  motivation 125

  performance 264

  rosters 253

  salaries 253–4

  temporary 252

volunteer   254
see also management; work
stakeholders and communication   474–83
State of the Environment (SoE)   319
streams   17–18
supply and demand   229–31
sustainability   264–5
  assessment   294
  considerations   298
  and environmental impacts   297–8
  infrastructure design   278–9
  lessons and principles   286–7
  need for   270–4
  performance assessment and monitoring   275–7
  policy and planning   274–5
  and protected areas   270, 271, 274–86
sustainable development   50–3
sustainable park management   280–2
sustainable tourism   275, 277
SWOT analysis   197

Tarra–Bulga National Park   468
Tasmania   38, 41, 42, 65, 66, 100–2, 155, 156, 444, 596–7
Tasmanian Seamounts Marine Reserve   547
Tasmanian Wilderness   94
technology
  bans   526
  and managing protected areas   179
terrestrial
  area for nature conservation   105
  fauna   28–33, 134
  invertebrates   33
  national reserve system   133–8
  protected areas   105–11, 133–8
  vegetation   23–7, 134, 135
Tertiary Period   5–6
timber industry   84
tourism
  and Aboriginal cultural heritage sites   363–4
  commercial operations   460
  definition   425
  ecotourism   426–7, 431, 442
  global   427–9
  management   442–63
  marine areas   544–8
  and minimising risk   459–60
  and protected areas   164, 165, 208–9, 382–3, 399–400, 422, 425–7, 429–42, 443
  sustainable   275, 277
  see also recreation; visitor(s)
Tower Hill (Victoria)   41

travel cost method and environments valuation   238–42
Triassic Period   7–8
Tri-National Wetlands Memorandum of Understanding   98
turtles   321

Uganda   185–6
Uluru–Kata Tjuta National Park   94, 500
United Nations   85–6
United Nations Conference on Environment and Development (UNCED)   50
United Nations Development Program (UNDP)   85
United Nations Educational, Scientific and Cultural Organisation (UNESCO)   85, 86, 94
United Nations Environment Program (UNEP)   85, 86
United Nations World Charter for Nature   80
United States   134
urban landscapes   19–20
user-pays
  fees   224–6
  justification for   227–8
  level of charges   228
  setting a price   228–37

Valley of the Giants, WA   455
valuation
  benefit–cost analysis   129, 211, 237–48
  contingent   238, 242–4
  environmental   237–48
  input–output analysis   243–5
value measurement   237–8
values
  direct use   80–1
  and environment   69–75, 78–84
  existence   82
  instrumental   80–4
  non-use   82, 220, 227
  protected areas   78–84
vandalism
  and Aboriginal cultural heritage sites   364, 385
  and historic sites   367
  and protected areas   383
vegetation
  management and fire   330
  plots and transects   330
  studies, long-term   175
  terrestrial   23–7, 135, 511
  types   134
  see also flora; weeds
Victoria   38–9, 43, 133, 142, 147–8, 151–4, 598–9
Victorian Environment Assessment Council Act 2001 (Vic)   68

visitor(s)
  and resource allocation   165
  limits of acceptable change   450
  use   164, 208–9, 382–3, 399–400, 422, 425–7, 438, 486–91
  see also tourism
Visitor Impact Management (VIM)   449, 450
volunteers   254
  and fauna surveys   318
von Mueller, Baron Ferdinand   38

Wakefield, Jessie   42
Walmsley, John   104
waste, solid   273, 285, 374
water
  catchments   332, 523–4
  consumption   283
  ground   333
  management   332–4
  potable   374
  quality   333
  and sustainability   270–1, 272
  tables   332–3
Water Act 1989 (Vic)   69
weather, extreme   418–21
web sites   xxix–xxx
Wee Jasper Gravillea (*Grevillea iaspicula*)   330
weeds   310, 382, 393–4, 399–400
Werribee Park   368
Western Australia   99, 134, 147, 149–51, 395, 600–1
Western Quoll (*Dasyurus geoffroii*)   395
Western Shield   395
Wet Tropics World Heritage Area (Queensland)   94, 97, 482–3, 500
wetlands   17, 94, 98–9, 511
Whale Bird (*Sterna fuscata*)   96
whale shark   548
White, Mary   13
White Box (*Eucalyptus albens*)   512
Wik   97
WildCountry   516
wilderness   155, 499
Wilderness Society   516

Wildlife Information and Rescue Service (WIRES)   322
wildlife conservation   69
wildlife incidents   417–18
Williams, Fred   81
Wilsons Promontory Marine National Park   142, 349
winds, extreme   418
Witjira National Park   506–7
Woodland Watch   530
woodlands   16–17, 25–6, 62, 511
Wollemi National Park   11
Wollemi Pine (*Wollemia nobilis*)   10, 11, 329, 397
work, defining nature of   252–3
workforce and conservation outcomes   256
workplace
  disputes   256
  safety and health   255
  skills   252–3
World Bank   85
World Commission on Environment and Development (WCED)   86
World Commission on Protected Areas (WCPA)   61, 87–8
World Conservation Monitoring Centre (WCMC)   88, 178
World Conservation Union   87
World Heritage   65, 87, 89, 90, 94, 98
World Heritage Convention   85, 86–7, 94
World Heritage Properties Conservation Act 1983   65
World Parks Congress   84, 88–9, 466, 572
World Summit on Sustainable Development   51, 52–3
World Tourism Organisation (WTO)   428
World Travel and Tourism Council (WTTC)   427–8
World Wide Fund for Nature (WWF)   87, 98
World Wildlife Fund (WWF)   87
Wright, Judith   42, 81
writers and protect areas   81
writing and the use of clear English   263–4

yellow-flowered *Lechenaultia*   171

zinc toxicity   385
zoning   214–16